An Integrated Framework
for Structural Geology

An Integrated Framework for Structural Geology

Kinematics, Dynamics, and Rheology of Deformed Rocks

Steven Wojtal
Oberlin College, Oberlin, Ohio, US

Tom Blenkinsop
Cardiff University, Cardiff, UK

Basil Tikoff
University of Wisconsin-Madison, Wisconsin, US

This edition first published 2022

Registered Office(s)
John Wiley & Sons, Inc., 111 River Street, Hoboken, NJ 07030, USA
John Wiley & Sons Ltd, The Atrium, Southern Gate, Chichester, West Sussex, PO19 8SQ, UK

Editorial Office
The Atrium, Southern Gate, Chichester, West Sussex, PO19 8SQ, UK

For details of our global editorial offices, customer services, and more information about Wiley products visit us at www.wiley.com.

Wiley also publishes its books in a variety of electronic formats and by print-on-demand. Some content that appears in standard print versions of this book may not be available in other formats.

Library of Congress Cataloging-in-Publication Data applied for
Paperback ISBN: 9781405106849

Cover Design: Wiley
Cover Images: Courtesy of Steven Wojtal/Photograph of a polished slab and crossed-nicols photomicrograph of S-C mylonite from the South Amorican shear zone. Slab is ~15 cm across; field of view of photomicrograph is ~4 cm across. See D. Berthé, P. Choukroune, & P. Jegouzo. *Journal of Structural Geology* **1**, 31-42

Set in 9.5/12.5pt STIXTwoText by Straive, Pondicherry, India

SKY6F1CCAC2-C285-49A7-BC5D-9664C2F1B5BB_062122

Contents

Acknowledgements

We are grateful to Ellen M. Nelson and Paul K. Wojtal for careful figure drafting. The following contributed material or images for figures: Alex Lusk (Fig. 9.4), Wolfgang Maier (Fig. 3.14), Michael Rubenach (Figs. 3.2c, 3.13a, 3.15b, 3.17a, 3.22), Chris Tulley (Figs. 3.15c, 3.23), and Rachel Wells (Figs. 9.20, 9.21).

We sincerely thank Cat Bent, Jack Hoehn, and Alex Lusk for thorough reviews of drafts of chapters. Their thoughtful comments led to significant improvements in the presentation. Deanna A. Flores, Kyrsten L. Johnston, Ellen M. Nelson, Claire Ruggles, and Eneas Torres Andrade proofread the entire book, for which we are appreciative. We thank Mandy Collison, Frank Otmar Weinreich, and especially Shiji Sreejish at Wiley for their patience and their collective efforts to see this project to completion.

All three authors acknowledge the patience and support of their families, particularly Wendy, Jessie, and Sara.

Website

This book is accompanied by a website which includes color versions of many figures from the book and additional resources.

https://geoscience.wisc.edu/structuralgeologybook/

1

A Framework for Structural Geology

1.1 Introduction

Structural geology is the branch of earth sciences that focuses on understanding the processes by which geological materials deform. Structural geologists use the rock record to study naturally deformed rocks, an endeavor that relies upon both fieldwork and, to an increasing degree, the use of microstructural observations to characterize deformed rocks. Other academic disciplines – such as branches of geophysics, engineering, materials science, and physics – share the goal of characterizing and understanding how natural materials deform. Those fields focus on laboratory deformation experiments or theoretical models of material behavior.

1.1.1 Deformation

The term *deformation* encompasses *changes of position* (movement) and *changes of shape* (distortion). Natural deformation of rock produces a variety of rock *structures* that include fractures, faults, folds, deformation fabrics (e.g. foliations and lineations), shear zones, and microstructures. Structural geologists are interested in how to interpret rock structures and microstructures in terms of geometry, kinematics (the motion associated with deformation), dynamics (the forces involved in deformation), and the material behavior of rocks undergoing deformation.

1.1.2 Empirical vs. Theoretical Approaches

Three critical components compose the framework for structural geology: (1) The **three-dimensional geometry** of all geological structures and microstructures in the region; (2) the exact path that each part of the material followed as the individual structures formed and microstructures developed (the **kinematic evolution**); and (3) how the rock structures and microstructures evolved in terms of the absolute rates at which component parts moved, the forces that operated on the rocks, and the deformation-related physical properties of the rocks during deformation (the **deformation dynamics**).

The terms *kinematics* and *dynamics* will be familiar to students of physics. For those unfamiliar with the terms, kinematics refers to descriptions of the motion of parts of a system, whereas dynamics considers how systems respond to forces and boundary conditions. Kinematics and dynamics carry no connotation of what sort of entities are moving or what type of system is under consideration. In the physical sciences, the term *mechanics* refers to the field of study that examines how materials respond, i.e. move or change, in response to forces or displacements imposed on them. Combining these concepts, we see that a branch of science that is highly appropriate to structural geology is the

An Integrated Framework for Structural Geology: Kinematics, Dynamics, and Rheology of Deformed Rocks, First Edition. Steven Wojtal, Tom Blenkinsop, and Basil Tikoff.

mechanics of continuous media (e.g. continuum mechanics). We need to introduce one final term – *rheology*, which is often used as shorthand for the mechanics of continuous media. Rheology refers to the study of the relations between the displacements or velocities observed within a deforming material and the forces or stresses that act upon that material. We recommend reading a short recounting of an after-dinner speech by Marcus Reiner, in which he outlined his role in coining the term "rheology" (Reiner 1964).

The Deborah Number

(*Physics Today*, January 1964, p. 62/with permission from AIP Publishing)

From an after-dinner talk at the 4th Int. Congress on Rheology by Marcus Reiner, Israel Institute of Technology

In 1928 I came from Palestine to Easton, PA, to assist Eugene Cook Bingham at the birth of Rheology. I felt strangely at home. There was Bethlehem quite near, there was a river Jordan, and a village called Little Egypt. The situation was, however, also slightly confusing. To go from Bethlehem to Egypt, one had to cross the river Jordan, a topographic feature which did not conform to the original. Then there were, here, places such as Allentown, to which there was no analogy. And this could lead to strange situations, such as when a girl at school was asked where Christ was born and replied, "In Allentown." When corrected by, "No, in Bethlehem," she remarked, "Well I knew it was somewhere around here."

In Palestine I was working as a civil engineer doing science as a hobby. In 1920 a chemist had asked my help in the problem of the flow of a plastic material through a tube. I solved the problem and derived what is now known as the Buckingham-Reiner equation, Buckingham at the US National Bureau of Standards having derived the equation before. When Bingham learned of my work, he invited me to Lafayette College.

When I arrived, Bingham said to me, "Here you, a civil engineer, and I, a chemist, are working together at joint problems. With the development of colloid chemistry, such a situation will be more and more common. We therefore must establish a branch of physics where such problems will be dealt with."

I said, "This branch of physics already exists; it is called the mechanics of continuous media, or mechanics of continua."

"No, this will not do," Bingham replied. "Such a designation will frighten away the chemists."

So he consulted a professor of classical languages and arrived at the designation of Rheology, taking as the motto of the subject Heraclitus' παυτα ρει or "everything flows."

Rheology has become a well-known branch of physics, but most typists think it is a misprint for theology. I constantly receive mail addressed to the Theological Laboratory of the Israeli Institute of Technology and, on the occasion of the Second International Congress at Oxford 10 years ago, there was a special coach in the train at Paddington Station reserved for members of the Theological Congress. This seems ridiculous, but there is some relation between rheology and theology and on this, I want to say a few words.

Heraclitus' "everything flows" was not entirely satisfactory. Were we to disregard the solid and deal with fluids only? There are solids in rheology, even if they show relaxation of stress and consequently creep. The way out of this difficulty had been shown by the Prophetess Deborah even before Heraclitus. In her famous song after the victory over the Philistines, she sang, "The mountains flowed before the Lord." When, over 300 years ago, the Bible was translated into English, the translators, who had never heard of Heraclitus, translated the passage as "The mountains *melted* before the Lord" – and so it stands in the authorized version. But Deborah knew two things. First, that the mountains flow, as everything flows. But, secondly, that they flowed before the Lord, and not before man, for the simple reason that man in his short lifetime cannot see them flowing, while the *time of observation of God is infinite.* We may therefore define the non-dimensional Deborah Number:

$$\boldsymbol{D} = \frac{\text{Time of relaxation}}{\text{Time of observation}}$$

The difference between solids and fluids is then defined by the magnitude of ***D***. If your time of observation is very large, or conversely if the time of relaxation of the material under observation is very small, you see the material flowing. On the other hand, if the time of relaxation of the material is larger than your time of observation, the material, for all practical purposes is a solid. In problems of industrial design, you may substitute the *time of service* for the time of observation. When designing a concrete bridge you make up your mind to decide how long you expect it to serve, and then compare this time interval with the relaxation time for concrete.

It therefore appears that the Deborah Number is destined to become the fundamental number of rheology, bringing solids and fluids under a common concept, and leaving Heraclitus' παυτα ρει as a special case for infinite time of observation, or infinitely small time of relaxation. The greater the Deborah Number, the more solid the material; the smaller the Deborah Number, the more fluid it is.

There is a story they tell about two students of theology, who were praising the Almighty God. Said one, "For God, 1000 years are like a minute. And as he is the Creator of all, 1000 dollars are for Him like a cent." Said the other, "Wonderful, next time I pray to God, I shall pray, 'God, give me a cent.'" Said the first: "What will it help you? He will say, 'Wait a minute.'"

The man did not take care of the difference between God's and his own time scale. This then is the connection between Theology and Rheology. In every problem of rheology, *make sure you use the right Deborah Number.*

Structural geologists, like all scientists, use the scientific method of developing ideas into testable *hypotheses*, which contribute to the construction of *theories*. Theories comprise hypotheses tested against data, but they also include deductions from tested hypotheses, mathematical and numerical analyses that extrapolate or interpolate data, predictions derived from models, etc. From plate tectonics theory, we infer that mountain belts may result from continent-continent collisions. Kinematic evidence supports the deduction that the Alpine mountain chain of Eurasia formed when continental landmasses on the European plate, on several small microplates, and on the African plate collided. *Models* contribute to our understanding by enabling us to make predictions to be tested against observations and data. A model can be formulated conceptually, mathematically (an analytical or numerical model), or physically (an analogue model). Many elegant numerical and physical models of the Alps mountain range have been made, which integrate forces and material behavior to provide insights into the dynamics of the Alps.

How any individual structural geologist approaches geological structures via the scientific method depends on their viewpoint and the tools that they use. The two end-member approaches are **empirical** and **theoretical**. Empiricism is the viewpoint that knowledge derives from experience. Empirical scientists seek understanding of phenomena by observing and measuring them. The theoretical approach is the viewpoint that derives insight from theories, including models. Thus, a theoretician will typically seek understanding by producing models and testing these models against the observation. Both are legitimate scientific viewpoints and rarely does any scientist use just one approach. Still, there can be tension between the empirical and the theoretical viewpoints, which has, more than once, produced severe scientific disagreements. Yet, it is also this tension that moves the field forward. The distinction between empirical and theoretical approaches is very similar to the distinction between *inductive* and *deductive* methods in science: induction proceeds from examples to general rules or principles, while the deductive approach is to use rules to understand examples.

The difference between the empirical and theoretical viewpoints largely determines how one approaches structural geology (Figure 1.1). Since the empirical viewpoint is based on observation, the most basic observation is that of the three-dimensional geometry of structures and structural fabrics. For example, a three-dimensional exposure of a fold is shown in Figure 1.2. This fold occurs on the beach in Bude, England, and was deformed in the Hercynian orogeny about 250 Ma ago (1 Ma = 1 Mega annum = 10^6 years). One can unambiguously describe the geometry of the fold. However, to determine the kinematic evolution of the fold, it is necessary to make an **inference**. An **inference** is the reasoning involved in reaching a conclusion on the basis of circumstantial evidence and/or prior conclusions, rather than on the basis of direct observation. For instance, structural geologists who create scaled experimental models of folds with putty can use the observed three-dimensional geometry of the fold; the arrangement of faults, fractures, and fabrics in the fold; and their experimental knowledge to *infer* how the fold formed (its kinematic evolution: how the rock particles moved from their original to final positions). Determining the forces involved in the fold

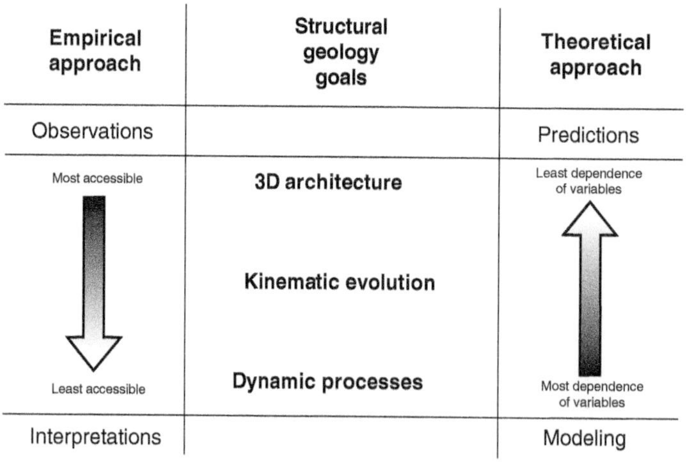

Empirical approach	Structural geology goals	Theoretical approach
Observations		Predictions
Most accessible	3D architecture	Least dependence of variables
↓	Kinematic evolution	↑
Least accessible	Dynamic processes	Most dependence of variables
Interpretations		Modeling

Figure 1.1 The empirical vs. theoretical approach. Details are given in the text.

Figure 1.2 Fold exposed in three dimensions. The height of the fold is approximately 1 m.

(dynamic processes) requires an even greater level of inference. The forces that caused the folds to form have completely ceased to exist – they operated 250 Ma ago – and completely different forces are transmitted through these rocks at present. Making direct observations of full kinematic histories of deformed rocks or of dynamic processes in structural geology is impossible, although one can make reasonable inferences about the kinematic evolution and thereby approach dynamic analysis.

A theoretician would approach the formation of the fold in Bude differently from an empiricist (Figure 1.1). The starting point for theoretical approaches is often a model. A model, in a scientific sense, is a mathematical description of a natural entity or process. Due to the complexity of natural entities and processes, models are usually abstracted or simplified representations of the entity or process. This approach requires understanding the fundamental physics and chemistry behind the process. Thus, a theoretician might *start* with dynamic analysis, rather than with observation. The outcome of the model is a **prediction**. Theoretical models typically predict a particular kinematic evolution because the forces are directly linked to motions through rules established to describe an inferred type of material behavior. The full development of the structure requires a time-wise integration of the kinematics. In cases where this approach is possible, particular characteristics of the structure are derived predictions from the modeling. For example, a dynamic model might predict the hinge shape for the fold, or a particular ratio between the fold wavelength and thickness of the folded layers. Because the modeling is directed at understanding the fold in Bude (Figure 1.2), no aspects of the three-dimensional geometry of the actual fold are introduced in the modeling. For the model to be useful, it must make a prediction regarding the three-dimensional geometry of the fold, which the geologist can compare against the fold's observed three-dimensional geometry.

Both empirical and theoretical approaches are useful in structural geology, as well as in most other sciences. Further, many structural geologists use both approaches, according to which approach will solve the particular problem that is faced. The only caveat is that geology is historical science. Any historical science will inherently rely more on the empirical rather than the theoretical approach because it endeavors to understand observed developmental sequences rather than impose deduced ones (see Tikoff et al. 2013). The geological time scale, for instance, is entirely the result of the empirical approach.

1.1.3 Continuum Mechanics and its Applicability to Structural Geology

The concepts of continuum mechanics regularly are integral components of the framework for structural geology. Continuum mechanics is the study of the deformation of materials that are continuous (i.e. space-filling), not composed of discrete, individual objects or particles in space. Because the continuum mechanics approach is both quantitative and theoretical, it is not typically covered in undergraduate structural geology courses. The wonderful book "Stress and Strain: Basic Concepts of Continuum Mechanics for Geologists" by Means (1976) helped introduce the continuum mechanics approach to a broader audience when it appeared almost 50 years ago, and it is still a valuable supplement to undergraduate structural geology texts. Means' book continues to exemplify the power of applying continuum mechanics to structural geology, although new advances in structural geology have occurred in the past few decades that are not covered. With this book, we endeavor to update and, to a degree, extend the coverage of the continuum mechanics topics elegantly presented in Means' "Stress and Strain."

We do not intend to encourage a wholly theoretical approach to structural geology. We are convinced that a solid grounding in continuum mechanics helps in understanding all aspects of rock deformation. For example, the development of plate tectonics in the 1960s and 1970s provided a rational framework for understanding the character of and connections among different geological provinces. Lithospheric plates might consist of either or both continental and oceanic crust, but it is the divergence, convergence, and transcurrence across plate boundaries that lead to the rock deformation responsible for folds, faults, fabrics, etc. Quantitative understanding of deformation across plate boundaries is essential to relate the rock structures to spatial changes of displacements or velocities measured using traditional surveying techniques, global positioning systems, or laser interferometry. Traditionally, structural geology texts emphasize one aspect of deformation – the distortion of rock masses and the development of rock structures. We expand this view to consider the formation of rock structures with regard to displacements, velocities, rheology, and deformation mechanisms.

The second focus of this book is rheology, which relates gradients of displacements and velocities of rock masses to the forces and stresses acting on them. Studies of rheology are a prerequisite to defining where and how much energy is dissipated in these global processes, and they may contribute to understanding what caused historical variations in plate geometries and rates of movement. A quantitative understanding of deformation mechanisms and rheological behavior may enable geologists to understand why the lithosphere is divided into relatively undeformed plates and deforming plate boundary zones, what factors are responsible for the precise location of the plate boundaries, what mechanism or mechanisms contribute to driving plate motion, and how lithospheric movements and plate boundary deformation relate to deeper Earth processes.

1.1.4 How to use this Book

1.1.4.1 Aims of the Book

Our aim with this book is to examine the core principles of deformation in ways that provide a quantitative foundation for developing a framework for understanding structural geology. Undergraduate structural geology texts focus on the development of rock structures. They typically include thorough descriptions and illustrations of macroscopic rock structures, outline variations in the geometry of those structures, and characterize to a greater or lesser degree of detail the grain-scale or microscopic fabrics that develop in deformed rocks. Most structural geology texts also introduce the ways that geologists quantitatively measure rock deformation, specify the forces responsible for deformation, and relate the rock deformation to the driving forces. Graduate-level monographs on structural geology typically address a subset of these topics, focusing, for example, on describing and quantifying in depth the geometry and evolution of structures or deformation fabrics, *or* on outlining models that material scientists use to relate rock deformation to the forces driving deformation.

This book is intended to bridge the gap between typical undergraduate texts and graduate-level monographs. We presume a level of familiarity with geological structures and endeavor to augment students' understanding by examining the microstructures associated with them. We then use the understanding of geological structures and microstructures to examine in detail continuum mechanics principles and practices that contribute to our understanding of tectonic processes.

1.1.4.2 A Word about Words

Structural geologists, like most other scientists, use scientific terminology to communicate as precisely as possible with each other. In this book, we have emphasized new words in two different ways throughout the book. **Bold** denotes critically important words, the first time they are used or when they are defined. We have used *italics* for words for which a student should know about, but do not necessarily need an exact working definition.

1.1.4.3 Organization of the Book

In Chapter 2, we review the geometry and general characteristics of rock structures. Chapter 3 describes the geometry and characteristics of deformation microstructures. These two chapters are intended to ensure a standard level of understanding of deformation structures and microstructures, and they describe the context for our focus on continuum mechanics. The rest of the book is organized to separate observations (that scientists make on rocks) from inferences (that they derive from those observations). This approach follows the scientific methodology used by many practitioners of structural geology. Thus, Chapters 4 and 5 address what structural geologists observe and measure – displacements and velocities, displacement and velocity fields, and gradients of displacement and velocity fields. These chapters are most directly related to the structures and microstructures that one observes in deformed rocks, and they enumerate what rock structures and microstructures tell us about how masses of rock have moved relative to each other. Chapter 6 examines the forces, pressures, and stresses within Earth that cause movements. We introduce and use well-established conceptualizations that guide the thinking of geologists in assessing how forces act on rocks and how spatial variations in forces impact rock masses. Chapter 7 addresses rheology, the study of how displacements and/or strains relate to stresses. This chapter examines the mathematical relationships developed to describe the different ways that particles move in response to driving forces. Chapter 8 considers the various mechanisms by which minerals and rocks deform and how variations in physical conditions impact which mechanisms predominate in accommodating deformation. Thus, Chapters 6–8 consider in detail deductions and inferences commonly utilized in analyzing the deformation of rocks. A final chapter (Chapter 9) uses selected case studies to illustrate situations where geologists have been able to derive displacement fields and strains from rock structures, infer rock rheology by using laboratory and theoretical analyses of fabrics and deformation mechanisms, and draw inferences on the mechanical behavior of rock masses.

Most undergraduate structural geology textbooks are focused on a geometric description of geological structures, and are inherently more empirical. Two chapters in this book follow this empirical approach. Chapter 2 briefly reviews the characteristics of macroscopic structures, with the aim of reviewing relevant terminology on fractures, faults, folds, deformation fabrics, and shear zones. Chapter 3 characterizes the geometries in a thin section of the most common deformation microstructures in detail. This more substantial discussion of microstructures is included for two reasons. First, microstructures are covered only slightly within many typical structural geology texts because of the emphasis on field observations. Second, distinguishing the particular mechanism or mechanisms by which a rock deformed hinges in many instances on being able to document a specific combination of microstructures. Thus, in order to accomplish the three goals outlined in the previous section, one must be able to recognize an array of deformation microstructures. The language associated with microstructural analysis is often a mixture of observation, inference, and interpretation. We hope that the organization of this book, with an early chapter (Chapter 3) outlining the geometry and character of microstructures and a later chapter (Chapter 8) addressing deformation mechanisms, will lessen the tendency to conflate what is observed in a thin section with what can be inferred or interpreted with the knowledge of rheology. Readers with interests in a broader understanding of microstructures are referred to "A Practical Guide to Rock Microstructure" by R. H. Vernon (2004), and those interested in exploring the role of microstructure analysis in structural geology are referred to "Microtectonics" by C. W. Passchier and R. A. J. Trouw (2005).

Each of Chapters 4 through 8 has two sections: Part A – Conceptual Foundation and Part B – Comprehensive Treatment. The main distinction between the Conceptual Foundation and Comprehensive Treatment sections is the level of mathematical rigor presented. The Conceptual Foundation section uses algebraic equations, while the Comprehensive Treatment section introduces and uses vectors and matrices, which are the basis of continuum mechanics approaches. Several chapters also have Appendices that outline in some detail specific mathematical approaches. The Part A and Part B sections are written in parallel to encourage students to read the Comprehensive Treatment section, with its more advanced mathematics, once the basic understanding of the concepts is developed in the Conceptual Foundation. The parallel structure of the sections means that the two sections present major topics in the same order and, to the extent possible, major divisions in the two sections have similar titles.

Happy reading!

References

Means, W.D. (1976). *Stress and Strain: Basic Concepts of Continuum Mechanics for Geologists*. New York: Springer-Verlag.

Passchier, C.W. and Trouw, R.A.J. (2005). *Microtectonics*, 2e. Berlin: Springer.

Reiner, M. (1964). The Deborah number. *Physics Today* **17**: 42.

Tikoff, B., Blenkinsop, T., Kruckenberg, S.C., Morgan, S., Newman, J., and Wojtal, S. (2013). A perspective on the emergence of modern structural geology: Celebrating the feedbacks between historical-based and process-based approaches. In: *The Web of Geological Sciences: Advances, Impacts, and Interactions: Geological Society of America Special Paper*, vol. 500 (ed. M.E. Bickford). Geological Society of America.

Vernon, R.H. (2004). *A Practical Guide to Rock Microstructure*. Cambridge: Cambridge University Press.

2

Structures Produced by Deformation

2.1 Geological Structures

The deformation of rocks gives rise to five main types of geological structures: **structural fabrics**, **folds and boudinage**, **fractures and stylolites**, **faults and fault zones**, and **shear zones**. These structures, and the deformation that gives rise to them, can occur naturally at scales ranging from the whole lithosphere to the microscopic. The aim of this book is to examine the general principles of deformation. So, we begin by reviewing the characteristic geometries of the different types of geological structures. We postpone discussing the macroscopic and microscopic processes by which the structures form until much later in the book (mainly in Chapters 7 and 8); the intervening chapters present concepts critical to observing, characterizing, and interpreting deformation.

We begin by introducing a few terms regularly used to characterize geological structures and/or their component parts in hand samples and outcrops. **Planar** refers to a single, locally planar surface. In contrast, **tabular** refers to a locally planar zone with a finite thickness. For example, some faults are planar, but all shear zones are tabular. **Competence** is a useful field term that refers to the tendency of the material to resist deformation (or deform at slow strain rates for a given set of conditions). Competent layers often fold or boudinage during deformation. **Incompetence** is the opposite of competence. The relative competence of adjacent units – termed **competency contrast** – is responsible for many of the observed geological structures in the field. Finally, an **enveloping surface** defines the extent of geological features. Often, a zone of tabular deformation is delineated by two enveloping surfaces, such as for a folded layer or a series of *en echelon* veins.

2.1.1 Structural Fabrics

Geologists use the term *fabric* to denote the geometric configuration, i.e. orientation and distribution, of the component parts of a rock. The component parts may be individual mineral grains, rock fragments, aggregates of mineral grains and rocks, or collections of structures. A configuration of component parts that is apparent in hand sample or outcrop is sometimes denoted a *macrofabric*, but most geologists drop the prefix and simply refer to the fabric in the rock. Fabrics apparent at the scale of a thin section are sometimes distinguished as *microfabrics*. A *primary* or *original* fabric or microfabric is a product of the processes by which the rock formed, for example, bedding in a sedimentary rock. Geologists regularly interpret igneous or sedimentary fabrics or microfabrics in rocks in the context of their formation. Structural fabrics and microfabrics are those for which deformation is a key agent responsible for rearranging rock's component parts.

An Integrated Framework for Structural Geology: Kinematics, Dynamics, and Rheology of Deformed Rocks,
First Edition. Steven Wojtal, Tom Blenkinsop, and Basil Tikoff.

Because they are a product of processes that act on rocks after their formation, they are sometimes called *secondary* fabrics or microfabrics.

Structural fabrics are defined by a preferred orientation or alignment of: (1) *deformed markers*, such as sedimentary clasts or fossils; (2) *tabular* or *elongate mineral grains*; (3) *aggregates of minerals*; and/or (4) *geological structures* (Figure 2.1). Fabrics that affect all parts of a rock equally are *penetrative*, whereas fabrics that affect only some portion of a rock are *non-penetrative*. Geologists distinguish two end-member types of fabric in deformed rocks: (1) *foliations* or planar fabrics and (2) *lineations* or linear fabrics. The name applied to a foliation and/or lineation in a rock depends on the rock's composition and conditions of deformation. For example, *cleavage* is the type of foliation found in unmetamorphosed to weakly metamorphosed rocks. In cleaved rocks, the aligned deformed markers or mineral grains are not visible to the naked eye. Some rocks exhibit *slaty cleavage*, i.e. foliation that is *continuous* or penetrative at the scale of a thin section. Other rocks exhibit *domainal* or *spaced cleavage*, where volumes of rock with no apparent cleavage or weakly developed cleavage separate volumes where the cleavage is well-developed. *Schistosity* is a type of foliation defined by planar mineral grains, typically sheet silicates or flattened quartz grains, that are visible to the naked eye. The tabular or platy minerals in schists may be disseminated throughout a rock, or they may be found in alternating domains with greater or lesser concentrations. *Gneissic layering* is a planar fabric defined by compositional banding in coarse-grained metamorphic rocks; the individual mineral grains within bands need not themselves have shapes that define a preferred orientation. There are also

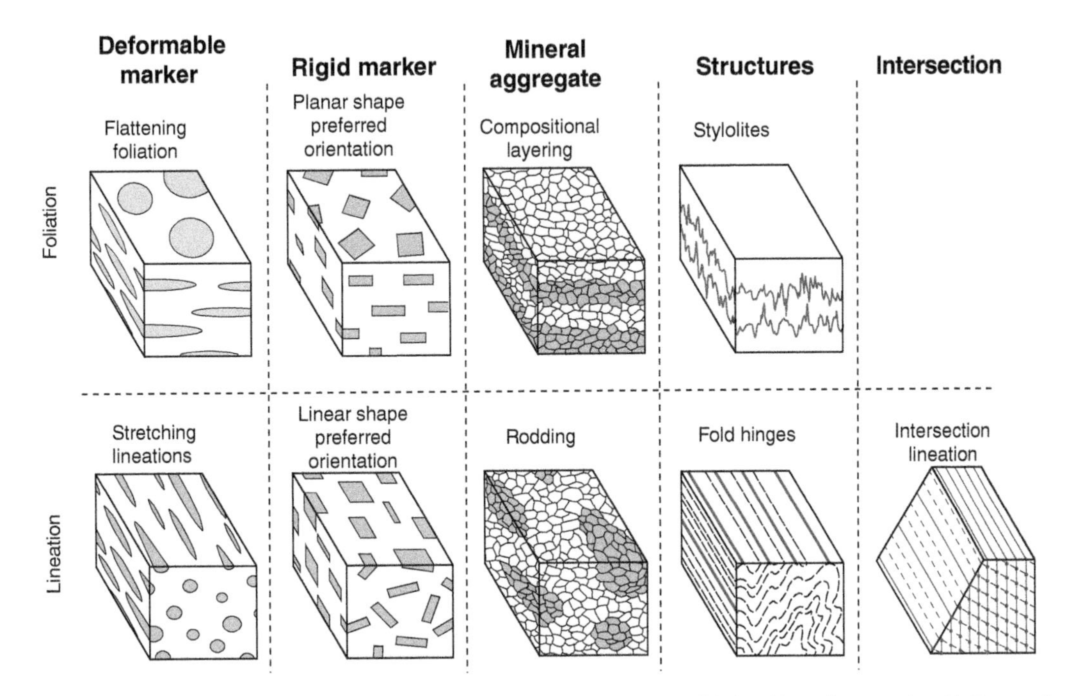

Figure 2.1 Structural fabrics. Foliations are locally planar secondary fabrics. Lineations are locally linear planar fabrics. Both are defined by a preferred alignment of different types of fabric elements, including deformable markers, rigid markers, mineral aggregates, rock structures, or the intersections between sets of surfaces.

different types of lineation (Figure 2.1). A rock in which elongate or needle-shaped minerals are preferentially oriented parallel to a single direction possesses a *mineral lineation*. If the mineral grains' elongate shape is due to deformation, the rock possesses a *stretching lineation*, and if the aligned elements are aggregates of minerals, the rock exhibits *rodding*. The intersection of surfaces or planar fabrics with different orientations in a rock (e.g. bedding and cleavage) creates a different sort of lineation, which is not penetrative. Rocks with a dominantly planar fabric or foliation are *S-tectonites*, and those with dominantly linear fabric or lineation are *L-tectonites*. Rocks often exhibit both planar and linear fabrics. If foliation dominates, the rock is an *SL-tectonite*, and if lineation dominates, it is an *LS-tectonite*.

Fabrics are important because:

1) They constrain the conditions of deformation. For example, one can infer that cleavage formation occurred under low-grade metamorphic conditions because it is typically accompanied by the growth of diagnostic minerals indicative of low-to-moderate pressure and temperature conditions. In contrast, a rock with gneissic layering indicates the rock was deformed under higher metamorphic grade conditions, again constrained by the distinctive assemblage and chemical composition of its mineral constituents.
2) They determine the relative timing of deformation events. Overprinting of one fabric by another provides critical information about the historical development of an area, particularly in high-grade terrains.
3) They control the material properties (heterogeneity and anisotropy) of the rock. Rocks without fabrics are isotropic, whereas rocks with increasing fabric intensity are increasingly anisotropic. Thus, fabric exerts directional control on rock strength and other material properties, such as porosity and permeability associated with fluid flow.

2.1.2 Folds and Boudinage

Folds form when nearly planar surfaces, either primary surfaces like bedding or secondary surfaces like foliation surfaces, are deformed into zigzag or wave-like shapes. Folded surfaces are often the buckled boundaries of rock layers, such as sedimentary strata, tabular igneous bodies, or bands of distinctive metamorphic rocks, and the intrinsic physical properties of the layers determine how buckling proceeds and the characteristics of the resulting folds. The deflections that define the folds sometimes result from fault movement or igneous intrusions. **Mullions**, cuspate-lobate forms developed on the surfaces separating distinct layers, result from processes akin to those responsible for the buckling of layers and, therefore, reflect the properties of the layers. **Boudinage** is the systematic variation in the thickness and/or continuity of layers, caused by elongation of competent layers. Similar to both folding and mullions, the geometry of the boudinaged layers reflects the properties of the layers. We begin by focusing on the geometry of folded individual surfaces. We consider the geometry of folded layers, boudinage, and mullions at the end of this section.

Folds form when an original surface is shortened primarily in a single direction. This produces a series of subparallel *fold hinges*, segments where the curvature of the folded surface is greatest, separated by *fold limbs*, segments where the folded surface has less curvature (Figure 2.2a). *Fold hinge lines*, lines connecting the points on the surface with the greatest curvature, are typically curved, although often curvature of the folded layer measured parallel to the hinge line is much lower than the curvature measured across the hinge line (Figure 2.2b). In segments of folded surfaces where the fold hinge line is approximately a straight line (often called the *fold axis*), geologists focus on the shape of the fold in a section plane, called the *profile plane*, taken perpendicular to the hinge line (Figure 2.2c).

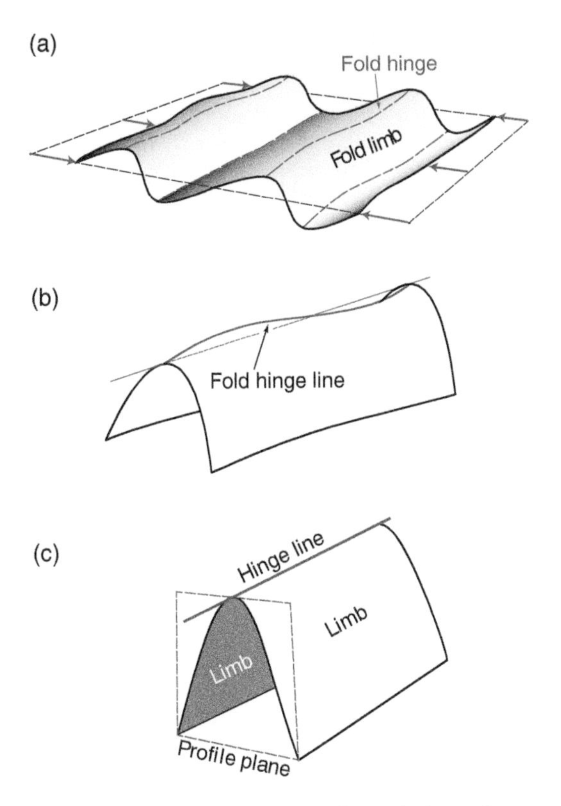

Figure 2.2 (a) A folded surface formed when a layer is shortened in a single direction. The *fold hinge lines* connect points of high curvature on the folded layer and separate *fold limbs*, regions of little curvature. (b) In most folds, the hinge line is curved, not straight. (c) For limited regions of many folds, the hinge line approximates a straight line and a preferred image of the folded layer is its trace on a *profile plane* perpendicular to the hinge line.

The profile shapes of folded surfaces provide information on the magnitude and kinematics of deformation. Two characteristics of the profile shape of a fold relate to the magnitude of shortening of the surface (Figure 2.3a): (1) the value of the angle between adjacent limbs of a folded surface and (2) whether, for a particular value of *inter-limb angle*, the fold surface has a rounded shape or is angular, i.e. has relatively long, straight fold limbs and a localized, highly curved fold hinge. Fold *symmetry* or *asymmetry* is an important clue to deformation kinematics (Figure 2.3b). Symmetric folds, where opposing fold limbs have approximately equal lengths, suggest shortening without significant shearing parallel to the surface. Asymmetric folds, with S or Z profile shapes (Figure 2.3c), have an associated sense of shear or *vergence*.

Only rarely does one find a single-folded surface. More commonly it is collections of subparallel surfaces that are folded. The *fold hinge surface* (*fold axial surface* or *axial plane* for some workers) is an imaginary surface that contains the hinge lines of adjacent folded surfaces (Figure 2.4a). The normal to the fold hinge surface defines a local direction of maximum shortening for the folding. So, geologists working in folded rocks typically measure the orientations of hinge surfaces and hinge lines to constrain the kinematics of deformation (Figure 2.4b).

Determining the orientations of the fold hinge lines and fold hinge surfaces in folded layers is an important first step toward understanding the kinematics of the deformation responsible for folding. In addition, geologists determine the shapes of bounding surfaces of folded layers, which define whether and how layer thickness varies with position about a fold. Structural geologists sometimes use two idealizations as benchmarks for the shapes of folded layers (Figure 2.5a): (1) *parallel folds*, where the thickness of the folded layer measured perpendicular to layer dip is the same at all limb dips; and (2) *similar folds*, where the thickness of the folded layer measured parallel to the fold hinge surface is the same at all limb dips. Parallel and similar folds are characterized by dip isogons (Figure 2.5). In general, parallel folds form in more competent units and similar folds form in less competent units. A second scheme for characterizing the shapes of folded layers utilizes a geometric element known as a *dip isogon*, which connects points with equal dip on the inner and outer arcs of a folded layer, to distinguish between different classes

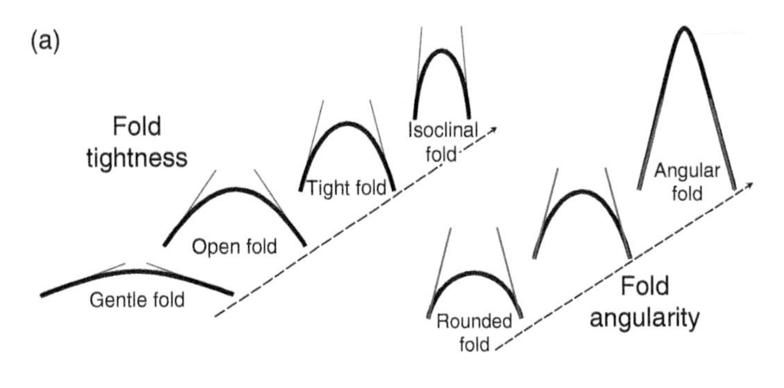

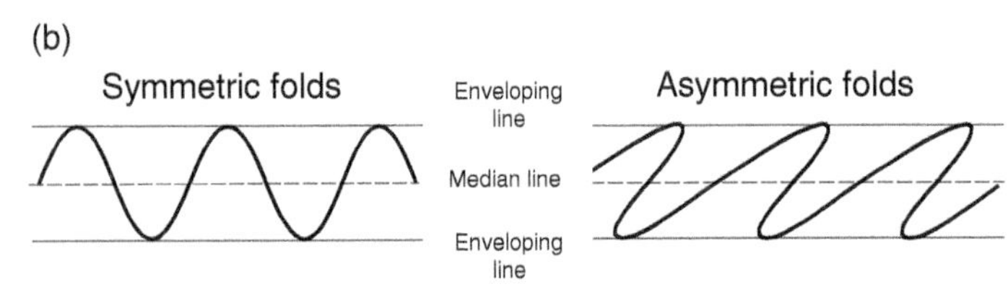

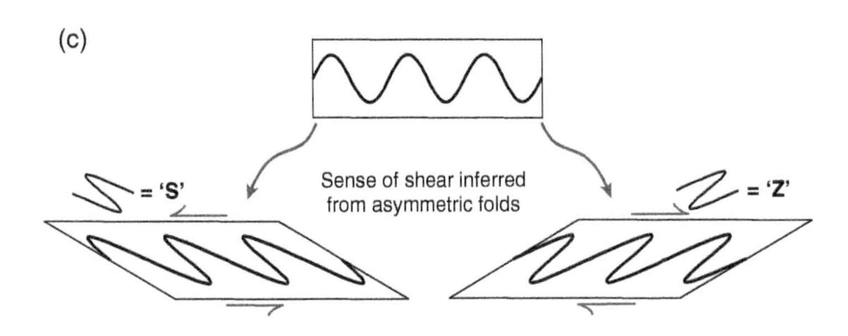

Figure 2.3 Fold kinematics. (a) Fold *tightness* defined by interlimb angle. Folds are *gentle* if interlimb angle ≥120°, *open* if 120° > interlimb angle > 70°, *close* if 70° > interlimb angle > 30°, *tight* if 30° > interlimb angle > 10°, and isoclinal if 10° ≥ interlimb angle. Fold *angularity* depends on the degree to which fold curvature is localized. (b) Comparison of *symmetric* and *asymmetric* folds. (c) Shearing superposed on existing symmetric folds generates asymmetric folds. *S*-shaped asymmetric folds imply top-to-the-left shearing, whereas *Z*-shaped asymmetric folds imply top-to-the-right shearing.

of folds. There are three general classes of fold shapes, with one class divisible into three subclasses (Figure 2.5b). By combining observations of fold symmetry with observations of the shapes of folded layers, geologists can draw inferences on the fundamental physical properties of the layers relative to their surroundings (their relative *competence*, or resistance to deformation) (Figure 2.6).

Shortening parallel to the boundary separating a competent layer from an incompetent layer sometimes distorts the boundary to form a *cuspate-lobate* form called *mullions* (Figure 2.7, top center). The cusps "point" toward the competent layer, and the incompetent layer often develops more prominent foliation. If mullions develop on both bounding surfaces of a layer shortened parallel to its length, the deformed layer may appear to be folded with lobate outer arcs and cuspate inner arc if the cusps and lobes are not aligned (Figure 2.7, upper left, upper layer). Alternatively, the deformed layer may appear to be a collection of bulbous forms if the cusps and lobes are aligned (Figure 2.7, upper left, lower layer).

Boudinage is a structure that commonly forms when relatively competent rock layers or segments of layers embedded between relatively incompetent layers are shortened across layering and lengthened parallel to layering (Figure 2.7, bottom center). Boudinage, like folds, may be symmetric (Figure 2.7, bottom left) or asymmetric (Figure 2.7, bottom right). Symmetric boudins indicate the directions of elongation and shortening;

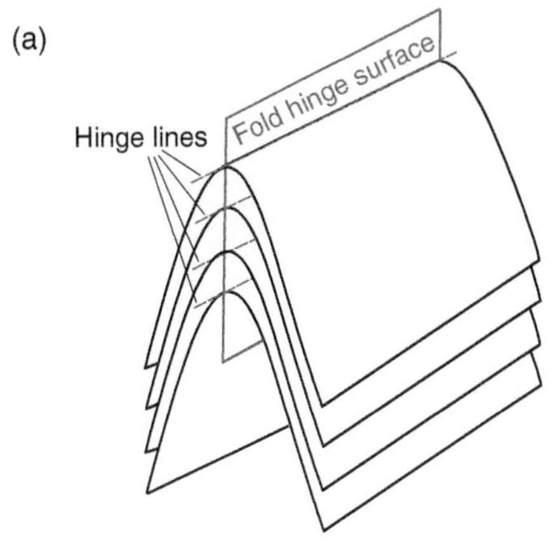

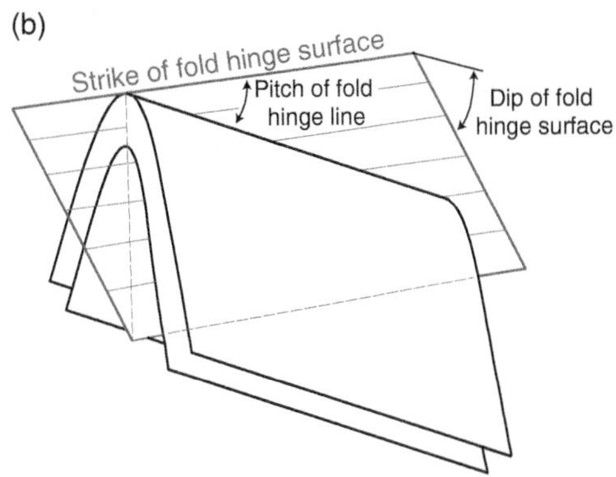

Figure 2.4 (a) Hinge lines drawn on adjacent folded surfaces together define the *fold hinge surface*. (b) To define the orientation of a fold, one must specify the strike and dip of the fold hinge surface and either the trend and plunge of the fold hinge line or its pitch.

comparing the original and final lengths of symmetric boudins gives a lower bound for the amount of elongation. Asymmetric boudins reflect the sense and magnitude of shear in the plane of boudinage. Differences in the competence of the boudinaged layer relative to its surrounding layers give rise to a variety of forms for boudinaged layers (Figure 2.8). Layers that are slightly to moderately more competent than the surrounding rock typically will develop a *pinch-and-swell structure*. Layers that are significantly more competent than the surrounding rock elongate by forming *boudins* with barrel-shaped or even rectangular cross sections. *Internal boudinage* is characterized by foliation surfaces that diverge and converge at regular intervals. Finally, any of these types of boudinage – pinch-and-swell, boudinage, or internal boudinage – may form in settings where layers are lengthened in two mutually perpendicular directions, giving rise to *chocolate-tablet boudinage*.

2.1.3 Fractures and Stylolites

Fractures are planar or tabular discontinuities within rock masses. Joints, extension fractures, and shear fractures (which are a type of small fault) are all examples of different types of fractures distinguished by the relative movement of the rock on either side of the fracture (Figure 2.9). A *joint* is a planar to sub-planar surface within a rock mass across which there is no cohesion and no visible offset or evidence for movement. Joints are often the locus of fluid flow through the rock. Those fluids sometimes alter the fracture walls, creating a tabular zone of rock whose character is distinct from the surrounding rock. *Extension fractures*, like joints, are generally planar discontinuities in a rock. In the case of an extension fracture, the rock masses separated by the discontinuity have moved away from each other, perpendicular to the fracture surface. Some extension fractures are filled by secondary minerals, creating a tabular volume of rock known as an *extension vein*. A *shear fracture* is a discontinuity surface within a rock across which there is displacement parallel to the surface. The movement of rock masses separated by fractures is not restricted to directions perpendicular or parallel to the fracture surface. Relative movements of opposing sides of a fracture with finite components parallel and perpendicular to the fracture surface generate *extensional shear fractures*.

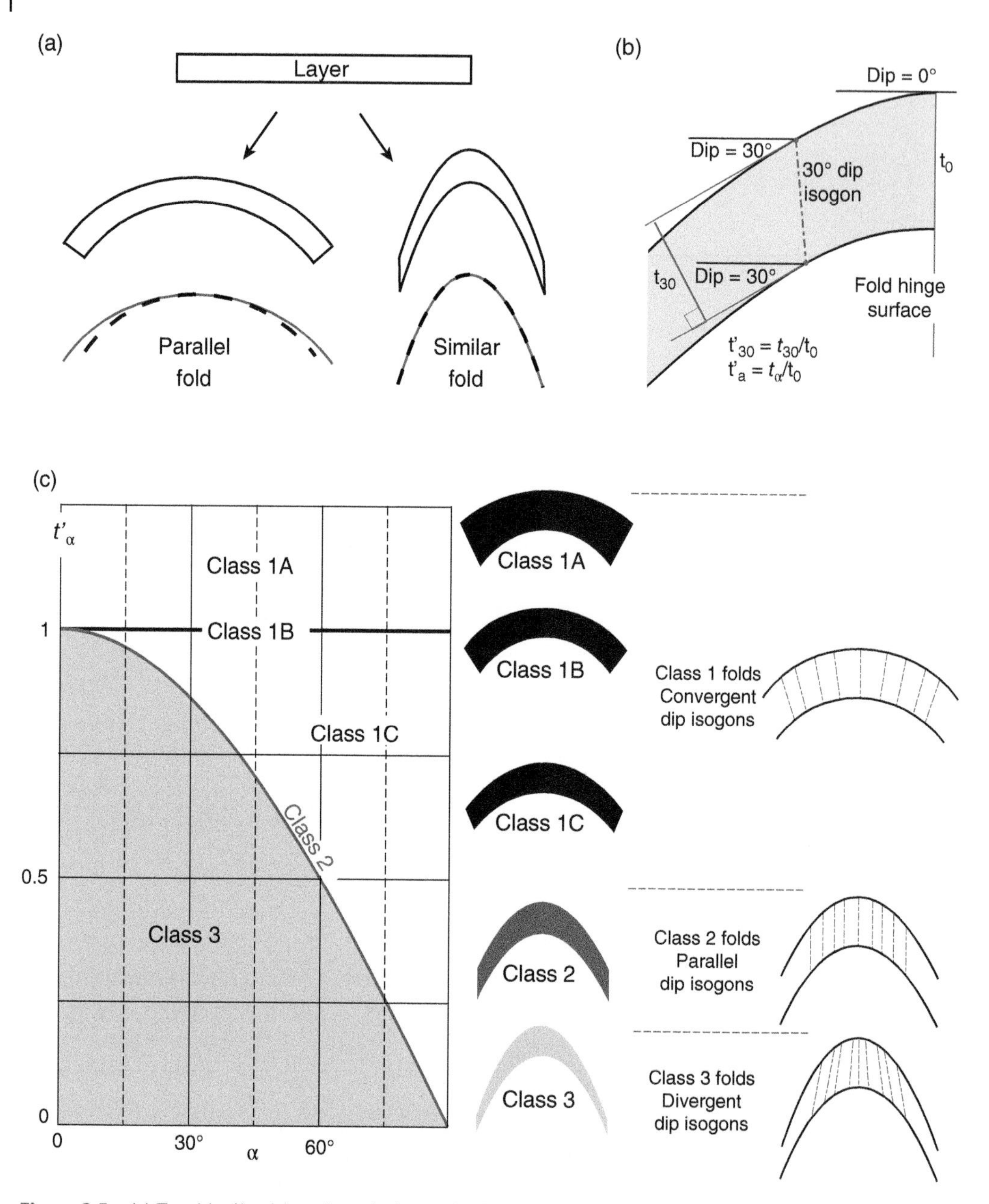

Figure 2.5 (a) Two idealized, benchmark shapes for folded layers. In *parallel folds*, the layer thickness is equal at all limb dips, although the inner (dashed black line) and outer arcs (red line) of the layer have different shapes. In *similar folds*, layer thickness varies with limb dip, but inner and outer arcs have identical shapes. (b) A *dip isogon* connects points with equal dip on the inner and outer arcs of a folded layer. t'_α gives the thickness of a layer at limb dip α compared to thickness at the hinge. (c) Fold shape classes defined on the basis of t'_α vs. α values and on the basis of dip isogon patterns. See text for explanation.

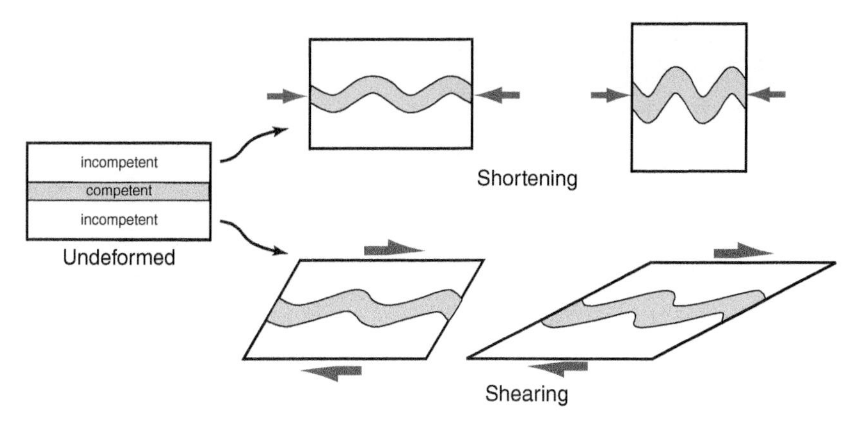

Figure 2.6 Layers of relatively competent rock embedded in less competent rock may form folds when the entire package is shortened parallel to the layering and elongated perpendicular to the general trend of the layering, or when sheared parallel to the general trend of the layering. Shortening is associated with symmetric folds, whereas shearing is associated with asymmetric folds.

Some geologists consider that *stylolites* and *slickolites* are also types of fractures across which the rock masses move toward each other. Stylolites and slickolites are planar to tabular discontinuities, and their displacement patterns share some characteristics with the displacement patterns associated with other extensional and shear fractures. For this reason, we have included drawings of them in Figure 2.9. The mechanisms by which stylolites and slickolites form are sufficiently different from those responsible for the formation of joints, extension fractures, and shear fractures that we consider stylolites as distinct from those types of fractures.

2.1.4 Faults and Fault Zones

Faults and *fault zones* are planar or tabular zones of discontinuity across which the rock masses have moved past each other parallel to the surface or zone boundaries (Figure 2.10a). Similar relative movement occurs across *shear zones*, described and discussed in detail in Section 2.1.5. Some faults are *slip surfaces*, i.e. discrete, outcrop-scale, individual surfaces that display distinctive surface markings and offset features in the wall rocks. In other cases, the displacement of one rock mass past another occurs within a localized, tabular region of discontinuous deformation – a *fault zone*. Rocks within the fault zone (*fault rocks*, see next) *and* rocks adjacent to the fault zone are commonly more deformed than rocks farther from the fault. Closely spaced joints or fractures, fractured rock, arrays of stylolites and/or veins, zones of alteration, structural fabrics, and folded layers often affect the rocks in proximity to a fault (the *wall rocks*) (Figure 2.10b). The zone of localized deformation associated with a fault is sometimes called a *damage zone*. Deformation within the damage zone may occur as the fault forms and/or after the fault is a through-going feature as the opposing walls move past each other.

Some geologists use the term *deformation zone* to refer to any tabular zone in which localized deformation results in relative movement of the wall rocks mainly parallel to the zone boundaries (Figure 2.10a). Thus, fault zones consisting of several anastomosing faults, shear fractures, and/or granulation zones are one type of deformation zone.

(a)

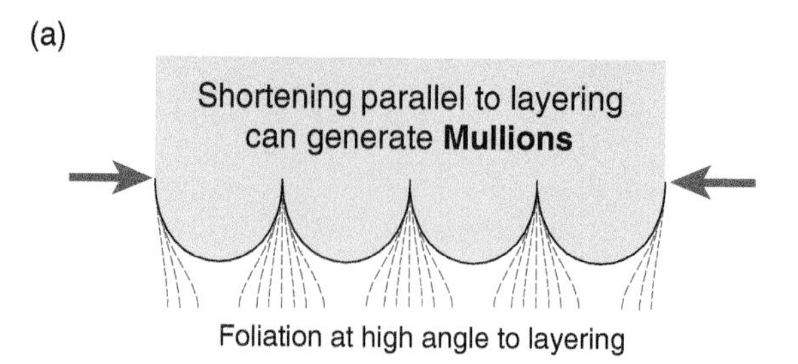

(b)

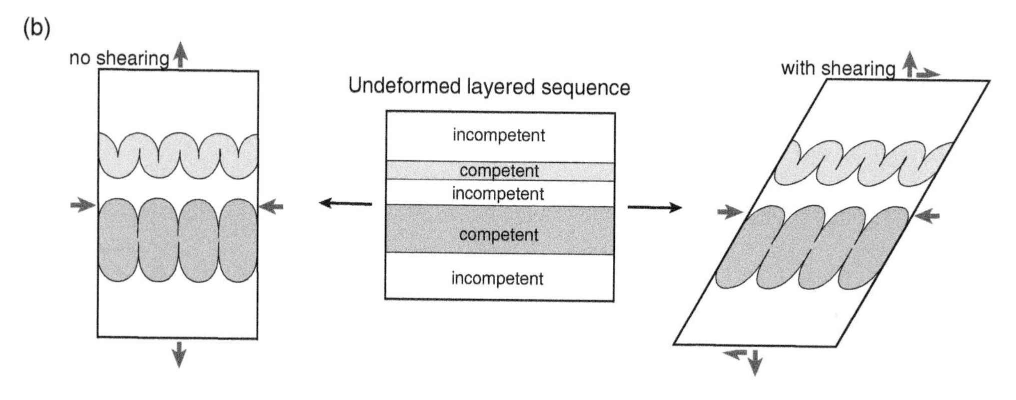

(c)

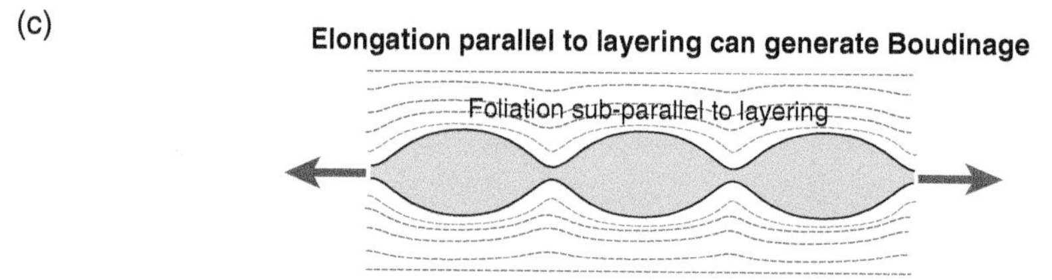

(d)

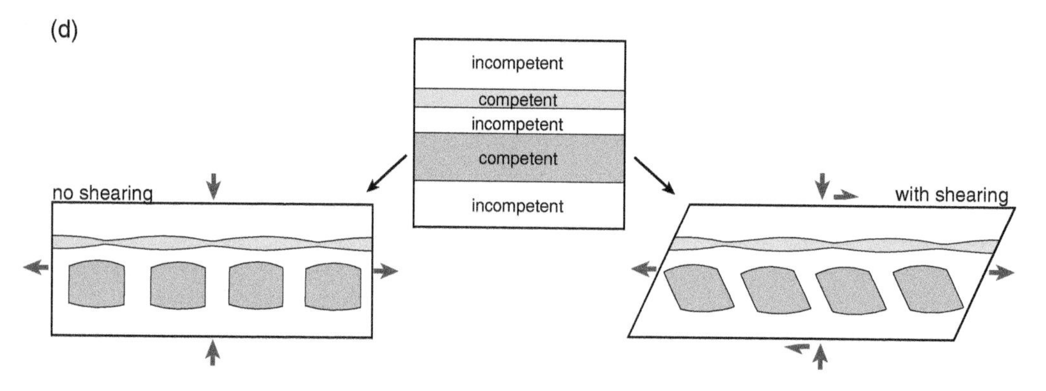

Figure 2.7 (a) Shortening parallel to the boundary separating a more competent material from a less competent material may deform the boundary in a series of cusps and lobes known as *mullions*. (b) The cusps of mullions developed on opposing boundaries of a layer may be offset, as in the light gray layer, or aligned, as in the dark gray layer. Concurrent shearing will give rise to asymmetric forms. (c) Shortening perpendicular to layering and elongation parallel to it regularly develops *boudinage*. (d) Elongated more competent layers may generate *pinch-and-swell structure*, as in the light gray layer, or a series of barrel-shaped boudins, as in the dark gray layer. Concurrent shearing will give rise to asymmetric forms.

Name	Description	Illustration
Pinch-and-swell structure	Boundaries of layers diverge and converge at regular intervals, producing linear regions where the layers are thinner (**pinches**) & thicker (**swells**)	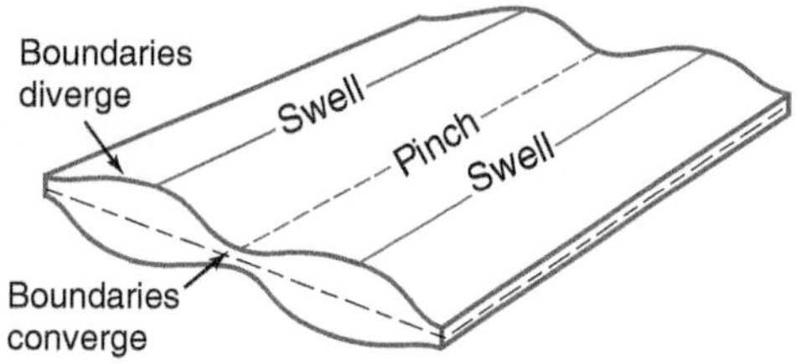 Boundaries diverge; Swell; Pinch; Swell; Boundaries converge
Boudinage	Two or more elongate bodies with rectangular, barrel, or elliptical shapes in cross-section (**boudins**) define a layer or segment of a layer	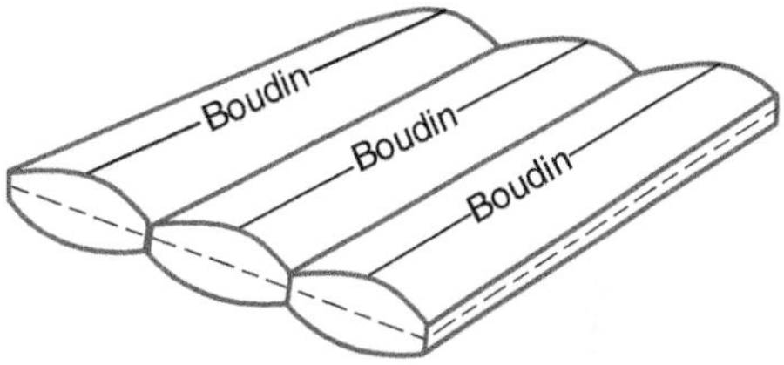 Boudin; Boudin; Boudin
Internal boudinage	Foliation surfaces regularly diverge and converge to define boudins within the foliation	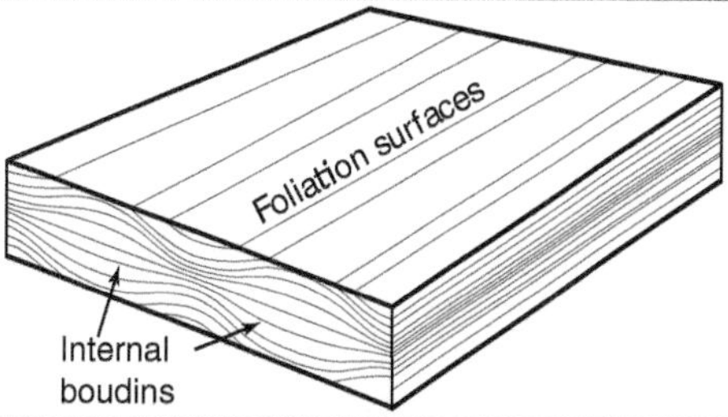 Foliation surfaces; Internal boudins
Chocolate-tablet boudinage	Numerous lens-or pillow-shaped bodies (**tablets**) define a layer or segment of a layer.	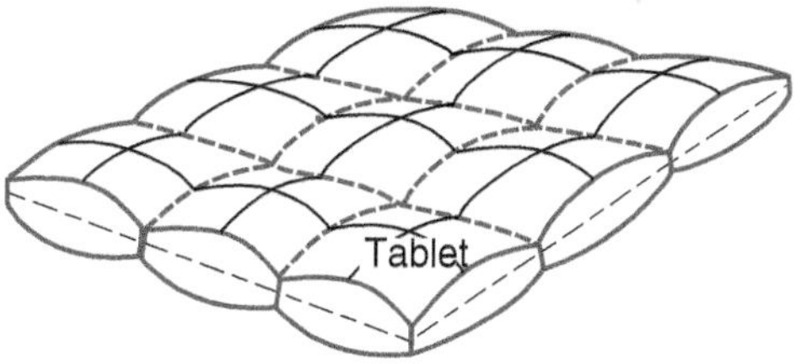 Tablet

Figure 2.8 Diagram illustrating the different types of boudinage.

In fault zones, the deformation is predominantly *discontinuous*, i.e. original features are cut by and displaced across discrete surfaces. Shear zones are deformation zones in which deformation is predominantly *continuous*, i.e. original features are stretched, shortened, or sheared but not cut by or offset across macroscopic fractures or slip surfaces. As shown in Figure 2.10a, fault zones and shear zones are in one sense end-member deformation zone types. Close examination of most deformation zones shows a combination of discontinuous and continuous deformations.

Geologists determine the relative movement (slip sense) on faults and across fault zones (their

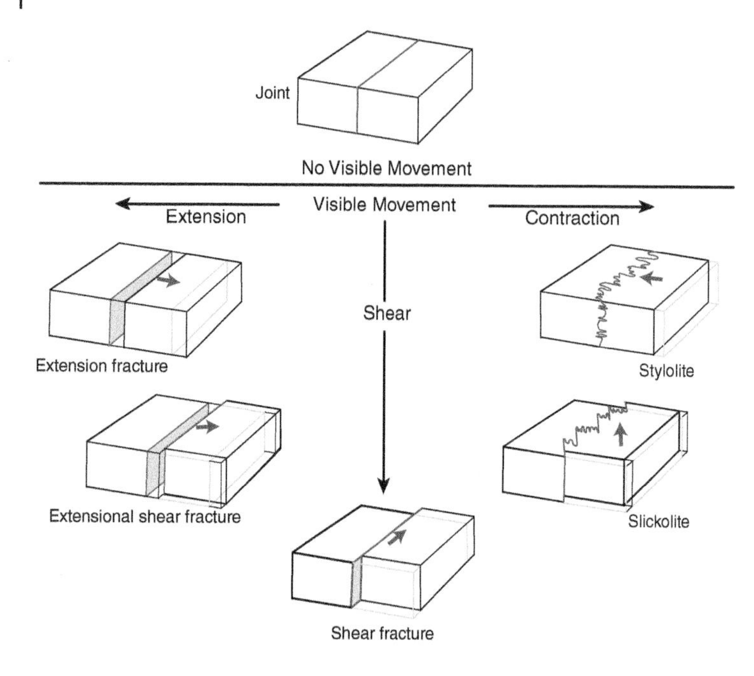

Figure 2.9 Different types of discontinuity surfaces (in red) cutting blocks of rock. At the top, a *joint* transects the block; there is no movement across the surface visible to the unaided eye. Diagrams beneath the line illustrate different types of relative movement of the blocks separated by the discontinuity surface. The walls of an *extension fracture* moved directly away from each other, whereas the walls of a *shear fracture* moved past each other, parallel to the discontinuity surface. The walls separated by an *extensional shear fracture* have finite movement components perpendicular and parallel to the fracture surface. Walls moved directly toward each other across a *stylolite* and obliquely toward each other across a *slickolite*.

slip sense) by: (1) observing offset markers; (2) recognizing characteristic geometries for the fractures, stylolites, and veins within the fault zone; and (3) examining the geometry of folds in wall rocks. Slip surfaces, whether they occur in isolation or within a fault zone, commonly are marked by: (1) *slickensides*, which are polished fault surfaces, and (2) *slickenlines*, which may be either striations, scratches, and grooves or coatings of mineral fibers, called *slickenfibers*, on the surface. Experimental data on and theoretical analyses of the origin of the different types of slickenlines indicate that they form parallel to the direction of relative movement on a fault. In many cases, the characteristics of slickenlines indicate the slip sense on a fault. In addition to slickensides and slickenlines, slip surfaces can display distinctive steps on surfaces that typically are orthogonal to fault movement. In some instances, these features, known as *congruous* or *incongruous steps* (Figure 2.11), also indicate slip sense. Thus, structures in the damage zone and fault-surface markings together contribute to understanding fault kinematics by indicating the direction of the relative motion of the wall rocks.

By combining information on offsets across faults, the geometry of fractures and folds in wall rocks, and the variety of surface markings on faults, geologists regularly assign faults to different classes depending upon their kinematics (Figure 2.12). *Dip slip faults* have fault slips parallel to the dip of the fault surface. Slip is *normal* or *reverse* depending on whether the hanging wall moved down or up relative to the footwall. *Strike slip faults* have fault slips parallel to the strike of the fault. Slip is *sinistral* (*left-lateral*) or *dextral* (*right-lateral*), depending upon whether once-continuous features are found to the right or to the left as one crosses the fault. *Oblique slip faults* have finite components of slip parallel to both fault dip *and* fault strike.

Distinctive rock types, known as *fault rocks*, occur within many fault zones. Fault rocks are altered versions of the sedimentary, igneous, or metamorphic rocks that compose the walls of

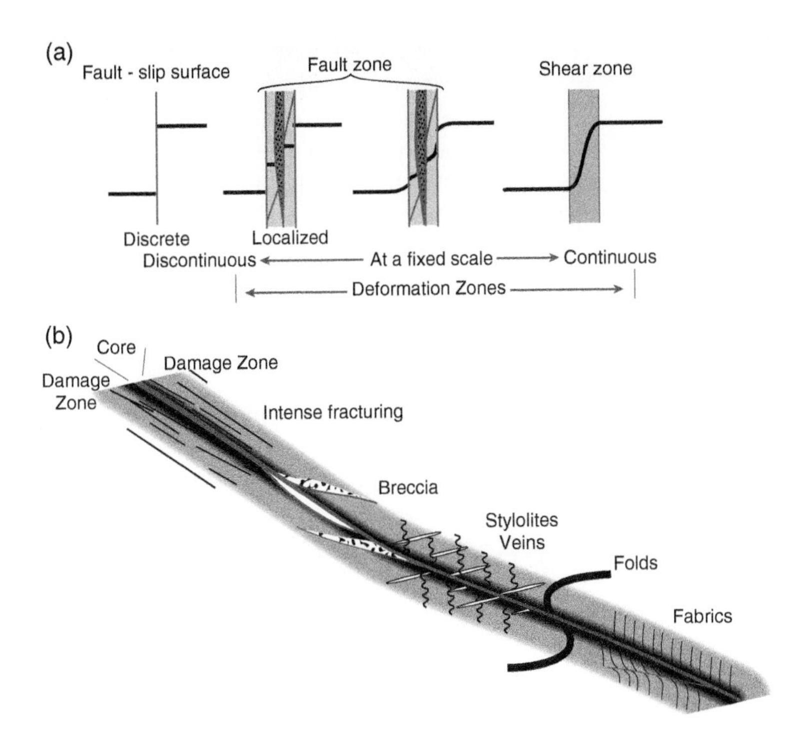

Figure 2.10 (a) A planar marker (in black) offset across different types of fault zones and deformation zones. At left, a *slip surface* is a discrete surface across which the marker is offset. More commonly, offset occurs across zones with finite thickness – *deformation zones*, depicted to the right. A *fault zone* is a localized zone of discontinuous deformation, and a *shear zone* is localized continuous deformation. Different types of deformation zones have different combinations of discontinuous and continuous deformation at a specified scale. (b) Features in wall rocks used to infer the existence of a fault include zones of intense fracturing, breccia, stylolites and veins, folds, and fabrics. These features are concentrated in the *damage zone* around faults.

the fault. In some cases, the alteration of fault rocks may be so complete that we cannot determine precisely the character of the *protolith* (starting material). More typically, fault rocks consist of angular to subangular fragments of the protolith, surrounded by finer-grained matrix and/or by infilled material precipitated from fluids. We call the fault rocks *gouge* if the aggregate of rock and mineral fragments found within a fault zone is *incohesive*, i.e. can be disaggregated in the hand or by a penknife. Gouges are often composed largely of clay-sized fragments and may be composed dominantly of phyllosilicate minerals. The phyllosilicate minerals in gouges may be preferentially aligned, imparting a foliation to the gouge. Clasts in gouges are generally small (<5 mm) and form small proportions of the rock. Fault rocks belonging to the *cataclasite* series (Figure 2.13) are cohesive rocks in which rock and mineral fracturing has played a dominant role in the initial stages of wall rock alteration. We assign different names to fault rocks in the cataclasite series based on the relative proportions of fine-grained matrix and coarser

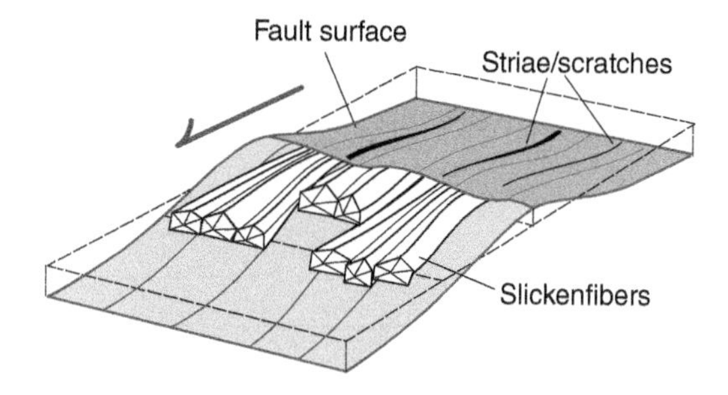

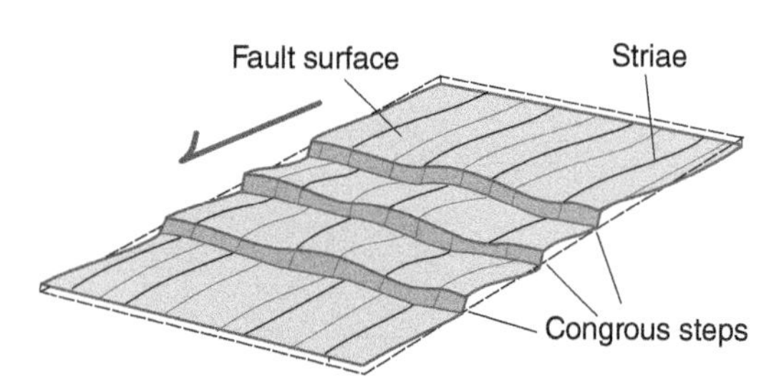

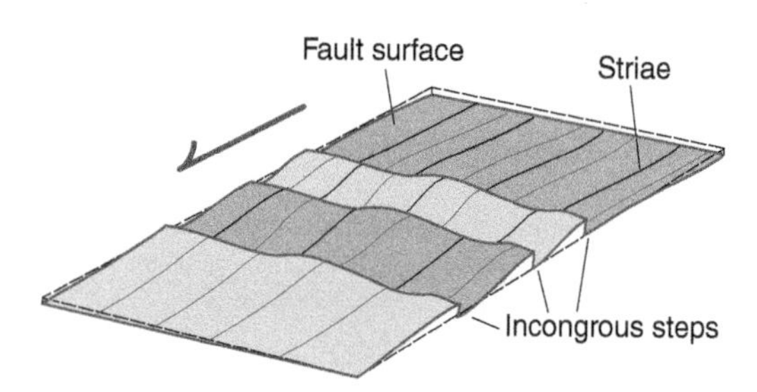

Figure 2.11 Fault surface markings include striae, scratches, slickenfibers, congruous steps, and incongruous steps.

fragments in the rock. If fault rocks have fragments greater than ~5 mm across, we call them *breccias* regardless of whether they are cohesive or incohesive.

Pseudotachylite is a fine-grained to very fine-grained rock with a dark matrix composed of glass or devitrified glass. Typically, pseudotachylite occurs along narrow planar fractures and also in triangular-shaped veins adjacent to the fractures, with the peak of the triangle pointing away from the planar fractures. Pseudotachylite veins often have a finer-grained, darker-colored rims, where the veins contact the adjacent rock. In many cases, slight color variations outline folds, swirls, and other structures within the veins. Pseudotachylite veins may contain vesicles, amygdales (amygdules in North America), and/or rounded fragments of the host rock.

2.1.5 Shear Zones

Shear zones are tabular regions possessing concentrations of one or more types of geological structure. As we outline further below, an individual shear zone may exhibit a distinctive, localized pattern of foliation, lineation, fractures, faults or fault zones, and folds or boudinage. These structures accommodate the displacement of a rock mass on one side of the tabular region relative to the rock mass on the opposite side of the tabular region.

Shear zones are regions: (1) with roughly parallel sides in which deformation intensity is greater than that in the surrounding rock mass, (2) across which rock on one side of the zone has moved relative to rock on the opposite side, and (3) in which deformation is mainly continuous at some defined scale (Figure 2.14a). In addition to shearing parallel to the zone boundaries, rock may be shortened or lengthened parallel to and normal to the boundaries. In order to maintain compatibility between rock within and outside of the shear zone, the rock outside of the shear zone may be distorted (Figure 2.14a). Figure 2.14a illustrates only situations with relatively simple wall rock deformation, i.e. there is no change in the lengths of lines perpendicular to the direction of shearing within the shear zone. This condition is not met in every shear zone. The tabular zone of shearing itself may be shortened in two directions as it thickens, shortened in one direction and extended in another as it thickens or thins, and extended in two directions as it thins.

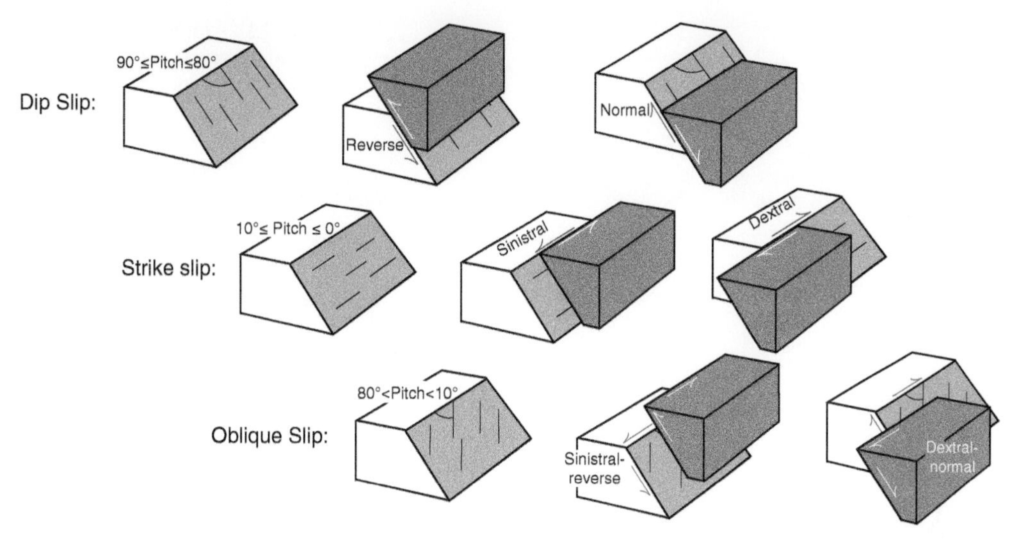

Figure 2.12 Diagrams indicating the kinematics of dip slip, strike slip, and oblique slip faults.

Figure 2.13 The main types of fault rocks.

	Gouge	**Cataclasite series**	**Breccia**
Character	Incohesive	Cohesive	Incohesive/cohesive
Composition	Clays commonly abundant	Matrix % Proto-cataclasite 50% Cataclasite 90% Ultra-cataclasite	Fragments > 5 mm
		Pseudotachylyte: glass/devitrified glass matrix	

Viewed perpendicular to the shearing direction, an idealized shear zone will have a gradient in deformation across the zone and no gradients in deformation parallel to the zone (Figure 2.13b-1). A variety of structures may accommodate the shearing deformation within the zone. In unmetamorphosed to weakly metamorphosed rocks, shearing sometimes is accommodated by arrays of stylolites and veins (Figure. 2.14b-2). In rocks deforming at conditions under which mineral grains deform or new minerals grow, shearing is apparent in deflected *S surfaces* or foliation (Figure 2.14b-3). In many instances, foliation formation in the shear zone is accompanied by the development of shearing surfaces oriented roughly parallel to the boundaries of the shear zone (Figure 2.14b-4). The shearing surfaces are known as *C surfaces* (where the *C* refers to *cisaillement; cisaille* is a French word that refers to the shearing or clipping of metal and the surfaces shear or clip the foliation surfaces). In some shear zones, another set of shearing surfaces, known as *C' surfaces*, develop (Figure 2.14b-5). Distinctive microstructures that are useful in determining the sense of shearing within the zone are described in Chapter 3.

There are distinctive types of rocks typically associated with shear zones. These rocks are generally analogous to the rocks associated with faults

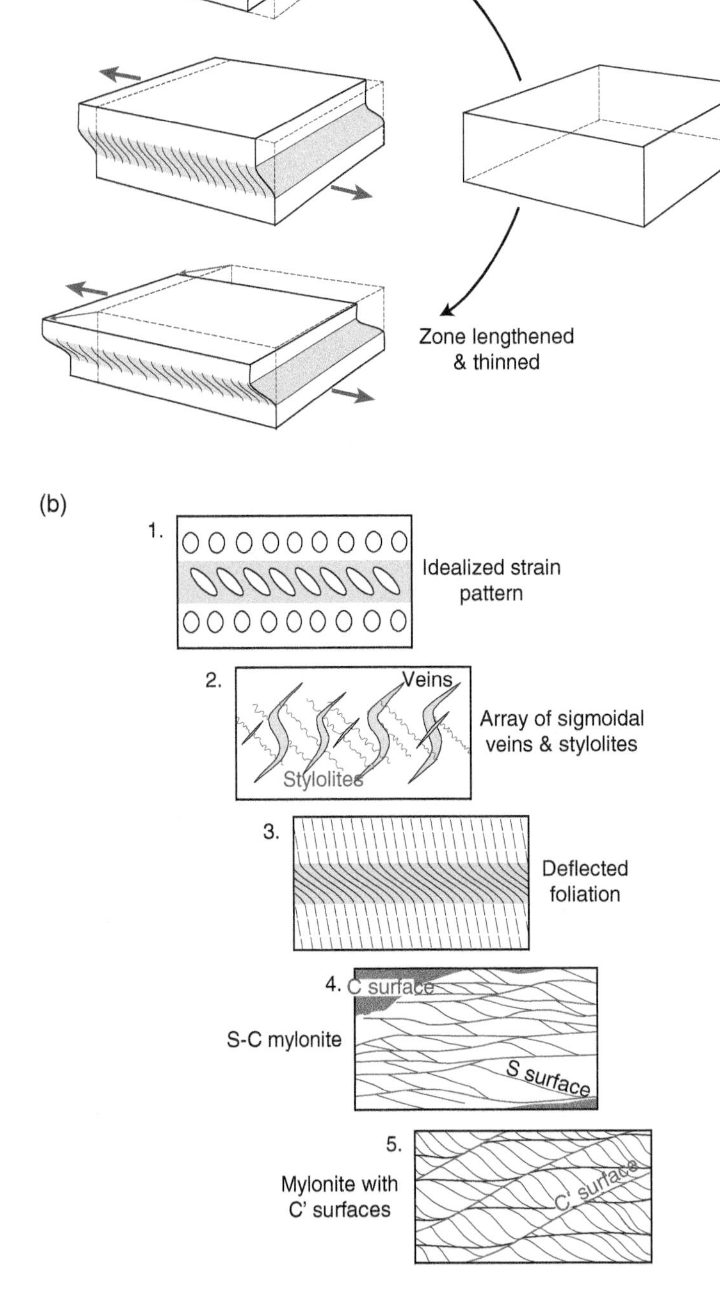

Figure 2.14 (a) Shear zones accommodate the displacement of one mass of rock past another by deformation localized within a tabular zone. Shearing may be accompanied by shortening parallel to the length of the zone with concomitant thickening of the shear zone, or may be accompanied by extension parallel to the length of the zone with concomitant thinning of the shear zone. (b) 1. The idealized deformation pattern in a top to left shear zone, viewed perpendicular to the shearing direction. 2. Shearing may be accommodated by arrays of stylolites and veins. 3. Deflected foliation surfaces (*S* surfaces) across a shear zone. 4. *C*-surfaces or shearing surfaces in a shear zone. 5. A second set of shearing surfaces, *C′*-surfaces, in a complex shear zone.

and fault zones, and like fault rocks, these rocks are found in tabular zones exhibiting evidence of localized shearing, that is, shear zones. *Mylonites* are cohesive, foliated rocks consisting of between 50 and 90% of mineral grains that are too fine to be seen without the aid of a hand lens. Mylonites often also possess a prominent lineation. The Greek root for the term mylonite is *mylos*, meaning "to mill." This terminology is a reflection that these rocks were originally thought to be formed by "grinding" a parent rock to generate fine grain size; we now know that mylonites are a product of recrystallization, not granulation. The visible mineral grains in mylonites may themselves contribute to the foliation and lineation in the rock, or those mineral grains may be relatively equant. Two subclasses of mylonitic rock types are distinguished on the basis of the relative volume percentage of fine mineral grains. More than 50% of a *protomylonite* consists of visible, identifiable mineral grains, and less than 10% of an *ultramylonite* consists of visible mineral grains. Some zones of localized shearing possess a mixture of non-foliated rocks from the cataclasite series and rocks from the mylonite series. In some instances, pseudotachylites also occur in mylonitic rocks, with or without rocks from the cataclasite series.

2.2 Additional Considerations

The different products of deformation typically occur together. For example, slip on a fault often generates folds or fabrics in rocks adjacent to the fault. Thus, a thorough examination of rock deformation requires an understanding of the characteristics of geological structures as well as an understanding of how deformation fabrics develop. *Crosscutting* deformation occurs when one feature (e.g. a fault) spatially cuts after another deformational feature (e.g. fold). These spatial relations are typically interpreted as a temporal development (e.g. the fault formed after the fold). *Overprinting* deformation is used when a younger deformation, often a fabric, disrupts or obliterates the geological structures of an earlier deformation.

In orogenic belts, it is also common to see more than one type of structure in rocks because they have been deformed more than once. As examples, either folds may be cut by faults or shear zones may be folded. Relations between structures, such as crosscutting or overprinting deformation, allow the sequence of events to be established. Each generation of planar or linear structures is labeled by its relative timing: thus S_1 and L_1 will have formed in the first deformation D_1 and be overprinted by S_2, L_2, etc. Folds formed in the D_1 will be F_1.

The types of structures that form in rocks are affected by three main factors: (1) the rock type, (2) the nature of the deformation (contraction, elongation, etc.), and (3) the stress or strain rate, temperature, and fluid content of the rock at the time of deformation. The effects of stress/strain rate, temperature, and fluid content are more difficult to work out. Characterizing their effects is an important theme of the book, which we address in the final three chapters.

3

Microstructures

3.1 Introduction

3.1.1 Overview

Microstructures are features of rocks that are visible at a microscopic scale. Most mesoscopic structures, including fractures, faults, stylolites, veins, folds, and boudinage, have microscopic counterparts. These are sometimes referred to with micro- as a prefix (e.g. microfault). However, we omit the prefix in this chapter because all the features described are microscopic. Other microstructures described in the chapter are visible only under the microscope. These include **undulatory extinction**, **deformation bands**, **deformation lamellae**, **subgrains**, **grain boundary bulges**, **deformation twins**, and **crystallographic preferred orientations (CPOs)**.

The study of structural geology on a microscopic scale began in the mid-nineteenth century when the technique of making thin sections of rocks was invented by H. C. Sorby. Note, however, that Robert Hooke used a microscope to examine fossils in the late seventeenth century. Microstructural studies constitute a large part of structural geology today. Any comprehensive structural study should include microstructural analysis because it can reveal aspects of a deformation history that are not evident in studies conducted on larger scales. For example, microstructures can give quantitative information about temperature, stress, and strain rate during deformation. Using microstructures to understand deformation conditions is a major application of microstructural research.

More fundamentally, microstructures can provide evidence in naturally deformed rocks that allows the operative deformation mechanisms to be identified. This link to deformation mechanisms is explored in Chapter 8. Knowledge of the deformation mechanisms, in turn, informs about the rheology during deformation, a topic covered in Chapter 7. The link between microstructures, deformation mechanisms, and rheology is established by comparing natural microstructures with those produced in experiments in the laboratory under controlled conditions.

This chapter presents an objective vocabulary and classification for microstructures. This microstructure vocabulary is necessary to link experiments and naturally deformed rocks correctly. Our understanding of microstructures owes a great deal to disciplines such as materials science that are outside mainstream structural geology. This situation has given rise to some confusion about microstructural terminology, and, consequently, there is a tendency to describe microstructures by inferred processes. It may also be for this reason that there is no generally accepted way to classify microstructures, which is a daunting barrier to the student beginning to learn about them.

An Integrated Framework for Structural Geology: Kinematics, Dynamics, and Rheology of Deformed Rocks,
First Edition. Steven Wojtal, Tom Blenkinsop, and Basil Tikoff.

Most microstructural studies made with microscopes or microprobes use a *thin section* of the rock. An important aspect of this method of study is that an individual thin section is a two-dimensional slice through a rock. The geometry of a structure expressed in one particular thin section may be radically different from a thin section with a different orientation from the same rock. For this reason, we advocate orienting samples for microstructural studies and cutting several thin sections in different orientations for any sample. Samples can be cut in a geographic reference frame, a fabric reference frame, or both (see Section 3.13). Three mutually perpendicular sections are often useful. This means that samples collected for microstructural studies may need to be somewhat larger than usual. Having thin sections of multiple orientations from samples partially overcomes the inherent limitation of the two-dimensional nature of thin sections.

3.1.2 Framework

At the microscopic scale, the geometry and other properties of grain boundaries have a fundamental effect on rock deformation. We, therefore, distinguish between microstructures that occur within grains (**intragrain**), those that include boundaries between several adjacent grains (**intergrain**), and those that involve many grains (**multigrain**) (Figure 3.1). In the context of holding the rate of deformation, fluid pressure, and rock-type constant, Figure 3.1 can also be read approximately as recording a transition from microstructures seen at low to high temperature from left to right. Figure 3.1 is, therefore, an outline of a classification scheme for microstructures. The organization of the subsections of this chapter follows this diagram. In general, we discuss microstructures as they occur at increasing metamorphic grades (increasingly high-temperature conditions). Thus, the first section (fractures) discusses features that typically occur in the upper crust while later sections (e.g. porphyroblasts) discuss features that need relatively high-temperature conditions. The wedges across the bottom of Figure 3.1 indicate the approximate relative importance of three basic classes of deformation mechanism in the origin of the microstructures.

The last two sections of the chapter are different. The first one addresses a group of distinctive microstructures, all characterized by asymmetry, that can occur over a wide range of temperature conditions. These **shear sense indicators** are particularly useful for determining the sense of shearing (i.e. top to right or top to left). The last section (Section 3.13) talks about how to uniquely orient your thin section in three-dimensional space.

3.1.3 Imaging of Microstructures

This chapter focuses mainly on petrographic (optical) observations, aiming to provide a diagnostic guide to the identification of the most common types of microstructure. Microstructural observations using the petrographic microscope, however, are complemented by observations from scanning electron microscopes (SEMs), transmission electron microscopes (TEMs), the electron microprobe, X-ray and neutron diffractometers, and computer-assisted tomography of minerals and rock specimens. The SEM is a particularly versatile tool that is increasingly essential in microstructural studies. Scanning electron microscopy can provide exceptionally detailed images of surfaces, mineral compositions by secondary electron imaging, details of mineral chemistry by wavelength and energy dispersive spectroscopy and cathodoluminescence (CL), and CPOs by electron backscatter diffraction (EBSD).

CL provides an image of the light emitted when the sample is energized by an electron beam. The light emitted under the electron beam is sensitive to mineralogy (including mineral chemistry) and the thermal/mechanical history of the sample.

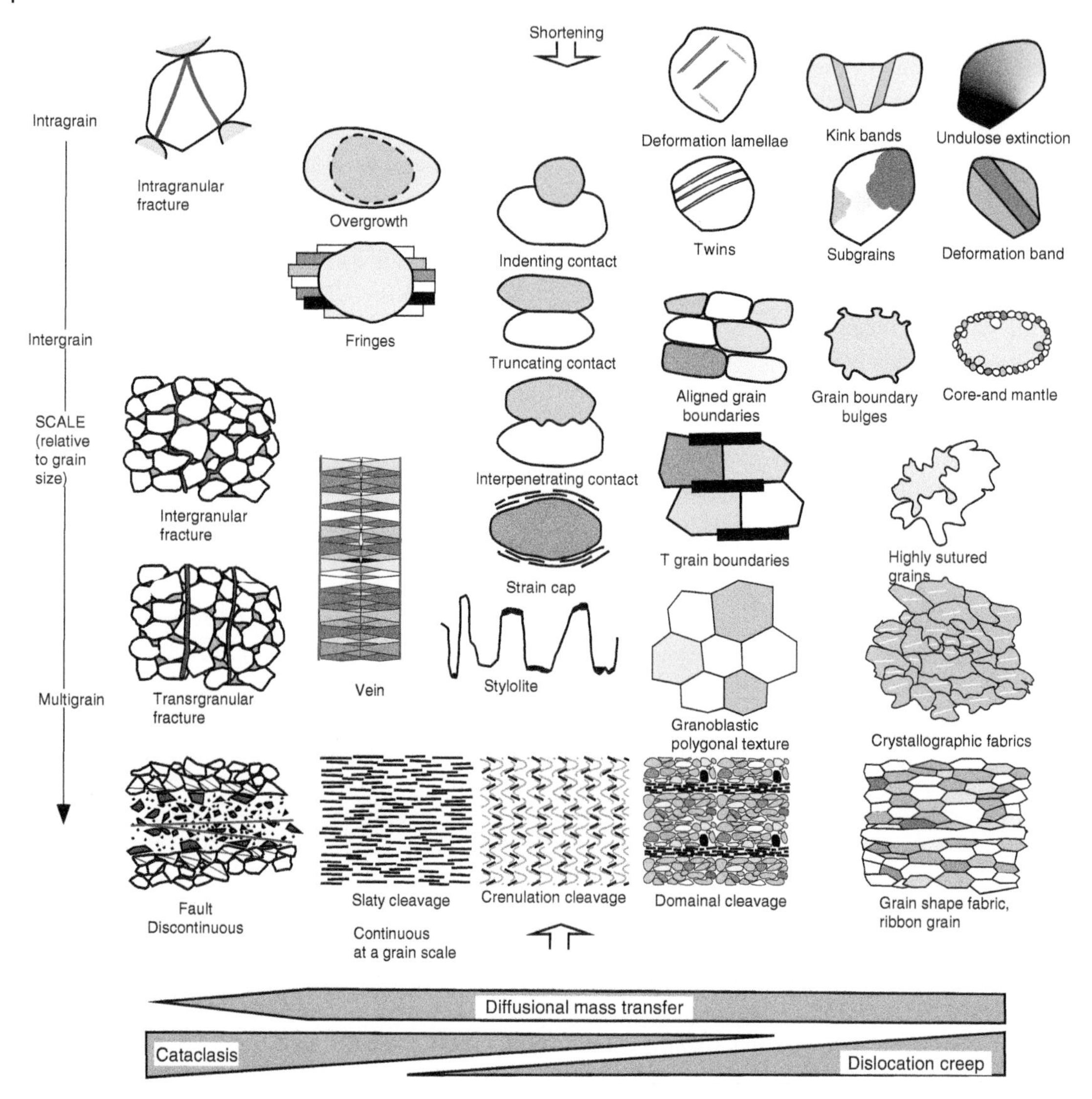

Figure 3.1 A framework for analyzing secondary microstructures. Microstructures within a grain (intragrain) are at the top, microstructures between adjacent grains (intergrain) are in the center, and microstructures involving many grains (multigrain) are at the bottom. Red lines indicate fractures. Microstructures are drawn in an orientation that shows vertical shortening. This diagram excludes asymmetric microstructures used for shear sense indicators, which are shown in Figure 3.24.

3.2 Fractures

Fractures are one of the most common microstructures in rocks of the upper crust. They are commonly referred to as cracks (or microcracks). Fractures contain a wealth of tectonic information: (1) fracture geometry constrains the orientation and magnitudes of stresses, and (2) the phase relations of the inclusion contents provide information on the conditions of fracture formation.

Fractures are also significant because they: (1) provide evidence for deformation events that may not be readily discernible on larger scales, and (2) have a large influence on the physical properties of rocks. For example, the presence of aligned fractures causes seismic waves to travel at different velocities in different directions within rocks of the upper crust, an effect known as seismic anisotropy. Consequently, the directions of seismic anisotropy can be used to assess the orientations of the fractures.

Fractures may be seen in thin sections as lines or thin planes that fragment grains (Figure 3.2a). Under close examination, linear arrays of inclusions occur along many fractures. The inclusions often contain fluids; these **fluid inclusions** are small pockets of liquid with vapor or solids, which were included when the fracture was cemented. Aligned collections of fluid inclusions are called **fluid inclusion planes** (**FIPs**; Figure 3.2a). Many fractures have fillings or cements, which may be optically continuous with the grains,

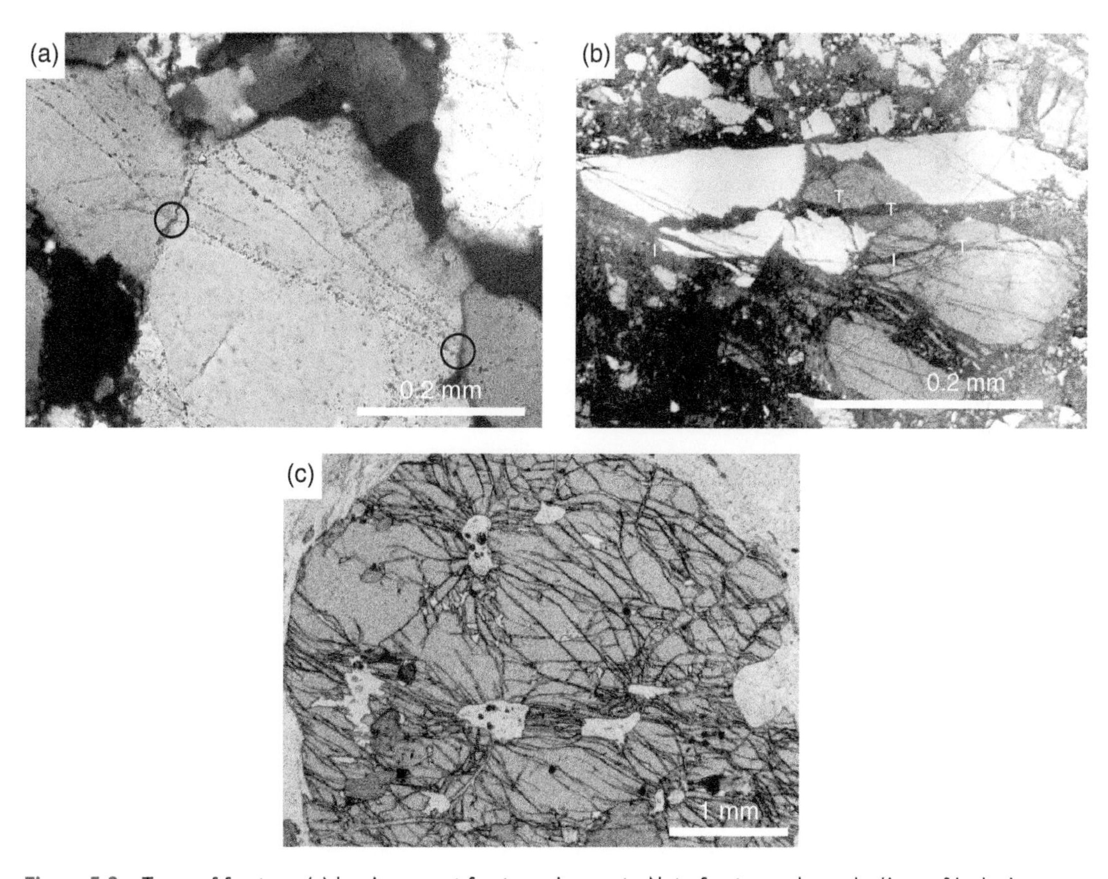

Figure 3.2 Types of fracture. (a) Impingement fractures in quartz. Note fractures shown by lines of inclusions connecting contact points (in circles) between quartz grains. These fractures are marked by fluid inclusions. XPL: North Spain. (b) Fractures in quartz revealed by cathodoluminescence. Intragranular and transgranular fractures (I and T, respectively) show as non-luminescing lines. CL: Isle of Skye, Scotland. (c) Inclusion microfractures in garnet (medium gray) around inclusions of coesite (light gray) inverted to quartz. PPL: Dora Meira, Italy. XPL, cross-polarized light; CL, cathodoluminescence; PPL, plane-polarized light.

rendering the fractures hard to see. Such fillings can be revealed by CL, which reveals fractures with fillings (Figure 3.2b).

Fractures are shown on the left-hand side of Figure 3.1 by red lines. They occur on all scales, and it is important to note whether fractures are confined to single grains (*intragranular*), occur around grain boundaries (*intergranular or circumgranular*), or cross two or more grains (*transgranular*) (Figures 3.1 and 3.2b). These distinctions have significance in determining the mechanisms that caused fracturing. Intragranular fractures commonly connect points of contact between grains, and may, therefore, have a range of orientations. Transgranular fractures, which cross several grains, are commonly oriented in one or more *sets* consisting of parallel to subparallel fractures (Figure 3.2b). The true orientation of the fractures in three dimensions can be ascertained through measurements on the universal stage or by combining measurements from two or more thin sections in different orientations, preferably as nearly perpendicular to each other to improve accuracy.

Table 3.1 outlines associations that have significance for the origin of fractures; those associations are useful in classifying the observed features of fractures.

Not all these categories are exclusive: for example, Figure 3.2c shows an inclusion-related fracture that is also a transformation-related fracture.

3.3 Fault Rocks

Fault rocks (see Section 2.1.4) are distinctive types of deformed rocks found within fault zones that often show spectacular microstructures. For example, gouges commonly have a foliation

Table 3.1 Types of fractures and their identifying features.

Type	Feature	Origin
Impingement	Intragranular; connects contact points between grains	Grain contacts increase local stresses
Flaw-related	Intragranular; joined to flaws such as grain boundaries, pores	Flaws increase local stresses
Cleavage	Intragranular; along cleavage planes	Cleavage planes are inherently weak
Dislocation-related	Intragranular; on subgrain or kink boundaries	Dislocations increase stress
Fault-related	All types; adjacent to a fault	Stresses can be concentrated around faults during fault propagation or displacement
Inclusion-related	Intragranular; confined to one mineral enclosed by different minerals, or surrounding or linking inclusions of one mineral in another	Stresses build up around inclusions because they respond differently to applied stress or temperature than their host
Transformation-related	Intragranular and transgranular; surrounding phases that have transformed to other phases	Transformations increase stress due to changes in volume

defined by the phyllosilicates or clays and regularly oriented shear surfaces known as Riedel shears (Figure 3.3), which can be used to determine slip sense as outlined in Section 3.12. Cataclasites and ultracataclasites are distinctive in thin section (Figure 3.4). The clasts of cataclasites are commonly angular with a range of grain sizes. They may have a very high density of microfracturing, accompanied by hydrothermal alteration and cementation. Ultracataclasites (Figure 3.4c) by definition have less than 10% fragments; they generally require electron microscopy

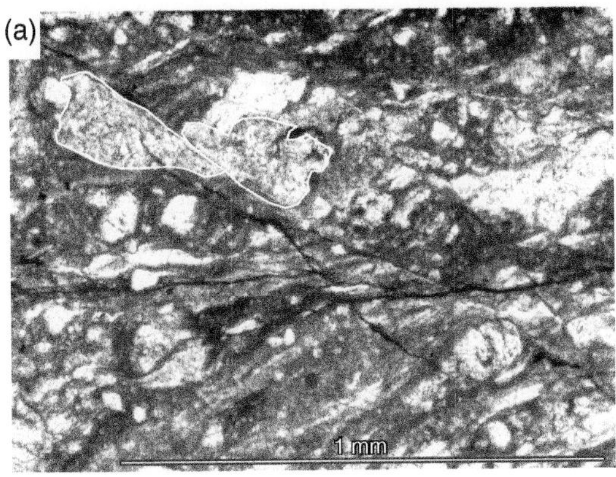

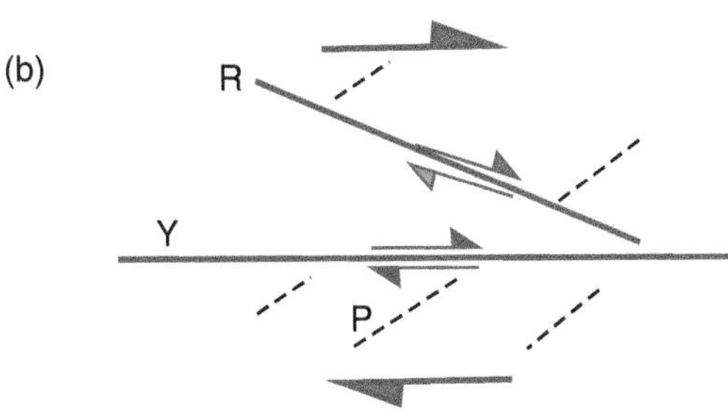

Figure 3.3 Gouge zone microstructures in a dextral gouge zone. (a) Photomicrograph. (b) Interpretation. The Y shear is in the center of the gouge zone and parallel to its margins. The Riedel shear (R) is inclined to the Y shear. It displaces a grain (outlined in white in (a)) in a dextral sense. The P foliation is inclined to the Y shear in the opposite direction and forms a penetrative foliation. PPL: Mount Isa, Australia. PPL, plane-polarized light.

to distinguish the components of the fine-grained material, which may be precipitated cements or finely crushed protolith.

Many fault surfaces are coated with fibrous minerals (crystals that are long in comparison to their width) that have grown on the fault surface in the direction of fault slip through precipitation from solution. Such *slickenfibers* are commonly made of quartz, calcite, chlorite, serpentine, or amphiboles, depending on host rock mineralogy, solution chemistry, and physical conditions of growth. Slickenfibers have similar microstructures to vein-filling fibers described in the next section.

Faults are rarely filled with pseudotachylites, occurring within narrow planar zones and triangular-shaped veins that join the planar zones. Pseudotachylites commonly consist of angular to rounded fragments surrounded by glass (isotropic and commonly colored orange, brown, or green) or devitrified glass (also commonly colored but with very low birefringence) (Figure 3.5). The glass may show devitrification textures such as spherulites (radiating bunches of crystal fibers) or perlitic texture (curved fractures), and evidence for flow in the form of folded banding (Figure 3.5a) or size sorting of fragments (Figure 3.5b). The edges of pseudotachylite veins may be embayed where they are adjacent to phyllosilicate minerals in the host rock.

Such pseudotachylite forms by: (1) melting rock immediately adjacent to the fault surface due to frictional heating associated with seismic slip, and (2) sudden chilling of the melt layer. Chilling results in the formation of glass (also evidenced by subsequent devitrification textures) and preferential melting of minerals such as phyllosilicates, resulting in embayments on vein margins. Because pseudotachylites are very fine-grained, they may be confused with a number of other rocks, including veins of hydrothermal minerals such as tourmaline and chlorite, fine-grained igneous dikes or ultracataclasite.

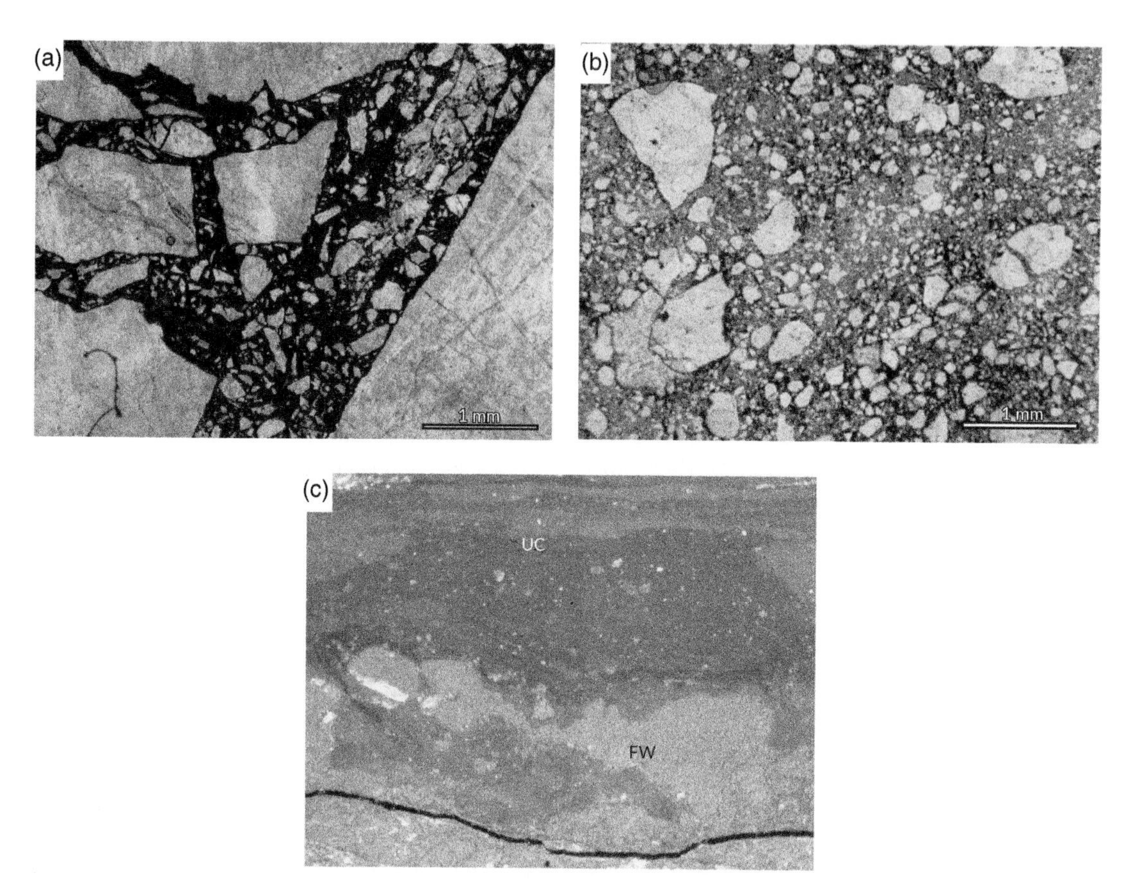

Figure 3.4 Cataclasites and ultracataclasites (a) Cataclasite. Angular fragments with a range of sizes derived from the quartzite wall rocks, surrounded by an infill of opaque hematite. PPL: Mount Isa, Australia. (b) Cataclasite. Note typical large range of poorly sorted, angular to subangular clasts in a dark matrix. PPL: Mount Isa, Australia. (c) Ultracataclasite. Fragments of detrital quartz, carbonate, and shale, some cut by calcite veins, in a matrix of very fine calcite. UC – mainly carbonate ultracataclasite, FW – shale in the footwall to the fault zone. XPL: Copper Creek thrust, Tennessee, USA. PPL, plane-polarized light; XPL, cross-polarized light.

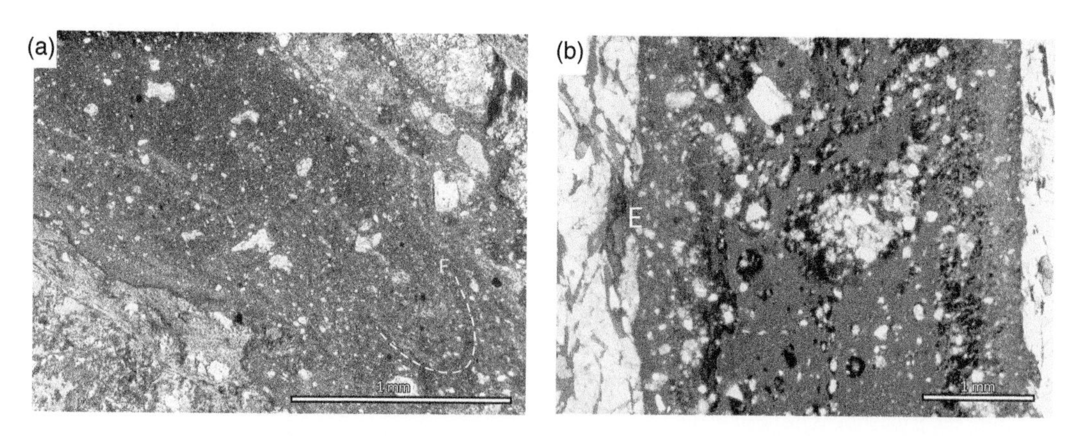

Figure 3.5 Pseudotachylite. Both views of narrow pseudotachylite veins show subrounded wall rock fragments (light in color) surrounded by opaque glass. (a) Flow banding (dashed white line labeled F) defined by slight contrasts in tone. PPL: Limpopo Belt, Zimbabwe. (b) E indicates an embayment on the edge of the vein where biotite in the host rock has melted. Larger fragments have concentrated in the vein center, and there appears to be a distinct margin (possibly due to chilling) on both sides of the vein. PPL: Zambezi Belt, Zimbabwe. PPL, plane-polarized light.

3.4 Overgrowths, Pressure Shadows and Fringes, and Veins

Overgrowths, pressure shadows and fringes, and veins are formed by minerals precipitated out of solutions, and they are quite common in rocks of relatively low metamorphic grade. These microstructures range from intergrain to multigrain in scale (Figure 3.1, left center). **Overgrowths** are rims of the same mineral composition as the grain that they surround (Figure 3.1). Overgrowths are commonly in crystallographic continuity with their core, i.e. the composition and lattice orientation of the overgrowth match those of the host. For this reason, it is sometimes difficult to distinguish overgrowths from the core. In some cases, the boundary of the host grain is marked by a line of small inclusions (Figure 3.6a). Overgrowths commonly have a slightly different chemical composition and so exhibit different luminescence from their host grains, allowing them to be distinguished well by CL. Pore spaces around grains can be completely occluded by material precipitated around grains, producing an **infill** or **cement**, which does not have to be the same composition as the grains (Figure 3.6b).

Mineral precipitates restricted to particular domains around grains or larger objects are **pressure shadows** or **pressure fringes** if they have fibrous habits (Figures 3.1 and 3.7). Fibrous pressure fringes of mica are sometimes called mica beards (Figure 3.7b). Important descriptive aspects of pressure shadows and fringes are: (1) their mineralogy, (2) the orientation of the whole shadow/fringe with respect to the surface of the object, (3) the orientations of fibers with respect to the surface of the object, and (4) the width of fibers and how they change with distance from the object. These are all critical features in distinguishing how the pressure shadows/fringes have grown, as discussed later.

Veins have similarities to overgrowths and pressure shadows/fringes because they consist of minerals that have grown (by either infill or replacement) in a host rock (Figure 3.1). Details of vein fillings, which are important for understanding their kinematic significance, are commonly best observed microscopically. Vein fillings can be classified according to Table 3.2, and some of these are illustrated in Figure 3.8. We distinguish between massive, equant or blocky, euhedral, laminated,

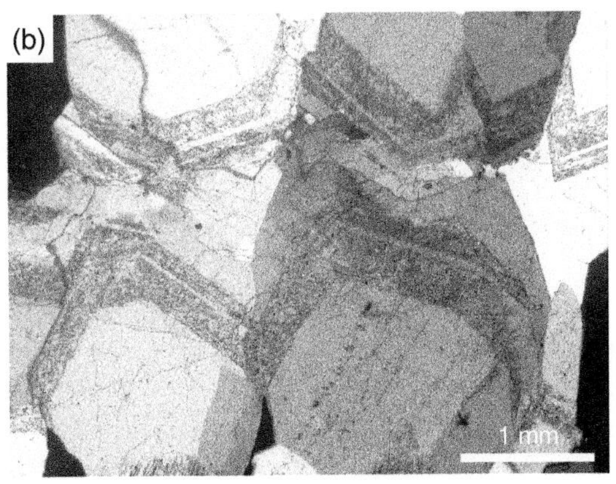

Figure 3.6 Overgrowths and infill. (a) Quartz overgrowths rendered visible by grain rim inclusions. XPL: North Spain. (b) Quartz infill around prismatic quartz crystal shapes defined by lines of primary fluid inclusions. XPL: Mount Isa, Australia. XPL, cross-polarized light.

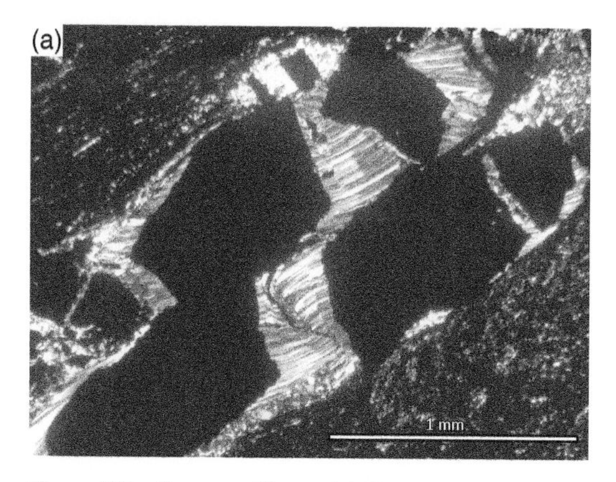

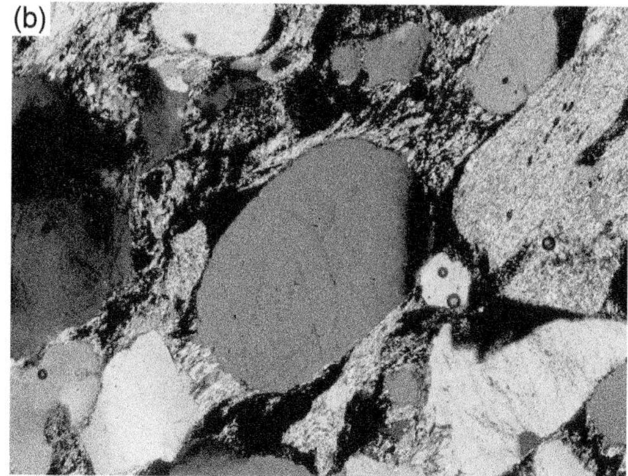

Figure 3.7 Pressure fringes. (a) Fine intergrowths of quartz and chlorite around rectangular opaque pyrite. XPL: Dalny mine, Zimbabwe. (b) Mica beards at the ends of the central quartz grain. XPL: Blue Ridge Province, Maryland, USA. In both cases, lengthening can be inferred in the upper right-lower left direction. XPL, cross-polarized light.

Table 3.2 Vein-filling textures.

Texture	Description	Interpretation
Massive	No particular structure	Single-stage filling
Equant/blocky	Crystals with equant/square shapes	Single-stage filling
Euhedral/idiomorphic	Crystals with typical habits and terminations, e.g. prismatic quartz	Growth into open space
Laminated	Planar bands subparallel to margins	Incremental growth ± shear
Fibrous	Elongated crystals	Incremental growth
Brecciated	Fragments surrounded by finer-grained matrix	Shear

fibrous, and brecciated vein fillings. Table 3.2 summarizes the differences between these types, and their interpretations, which are elaborated in Chapter 8.

In the context of vein fillings, **fibers** are interpreted to indicate that vein filling occurred progressively as the vein opened, so that the vein was mostly filled by the growing fibers as it widened. Fibers can be straight or curved, and may be perpendicular or oblique to the vein margins. Some fibers connect markers in the vein walls, such as grain fragments, that were contiguous in the rock before the vein formed. These fibers are likely to have grown in the direction that the vein opened, and they are referred to as *tracking fibers*. Straight tracking fibers indicate as single direction of vein opening (perpendicular or oblique to the vein margin); curved tracking fibers show that the opening direction changed during vein formation. If fibers show signs of deformation such as fractures or subgrains (see next), they have been deformed after vein formation, possibly obscuring information about the vein opening.

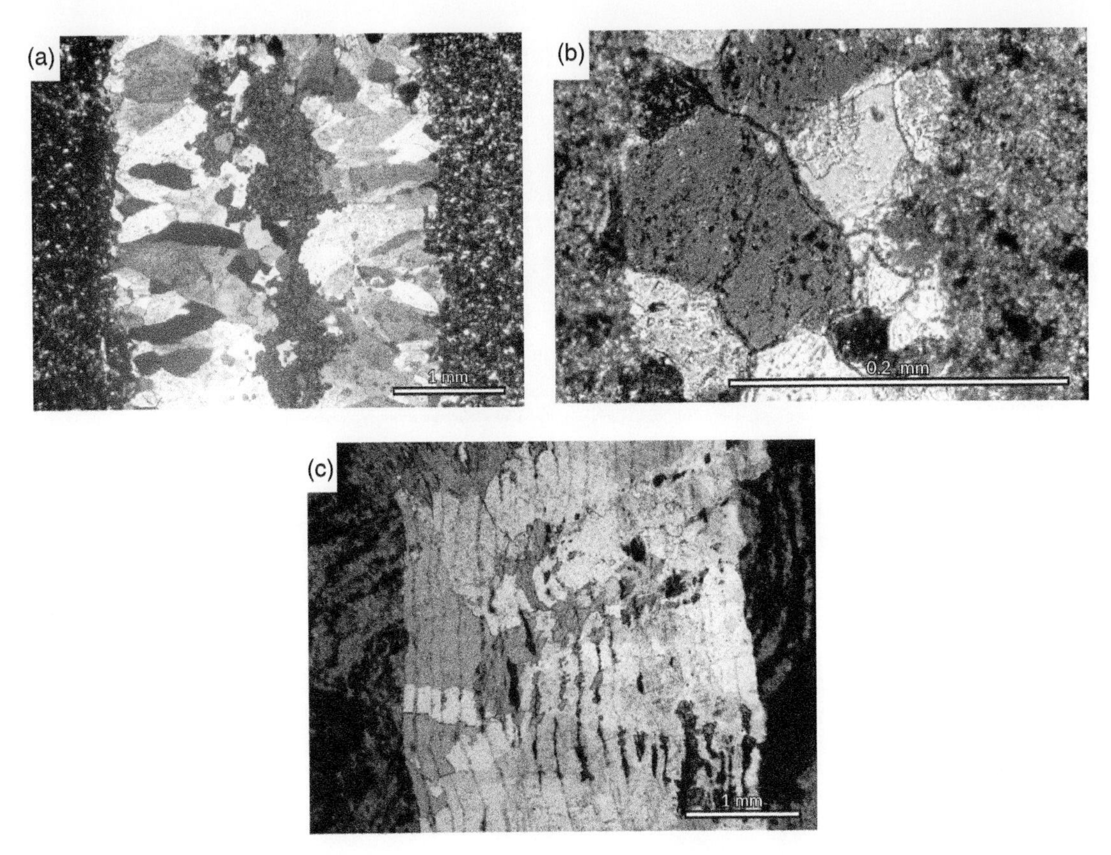

Figure 3.8 Vein-filling textures. (a) Massive chlorite in vein center, fibrous quartz at the edge of the vein. XPL: Century Mine, Australia. (b) Blocky calcite. PPL: Kimberley Block, Australia. (c) Laminated quartz. XPL: Bendigo mine, Australia. XPL, cross-polarized light; PPL, plane-polarized light.

Another basic observation about fibers is whether they have approximately constant width across veins, or whether they widen (Figure 3.9a). Widening is interpreted as a result of growth competition between adjacent fibers: the widening fibers have outgrown their neighbors, and the direction of widening shows the growth direction of the fiber. Growth directions in fibers with constant width may have varied. An important distinction is made between fibrous veins that have single fibers extending across the whole vein, and those in which there are two groups of fibers, on either side of a roughly central plane. It is also important to note whether fibers have a degree of symmetry about the center of the vein, or whether they are asymmetric. All these important observations have reasonably straightforward interpretations that are described in detail in Section 8B.1.1.2.

Not all fibers are tracking. *Non-tracking* fiber growth is determined by the orientation of the surface from which the fibers grew, rather than the opening direction of the vein. A common situation is that fibers grow perpendicular to the surface where they nucleate.

Some fibrous veins have an internal structure defined by inclusions, sometimes made of wall rock fragments or of similar minerals to those found in the wall rock. Such inclusions may be

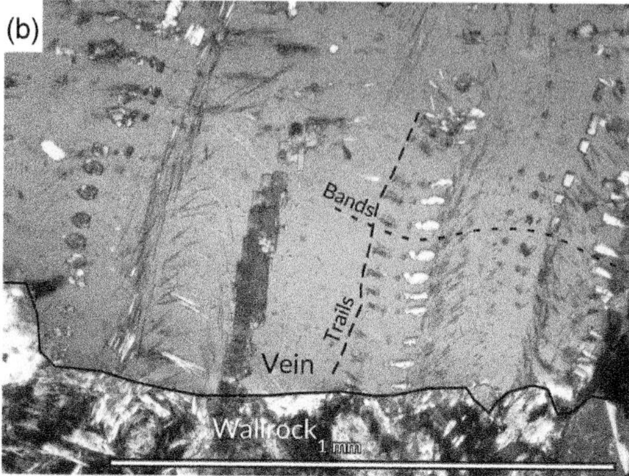

Figure 3.9 Vein-filling textures. (a) Quartz fibers widening toward the center of a vein. XPL: Mount Isa, Australia. (b) Inclusion trails and bands in quartz vein fill. The solid black line marks the boundary between the vein and the wall rock. Bands of inclusions (short dash line) are subparallel to the vein margin; inclusion trials (long dash line) are subperpendicular to the vein margin. XPL: Dalny mine, Zimbabwe. XPL, cross-polarized light.

Figure 3.10 Epithermal vein textures. Colloform banding in quartz. XPL: Mount Isa, Australia. XPL, cross-polarized light.

aligned both parallel to the vein margin (**inclusion bands**) and at some angle to it (**inclusion trails**) (Figure 3.9b).

Epithermal veins, formed less than 1 km below the surface by hydrothermal systems, have a variety of specific and distinctive textures, which are useful for identifying them and interpreting their formation (Figure 3.10). A more detailed classification for epithermal veins is given in Table 3.3.

Table 3.3 Epithermal quartz vein textures.

Texture	Description	Interpretation
Chalcedonic	Cryptocrystalline, waxy luster, fibrous	Primary growth, <180 °C
Saccharoidal	Appearance of sugar grains	Primary growth
Comb	Parallel crystals perpendicular to the vein wall	Primary growth away from vein margins
Zoned crystals	Alternating clear and milky bands, commonly in the euhedral form	Open space growth under changing conditions/fluids
Colloform	Fine bands with a kidney-like surface and radiating internal structure (reniform)	Open space growth from silica gel: below the chalcedonic zone
Crustiform	Bands parallel to vein walls defined by differences in mineralogy, texture, or color	Open space growth in fluctuating conditions/fluids; below the chalcedonic zone
Moss	Fine botryoidal (shaped like grape clusters) aggregates	Recrystallization from a silica gel texture
Microplumose	Splintery or feathery appearance within individual quartz crystals	Recrystallization from chalcedony/ amorphous silica
Mold	Impression left by a soluble phase	Replacement
Bladed	Quartz aggregates in knife-like shapes	Replacement, commonly of carbonate

3.5 Indenting, Truncating and Interpenetrating Grain Contacts, Strain Caps, and Stylolites

This set of dominantly intergrain microstructures shows clear evidence that material has been removed along grain boundaries (Figure 3.1) or from larger volumes of rock. Where a grain boundary with a high curvature is in contact with a boundary having a lower curvature, it is common to see the smaller grain *indenting* the boundary of the larger one. Generally, the contact is concave toward the grain with the larger curvature (e.g. contact between grains 1 and 2 in Figure 3.11a), implying that more of the larger grain has been removed. Where the two grains are of similar size, their contact is usually approximately planar, indicating similar amounts of removal from each grain (e.g. contact between grains 3 and 4 in Figure 3.11a); the contact can be called *truncating*. Grain boundaries comprising multiple protuberances of one grain to the other can be called *interpenetrating* (Figure 3.11b).

Strain caps consist of domains around relatively rigid objects enriched in phyllosilicates or other insoluble minerals (Figure 3.11c). Phyllosilicates in strain caps are strongly foliated and conform to, or "wrap around," the margins of the object.

Stylolites are irregular, commonly wave-like surfaces between aggregates of grains. High concentrations of minerals such as iron oxides, phyllosilicates, and graphite are common along stylolites (Figure 3.12). The protrusions of one side of the stylolite into the other are sometimes referred to as teeth. Stylolites seen in thin sections are cross sections of a complex 3D surface: the

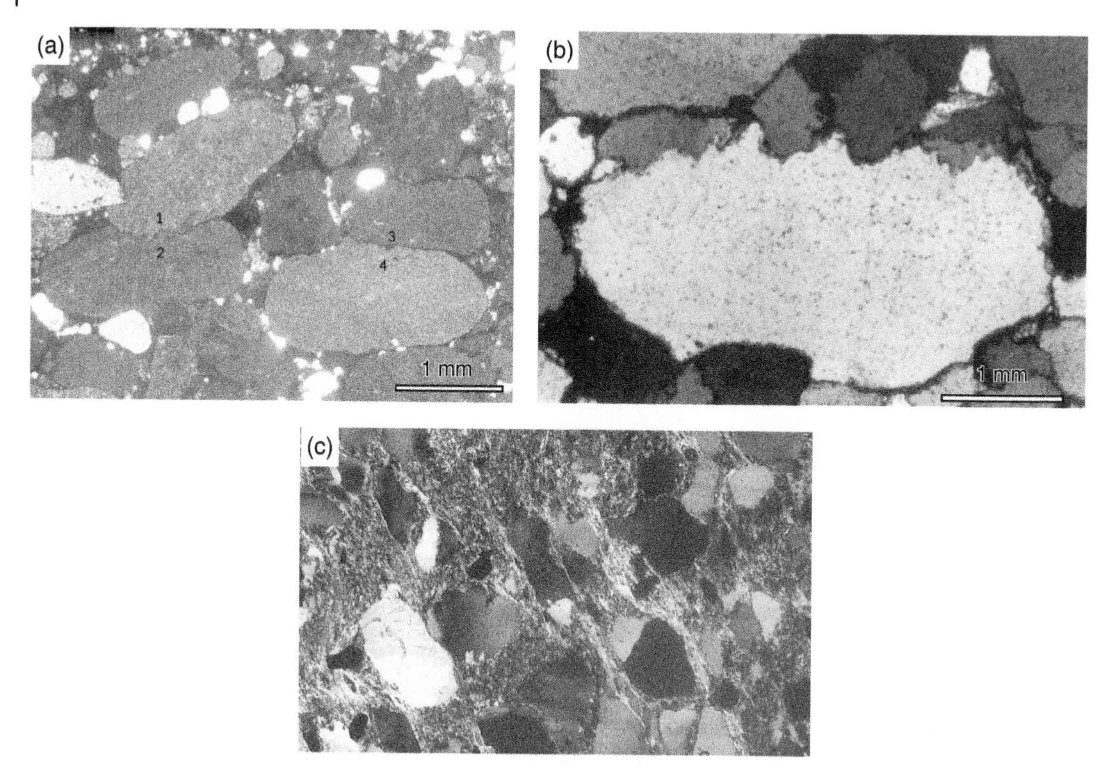

Figure 3.11 Features indicating pressure solution in quartz and lithic grains. (a) Grain contacts. The smaller grain 1 has indented grain 2; part of the contact is interpenetrating. Grains 3 and 4 have a truncating grain contact because they are of similar curvature. XPL: Kimberley Block, Australia. (b) Indenting grain contacts. The large central grain is indented by several smaller grains above and below. XPL: Cantabrian Zone, North Spain. (c) Strain caps of biotite around quartz clasts. XPL: Blue Ridge Province, Maryland, USA. XPL, cross-polarized light.

teeth-like structure seen in Figure 3.12 is a section through a surface of fingers and corresponding sockets in 3D.

Mechanisms for the formation of microstructures in Sections 3.4 and 3.5 are considered in Sections 8A.3 and 8B.2.

3.6 Aligned Grain Boundaries, T Grain Boundaries, and Foam Texture

Grain boundaries are surfaces along which one grain can slide past its neighbor. However, grain displacements are typically limited unless several grain boundaries line up. Such a configuration of **aligned grain boundaries** is a distinctive microstructure (Figure 3.13a). In some cases, the grain boundaries may be aligned in two directions, giving rise to a microstructure consisting of diamond-shaped grains.

In higher-grade metamorphic rocks containing quartz and micas, there is commonly a close relationship between the quartz grain boundaries and the mica grains: elongate quartz grains have longer sides parallel to the well-aligned micas and shorter sides perpendicular to these boundaries (Figure 3.13a). The long axes of the quartz and mica grains are generally parallel, but quartz-quartz-mica triple grain junctions have the quartz–quartz boundary at a high angle to the mica long axis. These are **T grain boundaries**. The micas appear to be controlling the orientation of the quartz grain boundaries due to an effect called *pinning*.

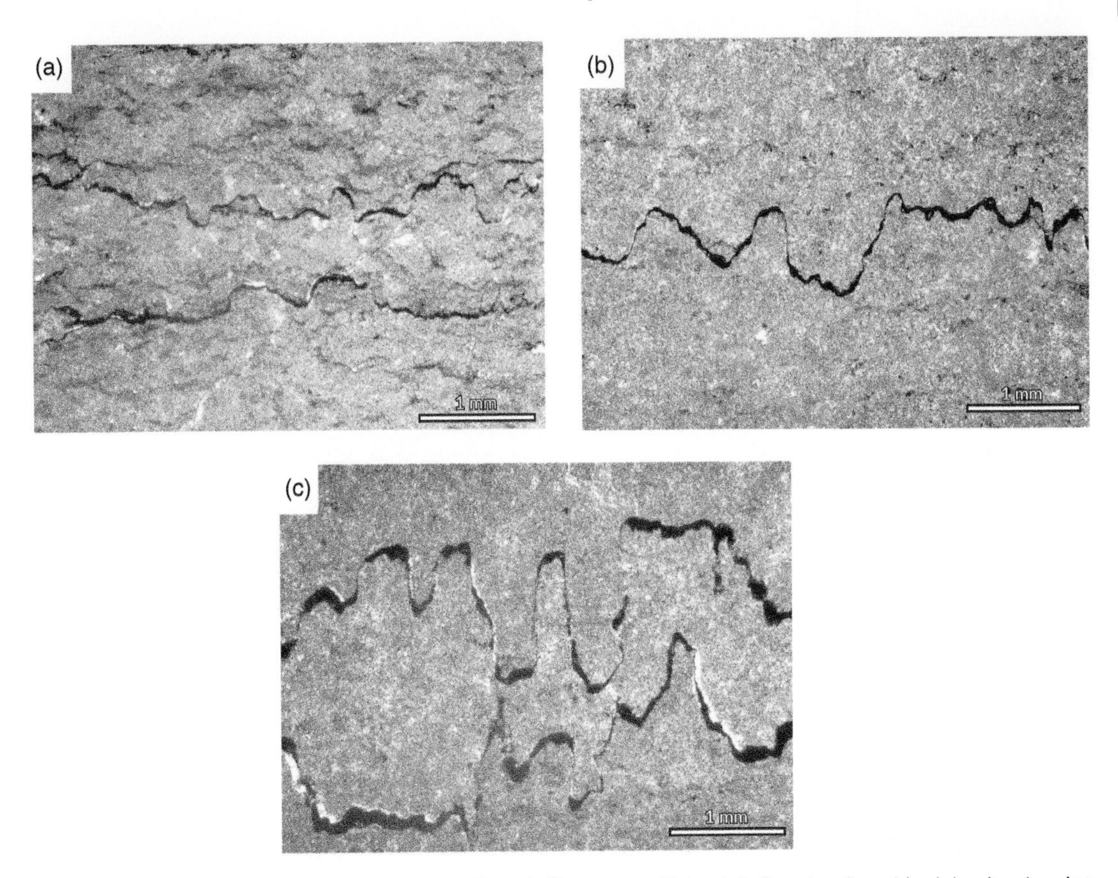

Figure 3.12 Microstylolites. (a–c) Sequence of teeth-like or crenellate stylolites developed in dolomite showing an increasing amount of volume loss, as indicated by the amplitude of the stylolite. PPL: Kimberley Block, Australia. PPL, plane-polarized light.

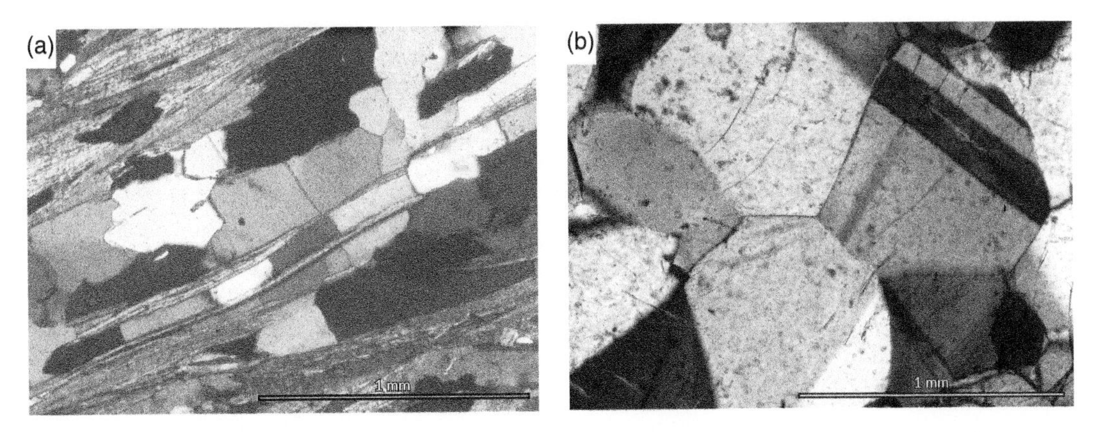

Figure 3.13 Aligned grain boundaries, T grain boundaries, and foam texture. (a) Aligned and T grain boundaries. Grain shape fabric of elongate quartz grains and biotite. Quartz grain boundaries are generally either parallel to adjacent biotite grains (aligned), or perpendicular, forming T shapes. XPL: Sample from below the Main Central Thrust, Annapurna, Nepal. (b) Foam texture between plagioclase grains. Note the 120° triple junction at the center of the picture. XPL: Ontario, Canada. XPL, cross-polarized light.

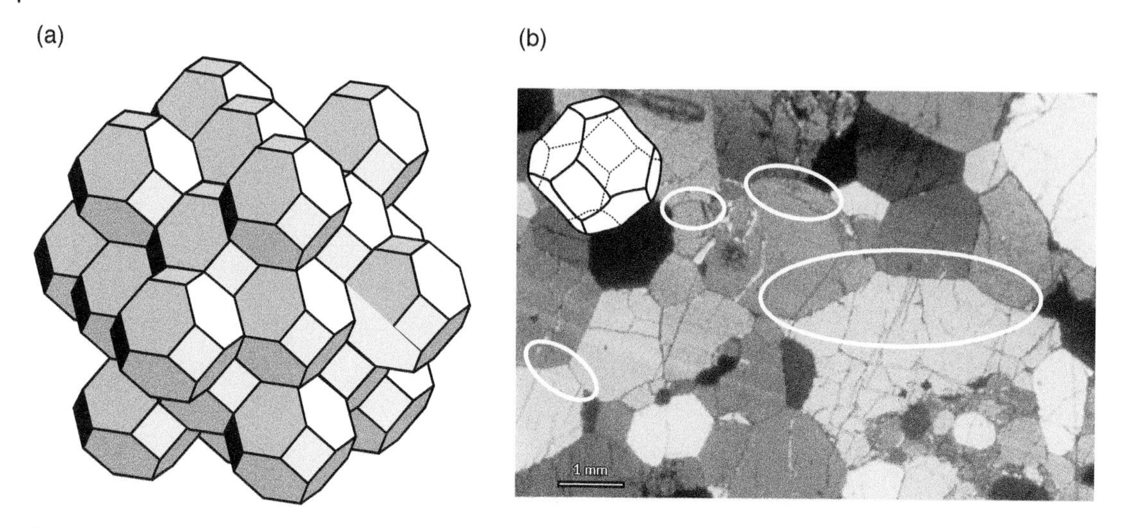

Figure 3.14 Granoblastic polygonal or foam texture. (a) Truncated octahedrons tessellate in three dimensions. (b) Granoblastic polygonal olivine crystals with curved grain boundaries, highlighted by the ellipses. The black outline shows how the truncated octahedron is modified by curved edges. XPL: Bushveld intrusion, South Africa. XPL, cross-polarized light.

Rocks that have experienced high metamorphic grades sometimes have very little microscopic fabric. Rather, mineral grains are equant and polygonal shapes with triple grain boundaries that intersect at approximately 120° in a section. This is known as **granoblastic polygonal** or **foam texture** (Figures 3.13b and 3.14). The section is likely to represent a cut through a tessellating 3D geometry consisting of truncated octahedra (Figure 3.14a). Both T grain boundaries and foam texture form in rocks to minimize surface energy. The truncated octahedron is the space-filling shape that minimizes surface energy. However, the angles between all edges are not equal in a truncated octahedron. In order to equalize these angles and minimize surface energy, the edges between the two sides may bend. Such curved grain boundaries around the roughly polygonal grains are a quite distinctive feature of well-developed granoblastic polygonal textures (Figure 3.14b).

3.7 Undulose Extinction, Subgrains, Deformation and Kink Bands, Deformation Lamellae, Grain Boundary Bulges, and Core-and-Mantle Microstructure

This section describes a variety of related intragrain microstructures, in which the crystal lattice is deformed but not fractured (top right of Figure 3.1). The presence of these features indicates deformation. **Undulose extinction** is a smooth variation in the extinction position of a single grain when examined in cross-polarized light (Figure 3.15a). It is a very common feature of deformed rocks, especially in quartz and feldspars. The variation in extinction position indicates a variation in the orientation of the crystal lattice of the grain, which, in turn, affects the polarizing

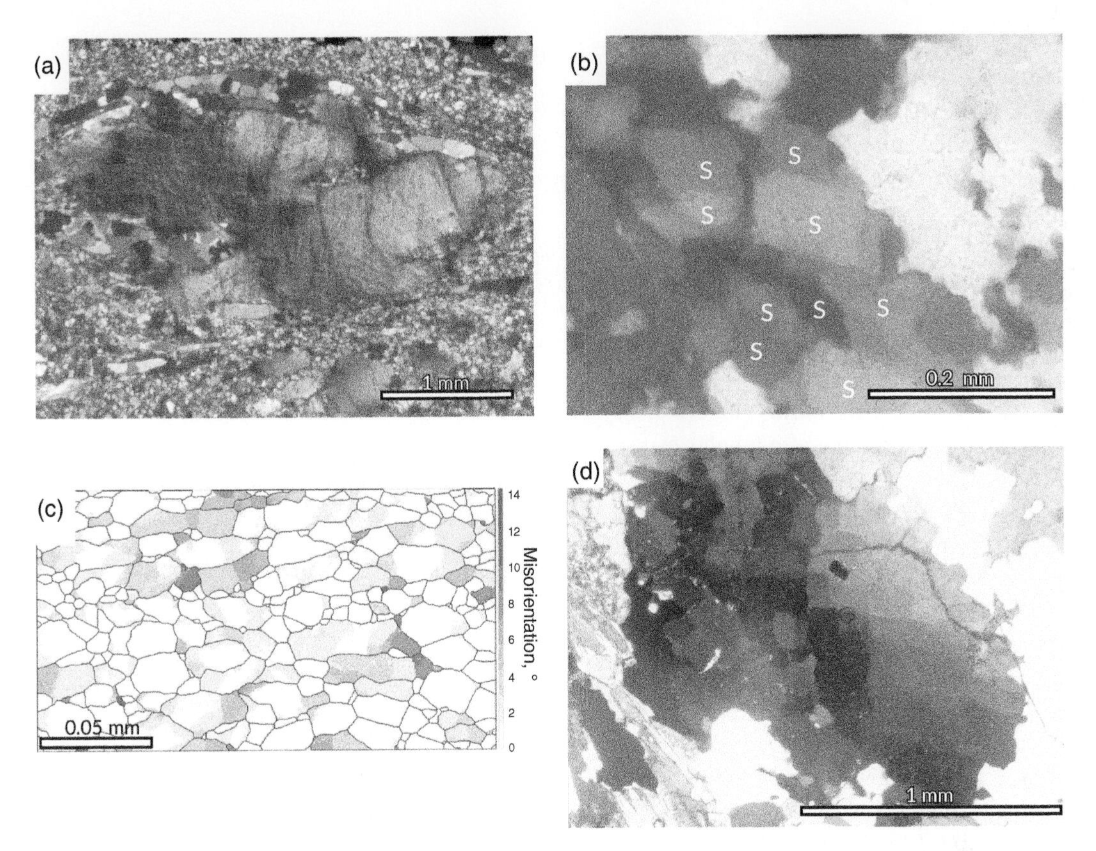

Figure 3.15 Features showing the initial stages of deformation by intracrystalline plasticity. (a) Undulatory extinction in perthite porphyroclast in a mylonite. XPL: Limpopo Belt, Zimbabwe. (b) Subgrains (s) can be identified as patches of slight variations in extinction position, showing slight variations in crystallographic orientation. XPL: Aureole of the Ardara pluton, Donegal, N. Ireland. (c) Misorientation map of grains, in which the color indicates the misorientation relative to the mean orientation of the grain, thus revealing the intragranular lattice strains. EBSD map. Quartzite, Tsunanose River, Eastern Kyushu, Japan. (d) Chessboard subgrains: approximately rectangular subgrains. XPL: Mount Isa, Australia. XPL, cross-polarized light; EBSD, electron backscatter diffraction.

direction of transmitted light and thus the extinction position of different parts of the grain.

In moderately deformed grains, these variations in extinction position and lattice orientation are commonly localized into **subgrains**: intracrystalline domains where the lattice orientation may vary by up to 10° from the rest of the grain (Figure 3.15b). This upper limit to the lattice misorientation for subgrains distinguishes a subgrain that forms within grains from two separate grains. A useful way to characterize microstructure in these rocks is to make maps of the misorientation of grains (Figure 3.15c). These misorientation maps require knowing the full crystallographic orientation in 3D for each grain and subgrain, which is commonly measured by EBSD studies.

Subgrains may have relatively planar walls in crystallographically controlled directions. In quartz, the basal (perpendicular to the sides of the hexagonal crystal shape seen in large quartz crystal) and prism (parallel to the sides of the hexagonal crystal shape) planes are common subgrain boundaries. If both directions are developed, the resulting effect is a pattern of square or rectangular domains of contrasting extinction, which is called a *chessboard* pattern of subgrains (Figure 3.15d).

Some subgrains have distinctly tabular shapes: these define **deformation bands**. If the subgrain boundaries are sharply defined and straight, they can be referred to as **kink bands** (Figure 3.16a). **Deformation lamellae** are also tabular features but restricted to very small widths: typically no more than 0.5–10 μm wide. They are generally visible due to their slightly different extinction and/or by their slightly different relief (index of refraction) compared to the host grain (Figure 3.16b). Deformation lamellae are often very closely spaced (on the order of their width), and they are commonly slightly curved. They can be pervasive across a whole grain or localized into smaller domains. They are seen best under high magnification with a small substage diaphragm aperture. Deformation lamellae form parallel to crystallographic planes (for example, parallel to the rhomb planes in quartz), and they may occur in more than one set in a grain.

The shapes of grain boundaries may be highly irregular and convoluted such as those in Figure 3.17a. These bulbous domains can be described as **grain boundary bulges**, and they are considered to form by the movement of grain boundaries through adjacent grains, and a process called *grain boundary migration* (Section 8A.4.4.1). The result is an interlocked or **sutured** grain boundary.

Small grains in the matrix may have a close spatial and sometimes lattice relationship to larger grains suggesting that they are derived from the larger grains; they are called **new grains** or neoformed grains. The **core-and-mantle** structure is a particularly distinctive example of this pattern in which the porphyroclasts are surrounded by a rim of smaller new grains (Figure 3.17b).

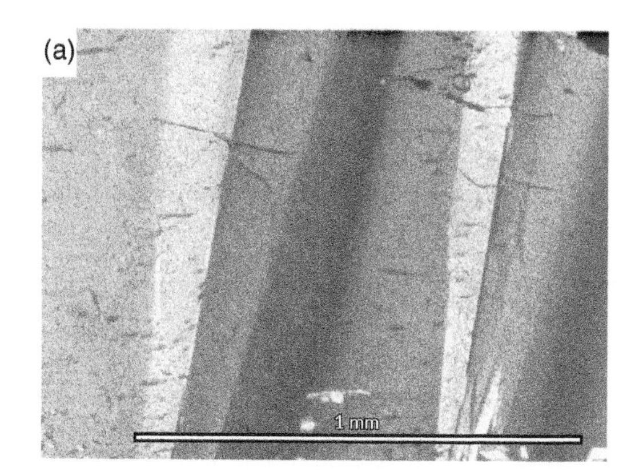

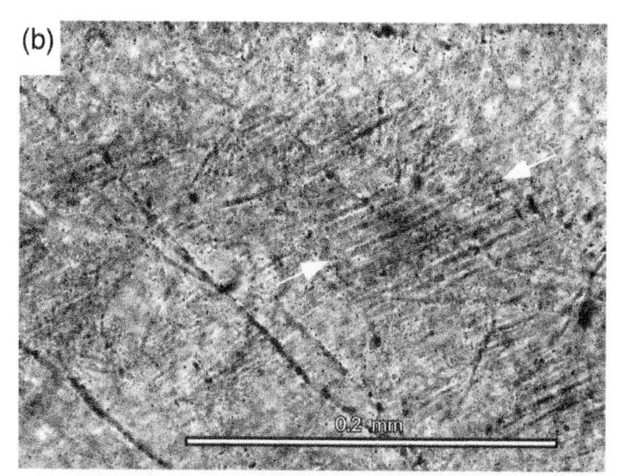

Figure 3.16 Features showing advanced stages of deformation by intracrystalline plasticity. (a) Kink bands in kyanite: sharp boundaries between bands of different extinction position. XPL. (b) Deformation lamellae. The narrow short zones of slight contrast in relief (e.g. between arrows) are typical deformation lamellae. Note that they are slightly curved. PPL: Mount Isa, Australia. (c) Twins. Deformation twins in calcite. Note that individual twins are curved. PPL. XPL, cross-polarized light; PPL, plane-polarized light.

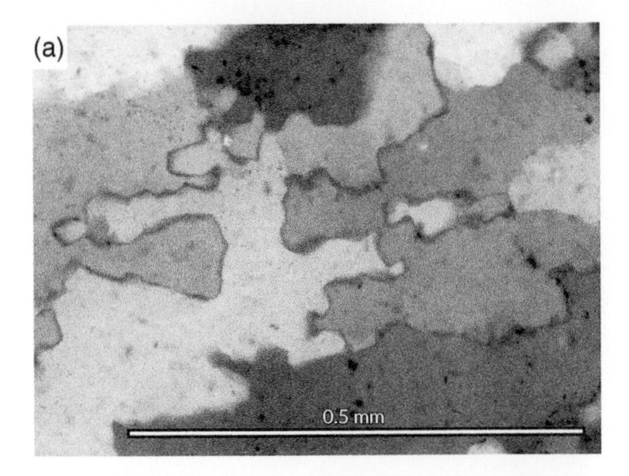

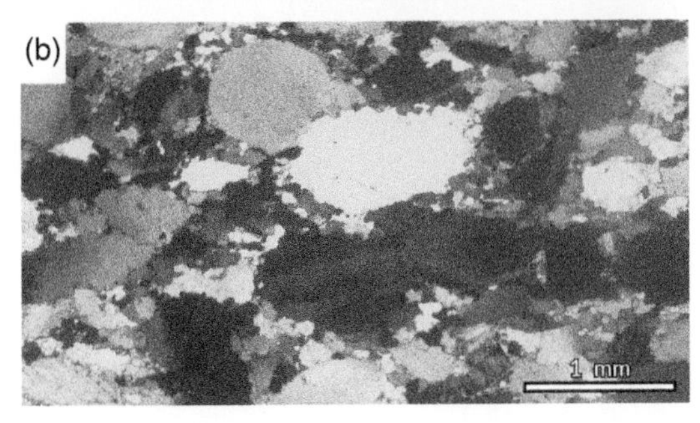

Figure 3.17 (a) Grain boundary bulges. XPL: Aureole of the Ardara pluton, Donegal, N. Ireland. Sample from the collection of Mike Rubenach. (b) Core and mantle structure. Small, recrystallized grains around the margins of large quartz grains. XPL: Blue Ridge Province, Maryland, USA. XPL, cross-polarized light.

Deformation mechanisms for microstructures in this section are considered in Sections 8A.4 and 8B.3.

3.8 Deformation Twins

Twins or *twin lamellae* are intracrystalline domains in which the lattice orientation of the host grain has been rotated relative to the untwinned grain, like subgrains. However, twins differ from subgrains in that the twin has a fixed angular relationship with the host. As we have seen earlier, subgrains may have variable misorientations. Twins are recognized under the microscope as straight zones with different extinction positions from the host, having sharp boundaries parallel to crystallographic directions. Twins can be distinguished from tabular subgrains because there are usually several parallel twins extending across the whole grain, unlike tabular subgrains, which tend to form only one domain within a grain.

Two types of twins can be distinguished. **Growth twins** are commonly straight and have an even width across the grain. These are primary microstructures, formed when the mineral grows. **Deformation twins** can be recognized because they may be bent and have a variable thickness along their length (Figure 3.16c). These twins are intragrain, secondary microstructures (Figure 3.1) that form during deformation. Mechanisms of twinning are considered in Sections 8A.4.9.1 and 8B.3.8.

3.9 Grain Shape Fabrics, Ribbon Grains, and Gneissic Banding

Deformed rocks commonly have parallel, inequant grains that define a **grain shape fabric** (Figure 3.18), which may be visible at mesoscopic or microscopic scales. These fabrics are by definition multigrain (Figure 3.1). If the grains are platy and parallel to each other, the fabric is a foliation; if they are rod-like and parallel to each other, it is a lineation. The distinction between foliations and lineations cannot be made on the basis of a single thin section because a section parallel to a lineation will look very similar to a section perpendicular to a foliation. A thin section perpendicular to a pure linear fabric, however, will not show any preferred orientation at all. Figure 3.18 shows sections parallel (a), oblique (b), and perpendicular (c) to a dominantly linear fabric of actinolite needles. Sections (a) and (b) both exhibit grain shape alignments, which are strong in (a) and moderate in (b). There is very little fabric apparent in (c) due to the weak alignment of the intermediate axes of

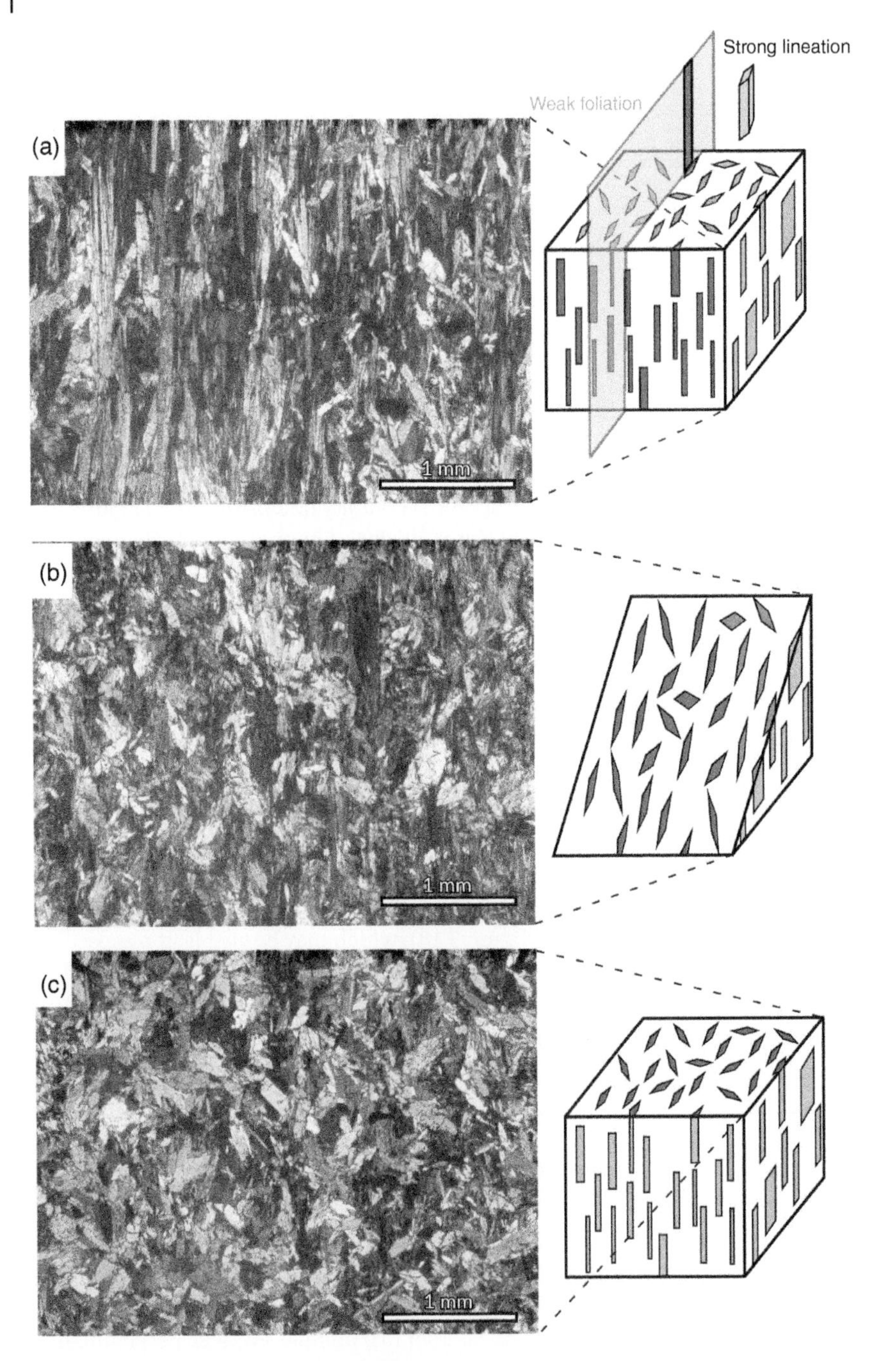

Figure 3.18 Sections through a dominantly linear fabric of actinolite needles. The block diagram shows the fabric in three dimensions. (a) Parallel to lineation: strong fabric. (b) Oblique to the lineation: moderate fabric. (c) Perpendicular to the lineation: fabric is very weak. Sections (a) and (b) would look the same as sections in a similar orientation through a planar fabric. Photomicrographs are all PPL: Arcturus, Zimbabwe. PPL, plane-polarized light.

the prismatic actinolite crystals. The intensity of a grain shape fabric observed in a thin section, therefore, depends on the type of fabric and the orientation of the thin section, as well as the deformation experienced by the rock.

Fabrics can be composed of almost any mineral, and at the scale of the thin section, they can be **pervasive** or **penetrative**, i.e. affecting all parts of a rock equally (Figure 3.19a), to **domainal** (Figure 3.19b, 3.20), i.e. affecting only particular regions or domains in the rock. The grain size of the mineral defining the fabric generally increases with metamorphic grade. In polymineralic rocks, the alignment is usually stronger in one of the minerals, such as micas in a quartz-mica schist (Figure 3.20).

Thin section observations are particularly useful for defining age relationships between fabrics. One of the clearest relationships to observe is the crenulation, or microscale folding, of an earlier fabric to form a later fabric (Figure 3.19c). This microstructure typically occurs in pelites (metamorphosed mudstones) dominated by quartz and phyllosilicates.

Ribbon grains are monomineralic layers, approximately one grain wide, which record intense fabric formation during deformation at higher metamorphic grades. Quartz and carbonates commonly form ribbon grains. Two types of ribbon grain can be distinguished: (1) *polycrystalline ribbon* grains (Figure 3.21a) consist of several small, typically tabular crystals that occupy the width of the ribbon and commonly show internal strain features such as undulatory extinction or subgrains; and (2) *monocrystalline ribbon grains*, which are single crystals (Figure 3.21b).

Ribbon grains can be seen as an intermediate microstructure between foliations and **gneissic layering**, or banding, also called gneissose structure. Rocks exhibiting gneissic layering consist of layers of differing composition, typically on a mm to cm scale, which is formed in higher-grade metamorphic rocks. The formation of gneissic banding

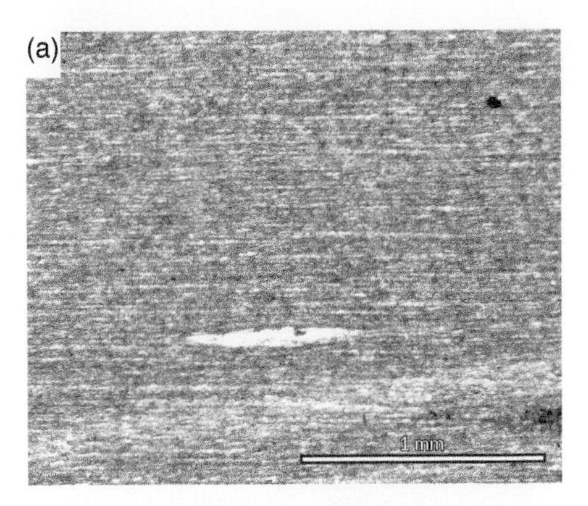

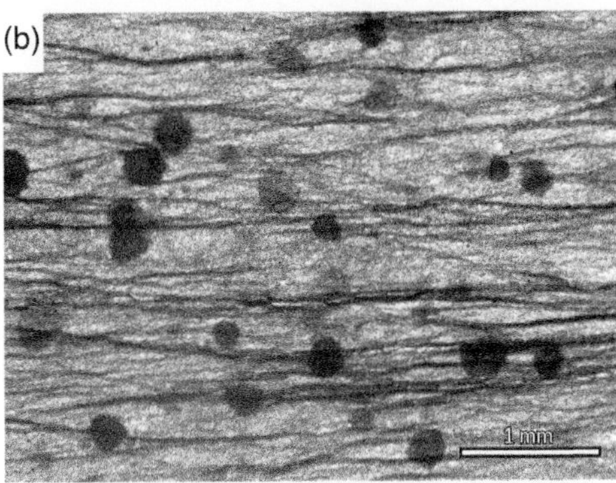

Figure 3.19 Types of cleavage. (a) Slaty cleavage. This cleavage is penetrative and homogeneous at a mm scale. PPL: Bendigo, Australia. (b) Spaced or domainal in argillaceous limestone. Domains are distinct at a mm scale. XPL: Valley and Ridge Province, Virginia, USA. (c) Crenulation cleavage. Q (dominantly quartz) and P (dominantly mica) domains are distinct at a mm scale. PPL: Appalachian Mountains, USA. XPL, cross-polarized light; PPL, plane-polarized light.

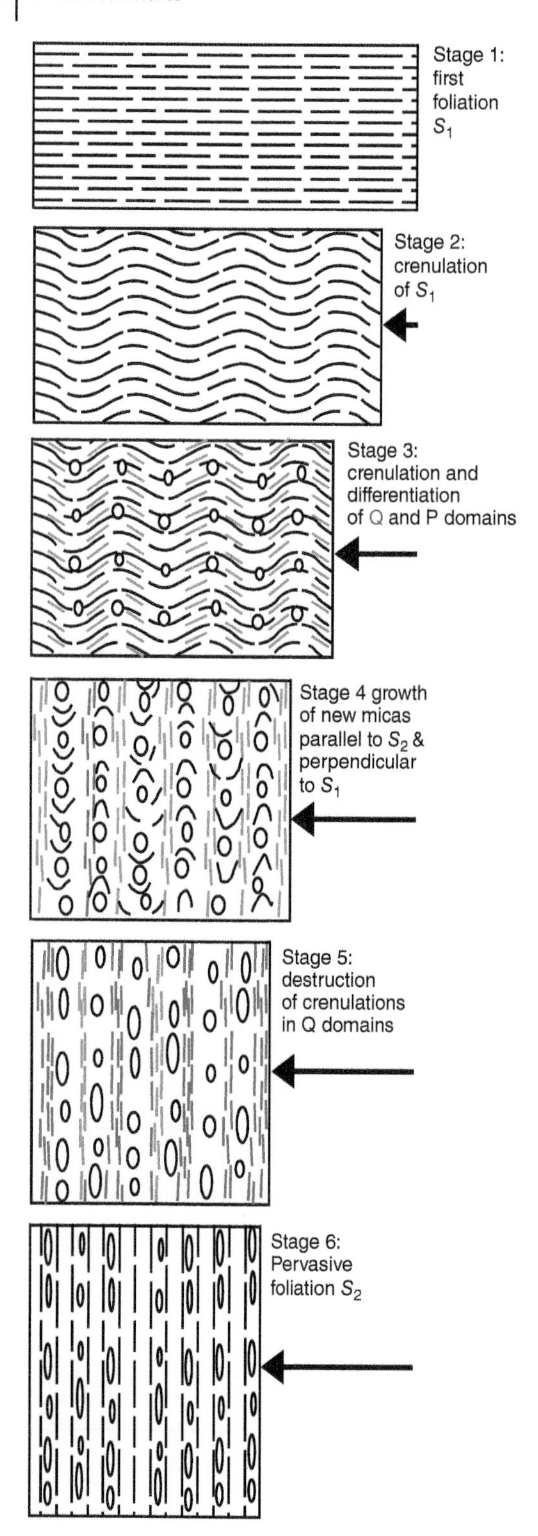

Figure 3.20 Six stages can be defined in the progression from a first foliation S_1 to a completely new foliation S_2. The second foliation is a **crenulation cleavage** if the grain size of the rocks is not visible to the unaided eye. In stage 2, the first foliation (S_1) is folded. Stage 3 is marked by the concentration of micas on the limbs of the crenulations (folds). At this stage, the limbs of the folds define P domains (P for phyllosilicate), and the hinges of the folds are identified as Q domains (because quartz, the other most common phase, is concentrated there). In stage 4, new phyllosilicates start to grow parallel to the hinge surfaces of the folds. In the fifth stage, a domainal fabric of P and Q domains in the orientation of the new foliation S_2 is created. The final stage is the pervasive development of this fabric, in which all trace of the original fabric is completely lost. Source: Modified from Bell and Rubenach (1983).

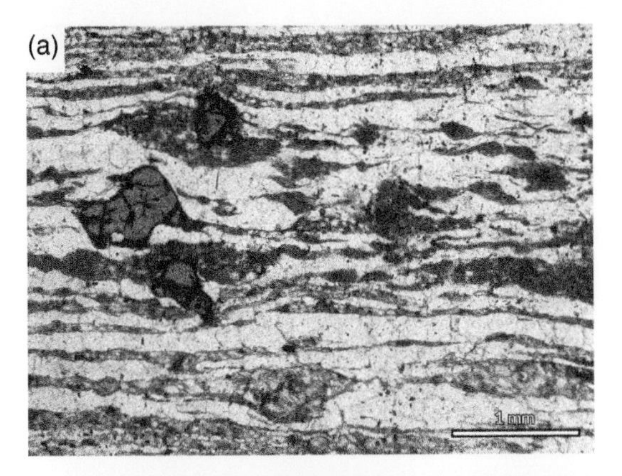

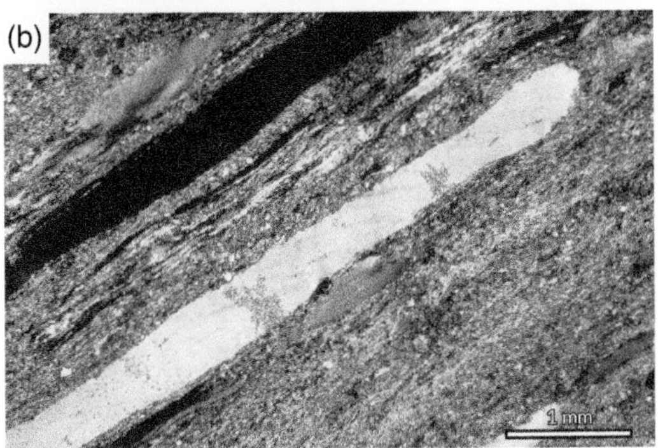

Figure 3.21 Ribbon grains. (a) Polycrystalline quartz forms ribbon grains around garnet and feldspar. PPL: Limpopo Belt, Zimbabwe. (b) Monocrystalline quartz. This ribbon grain has been formed by extreme flattening of a quartz clast. XPL: Blue Ridge Province, Tennessee, USA. PPL, plane-polarized light; XPL, cross-polarized light.

is not well understood, but it is clear that several processes can be involved, most of which are associated with deformation. Features such as isolated isoclinal fold hinges, known as *rootless* isoclinal folds, form within gneissic layers because these rocks are so highly strained that the limbs were attenuated. Yet, even in these high-strain settings, a grain shape fabric may not be visible in the different layers, even microscopically. The initial microstructure of the protolith plays a role in the formation of some gneissose banding. Other types of gneissic banding are due to veining and, in migmatites, to layers of partial melt.

3.10 Porphyroblasts

Porphyroblasts are large grains of metamorphic minerals surrounded by smaller grains, which are common in metamorphic rocks and which have grown by metamorphic reactions from the minerals in the primary rock. Some metamorphic minerals are characteristically porphyroblastic, for example, garnet, andalusite, staurolite, and cordierite. Other metamorphic minerals that may appear as either porphyroblasts or matrix phases are micas, feldspars, and amphiboles. The chemical composition of porphyroblasts may change considerably from the center or core to their rim, yielding *zoned* porphyroblasts. Porphyroblasts commonly contain inclusions of other minerals (Figure 3.22), which may be the same size, phase, and composition as the matrix outside the porphyroblast, or may be different in all these respects. The density and type of inclusions commonly vary from core to rim: for example, an inclusion-rich core can be surrounded by an inclusion-free rim.

Porphyroblast inclusions may define a fabric inside the porphyroblast, called an **internal fabric** (intragrain), and given the symbol S_i to distinguish it from the **external fabric** (multigrain) of the matrix around the porphyroblast (S_e). Internal fabrics may be defined by anisotropy in their distribution (i.e. the centers of the inclusions are aligned), anisotropy in shape, or both. A line formed by tracing out the inclusion anisotropy is called an **inclusion trail** (the same term that is used for a similar feature in veins: Section 3.4). Porphyroblasts can contain zones with different inclusion geometries.

The relations between the internal and external fabrics have particular significance for microstructural studies. A simple classification of inclusion geometries and their relationships to the matrix is given in Table 3.4. Three critical observations to make are:

1) Do the inclusions define a fabric (S_i) or are they unoriented/unaligned?
2) If S_i exists, is it continuous or discontinuous with S_e?
3) If S_i exists, does it have the same or a variable orientation with respect to S_e?

Inclusion trails defining S_i may curve continuously (e.g. Figure 3.22a, b) or be relatively straight. Curved trails are sometimes spectacular, curving around the center of the porphyroblast by more than 360°. Spiral trails that curve by more than 180° are called *snowball structures* or trails (Figure 3.22b), and have only been described from garnets. The trails themselves may be domainal, consisting of portions that are rich in inclusions such as micas, and other portions that are mainly quartz. Careful study of inclusion trails shows that many consist of relatively straight portions of trails that are truncated or deflected against other relatively linear portions of trails.

The three-dimensional shape of inclusion trails has been studied in detail, both from individual garnets by computer-assisted tomography and by looking at numerous examples of garnets in multiply-oriented thin sections. Results of these studies show that inclusion trails can have complex shapes, and a single geometry may appear quite different in different sections.

The patterns of inclusions described earlier are understood in terms of the relationships between episodes of porphyroblast growth, which traps inclusions from the matrix, and changes of orientation between the porphyroblasts and the matrix. Unoriented inclusions indicate that the porphyroblast grew before any deformation: such porphyroblasts can be called **pretectonic**. If the inclusions define a fabric that is discontinuous with S_e, then a first event of deformation and porphyroblast

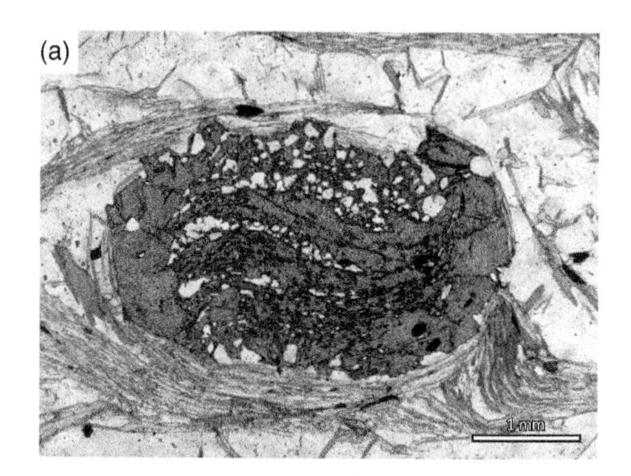

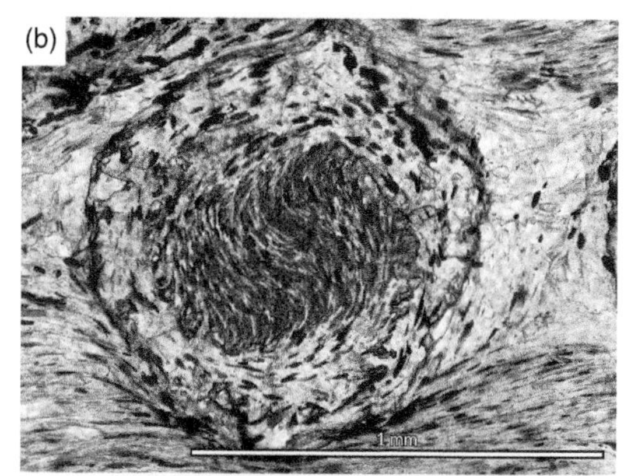

Figure 3.22 Relationships between porphyroblasts and inclusions. (a) Intertectonic porphyroblast of garnet: S_i is discontinuous with S_e. PPL. (b) Syntectonic porphyroblast: S_i continuous with S_e, but has variable orientations. PPL: Sample AN29, below the Main Central Thrust, Annapurna, Nepal. (c) Post-tectonic porphyroblast: a few inclusions within the larger porphyroblast align with the matrix. PPL: Sample SI597. PPL, plane-polarized light.

Table 3.4 Relationships between inclusion geometry and matrix fabric S_e.

Inclusions	Continuity of S_i with S_e	Orientation of S_i relative to S_e	Classification	Figure
No fabric			Pretectonic	
Define S_i	Discontinuous	Different	Intertectonic	3.23a
Define S_i	Continuous	Variable	Syntectonic	3.23b
Define S_i	Continuous	The same	Posttectonic	3.23c

growth was followed by a second deformation after porphyroblast growth. Such porphyroblasts can be called **intertectonic** (Figure 3.22a). If S_i is continuous with S_e but has different orientations, it can be inferred that the porphyroblast grew during deformation and that there was a change in orientation between the porphyroblast and the matrix during the deformation event. These porphyroblasts are called **syntectonic** (Figure 3.22b). Finally, if the porphyroblast has S_i continuous with S_e, this suggests that the porphyroblast grew after deformation: it is **posttectonic** (Figure 3.22c). Curved inclusion trails indicate a progressive relative rotation of the porphyroblast and the matrix foliation.

3.11 Crystallographic Fabrics (Crystallographic Preferred Orientations)

In many deformed rocks, the crystal lattices of different grains of the same mineral are nearly parallel to each other (Figure 3.1): this is known as a crystallographic preferred orientation (CPO) or a lattice-preferred orientation (LPO). The crystal lattices are also at constant angles to other fabric elements such as foliation and lineation. CPOs may not be obvious from a cursory thin section examination, but strong hints for their presence may be the subparallel extinction positions of many grains, or common patterns of birefringence or pleochroism. They can only be observed in multiple grains, and CPOs may only be detectable after measuring hundreds of grains.

More detailed examination of such CPOs requires the determination of the crystallographic orientation of many grains. This evaluation can be done in multiple ways, including using the universal stage with an optical microscope, using X-ray diffraction patterns from EBSD measurements in an SEM, or from automated techniques using photographs of thin sections taken in different orientations in cross-polarized light.

Crystallographic orientations are commonly represented on stereoplot in which the positions of crystallographic axes, such as the *c*-axis of quartz, are shown relative to the orientations of fabric (lineation, foliation) (Figure 3.23a). This method has the disadvantage that each crystallographic axis requires a separate stereoplot. Equivalently, the fabric elements can be plotted relative to the crystallographic elements on an *inverse pole figure* (Figure 3.23b). Neither of these methods preserves the spatial context of the grain orientation. Rather, a map of grain orientations is made using symbols and/or colors to show crystallographic orientations (the *AVA* or *Aschenverteilungsanalyse* or *axial-distribution* or *Angle vs. Azimuth diagram*) (Figure 3.23c, d). Mechanisms for the formation of CPOs are discussed in Section 8A.4.

3.12 Shear Sense Indicators, Mylonites, and Porphyroclasts

Microstructural studies are commonly aimed at finding the shear sense in zones of deformation including both fault and shear zones. Microstructures are of particular value because they can reveal details of the kinematics that are not visible

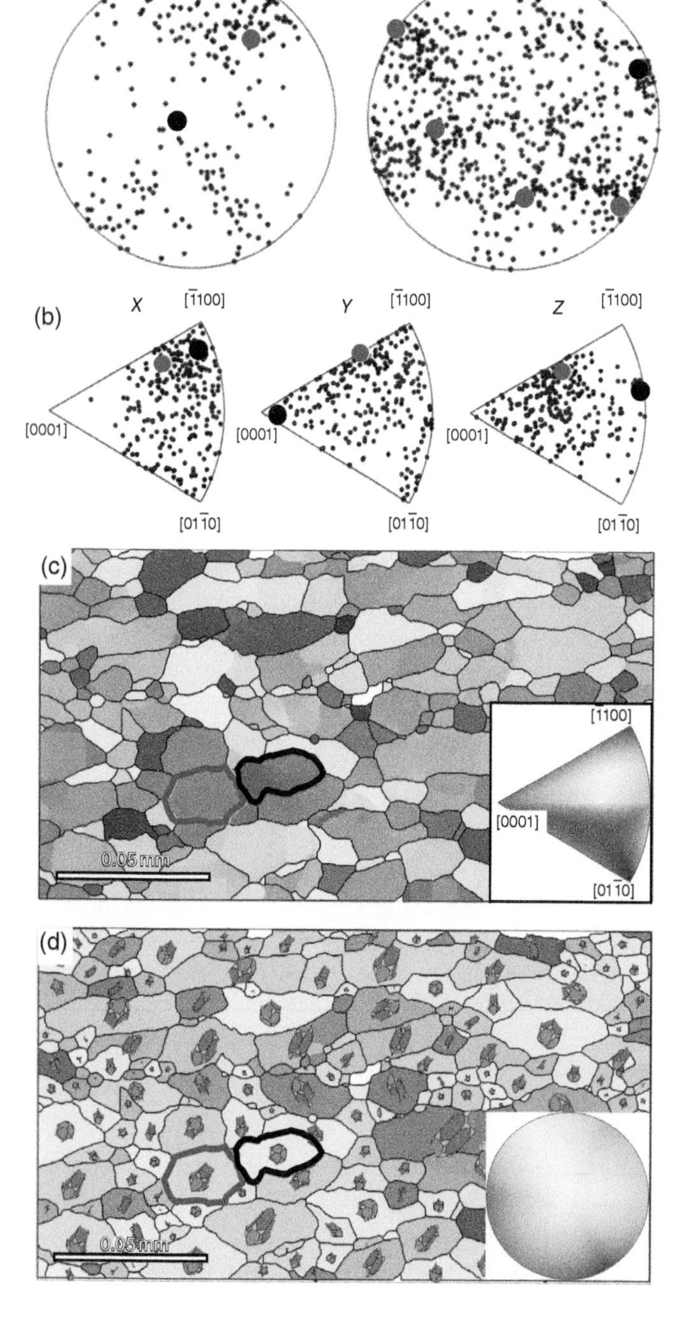

Figure 3.23 Different representations of crystallographic preferred orientations in quartzite. Red and black large data points in (a) and (b) correspond to the red and black outlined grains in (c) and (d). (a) Conventional crystallographic axes plotted on a lower hemisphere, equal-area stereographic projection. The labeled crystallographic axes are plotted relative to the conventional orientation of the fabric axes with *X* horizontal, EW, and the *XZ* plane vertical EW. (b) Inverse pole figures showing the orientation of the fabric axes *X*, *Y*, and *Z* relative to the crystallographic axes of quartz. (c) An orientation map with grey scale according to the inverse pole figure key, in which grey scale shows the orientation of the *Y* axis for each grain relative to the crystallographic axes. The inset shows how the orientation of the *Y* axis is represented by the grey scale. (d) An AVA diagram, which shows the orientation of the *c*-axis by grey scale as indicated by the equal area, lower hemisphere colored stereoplot. The orientation of the quartz crystals is also shown schematically in each grain. All data sets were obtained from quartzite, Tsunanose river, Kyushu, Japan, and plots were constructed using the MTEX toolbox (Modified from Bachmann et al. 2011).

on the outcrop or larger scales. The key feature of a microstructure that allows shear sense to be determined is that it is asymmetric with respect to a plane through the center of the microstructure and parallel to the fault or shear zone. Microstructures used for shear sense determination are, in fact, the asymmetric equivalent of microstructures that have been described in previous sections of this chapter. Figure 3.24 shows some of the more commonly used shear sense indicators, arranged with respect to grain size and temperature as shown in Figure 3.1.

When we make microstructural observations, we usually examine small samples that we have taken from the field, under the optical or electron microscope. For simplicity of description, it is useful to refer to the sense of shear in the framework of the sample (*sample frame of reference*), temporarily overlooking the three-dimensional geographical context of the sample in the shear zone.

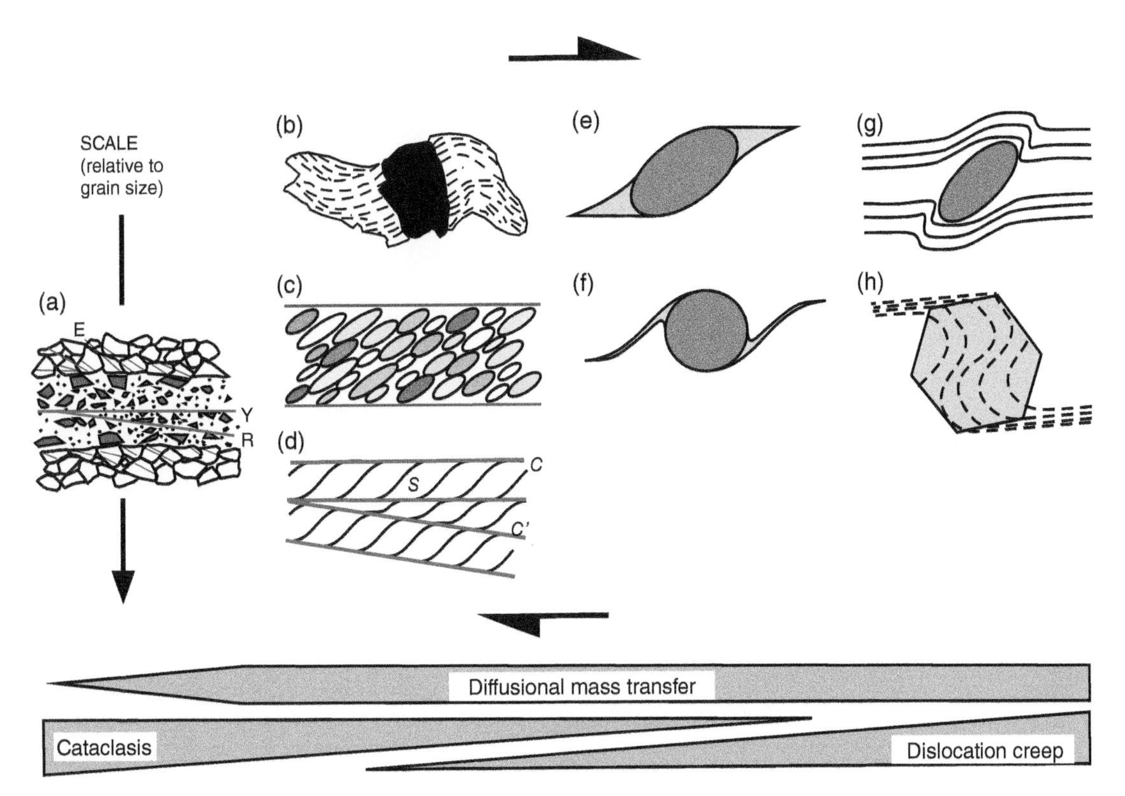

Figure 3.24 Summary of asymmetric microstructures that can be used for shear sense determination. All microstructures show a dextral shear sense. Scale varies from single grains at the top to multigrain at the bottom. Generalized dominant deformation mechanisms shown along the bottom of the figure. (a) Fault zone microstructures, including extensional fractures (E), Riedel shear (R), and a Y shear, parallel to the fault surface. (b) Asymmetric pressure fringes. (c) Oblique grain shape fabrics. (d) *S*, *C*, and *C′* fabrics. (e) Sigma porphyroclast. (f) Delta porphyroclast. (g) Rolling structures. (h) Curved inclusion trails in porphyroblasts.

Conventionally, we look down the microscope at a thin section that is oriented with the shear plane right-left in the field of view. The shear sense is described as dextral or sinistral across this plane using the same convention as for faults: if the side of the sample opposite to the observer moves to the right, this is dextral and vice versa for sinistral. Alternatively, the shear sense can be described as clockwise (dextral) or anticlockwise (sinistral). These terms can be understood if they are imagined as the rolling action induced by the shear on an axle within the shear zone.

Dextral and sinistral are used for convenience in describing two-dimensional images, and do not imply that the movement is strike-slip in a *geographic* frame of reference. To understand the implication of the shear sense observed in the sample frame of reference for shear sense in the geographic frame of reference, we need to consider how the sample was collected and oriented in the field: this is described in Section 3.13.

Fault zones contain gouges or cataclasites and, less commonly, pseudotachylites. Shear sense can be determined from Riedel shears (Figure 3.3b), or from extension fractures, which are commonly oriented about 45° to the fault plane (Figure 3.24a). Shear zones often contain or are completely composed of **mylonites**, which typically consist of finer-grained mineral components known as the **matrix** surrounding coarser grains known as **porphyroclasts**. Mylonites can be subdivided into protomylonites (0–50% matrix), mylonites *sensu stricto* (50–90% matrix), and ultramylonites (90–100% fine-grained material) (Figure 3.25). Both the matrix and the porphyroclasts have been derived from a protolith by intense deformation, and mylonites commonly have strong planar and/or linear fabrics. Porphyroclasts may have internal deformation features such as microfractures, undulatory extinction, and subgrains. Matrix grains may wrap around the porphyroclast (Figure 3.24g), and in some instances, collections of small grains derived from the porphyroclast extend as a **tail** into the matrix (Figure 3.24e, f). Domainal mineral growths, such as pressure shadows or fringes, are also common around porphyroclasts (Figure 3.24b).

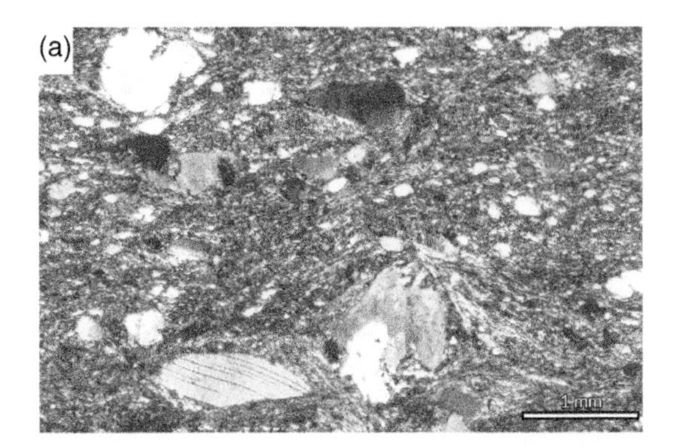

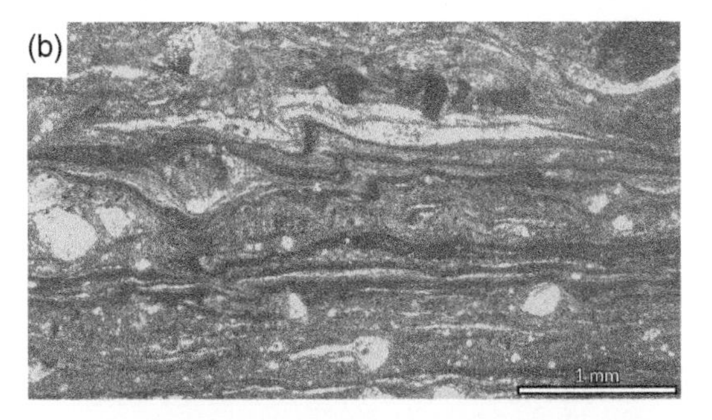

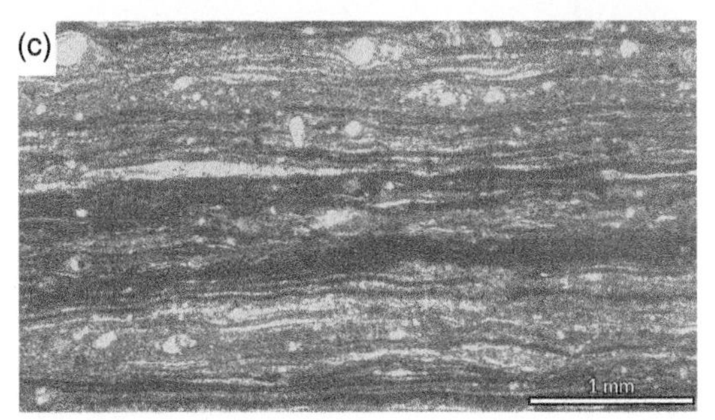

Figure 3.25 Typical microstructures of mylonites, showing porphyroclasts (larger fragments) in matrix. (a) Protomylonite. XPL: South Armorican shear zone, Brittany, France. (b) Mylonite. XPL. (c) Ultramylonite. XPL. Both (b) and (c) are sections of rounded cobbles from Pleistocene till in Ohio, USA, presumedly eroded from Canadian Shield in Ontario, Canada. XPL, cross-polarized light.

3.12.1 Asymmetric Pressure Shadows and Fringes

Pressure shadows and fringes (fibrous shadows) are formed as new mineral growths around a *core object*. Asymmetric pressure shadows/ fringes may be excellent shear sense indicators (Figure 3.24b), although they may have very complicated geometries when examined in detail (Figure 3.26). We distinguish **simple** types of pressure shadow (Figure 3.27a–c) where the shapes of the shadow and the fibers are regular, from **complex** types (Figure 3.27d, e) where the shadow and fiber shapes are generally curved and/or segmented.

The shapes of pressure shadows depend on: (1) the type of deformation in the matrix, and (2) the shape and orientation of the core object. The contrast between the shadow shapes in Figure 3.27b, c illustrates how the orientation of the core object can have a strong influence on the shape of the shadow. The shapes of fibers in pressure fringes depend on both these variables and how they grow. Some types of fibers always grow perpendicular to the faces of the core object (Figure 3.27a, c), while the growth of other types depends on strain in the matrix with respect to the core object, as shown in Figure 3.27b. These are called *face-controlled* and *displacement-controlled fiber* growth, respectively.

Two aspects of the pressure shadow/fringe geometry can be used to deduce shear sense: (1) the external shape of the pressure shadow, and (2) internal fibrous growths within pressure fringes. A reliable way to assess the asymmetry and sense of shear for simple shapes is to construct a reference line through the center of the core object parallel to the shear plane or foliation (Figure 3.27a) and a normal, perpendicular to the shear plane, through the center of the core object. These two lines define four quadrants 1–4. An asymmetric pressure shadow will have a greater proportion of its area in two opposite quadrants

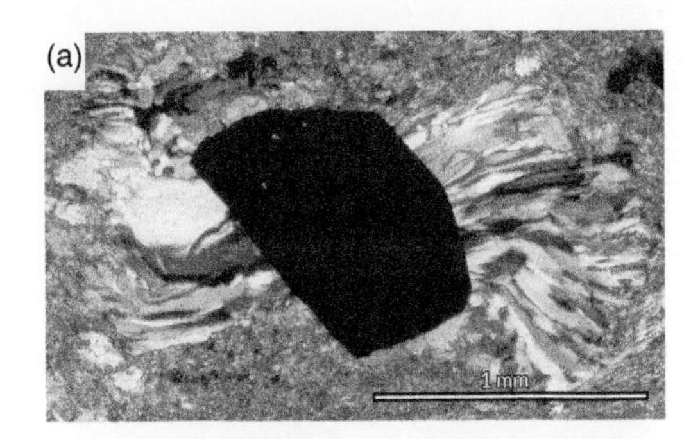

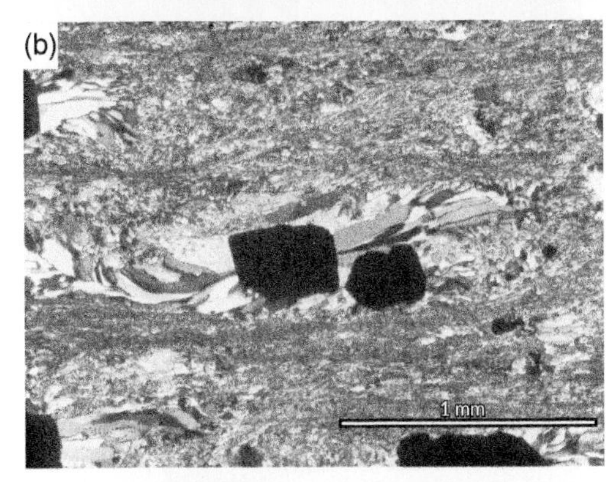

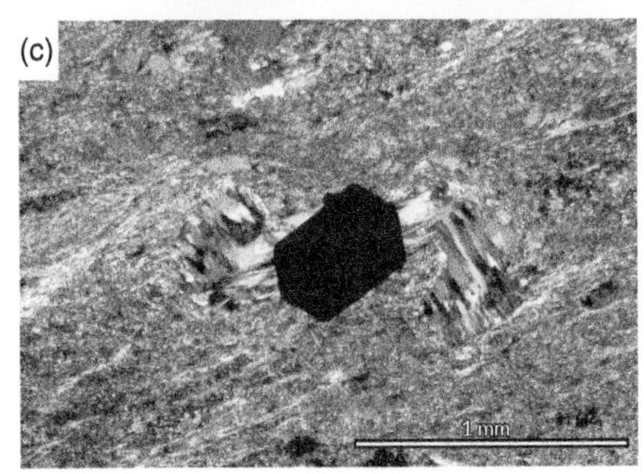

Figure 3.26 Pressure fringes. Fibers in the fringes directly adjacent to the pyrites grow perpendicular to their faces. These complex pressure fringes of quartz around pyrite polygons all have an S shape, suggesting dextral shear. The main differences between a, b and c relate to the shape of the pyrite polygon. a) Irregular section through pyrite with pressure fringes of quartz b) Complex pressure fringes around two adjacent pyrite grains c) Symmetrical, six-sided section of pyrite has similar pressure fringes on both sides. XPL: Yilgarn, Australia. XPL, cross-polarized light.

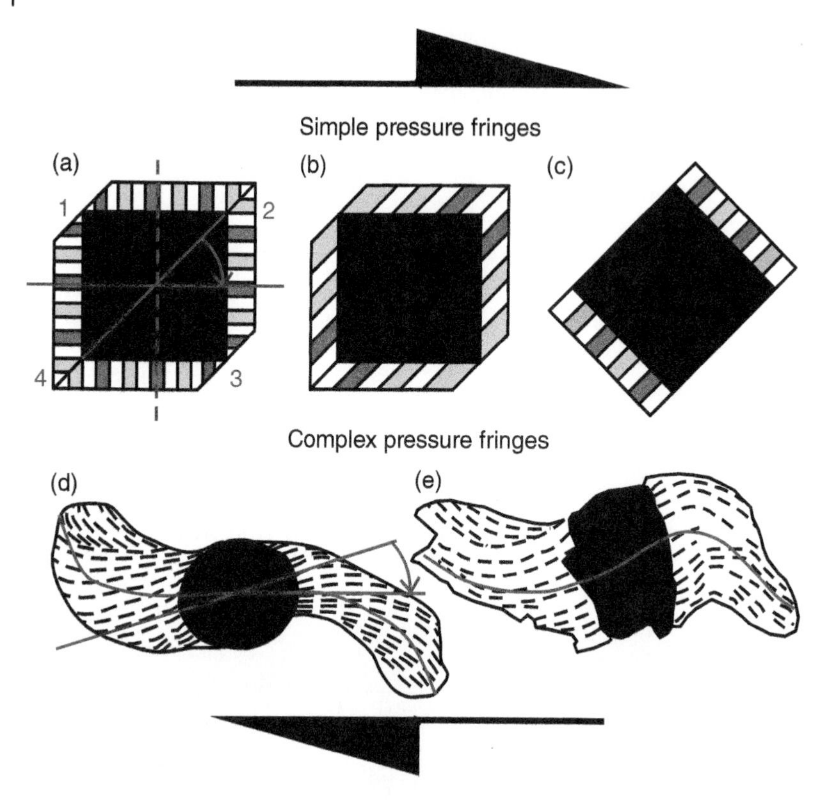

Figure 3.27 Geometry and use of pressure shadows as shear sense indicators. The black shape is a core object; the pressure shadows are indicated by fibers in gray or white colors, or by the dashed black line. All pressure shadows show a dextral sense of shear. (a–c) Simple types. In (a, c), the orientation of fibers is perpendicular to the object face, but in (b), the fiber growth direction is determined by strain in the matrix. To determine the shear sense from simple types, a reference line is constructed through the object parallel to the shear plane (red line), and the perpendicular line (red dashed line) is constructed, defining the four labeled quadrants. There is more pressure shadow in quadrants 2 and 4 than in quadrants 1 and 3. The shear sense is given by the clockwise rotation (red arc) from quadrants 2/4 to the reference plane. (d) and (e) are complex types. The shear sense is given by the shape of the shadow, defined by a line through its geometric center. The red curve has an S shape, giving a dextral sense of shear. (c) Pyrite framboid, Leonora, Australia, traced from Koehn et al. (2000). (d) Pyrite. Pyrenees, France, modified from Koehn et al. (2003).

than in the alternate pair (e.g. in Figure 3.27a–c, quadrants 2 and 4 have a greater proportion of the pressure shadow than 1 and 3). The shear sense is given by the vergence (sense of rotation) from these quadrants to the shear plane (Figure 3.27a).

Complex pressure fringes can have very intricate shapes that are affected by the shape of the core object, the growth mechanism of the fringe, rotation of the core object, and rotation and deformation of the pressure fringe. However, there is a simple rule that works for all types of pressure fringes to deduce shear sense. Examine the fringe adjacent to the core object. The fibers will point into one pair of opposite quadrants as defined earlier. The vergence from those quadrants to the shear plane will give the shear sense (Figure 3.27d). Additionally, the curved nature of many pressure fringes can be used. The center of the fringe will define a smooth curve through the center of the core object. The shape of this curve can be either

S or Z. S shapes imply dextral (clockwise) shear and Z shapes imply sinistral (counterclockwise) shear (Figure 3.27d, e).

3.12.2 Foliation Obliquity and Curvature

Many shear zones have a foliation that is oblique to their margins and the shear plane (Figures 3.24c, 3.28a, and 3.29a). If this foliation is the only fabric in the shear zone, the shear sense determination is straightforward: the shear sense is given by the rotation through the smaller angle that is needed to bring the oblique foliation parallel to the shear plane (Figure 3.29a). If the foliation in the shear zone is curved, the rotation of the curvature toward the center of the shear zone is in the sense of shear (Figures 3.28b and 3.29b).

3.12.3 *SC*, *SC′*, and *SCC′* Fabrics

Many shear zones in a strongly foliated rock have two planar fabrics related to shear (Figures 3.24d, 3.28b, c). Typically, one of the fabrics is a penetrative schistosity that has a sigmoidal shape (***S* foliation**) and the other is a planar zone of intense shearing (***C*** or ***C′* surface**) that is spaced at discrete and approximately constant intervals. *C* surfaces are parallel to the shear plane (Fig. 3.29c), while *C′* surfaces are slightly oblique in the opposite direction to the *S* foliation (Figures 3.28c and 3.29d). *C* and *C′* planes are also known as *shear band cleavages*, and *C′* planes have been called *extensional crenulation cleavage.*

Shear zones may form with only the *S* foliation, with *S* and *C* surfaces, with *S* and *C′* surfaces, or with *S*, *C*, and *C′* surfaces, defining ***SC***, ***SC′***, and ***SCC′*** fabrics, respectively. These fabrics are very reliable indicators of the sense of shear in mylonite zones. The curvature of the *S* foliation can be used independently of its obliquity, as described earlier (Section 3.12.2) (Figure 3.29b, c). Alternatively, the obliquity of the *S* foliation to the *C* or *C′* surfaces can also be used as described earlier (Section 3.12.2)

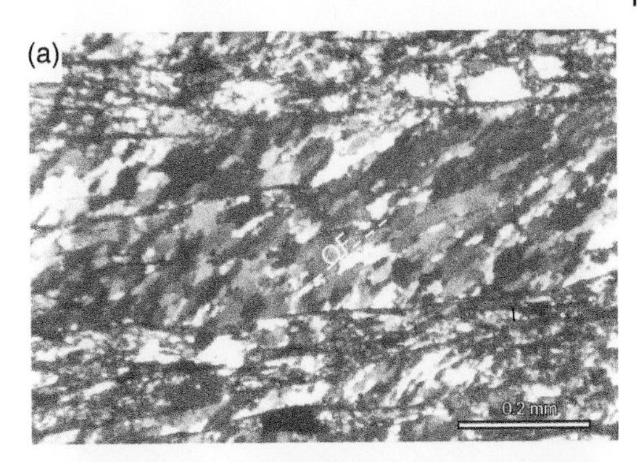

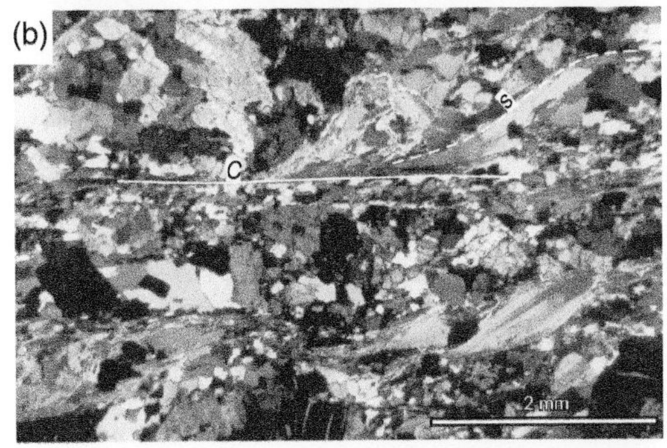

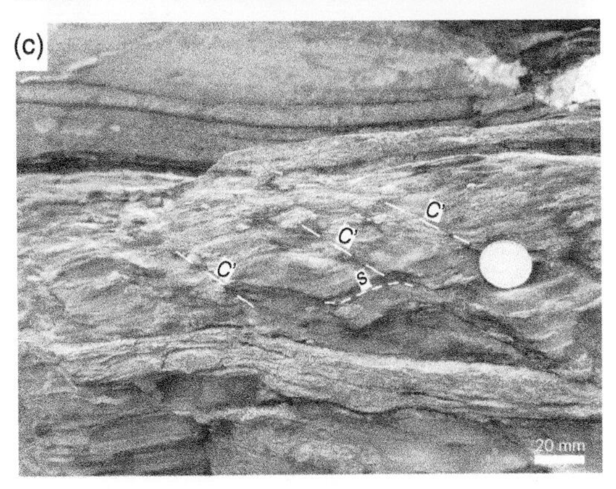

Figure 3.28 Oblique foliation, *SC*, and *SC′* fabrics. (a) Oblique foliation, OF. The dextral shear sense given by clockwise rotation onto the shear plane (cf. Figure 3.29a). XPL: Zivuku, Zimbabwe. (b) *SC* fabric. XPL: South Armorican shear zone, Brittany, France. Dextral shear sense given by clockwise rotation of *S* onto *C* (cf. Figure 3.29a) and curvature of *S* in the vicinity of *C* (cf. Figure 3.29b). (c) *SC′* fabric. Dextral shear sense given by obliquity of *S* surfaces to *C′* and by the curvature of *S* relative to *C′* (cf. Figure 3.29d). Cangas de Foz, North Spain. XPL, cross-polarized light.

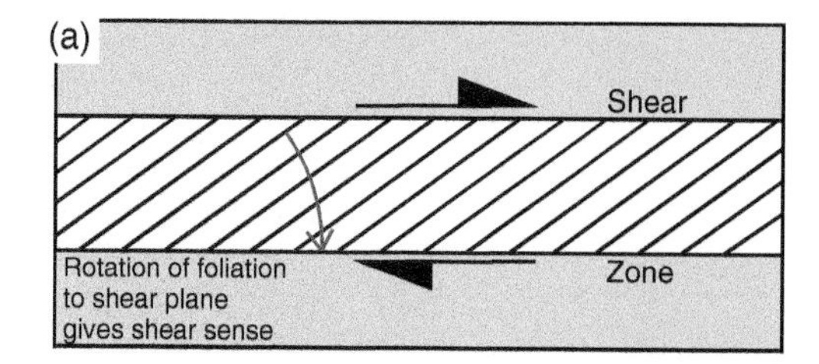

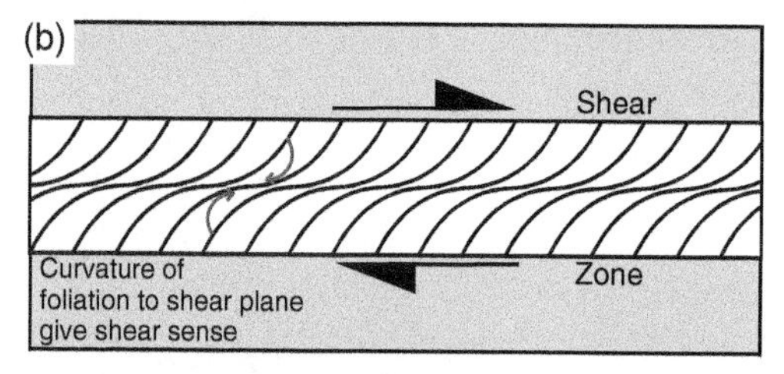

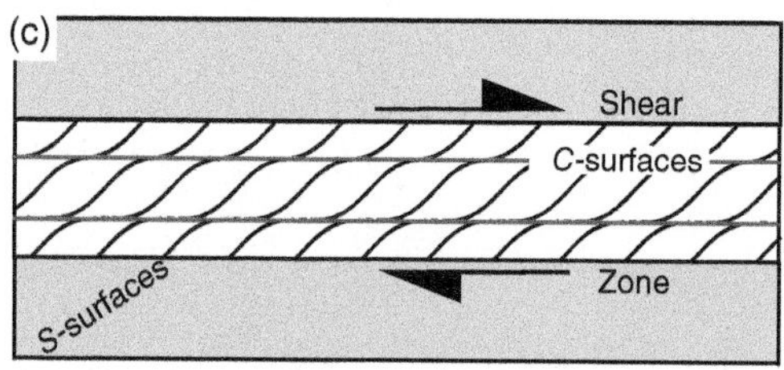

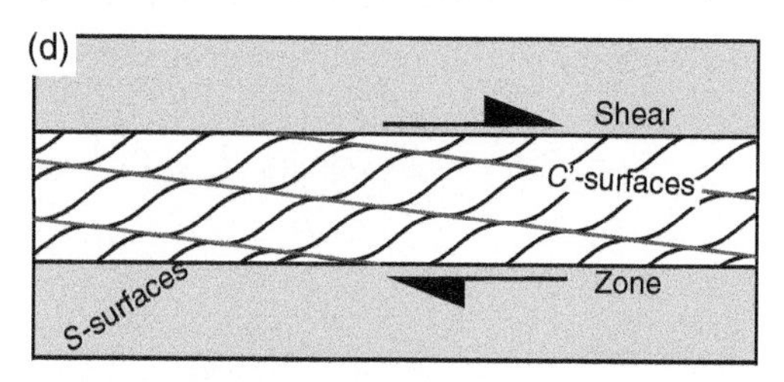

Figure 3.29 Shear sense determination from oblique and curved foliations and *SC* and *SC* fabrics in shear zones. The shear zone is the white area in each diagram between the gray wall rocks. Black lines indicate fabrics within the shear zone. (a) Oblique foliation. The red arrow shows a clockwise rotation from the oblique foliation onto the edge of the shear zone. This geometry implies a dextral sense of shear. (b) Curved foliation. The foliation curves in a clockwise sense toward the center of the shear zone (red arrow), also implying dextral shear. (c) *SC* fabric. The curved, more pervasive surfaces (black) are the *S* surfaces. They indicate a dextral sense of shear by the same criterion used in (b). The less curved *C* surfaces approximately parallel to the boundaries of the zone (d) *SC* fabric. This micrograph is similar to (c) except that the *C* surfaces (red) are oblique to the shear zone. The dextral sense of shear can be seen from the curvature of the *S* surfaces into the *C* surfaces.

(Figure 3.29c, d). This asymmetry is particularly evident where the *S* surfaces curve into the *C* or *C′* surfaces (Figure 3.29c, d). Small offsets on *C* and *C′* surfaces can sometimes be seen on grains. The obliquity of the *C′* surfaces, which is in the opposite direction to the *S* foliation, can be used (Figure 3.29d).

3.12.4 Porphyroclast Systems

A porphyroclast system consists of a **porphyroclast** (a clast that is large relative to its matrix) and a **tail** or **wing**, which has been derived from the clast by grain-size reduction, possibly with metamorphic reactions (Figures 3.24e, f, and 3.30). Porphyroclast systems are also called *mantled porphyroclasts*. Porphyroclast systems have very varied geometries that are best described in two dimensions relative to a reference plane through the center of the porphyroclast and parallel to the trace of the shear plane (Figure 3.31). If the boundaries of the shear zone are apparent, they define the orientation of the shear plane. A *C* foliation defines the shear plane in many cases. This assumption allows the distinction between five different types of two-dimensional geometry: θ, ϕ, σ, δ, and **complex types** (Figure 3.31). Porphyroclast systems that are symmetrical about the reference plane cannot be used for shear sense indicators: this applies to θ and ϕ types, which differ in the presence of elongate tails on ϕ types.

The three asymmetric types, σ, δ, and complex, are distinguished from each other by the shape of the tail in the vicinity of the porphyroclast. σ types usually have one relatively straight side and one that is curved (Figure 3.31 row a). δ types (Figure 3.31 rows c and d) have tails that have a pronounced curvature adjacent to the porphyroclast, and complex types (Figure 3.31 row e) have more than one distinct tail. Row (b) in Figure 3.32 shows porphyroclasts that are intermediate between σ and δ types.

It is important to realize that real porphyroclast systems have much more variety in shapes than

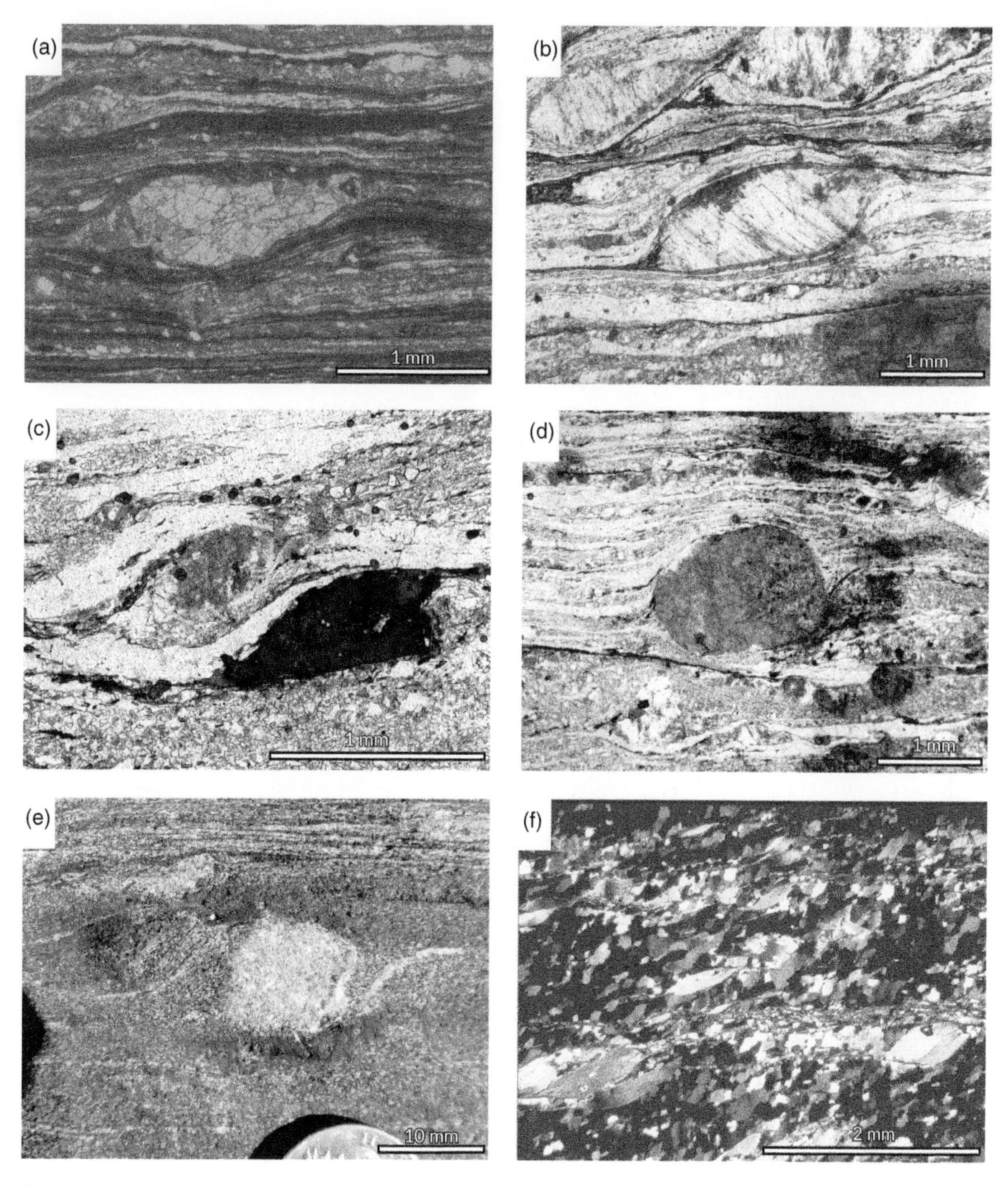

Figure 3.30 σ, clasts, δ clasts, and mica fish, showing some of the variety of possible shapes for porphyroclasts. All photomicrographs show a dextral sense of shear. (a) A sigma clast of clinopyroxene. The left stepping of the tails gives the dextral shear sense (see Figures 3.31 and 3.32a). XPL: Grenville Belt, Canada. (b) A sigma clast of feldspar with a left-stepping tail indicating dextral shear (see Figures 3.31a and 3.32a). PPL: Limpopo Belt, Zimbabwe. (c) Two sigma clasts (feldspar and an opaque mineral). Both show a clear left step and therefore dextral shear sense (see Figures 3.31a, b, and 3.32a). PPL: Mount Isa, Australia. (d) A delta clast of feldspar. The dark narrow tail curves in an anticlockwise sense as the porphyroclast is approached, indicating a clockwise rotation of the clast and a dextral shear sense. PPL: Mount Isa, Australia (see Figures 3.31c and 3.32b). (e) A delta clast of feldspar, showing similar anticlockwise rotation of tail toward clast (see Figures 3.31c and 3.32b). Mount Isa Inlier, Australia. (f) Mica fish. Rotating the long axes of the fish to the horizontal foliation suggests a dextral shear (see Figure 3.31a, b). XPL: South Armorican Shear Zone, Brittany, France. XPL, cross-polarized light; PPL, plane-polarized light.

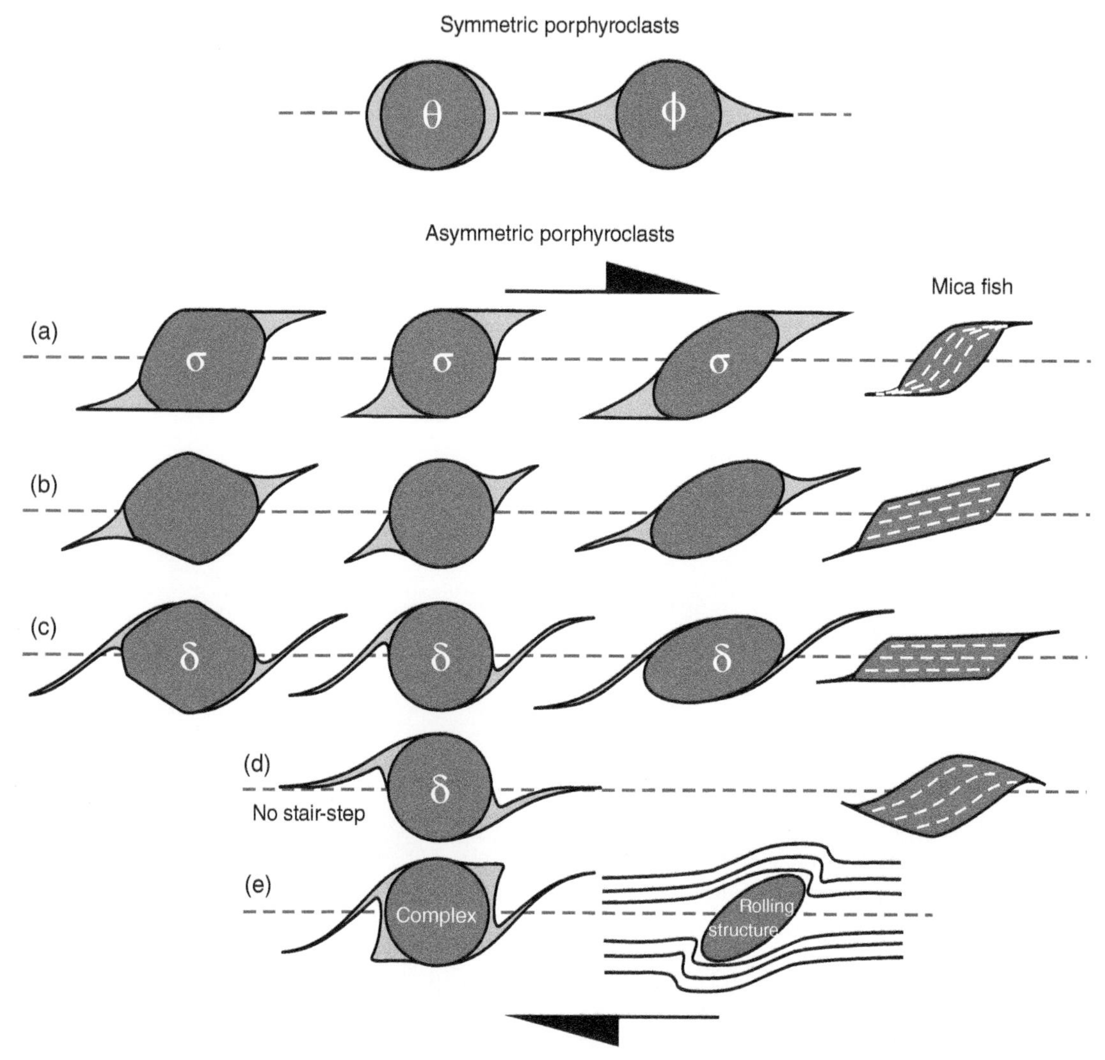

Figure 3.31 Types of porphyroclast system. In all diagrams, the plane through the center of the porphyroclast (red dashed line) defines a reference plane parallel to the sides of the shear zone. The top panel shows the two symmetric types, θ and ϕ. These are distinguished by the presence of an elongate tail in the case of ϕ types. The lower panels show: (a) σ types; (b) σ to δ types; (c) δ types (stairstepping); (d) δ types (no stairstepping); and (e) complex types. The three columns at left show how each type of tail looks with different shapes of the core porphyroclast. The right column, labeled "fish," shows analogous structures that are developed in single large grains of mica, amphibole, etc.

can be represented in a simple diagram. This variety depends on:

1) the type of flow in the matrix around the porphyroclast;
2) the initial shape of the porphyroclast (the columns in Figure 3.31 illustrate some ways in which the porphyroclast shape may change the shape of the system);

3) the deformation (including rotation) of the porphyroclast and its tail;
4) the degree of cohesion between the porphyroclast, its tail, and the matrix; and
5) the orientation of the section through the porphyroclast. For example, ϕ porphyroclast systems can occur in orthogonal sections through δ porphyroclast systems.

Due to the complexity of natural porphyroclast systems, it is useful to have a few simple guidelines to using porphyroclast systems as shear sense indicators. The most reliable method of using either of the asymmetric porphyroclasts (σ and δ types) is the *stairstepping rule* for tails. If the tails on either side of the porphyroclast are on opposite sides of the reference plane, they are said to be stairstep. This geometry applies to rows a–c and e of Figure 3.31. The sense of a stairstep is determined in the same way as for *en echelon* fractures. Imagine walking toward a giant porphyroclast along one tail. In order to cross the center of the porphyroclast and to continue walking along the other tail, it will be necessary to step to the left or right, giving the sense of step (Figure 3.32a). A reliable rule for determining the shear sense is that it is the opposite to the sense of step. In Figure 3.31, all the stairstepping tails step to the left, and the sense of shear is right-lateral or dextral. An asymmetric tail does not necessarily imply a stairstep geometry. Figure 3.31 row d shows an asymmetric δ-type porphyroclast that does not stairstep. It is nevertheless possible to deduce the sense of shear using the guideline that is described next.

The sense of curvature of the tail of a δ porphyroclast is a reliable shear sense indicator. Imagine walking along a δ tail toward the porphyroclast. If you were to walk along the center line (commonly quite easy to define since δ tails are usually narrow), your path would curve as you approached the porphyroclast (Figure 3.32b). If the center line of the tail curves to the left, the sense of shear is right-lateral or dextral (another way of expressing this is to say that the tail has an anticlockwise curvature as the porphyroclast is approached, implying a clockwise sense of rotation). All the δ tails shown in Figure 3.31 curve to the left as the porphyroclast is approached, and imply the dextral sense of shear shown in the figure. The *curvature rule* for shear sense is that the shear sense is the opposite to the sense of curvature of δ tails, i.e. anticlockwise curvature means clockwise shear sense and vice versa.

Structures in compositional layering that is deflected around porphyroclasts are sometimes called **rolling structures**, or quarter folds (Figures 3.24g and 3.31e). They can be regarded as a type of asymmetric fold, and the asymmetry of these types of folds can be a reliable shear sense indicator (Figure 3.32c). S and Z folds, respectively, form in sinistral and dextral shear.

Fish are elongate porphyroclasts with long axes inclined toward the shear direction in the same way as an oblique foliation (Figure 3.30f). They are most commonly made of mica, but porphyroclasts with similar shapes are also known as amphiboles, kyanite, and feldspar. They may have tails of recrystallized grains or grain fragments. The strong basal cleavage planes of mica in mica fish are oriented within 20° of the fish long axis, which may be inclined by up to 30° to the shear plane. There are close analogies between the shapes of fish and porphyroclasts, which are shown in the right-hand column of Figure 3.31. Fish can be used as reliable shear sense indicators by observing the sense of rotation implied from the long axis of the porphyroclast to the shear zone: this will be the same as the shear sense.

3.12.5 Precautions with Shear Sense Determination

One of the most basic precautions in determining shear sense from a section is that **only** truly asymmetric porphyroclasts must be used. Although

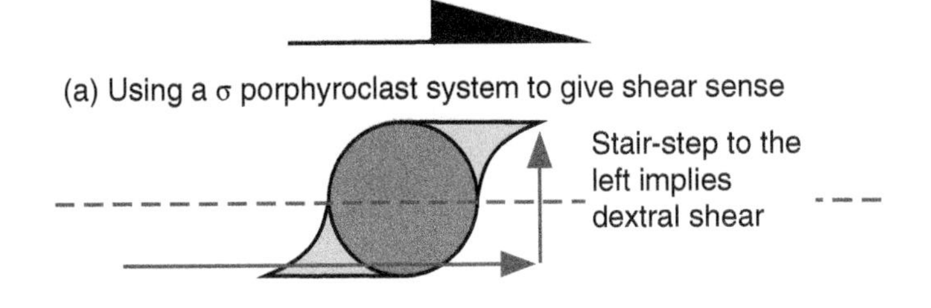

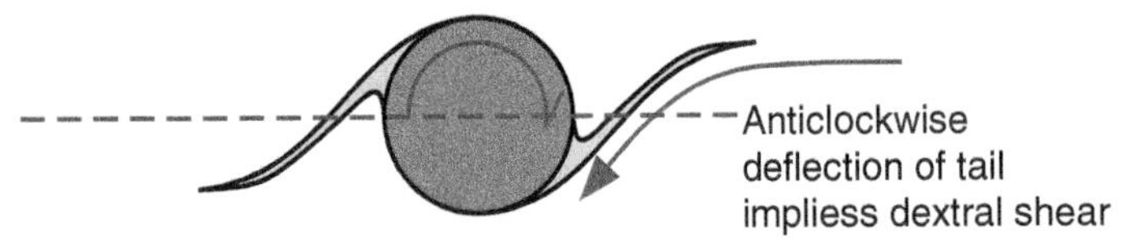

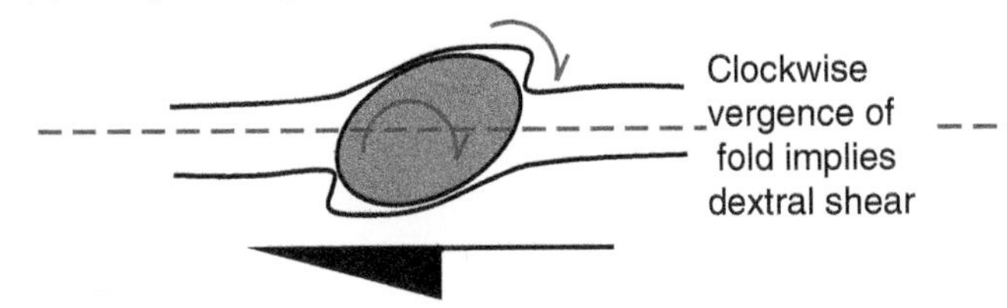

Figure 3.32 How to use porphyroclast systems to deduce shear sense. (a) σ clasts. The sense of stairstep (up to the right) implies a dextral shear sense. (b) δ clasts. The deflection of the tail as the porphyroclast is approached is to the left: this geometry indicates clockwise rotation of the clast and dextral shear. (c) A rolling structure. The folds in the layers around the clast have a clockwise vergence, indicating dextral shear.

this sounds trivial, it can be difficult to be sure that a shape is genuinely asymmetric, especially if the surface on which it is examined is not truly planar. Porphyroclasts that are very close to each other can also cause problems with shear sense determination; interactions between nearby porphyroclasts may produce local complex flow patterns that differ from the bulk shear.

It is possible that the sense of rotation of porphyroclasts may change in a single shear zone in a single deformation, depending on their initial orientation, shape, and type of flow. Therefore, opposite shear senses can be expected in a shear zone. A full consideration of all these factors is needed for any shear zone: several techniques can be employed for such a *vorticity analysis*. The more circular and the more isolated porphyroclasts are the better indicators of bulk flow. Complications may occur due to reactivation, possibly in the opposite sense to the initial shear.

The following methods of shear sense determination have been shown to be unreliable in particular instances: (1) Curved or truncated inclusion trails in porphyroblasts; (2) The asymmetry of crystallographic fabrics, especially in quartz; and (3) Fold asymmetry in shear zones. Using multiple independent shear sense indicators together can overcome these difficulties.

3.13 Collecting Oriented Samples and Relating Sample to Geographic Frames of Reference

Microstructural observation requires the use of thin sections of rocks attached to microscopic slides. The first step in making a thin section is cutting a **billet**, a rectangular block, from the rock sample. The thin section is made by gluing the thin section glass to one side of the billet and cutting the rest of the billet away from a thin slice of rock glued to the glass slide. The thin slice of rock is then polished to a uniform thickness. *Standard thin sections* are ground to a thickness of 0.03 mm = 30 μm thick, and *ultrathin sections* are ground to a thickness of 0.01 mm = 10 μm. The lengths and widths of a typical thin section are different in North America and Europe.

Any image of a microscopic object is a **micrograph** or microphotograph. The use of micrographs to interpret the geometry and kinematics of geological structures requires that the thin sections are correctly oriented. There are two basic approaches to orienting thin sections, using: (1) geographic-based orientation, or (2) fabric-based orientation. Geographic-based orientations are less common but more straightforward. Here the billet is cut with its long or short edges parallel to known geographic directions, typically the cardinal directions (NS and EW), and with its faces vertical or horizontal. This approach is used when one wants to preserve spatial information in a sample that lacks a fabric, such as might occur for an unfoliated granitic rock.

Most structural analyses are conducted on samples that are oriented relative to the fabric. If a rock has a foliation and lineation, workers typically use an *X*–*Y*–*Z* system of orientations where the *X*, *Y*, and *Z* directions are **fabric axes** oriented parallel to the finite strain axes S_1, S_2, and S_3, respectively. In a fabric sense: (1) *X* is parallel to the lineation and lies in the foliation plane, (2) *Y* is perpendicular to the lineation and lies in the foliation plane, and (3) *Z* is the pole to the foliation plane (i.e. perpendicular to foliation) (Figure 3.33a). The axis of rotation is called the vorticity vector; for fabrics recording simple shear, the vorticity vector is assumed to be in the *Y* direction. Structural geologists most commonly cut thin section parallel to the *XZ* plane, i.e. perpendicular to foliation and parallel to lineation. If deformation is accumulated by simple shear, the

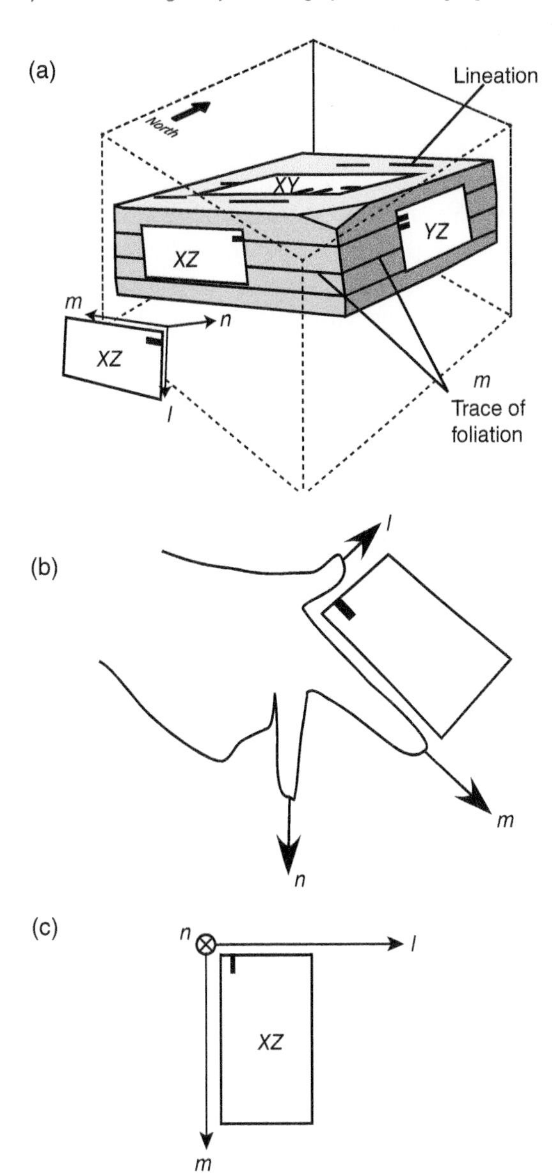

Figure 3.33 (a) The orientation of three typical thin sections cut relative to the fabric axes *X*, *Y*, and *Z* (sample frame of reference). *X* is the lineation, *XY* is the foliation plane. (b) The coordinate axes for a thin section *l*, *m*, and *n* form a right-hand system. (c) An *XZ* section looking from above in the direction of the positive *n* axis.

XZ section is perpendicular to the vorticity vector and should contain the maximum asymmetry (note: this assumption is not generally valid).

Geologists typically note the orientation by putting a series of notches on the edge of the thin section. The notches are made at the edge of the billets, which then are automatically transferred to the thin section during section making. There is currently no agreed convention for the notches relative to the fabrics. In the absence of agreement, it is difficult for any individual to determine the original geographic orientation of another geologist's thin section. If all one knows is the orientation of the fabric and the orientation of the fabric in the thin section, there are four possibilities for orienting every thin section in geographic space. Consider the case of a section of a rock with: (1) NS-oriented fabric with a horizontal lineation in a vertical strike-slip shear zone, (2) a notch in the *X* direction, and (3) notes indicating that the thin section is an *XZ*-oriented section. The notch denoting the *X* direction could be on the N or S end of the sample and slide. The section could be right-side-up (so geologists look down onto the plane of the thin section) or upside down (so geologists look up into the plane of the thin section). There is insufficient information to distinguish between the four possible orientations. This ambiguity is problematic because a dextral shear zone will appear to have a sinistral shear sense if one looks upward into the thin section, and vice versa. In general, one wants to look down (vertically) onto a thin section. However, even if a thin section is oriented so that one is looking down at it, there is twofold geographical ambiguity.

Tikoff et al. (2019) proposed a method to orient a thin section and to notch it relative to a sample's fabric in a manner that retains the thin section's geographic information. The following description is a simplified version of that proposed method. The core concept is that one selects a corner of the thin section as the origin of an orthogonal, 3D coordinate frame (more about this next). Three coordinate axes l, m, n emanate from that corner and conform to a right-hand convention scheme in which the third axis n points down into the rock mass. In order to visualize these coordinate axes, using your right hand, orient your thumb, index finger, and middle finger so they are mutually perpendicular (Figure 3.33b, c). By convention, one's thumb points in the l-direction, one's forefinger points in the m-direction, and one's middle finger points in the n-direction. The l- and m-axes are parallel to the plane of the glass. To orient the thin section in space unambiguously, one simply needs to specify quantitatively the orientations of those three vectors. Because structural geologists are always looking into Earth, we prefer that the downward-pointing ends of the l, m, and n coordinate axes have positive values. This occurs if the reference corner is the highest elevation or topmost corner of the thin section.

We take each of the l-, m-, and n-axes to be unit vectors, i.e. each having a magnitude of one (1). In order to define the geographic orientation of the thin section, we need to specify the orientations of the l-, m-, and n-axes relative to a geographic coordinate frame defined by a horizontal line pointing north (N = a direction having an azimuth of 000), a horizontal line pointing east (E = a direction have an azimuth of 090), and D = a vertical line pointing down. Because we want down to be positive and we want a right-hand coordinate system, we specify that N corresponds to the thumb on a right hand, E corresponds to the forefinger on a right hand, and D corresponds to the downward-pointing middle finger on a right hand (Figure 3.34a). The orientation of a unit vector parallel to the m-axis in the N–E–D coordinate system is given by the three angles: (1) α_m = the angle between the positive ends of N and m; (2) β_m = the angle between the positive ends of E and m; and (3) γ_m = the angle between the positive ends of D and m. Figure 3.34b shows that one can

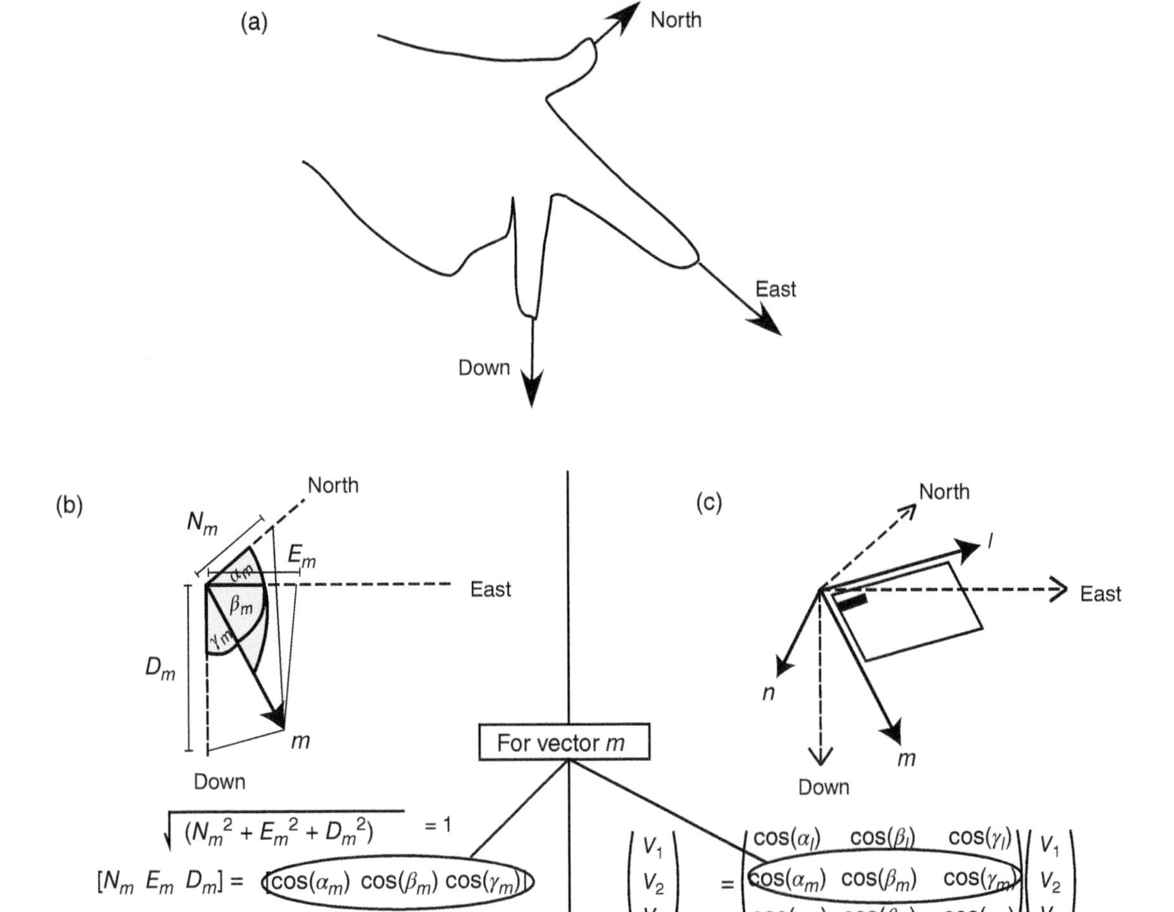

Figure 3.34 (a) The geographic axes North, East, Down constitute a right-hand coordinate framework (geographic frame of reference). (b) The angles α_m, β_m, and γ_m between m and North East Down axes. (c) The nine angles between *l, m, n* and N, E, D define a matrix that can be used to rotate any vector from the thin section coordinate framework to the geographic coordinate framework.

specify the unit vector *m* using three components in the *N*–*E*–*D* coordinate frame:

$$N_m = \cos(\alpha_m) \quad E_m = \cos(\beta_m) \quad D_m = \cos(\gamma_m). \tag{3.1}$$

A comparable analysis for the *l*- and *n*-axes yields

$$N_l = \cos(\alpha_l) \quad E_l = \cos(\beta_l) \quad D_l = \cos(\gamma_l) \tag{3.2}$$

and

$$N_n = \cos(\alpha_n) \quad E_n = \cos(\beta_n) \quad D_n = \cos(\gamma_n). \tag{3.3}$$

In relations (3.1), (3.2), and (3.3), the N_i, E_i, and D_i values are the components of the unit vectors parallel to the *N*-, *E*-, and *D*-axes, respectively. Recording the values of the three α angles, the three β angles, and the three γ angles, or three N_i

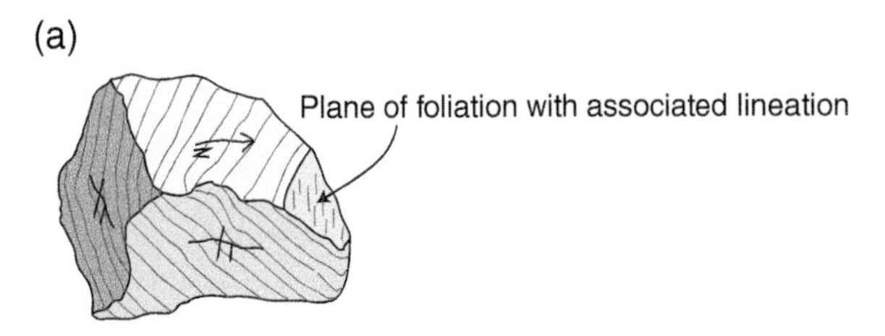

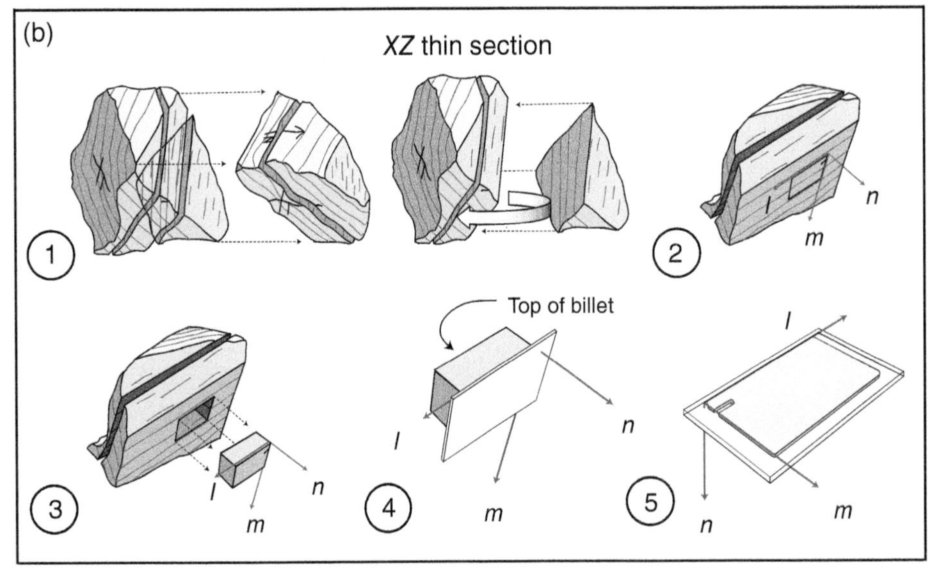

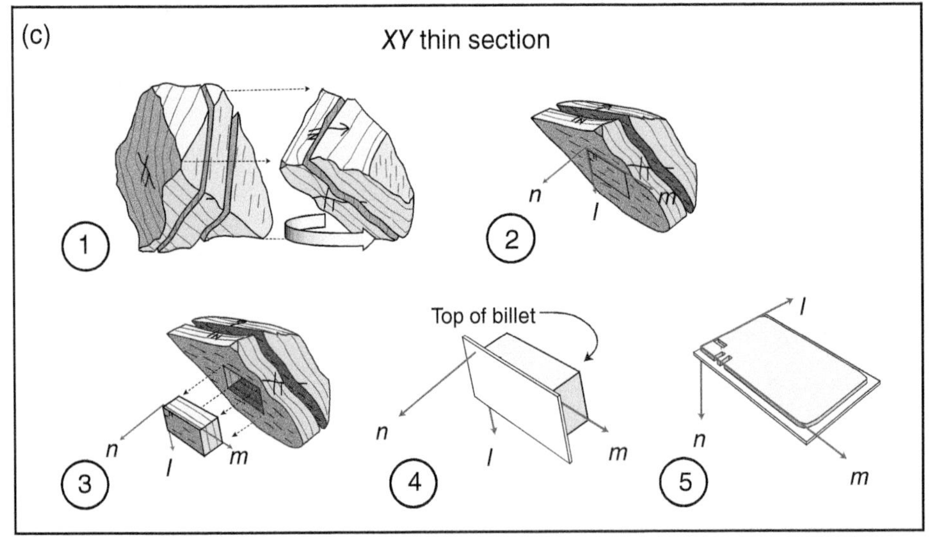

Figure 3.35 From oriented hand specimen to thin section, showing the notching system. (a) Oriented hand specimen. Strikes and dips of marked faces show their geographic orientation. (b) Cutting and notching an *XZ* thin section. (c) Cutting and notching an *XY* section. In both cases, the thin section is looked at from the top, i.e. *n* points downward.

values, the three E_i values, and the three D_i values, will specify the orientation of the thin section in geographic space (Figure 3.34c).

It is not, in fact, necessary to position the origin of the l–m–n coordinate frame at the highest corner of the thin section. As long as the n-axis points down and the coordinate frame is right-handed, the magnitudes of the l and m components will preserve the appropriate geographic orientation. The N_l, E_n, D_l, and D_n values may, in those instances, have negative values, but they will give the appropriate relative location of elements in the thin section.

Tikoff et al. (2019) also proposed a convention scheme for notching the billet (Figure 3.35): (1) the X orientation has a single notch, (2) the Y orientation has a double notch, and (3) there is no notch in the Z direction. This arrangement has the advantage of a single notch on the common XZ thin section. If the billet is notched in a manner that the notches are closest to a reference corner, and the reference corner is the highest point on the thin section, the notches always are on the uphill end of a thin section placed back in real space. An example of how to cut billets from a rock sample, to ensure that you are looking down onto the rock sample, is given in Figure 3.35.

There are cases where the notching scheme as outlined earlier is ambiguous, e.g. if a rock has a vertical fabric with horizontal lineations and the foliation strikes exactly EW or NS. These issues are discussed in Tikoff et al. (2019). The orientation of the notches does not really matter, provided that one identifies a reference corner and provides l-, m-, and n-axis orientations for the thin section. If this information is known, one can reconstruct the thin section orientation to either a known fabric or geographic coordinate system.

References

Bachmann, F., Hielscher, R., and Schaeben, H. (2011). Grain detection from 2d and 3d EBSD data – specification of the MTEX algorithm. *Ultramicroscopy* 111: 1720, 2011–1733.

Bell, T.H. and Rubenach, M.J. (1983). Sequential porphyroblast growth and crenulation cleavage development during progressive deformation. *Tectonophysics 92*: 171–194. https://doi.org/10.1016/0040-1951(83)90089-6.

Koehn, D., Hilgers, C., Bons, P.D., and Passchier, C.W. (2000). Numerical simulation of fibre growth in antitaxial strain fringes. *Journal of Structural Geology 22*: 1311–1324. https://doi.org/10.1016/S0191-8141(00)00039-0.

Koehn, D., Bons, P.D., and Passchier, C.W. (2003). Development of antitaxial strain fringes during non-coaxial deformation: an experimental study. *Journal of Structural Geology 25*: 263–275. https://doi.org/10.1016/S0191-8141(02)00022-6.

Tikoff, B., Chatzaras, V., Newman, J., and Roberts, N.M. (2019). Big data in microstructure analysis: building a universal orientation system for thin sections. *Journal of Structural Geology 125*: 226–234. https://doi.org/10.1016/j.jsg.2018.09.019.

4

Displacements

4.1 Overview

Displacements are a fundamental and observable property of rock deformation. A **displacement** describes the movement of a particle from an initial position to a final position. In order to define a displacement, we must specify both the direction and distance or magnitude that the particle moved. Since displacements have both magnitude and direction, we represent them mathematically as **vectors**. In cases where structural geologists can determine a series of positions that a particle occupied over a long time interval, we distinguish **displacement path**, the route through space followed by a particle, and **displacement history**, the timing of when particles occupied different positions along the displacement path. If we know the times at which a particle occupied two positions, we can calculate its **displacement rate** (or **velocity**) by dividing the displacement during that time interval by the length of the time interval. In studies of deformation that occurred long ago, the limited precision of dating techniques means that we can only calculate average displacement rates over millennia. Studies of neotectonic deformation typically measure displacement rates that have prevailed over time intervals of decades to centuries. Long-term average displacement rates may differ from short-period rates. For example, average displacement rates determined for movement on individual faults typically are millimeters per year to centimeters per year, but during earthquakes, displacement rates can be meters per second.

The collection of displacement vectors for the particles originally found at different positions defines a **displacement field**. The displacement field, relating the initial configuration to the final state of a material body, defines the **deformation** of any rock mass. In order to gain insight into deformations, we sometimes envision that displacement fields consist of three components: (1) **Translation**; (2) **Rotation**; and (3) **Pure strain**. Structural geologists often consider two different combinations of rotation and pure strain: **pure shear** and **simple shear** displacement fields. We can distinguish pure shear and simple shear displacement fields by examining the displacement paths of several particles.

4.2 Chapter Organization

This chapter has three sections: (1) Conceptual Foundation (Section A); (2) Comprehensive Treatment (Section B); and (3) Appendices. The distinction between the Conceptual Foundation and Comprehensive Treatment sections is the

An Integrated Framework for Structural Geology: Kinematics, Dynamics, and Rheology of Deformed Rocks,
First Edition. Steven Wojtal, Tom Blenkinsop, and Basil Tikoff.

level of mathematical rigor presented. The Conceptual Foundation section uses algebraic equations, while the Comprehensive Treatment section introduces and uses vectors and matrices, which are the basis of continuum mechanics approaches. The sections are written in parallel to encourage students to read the Comprehensive Treatment section, with its more advanced mathematics, once the basic understanding of the concepts is developed in the Conceptual Foundation. The parallel structure of the sections means that they have the same overall organization. Major divisions in the two sections have the same title and present the same material in the same order. Thus, Sections 4A.2 and 4B.2, both titled Particle Paths and Velocities, present the same concepts with different levels of mathematical support. Although major divisions cover the same topics, the material presented in one paragraph in the Conceptual Foundation often requires several paragraphs, and sometimes multiple subsections, to be addressed fully in the Comprehensive Treatment section. The Appendices present a brief overview of vector and matrix operations, which are utilized in the Comprehensive Treatment section of this chapter. The same organization is used for the next four chapters. Table 4.1 gives commonly used symbols in this chapter.

Table 4.1 Symbols used in this chapter.

Scalars (magnitude only)	
X, Y, Z	Coordinates of initial positions
x, y, z	Coordinates of final positions
u	displacement
v	velocity
t	time
du_x/dX	Gradient of u_x in the X direction
Vectors (magnitude and direction)	
$\mathbf{X, Y, Z}$	Initial position vector
$\mathbf{x, y, z}$	Final position vector
$\mathbf{u}$	Displacement, with components u_x, u_y, u_z
$\mathbf{v}$	Velocity, with components v_x, v_y, v_z
$\mathbf{i, j, k}$	Vectors parallel to cartesian coordinate axes 1, 2, 3
Tensors	
J	Displacement Gradient Tensor

N.B. Magnitudes of vectors, which are scalar quantities, are given by the same symbols are the vectors but not in bold. Tensor components are also not in bold.

4A Displacements: Conceptual Foundation

4A.1 Specifying Displacements or Individual Particles

4A.1.1 Basic Ideas

In the fault images shown in Figure 4.1, distinctive strata on the left side abut against the fault, and their displaced continuations on the right side (hanging wall) have moved down with respect to the strata on the left side (footwall). Thus, the fault is a normal fault. The term *net slip* describes the movement of particles on one side of a fault relative to particles on the opposite side of the fault. The letter *p* in Figure 4.1b identifies the location where the top of the hanging wall segment of layer A intersects the fault trace. The letter *q* in Figure 4.1b identifies the location where the top of the footwall segment of layer A intersects the fault trace. The line *pq* is parallel to striations on the fault surface. We infer that line *pq* defines the net slip on this fault.

The net slip on this or on any fault is an example of a **displacement**. Since the direction of fault slip is parallel to the plane of the image in Figure 4.1b, the measured offset of the marker layer in the plane of the section is precisely equal to the length of the net slip vector for that fault. To characterize the displacement across this fault, we must know both the magnitude and direction of fault slip. The magnitude is the measured offset of the marker

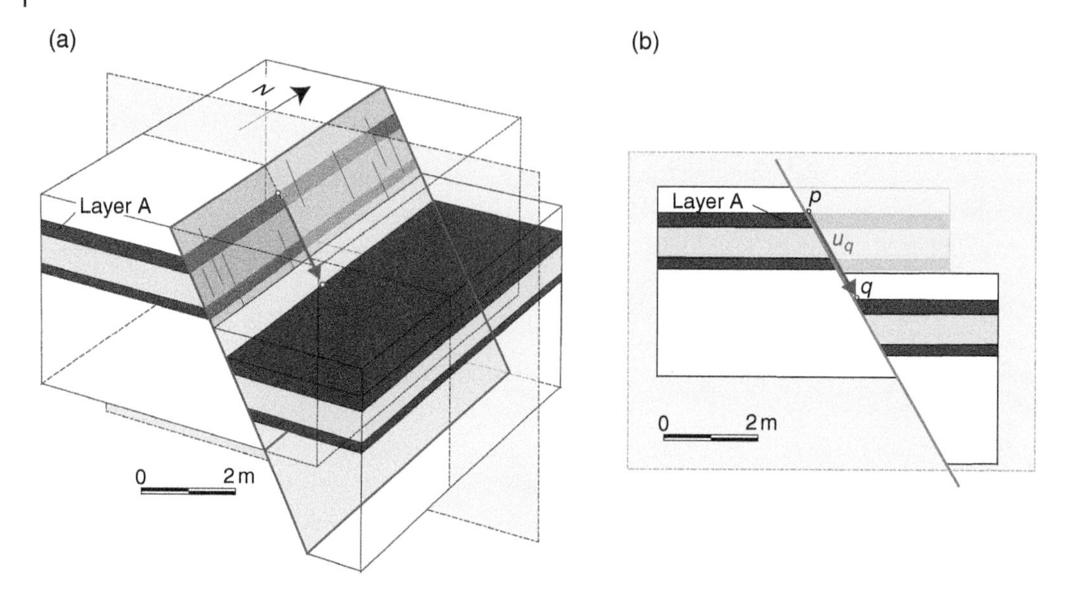

Figure 4.1 (a) Block diagram of layers cut and offset by a fault with NS strike and 60° east dip. The transparent gray plane is the plane of section in (b), and small red circles identify formerly adjacent points in hanging wall and footwall blocks. (b) Vertical section showing offset layers. $\boldsymbol{u}_q$ indicates the displacement of point q.

layer, 2.1 m, and the direction is given by dip and dip direction of the fault, 60 → 090 for the north-south (NS)-striking fault that dips 60° to the east.

The net slip vector connects the *initial* (or original) position of a particle to its *final* position and indicates the displacement of that particle. Notice that we distinguish between *particles* (which are small, identifiable pieces of rock, such as a particular mineral grain) and *positions* (which are locations or points in space). Faulting is just one straightforward geological example of the movement of one or more particles during a particular time interval. All of the familiar geological structures (faults, folds, fabrics, etc.) result from the movements of particles from initial positions to different final positions.

Consider the particle with position coordinates (X, Y) prior to deformation (Figure 4.2). The pre-deformation position coordinates define the *initial position* of a particle and provide a way to identify or name that particle. After deformation, the particle has moved to coordinates (x, y), which define its *final position*. That is, the particle that once occupied position (X, Y) has now moved to the position (x, y).

You will also recognize that we have imposed a reference frame consisting of an origin and coordinate axes in the vicinity of the fault. The location of the origin is arbitrary, but we often position it within a part of the system that has not moved. In this case, we assume that the origin is fixed and particles move with respect to it. Likewise, we have arbitrarily defined two-dimensional Cartesian coordinate axes, that is two orthogonal directions in space that are x and y axes with equal unit lengths along them. We determine both initial position coordinates (X, Y) and final position coordinates (x, y) with respect to the same coordinate axes.

The line between the initial position (X, Y) and the final position (x, y) of the particle is its **displacement vector** (Figure 4.2). We typically designate a displacement vector with a directed line segment or arrow, with the arrowhead pointing toward the final position of the particle. It is also

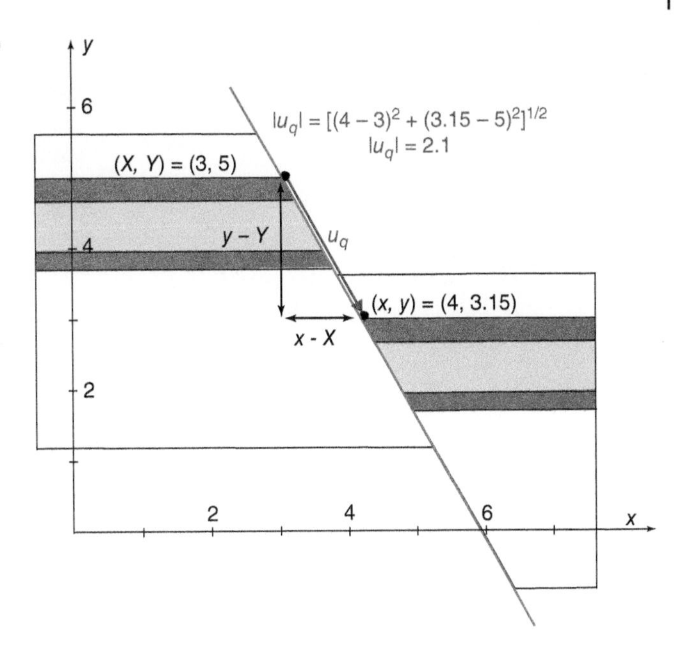

Figure 4.2 Vertical section from Figure 4.1b with Cartesian coordinate axes superposed. The offset of layer A given by vector $\boldsymbol{u_q}$.

possible to calculate algebraically the vector by determining the change in x and y coordinates of the particle. Thus, each displacement vector is defined by two values, u_X and u_Y, where u_X is the magnitude of change in position along the x axis and u_Y is the magnitude of change in position along the y axis. We can represent the displacement vector using a notation analogous to that used to represent a position in space, that is displacement vector $= (u_X, u_Y)$. We call u_X and u_Y the *components* of the displacement vector. The values of the components of the displacement vector are:

$$u_X = x - X \tag{4.1a}$$

$$u_Y = y - Y. \tag{4.1b}$$

The magnitude of the displacement vector is given by the equation:

$$\text{Magnitude} = \sqrt{\left[(x-X)^2 + (y-Y)^2\right]} = \left[(x-X)^2 + (y-Y)^2\right]^{1/2}. \tag{4.2}$$

This equation is just the Pythagorean theorem, with the change in position along the x axis and change in position along the y axis forming the two sides of a right triangle.

4A.1.2 Geological Example

We now apply this approach to the fault shown in Figures 4.1 and 4.2. We position the origin of the coordinate axes at the lower, left side corner of the image; each axis has a 1 m unit length. The original position of point q is located at $(X, Y) = (3, 5)$, and the final position of point q is located at $(x, y) = (4, 3.15)$. Measuring directly along the line connecting (X, Y) and (x, y), we find the net slip $\boldsymbol{u_q}$ to be 2.1 m, with 1 m of movement in the positive x direction ($u_X = 1$) and 1.85 m of movement in the negative y direction ($u_Y = -1.85$). The displacement is given by the vector $(1, -1.85)$. The magnitude of the vector can be calculated using Equation (4.2):

$$\text{Magnitude } \boldsymbol{u_q} = |\boldsymbol{u_q}| = \sqrt{\left[(4-3)^2 + (3.15-5)^2\right]} = \sqrt{4.4225} = 2.1\text{m}.$$

The calculated magnitude makes sense, of course; it is exactly the length measured directly on the image.

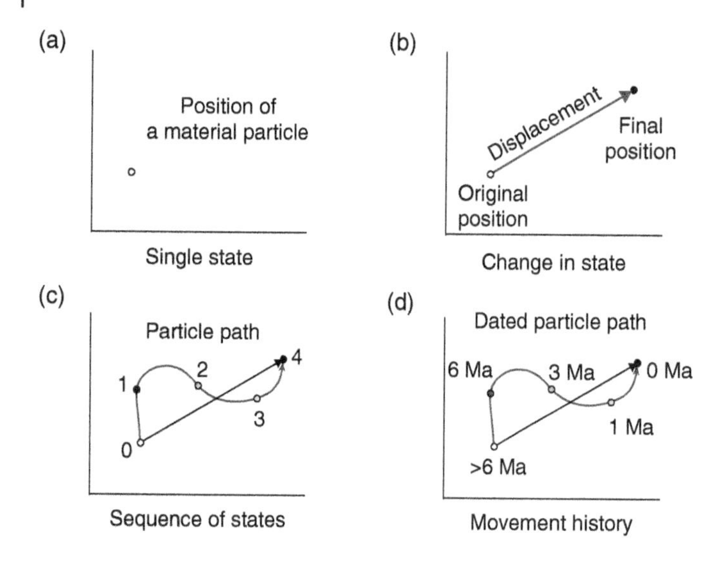

Figure 4.3 (a) Diagram showing the initial position of a material particle. (b) Diagram showing initial and final positions of the particle. The displacement of the particle relates to its initial and final positions. (c) The red curve shows the path followed by the material particle, with 1, 2, 3, and 4 indicating successive positions along the path. The displacement need not parallel the path followed by the particle. (d) Determining when the particle occupied different positions defines its movement history.

This treatment shows that determining the displacement of a particle is straightforward – if one can ascertain the original and final positions of a particle.

4A.2 Particle Paths and Velocities

4A.2.1 Particle Paths

4A.2.1.1 Basic Ideas

The movements of particles required to form geological structures, including faults, do not occur during a single instant of time. Rather, geological structures develop by increments over time. The net slip we observe on a fault, for example, may occur through multiple seismic events, through creeping motion over time, or some combination of seismic events and creeping motion. The concept of *incremental displacement* enables geologists to separate the total displacement into numerous, specific movements, each of which is a fractional portion of the overall displacement. The net slip on a fault or fault displacement considered in Section 4A.1 can therefore be subdivided into multiple increments.

Structural geologists have specific terms for the time-dependent development of geological structures. The position of a particle – either the original position (X, Y) or the final position (x, y) discussed here – describes the instantaneous condition of a system. This condition is known as a *single state* (Figure 4.3a). Any displacement describes a *change in state* (Figure 4.3b), typically the difference between the initial and final positions. A **path** (Figure 4.3c) defines a continuous series of states followed by an individual particle. Because an individual particle may move in different directions during successive intervals, the path that a particle follows need not coincide with the total displacement of that particle at any time. One way to characterize the path that a particle follows is to define incremental displacements for successive intervals from the start to the finish of deformation. In order to characterize fully the movement of a particle, one needs to define *when* the particle occupied different positions along its path, that is one must specify a **dated path** or movement history (Figure 4.3d).

4A.2.1.2 Geological Example

It is increasingly common for geologists to establish the paths that particles followed as part of their investigation of natural deformations. In a few settings, geologists have managed to reconstruct dated paths. Figure 4.4a is a generalized map of one such setting, the Wallace Creek area of the San Andreas fault in central California, United States. The upper reaches of the modern channel of Wallace Creek, an ephemeral stream that drains higher elevations to the north-east (NE) of the fault trace, flows roughly perpendicular to topographic contours until it reaches the San Andreas fault. The channel is offset ~100 m to the north-west (NW), parallel to topographic contours, as it crosses the fault before it turns to again flow down the slope to an alluvial fan. Along the base of this channel is a continuous string of alluvial deposits; plant material extracted from sediments near the bottom of these deposits yield a radiocarbon age of ~3700 years before present. Also visible to the south-west (SW) of the fault trace are the remnants of an older, now-abandoned channel that: (1) runs parallel to the trace of the fault for more than 100 m before turning to run down the slope to an alluvial fan; and (2) is filled by alluvium with plant material older than 3700 years old and younger than ~6000 years old. Deposits with composition and age comparable to this 3700–6000-year-old alluvium are preserved along the walls of the upstream portion of Wallace Creek. The spatial distribution of the channels and the deposits found in them are consistent with the sequence of states shown in Figure 4.4b. From the ages determined for the deposits, and the configurations of the channels in which they accumulated, one can estimate the path followed by a particle that would have been near the base of a ~6000-year-old alluvial channel on the SW side of the fault (Figure 4.4c and d). For the Wallace Creek segment of the San Andreas fault, incremental displacement vectors determined for particles near the fault are parallel to the total displacement vectors. The straight line parallel to the San Andreas fault in Figure 4.4c and d defines the movement history for particles on the NW side of the fault.

4A.2.2 Velocities

4A.2.2.1 Basic Ideas

The **velocity** of a particle is the change in its position divided by the time required to move from one position to the other. The change in the position of a particle is, of course, the displacement of that particle. Stated another way, then, the velocity of the particle is its displacement (final position–initial position) divided by the length of time over which it moved (final time [t_f] – initial time [t_i]). Written as an equation, we have:

$$\text{velocity} = \text{displacement} / \text{elapsed time}. \tag{4.3}$$

Since the displacement has both magnitude and direction, the velocity will have both magnitude and direction. Thus, a velocity in our two-dimensional example has two components (v_X, v_Y) where

$$v_X = (x - X)/(t_f - t_i) \tag{4.4a}$$

$$v_Y = (y - Y)/(t_f - t_i). \tag{4.4b}$$

The total velocity is given by

$$\text{velocity} = \left[(x - X)/(t_f - t_i), (y - Y)/(t_f - t_i)\right]. \tag{4.4c}$$

The direction of the velocity is exactly the direction of the displacement. The magnitude of the velocity is the magnitude of the displacement divided by the elapsed time:

$$\text{magnitude of velocity} = \text{magnitude of displacement} / \text{elapsed time} \tag{4.5a}$$

or

$$\text{magnitude of velocity} = \left[(x - X)^2 + (y - Y)^2\right]^{1/2} \Big/ (t_f - t_i). \tag{4.5b}$$

Because the standard unit of length is meters (so displacements are measured in meters) and the standard unit of time is seconds, velocities are sometimes given as m/s. The movements of particles during natural deformations are so slow,

(a)

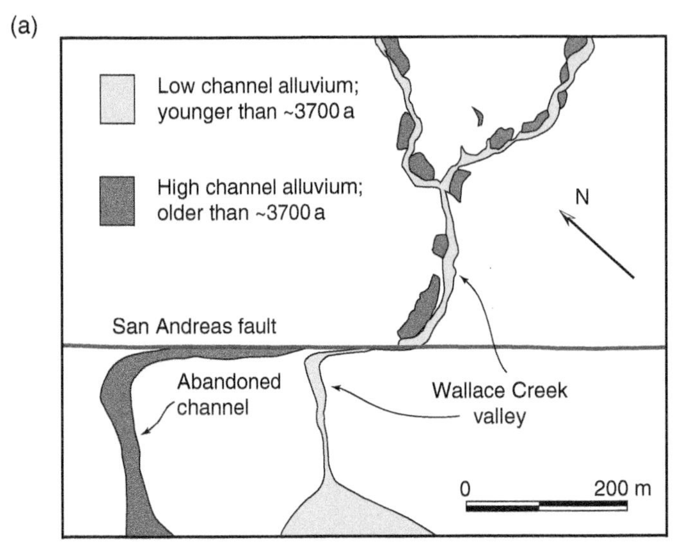

(b)

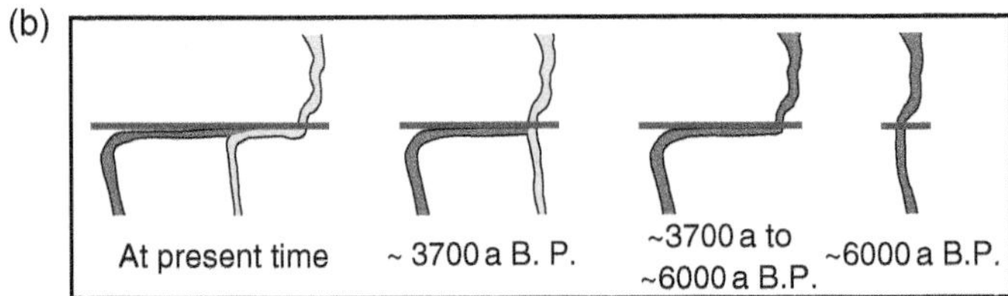

(c)

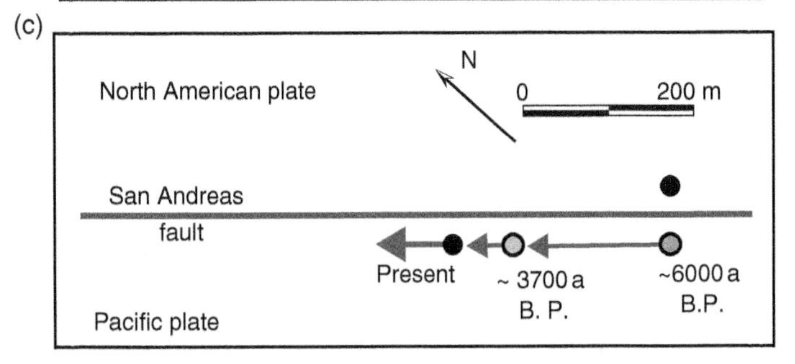

(d)

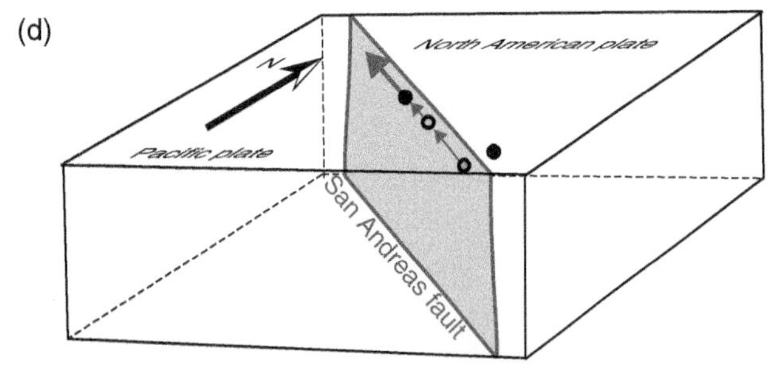

Figure 4.4 (a) Sketch map of a segment of the San Andreas strike-slip fault in central California, United States. Wallace Creek is deflected as it crosses the fault, and the downstream portion of its channel is offset by slip on the fault. An older, now-abandoned channel is also visible at this location. (b) Series of sketches illustrating different times in the development of the geomorphologic features seen at Wallace Creek. *Source:* (a, b) Modified from Sieh and Jahns (1984). (c) Sketch map showing the incremental displacements of the Pacific plate relative to the North American plate. (d) Block diagram of (c).

however, that structural geologists often calculate velocities in cm/a or mm/a (where the abbreviation "a" stands for *annum*, Latin for year). Finally, it is important to recognize that in calculating a velocity from a displacement, one is implicitly assuming that the displacement rate has been constant during the time interval in which the displacement occurred.

We first apply these ideas to the hypothetical fault given in Figure 4.1. If we assume that the fault moved at a constant rate for 1 million years, we can calculate the velocity. The magnitude of slip is 2.1 m and the elapsed time is 1,000,000 (10^6) years. Calculating the magnitude of the average velocity, we find:

$$\begin{aligned}\text{average velocity} &= 2.1\,\text{m}/10^6\,\text{a}\\ &= 2.1\times10^{-6}\,\text{m/a}.\end{aligned}$$

4A.2.2.2 Geological Example

We return to the Wallace Creek example (Figure 4.4) to explore further the concept of velocities. We must first define a frame of reference. We consider that the upstream portion of the creek (rock to the NE of the fault) is stationary and that the downstream portion of the creek (rock to the SW of the fault) moves past the stationary NE side of the fault. The downstream portion of Wallace Creek moved ~125 m relative to the upstream portion over the last ~3700 a. Dividing the distance moved by the time interval yields an average velocity of ~3.4×10^{-3} m/a or ~34 mm/a over the last 3700 years. Using the older, ~6000-year-old, geomorphologic features offset by the San Andreas fault at Wallace Creek, comparable calculations yield average velocities as high as 63 mm/a. This suggests that the rate of movement across this fault has not been steady. Historic observations emphasize that the relative movement of the two sides of this fault is highly episodic. Careful monitoring shows that there has been no movement for more than 150 years. The last slip increment on this segment of the San Andreas fault, which was 5.3–7.9 m of right-lateral movement, occurred within a few minutes on the morning of 9 January 1857. The movement across this fault is, then, definitely unsteady. By studying disrupted layers or abrupt changes in the type of sediment accumulating at individual locations along this fault, investigators have inferred that slip episodes have occurred repeatedly on this segment of the San Andreas fault. The time interval between different slip episodes ranges from 150 to 450 years.

It is only in a few settings that geologists have been able to estimate velocities using solely geologic observations. Moreover, geologists often must accept sizable uncertainties in the absolute positions occupied by particles and especially in the time at which the particles occupied those positions. The paths reconstructed often indicate that particles moved appreciable distances, but it is rare to have more than a few well-defined and precisely dated positions along a particle path.

With the widespread use of the global positioning system (GPS) in the past two decades, however, velocities are now much more commonly reported in the geological literature. Geologists are now capable of determining the positions of particles at Earth's surface at well-defined times with high precision. Even though GPS receivers determine the positions (single states) of stations at successive times, geologists almost always report GPS data as changes in position over time, that is as velocities. GPS data define the current velocities of plates relative to each other. In active plate margins where the crust is currently undergoing deformation, GPS data are precise enough to enable geologists to recognize velocities as slow as ~2 mm/a. Through a widespread collection of GPS data, geologists have seen significant growth in the number of well-defined particle velocities and short-term movement histories for particles (Figure 4.5).

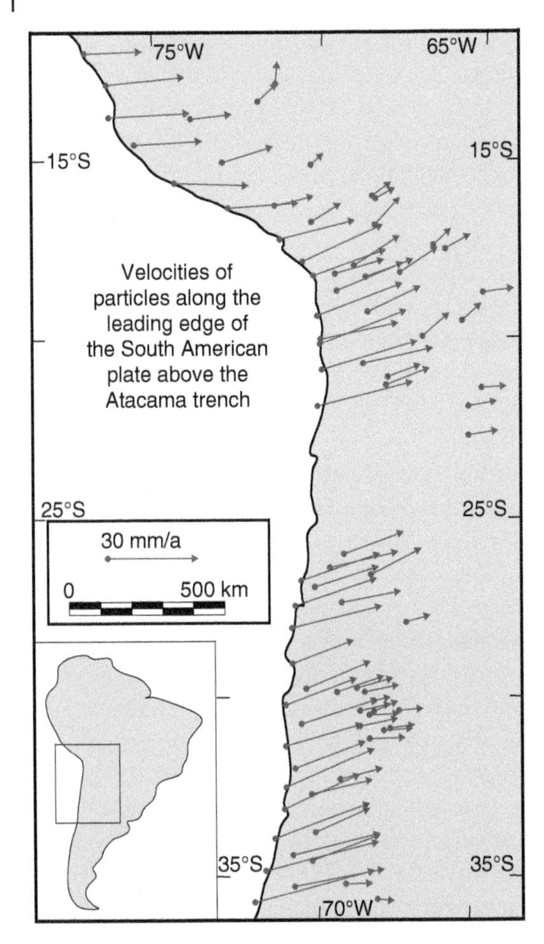

Figure 4.5 Vectors denoting the velocities of particles above the Atacama trench relative to a fixed South American craton. *Source:* Modified from Allmendinger et al. (2009).

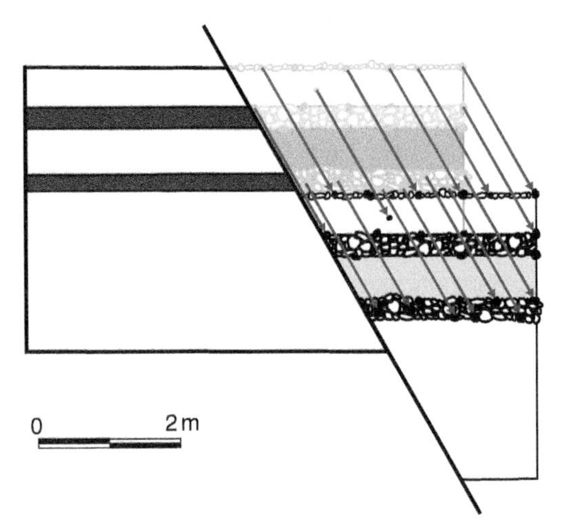

Figure 4.6 Vertical section of the normal fault from Figure 4.1 showing displacement vectors for some identifiable detrital particles. The magnitudes and directions of the displacements of all particles are the same.

4A.3 Displacements of Collections of Particles – Displacement Fields

4A.3.1 Displacement Fields

We return to Figure 4.1, which shows the net slip along a normal fault, to illustrate the concept of displacement fields. The displacement of an individual particle refers to the change in the state of that single particle, that is the movement of the particle from one position to another position. Thus, the line *pq*, which connects the initial and final positions of a single, identifiable particle, is that particle's displacement. Clearly, all particles on the hanging wall side of the fault have moved relative to the footwall. In this system, like many other systems where there are many particles, we can picture a group of arrows, each depicting the displacement vector of one of the many particles. This group of displacement vectors (Figure 4.6) is a **displacement field**. In the situation depicted in Figure 4.6, the displacement vectors of each of the many particles are the same. All of the displacement vectors have the same u_X and u_Y values. This need not be true of a displacement field. In most displacement fields, the displacement vectors for different particles are neither parallel nor of the same magnitude (Figure 4.7). Each displacement vector in a displacement field must only point in the direction of a particle's displacement and be proportional to the magnitude of that displacement.

As is the case for individual particles, the positions of several particles that are related to each

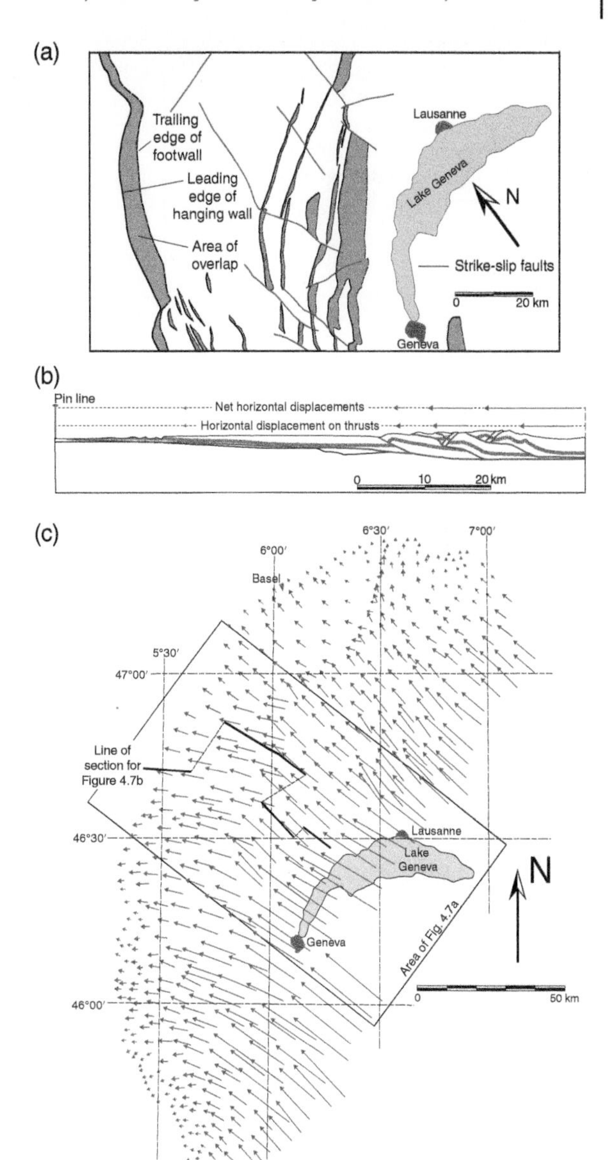

Figure 4.7 (a) Map of a portion of the Jura Mountains of France and Switzerland showing the estimated overlap of strata due to thrusting (solid line = leading edge of thrust hanging wall; dashed line = trailing edge of thrust footwall; red shading = area of duplicated strata) and separation across major strike-slip faults. (b) Schematic cross section showing inferred horizontal displacements at different positions across the Jura Mountains along the profile indicated in (c). (c) Plot of displacement vectors for different material particles in the Jura Mountains inferred from offsets on thrust and strike-slip faults and structural relief of folds. Note that the magnitude and direction of displacement vectors vary with position; this is a nonuniform displacement field. *Source:* Modified from Affolter and Gratier (2004).

other define a single state of a material object. In Figure 4.8a, several particles collectively define the square. The displacement field, that is the displacements of those several particles, records a change in state for the material object; if different particles experience different displacements the shape of the object will change. Thus, in Figure 4.8b, the material particles now define a rectangle. Defining the intermediate positions of the several particles, that is defining a sequence of states, enables one to specify the paths these related particles followed during deformation (Figure 4.8c). To define the movement history for the object, we must specify dated

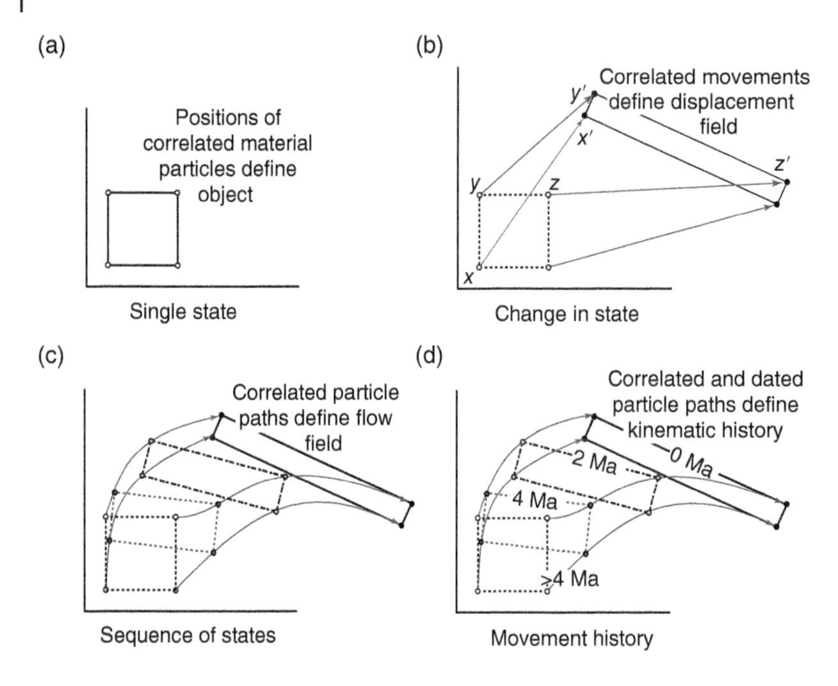

Figure 4.8 (a) Particles that are related to each other define a material object; here four related particles outline a square object. (b) A displacement field describes the overall movements of the related particles. Differences in the displacements of individual particles transform the square object into a rectangular object. In order to describe completely the deformation kinematics, we must determine paths followed by the individual particles shown in (c) and movement histories for the particles shown in (d).

paths for the related particles that define the object (Figure 4.8d).

4A.3.2 Uniform vs. Nonuniform and Distributed vs. Discrete Displacement Fields

There are two types of displacement fields: uniform and nonuniform. In *uniform displacement fields*, the displacement vectors of all particles within an area of focus are parallel to each other and have equal magnitudes. Because all particles experience the same displacement, the relative positions of particles do not change during or after uniform displacement. The displacement field for particles in the hanging wall in Figure 4.6 is an example of a uniform displacement field. We call this type of displacement field *translation* because it entails the movement of a collection of particles without rotation or distortion. Uniform displacement fields exist only for limited collections of particles. If we enlarge our field of view to include the footwall of the fault, the displacements of all individual particles are no longer parallel, nor do they have the same magnitude. A displacement field in which different particles have displacement vectors with either different magnitudes or different directions is a *nonuniform* displacement field. In nonuniform displacement fields, displacements vary with the position (either initial or final position) of particles (e.g. Figure 4.7).

There are two different ways that the displacement of particles can change with position in a nonuniform displacement field. In some cases, the displacement of neighboring particles are markedly different, that is displacements jump abruptly from one value to a different value across a boundary. The fault in Figure 4.6 is such a boundary; it separates particles that have moved from particles that have not. We use the term **discrete** to describe displacement fields that exhibit abrupt, step-like changes in displacement (Figure 4.1). The displacement field associated with this and other faults, which cut and offset preexisting features like marker layers, is a discrete displacement field. Individual fractures, veins, and stylolites are other structures responsible for discrete displacement fields. When the displacements of neighboring particles vary only slightly, changes in the direction or magnitude of displacement occur gradually with changes in the positions of particles. We use the term **distributed** to describe displacement fields where displacements change smoothly from each particle to its neighbor. In distributed displacement fields, marker layers are reoriented and/or distorted but exhibit no breaks or gaps. Distributed displacement fields are associated with individual shear zones (Figure 4.9), folds, pinch and swell structure, and many deformation fabrics. The examples we list are just a few of the almost infinite number of different displacement fields. The distinction between discrete and distributed deformation does depend on the scale at which we examine structures. For a convenient, standard scale to use in the field, we recommend examining on a scale of centimeters, as is shown in Figure 4.9.

Examinations of distributed displacement fields inherently lead us to consider how to quantify variations in displacement with the position. We use the notion of a **displacement gradient** to characterize how displacements change with changes in the positions of the particles. For example, moving from left to right across Figure 4.9b, the magnitude of the displacements of material particles along the margin of the dike increase as magnitudes of the x components of the particles' positions along the periphery of the dike increase. We show how to calculate displacement gradients in the Comprehensive Treatment section of this chapter; at this point, it is important to recognize that displacement gradients give rise to reorientations and distortions of collections of particles.

4A.3.3 Classes of Displacement Fields

In nonuniform displacement fields, the relative positions of particles change because their displacements are not identical. We distinguish three end-member types of relative movement for pairs of particles. The relative movement of two particles is **convergent** if the two particles move toward each other along the line that connects them (Figure 4.10a). The relative movement of two particles is **divergent** if the particles move away from each other along the line that connects them (Figure 4.10b). The relative movement of two particles is **transcurrent** if the particles move past each other along directions perpendicular to the line that connects them (Figure 4.10c).

You may recall seeing the terms convergent, divergent, and transcurrent used to describe the boundaries of the tectonic plates. Plate tectonics is a kinematic description of plate motions based on the displacement of one plate relative to another. The relative motion is observed by "tagging" two particles on the two different plates at opposite ends of *any* line segment drawn perpendicular to that plate boundary segment (Figure 4.11). If the two particles at the end of this perpendicular line experience relative motion toward each other, we say that the plate boundary segment is convergent. Likewise, the plate boundary segment is divergent or transcurrent, respectively, if the particles move away from each other or move past each other.

Geodetic measurements give the displacement per year of particles at different positions in the vicinity of plate boundaries. End-member examples include the divergent plate boundary in Iceland (Figure 4.12a), the convergent plate boundaries associated with subduction (Figure 4.5; western

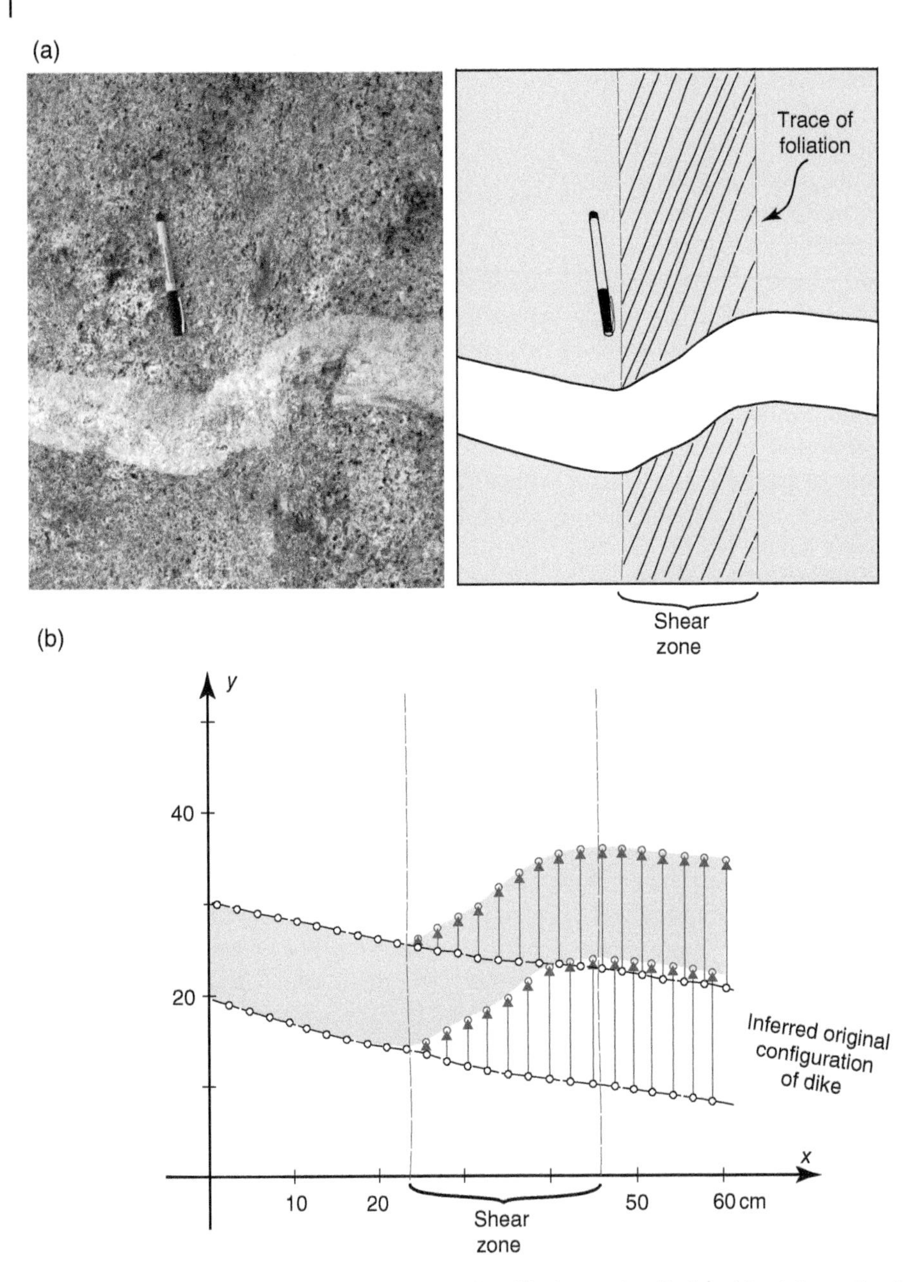

Figure 4.9 (a) Photograph and sketch of an aplite dike in granite displaced by deformation in a shear zone (the odd perspective in the photo arises because the image was rotated 90° counterclockwise); foliation traces indicated in the sketch show that shearing is concentrated within narrow, planar-sided zone denoted by dashed red lines. (b) Plot of the same field of view from (a), showing that displacements of particles change as a function of position. Thus, displacement gradients exist between dashed red lines.

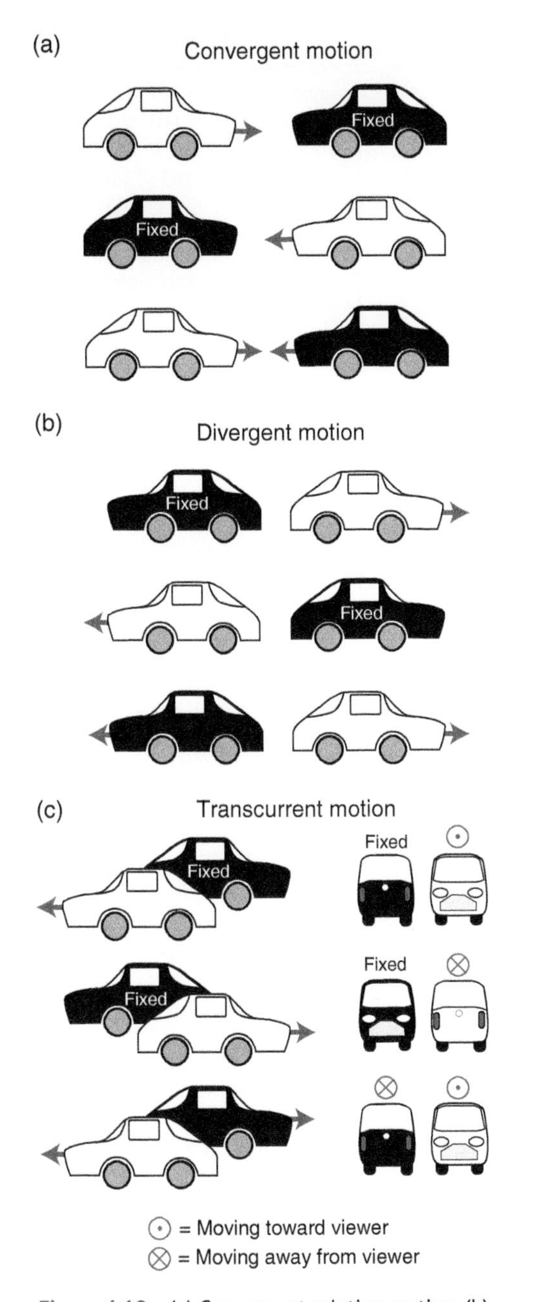

Figure 4.10 (a) Convergent relative motion. (b) Divergent relative motion. (c) Transcurrent relative motion (as shown, the circle with the dot indicates movement out of the plane of the diagram and the circle with "×" indicates movement into the plane of the diagram).

South America) or collision (Figure 4.12b; Tibet-India suture), and the transcurrent plate boundary of central California (Figure 4.12c).

4A.4 Components of Displacement Fields: Translation, Rotation, and Pure Strain

A **deformation** is a change in the position of particles. Because the displacement field relates the initial configuration of the collection of particles that define a rock mass to their final state, it defines the **deformation** of that rock mass. If there is no displacement field (all particles occupy the same positions before and after a particular time interval), there is no deformation. In **homogeneous deformations**: (1) any collection of particles that defined a straight line before deformation define a straight line after deformation; (2) any two lines that were parallel before deformation remain parallel after deformation; and (3) the ratio of original lengths of portions of a line to their deformed lengths is constant along all directions through the collection of particles (Figure 4.13a). Uniform displacement fields, as defined here, are a special type of homogeneous deformation. In **heterogeneous** (or **inhomogeneous**) **deformations**: (1) collections of particles that define straight lines before deformation generally define curved lines after deformation; (2) most pairs of lines that were parallel before deformation are no longer parallel after deformation; and (3) the ratio of original lengths of portions of a line to their deformed length will vary along most directions through the collection of particles (Figure 4.13b). Most homogeneous deformations, and all heterogeneous deformations, have nonuniform displacement fields.

We divide displacement fields responsible for homogeneous deformations into three different components: translation, rotation, and pure strain (Figure 4.14). Each of the three components

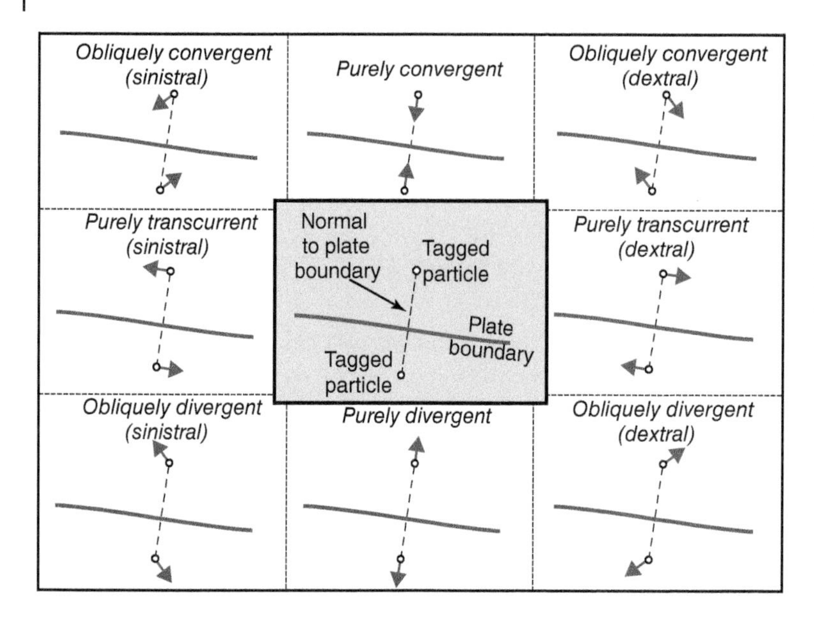

Figure 4.11 We use two tagged particles at opposite ends of a line normal to a plate boundary (see shaded box at the center of the figure) to define the type of plate movement. Different types of relative movement of those particles are illustrated in the unshaded squares. See text for explanation.

correlates directly to physical features or effects that we can observe.

The translation component is defined by a vector whose length and magnitude describes precisely the displacement of at least one of the particles within a mass and whose length and magnitude describes approximately the displacement of all other particles in the mass. We can envision the translation as the displacement of the center of mass of a collection of particles that experience nonuniform displacements. If a displacement field consists of *only* the translation component, it will be a uniform displacement field where the individual displacement vectors for all particles are parallel to each other and have equal magnitudes. In this case (Figure 4.14a), we can describe the displacement vector components of all material particles using

$$u_X = a \tag{4.6a}$$

$$u_Y = b, \tag{4.6b}$$

where a and b are constants. a is the magnitude of the translation vector component in the x direction, and b is the magnitude of the translation vector component in the y direction. Notice that a translation-only displacement field moves an object in space without rotating or distorting it.

The translation is sometimes a good approximation for the overall displacement field for a collection of particles offset by a fault (Figure 4.1). For example, the San Andreas fault dissected the Neenach–Pinnacles volcanic complex, which had formed shortly before slip commenced on the fault. Fault slip separated this complex into the Pinnacles Volcanic Formation now found in central California near the town of Hollister and the Neenach Formation in southern California (Figure 4.15). Although there is some distortion and rotation of the collections of particles that compose these bodies, the total displacement field for them is dominated by a bulk translation vector with a length of ~315 km and an azimuth of 130° or 310°, depending upon which side of the fault you are standing (i.e. your reference frame).

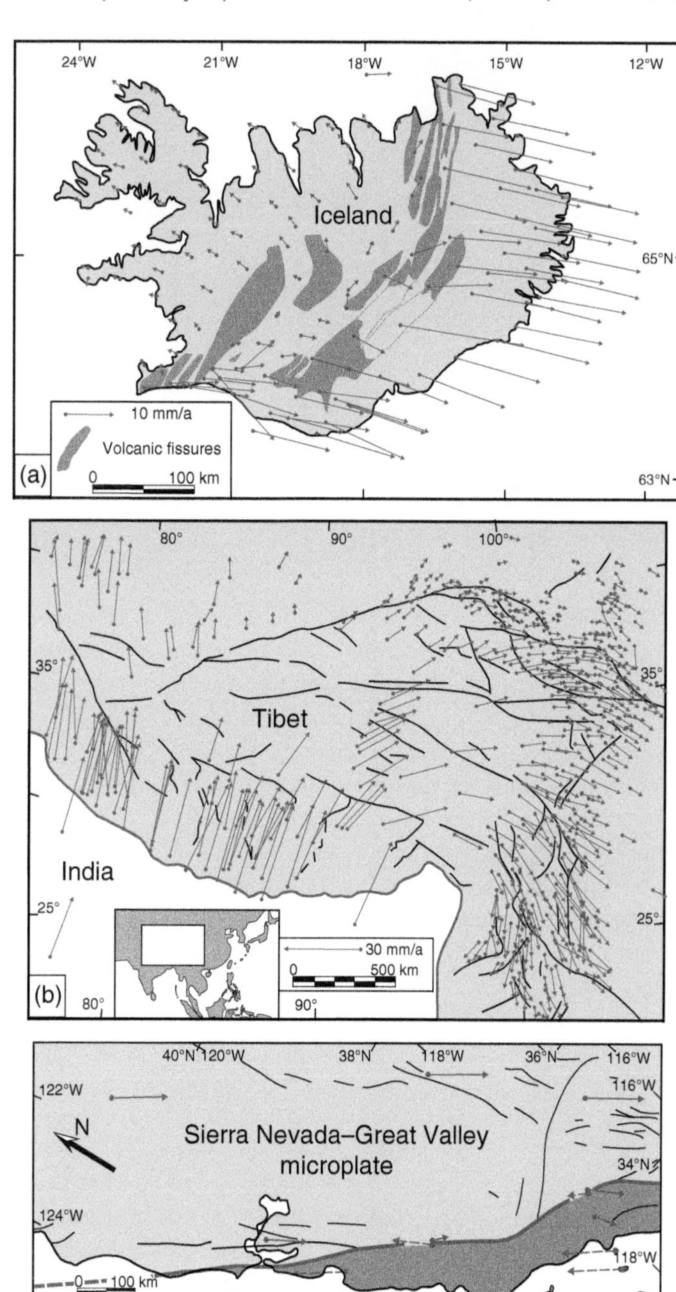

Figure 4.12 Vectors denoting the velocities of particles near different types of plate boundaries. (a) Vectors denote the movement of particles relative to fixed North America. Particles east of the volcanic fissures are moving away from North America, identifying the location of the divergent plate boundary in central Iceland. *Source:* Modified from Árnadóttir et al. (2009. (b) Vectors denoting movement of particles (relative to fixed Eurasia) in the vicinity of the India–Tibet suture (bold red line). *Source:* Modified from Allmendinger et al. 2009. Material particles near the suture and in the western portion of Tibet indicate convergent relative movements, whereas those east of Tibet and of the suture exhibit transcurrent relative movements. (c) The San Andreas fault in central California separates the Sierra Nevada–Great Valley microplate (in lighter gray) from the Pacific plate (in darker gray). Solid arrows denote velocities of the Sierra Nevada-Great Valley microplate (in lighter gray), and dashed arrows denote velocities of the Pacific plate; data from DeMets.

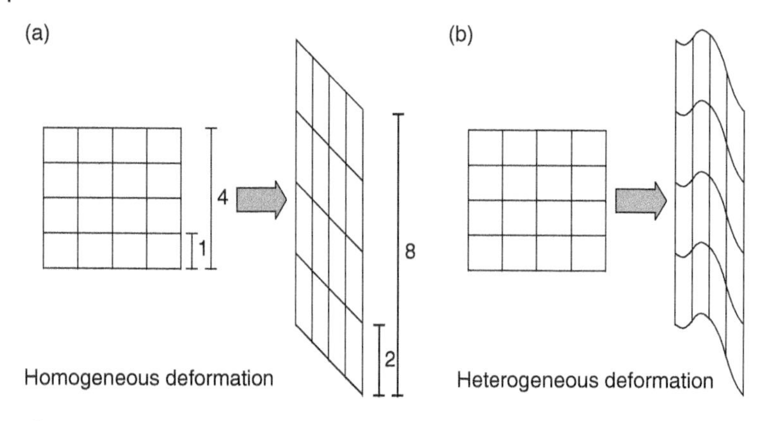

Figure 4.13 Examples of homogeneous (a) and heterogeneous (or inhomogeneous) (b) deformation. (a) In homogeneous deformation, lines that were originally straight and parallel remain straight and parallel after deformation. Also, deformation in a small area is the same as deformation of the whole area. (b) In heterogeneous deformation, these conditions do not apply.

The other two components of uniform displacement fields, the rotation and pure strain components, exist only if displacements vary with position, that is only if displacement gradients have nonzero values. Because different particles experience different displacements during rotation and/or pure strain, these two components are more complex than the translational component. A displacement field associated with **rotation** alone is shown diagrammatically in Figure 4.14b. The magnitude of a particle's displacement vector depends on its initial position:

$$u_X = (\cos\phi - 1)\cdot X + (\sin\phi)\cdot Y \quad (4.7a)$$

$$u_Y = -(\sin\phi)\cdot X + (\cos\phi - 1)\cdot Y \quad (4.7b)$$

where X and Y are the particle's position coordinates and ϕ is the counterclockwise angle through which the collection of particles rotates.

As with all mathematical equations, it is worthwhile to work through some cases for which one already has an intuitive feel for the answer. If there is no rotation, then $\phi = 0$. Since $\cos 0 = 1$ and $\sin 0 = 0$, then $u_X = 0\cdot X + 0\cdot Y = 0$, and $u_Y = -0\cdot X + 0\cdot Y = 0$. Thus, if $\phi = 0$, then the rotation-only displacement field is null and there is no deformation. Consider next the case of $\phi = 45°$ (shown in Figure 4.14b). Since $\cos 45° = 0.707$ and $\sin 45° = 0.707$, then $u_X = -(0.293)\cdot X + (0.707)\cdot Y$, and $u_Y = -(0.707)\cdot X + (-0.293)\cdot Y$. For this case, the magnitude of displacement for particles that lie on either the x or the y axis is directly proportional to their distance from the origin. That is, the displacements of particles originally at positions $(X, Y) = (0, 1)$ or $(X, Y) = (1, 0)$ are half the displacements of particles originally at positions $(X, Y) = (0, 2)$ or $(X, Y) = (2, 0)$. This proportion relationship holds true for any line emanating from the origin. Consequently, all material lines remain straight and rotate about the origin.

The tilting of geological strata can be an example of a rotation-only displacement field (Figure 4.16). Originally horizontal sedimentary strata are often rotated to moderately to steeply dipping positions, yielding geological structures known as "hogbacks" or "flatirons." In some geological provinces, this tilting occurs even though the sedimentary strata have not been translated far from where they were deposited and show very little internal distortion. The displacement field for particles that

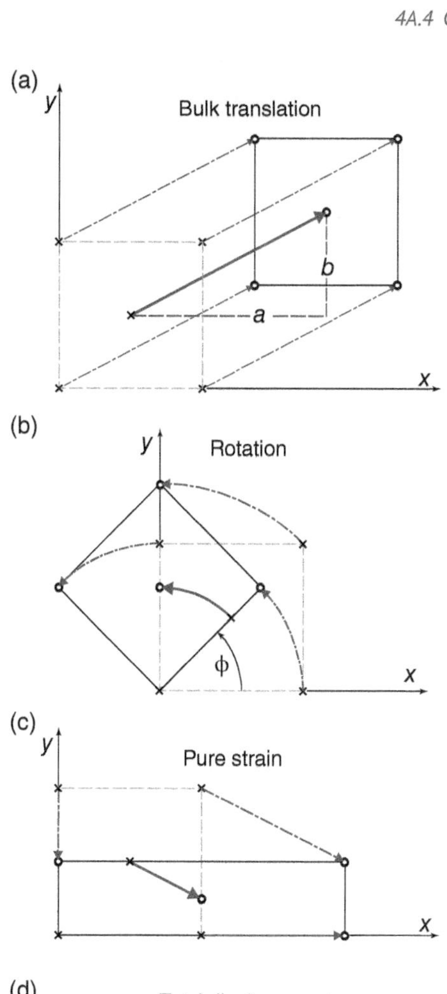

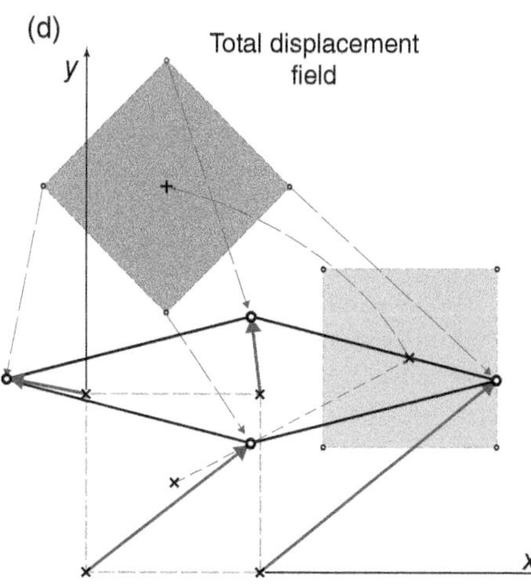

compose sedimentary strata in such tilted strata is dominated by a rigid rotation component.

The magnitudes of displacement vectors in the **pure strain** component of a homogeneous displacement field (Figure 4.14c) also depend on the initial position of the particles:

$$u_X = (A-1)\cdot X + 0\cdot Y \tag{4.8a}$$

$$u_Y = 0\cdot X + (D-1)\cdot Y \tag{4.8b}$$

where A and D are constants. The values A and D reflect displacements proportional to a particle's position along the x- and y-coordinate axes, respectively. If $A < 1$, particles along lines parallel to the x axis move toward the y axis an amount proportional to their original position. This causes any two particles on any line parallel to the x axis to converge. If $A > 1$, particles along lines parallel to the x axis move away from the y axis an amount proportional to their original position. This causes particles along lines parallel to the x axis to diverge. If $A = 1$, then $u_X = 0\cdot X + 0\cdot Y$, and there is no movement in the x-direction. Similar relations hold for D: $D < 1$ causes convergent movement in the y axis direction, $D > 1$ causes divergent movement in the y axis direction, and $D = 1$ results in no movement in the y axis direction. The divergent and/or convergent movement of particles results in the distortion of collections of particles. In Figure 4.14c, $A = 2$ and $D = ½$, so particles move away from the origin along the x axis and toward the origin along the y axis.

Familiar examples of pure strain-only displacement fields are not common. Pushing straight

Figure 4.14 Diagrams illustrating (a) a bulk translation; (b) a rigid rotation, and (c) a pure strain. In each case, dashed lines depict the initial state, solid black lines depict the final state, and red arrows indicate the movements of the corners and center of the original square. (d) Combining a bulk translation component, rigid rotation components, and pure strain component accounts for changes in position, orientation, and shape.

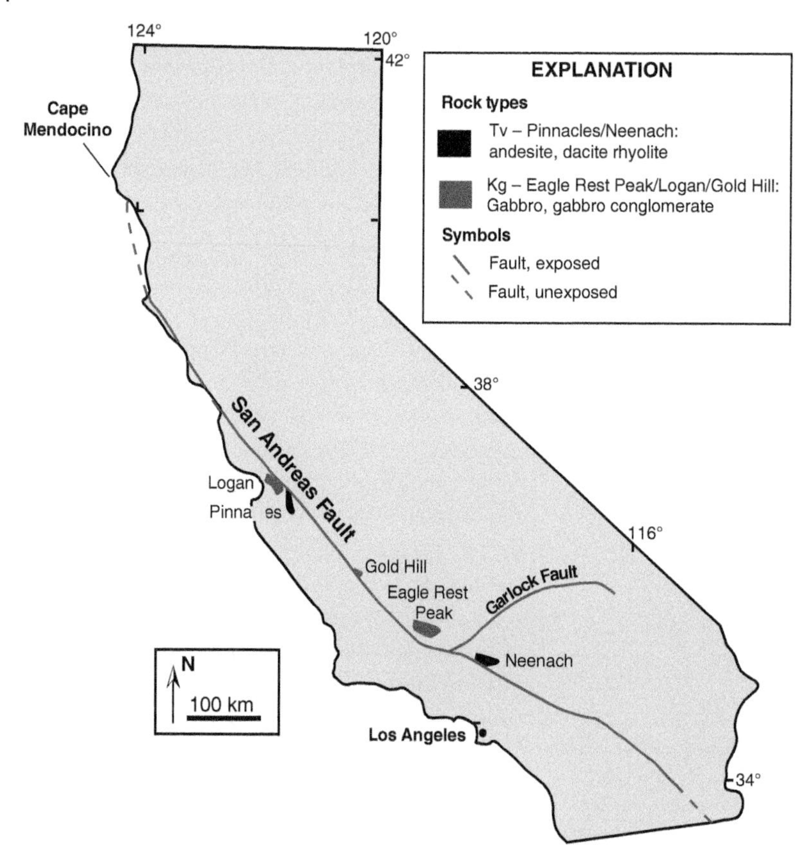

Figure 4.15 Map showing piercing points on different sides of San Andreas fault. The igneous bodies are offset by approximately 315 km. The Gold Hill gabbro is caught between different strands of the fault and records about half of the offset.

down on a sponge is a one-dimensional example. Drawing a small circle on the outside of a large balloon and then watching the circle increase in size as the balloon inflates approximates a two-dimensional situation. In detail, however, the inflating balloon case is not a perfect analogy – the circle increases in size, but the material inside the circle does not remain in the same plane as the circle and the balloon material itself thins as the balloon inflates.

All of the geological examples given earlier are exceptional cases, in which one component of the displacement field can be isolated. In general, rock masses experience total displacement fields that consist of combinations of translation, rotation, and distortion. We can simply add together the displacement components given by Equations (4.6)–(4.8) to determine the total displacement field (see Figure 4.14d). Consider also an example from the Sierra Nevada batholith in California (Figure 4.17a). In this case, the NE side of a relatively young pluton is interpreted to have "ballooned" or forcefully expanded, pushing the existing rock out of the way. Because we can infer the position and orientation of the surrounding sedimentary layers prior to intrusion, we can

(a)

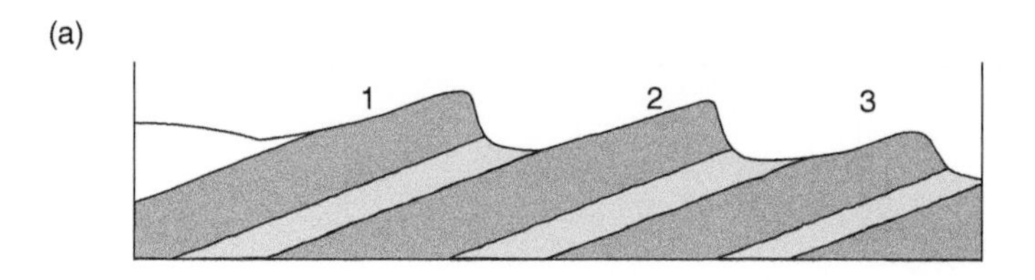

(b)

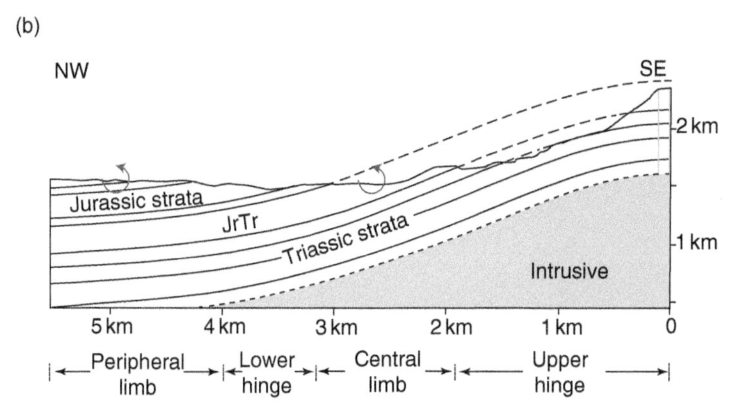

Figure 4.16 (a) Sketch of hogbacks (1, 2, and 3) developed on tilted strata consisting of more resistant (darker gray) and less resistant (lighter gray) units. (b) Cross section of distorted sedimentary strata above an igneous intrusion in the Henry Mountains, Utah, United States. *Source:* After Jackson and Pollard (1988). Deformation of strata in the peripheral limbs consists primarily of solid body rotation, and deformation of strata in the central limb is primarily uplift and solid body rotation.

resolve the components of translation, rotation, and pure strain in the deformation of the wall rock of this intrusion (Figure 4.17b and c).

4A.5 Idealized, Two-Dimensional Displacement Fields

Structural geologists often correlate the displacement field associated with a natural deformation to one of two idealized displacement fields, **simple shear** or **pure shear** displacement fields. Both simple shear and pure shear displacement fields give rise to homogeneous deformations. Simple shear and pure shear displacement fields are widely invoked because they closely approximate two common but distinctive types of homogeneous deformations found in nature. For example, settings where one rock mass moves past another, such as fault zones or shear zones, often exhibit dominantly bulk simple shear displacement fields. Similarly, settings where one rock mass converges to or moves toward another, such as fold belts or slate belts, or where one rock mass diverges from or moves away from another, such as zones of crustal extension, often exhibit bulk pure shear displacement fields. Distinguishing these two idealized displacement fields requires knowledge of the displacement paths of particles (Figure 4.3c). In this sense, determining whether a deformation conforms to a simple shear or pure shear displacement field requires more information than is needed to resolve a displacement field into its translation, rotation, and pure strain components, which only requires knowledge of the final and initial positions of particles.

Finally, it is important to take care not to confuse *pure strain* and *pure shear* despite the similarity of

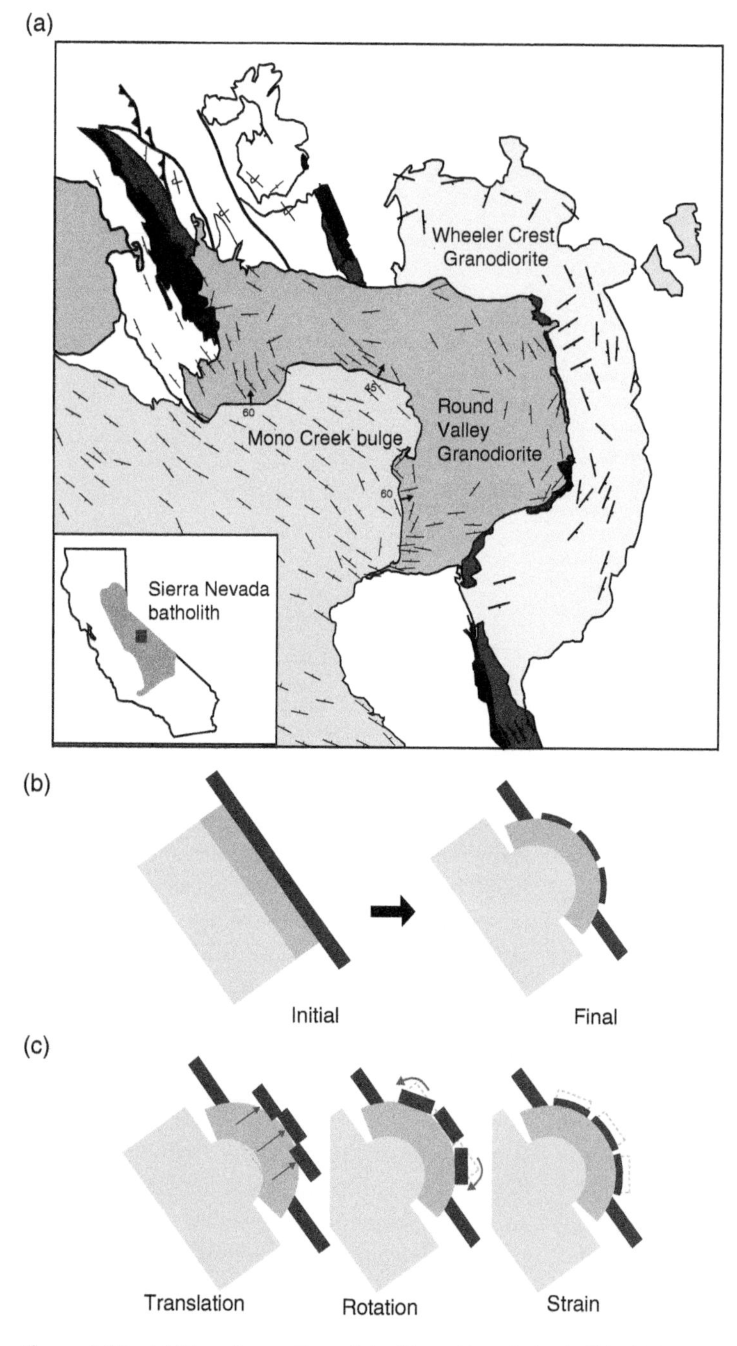

Figure 4.17 (a) Map of a portion of the Sierra Nevada batholith. (b) Two-dimensional rendering of the total displacement field for metasedimentary rocks east of the batholith. (c) Factoring the total displacement field in translation, rotation, and pure strain components.

the terms. Pure shear is just one example of a two-dimensional pure strain deformation. There are many three-dimensional deformations that are also pure strain.

4A.5.1 Simple Shear

Simple shear is a distinctive type of pure strain in which the displacement field, that is the displacements of material particles as a function of their positions, is quite simple (Figure 4.18). An example of simple shear is given by the equations:

$$u_X = 0 \cdot X + \gamma \cdot Y \tag{4.9a}$$

$$u_Y = 0 \cdot X + 0 \cdot Y \tag{4.9b}$$

in which γ is a constant known as the shear strain. In this simple shear displacement field, all displacement vectors are parallel to each other (and in this case parallel to the x axis). Different material particles generally have different displacement magnitudes, and the magnitude of any particular particle's displacement depends solely on the y component of the particle's original position. Because of this relationship, all particles with equal y component values have equal displacement magnitudes. If the value of $\gamma = 1$, the offset of any point is exactly equal to its distance from the x axis. Simple shear can occur at any orientation in space; to use Equations (4.9) we just have to reorient the coordinate axes so that the x axis parallels the displacement vectors.

One way to visualize the displacements associated with simple shear is to imagine the movement of a group of playing cards (Figure 4.18c). In depicting simple shear, we allow individual cards to move only in one direction (Figure 4.18c). Because the cards cannot stretch, every particle on every card must have the same displacement as every other particle. Particles within some naturally deformed shear zones exhibit displacements that closely resemble simple shear displacement fields (e.g. Figure 4.9).

Any displacement field, such as simple shear, technically only defines a change in state, the difference between an initial and final position. In contrast, *simple shearing* can also be thought of as a deformation path; we use the suffix "-ing" to indicate that the terms describe a displacement path, that is a series of small steps that denote the

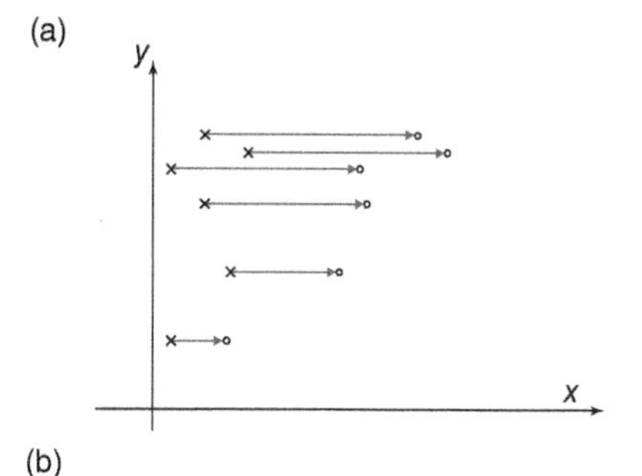

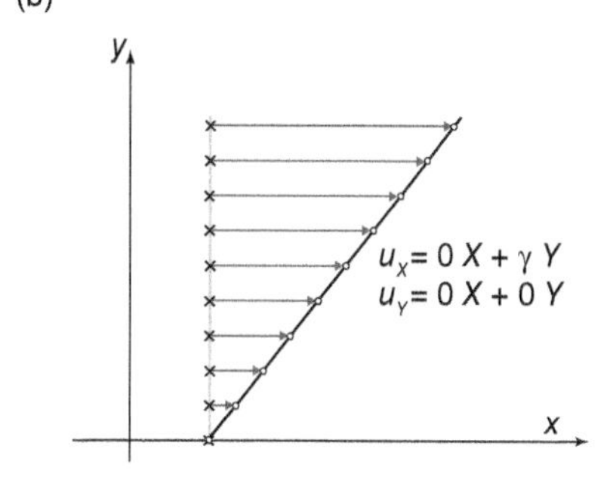

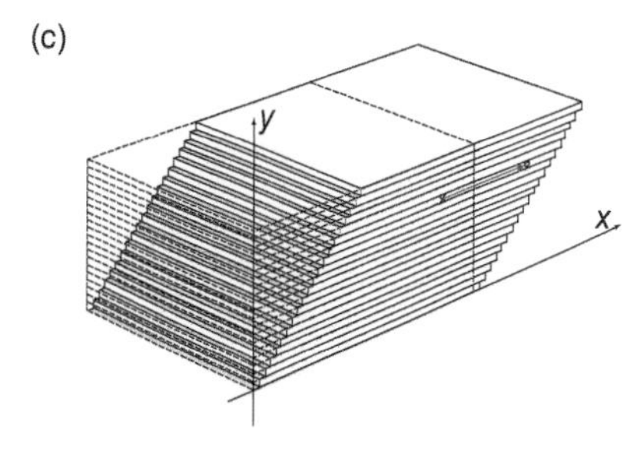

Figure 4.18 (a) In two-dimensional, simple shearing displacement fields, particles are displaced parallel to a particular direction (x = initial position of a particle; o = final position of the particle), and the magnitude of the displacement of particles is proportional to their distance from a particular line, here the x axis. (b) A collection of particles originally along a line orthogonal to the shear direction define a line inclined to the shear direction after deformation. (c) The displacement field is like that observed by shearing a deck of cards.

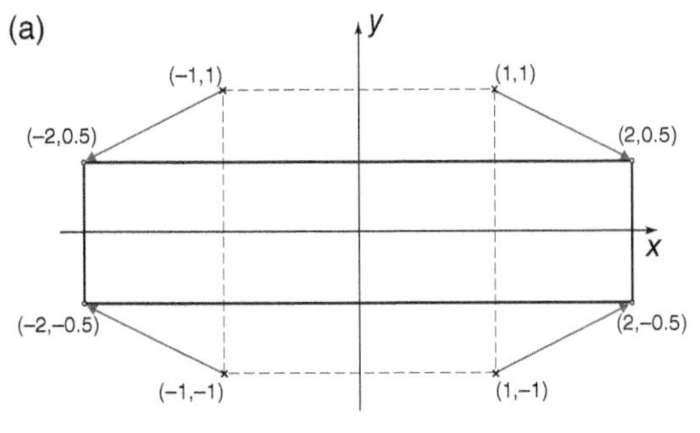

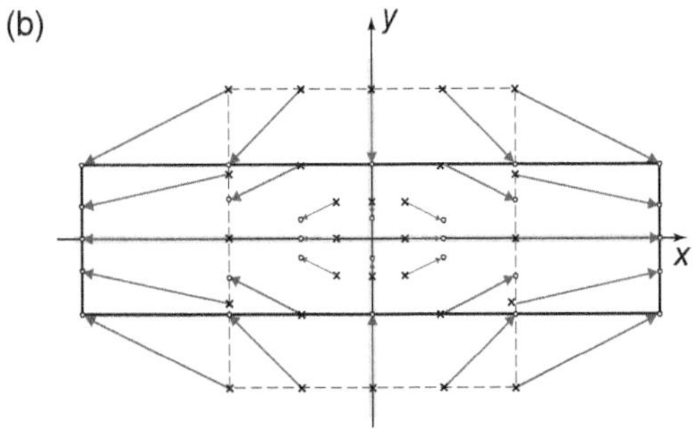

Figure 4.19 (a) Particles defining a square deformed to a rectangle by two-dimensional pure shearing displacement field with divergence along the x axis and convergence along the y axis. (b) Displacements of selected particles resulting from the same two-dimensional pure shearing displacement field.

progression from initial state to the final state. For simple shear, the path of any particle follows exactly the line of the particle's finite displacement vector.

4A.5.2 Pure Shear

Pure shear is a second, distinctive type of pure strain in which displacements are defined by:

$$u_X = (A-1)\cdot X + 0\cdot Y \quad (4.10a)$$

$$u_Y = 0\cdot X + (D-1)\cdot Y \quad (4.10b)$$

in which A and D are constants. At first impression, it might seem that pure shear, with two constants (A and D) to be defined, is less restrictive than simple shear, which has just one constant (γ) to be defined. In fact, in pure shear, the amount of divergence (or convergence) along the x axis must be exactly matched by convergence (or divergence) along the y axis. Mathematically, this means that:

$$A = 1/D$$

and/or

$$A\cdot D = 1.$$

If the convergence along one axis is matched by divergence along that other axis, there is no area loss of any collection of particles deformed by pure shear. For example, in Figure 4.19a, particles at (1, 1), (1, −1), (−1, −1) and (−1, 1) define the corners of a square whose sides are all 2 units long. The area enclosed by the lines connecting these particles is the product of the lengths of the sides, that is 2 × 2 = 4. Using the displacement field defined by Equation (4.10) with $A = 2$ and $D = 0.5$, we can deform that square in pure shear by displacing the corners to new positions. After deformation, the four particles now occupy positions (2, 0.5), (2, −0.5), (−2, −0.5), and (−2, 0.5) and define a rectangle whose length parallel to the x axis is 4 and whose height parallel to the y axis is 1. The area enclosed by the lines connecting the final positions of these points also has an area of 4.

The displacement field of pure shear is shown in Figure 4.19b. Particles whose original positions lie on the coordinate axes (denoted by gray shading) experience only convergence or divergence. Particles whose original positions do not fall on the coordinate axes, such as those defining the corners of the dashed square, experience a component of convergence in the y coordinate axis direction and a component of divergence in the x coordinate axis direction. Alternatively stated, these particles move toward the x axis and away from the y axis.

For particles whose original positions fall on the two coordinate axes, their displacement path is a straight line coincident with their finite displacement. For particles whose original positions are not on either axis, their pure shearing displacement path is fundamentally different from the straight lines that define their finite displacements. In *pure shearing* (we again use the suffix "-ing" to refer to the displacement path rather than the displacement field), particles follow hyperbolic (curved) trajectories. The paths followed by different particles are, however, defined by the values of A and D in Equation (4.10).

4A.6 Idealized, Three-Dimensional Displacement Fields

The simple shear and pure shear displacement fields described in the previous section define particles motions in three dimensions. They are just special cases of three-dimensional motion in which displacements only vary as functions of the x and y coordinates of points. As a result, particles whose original positions fall in the x–y plane remain in that plane. In both simple shear and pure shear, there is no motion perpendicular to any plane parallel to the x–y plane. For this reason, we call simple shear and pure shear **plane strain** deformations because all displacements (and strains) occur in a plane (e.g. Figure 4.20a).

We can envision, however, displacement fields in which divergence in one coordinate direction results in equal convergence in the other two coordinate directions: this case is **purely constrictional shear** (Figure 4.20b). In contrast, **purely flattening shear** occurs when convergence occurs in one coordinate direction is compensated by equal divergence in the other two coordinate directions (Figure 4.20c). To depict either of these idealized, three-dimensional displacement fields mathematically, it is necessary to introduce a few more variables. For three-dimensional displacement fields, there is a third component of the displacement of any material particle (u_Z), which moves the particle in the z direction depending on its original Z position. For the case of the purely constrictional shear and purely flattening shear, the equations are given by:

$$u_X = (A-1)\cdot X \tag{4.11a}$$

$$u_Y = (D-1)\cdot Y \tag{4.11b}$$

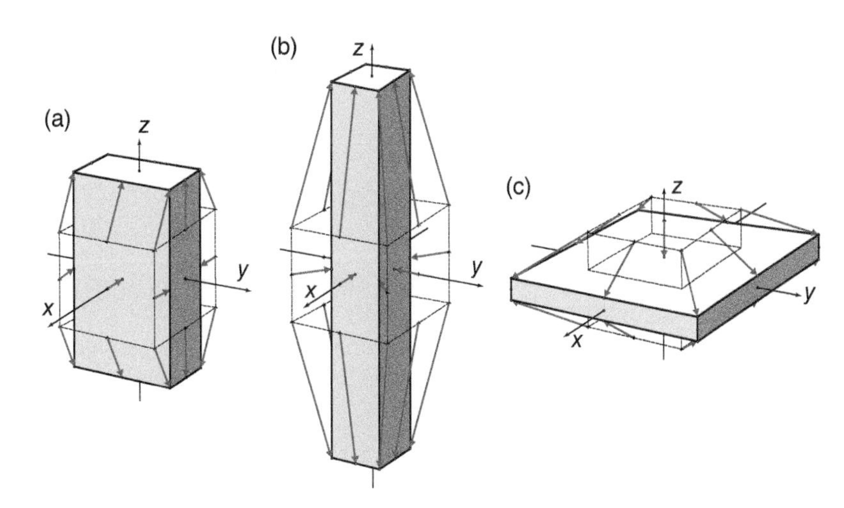

Figure 4.20 Three-dimensional orthogonal convergence and divergence displacement fields. (a) Three-dimensional pure shearing displacement field. (b) Purely constrictional displacement field. (c) Purely flattening displacement field.

$$u_Z = (E-1) \cdot Z. \quad (4.11c)$$

Once again, A, D, and E are constants. In order for the volume occupied by a collection of particles to remain constant,

$$A \cdot D \cdot E = 1.$$

For the case of purely constrictional shear, one of the three (A, D, E) variables is equal to or greater than 1 and the other two are less than one and equal to each other. For the case of Figure 4.20b, $E > 1$ and $A = D < 1$. For the case of purely flattening shear, two of the variables are equal to each other and greater than 1 and the third is equal to or less than 1. As shown in Figure 4.20c, $E < 1$ and $A = D > 1$.

In purely constrictional shearing and purely flattening shearing (where we again use the "-ing" to denote deformation path), particles whose original positions fall on one of the three coordinate axes will move directly toward or away from the origin on those coordinate axes, whereas particles whose original positions do not lie on the axes follow curved (hyperbolic) paths. As such, they share many attributes with pure shearing. Pure shearing, purely constrictional shearing, and purely flattening shearing are all examples of coaxial deformation paths; they are also all examples of pure strain.

4A.7 Summary

There are several fundamental facts concerning displacements and displacement fields:

1) In order to characterize the displacement of a particle, structural geologists must specify its magnitude and direction in space; mathematically we use vectors to denote displacements. A geological example of displacement is the net slip on a fault.
2) Structural geologists distinguish between displacement (the change in state between an initial and final position of a particle), displacement path (the route of a particle through space), and a displacement history (the timing of points along the path).
3) In nearly all geological settings, the displacements of different particles vary with their spatial position; the description of the variation of displacements is called a displacement field.
4) Any two particles within a displacement field can move toward each other along the line that connects them (converge or undergo convergent motion), move away from each other along the line that connects them (diverge or undergo divergent motion), or move past each other along paths perpendicular to the line that connects them (undergo transcurrent motion).
5) Deformation is characterized by the displacement field.
6) Any homogeneous displacement field consists of components of translation, rotation, and strain.
7) Simple shear and pure shear are two idealized examples of two-dimensional displacement fields.
8) Purely constrictional shear and purely flattening shear are two idealized examples of three-dimensional displacement fields.

4B Displacements: Comprehensive Treatment

4B.1 Specifying Displacements for Individual Particles

4B.1.1 Defining Vector Quantities

In the Conceptual Foundation section, we introduced the idea of using vectors to represent displacements and used algebraic equations to describe the components of displacements in a few straightforward displacement fields. Many structural geologists, and most engineers and geophysicists, use a more rigorous mathematical foundation that draws upon concepts from tensor analysis to examine vectors and vector fields. In this Comprehensive Treatment section, we explore more thoroughly the properties of vectors, introduce a widely used notation scheme to represent vectors, and examine how to use vector and tensor notation to perform mathematical operations on vector quantities.

We begin by distinguishing *scalar quantities* or *scalars* from vector quantities or vectors. A scalar quantity is any measurable quantity that we can specify using a single number. Examples of scalar quantities are length, width, volume, mass, density, and temperature. One needs more information to specify other measurable quantities. Among those other measurable quantities is a class called *vector quantities* or *vectors*. Vectors have both **magnitude** (or length) and **orientation** (or attitude). Fault slip is a vector quantity. Knowing only one of those two pieces of information, either the magnitude or the orientation of the fault slip, is not enough to enable us to define the displacement of a particle; both are required. In the Conceptual Foundation section, we used directed line segments or arrows to represent vector quantities (e.g. Figures 4.2b and 4.3b); we will continue to use directed line segments to represent vectors graphically. In the text and equations given in this section, we follow a convention used in many other texts and in some geology journals and use either normal or italic typefaces to denote scalar quantities and the combination of bold and italic typeface to denote vector quantities. This convention is used in many other texts and in some geology journals.

We begin to develop the mathematical foundation for vectors by examining a geological example – the displacement across the Superstition Hills fault, a strike-slip fault in southern California. Figure 4.21a is a topographic map of a V-shaped, erosional gully cut and offset by slip on the Superstition Hills fault. The collection of points at the base of this small valley once defined a continuous linear feature that was cut into two lines by fault slip. The upstream segment of this once-continuous line at the base of the V-shaped gully pierces or intersects the fault zone at point a, indicated by the filled circle. The downstream segment of this once-continuous line pierces or intersects the fault zone at point b, indicating the open circle. Prior to fault slip, the particles now found at a and b were immediately adjacent to each other (Figure 4.21b). Fault slip has displaced these particles relative to each other.

One can envision that the fault slip occurred in one of three ways: (1) the particle now at a moved away from a stationary particle at b (Figure 4.21c); (2) the particle now at b moved away from a stationary particle at a (Figure 4.21d); or (3) the particles now at both a and b moved away from a third position (Figure 4.21e). If possibility (1) described what actually happened, we could indicate that by using the letter A to identify the original position to the particle now at a (Figure 4.21c). If possibility (2) were the case, we could indicate that by using the letter B to identify the original position of the particle now at b (Figure 4.21d). In either of those instances, we would use position coordinates (X, Y) to denote the initial position of the particle and position coordinates (x, y) to denote its final position (Figure 4.21e). Note that each of a point's position coordinates is a scalar quantity.

Because we cannot determine which of these cases pertains, we use different, lower case letters to identify the particles at different positions. We will follow this convention throughout the remainder of this chapter. We show next that all three cases yield equivalent relative movement vectors, meaning that our uncertainty of knowing which of these situations occurred does not prevent us from determining the direction and magnitude of the displacements of the particles currently at a and b. Still, we must define a **reference frame** to proceed, which typically involves

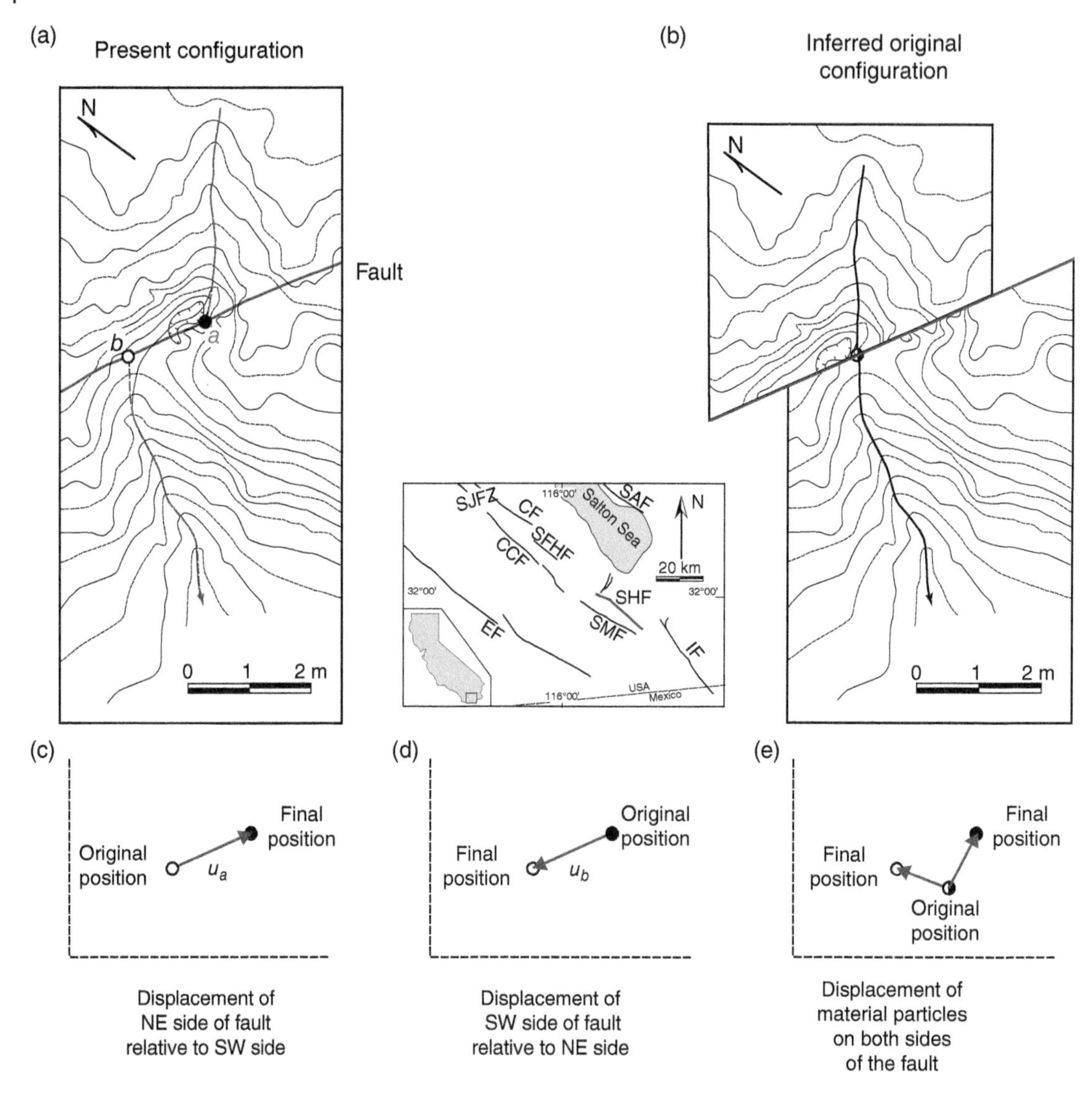

Figure 4.21 (a) Map of an erosional gully cut and offset by a portion of the Superstition Hills dextral, strike-slip fault in southern California (see inset for fault location). a and b are the locations where the upstream and downstream segments of the gully pierce the fault. (b) Inferred pre-faulting configuration of the erosional gully. *Source:* (a) and (b) Modified from Lindvall et al. (1989). (c) $\boldsymbol{u}_a$ gives the displacement of a relative to b. (d) $\boldsymbol{u}_b$ gives the displacement of b relative to a. (e) Both a and b may have been displaced from a separate third position.

envisioning that some of the particles in our area of concern remained stationary. For this case, we envision that particles on the SW side of the fault remained stationary.

4B.1.2 Types of Vectors

Figure 4.22 is an enlargement of a portion of the map of the faulted gully in Figure 4.21 showing a portion of the fault, the offset rill, and the positions

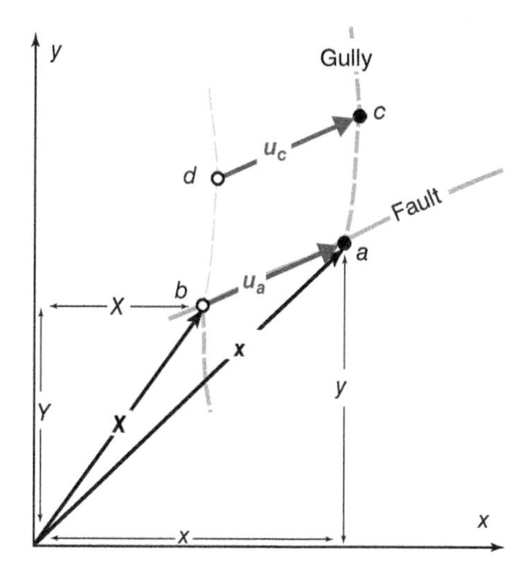

Figure 4.22 Comparison of the displacements of different portions of the offset gully from Figure 4.21. See the text for more explanation.

of particles a and b. Superposed on this image is a set of Cartesian coordinate axes with the origin on the SW side of the fault. A directed line segment from the origin to the position of particle a has a different magnitude and orientation than the vector from the origin to the position of particle b. Because no two vectors starting at the origin and ending at distinct locations are equal (have equal magnitudes *and* directions), each position in space has a unique vector tying it to the origin. This one-to-one relationship between positions and vectors means that we can identify any position in space by its associated vector: each position lies at the head of one vector that starts at the origin of a coordinate frame. Any vector whose tail coincides with the origin of a coordinate frame and whose head coincides with any specific position is a **position vector** (Figure 4.22). Position vectors belong to a class of vectors called *bound vectors* because all of the vectors defining positions with respect to the origin of a particular coordinate frame have their tails bound to, tied to, or emanating from a single location in space – the origin of the coordinate frame.

We can use the position vector to a particular point in space as a way to identify, or "tag," the individual particle that occupies that position at a particular time. For example, the position vector $\boldsymbol{X}$ (=X, Y) in Figure 4.22 "tags" or identifies the inferred original position of particle a (immediately adjacent to particle b). The position vector $\boldsymbol{x}$ (=x, y) "tags" or identifies the current position of particle a.

In Figures 4.21c and 4.22, the vector $\boldsymbol{u}_{\mathbf{a}}$ connects the position of particle b to the position of particle a. Since particle a was originally immediately adjacent to particle b, the vector $\boldsymbol{u}_{\mathbf{a}}$ depicts the **displacement** of particle a relative to a stationary particle b. Similarly, vector $\boldsymbol{u}_{\mathbf{b}}$ in Figure 4.21d depicts the **displacement** of particle b relative to a stationary particle a.

Displacement vectors belong to a general class called *free vectors*. Free vectors do not necessarily start at the origin. Thus, these vectors are "free" to emanate from any point in space, such as point b or point d in Figure 4.22. Further, two or more free vectors emanating from different points can have identical magnitudes and directions. Displacements are good examples of free vectors because the tail of the displacement vector for any individual particle must emanate from the original position of that particle (not the origin). Further, two or more particles may experience the same movement, in which case they will have identical displacement vectors.

To return to the geological example (Figures 4.21 and 4.22), the displacement vector $\boldsymbol{u}_{\mathbf{a}}$ with its tail at the original position of particle a (immediately adjacent to the current position of particle b) and its head at the final position of particle a describes the displacement of particle a relative to particle b. Thus, displacement vector $\boldsymbol{u}_{\mathbf{a}}$ is a free vector. Additionally, multiple points can have the same displacement. For example, the displacement vector for particle c ($\boldsymbol{u}_{\mathbf{c}}$) is identical to the displacement vector for particle a ($\boldsymbol{u}_{\mathbf{a}}$), making these both free vectors.

4B.1.3 Relating Position and Displacement Vectors

Figure 4.23a shows two of many potential routes from the origin to the current location of particle a. The most direct route is to follow the position vector $\boldsymbol{x}$ directly from the origin to the location of the particle a. An alternative route is to follow the position vector $\boldsymbol{X}$ to the original position of particle a and then to follow the displacement vector $\boldsymbol{u}_a$ to the current location of this particle. We can write a simple vector sum that shows that these two routes are equivalent (see Appendix 4-I for a description of vector addition).

The original position $\boldsymbol{X}$, final position $\boldsymbol{x}$, and displacement vector $\boldsymbol{u}_a$ are all related by a simple mathematical equation:

$$\boldsymbol{X} + \boldsymbol{u}_a = \boldsymbol{x}. \tag{4.12}$$

(a)

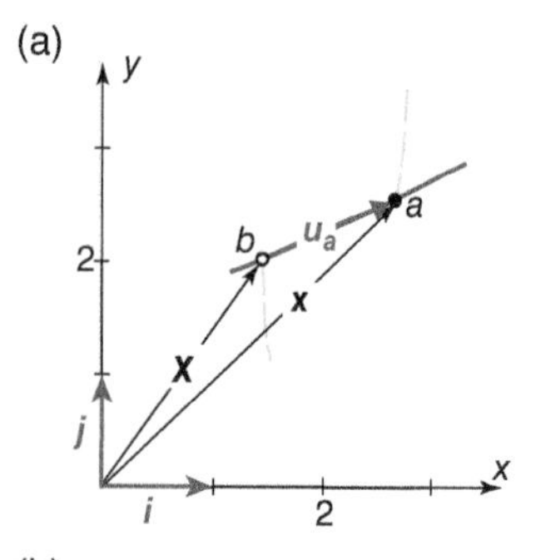

(b)

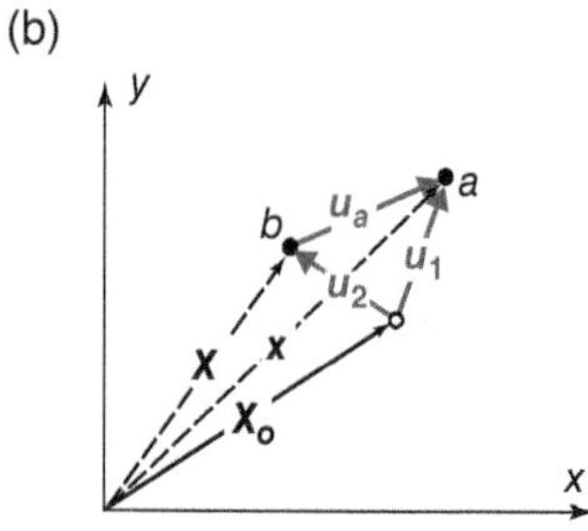

Figure 4.23 (a) Diagram relating the displacement vector of a particle to its original and final position vectors. Vectors $\boldsymbol{i}$ and $\boldsymbol{j}$ are the unit vectors parallel to the x and y coordinate axes. (b) The displacement of one particle relative to another may be due to the movement of both particles moving away from their original position, denoted by $\boldsymbol{X}_o$.

Rearranging this equation yields:

$$\boldsymbol{u}_a = \boldsymbol{x} - \boldsymbol{X}. \tag{4.13}$$

Equation (4.13) provides an explicit definition of the displacement of particle a: the displacement is the vector difference between the position vector for the current position of a particle and the position vector for its original position. This definition holds for any point we choose. If we elected to focus instead on the movement of particle b relative to a fixed particle a, we know that its displacement would be $\boldsymbol{u}_b$. Written in the form of a vector difference we have:

$$\boldsymbol{u}_b = \boldsymbol{X} - \boldsymbol{x} = -(\boldsymbol{x} - \boldsymbol{X}) = -\boldsymbol{u}_a. \tag{4.14}$$

This equation states – in a formal way – what you can observe by inspecting Figure. 4.21c and d: the displacement of particle b relative to particle a has the same magnitude but the opposite direction of the displacement of particle a relative to particle b.

In this example, Equation (4.13) enabled us to calculate the displacement vector $\boldsymbol{u}_a$ from the two position vectors ($\boldsymbol{x}$ and $\boldsymbol{X}$). Note also that Equation (4.12) enables one to determine the final position ($\boldsymbol{x}$) of a particle if one knows its initial position and its displacement.

These relations also allow us to address explicitly the frame of reference issue raised in our earlier discussion of the geological example. We now return to the situation illustrated in Figure 4.21e, where both particles a and b moved to their current locations from a third, intermediate position. The movements of particles a and b must yield a difference in final positions equivalent to that created by the movement of only one particle relative to a fixed other particle. In Figure 4.23b, $\boldsymbol{X}_o$ is the original position vector for both particles a and b. Particle a experiences displacement $\boldsymbol{u}_1$ and particle b experiences displacement $\boldsymbol{u}_2$. We note that $\boldsymbol{X}_o + \boldsymbol{u}_2 = \boldsymbol{X}$ and $\boldsymbol{X}_o + \boldsymbol{u}_1 = \boldsymbol{x}$. Thus, $\boldsymbol{u}_2 = \boldsymbol{X} - \boldsymbol{X}_o$ and $\boldsymbol{u}_1 = \boldsymbol{x} - \boldsymbol{X}_o$.

The vector difference between the two displacement vectors is

$$\begin{aligned} u_1 - u_2 &= (x - X_o) - (X - X_o) \\ &= x - X - X_o + X_o = x - X = u_a \end{aligned}$$

or

$$\begin{aligned} u_2 - u_1 &= (X - X_o) - (x - X_o) \\ &= X - x - X_o + X_o = X - x = u_b. \end{aligned}$$

Thus, the situation where both particles a and b moved is precisely equivalent to the situation where one particle moved relative to a second, stationary particle.

4B.1.4 Characterizing Vector Quantities

One way to characterize a vector quantity is to specify its direction and magnitude (Figure 4.24a). Typically, we use a Cartesian coordinate system, consisting of two orthogonal axes: the x and y axes (some texts denote the two coordinate axes x_1 and x_2). We envision that there is a unit vector, a vector whose length is one unit, parallel to each axis: $\boldsymbol{i}$ parallel to the x axis and $\boldsymbol{j}$ parallel to the y axis. Figure 4.24b shows that the vector $\boldsymbol{X}$ is the *resultant* (or vector sum) of *component vectors* parallel to the two coordinate axes. Each component vector is, in turn, a scalar multiple (see Appendix 4-I) of the unit vector parallel to the corresponding coordinate axis:

$$\boldsymbol{X} = X\boldsymbol{i} + Y\boldsymbol{j}. \tag{4.15}$$

Returning to Figure 4.23a and taking the lengths of the (east-west) EW unit vector $\boldsymbol{i}$ and the NS unit vector $\boldsymbol{j}$ to be 1 m, we have $\boldsymbol{X} = 1.4\boldsymbol{i} + 2.0\boldsymbol{j}$ and $\boldsymbol{x} = 2.7\boldsymbol{i} + 2.5\boldsymbol{j}$. We can use the values of the coefficients of the component vectors to determine the components of the vector $\boldsymbol{u}_a$, which gives the displacement of particle a relative to a fixed particle b:

$$\begin{aligned} \boldsymbol{u}_a = \boldsymbol{x} - \boldsymbol{X} &= [x\boldsymbol{i} + y\boldsymbol{j}] - [X\boldsymbol{i} + Y\boldsymbol{j}] \\ &= [x - X]\boldsymbol{i} + [y - Y]\boldsymbol{j} = u_X\boldsymbol{i} + u_Y\boldsymbol{j} \end{aligned} \tag{4.16}$$

(a)

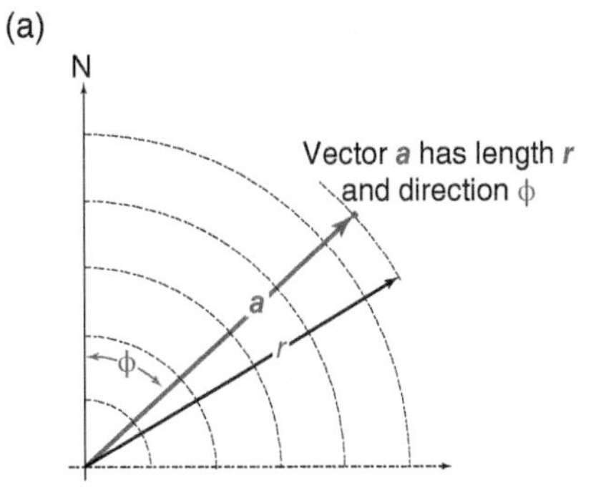

(b)

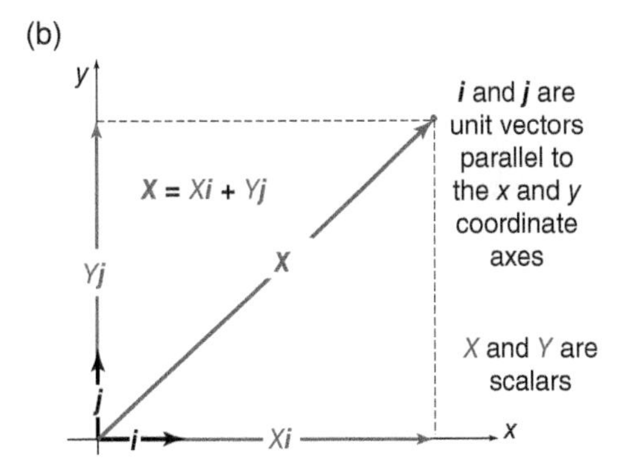

Figure 4.24 (a) One way to envision a vector quantity is to define the vector's length r and the angle ϕ it makes with an arbitrary reference direction. (b) Alternatively, one can envision that a vector is the sum of components parallel to Cartesian coordinate axes. Each component is a scalar multiple of a unit vector parallel to one of the coordinate axes. Here $\boldsymbol{i}$ is the unit vector parallel to x, and $\boldsymbol{j}$ is the unit vector parallel to y.

where $u_X = x - X$ and $u_Y = y - Y$. Substituting the values of the coefficients of the component vectors of the two position yields $u_X = x - X = 2.7 - 1.4 = 1.3$ and $u_Y = y - Y = 2.5 - 2.0 = 0.5$. Thus,

$$\boldsymbol{u}_a = 1.3\boldsymbol{i} + 0.5\boldsymbol{j}.$$

As long as the unit vectors $\boldsymbol{i}$ and $\boldsymbol{j}$ have the same orientations in space, the displacement we calculate for particle a does not depend upon the location of the origin of the coordinate system. In Figure 4.25, the lengths of EW unit vector $\boldsymbol{i}$ and the NS unit vector $\boldsymbol{j}$ are again 1 m. Taking the position of the coordinate frame origin in Figure 4.25

(on the NE side of the fault), $\boldsymbol{X} = -2.8\boldsymbol{i} - 1.8\boldsymbol{j}$ and $\boldsymbol{x} = -1.5\boldsymbol{i} - 1.3\boldsymbol{j}$. Using these values to determine the coefficients of the vector $\boldsymbol{u}_\mathbf{a}$ yields $u_X = x - X = -1.5 - (-2.8) = 1.3$ and $u_Y = y - Y = -1.3 - (-1.8) = 0.5$. We again have

$$\boldsymbol{u}_a = 1.3\boldsymbol{i} + 0.5\boldsymbol{j}.$$

We return to Superstition Hills fault (Figure 4.21) to illustrate these concepts. In Figure 4.21c, $\boldsymbol{u}_\mathbf{a}$ is

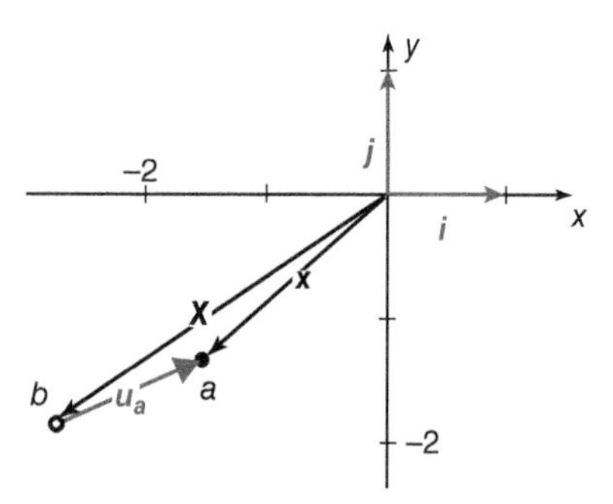

Figure 4.25 The displacement of one particle relative to another is the same regardless of the relative position of the origin of the coordinate axes used to determine its components.

the vector denoting the movement of particle a relative to a particle b. Since the original position of particle a was immediately adjacent to particle b, the vector $\boldsymbol{u}_\mathbf{a}$ gives the displacement of particle a relative to particle b. Similarly, vector $\boldsymbol{u}_\mathbf{b}$ in Figure 4.21d gives the displacement of particle b relative to particle a.

By identifying points that were originally adjacent and are no longer adjacent in inactive structures, we can also determine movements that occurred in the distant past, which we cannot possibly have observed. Figure 4.26a is an example of such an inactive structure. Approximately horizontal limestone beds are offset across a discrete fault (or slip surface). In principle, we can find two particles, particle p on the footwall side of the fault and particle q on the hanging wall side of the fault that were adjacent prior to fault slip. Figure 4.26b shows two such particles. Particle p lies on the footwall trace of the bottom of bed A. Particle q lies on the hanging wall trace of the bottom surface of bed A. The line pq is parallel to surface markings on the fault surface. Fault slip either has displaced particle p relative

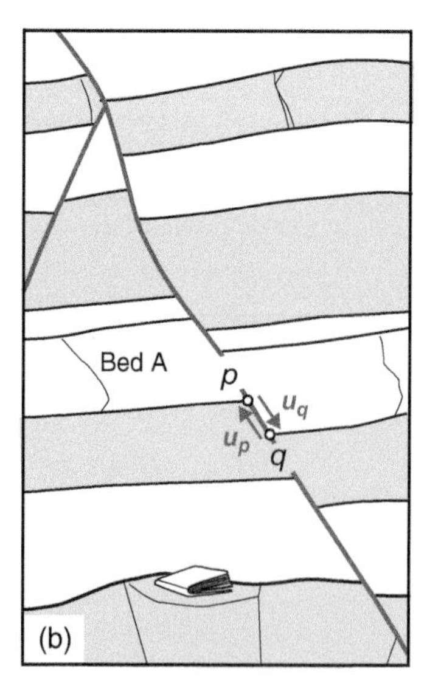

Figure 4.26 (a) Photograph of a fault cutting subhorizontal limestone beds. (b) Drawing illustrating the relative movement of the hanging wall and footwall of the fault in (a).

to a stationary particle q, has displaced particle q relative to a stationary particle p, or has displaced both particles p and q away from a common original position. The vector $\boldsymbol{u}_\mathbf{q}$ gives the displacement of particle q relative to particle p. In this example, this vector is a plunging line. Like this illustrated case, we can envision a vector $\boldsymbol{u}_\mathbf{p}$ that gives the displacement of particle p relative to particle q. The vector $\boldsymbol{u}_\mathbf{p}$ has the same length as $\boldsymbol{u}_\mathbf{q}$ but points in the opposite direction, that is it is directed upward.

4B.2 Particle Paths and Velocities

4B.2.1 Incremental Displacements for Particles

Just as the displacement of a particle during deformation need not coincide with the particle's path, the slip vector for any two formerly adjacent particles separated by a fault need not indicate the path that these particles followed as they moved away from each other. One reason for this discrepancy is that the slip we observe on any fault is not likely to have accrued in a single episode. The finite slip we observe typically occurs in numerous increments over long time intervals; this observation is readily apparent for the kilometer-scale displacements determined for the largest natural faults (Figure 4.4), but similarly true for many small-offset faults. Moreover, there is necessarily a difference between the *path* that particles follow and their *finite* (or net) displacement vectors if fault surfaces are curved. Particles must follow curved paths if the rock adjacent to a fault is to remain in contact with a curved fault surface.

Figure 4.27 illustrates how the finite displacement we measure across a fault is the result of numerous small-displacement *increments.* In Figure 4.27a, the points labeled b_1, b_2, b_3, etc. show the successive positions of particle b, which was initially adjacent to particle a, as strike-parallel slip accumulates on a hypothetical planar fault

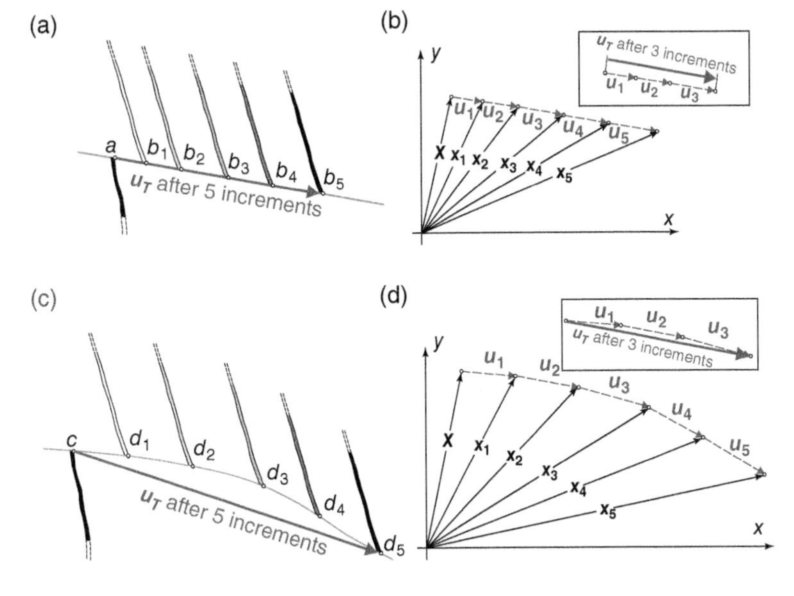

Figure 4.27 (a) Successive positions of a particle b displaced by a planar fault. (b) Diagram relating successive position vectors for particle b to its incremental and total displacements. (c) Successive positions of a particle d displaced on a curved fault. (d) Diagram relating successive position vectors for particle d to its incremental and total displacements.

cutting a mineral-filled vein. In Figure 4.27b, $\boldsymbol{X}(b)$ is the position vector indicating the initial position of particle b. Vectors $\boldsymbol{x}_1(b)$, $\boldsymbol{x}_2(b)$, $\boldsymbol{x}_3(b)$, etc. are position vectors corresponding to the successive positions of particle b. From Figure 4.27b, we see that the displacement of particle b during the first increment is

$$\boldsymbol{u}_1(b)=\boldsymbol{x}_1(b)-\boldsymbol{X}(b).$$

The displacements during successive increments, given by the **incremental displacement vectors** $\boldsymbol{u}_2(b)$, $\boldsymbol{u}_3(b)$, etc. are the vector differences between the position vectors marking the location of particle b (Figure 4.27b). Thus,

$$\boldsymbol{u}_2(b)=\boldsymbol{x}_2(b)-\boldsymbol{x}_1(b)$$

and

$$\boldsymbol{u}_3(b)=\boldsymbol{x}_3(b)-\boldsymbol{x}_2(b).$$

Inspection of the inset in Figure 4.27b shows that the total displacement after three increments is the vector sum of the three incremental displacement vectors:

$$\boldsymbol{u}_T(b)=\boldsymbol{u}_1(b)+\boldsymbol{u}_2(b)+\boldsymbol{u}_3(b).$$

In algebraic form, we have

$$\begin{aligned}\boldsymbol{u}_T(b)&=\boldsymbol{u}_1(b)+\boldsymbol{u}_2(b)+\boldsymbol{u}_3(b)\\&=\left[\boldsymbol{x}_1(b)-\boldsymbol{X}(b)\right]+\left[\boldsymbol{x}_2(b)-\boldsymbol{x}_1(b)\right]\\&\quad+\left(\boldsymbol{x}_3(b)-\boldsymbol{x}_2(b)\right]\\&=\boldsymbol{x}_3(b)-\boldsymbol{X}(b),\end{aligned}$$

showing that our graphical and algebraic analyses are equivalent.

Similarly, we can represent the displacement after five increments as the vector sum of the first five incremental displacement vectors or the vector difference between the position vector for the particle b at that time and the position vector for the initial position of particle b. Because all of the incremental displacement vectors in Figure 4.27a have the same direction, the total displacement vector is a scalar multiple of the incremental displacement vectors. Figure 4.27c shows successive locations of particle d, which was originally adjacent to particle c, as slip accumulates on a curved fault surface. From Figure 4.27d, we see that the displacements of particle d during the successive increments are

$$\boldsymbol{u}_1(d)=\boldsymbol{x}_1(d)-\boldsymbol{X}(d),$$

$$\boldsymbol{u}_2(d)=\boldsymbol{x}_2(d)-\boldsymbol{x}_1(d),$$

and

$$\boldsymbol{u}_3(d)=\boldsymbol{x}_3(d)-\boldsymbol{x}_2(d).$$

The inset in Figure 4.27d again shows that the total displacement of particle d after three increments is the vector sum of the three incremental displacement vectors:

$$\boldsymbol{u}_T(d)=\boldsymbol{u}_1(d)+\boldsymbol{u}_2(b)+\boldsymbol{u}_3(d)$$

or

$$\begin{aligned}\boldsymbol{u}_T(d)&=\boldsymbol{u}_1(d)+\boldsymbol{u}_2(d)+\boldsymbol{u}_3(d)\\&=\boldsymbol{x}_3(d)-\boldsymbol{X}(d).\end{aligned}$$

Thus, even though the individual incremental displacement vectors are not parallel, the total displacement at any instant is the vector sum of those incremental displacement vectors.

4B.2.2 Particle Paths and Movement Histories

In Figure 4.27b and d, the series of positions occupied by the particles together define the **path** that particle followed over time. If a geologist is able to determine a series of closely spaced successive positions for a particle, she or he can define a relatively smooth curve that indicates the path followed by that particle. If she or he can determine when a particle occupied different positions along the particle path, the geologist has ascertained the **movement history** for that particle.

An example of movement history is provided by a neotectonic study conducted in New Zealand. Figure 4.28a is a sketch map of the Gaunt Creek exposure of the Alpine Fault, which separates the Australian and Pacific plates, on the South Island of New Zealand. In a series of outcrops, mylonites in the hanging wall (SE side) of the fault are thrust over three distinctive types of poorly consolidated sedimentary deposits. The oldest of these are fluvioglacial gravels deposited at the end of the last glacial maximum, probably about 20,000 a before present (B.P.). Next is a collection of talus and slump deposits containing fragments of cataclasite that, in the modern setting, disintegrate within tens of meters of the cataclasite exposures. A fragment of wood extracted from the base of these talus and slump deposits is ~12,650 a old. The youngest deposits are a series of outwash gravels with ~10,300 a old wood fragments preserved near the base of the series. Figure 4.28b shows (at left) reconstructions of five stages in the development of the structures seen in the Gaunt Creek exposures along with (at right) a schematic rendering of the successive positions of a particle currently in the mylonites just above the younger gravels. Because the slickenlines and other indicators of the direction of slip on the Alpine Fault are approximately parallel to the direction of relative motion for the Australian and Pacific plates, we can infer that the successive positions of a particle in the mylonites define the curved path indicated in Figure 4.28c. This curved path defines the movement history for the mylonites adjacent to the Alpine Fault.

4B.2.3 Dated Particle Paths, Instantaneous Movement Directions, and Velocities

The **velocity** of a particle is the change in its position divided by the time required to move from one position to the other. The change in position of a particle is a vector quantity, so the velocity too must be a vector quantity:

$$v = (X - x)/(t_f - t_0). \tag{4.17}$$

A velocity vector $\boldsymbol{v}$ is a directed line segment whose orientation gives the direction in which a particle is moving or has moved and whose magnitude corresponds to the distance per unit time or speed of the particle. The magnitude of a velocity vector has dimensions of length divided by time, such as mm/a or m/s.

If we know the time at which a particle occupied one position along its particle path and the time that the particle occupied a second position along that path, we can calculate the average velocity of the particle during that time interval. If the time interval is long or if the particle path is curved, the average velocity may not be representative of the movement at different times during that interval. Consider the movement of a particle along a curved, dated particle path (Figure 4.29). At time t_0, the original position vector for the particle is $\boldsymbol{X}$, and at time t_f, the final position vector of the particle is $\boldsymbol{x}$. The overall velocity vector, $\boldsymbol{v}_{\mathbf{av}}$, is a relatively crude indicator of the movement of the particle.

Even if a particle follows a straight-line path, it need not travel at a steady speed along that path. If we calculate velocities $\boldsymbol{v}_{\mathbf{inc}}$ using displacement increments accrued during shorter time intervals (Figure 4.29b and c), the velocity vector more accurately portrays the instantaneous direction of movement of the particle and may more closely approximate the true speed at which the particle moved during that time interval. Figure 4.29c shows that the tangent at any point to a curved particle path defines the instantaneous direction in which that particle moved when it occupied that position.

It is useful to reconsider the Alpine Fault example (Figure 4.28), to distinguish between the velocity (or instantaneous displacement) of a particle and the total displacement of that

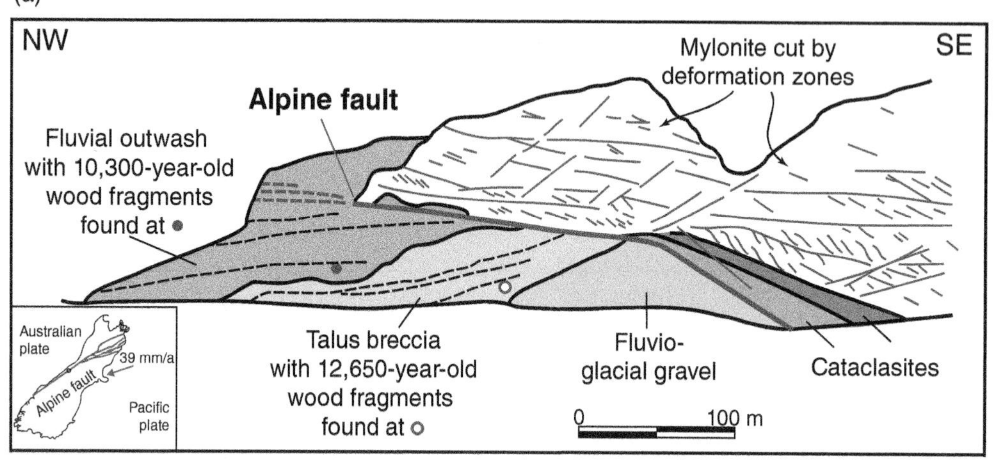

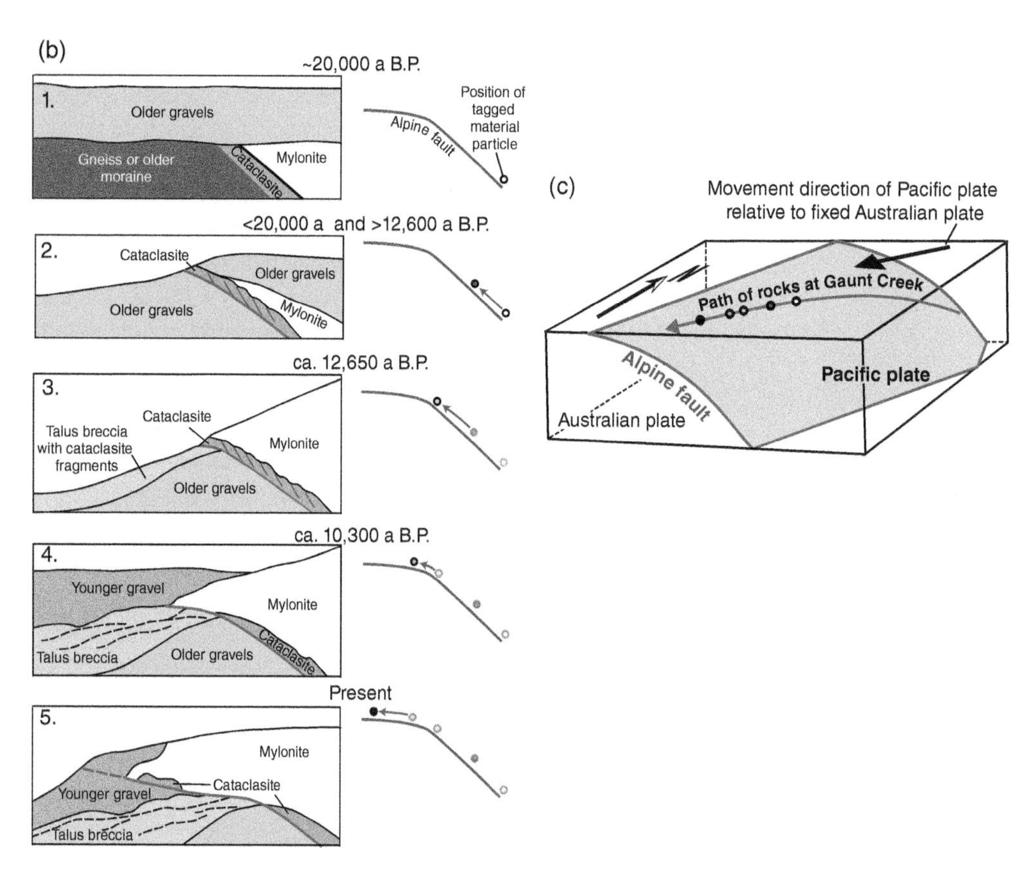

Figure 4.28 (a) Sketch of the Gaunt Creek exposure of the Alpine Fault in New Zealand (see inset map for location). At this exposure, mylonites formed at depth along the fault are thrust over unconsolidated fluvioglacial deposits, talus, and fluvial outwash deposits. (b) Series of sketches showing the temporal development of the structure depicted in (a). 1. Inferred geometry post-deposition of the older gravels. 2.–5. Successive stages after deformed cataclasite and mylonite were thrust over older gravels. At right, small circles show successive positions of the mylonites now exposed at the surface above the unconsolidated sediments. *Source:* (a) and (b) modified from Cooper and Norris (1994). (c) Block diagram showing schematically the curved path followed by the hanging wall mylonites.

(a)

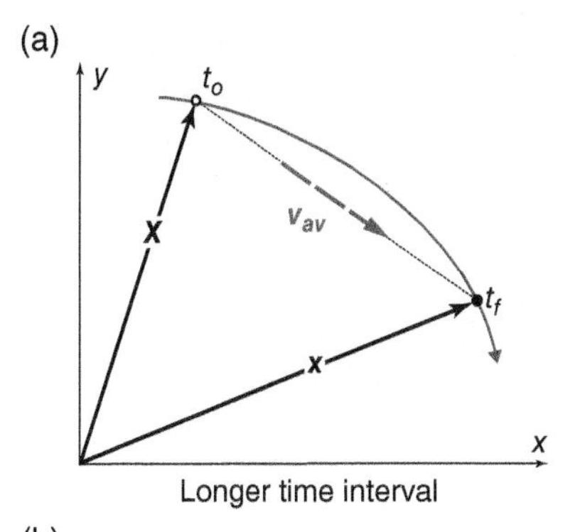

(b)

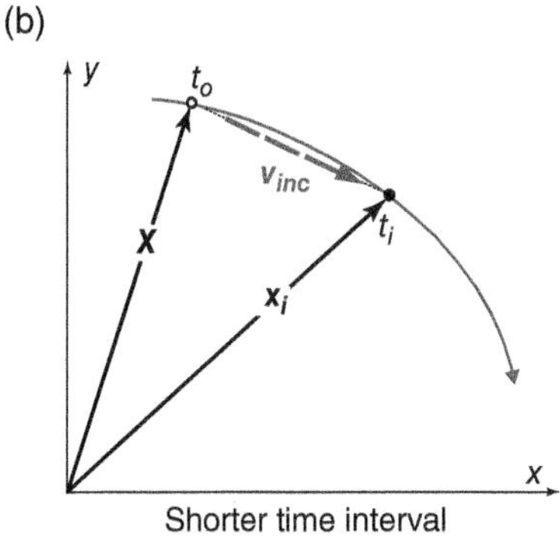

(c)

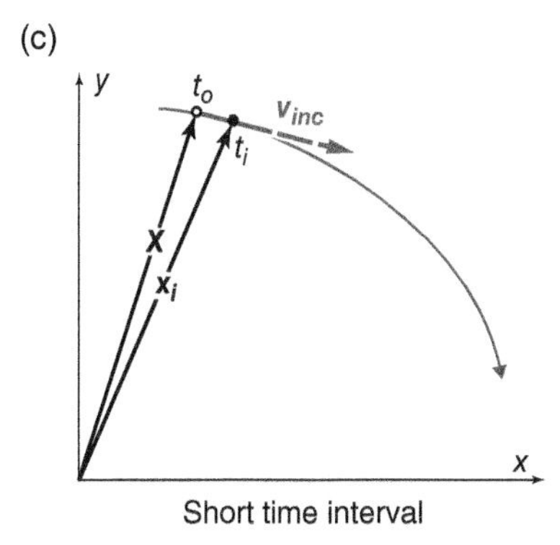

Figure 4.29 (a) For longer time intervals, the average velocity of a particle is unlikely to parallel the path followed by the particle. (b) For shorter time intervals, the incremental velocity is more nearly parallel to the particle path. (c) For short time intervals, the incremental velocity is likely to be tangent to the path the particle follows.

particle. At the Gaunt Creek exposures, mylonites in the hanging wall were thrust ~180 and ~110 m over the underlying unconsolidated sediments during the last ~12,650 and 10,300 a, respectively. These values are consistent with average thrust velocities of 11–14 mm/a. Using the minimum offsets parallel to slickenlines on the Alpine Fault needed to account for the observed magnitude of overthrusting, average particle velocities are 18–24 mm/a. The Alpine Fault, like the San Andreas fault, is a major geologic feature inferred to demarcate the boundary between two lithospheric plates. The average velocities calculated for particles near the faults are nearly equal to the velocities inferred for the relative movement of the lithospheric plates in both instances. Velocities of tens of millimeters per year are near the high end of the velocities of particles in natural deformations. The relative movements of particles in other settings, such as those in which folds or foliations form or those with smaller offset faults or shear zones, are likely to be slower than those observed in these plate-boundary settings.

4B.3 Displacements of Collections of Particles – Displacement Fields

4B.3.1 Concept of a Displacement Field

Figure 4.30 is a view from a different perspective of the erosional gully cut and offset by slip on the Superstition Hills fault in southern California (cf. Figures 4.21 and 4.22). We have omitted the topographic contours to highlight the linear gully, and we have superimposed on the field of view a set of Cartesian coordinate axes whose *y* axis is parallel to the fault. Particles *a* through *h* are a related collection of tagged or identified particles that demarcate the base of the gully prior to fault slip. Fault slip did not move particles *a* through *d* relative to an origin arbitrarily placed on the south side of the fault. The position vectors identifying the original positions of these particles are identical to the position vectors identifying their final positions (this is shown for particle *a* in Figure 4.30). Fault

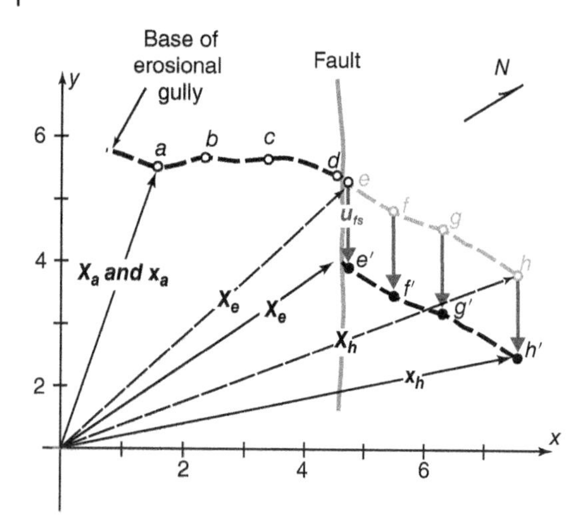

Figure 4.30 Plot showing displacement of particles along a transect across the Superstition Hills dextral strike-slip fault. The field of view is the same as Figure 4.22 but rotated 90° clockwise. Note that particle displacement is a function of position coordinate X.

slip has, however, moved particles e through h relative to the origin. Their new positions are e' through h', respectively. For the deformation associated with fault slip within the field of view, the displacement vector is the same for each of those particles. That is $\mathbf{x_i} - \mathbf{X_i} = \mathbf{u_{fs}}$ for $i = e, f, g$, or h.

To summarize the situation illustrated in Figure 4.30, the displacement vectors of particles is *a function of their original position*. For particles a through d, $\mathbf{u}(a) = \mathbf{u}(b) = \mathbf{u}(c) = \mathbf{u}(d) = 0$, whereas for particles e through h, $\mathbf{u}(e) = \mathbf{u}(f) = \mathbf{u}(g) = \mathbf{u}(h) = \mathbf{u_{fs}} = -1.4\mathbf{j}$, a displacement of ~1.4 m in a $-y$ direction. In fact for this field of view, as a result of the orientation of our Cartesian coordinate frame, we can state that $\mathbf{u}(\mathbf{X}) = 0$ provided that the x component of the position vector, X, is less than 4.6 m and $\mathbf{u}(\mathbf{X}) = \mathbf{u_{fs}}$ if $X \geq 4.6$ m.

This example introduces the idea of tracking the displacements of a collection of several, related particles, which together constitute a material region or define a material object, to understand a deformation. Geologists rely extensively on studies of the movements of related particles to understand deformations, and routinely envision deformations by considering that the displacement vectors of many related particles constitute a **displacement field**. Movement on this fault is an example of a discrete displacement field, where displacement values change abruptly or in a step-like manner with changes in position.

We can use the planar-sided shear zone cutting across and offsetting a dike (see Figure 4.9) to examine another displacement field. In Figure 4.31, we show a drawing of the distorted dike superposed on a set of Cartesian coordinate axes. Particles a through i are a collection of related particles that define the inferred position of the side of the dike prior to the formation of the shear zone. Deformation has not altered the positions of particles a and b relative to the origin; the position vectors identifying the original positions of these particles are identical to the position vectors identifying their final positions (this is shown for particle b in Figure 4.31). Deformation has displaced particles c through i relative to the origin. Their new positions are c' through i', respectively. Deformation associated with the shear zone has displaced

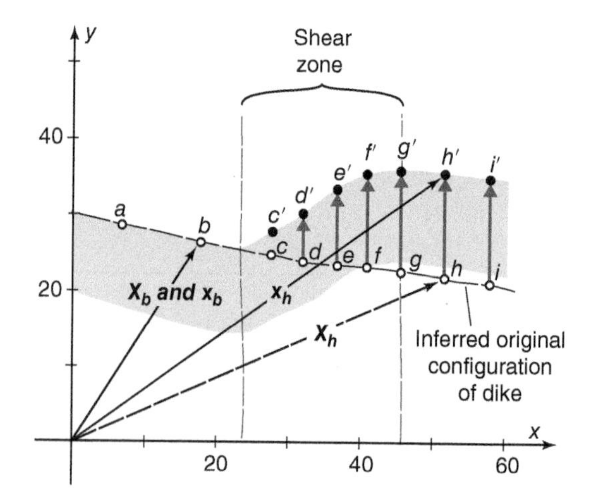

Figure 4.31 Same field of view as Figure 4.9b showing how displacements of particles relate to their original and final position vectors. Note that displacements vary as a function of position coordinate X.

particles c through i in the y direction (Figure 4.31). Particles c to g have successively larger magnitude displacements; particles g, h, and i have the same displacement = $\boldsymbol{u}_{\mathbf{sz}}$ = 14$\boldsymbol{j}$. Within the field of view, then, the displacement magnitude varies with position. For particles with $X < \sim 23$ cm, the displacement magnitude is zero. For particles with $X > \sim 47$ cm, displacement = $\boldsymbol{u}_{\mathbf{sz}}$, a displacement of about 14 cm in the y direction. For particles with 23 cm < X < 47 cm, the magnitude of the displacement in the y direction increases as X increases. The displacement field associated with this shear zone is an example of a distributed displacement field, where displacements change gradually or smoothly with changes in position.

4B.3.2 Field Quantities

You probably have some experience working with field quantities, although you may not have thought about them as such. Geologists know that their compass needle points toward the north magnetic pole regardless of where they are on Earth. We use the idea of the magnetic field, which defines the orientation and magnitude of the magnetic force lines with which the compass needle aligns, to describe how magnetism varies at different locations in space. A displacement field is simply a way to visualize collectively the displacements of individual particles at different positions. In the earlier geological example, we considered the displacement of related particles as a function of particles' initial positions; this formulation envisions the particles as moving from their *initial positions* to their *final positions*. We can represent the idea that the displacement of particles is a function of their initial position in the following way:

$$\boldsymbol{u} = f(\boldsymbol{X}). \tag{4.18}$$

An equivalent way to represent this functional relationship is to write equations giving the different components of the displacements of particles as a function of their initial position coordinates:

$$u_X = f_X(X,Y,Z) \tag{4.18a}$$

$$u_Y = f_Y(X,Y,Z) \tag{4.18b}$$

$$u_Z = f_Z(X,Y,Z). \tag{4.18c}$$

We can use Equation (4.18) to analyze the situations illustrated in Figures 4.30 and 4.31. In both cases, the displacements $\boldsymbol{u}_{\mathbf{fs}}$ and $\boldsymbol{u}_{\mathbf{sz}}$ are parallel to the y axes, meaning that $u_X = u_Z = 0$ for all values of X and Y (and Z) in both examples. Furthermore, the displacement values do not change with Y in either case. Thus $\boldsymbol{u} = u_y\,\boldsymbol{j}$ where $u_Y = f_Y(X)$ describes all variations of displacement with position. For Figure 4.30, $u_Y = f_Y(X)$ is a step function:

$$u_Y = 0 \quad \text{for} \quad X < 4.6\text{m},$$

and

$$u_Y = -1.4\text{m} \quad \text{for} \quad X \geq 4.6\text{m}.$$

For Figure 4.31, $u_Y = f_Y(X)$ has three distinct regions:

$$u_Y = 0 \quad \text{for} \quad X < 23\text{cm},$$

$$u_Y = f_Y(X) \approx 0.583(X-23) \quad \text{for} \quad 23\text{cm} \leq X \leq 47\text{cm},$$

and

$$u_Y = 14\text{cm} \quad \text{for} \quad X > 43\text{cm}.$$

4B.3.3 Gradients of the Displacement Field: Discrete and Distributed Deformation

We use the concept of a **gradient** to evaluate variations in displacement with differences in position. Most geologists are familiar with the geothermal gradient, which quantifies the change in temperature with depth in Earth, or pressure gradients, where the hydrostatic pressure changes with changes in position in a flowing fluid, such as within an aquifer. A **displacement gradient** quantifies how the displacement experienced by different particles changes with changes in the positions of the particles. In this chapter, we will focus on determining gradients with respect to the initial or original positions of particles, but it is also possible to define gradients relative to the final or current positions of particles. In this initial discussion of displacement gradients, we will use $\Delta \boldsymbol{u}/\Delta \boldsymbol{x}$ to denote the difference in the displacement of two particles a finite distance apart. Many mathematics texts use a similar notation $\Delta F/\Delta x$ to work toward defining the derivative a function $F(x)$, that is $\Delta F/\Delta x \approx dF/dx$. Defining the derivative of function where both the function ($\boldsymbol{u}$) and its argument ($\boldsymbol{X}$) are vectors is not straightforward. This task is easier to accomplish by considering the functional form of the components of $\boldsymbol{u}$, and this is what we will do in the following chapters. In our fault and shear zone examples (Figures 4.30 and 4.31, respectively), the displacement gradient, $\Delta \boldsymbol{u}/\Delta \boldsymbol{X}$, reduces to $\Delta u_Y/\Delta X$ ($\approx du_Y/dX$) because the displacements of all particles have $u_x = u_z = 0$ and because the only nonzero component of the displacement, u_y, does not vary with Y or Z. In the case of the Superstition Hills fault as illustrated in Figure 4.30, we have

$$\Delta u_Y/\Delta X = 0 \text{ for } X < 4.6\text{m and for } X > 4.6\text{m},$$

and

$$\Delta u_Y/\Delta X \text{ is undefined } \left(\Delta u_Y \text{ is finite, but } \Delta X = 0\right) \text{ when } X = 4.6\text{m}.$$

Stated in words, the magnitude of the displacement is constant across broad areas but jumps abruptly from one value to another along the line $X \approx 4.6$ m. In the case of the shear zone in Figure 4.31, we have

$$\Delta u_Y/\Delta X = 0 \text{ for } X < 23\text{cm and for } X > 47\text{cm}$$

$$\Delta u_Y/\Delta X \approx 0.583 \text{ for } 23\text{cm} \le X \le 47\text{cm}.$$

Stated in words, the magnitude of the displacement is constant across broad areas but changes regularly and smoothly within the zone between the lines $X = 23$ cm and $X = 47$ cm. It is no coincidence that changes in the magnitude of the displacement correlate with geologic structures – nonzero gradients in the displacement field are responsible for the geologic structures and fabrics we see in deformed rocks.

It is important to recognize that the **scale** at which we observe determines whether a particular deformation is discrete or distributed. For example, what appears to be a discrete slip surface accommodating an abrupt jump in displacement might be a very thin shear zone in which each particle is displaced slightly differently from its neighbors. Similarly, the apparently smooth change in displacement distributed across what appears to be a shear zone might occur on numerous discrete slip surfaces. For this reason, it is necessary to attach a scale-related modifier to the terms discrete and distributed, such as "discrete at the outcrop scale" or "distributed at the hand-sample scale."

4B.3.4 Idealized Versus True Gradients of the Displacement Field

In the fault and the shear zone examples considered above, displacement values change mainly along a single linear direction within the areas on which we focused. In both cases, we chose to orient our Cartesian coordinate axes so that the x axes were parallel to that one direction. By constructing a profile of the displacements of particles along that one direction (parallel to the x axes in the two examples considered), we could document the variation in displacement along that one *dimension*. Within the field of view of either figure, the displacement field exhibits a *one-dimensional displacement gradient*. Geologists often infer that an idealized displacement field with finite gradients in only one direction closely approximates the true displacement field within an area of interest.

In fact, the finite displacement observed on any fault must eventually diminish to zero both along

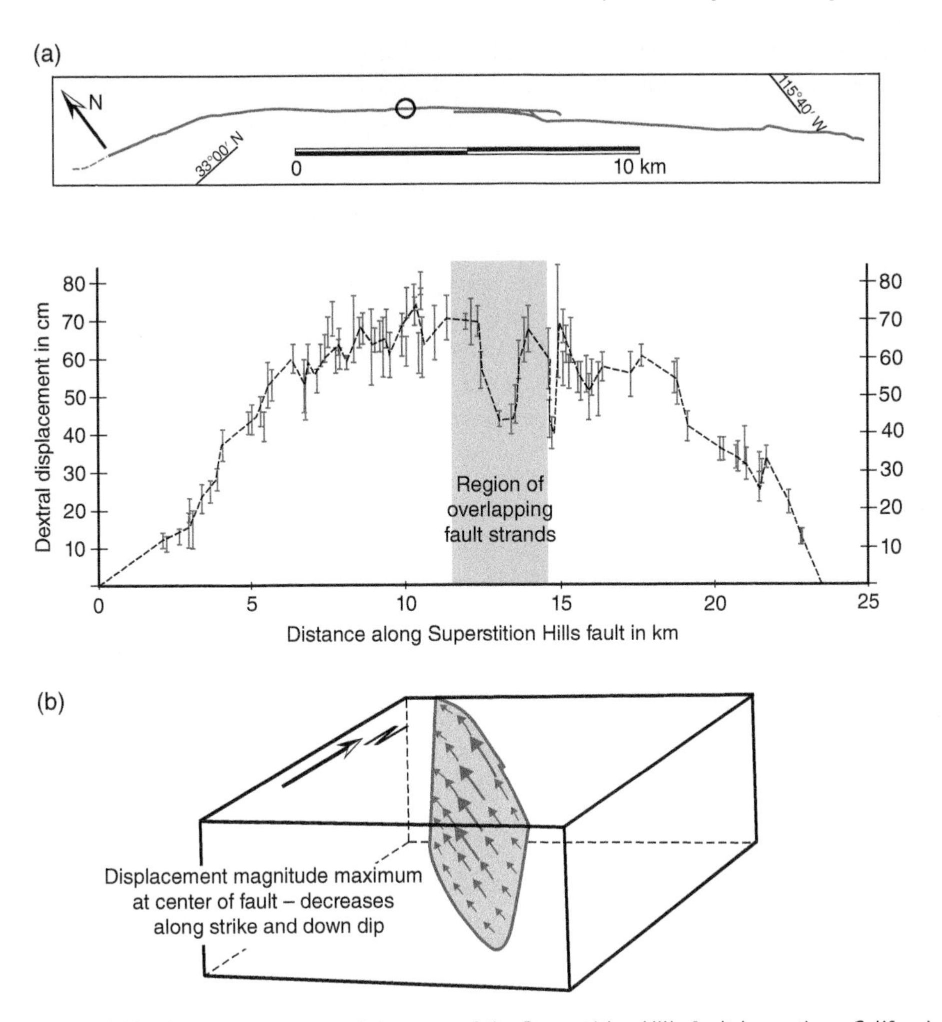

Figure 4.32 (a) Across top, map of the trace of the Superstition Hills fault in southern California. Below, plot of dextral displacement associated with the M6.6 1987 earthquake measured at different positions along the length of this fault. *Source:* Modified from Lindvall et al. (1989). (b) Schematic representation of the magnitude of displacement on this strike-slip fault during a slip event, with displacement decreasing from the maximum at the center of fault to zero along the tip line of the fault.

the strike and the dip. Stated in different terms, the actual displacement fields associated with the fault cannot have zero displacement gradients in the plane of the fault. Figure 4.32a shows the measured magnitude of dextral displacement associated with a 1987 M6.6 earthquake on the Superstition Hills fault. For several km along the fault NW and south-east (SE) of the location of the offset gully depicted in Figure 4.30, the magnitude of the earthquake-associated displacement is roughly constant at 65–70 cm (about one half of the offset of the gully shown in Figures 4.21 and 4.22 occurred during this earthquake). Eventually, however, the measured magnitude of the displacement decreases along the strike of the fault to zero at the surface terminations of the fault. Thus, in the true displacement field associated with the faulting, displacement magnitudes vary in the strike direction, which is parallel to the displacement vectors and perpendicular to the one direction specified in the previous section to have a non-zero displacement gradient. The displacement magnitude must also decrease to zero down the dip of the fault, for this fault does not divide Earth into two parts. Thus, the true displacement field has nonzero gradients in the along-strike and along-dip directions (Figure 4.32b).

4B.4 The Displacement Gradient Tensor – Relating Position and Displacement Vectors

Earlier in the Comprehensive Treatment section, we distinguished *scalar quantities* or *scalars* from *vector quantities* or *vectors*. There is another class of quantities, one of which is a quantity that characterizes how displacements change with position in a deformation, that possess a greater number of independent components than any vector. We call these quantities that exhibit a level of complexity beyond vectors *tensor quantities* or *tensors* (precise, formal definitions of tensors are abstruse and seem, at first glance, to ask you to engage in circular reasoning; we prefer, then, to introduce tensors and tensor properties through examples). To depict the tensors encountered in analyses of deformation, we use matrices. In two-dimensional settings, the relevant matrices are 2×2 arrays of numbers, such as

$$\mathbf{A} = \begin{vmatrix} a_{11} & a_{12} \\ a_{21} & a_{22} \end{vmatrix}$$

where the matrix components a_{11}, a_{12}, etc. will have different values depending upon the type of matrix and the particulars of a specific deformation. In three-dimensional analyses of deformation, the relevant matrices are 3×3 arrays of numbers:

$$\mathbf{B} = \begin{vmatrix} b_{11} & b_{12} & b_{13} \\ b_{21} & b_{22} & b_{23} \\ b_{31} & b_{32} & b_{33} \end{vmatrix}$$

Appendix 4-II describes in detail how to perform the straightforward mathematical operations we need using matrices.

One way to understand the tensors used to analyze deformation is to recognize that they are mathematical entities that relate two different vector quantities. The tensor quantity we consider here, the **displacement gradient tensor**, relates the initial position vector of a particle to its displacement vector. In this chapter, we consider only situations where displacements are linear functions of position coordinates. In those instances, the components of the displacement gradient tensor are the coefficients of linear equations relating particle displacement vector components to particle position components. With the stipulation that the displacement vector components are linear functions of the position vector components, $\Delta u_X/\Delta X = du_X/dX$, $\Delta u_X/\Delta Y = du_X/dY$, $\Delta u_Y/\Delta X = du_Y/dX$, and $\Delta u_Y/\Delta Y = du_Y/dY$. We, therefore, use the notation du_X/dX instead of $\Delta u_X/\Delta X$.

In Chapter 5, we consider a second tensor, the **position gradient tensor**, which relates the initial position vector of a particle to its final position vector. In the examples presented in Chapters 4 and 5, the components of a particle's final position vector are also linear functions of its initial position coordinates. Thus, for these examples, the components of the position gradient tensor are the coefficients of linear equations that relate a particle's final position vector components to its initial position vector components. In Chapter 5, we show that some components of the displacement gradient tensor and the position gradient tensor denote meaningful or measurable physical quantities. This is not true of all tensor quantities. Even though a matrix that represents a tensor quantity relates two vector quantities in a meaningful way, the individual components may not denote easily visualized, physically meaningful, or measurable quantities.

One property shared by scalars, vectors, and tensors is that they quantify properties that exist *independently* of the orientation of the coordinate frame in which we define them. The magnitudes of the individual components of the matrix depicting a tensor are different when referred to coordinate axes with different orientations, but the essential character of the tensor does not change. We will see that vector and tensor components referred to any one coordinate orientation can be simply related to the components referred to any other coordinate frame. Thus, we consider ways to determine "unknown" vector or tensor components in a new coordinate frame as functions of "known" vector or tensor components in the original coordinate frame. In the text and equations presented in this section, we follow a convention used in many other texts and in some geology journals to use bold, upper case letters to denote tensor quantities. Before this introduction delves too deeply into abstraction, we turn to some examples to demonstrate how tensors and tensor notation are straightforward and useful.

4B.4.1 Components of Displacement Fields: Translation, Rotation, and Pure Strain

In the Conceptual Foundation, we showed that any displacement field can be divided into components of translation, rotation, and distortion/dilation. Although most geological deformations involved a combination of terms, mathematics are much easier to understand if we treat these components individually.

4B.4.2 Translation Displacement Fields

In a two-dimensional, **translation displacement field**, all particles have identical displacement vectors. Thus, the displacement vectors for all particles within a specific area are: (1) are parallel to each other; and (2) have magnitudes that do not vary regardless of the direction one follows across the field. In mathematical terms, these two conditions mean that all four components of the displacement gradient vanish, that is $\mathrm{d}u_X/\mathrm{d}X = \mathrm{d}u_Y/\mathrm{d}X = \mathrm{d}u_X/\mathrm{d}Y = \mathrm{d}u_Y/\mathrm{d}Y = 0$, and

$$u_X = k \tag{4.19a}$$

$$u_Y = k' \tag{4.19b}$$

where k and k' are constants. It is possible for $k = k' = 0$ (if particles are not displaced at all), for $k \neq 0$ and $k' = 0$ (Figure 4.33a), for $k = 0$ and $k' \neq 0$ (Figure 4.33b), or for $k \neq 0$ and $k' \neq 0$ (Figure 4.33c). Because all particles in the field of view experience the same movement, there is no movement of particles relative to each other. Consequently, in a uniform translation displacement field, an object is neither reoriented or distorted.

4B.4.3 Rigid Rotation Displacement Fields

A two-dimensional, **rigid rotation displacement field** is a displacement field in which particles do not change their relative positions yet all particles (except those on the axis of rotation) move. In Figure 4.34a, we illustrate a rigid rotation displacement field in which the particle positioned at the origin of a Cartesian coordinate frame is stationary. All other particles are displaced an amount so that there is an angle ϕ between the position vector to the particle's original position ($\boldsymbol{X}$) and the position vector to the particle's current position ($\boldsymbol{x}$). Figure 4.34b shows the displacement $\boldsymbol{u} = \boldsymbol{x} - \boldsymbol{X}$ for an arbitrary particle. The initial position vector for this particle, $\boldsymbol{X} = X\boldsymbol{i} + Y\boldsymbol{j}$, makes an angle of θ with the x axis. The components of this position vector are, then, $X = |\boldsymbol{X}|\cos\theta$ and $Y = |\boldsymbol{X}|\sin\theta$, where $|\boldsymbol{X}|$ is the magnitude of the vector $\boldsymbol{X}$. The final position vector for this particle, $\boldsymbol{x} = x\boldsymbol{i} + y\boldsymbol{j}$, makes an angle of $(\theta + \phi)$ with the x axis. The components of this position vector are: $x = |\boldsymbol{x}|\cos(\theta + \phi)$ and $y = |\boldsymbol{x}|\sin(\theta + \phi)$, where $|\boldsymbol{x}|$ is the length of the

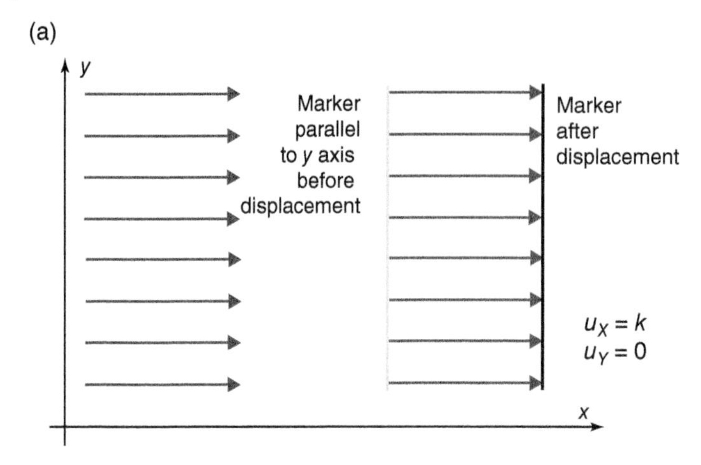

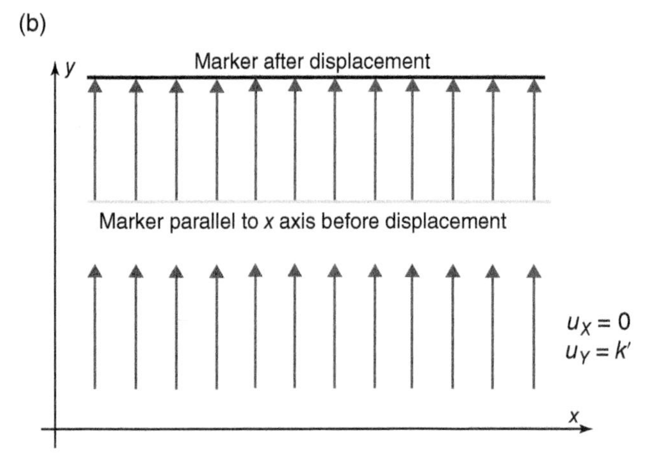

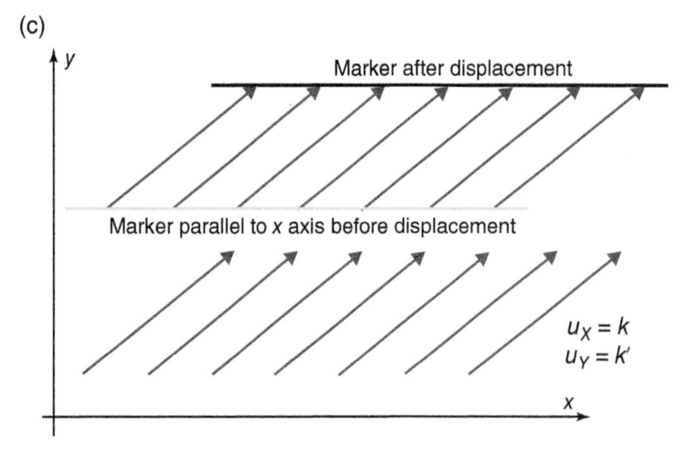

Figure 4.33 Uniform translation displacement fields. (a) Translation parallel to the x axis. (b) Translation parallel to the y axis. (c) Translation oblique to the coordinate axes.

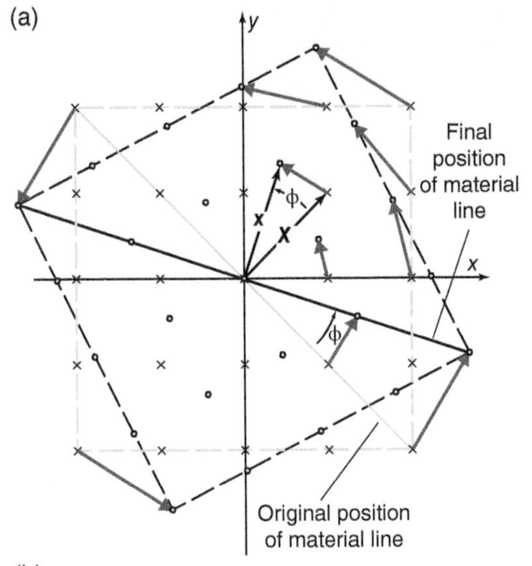

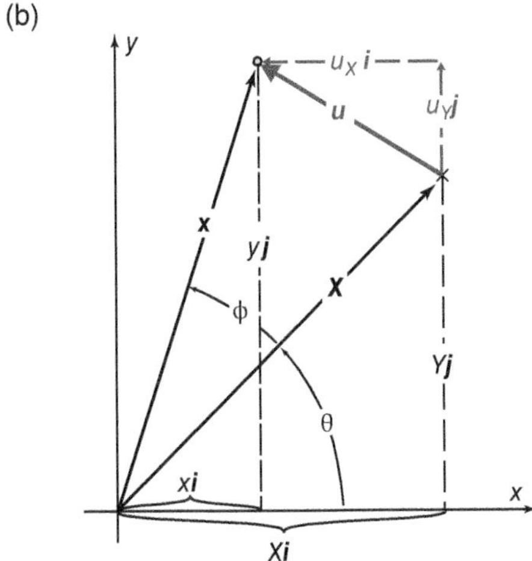

Figure 4.34 (a) Rigid body rotation displacement field. All particles displaced in a manner that conforms with a counterclockwise rotation of their position vector through an angle ϕ. (b) Diagram relating the original position coordinates X and Y to the final position coordinates x and y.

vector $\boldsymbol{x}$. Thus, the components of the displacement vector are:

$$u_X = x - X = |\boldsymbol{x}|\cos(\theta+\phi) - |\boldsymbol{X}|\cos\theta$$

$$u_Y = y - Y = |\boldsymbol{x}|\sin(\theta+\phi) - |\boldsymbol{X}|\sin\theta.$$

Using the trigonometric formula for the sine and cosine of the sum of two angles, we have

$$u_X = |\boldsymbol{x}|\left(\cos\theta\cos\phi - \sin\theta\sin\phi\right) - |\boldsymbol{X}|\cos\theta$$

$$u_Y = |\boldsymbol{x}|\left(\sin\theta\cos\phi + \cos\theta\sin\phi\right) - |\boldsymbol{X}|\sin\theta.$$

Now, $|\boldsymbol{X}| = |\boldsymbol{x}|$ because rotation about the origin does not change the length of the particle's position vector. Thus,

$$\begin{aligned} u_X &= |\boldsymbol{X}|\left(\cos\theta\cos\phi - \sin\theta\sin\phi\right) - |\boldsymbol{X}|\cos\theta \\ &= |\boldsymbol{X}|\cos\theta\cos\phi - |\boldsymbol{X}|\cos\theta - |\boldsymbol{X}|\sin\theta\sin\phi \end{aligned}$$

$$\begin{aligned} u_Y &= |\boldsymbol{X}|\left(\sin\theta\cos\phi + \cos\theta\sin\phi\right) - |\boldsymbol{X}|\sin\theta \\ &= |\boldsymbol{X}|\sin\theta\cos\phi - |\boldsymbol{X}|\sin\theta + |\boldsymbol{X}|\cos\theta\sin\phi. \end{aligned}$$

Substituting X for $|\boldsymbol{X}|\cos\theta$ and Y for $|\boldsymbol{X}|\sin\theta$ yields two expressions for the components of the displacement of a particle as a function of its original position coordinates.

$$u_X = X\left(\cos\phi - 1\right) - Y\sin\phi \tag{4.20a}$$

$$u_Y = X\sin\phi + Y\left(\cos\phi - 1\right). \tag{4.20b}$$

Figure 4.34a shows this result graphically; the displacements vary with position. Thus, the displacement gradients must have finite values. From Equations (4.20a) and (4.20b), we can see that $du_X/dX = (\cos\phi - 1)$, $du_Y/dX = \sin\phi$, $du_X/dY = -\sin\phi$, and $du_Y/dY = (\cos\phi - 1)$, confirming that the gradients of the displacement field are nonzero for all values of ϕ other than 0°, 360°, or multiples of 360°.

4B.4.4 Pure Strain Displacement Fields

The pure strain component changes the size (dilates), changes the shape (distorts), or changes both the size and shape of (both dilates and distorts) a material object. In two dimensions, a pure strain displacement field only allows movement along the x and y axes, such that either $du_X/dX \neq 0$ or $du_Y/dY \neq 0$. Any displacement parallel to the y axis that depends on (or is a function of) the

x position coordinate (du_Y/dX) or any displacement parallel to the x axis that depends on (or is a function of) the y position coordinate (du_X/dY) induces a component of rotation, which is not allowed in pure strain. Thus, $du_Y/dX = du_X/dY = 0$.

For pure strain displacement fields, then, the displacements of a particle are a function of its original position coordinates.

$$u_X = \left[du_X/dX\right]\cdot X + \left[du_X/dY\right]\cdot Y = \left[du_X/dX\right]\cdot X + 0 = \left[du_X/dX\right]\cdot X \tag{4.21a}$$

$$u_Y = \left[du_Y/dX\right]\cdot X + \left[du_Y/dY\right]\cdot Y = 0 + \left[du_Y/dY\right]\cdot Y = \left[du_Y/dY\right]\cdot Y. \tag{4.21b}$$

Comparing these equations to Equations (4.10) in the Conceptual Foundations section, we see that $du_X/dX = A - 1$ and $du_Y/dY = D - 1$, where A and D were constants such that $A \cdot D = 1$. In such a case, the displacement components take the form:

$$u_X = (A-1)\cdot X \tag{4.21a1}$$

$$u_Y = (D-1)\cdot Y. \tag{4.21b1}$$

4B.4.5 Total Displacement Fields

A displacement field that includes translation, rotation and pure strain will have displacement vector components that are simply the sum of Equations (4.19)–(4.21), that is

$$u_X = (A + \cos\phi - 2)\cdot X - (\sin\phi)\cdot Y + k \tag{4.22a}$$

$$u_Y = (\sin\phi)\cdot X + (D + \cos\phi - 2)\cdot Y + k'. \tag{4.22b}$$

Note that coefficients of X and Y in these two linear equations, which are also the components of the displacement gradient tensor that relates a particle's original position vector to its displacement vector, are likely to be nonzero if there is any rotation and pure strain. In the idealized situations, we describe and illustrate in the following section, we consider situations where unique combinations of rotation and pure strain result in displacement gradient values with especially straightforward values.

4B.4.6 Using Displacement Gradient Matrices to Represent Displacement Fields

In Equations (4.19) through (4.22), the components of a particle's displacement vector, u_X and u_Y, are linear functions of X and Y, the coordinates of that particle's original position vector:

$$u_X = K \cdot X + L \cdot Y$$

$$u_Y = M \cdot X + N \cdot Y.$$

Matrix notation (see Appendix 4-II) is a convenient way to represent such a system of linear equations. In order to use matrix notation, we envision the components of the displacement vector u_X and u_Y as elements of one 2 × 1 *column matrix* and the components of the original position vector X and Y as elements of a second 2 × 1 column matrix. Similarly, the coefficients of the system of linear equations, K, L, M, and N, are elements of a 2 × 2 square matrix $\boldsymbol{J}$. The column vector of a particle's displacement vector is the product of the square matrix and the particle's original-position column vector:

$$\begin{vmatrix} u_X \\ u_Y \end{vmatrix} = \begin{vmatrix} K & L \\ M & N \end{vmatrix} \begin{vmatrix} X \\ Y \end{vmatrix} \tag{4.23a}$$

Matrix multiplication dictates that the element in the first row of the product matrix (u_X) is the sum of two products: (1) the first-row, first-column element in the square matrix (K) times the first-row element of the second-column matrix (X); and (2) the second-column, first-row element of the square matrix (L) times the second-row element of the second-column matrix (Y). The second-row element of the product matrix (u_Y) is also the sum of two products: (1) the first-row, first-column element in the square matrix (M) times the first-row element of the second column matrix (X); and (2) the second-row, second-column element of the square matrix (N) times the second-row element of the second-column matrix (Y). Substituting J_{11} for K, J_{12} for L, J_{21} for M, and J_{22} for N, Equation (4.23a) becomes

$$\begin{vmatrix} u_X \\ u_Y \end{vmatrix} = \begin{vmatrix} J_{11} & J_{12} \\ J_{21} & J_{22} \end{vmatrix} \begin{vmatrix} X \\ Y \end{vmatrix} \tag{4.23b}$$

Using matrix shorthand, this is

$$\boldsymbol{u} = \mathbf{J} \cdot \boldsymbol{X} = \left[\mathrm{J}\right] \cdot \boldsymbol{X} \tag{4.23c}$$

where $\boldsymbol{u}$ is a particle's displacement vector, $\mathbf{J}$ (sometimes denoted [J]) is the 2 × 2 matrix, and $\boldsymbol{X}$ is a particle's position vector in the undeformed state.

The elements J_{ij} in this 2 × 2 matrix are:

$$J_{11} = \mathrm{d}u_X/\mathrm{d}X, \quad J_{12} = \mathrm{d}u_X/\mathrm{d}Y,$$

$$J_{21} = \mathrm{d}u_Y/\mathrm{d}X, \quad \text{and} \quad J_{22} = \mathrm{d}u_Y/\mathrm{d}Y.$$

In Equation (4.23c), the matrix $\mathbf{J}$ (or [J]) is the 2 × 2 **displacement gradient matrix**. An alternative version of Equations (4.23) is

$$\begin{vmatrix} u_X \\ u_Y \end{vmatrix} = \begin{vmatrix} (\mathrm{d}u_X/\mathrm{d}X) & (\mathrm{d}u_X/\mathrm{d}Y) \\ (\mathrm{d}u_Y/\mathrm{d}X) & (\mathrm{d}u_Y/\mathrm{d}Y) \end{vmatrix} \begin{vmatrix} X \\ Y \end{vmatrix} \tag{4.23d}$$

A common alternative form of this equation results by replacing X by X_1 and Y by X_2. In that case, the matrix shorthand version of Equation (4.23d) is

$$\boldsymbol{u} = \left[\mathrm{d}u_i/\mathrm{d}X_j\right] \cdot \boldsymbol{X} \tag{4.23e}$$

where each element $\mathrm{d}ui/\mathrm{d}Xj$ of the displacement gradient matrix is the change in the value of the ith row of the original position vector with respect to the change in the jth row of the original position column vector. Matrix notation does not simplify the representation of displacement vector components in the examples we consider here. We will see later that matrix representations provide significant insight into calculations related to displacement gradients, position gradients and other meaningful physical measurements. We introduce matrices and matrix notation here so that you will understand matrix algebra and can focus on its implications in later analyses.

4B.5 Idealized, Two-dimensional Displacement Fields

4B.5.1 Simple Shear Displacement Fields

A **simple shear displacement field** (Figure 4.35) is an idealized displacement field that resembles the displacement pattern associated with shear zones (cf. Figures 2.14, 4.9, 4.18, and 4.31). In a two-dimensional simple shear displacement field, all particles are displaced in a single direction. As a result, the displacement vectors for all particles within the area of interest: (1) are parallel to each other; (2) have magnitudes that do not vary in the direction parallel to the displacement vectors; and (3) have magnitudes that vary smoothly along any traverse perpendicular to the displacement vectors. For situations where the displacement vectors parallel either the x or y axes, the second and third conditions imply that either

$$u_X = 0 \tag{4.24a}$$

$$u_Y = f_Y(X) \tag{4.24b}$$

meaning $\mathrm{d}u_Y/\mathrm{d}X$ is nonzero and $\mathrm{d}u_Y/\mathrm{d}X = \mathrm{d}u_X/\mathrm{d}Y = \mathrm{d}u_Y/\mathrm{d}Y = 0$ (as in Figure 4.31) or

$$u_X = f_X(Y) \tag{4.24c}$$

$$u_Y = 0 \tag{4.24d}$$

meaning $\mathrm{d}u_X/\mathrm{d}Y$ is nonzero and $\mathrm{d}u_X/\mathrm{d}X = \mathrm{d}u_Y/\mathrm{d}X = \mathrm{d}u_Y/\mathrm{d}Y = 0$ (as in Figure 4.35).

Displacement fields that closely approximate two-dimensional, simple shear occur between two straight and parallel lines that define the boundaries of a shear zone. Outside the shear

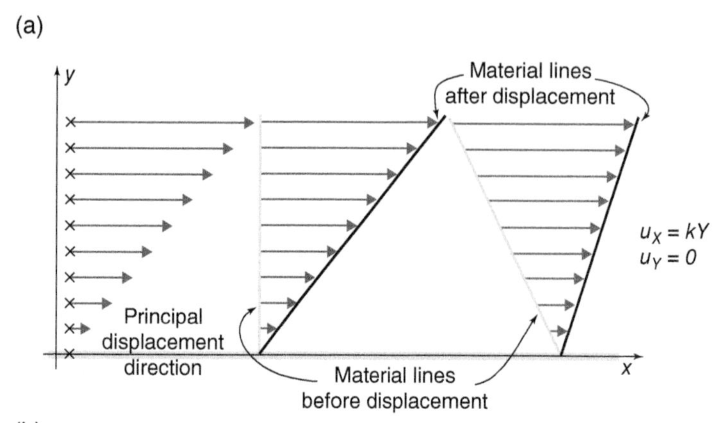

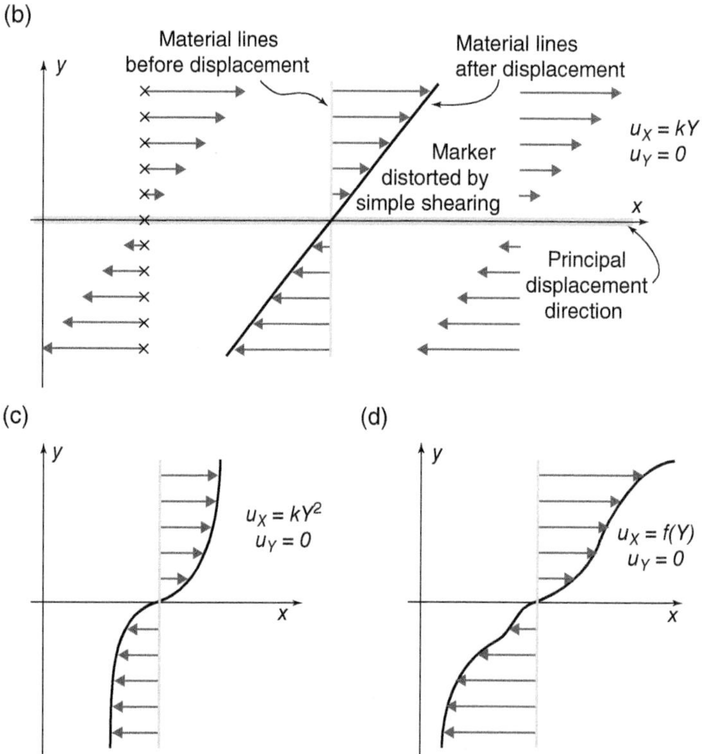

Figure 4.35 (a) Two-dimensional simple shearing displacement field with the x axis positioned along the bottom edge of the field. (b) Two-dimensional simple shearing displacement field with the x axis positioned along the center of the field. (c) and (d) Displacement profiles indicative of heterogeneous simple shearing.

zone, all displacement gradients are zero. Depending on the reference frame, particles outside the shear zone experience either experience no displacement or experience uniform translation. Within a zone of simple shear, we can envision the smooth variation in displacement magnitude in two different ways, depending on the reference frame. If one side of the shear zone is used as a reference frame, the magnitudes of the particle displacements increase smoothly from zero along the margin to a finite value along the opposite side of the zone (Figure 4.35a).

Alternatively, we can use the center of the shear zone as the reference frame. In this case, displacement magnitudes vary smoothly from zero to a positive value (with a magnitude equal to one half that reckoned relative to a fixed margin) on one side of the centerline to a negative finite value (also with a magnitude equal to one half that reckoned relative to a fixed margin) on the other side of the centerline (Figure 4.35b).

Note that regardless of its position within the area of interest, a collection of particles that define a material line oblique to the direction of zero displacement gradient (such as a line oblique to the x axis direction in Figure 4.35a and b) will have a different orientation after displacement. Thus, material lines will appear to "rotate" or "spin" toward the shearing direction. The reorientation of lines in this manner is *not*, however, the same thing as the rotation we described earlier. Because the particles defining these lines will either converge or diverge as the line reorients, these lines will also tend to shorten or elongate. For this reason, we refer to the *reorientation of material lines* not the rotation of material lines. All material lines originally parallel to the x axis retain their orientations (parallel to the x axis) and lengths, although those farther from the x axis experience greater total displacement.

In most displacement fields, we can identify special linear directions for which the collection of particles that define lines along those directions maintains a constant orientation throughout the movement. In recognition of this special property, we call these directions *eigendirections*. Mathematicians add the root "eigen" – meaning "own" in German – to terms in order to refer to properties that are "innate" or "idiosyncratic," that is are characteristic or unique. The particles that define a line parallel to such a unique direction retain that orientation. In all examples illustrated in Figure 4.35, the eigendirections are parallel to the x axis.

For a simple shearing parallel to the x axis, as long as the displacement gradient du_X/dY is constant (so that $u_X = \gamma Y$), we have a *homogeneous*, two-dimensional, simple shear displacement field. In many natural settings, we have displacement fields that approximate two-dimensional, simple shearing displacement fields, but the displacement gradient du_X/dY is not constant (Figure 4.35c and d). In those cases, we have an *inhomogeneous*, two-dimensional, simple shear displacement field.

4B.5.2 Uniaxial Convergence or Uniaxial Divergence Displacement Fields

Uniaxial convergence and uniaxial divergence are two-dimensional displacement fields in which all particles are displaced in a single direction. The displacement field again has no gradients along one direction. In this case, the displacement gradients are zero along the direction perpendicular to the direction the particles move. In such displacement fields, the displacement gradient component measured parallel to the direction in which particles move is either the greatest (positive) or smallest (negative) displacement gradient to exist in this displacement field. Negative displacement gradients correlate to converging particles (Figure 4.36a and b), while positive displacement gradients correlate to diverging particles (Figure 4.36c and d).

When particles move parallel to the x axis, convergence occurs if

$$u_X = f_X\left(X\right),$$

where

$$du_X/dX = df_X\left(X\right)/dX < 0 \qquad (4.25a)$$

$$u_Y = 0. \qquad (4.25b)$$

Figure 4.36a shows a case where $f_X(X) = X/k - X = (1/k - 1)X$, where k is a constant whose magnitude is greater than 1. Note that $f_X(X)$ is a linear function of X in which the magnitude of the coefficient of X always has a value between 0 and −1. These values result in particles moving toward, but not past, the y axis for all values of X. The gradients of the displacement field, in this case, are

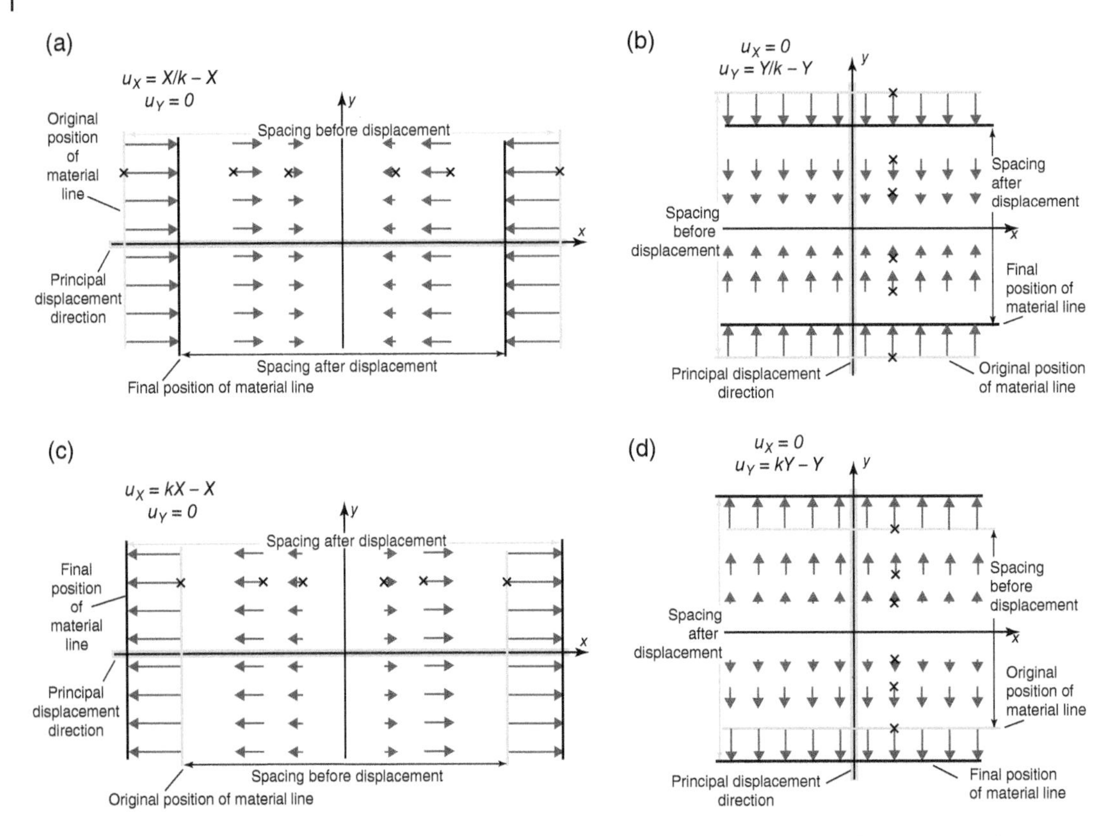

Figure 4.36 (a) Displacement field with uniaxial convergence parallel to the x axis. (b) Displacement field with uniaxial convergence parallel to the y axis. (c) Displacement field with uniaxial divergence parallel to the x axis. (d) Displacement field with uniaxial divergence parallel to the y axis.

$du_Y/dX = du_X/dY = du_Y/dY = 0$ and $du_X/dX = (1/k - 1) < 0$.

In the case where particles move parallel to the y axis, convergence occurs if

$$u_X = 0 \tag{4.25c}$$

$$u_Y = f_Y(Y), \text{ where } du_Y/dY = df_Y(Y)/dY < 0. \tag{4.25d}$$

Figure 4.36b shows the case where $f_Y(Y) = Y/k - Y$ where k is again a constant whose magnitude is greater than 1. The gradients of this displacement field are $du_X/dX = du_X/dY = du_Y/dX = 0$ and $du_Y/dY = (1/k - 1) < 0$. Because the convergence occurs only along one direction, we call this *uniaxial* convergence. Uniaxial convergence reduces the area in the plane occupied by any collection of particles. In reality, either some material must move out of the plane, or the material within the plane must experience an increase in density to accommodate these displacements.

In the case where particles move parallel to the x axis, uniaxial divergence occurs when

$$u_X = f_X(X), \text{ where } du_X/dX = df_X(X)/dX > 0 \tag{4.25e}$$

$$u_Y = 0. \tag{4.25f}$$

Figure 4.36c shows a case where $f_X(X) = k X - X = (k - 1) X$, with k is a constant whose magnitude is greater than 1. $f_X(X)$ is again a straightforward linear function of X. Here $(k-1)$, the coefficient of X, always has a value greater than 0. These values result in particles moving away from the y axis for all values of X. The gradients of the displacement field, in this case, are $du_Y/dX = du_X/dY = du_Y/dY = 0$ and $du_X/dX = (k-1) > 0$.

Uniaxial divergence for particles moving parallel to the y axis occurs when

$$u_X = 0 \tag{4.25g}$$

$$u_Y = f_Y(Y), \text{ where } du_Y/dY = df_Y(Y)/dY > 0. \tag{4.25h}$$

Figure 4.36d shows a case where $f_Y(Y) = kY - Y$ with k is a constant whose magnitude is greater than 1. The gradients of the displacement field in this case are $du_X/dX = du_X/dY = du_Y/dX = 0$ and $du_Y/dY = (k-1) > 0$. Uniaxial divergence increases the area in the plane occupied by any collection of particles, meaning that either material remains in the plane and experiences a decrease in its density or that some material has moved into this plane from positions off the plane.

In uniaxial convergence or uniaxial divergence displacement fields, material lines that are oriented parallel and perpendicular to the coordinate axes maintain their orientations during movement. The coordinate axis directions are, then, eigendirections. Consequently, one of the eigendirections is also the direction of a maximum or minimum displacement gradient. In uniaxial convergence, particles converge along one eigendirection (as in the x axis direction in Figure 4.36a or in the y axis direction in Figure 4.36b). In uniaxial divergence, particles diverge along one eigendirection (as in the x axis direction in Figure 4.36c or in the y axis direction in Figure 4.36d). These examples indicate another criterion by which one can recognize eigendirections: particles move directly toward or away from the origin along some eigendirections.

4B.5.3 Pure Shear Displacement Fields

Pure shear is a two-dimensional displacement field that is a combination of uniaxial convergence along one direction (the x or the y axis) and uniaxial divergence along a second direction at right angles (the y or x axis). We create a displacement field with convergence along one axis and divergence along the other by combining the appropriate versions of Equations (4.25):

$$u_X = f_X(X) + 0 = f_X(X) \tag{4.26a}$$

$$u_Y = 0 + f_Y(Y) = f_Y(Y) \tag{4.26b}$$

where either $du_X/dX = df_X(X)/dX < 0$ and $du_Y/dY = df_Y(Y)/dY > 0$ or $du_X/dX = df_X(X)/dX > 0$ and $du_Y/dY = df_Y(Y)/dY < 0$. **Pure shear displacement fields** are a subset of combined convergence-divergence displacement fields where $du_X/dX \neq 0$, $du_X/dY = du_Y/dX = 0$, and $(du_Y/dY + 1) = 1/(du_X/dX + 1)$. The condition that $(du_X/dX + 1)(du_Y/dY + 1) = 1$ ensures that any collection of particles occupies the same area in the plane before and after the movement. The equivalence of the area occupied by a collection of particles before and after movement is an essential condition of a pure shear displacement field. Note that pure shear records only the pure strain component of deformation; it has no component of solid-body rotation.

Figure 4.37a shows a pure shearing displacement field generated by adding the uniaxial convergence along the x axis shown in Figure 4.36a with the uniaxial divergence along the y axis shown in Figure 4.36d. The expressions for displacement as a function of position for pure shear displacement field are simple linear functions:

$$u_X = f_X(X) + 0 = (1/k - 1)X + 0 = X/k - X$$

$$u_Y = 0 + f_Y(Y) = 0 + (k-1)Y = kY - Y$$

where k is a constant greater than 1. One can also generate a pure shear displacement field by combining the uniaxial divergence along the x axis shown with the uniaxial convergence along the

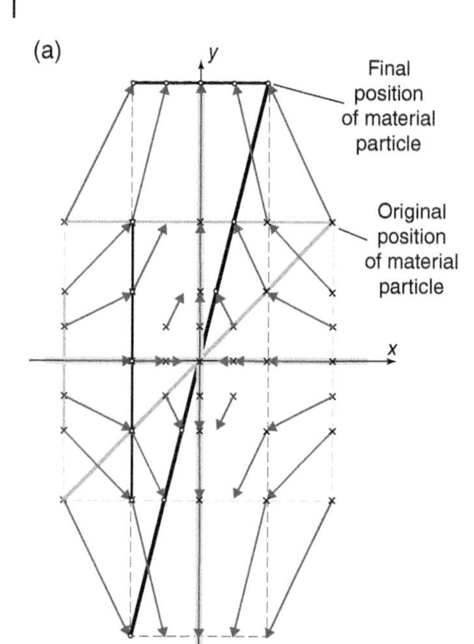

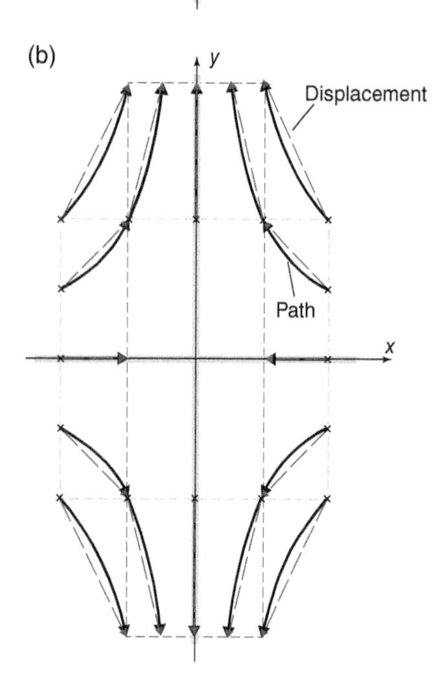

Figure 4.37 (a) Two-dimensional pure shearing displacement field with convergence along the x axis and divergence along the y axis. (b) Comparison of total displacements of particles from (a) with paths followed by particles.

y axis (cf. Figure 4.19). The expressions for displacement as a function of position for this pure shear displacement field are:

$$u_X = f_X(X) + 0 = (k-1)X + 0 = kX - X$$

$$u_Y = 0 + f_Y(Y) = 0 + (1/k - 1)Y = Y/k - Y.$$

In Figure 4.37a, each × marks the original position of a particle within a Cartesian coordinate frame. The arrows give the direction and magnitude of the displacement of that particle relative to the particle located at the origin of the coordinate frame. Particles whose original positions fall on either coordinate axis have relative displacements directed exactly toward or exactly away from the origin. Those particles, therefore, remain on the coordinate axis. The collection of particles along the axes – material lines coincident with the axes – do not change their orientation, and the successive incremental displacements of these particles are parallel. The coordinate axes are therefore eigendirections for these displacement fields. Particles converge along one eigendirection and diverge along the other. Therefore, any collection of particles that define a material line originally parallel to one of the coordinate axes remain parallel to the coordinate axes after movement. Two such material lines are shown in Figure 4.37a, with the original positions of the material lines in gray and the displaced positions in black.

Particles not originally located on one of the axes will move obliquely to the coordinate axes. Material lines originally inclined to the coordinate axes (like the inclined gray line on Figure 4.37a) are reoriented by the displacement field; reorientation brings the material line closer to being parallel with the coordinate axis experiencing divergence. Material lines originally inclined to the coordinate axes are never reoriented into parallelism with the axes, even for large displacements.

The *change of state* (displacement) of a particle differs distinctly from the *path* during pure shear. Figure 4.37b compares the displacements of selected particles from Figure 4.37a with the paths that those particles follow. Incremental displacements, relative to the origin, of particles not lying on the coordinate axes, are not parallel, meaning that the paths followed by those particles are curved.

4B.5.4 General Shear Displacement Fields

To generate the pure shear displacement field, we added together the expressions defining the displacements as a function of the original positions for convergence along one axis and divergence along the other axis. We can generate other displacement fields of interest by combining two or more algebraic expressions that define displacement values as functions of position. One such two-dimensional displacement field that is especially useful in geological contexts is a **general shearing displacement field**, which combines simple shearing and pure shearing. For simplicity, we will only consider the cases of simple shearing parallel to the x axis (Figure 4.38). The conditions for a general shearing displacement field are that $du_X/dX \neq 0$, $du_X/dY \neq 0$, $du_Y/dX = 0$ and $(du_X/dX + 1)(du_Y/dY + 1) = 1$. Adding the expressions for displacements as a function of position for simple shearing (given by Equations 4.24) to those for displacements as a function of position for pure shearing (given by Equations 4.26), we get the following expressions for displacements as a function of position:

$$u_X = f_X(X) + f_X(Y) = f_X(X,Y) \tag{4.27a}$$

$$u_Y = 0 + f_Y(Y). \tag{4.27b}$$

If the pure shearing component has divergence along the x axis and convergence along the y axis, the simplest forms for the functions $f_X(X,Y)$, and $f_Y(Y)$ are

$$u_X = k_1X - X + k_2Y$$

$$u_Y = Y/k_1 - Y.$$

Figure 4.38a illustrates the resulting displacement field. Dashed, thin, gray lines connect several particles to define a material object – a square – prior to displacement. The displacements experienced by these particles distort the square into a parallelogram. Particles positioned along the x axis remain on the axis but experience divergence. Similarly, particles defining lines parallel to the x axis before displacement (such as those defining the top and bottom sides of the square) define lines parallel to the x axis after displacement (the top and bottom sides of the parallelogram). Thus, the x axis direction is an eigendirection; particles experience divergence along this direction in this instance. There is also an eigendirection along which particles experience convergence; it is inclined from upper left to lower right relative to the horizontal x axis (Figure 4.38a and b). Material lines inclined to the eigendirections (such as the diagonals of the dashed square) are reoriented during displacement. As is true for two-dimensional pure shearing, the curved paths of particles are not generally parallel to their finite displacements.

If the pure shearing component has convergence along the x axis and divergence along the y axis, the simplest form for the functions $f_X(X,Y)$, and $f_Y(Y)$ is

$$u_X = X/k_1 - X + k_2Y$$

$$u_Y = k_1Y - Y.$$

Figure 4.38c illustrates the resulting displacement field. Convergence occurs along the eigendirection parallel to the x axis and divergence occurs along an eigendirection inclined from lower left to upper right relative to a horizontal x axis. Similar to the preceding case, the paths of particles are not generally parallel to their finite displacements (Figure 4.38d).

4B.6 Idealized, Three-Dimensional Displacement Fields

In natural settings, the arrays of vectors defining a displacement field are rarely confined to a single, two-dimensional plane. Rather, in displacement

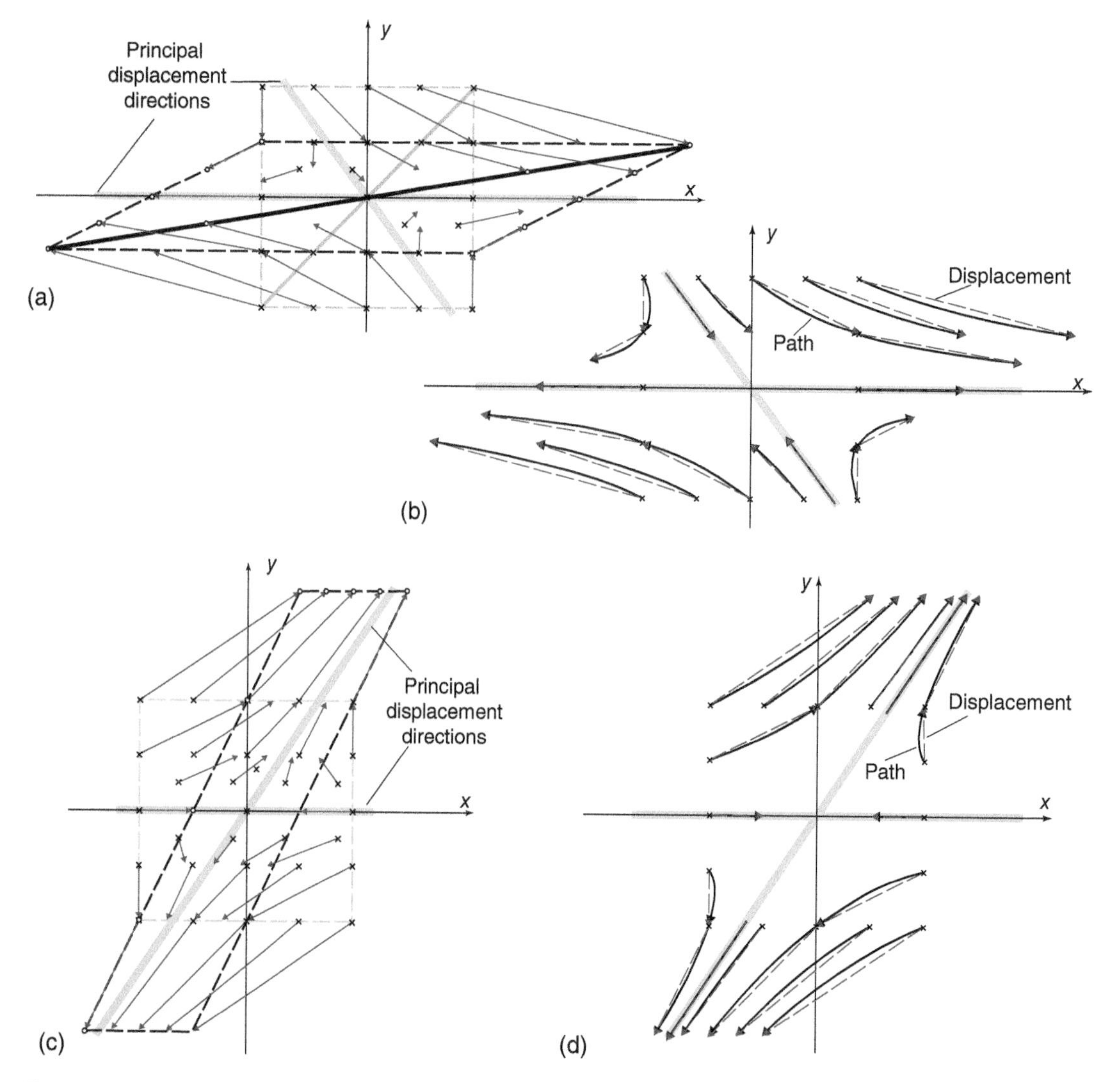

Figure 4.38 (a) General shearing displacement field generated by adding together simple shearing parallel to the x axis and pure shearing with divergence along the x axis and convergence along the y axis. (b) Comparison of displacements of particles from (a) with paths followed by particles. (c) General shearing displacement field generated by adding together simple shearing parallel to the x axis and pure shearing with convergence along the x axis and divergence along the y axis. (d) Comparison of displacements of particles from (c) with paths followed by particles.

fields associated with natural deformations, the displacement vectors at different positions typically have orientations and magnitudes that vary in three dimensions (Figure 4.39). We envision that three-dimensional deformations also consist of bulk translation, rigid body rotation, and pure strain components (Figure 4.40). There are three major complications of three-dimensional displacement fields, relative to two-dimensional displacement fields. First, there are nine components to the displacement field, as opposed to the four components in two dimensions. Second, there are generally three eigendirections. Third, we often encounter difficulty in depicting three-dimensional displacements on a two-dimensional page. We try to overcome these problems by starting with a simple shear, which was discussed earlier.

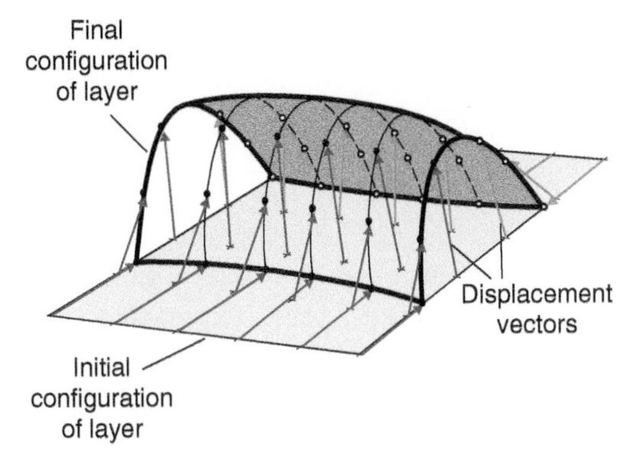

Figure 4.39 Displacement vectors of selected particles during the formation of a simple anticline.

4B.6.1 Three-Dimensional Simple Shear Displacement Fields.

Simple shear is plane strain (two-dimensional) deformation, yet it can be calculated mathematically and depicted visually in three dimensions. In both two- and three-dimensional simple shear, the displacement vectors for all particles are parallel to each other. The magnitudes of displacements do not vary in the direction parallel to the displacement vectors nor in a second direction perpendicular to the displacement direction. These two directions, the shearing direction and the line perpendicular to it along which displacements are constant, define an orientation for planes within which displacement magnitudes

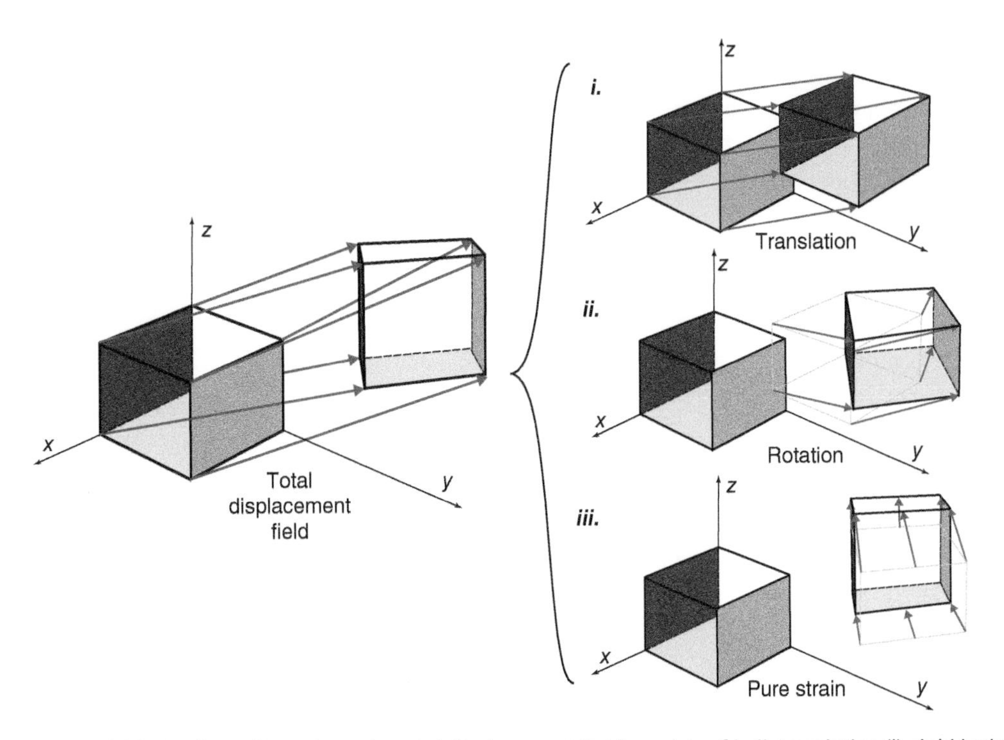

Figure 4.40 In three dimensions, the total displacement field consists of bulk translation (i), rigid body rotation (ii), and pure strain components (iii).

are constant (to use an analogy, this plane would be the actual computer card in Figure 4.18c). We call these planes **shear planes**. There are no relative displacements of particles within the shear plane, just as the particles in the computer card do not move relative to each other. Along the direction perpendicular to a shear plane (which is also perpendicular to the displacement direction), the magnitudes of displacements change smoothly.

To describe simple shear in three dimensions, we orient the Cartesian coordinate axes so that two of them lie within a shear plane, one parallel to and one perpendicular to the displacement direction. The third coordinate axis is, then, perpendicular to the shear plane. Particles lying within the shear plane containing the two coordinate axes do not move relative to the origin. Relative to any single shear plane, all particles on one side of that shear plane move in a single direction, and all particles on the other side of that shear plane move in the opposite direction (180° from the displacement direction on the first side). In Figure 4.41, particles on the x–y plane do not move relative to the origin of the coordinate frame, displacement vectors for all other particles are parallel to the x axis, and displacement magnitudes increase smoothly for particles at greater distances above or below the x–y plane. The forms of the displacement components for this three-dimensional simple shearing displacement field are:

$$u_X = f_X(X,Y,Z) = f_X(Z) \tag{4.28a}$$

$$u_Y = f_Y(X,Y,Z) = 0 \tag{4.28b}$$

$$u_Z = f_Z(X,Y,Z) = 0. \tag{4.28c}$$

du_X/dZ is nonzero, but $du_X/dX = du_X/dY = du_Y/dX = du_Y/dY = du_Y/dZ = du_Z/dX = du_Z/dY = du_Z/dZ = 0$. Displacement gradients vanish within any plane parallel to the x-y plane, but gradients are nonzero along any direction oblique to the x–y plane. Material lines originally parallel to the x–y plane are displaced but not reoriented. Thus, all material lines parallel to the x–y plane are eigendirections. Any material line oblique to the x–y plane is reoriented during simple shearing displacement. Any collection of material lines that define a material plane inclined to the shear plane are reoriented during shearing.

In three dimensions, there are five other geometries for three-dimensional simple shearing on

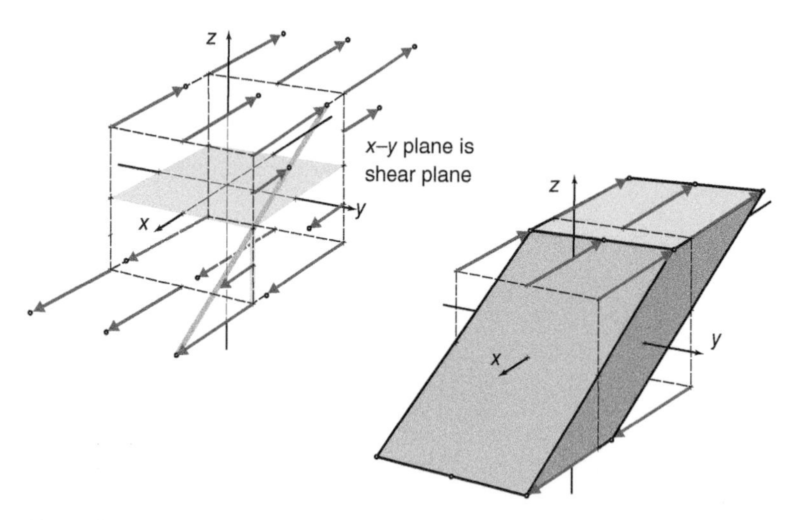

Figure 4.41 Three-dimensional simple shearing displacement field, with a shear plane parallel to the x–y plane and a shear direction parallel to the x axis.

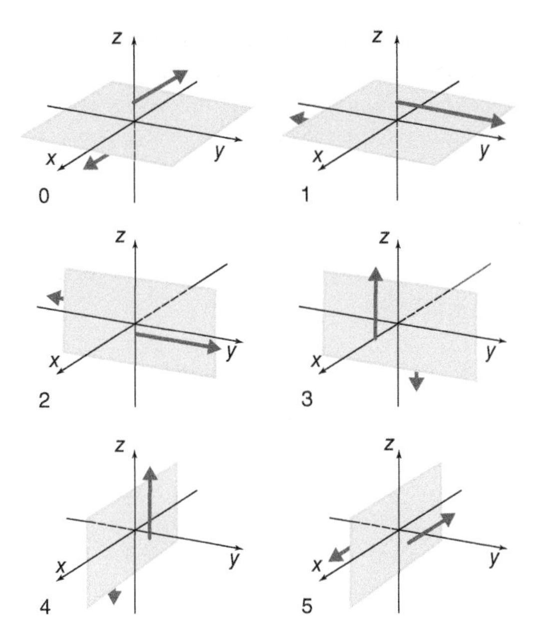

Figure 4.42 Schematic representations of three-dimensional simple shearing displacement fields. (0) Simple shearing displacements illustrated in Figure 4.41. (1) to (5) are alternative geometries for three-dimensional simple shearing fields. See text for explanation.

shear planes parallel to two coordinate axes (Figure 4.42): (1) shearing displacements parallel the y axis on shear planes parallel to the x–y plane; (2) shearing displacements parallel the y axis on shear planes parallel to the y–z plane; (3) shearing displacements parallel the z axis on shear planes parallel to the y–z plane; (4) shearing displacements parallel the z axis on shear planes parallel to the x–z plane; and (5) shearing displacements parallel the x axis on shear planes parallel to the x–z plane.

4B.6.2 Three-Dimensional Orthogonal Convergence and Divergence Displacement Fields

There are twelve three-dimensional displacement fields consisting only of combinations of convergence and divergence along different coordinate directions. Restricting the displacement field to orthogonal convergence and/or divergence along three orthogonal coordinate axes places the following conditions on the functional relationships for displacement components:

$$u_X = f_X(X,Y,Z) = f_X(X)$$
$$u_Y = f_Y(X,Y,Z) = f_Y(Y)$$
$$u_Z = f_Z(X,Y,Z) = f_Z(Z).$$

Convergence occurs along the x axis if $du_X/dX = df_X(X)/dX < 0$, along the y axis if $du_Y/dY = df_Y(Y)/dY < 0$, and along the z axis if $du_Z/dZ = df_Z(Z)/dZ < 0$. Divergence occurs along the x axis if $du_X/dX = df_X(X)/dX > 0$, along the y axis if $du_Y/dY = df_Y(Y)/dY > 0$, and along the z axis if $du_Z/dZ = df_Z(Z)/dZ > 0$. In displacement fields characterized by convergence (or divergence) along only one axis or along all three axes, the volume occupied by a collection of particles after displacement is less than (or greater than) the volume occupied by the same particles before displacement. The condition for constant volume is

$$(du_X/dX + 1)(du_Y/dY + 1)(du_Z/dZ + 1) = 1. \tag{4.29}$$

Three-dimensional displacement fields with orthogonal convergence and divergence and equal volume before and after displacement occur with: (1) convergence along one axis, divergence along a second axis, and neither convergence nor divergence along the third axis (plane strain); (2) with convergence along two axes and divergence along the third axis (constriction); or (3) with divergence along two axes and convergence along the third axis (flattening) (Figure 4.43).

4B.6.3 Pure Shearing Displacement Fields

In a **pure shearing displacement field**, there is convergence along one axis, divergence along a second axis, and neither convergence nor divergence along the third axis. With neither

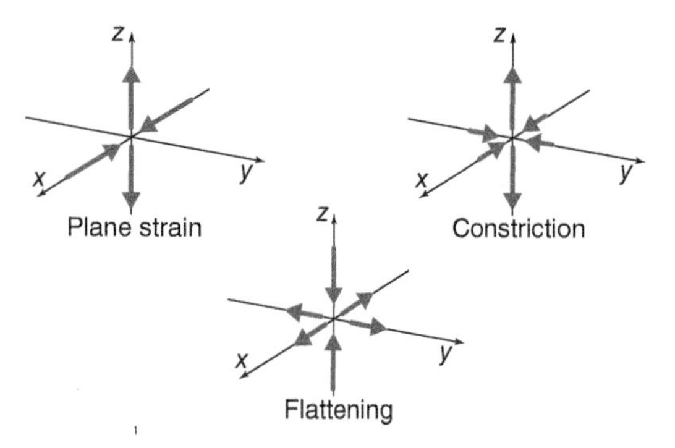

Figure 4.43 Schematic representations of pure shearing, constrictional, and flattening three-dimensional displacement fields.

convergence nor divergence along one axis, gradients of the displacement component parallel to that axis are zero. The condition that a collection of particles occupy equal volumes before and after displacement given in Equation (4.29) then reduces to the form defined for two-dimensional pure shear displacement fields in Equations (4.26). For the situation depicted in Figure 4.20a, with convergence along the x axis, neither convergence nor divergence along the y axis, and divergence along the z axis, the simplest form the displacement components can take is:

$$u_X = f_X(X) = (1/k - 1)X = X/k - X \quad (4.30a)$$

$$u_Y = 0 \quad (4.30b)$$

$$u_Z = f_Z(Z) = (k - 1)Z = kZ - Z \quad (4.30c)$$

where k is a constant whose magnitude is greater than 1. Note that $du_X/dX = 1/k - 1 < 0$ and $du_Z/dZ = k - 1 > 0$ and $(du_X/dX + 1)(du_Y/dY + 1)(du_Z/dZ + 1) = (1/k)\cdot 1\cdot k = 1$. In the x–z plane or any plane parallel to it, the pattern of displacements relative to the origin is the two-dimensional pure shearing displacement field depicted in Figure 4.37. As discussed earlier, there are two mutually perpendicular eigendirections in the x–z plane, lying along the x (convergence) and z (divergence) coordinate axes. Particles lying along the y axis do not move relative to the origin. Particles that fall along any line originally parallel to the y axis all experience identical displacements, meaning that the line retains its orientation parallel to the y axis. Thus, the y axis too is an eigendirection, so there are three mutually perpendicular eigendirections. Material lines initially inclined to the coordinate axes reorient during displacement, as do material planes inclined to the x–z plane.

4B.6.4 Constrictional Displacement Fields

Figure 4.20b shows a three-dimensional displacement field with convergence along the x and y axes and divergence along the z axis. We call this constant volume example a **purely constrictional displacement field** because: (1) any collection of particles defining an originally equant shape (such as the particles defining the dashed cube) define a *constricted* and *elongated* shape (the elongate prism) after displacement; and (2) particles undergo convergence at equal rates along all lines perpendicular to the z axis. The example shown in Figure 4.20b has the simplest form possible for the displacement components:

$$u_X = f_X(X) = (1/k - 1)X = X/k - X \quad (4.31a)$$

$$u_Y = f_Y(Y) = (1/k - 1)Y = Y/k - Y \quad (4.31b)$$

$$u_X = f_Z(Z) = (k_1 - 1)Z = k_1 Z - Z \quad (4.31c)$$

where k and k_1 are constants whose magnitudes are greater than 1. In order to conserve volume, $(du_X/dX + 1)(du_Y/dY + 1)(du_Z/dZ + 1) = (1/k)^2\cdot k_1 = 1$, meaning that $k_1 = k^2$. In a purely constrictional displacement field, all x–y planes exhibit identical patterns of displacement relative to the origin. In any single section, the pattern is similar to that shown in Figure 4.37a.

Incremental displacements relative to the origin are collinear for particles lying along the z axis. The z axis is, then, an eigendirection where particles experience divergence. Within the x–y plane, particles move uniformly toward the z axis. The lines defined by the intersection between the x–y plane and any vertical plane containing the z axis are eigendirections where particles experience convergence.

Generally, constrictional displacement fields differ from the purely constrictional displacement field illustrated in Figure 4.20b in that there is unequal convergence along the x and y axes. The displacement components for a generally constrictional displacement field are

$$u_X = f_X(X) = (1/k_1 - 1)X = X/k_1 - X \quad (4.32a)$$

$$u_Y = f_Y(Y) = (1/k_2 - 1)Y = Y/k_2 - Y \quad (4.32b)$$

$$u_Z = f_Z(Z) = (k_3 - 1)Z = k_3 Z - Z \quad (4.32c)$$

where k_1 and k_2 are both constants whose magnitudes are greater than 1, with $k_1 \neq k_2$. Since $(du_X/dX + 1)(du_Y/dY + 1)(du_Z/dZ + 1) = (1/k_1)\cdot(1/k_2)\cdot k_3 = 1$, $k_3 = k_1 \cdot k_2$. Comparing displacement fields with equal magnitudes for k_1 but smaller and smaller magnitudes for k_2, you would see constrictional displacement fields that resemble Figure 4.20b for $k_1 \approx k_2$ and weakly constrictional displacement fields that resemble Figure 4.20a for $k_1 >> k_2$.

The patterns of displacements relative to the origin within the x–y plane differ from that in the purely constrictional displacement field. Only along the x and y axes do particles move directly toward the origin, and thus these axes are eigendirections. One eigendirection experiences the greatest magnitude of convergence in the plane normal to the z axis, whereas the other eigendirection experiences the least magnitude of convergence in the plane normal to the z axis. Other particles in the x–y plane have successive incremental displacements that are not collinear.

4B.6.5 Flattening Displacement Fields

Figure 4.20c shows a three-dimensional displacement field with divergence along the x and y axes and convergence along the z axis. This constant volume deformation is called a **purely flattening displacement field** because any collection of particles defining an originally equant shape (such as the particles defining the dashed cube) defines a *flattened* shape (the tabular shape) after displacement. In a purely flattening displacement field, all planes that contain the z axis exhibit identical patterns of displacement vectors. The displacement field is given by the following equations:

$$u_X = f_X(X) = (k - 1)X = kX - X \quad (4.33a)$$

$$u_Y = fY_y(Y) = (k - 1)Y = kY - Y \quad (4.33b)$$

$$u_Z = f_Z(Z) = (1/k_1 - 1)Z = Z/k_1 - Z \quad (4.33c)$$

where k and k_1 are constants whose magnitudes are greater than 1. Once again $(du_X/dX + 1)(du_Y/dY + 1)(du_Z/dZ + 1) = k^2\cdot(1/k_1) = 1$, with $k_1 = k^2$.

Incremental displacements relative to the origin are collinear for particles lying along the z axis, making the z axis an eigendirection where particles experience convergence. Within the x–y plane, particles move uniformly away from the z axis; the successive incremental displacements relative to the origin of all particles in the x–y plane are collinear. Thus, every line parallel to the x–y plane is an eigendirection in which particles experience divergence.

General flattening displacement fields differ from the purely flattening displacement field in that there is unequal divergence along different directions perpendicular to the z axis. As a result, the x and y axes are eigendirections. One eigendirection experiences the greatest magnitude of divergence, whereas the other eigendirection experiences the least magnitude of divergence.

The displacement components for a general flattening displacement field are

$$u_X = f_X\left(X\right) = \left(k_1 - 1\right)X = k_1X - X \qquad (4.34a)$$

$$u_Y = f_Y\left(Y\right) = \left(k_2 - 1\right)Y = k_2Y - Y \qquad (4.34b)$$

$$u_Z = f_Z\left(Z\right) = \left(1/k_3 - 1\right)Z = Z/k_3 - Z \qquad (4.34c)$$

where k_1 and k_2 are constants whose magnitudes are greater than 1, with $k_1 \neq k_2$. Since $(du_X/dX + 1)(du_Y/dY + 1)(du_Z/dZ + 1) = k_1 \cdot k_2 \cdot (1/k_3) = 1$, $k_3 = k_1 \cdot k_2$. Displacement fields either resemble purely flattening displacement fields ($k_1 \approx k_2$) or pure shearing displacement fields ($k_1 >> k_2$) (Figure 4.20c).

4B.6.6 Three-Dimensional General Shearing Displacement Fields

Two-dimensional general shearing displacement fields result from combining simple shearing along one coordinate axis with pure shear (convergence along one coordinate axis and divergence along the other coordinate axis). By analogy, **three-dimensional general shear displacement fields** combine simple shearing with a three-dimensional displacement field of flattening, constriction, or plane strain (pure shear). The latter have orthogonal convergence and divergence parallel to coordinate axes. There are twelve different, end-member cases involving purely constructional, pure shear, and purely flattening displacement fields, but in different orientations with respect to the simple shearing. These cases, and others, will be described in more detail in the strain chapter (Comprehensive Treatment).

4B.7 Summary

Displacement vectors connect the initial and final positions of a particle during a particular time interval. The positions of particles at different times are mathematically described by bound vectors. The displacement vectors connecting original and final positions are free vectors. Application to geological studies requires the specification of a particular frame of reference. A displacement field describes the movement of a collection of particles from their initial positions to their final positions. The displacement field is characterized by a set of equations that specify displacement vector components as functions of components of the original position vector. Alternatively, we can use the values of the gradients of the displacement field to determine the components of a displacement gradient matrix. We can determine the displacement vector for a particle by pre-multiplying its initial position vector by the displacement gradient matrix. All displacement fields are all denoted by the form of their displacement gradient matrix, including the components of the displacement fields (translation, a rigid body rotation, and pure strain) and geologically relevant displacement fields (simple shear, pure shear, purely constructional shear, purely flattening shear)

Appendix 4-I: Vectors

4-I.1 Simple Mathematical Operations with Vectors

Due to their greater information content, the mathematical manipulation of vectors is more involved than the mathematical manipulation of scalars. Still, the mathematics of vector addition, vector subtraction, and some types of vector multiplication are straightforward. Knowing how to perform these simple operations will help you to understand and analyze displacements and eventually deformations more thoroughly.

Vector addition: In order to add one vector to another, place the tail of the directed line segment

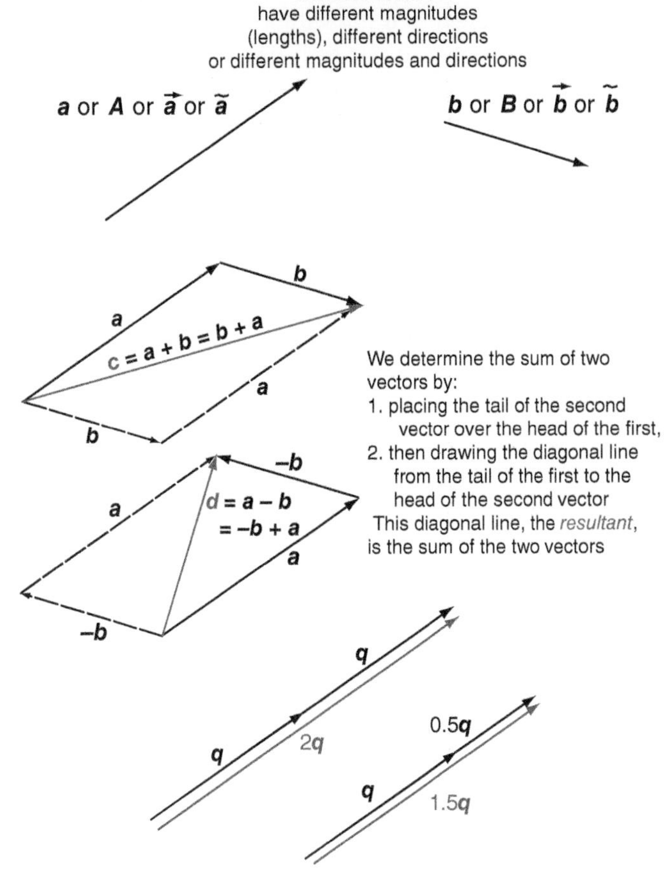

Figure 4A.1 Diagrams showing how directed line segments represent vectors and the reasoning behind vector addition, vector subtraction, and scalar multiplication of vectors.

representing the second vector adjacent to the head of the directed line segment representing the first vector (Figure 4A.1). The directed line segment from the tail of the first vector to the head of the second vector is the sum of the two vectors, called the *resultant*. The resultant $\mathbf{c} = \boldsymbol{a} + \mathbf{b}$ is also a diagonal of a parallelogram having the vectors $\boldsymbol{a}$ and $\boldsymbol{b}$ as its sides. Subtraction of vectors works just like the subtraction of scalars – one adds a "negative" quantity. A negative vector and a positive vector with equal magnitudes have the same lengths but point in the opposite directions. As an example, think of Cartesian coordinate axes, where the opposite directions denote positive and negative x or y values. In order to subtract one vector from another then, add a "negative" second vector to the first vector. One can envision the vector $\mathbf{d} = \boldsymbol{a} - \mathbf{b}$ as the resultant of the vectors $\boldsymbol{a}$ and $-\boldsymbol{b}$. Vector addition and subtraction are both commutative and associative. Thus, $\boldsymbol{a} + \mathbf{b} = \mathbf{b} + \boldsymbol{a}$ and $(\boldsymbol{a} + \mathbf{b}) + \mathbf{c} = \boldsymbol{a} + (\mathbf{b} + \mathbf{c})$.

Scalar multiplication: If we add the vector $\boldsymbol{q}$ to itself, we get a vector that points in the same direction as $\boldsymbol{q}$ yet has a length twice that of the vector $\boldsymbol{q}$. The vector $0.5\boldsymbol{q}$ has the same direction as $\boldsymbol{q}$ with a magnitude or length one half of the length of $\boldsymbol{q}$, and adding $0.5\boldsymbol{q}$ to $\boldsymbol{q}$ yields a vector pointing in the same direction as $\boldsymbol{q}$ but with a length 1.5 times the length of $\boldsymbol{q}$. In general, $n\boldsymbol{q}$, a **scalar multiple** of $\boldsymbol{q}$, has the same direction as the vector $\boldsymbol{q}$ and a magnitude n times the magnitude of $\boldsymbol{q}$ (Figure 4A.1).

4-I.2 Vector Magnitudes

We normally represent any vector as the sum of two (or three) component vectors, each of which is parallel to one of the two (or three) Cartesian coordinate axes. Further, we envision each component vector as a scalar multiple of a unit vector parallel to the corresponding coordinate axis. If $\boldsymbol{i}$, $\boldsymbol{j}$, and $\boldsymbol{k}$ are unit vectors parallel to the x, y, and z coordinate axes, respectively, we represent the vector $\boldsymbol{a}$ in component form as

$$\boldsymbol{a} = a_1\boldsymbol{i} + a_2\boldsymbol{j} + a_3\boldsymbol{k}, \tag{4A.1}$$

where a_1, a_2, and a_3 are scalar values. The magnitude of the length or magnitude of the vector $\boldsymbol{a}$, denoted $|\boldsymbol{a}|$, is

$$|\boldsymbol{a}| = \sqrt{\left(a_1^2 + a_2^2\right)} \tag{4A.2}$$

in two dimensions and

$$|\boldsymbol{a}| = \sqrt{\left(a_1^2 + a_2^2 + a_3^2\right)} \tag{4A.3}$$

in three dimensions.

The length or magnitude of the vector $\boldsymbol{b}$ is

$$|\boldsymbol{b}| = \sqrt{\left(b_1^2 + b_2^2 + b_3^2\right)}$$

If $\mathbf{a} + \mathbf{b} = \mathbf{c}$,

$$\begin{aligned} \boldsymbol{a}+\boldsymbol{b} &= \left(a_1\boldsymbol{i}+a_2\boldsymbol{j}+a_3\boldsymbol{k}\right)+\left(b_1\boldsymbol{i}+b_2\boldsymbol{j}+b_3\boldsymbol{k}\right) \\ &= \left(a_1+b_1\right)\boldsymbol{i}+\left(a_2+b_2\right)\boldsymbol{j}+\left(a_3+b_3\right)\boldsymbol{k} \\ &= c_1\boldsymbol{i}+c_2\boldsymbol{j}+c_3\boldsymbol{k}=\boldsymbol{c} \end{aligned}$$

where

$$\left(a_1+b_1\right)=c_1$$

$$\left(a_2+b_2\right)=c_2$$

$$\left(a_3+b_3\right)=c_3$$

Note that

$$|\boldsymbol{c}| = |\boldsymbol{a}+\boldsymbol{b}| = \sqrt{\left\{\left(a_1+b_1\right)^2+\left(a_2+b_2\right)^2+\left(a_3+b_3\right)^2\right\}}. \tag{4A.4}$$

In general

$$\begin{aligned} |\boldsymbol{a}+\boldsymbol{b}| &= \sqrt{\left\{\left(a_1+b_1\right)^2+\left(a_2+b_2\right)^2+\left(a_3+b_3\right)^2\right\}} \\ &\neq \sqrt{\left(a_1^2+a_2^2+a_3^2\right)}+\sqrt{\left(b_1^2+b_2^2+b_3^2\right)} \\ &= |\boldsymbol{a}|+|\boldsymbol{b}| \end{aligned}$$

The scalar multiple of a vector is the sum of the scalar multiples of its components. Thus,

$$\begin{aligned} n\boldsymbol{a} &= n\left(a_1\boldsymbol{i}+a_2\boldsymbol{j}+a_3\boldsymbol{k}\right) \\ &= na_1\boldsymbol{i}+na_2\boldsymbol{j}+na_3\boldsymbol{k} \end{aligned} \tag{4A.5}$$

Note that

$$\begin{aligned} |n\boldsymbol{a}| &= \sqrt{\left\{\left(na_1\right)^2+\left(na_2\right)^2+\left(na_3\right)^2\right.} \\ &= \sqrt{n^2\left(a_1^2+a_2^2+a_3^2\right)} \\ &= n\sqrt{\left(a_1^2+a_2^2+a_3^2\right)} = n|\boldsymbol{a}| \end{aligned}$$

4-I.3 Properties of Vector Quantities

In the same way that the mass of an object does not change if one uses different units to measure it, the unique direction and magnitude of any vector do not change even if we use coordinate frames with different orientations to characterize them. Because the unit vectors in different coordinate frames have different orientations, however, the magnitude of the scalar values required to multiply the unit vectors in those different coordinate frames must change as the orientation of the Cartesian coordinate axes changes. The magnitudes of the scalar values for the component vectors in different Cartesian coordinate axes are not independent of each other, however, because every set of component vectors must sum to yield the vector (Figure 4A.2a). We can easily demonstrate this (for the two-dimensional case) by using the known magnitudes of the components relative to "old," known Cartesian coordinate axes to calculate the magnitudes of the components relative to "new," desired Cartesian coordinate axes. We call mathematical "rule" by which we calculate components in the new Cartesian coordinate frame from the components in the old Cartesian coordinate frame the *transformation law* for vectors. Figure 4A.2a shows that a vector $\boldsymbol{X}$ has components $X\boldsymbol{i}$ parallel to the x axis and $Y\boldsymbol{j}$ parallel to the y axis and components $X'\boldsymbol{i}'$ parallel to the x' axis and $Y'\boldsymbol{j}'$ parallel to the y' axis. The transformation law provides a way to calculate the magnitudes of the new components, X' and Y', from the magnitudes of the old components, X and Y. If θ is the angle, measured counterclockwise, between a coordinate axis in the old frame and the corresponding axis in the new Cartesian coordinate frame, we can use simple trigonometry to relate the vector's components in the new (primed) coordinate frame to the components in the old (unprimed) coordinate frame (Figure 4A.2b, c):

$$X' = X\cos\theta + Y\sin\theta \tag{4A.6a}$$

$$Y' = -X\sin\theta + Y\cos\theta \tag{4A.6b}$$

In Appendix 4-II.6, we show an alternative way to express this relationship using the multiplication of matrices. In that formulation, the scalar values by which we multiply the unit vectors in the components of the vectors $\boldsymbol{X}$ and $\boldsymbol{X}'$ are the elements of 2×1 column matrices and that the trigonometric functions, called *direction cosines*, are the elements of a 2×2 *rotation matrix*.

4-I.4 Relating Magnitude and Orientation to Cartesian Coordinates

Throughout the chapter, we used Cartesian coordinate systems to specify vector quantities. One can also specify a vector by giving its magnitude

(a)

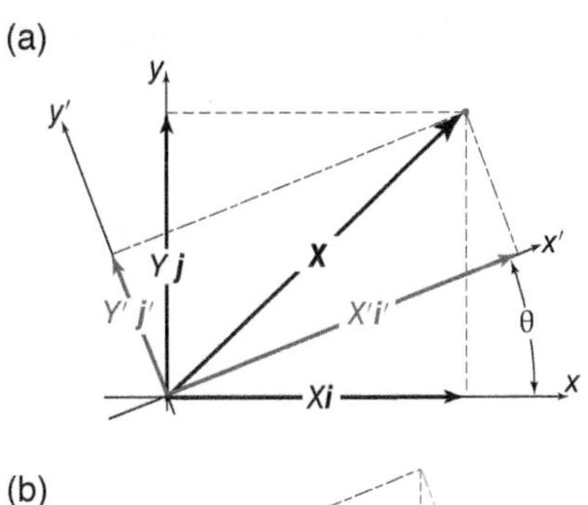

(b)

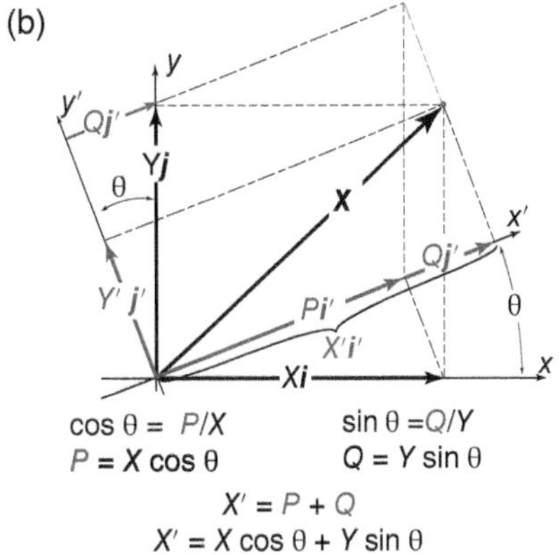

(c)

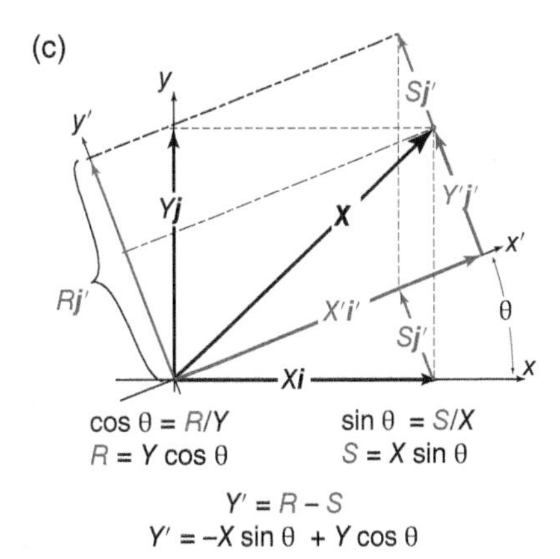

Figure 4A.2 (a) Any vector $\boldsymbol{X}$ will have different components relative to Cartesian coordinate axes with different orientations. (b) and (c) Beginning with the magnitudes of the components relative to one coordinate frame (X, Y) and the angle between the two coordinate frames (θ), one can determine the magnitudes of the coordinates relative to a second coordinate frame (X', Y').

and orientation relative to a reference direction. The two different ways of defining the vectors are interchangeable. We can, therefore, formally define ways to calculate the magnitude and orientation of a vector from its Cartesian coordinates, or to calculate the Cartesian coordinates of a vector from its magnitude and orientation.

We begin with vector $\boldsymbol{A}$ (Figure 4A.3). Its coordinates with respect to a Cartesian coordinate system with the x axis-aligned E–W (with the positive end pointed east), the y axis-aligned N–S (with the positive end pointed north), and the z axis-vertical (with the positive end pointed up) are

$$\boldsymbol{A} = a\boldsymbol{i} + b\boldsymbol{j} + c\boldsymbol{k}$$

(a)

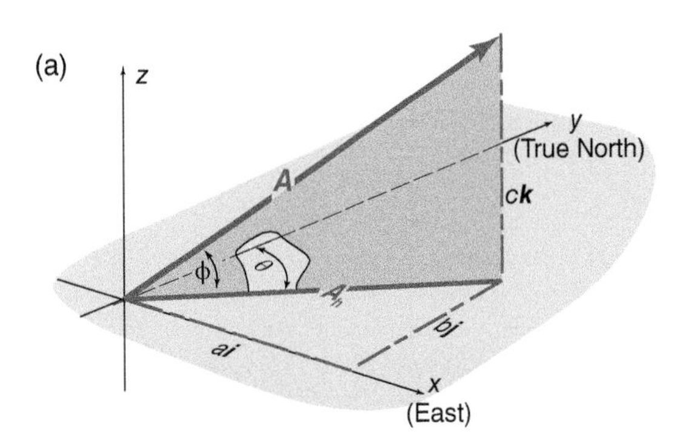

(b)

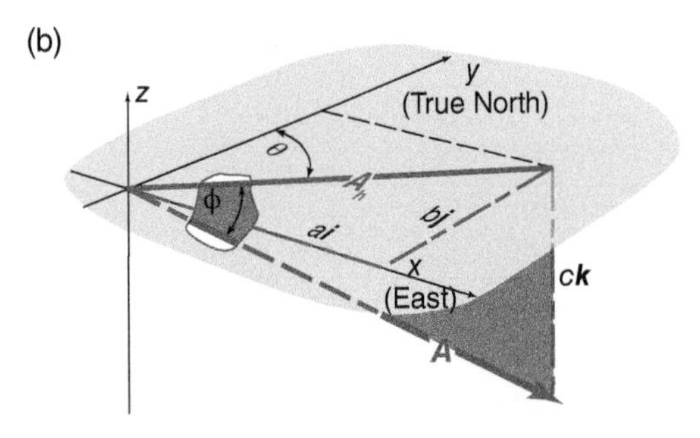

Figure 4A.3 Relationships in three dimensions between the orientation and magnitude of the vector A and its components parallel to EW, NS, and vertical Cartesian coordinate axes. In (a), the angle θ between the y axis and $\boldsymbol{A}_h$ is visible through the "window" in the vertical, dark gray plane. In (b), the angle ϕ between $\boldsymbol{A}_h$ and $\boldsymbol{A}$ is visible through the "hole" in the horizontal gray plane.

From Equation (4A.3), we know that the magnitude of $\boldsymbol{A} = |\boldsymbol{A}| = \sqrt{(a^2 + b^2 + c^2)}$. In Figure 4A.3, $\boldsymbol{A}_\mathbf{h}$ is the horizontal projection of vector $\boldsymbol{A}$, which is simply the sum of the x and y components of the vector. The magnitude of $\boldsymbol{A}_\mathbf{h} = |\boldsymbol{A}_\mathbf{h}| = \sqrt{(a^2 + b^2)}$. The azimuth of $\boldsymbol{A}_\mathbf{h}$ is the angle measured in the horizontal plane between true north, the y axis, and $\boldsymbol{A}_\mathbf{h}$. Inspecting Figure 4A.3, we see that

$$\tan\phi = |a\boldsymbol{i}| / |b\boldsymbol{j}| = a|\boldsymbol{i}| / b|\boldsymbol{j}| = a/b$$

and

$$\theta = \tan^{-1}(a/b) \qquad (4A.7)$$

If vector $\boldsymbol{A}$ is directed downward relative to the horizontal, the angle θ is the trend or azimuth of vector $\boldsymbol{A}$. If vector $\boldsymbol{A}$ is directed upward relative to the horizontal, the trend or azimuth of vector $\boldsymbol{A} = (\theta \pm 180°)$.

We use the lengths of $\boldsymbol{A}$ and $\boldsymbol{A}_\mathbf{h}$ to determine the inclination ϕ of vector $\boldsymbol{A}$.

$$\begin{aligned}\tan\phi &= |a\boldsymbol{i}| / |\boldsymbol{A}_h| = c|\boldsymbol{k}| / \sqrt{(a^2 + b^2)}\\ &= c/\sqrt{(a^2 + b^2)}\end{aligned}$$

$$\phi = \tan^{-1}\{c/\sqrt{(a^2 + b^2)}\} \qquad (4A.8)$$

The absolute value of the angle ϕ calculated from Equation (4A.8) gives the plunge of a line parallel to the vector $\boldsymbol{A}$. Since we generally reckon plunges to have positive values when measured downward from the horizontal, you need to take care to ensure that you determine the correct trend of the line for the sign of the z coordinate.

Beginning with the magnitude and orientation of a vector, we can readily determine the magnitudes of the components along Cartesian coordinate axes aligned parallel to the north, east, and vertical. Because vectors can point upward or downward relative to the horizontal, we will first envision that we are looking along the vector from its tail toward its head. If the line of sight is directed upwards relative to the horizontal, we consider the inclination to have a positive value. If the line of sight is directed downwards relative to the horizontal, we consider the inclination have a negative value. Note that this is different from the way that most geologists reckon inclinations, where a positive number denotes the angle measured downward from the horizontal.

We first determine the horizontal and vertical components of the vector. Taking a vertical section along the line of sight, we get a right triangle composed by the horizontal projection of the vector, the vertical component of the vector and the vector itself. The vector's inclination will be an acute angle directed upward or downward. From this triangle, we have

$$\sin\phi = |c\boldsymbol{k}|/|\boldsymbol{A}| = c|\boldsymbol{k}|/|\boldsymbol{A}| = c/|\boldsymbol{A}|$$

$$\cos\phi = |\boldsymbol{A}_h|/|\boldsymbol{A}|$$

Rearranging these equations, we have

$$c = |\boldsymbol{A}|\sin\phi \quad (4A.9)$$

$$|\boldsymbol{A}_h| = |\boldsymbol{A}|\cos\phi \quad (4A.10)$$

Note that c will be positive when ϕ is positive and negative when ϕ is negative. $|\boldsymbol{A}_h|$ will be positive regardless of the sign of ϕ.

From the magnitude of the horizontal component $|\boldsymbol{A}_h|$, we have

$$\sin\theta = |a\boldsymbol{i}|/|\boldsymbol{A}_h| = a|\boldsymbol{i}|/|\boldsymbol{A}_h| = a/|\boldsymbol{A}_h|$$

$$\cos\theta = |b\boldsymbol{j}|/|\boldsymbol{A}_h| = b|\boldsymbol{j}|/|\boldsymbol{A}_h| = b/|\boldsymbol{A}_h|$$

Rearranging, we have

$$a = |\boldsymbol{A}_h|\sin\theta \quad (4A.11)$$

$$b = |\boldsymbol{A}_h|\cos\theta \quad (4A.12)$$

4-I.5 Vector Products

Due to the differences between vectors and scalars, there are algebraic operations that apply uniquely to vector quantities.

The *dot product* or *scalar product* of two vectors $\boldsymbol{a}$ and $\boldsymbol{b}$ is

$$\boldsymbol{a}\cdot\boldsymbol{b} = |\boldsymbol{a}||\boldsymbol{b}|\cos\alpha \quad (4A.13)$$

where α = the angle between the two vectors. Scalar multiplication of vectors is commutative and distributive. Thus $\boldsymbol{a} \cdot \boldsymbol{b} = \boldsymbol{b} \cdot \boldsymbol{a}$ and $\boldsymbol{a} \cdot (\boldsymbol{b} + \boldsymbol{c}) = \boldsymbol{a} \cdot \boldsymbol{b} + \boldsymbol{a} \cdot \boldsymbol{c}$. When two vectors are perpendicular to each other (when $\alpha = 90°$), their scalar product is zero. Considering the unit vectors $\boldsymbol{i}, \boldsymbol{j}$, and $\boldsymbol{k}$,

$$\boldsymbol{i}\cdot\boldsymbol{i} = \boldsymbol{j}\cdot\boldsymbol{j} = \boldsymbol{k}\cdot\boldsymbol{k} = 1$$
$$\boldsymbol{i}\cdot\boldsymbol{j} = \boldsymbol{j}\cdot\boldsymbol{i} = \boldsymbol{i}\cdot\boldsymbol{k} = \boldsymbol{k}\cdot\boldsymbol{i} = \boldsymbol{j}\cdot\boldsymbol{k} = \boldsymbol{k}\cdot\boldsymbol{j} = 0.$$

Thus, the scalar product of two vectors in terms of their components is

$$\begin{aligned}\boldsymbol{a}\cdot\boldsymbol{b} &= (a_1\boldsymbol{i} + a_2\boldsymbol{j} + a_3\boldsymbol{k})\cdot(b_1\boldsymbol{i} + b_2\boldsymbol{j} + b_3\boldsymbol{k}) \\ &= (a_1\boldsymbol{i}\cdot b_1\boldsymbol{i}) + (a_1\boldsymbol{i}\cdot b_2\boldsymbol{j}) + (a_1\boldsymbol{i}\cdot b_3\boldsymbol{k}) \\ &\quad + (a_2\boldsymbol{j}\cdot b_1\boldsymbol{i}) + (a_2\boldsymbol{j}\cdot b_2\boldsymbol{j}) + (a_2\boldsymbol{j}\cdot b_3\boldsymbol{k}) \\ &\quad + (a_3\boldsymbol{k}\cdot b_1\boldsymbol{i}) + (a_3\boldsymbol{k}\cdot b_2\boldsymbol{j}) + (a_3\boldsymbol{k}\cdot b_3\boldsymbol{k}) \\ &= a_1b_1 + a_2b_2 + a_3b_3\end{aligned} \quad (4A.14)$$

The *vector product* or *cross product* of two vectors $\boldsymbol{a}$ and $\boldsymbol{b}$, denoted $\boldsymbol{a} \times \boldsymbol{b}$, generates a third vector, $\boldsymbol{c}$.

$$\boldsymbol{c} = \boldsymbol{a}\times\boldsymbol{b} \quad (4A.15)$$

$\boldsymbol{c}$, the cross product of $\boldsymbol{a}$ and $\boldsymbol{b}$: (1) is oriented perpendicular to the plane defined by $\boldsymbol{a}$ and $\boldsymbol{b}$; (2) is directed in the direction from which the smaller of the two angles from $\boldsymbol{a}$ to $\boldsymbol{b}$ appears to be counterclockwise; and (3) has a magnitude is given by $|\boldsymbol{a}||\boldsymbol{b}| \sin \alpha$. The vector product is not commutative, although $\boldsymbol{a} \times \boldsymbol{b} = -\boldsymbol{b} \times \boldsymbol{a}$. The vector product is distributive, so $\boldsymbol{a} \times (\boldsymbol{b} + \boldsymbol{c}) = \boldsymbol{a} \times \boldsymbol{b} + \boldsymbol{a} \times \boldsymbol{c}$. When two vectors are parallel, their cross product is zero. For the unit vectors $\boldsymbol{i}, \boldsymbol{j}$, and $\boldsymbol{k}$,

$$\boldsymbol{i}\times\boldsymbol{i} = \boldsymbol{j}\times\boldsymbol{j} = \boldsymbol{k}\times\boldsymbol{k} = 0$$

$$\boldsymbol{i}\times\boldsymbol{j} = \boldsymbol{k} \quad \boldsymbol{j}\times\boldsymbol{k} = \boldsymbol{i} \quad \boldsymbol{k}\times\boldsymbol{i} = \boldsymbol{j} \quad \boldsymbol{j}\times\boldsymbol{i} = -\boldsymbol{k}$$
$$\boldsymbol{k}\times\boldsymbol{j} = -\boldsymbol{i} \quad \boldsymbol{i}\times\boldsymbol{k} = -\boldsymbol{j}$$

Thus, the cross product of two vectors, written in terms of their components, is

$$\begin{aligned}
\boldsymbol{a}\times\boldsymbol{b} &= (a_1\boldsymbol{i}+a_2\boldsymbol{j}+a_3\boldsymbol{k})\times(b_1\boldsymbol{i}+b_2\boldsymbol{j}+b_3\boldsymbol{k}) \\
&= (a_1\boldsymbol{i}\times b_1\boldsymbol{i})+(a_1\boldsymbol{i}\times b_2\boldsymbol{j})+(a_1\boldsymbol{i}\times b_3\boldsymbol{k}) \\
&\quad +(a_2\boldsymbol{j}\times b_1\boldsymbol{i})+(a_2\boldsymbol{j}\times b_2\boldsymbol{j})+(a_2\boldsymbol{j}\times b_3\boldsymbol{k}) \\
&\quad +(a_3\boldsymbol{k}\times b_1\boldsymbol{i})+(a_3\boldsymbol{k}\times b_2\boldsymbol{j})+(a_3\boldsymbol{k}\times b_3\boldsymbol{k}) \\
&= a_1b_2\boldsymbol{k}-a_1b_3\boldsymbol{j}-a_2b_1\boldsymbol{k} \\
&\quad +a_2b_3\boldsymbol{i}+a_3b_1\boldsymbol{j}-a_3b_2\boldsymbol{i} \\
&= (a_2b_3-a_3b_2)\boldsymbol{i}+(a_3b_1-a_1b_3)\boldsymbol{j} \\
&\quad +(a_1b_2-a_2b_1)\boldsymbol{k}
\end{aligned} \tag{A.16}$$

Appendix 4-II: Matrix Operations

4-II.1 Defining Matrices

A *matrix* is a collection of symbols or numbers arranged in a rectangular array. Each *element* in a matrix, that is each symbol or number in the array, falls in a particular *row* and a particular *column* in the rectangular array. We describe the size of the array, and thus the size of the matrix, by giving the numbers of rows and columns there are in the rectangular array. An "$n \times m$" matrix (read as "n by m" matrix) has "n" rows and "m" columns. We use matrices because they are a convenient, shorthand way of representing several related numbers or symbols (the elements of the matrix) by a single symbol (the name of the matrix). Further, *matrix algebra*, which consists of a number of mathematical operations we can perform with matrices, provides a very convenient shorthand representation of systems of equations. We use *matrix algebra* to represent systems of linear equations, linear transformations such as the vector transformation law, and transformation laws for other quantities.

Different authors use different symbols to represent matrices. The symbols most often used are a single capital letter in square brackets = $[Z]$, a boldface capital = $\boldsymbol{Z}$, or a single capital letter with two subscripts = Z_{ij}. We will use a single boldface capital, such as $\mathbf{Z}$, to denote a matrix.

Different authors use different schemes to represent the elements in the array. We will use lower case letters, often with subscripts to indicate the row and column in which we find the element. Thus z_{23} is an element in the 2nd row and the 3rd column of matrix $\mathbf{Z}$. A shorthand notation for an element in the ith row and the jth column of matrix $\mathbf{Z}$ is z_{ij}.

If $\mathbf{Z}$ is a 3×4 matrix, it must have 3 rows and 4 columns.

$$\mathbf{Z} = \begin{vmatrix} 1 & 15 & 24 & b \\ 0 & \mathrm{c} & 24 & 17 \\ 34 & -1 & 7 & d \end{vmatrix} \leftarrow 3\,\text{rows} \qquad \begin{matrix} z_{11}=1; & z_{13}=24; & z_{22}=c; \\ z_{23}=24; & z_{31}=34; & z_{34}=d \end{matrix}$$
$$\uparrow 4\,\text{columns}$$

For square matrices (2×2, 3×3, or $n \times n$ matrices), the elements found along the diagonal that runs from the upper left corner to the lower right corner are the *principal diagonal* elements. The other elements are *off-diagonal* elements. A matrix composed of elements where $a_{ij} = a_{ji}$ (that is $a_{21} = a_{12}$, $a_{13} = a_{31}$, etc.) is a *symmetric* matrix. A matrix is composed of elements where $a_{ij} = -a_{ji}$ (that is $a_{21} = -a_{12}$, $a_{23} = -a_{32}$, etc.) is an *antisymmetric* or *skew-symmetric* matrix. Note that $a_{11} = -a_{11}$, $a_{22} = -a_{22}$, and $a_{33} = -a_{33}$ can be satisfied only if $a_{11} = a_{22} = a_{33} = 0$. If the off-diagonal elements of a matrix are all zero, the matrix is a *diagonal matrix*. If a diagonal matrix has $a_{11} = a_{22} = a_{33}$ = a single value, it is a *scalar matrix*. The identity matrix $\mathbf{I}$, which we define and illustrate next, is a special scalar matrix where $a_{11} = a_{22} = a_{33} = 1$.

4-II.2 Matrix Addition and Subtraction

We can add or subtract any two matrices of equal size, where the two matrices have equal numbers of rows *and* columns. We find the sum of two matrices of the same size by adding together the elements in corresponding positions and placing each sum in the corresponding site of the third matrix (the matrix sum). To find the difference between two matrices, subtract corresponding elements and place the result in the corresponding site. Thus, given two 2 × 3 matrices **P** and **Q**, both the sum of **P** and **Q** and the difference between **P** and **Q** must be 2 × 3 matrices. We have

$$\mathbf{P} = \begin{vmatrix} 15 & a & 17 \\ 0 & -4 & r \end{vmatrix} \qquad \mathbf{Q} = \begin{vmatrix} -7 & 12 & b \\ 6 & 4 & d \end{vmatrix}$$

$$\mathbf{P}+\mathbf{Q} = \begin{vmatrix} 8 & a+12 & 17+b \\ 6 & 0 & r+d \end{vmatrix} \quad \mathbf{P}-\mathbf{Q} = \begin{vmatrix} -7 & a-12 & 17-b \\ -6 & -8 & r-d \end{vmatrix}$$

(4A.17a and b)

4-II.3 Matrix Multiplication

Starting with two matrices where the number of columns in the first equals the number of rows in the second, we can multiply the first matrix times the second. Each element in any product matrix is the sum of the products of elements in the rows of the first matrix times elements in the columns of the second matrix. The row number of the product element corresponds to the row number of the elements from the first matrix, and the column number of the product element corresponds to the column number of the elements from the second matrix. For example, to calculate the element in the first row, first column position of the product matrix, multiply the first element in the first row of the first matrix times the first element in the first column of the second matrix, the second element in the first row of the first matrix times the second element in the first column of the second matrix, etc. Add together all of those products together and place the sum in the first row, first column position of the product matrix. Then, to calculate the element in the first row, second column position of the product matrix, take the first row of the first matrix and the second column of the second matrix. Multiply the first element in the row of the first matrix times the first element in the second column of the second matrix, the second element in the first row of the first column times the second element in the second column of the second matrix, etc. Add all the products together and place the sum in the first row, second column position of the product matrix. Repeat this process for each term in the product matrix. Matrix multiplication sounds immensely complicated when described in words, but the next examples show that it is a straightforward operation in reality.

Clearly, the number of columns in the first matrix must equal the number of rows in the second matrix in order for us to multiply them. The number of rows and columns in the resulting product is found by the following reasoning:

- Matrix **G** is "$m \times n$" or has "m" rows and "n" columns.
- Matrix **H** is "$n \times o$" or has "n" rows and "o" columns.
- The product of **G·H** will have "m" rows and "o" columns or will be "$m \times o$."

In principle or in practice, "m," "n," and "o" can have different values. In such cases, even though the matrix product **G·H** exists (because the number of columns in **G** equals the number of rows in **H**), we cannot multiply **H·G** because "m" ≠ "o."

For any square matrix **G**, we can define an *identity matrix* for multiplication, denoted **I**, such that **G·I** = **I·G** = **G**.

$$\text{For } 2\times 2 \text{ matrices } \mathbf{I} = \begin{vmatrix} 1 & 0 \\ 0 & 1 \end{vmatrix}$$

$$\text{For } 3\times 3 \text{ matrices } \mathbf{I} = \begin{vmatrix} 1 & 0 & 0 \\ 0 & 1 & 0 \\ 0 & 0 & 1 \end{vmatrix}$$

If two square matrices **A** and **B** exist such that

$$\mathbf{A}\times\mathbf{B} = \mathbf{B}\times\mathbf{A} = \mathbf{I}$$

B is the *inverse* of **A** and **A** is the inverse of **B**. We denote the inverse of matrix **A** as $\mathbf{A}^{-1}$. Not all square matrices possess inverses (if the *determinant* of a matrix, defined next, is zero, the matrix's inverse does not exist).

Most of the matrix multiplication used in structural geology entails:

1) Multiplying one “2 × 2” matrix by another, so the product is a “2 × 2” matrix
2) Multiplying one “3 × 3” matrix by another, so the product is a “3 × 3” matrix
3) Multiplying a “2 × 2” matrix times a “2 × 1” matrix, so the product is a “2 × 1” matrix
4) Multiplying a “3 × 3” matrix times a “3 × 1” matrix, so the product is a “3 × 1” matrix.

Examples are much easier to follow, so we illustrate next the four different matrix products used in this text.

4-II.3.1 Multiplying Two “2 × 2” Matrices

$$\mathbf{A}\times\mathbf{B}=\mathbf{C}$$

$$\mathbf{A}=\begin{vmatrix} a_{11} & a_{12} \\ a_{21} & a_{22} \end{vmatrix} \quad \mathbf{B}=\begin{vmatrix} b_{11} & b_{12} \\ b_{21} & b_{22} \end{vmatrix}$$

$$\begin{vmatrix} a_{11} & a_{12} \\ a_{21} & a_{22} \end{vmatrix} \cdot \begin{vmatrix} b_{11} & b_{12} \\ b_{21} & b_{22} \end{vmatrix} = \begin{vmatrix} a_{11}\ b_{11}+a_{12}\ b_{21} & a_{11}\ b_{12}+a_{12}\ b_{22} \\ a_{21}\ b_{11}+a_{22}\ b_{21} & a_{21}\ b_{12}+a_{22}\ b_{22} \end{vmatrix} = \begin{vmatrix} c_{11} & c_{12} \\ c_{21} & c_{22} \end{vmatrix} \tag{4A.18}$$

Thus,

$$c_{11}=a_{11}\ b_{11}+a_{12}\ b_{21} \tag{4A.18a}$$

$$c_{12}=a_{11}\ b_{12}+a_{12}\ b_{22} \tag{4A.18b}$$

$$c_{21}=a_{21}\ b_{11}+a_{22}\ b_{21} \tag{4A.18c}$$

$$c_{22}=a_{21}\ b_{12}+a_{22}\ b_{22} \tag{4A.18d}$$

In each product ($a_{11}\ b_{11}$, $a_{12}\ b_{22}$, etc.), the “inside” subscripts are always the same. The “outside” subscripts give the row and column of their ultimate position in matrix **C**. Thus,

"Outside" subscripts give the position in the matrix **C**, here **2**nd row, **1**st column
↑ ↑ ↑ ↑ ↑

Element in the **2**nd row, **1**st column of **C** → $\quad$ $\boldsymbol{c21=a21b11+a22b21}$

the same (both ones) ┘┘ ↓↓

the same (both twos)

4-II.3.2 Multiplying Two “3 × 3” Matrices

$$\mathbf{A}\cdot\mathbf{B}=\mathbf{C}$$

$$\mathbf{A}=\begin{vmatrix} a_{11} & a_{12} & a_{13} \\ a_{21} & a_{22} & a_{23} \\ a_{31} & a_{32} & a_{33} \end{vmatrix} \quad \mathbf{B}=\begin{vmatrix} b_{11} & b_{12} & b_{13} \\ b_{21} & b_{22} & b_{23} \\ b_{31} & b_{32} & b_{33} \end{vmatrix}$$

$$\begin{vmatrix} a_{11} & a_{12} & a_{13} \\ a_{21} & a_{22} & a_{23} \\ a_{31} & a_{32} & a_{33} \end{vmatrix} \cdot \begin{vmatrix} b_{11} & b_{12} & b_{13} \\ b_{21} & b_{22} & b_{23} \\ b_{31} & b_{32} & b_{33} \end{vmatrix}$$

$$=\begin{vmatrix} a_{11}b_{11}+a_{12}b_{21}+a_{13}b_{31} & a_{11}b_{12}+a_{12}b_{22}+a_{13}+b_{32} & a_{11}b_{13}+a_{12}b_{23}+a_{13}b_{33} \\ a_{21}b_{11}+a_{22}b_{21}+a_{23}b_{31} & a_{21}b_{12}+a_{22}b_{22}+a_{23}b_{32} & a_{21}b_{13}+a_{22}b_{23}+a_{23}b_{33} \\ a_{31}b_{11}+a_{32}b_{21}+a_{33}b_{31} & a_{31}b_{12}+a_{32}b_{22}+a_{33}b_{32} & a_{31}b_{13}+a_{32}b_{23}+a_{33}b_{33} \end{vmatrix} = \begin{vmatrix} c_{11} & c_{12} & c_{13} \\ c_{12} & c_{22} & c_{23} \\ c_{31} & c_{32} & c_{33} \end{vmatrix} \tag{4A.19}$$

Thus,

$$c_{11} = a_{11} b_{11} + a_{12} b_{21} + a_{13} b_{31} \quad (4A.19a)$$

$$c_{12} = a_{11} b_{12} + a_{12} b_{22} + a_{13} b_{32} \quad (4A.19b)$$

$$c_{13} = a_{11} b_{13} + a_{12} b_{23} + a_{13} b_{33} \quad (4A.19c)$$

$$c_{21} = a_{21} b_{11} + a_{22} b_{21} + a_{23} b_{31} \quad (4A.19d)$$

$$c_{22} = a_{21} b_{12} + a_{22} b_{22} + a_{23} b_{32} \quad (4A.19e)$$

$$c_{23} = a_{21} b_{13} + a_{22} b_{23} + a_{23} b_{33} \quad (4A.19f)$$

$$c_{31} = a_{31} b_{11} + a_{32} b_{21} + a_{33} b_{31} \quad (4A.19g)$$

$$c_{32} = a_{31} b_{12} + a_{32} b_{22} + a_{33} b_{32} \quad (4A.19h)$$

$$c_{33} = a_{31} b_{13} + a_{32} b_{23} + a_{33} b_{33} \quad (4A.19i)$$

Once again, in each product (a_{11} $b_{11,}$ a_{12} $b_{22,}$ etc.), the "inside" subscripts are always the same, and the "outside" subscripts give the row and column of their ultimate position in the product matrix **C**.

When multiplying two 2 × 2 (or two 3 × 3 matrices) **A** and **B**, both matrix products **C** = **A**·**B** and **D** = **B**·**A** are well defined. Matrix multiplication is generally not, however, commutative. Thus, normally **C** ≠ **D**. Matrix multiplication is commutative, of course, for **A**·**I** = **I**·**A** = **A** and is commutative (provided that a matrix's inverse exists) $\mathbf{A}\cdot\mathbf{A}^{-1} = \mathbf{A}^{-1}\cdot\mathbf{A} = \mathbf{I}$. Matrix multiplication is also commutative for diagonal matrices.

4-II.3.3 Multiplying a 2 × 2 Matrix Times a 2 × 1 Matrix

$$\mathbf{A} \cdot \boldsymbol{X} = \boldsymbol{x}$$

$$\mathbf{A} = \begin{vmatrix} a_{11} & a_{12} \\ a_{21} & a_{22} \end{vmatrix} \qquad \boldsymbol{X} = \begin{vmatrix} X_1 \\ X_2 \end{vmatrix}$$

$$\begin{vmatrix} a_{11} & a_{12} \\ a_{21} & a_{22} \end{vmatrix} \begin{vmatrix} X_1 \\ X_2 \end{vmatrix} = \begin{vmatrix} a_{11}X_1 + a_{12}X_2 \\ a_{21}X_1 + a_{22}X_2 \end{vmatrix} = \begin{vmatrix} x_1 \\ x_2 \end{vmatrix} \quad (4A.20)$$

or

$$x_1 = a_{11} X_1 + a_{12} X_2 \quad (4A.20a)$$

$$x_2 = a_{21} X_1 + a_{22} X_2 \quad (4A.20b)$$

4-II.3.4 Multiplying a 3 × 3 Matrix Times a 3 × 1 Matrix

$$\mathbf{A} \cdot \boldsymbol{X} = \boldsymbol{x}$$

$$\mathbf{A} = \begin{vmatrix} a_{11} & a_{12} & a_{13} \\ a_{21} & a_{22} & a_{23} \\ a_{31} & a_{32} & a_{33} \end{vmatrix} \qquad \boldsymbol{X} = \begin{vmatrix} X_1 \\ X_2 \\ X_3 \end{vmatrix}$$

$$\begin{vmatrix} a_{11} & a_{12} & a_{13} \\ a_{21} & a_{22} & a_{23} \\ a_{31} & a_{32} & a_{33} \end{vmatrix} \cdot \begin{vmatrix} X_1 \\ X_2 \\ X_2 \end{vmatrix} = \begin{vmatrix} a_{11}X_1 + a_{12}X_2 + a_{13}X_3 \\ a_{21}X_1 + a_{22}X_2 + a_{23}X_3 \\ a_{31}X_1 + a_{32}X_2 + a_{33}X_3 \end{vmatrix} = \begin{vmatrix} x_1 \\ x_2 \\ x_3 \end{vmatrix} \quad (4A.21)$$

or

$$x_1 = a_{11} X_1 + a_{12} X_2 + a_{13} X_3 \quad (4A.21a)$$

$$x_2 = a_{21} X_1 + a_{22} X_2 + a_{23} X_3 \quad (4A.21b)$$

$$x_3 = a_{31} X_1 + a_{32} X_2 + a_{33} X_3 \quad (4A.21c)$$

Comparing Equations (4A.6a and b) and (4A.20) indicates that we can represent the two-dimensional vector transformation law in matrix form by

$$\boldsymbol{x}' = \boldsymbol{\Omega} \cdot \boldsymbol{X} \quad (4A.22)$$

where

$$\boldsymbol{\Omega} = \begin{vmatrix} \cos\theta & \sin\theta \\ -\sin\theta & \cos\theta \end{vmatrix}$$

The components of the matrix **Ω** are determined by relative orientations of the new and old coordinate axes (Figure 4A.4). Each component is derived from the length of a vector parallel to one of the new coordinate axes relative to the length of its projection on one of the old coordinate axes (Figure 4A.4). A vector of length $|\boldsymbol{x}|$ parallel to the x' axis has a projection along the x axis of length $|\boldsymbol{x}|$ cos θ (c.f. Equation (4A.13)) and has a projection along the y axis of length $|\boldsymbol{x}|\cos(90 - \theta) = |\boldsymbol{x}|\sin\theta$. Because a vector of length $|\boldsymbol{y}|$ parallel to the positive end of the y' axis has a projection along the negative end of the x axis, its projection is $-|\boldsymbol{y}|\cos(90 - \theta) = -|\boldsymbol{y}|\sin\theta$. Some workers call the

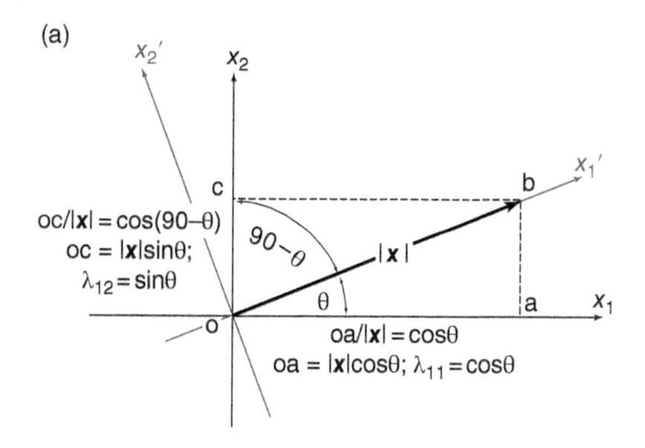

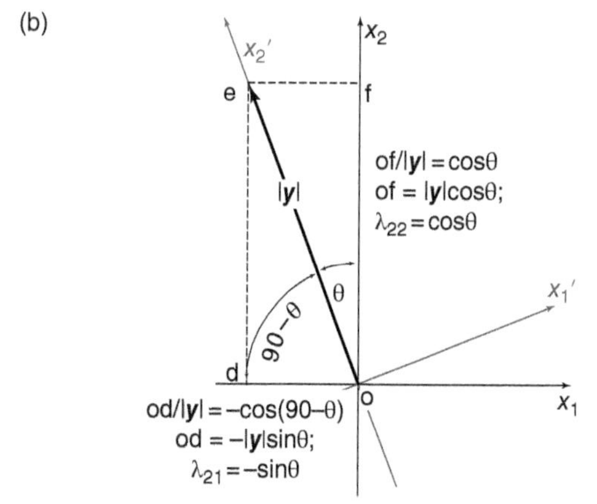

Figure 4A.4 (a) Diagram resolving vector ***x***, parallel to the x_1' axis, onto the x_1 and x_2 axes. (b) Diagram resolving vector ***y***, parallel to the x_2' axis, onto the x_1 and x_2 axes.

components of the matrix $\boldsymbol{\Omega}$ *direction cosines*, denoted by λ_{ij}. The subscript i identifies the new coordinate axis and the subscript j identifies the old coordinate axis related by the component. Thus, λ_{xx} (or λ_{11} if one labels the coordinate axes x_1 and x_2 instead of x and y) is the direction cosine relating the new x' and the old x axes, and λ_{yx} (or λ_{21} if one uses x_1 and x_2 instead of x and y) is the direction cosine relating the new y' and the old x axes.

4-II.3.5 Scalar Multiplication

One can multiply any general matrix $\mathbf{A}$ by a scalar quantity q. Each element in the matrix $q\cdot\mathbf{A}$, which is a scalar multiple of the matrix $\mathbf{A}$, is a scalar multiple of the corresponding element of $\mathbf{A}$. Thus,

$$q\cdot\mathbf{A}=\begin{vmatrix} q\cdot a_{11} & q\cdot a_{12} & q\cdot a_{13}\\ q\cdot a_{21} & q\cdot a_{22} & q\cdot a_{23}\\ q\cdot a_{31} & q\cdot a_{32} & q\cdot a_{33}\end{vmatrix} \quad (4A.23)$$

For the special case of a scalar multiple of the identity matrix $\mathbf{I}$, Equation (4A.23) becomes

$$q\cdot\mathbf{I}=\begin{vmatrix} q\cdot 1 & q\cdot 0 & q\cdot 0\\ q\cdot 0 & q\cdot 1 & q\cdot 0\\ q\cdot 0 & q\cdot 0 & q\cdot 1\end{vmatrix}=\begin{vmatrix} q & 0 & 0\\ 0 & q & 0\\ 0 & 0 & q\end{vmatrix}$$

The multiplication of matrix $\mathbf{A}$ by a scalar q is equivalent to multiplying the matrix A by a scalar matrix in which $a_{11}=a_{22}=a_{33}=q$.

$$q\mathbf{A}=\begin{vmatrix} q & 0 & 0\\ 0 & q & 0\\ 0 & 0 & q\end{vmatrix}\cdot\begin{vmatrix} a_{11} & a_{12} & a_{13}\\ a_{21} & a_{22} & a_{23}\\ a_{31} & a_{32} & a_{33}\end{vmatrix}$$
$$=\begin{vmatrix} qa_{11} & qa_{12} & qa_{13}\\ qa_{21} & qa_{22} & qa_{23}\\ qa_{31} & qa_{32} & qa_{33}\end{vmatrix}$$

4-II.4 Transpose of a Matrix

Beginning with any $m \times n$ matrix $\mathbf{P}$, one creates the transpose of that matrix, denoted $\mathbf{P}^{\mathrm{T}}$ by turning each row in matrix $\mathbf{P}$ into a column in matrix $\mathbf{P}^{\mathrm{T}}$. Thus, the component in the ith row and jth column in $\mathbf{P}$ (p_{ij}) becomes the component in the jth row and ith column in $\mathbf{P}^{\mathrm{T}}$ ($p_{ji}{}^{\mathrm{T}}$).

$$\mathbf{P}=\begin{vmatrix} p_{11} & p_{12}\\ p_{21} & p_{22}\\ p_{31} & p_{32}\end{vmatrix} \qquad \mathbf{P}^{\mathrm{T}}=\begin{vmatrix} p_{11} & p_{21} & p_{31}\\ p_{12} & p_{22} & p_{32}\end{vmatrix}$$

For a square matrix $\mathbf{A}$, this operation is equivalent to taking an element in the ith row and jth column of the original matrix and placing this element in the jth row and ith column of the transpose.

$$\mathbf{A}^{\mathrm{T}} = \begin{vmatrix} a_{11} & a_{12} & a_{13} \\ a_{21} & a_{22} & a_{23} \\ a_{31} & a_{32} & a_{33} \end{vmatrix}^{\mathrm{T}} = \begin{vmatrix} a_{11} & a_{21} & a_{31} \\ a_{12} & a_{22} & a_{32} \\ a_{13} & a_{23} & a_{33} \end{vmatrix} \quad (4A.24)$$

For square matrices, one can envision the transpose operation as reflection across the matrix's principal diagonal. Earlier, we defined symmetric and antisymmetric matrices. For any symmetric matrix,

$$\mathbf{S} = \mathbf{S}^{\mathrm{T}} \quad (4A.25)$$

For any antisymmetric matrix,

$$\mathbf{S} = -\mathbf{S}^{\mathrm{T}} \quad (4A.26)$$

The transpose of a matrix product is the product in reverse order of the transposes of the two matrices. Thus,

$$(\mathbf{A} \cdot \mathbf{B})^{\mathrm{T}} = \mathbf{B}^{\mathrm{T}} \cdot \mathbf{A}^{\mathrm{T}} \quad (4A.27)$$

Further, given any general square matrix $\mathbf{A}$, one can easily show that

$$\mathbf{A} = {}^{1}\!/_{2}\left(\mathbf{A} + \mathbf{A}^{\mathrm{T}}\right) + {}^{1}\!/_{2}\left(\mathbf{A} - \mathbf{A}^{\mathrm{T}}\right) \quad (4A.28)$$

The matrix ${}^{1}\!/_{2}(\mathbf{A} + \mathbf{A}^{\mathrm{T}})$ is symmetric, and the matrix ${}^{1}\!/_{2}(\mathbf{A} - \mathbf{A}^{\mathrm{T}})$ is antisymmetric.

4-II.5 Determinant of a Square Matrix

Some interesting properties of a general square matrix $\mathbf{A}$ depend upon the value of the *determinant* of the matrix, denoted $|\mathbf{A}|$ or det $\mathbf{A}$. For example, if the determinant of a square matrix $\mathbf{A}$ is nonzero, the inverse of the matrix $\mathbf{A}$ ($=\mathbf{A}^{-1}$) exists. Similarly, the ratio of the volume of an object to the volume of its deformed equivalent depends upon the determinant of the matrix that transforms position vectors for particles in the undeformed object to the position vectors of the same particles in the deformed object. One can find general formulae to calculate the determinant of an $n \times n$ matrix in mathematics texts. For our purposes, we need only to calculate the determinants of 2×2 or 3×3 square matrices. For a 2×2 matrix $\mathbf{A}$,

$$\det \mathbf{A} = |\mathbf{A}| = \left(a_{11}a_{22} - a_{12}a_{21}\right) \quad (4A.29a)$$

For a 3×3 matrix $\mathbf{A}$,

$$\begin{aligned} \det \mathbf{A} = |\mathbf{A}| &= a_{11}\left(a_{22}a_{33} - a_{23}a_{32}\right) \\ &\quad - a_{12}\left(a_{21}a_{33} - a_{23}a_{31}\right) \\ &\quad + a_{13}\left(a_{21}a_{32} - a_{22}a_{31}\right) \end{aligned} \quad (4A.29b)$$

It is easy to show that

$$\det \mathbf{A} = \det \mathbf{A}^{\mathrm{T}} \quad (4A.30)$$

Using the definition of the determinant of a 3×3 matrix, one can write a simple way to calculate any vector cross product. Taking, for example, the product $\boldsymbol{c} = \boldsymbol{a} \times \boldsymbol{b}$,

$$\begin{aligned} \boldsymbol{c} = \det \begin{vmatrix} \boldsymbol{i} & \boldsymbol{j} & \boldsymbol{k} \\ a_1 & a_2 & a_3 \\ b_1 & b_2 & b_3 \end{vmatrix} &= \left(a_2b_3 - a_3b_2\right)\boldsymbol{i} \\ &\quad - \left(a_1b_3 - a_3b_1\right)\boldsymbol{j} + \left(a_1b_2 - a_2b_1\right)\boldsymbol{k} \end{aligned} \quad (4A.31)$$

4-II.6 Inverse of a Square Matrix

If the determinant of a matrix $\mathbf{A}$ is nonzero, the inverse of the matrix exists. Mathematics texts provide general formulae to calculate the components of the inverse of an invertible matrix. For the case of a 2×2 matrix for which $\det \mathbf{A} = |\mathbf{A}| \neq 0$,

$$\mathbf{A}^{-1} = 1/\left(\det \mathbf{A}\right) \cdot \begin{vmatrix} a_{22} & -a_{12} \\ a_{21} & a_{11} \end{vmatrix} \quad (4A.32a)$$

$$\mathbf{A}^{-1} = \begin{vmatrix} a_{22}/(a_{11}a_{22} - a_{12}a_{21}) & -a_{12}/(a_{11}a_{22} - a_{12}a_{21}) \\ -a_{21}/(a_{11}a_{22} - a_{12}a_{21}) & a_{11}/(a_{11}a_{22} - a_{12}a_{21}) \end{vmatrix} \tag{4A.32b}$$

For the case of a 3 × 3 matrix for which det $\mathbf{A} = |\mathbf{A}| \neq 0$,

$$\mathbf{A}^{-1} = 1/(\det \mathbf{A}) \cdot \begin{vmatrix} (a_{22}a_{23} - a_{23}a_{32}) & -(a_{12}a_{33} - a_{13}a_{32}) & (a_{12}a_{23} - a_{13}a_{22}) \\ -(a_{21}a_{33} - a_{23}a_{31}) & (a_{11}a_{33} - a_{13}a_{31}) & -(a_{11}a_{23} - a_{21}a_{13}) \\ (a_{21}a_{32} - a_{31}a_{22}) & -(a_{11}a_{32} - a_{12}a_{31}) & -(a_{11}a_{32} - a_{12}a_{31}) \end{vmatrix} \tag{4A.33a}$$

$$\mathbf{A}^{-1} = \begin{vmatrix} (a_{22}a_{23} - a_{23}a_{32})/|\mathbf{A}| & -(a_{12}a_{33} - a_{13}a_{32})/|\mathbf{A}| & (a_{12}a_{23} - a_{13}a_{22})/|\mathbf{A}| \\ -(a_{21}a_{33} - a_{23}a_{31})/|\mathbf{A}| & (a_{11}a_{33} - a_{13}a_{31})/|\mathbf{A}| & -(a_{11}a_{23} - a_{21}a_{13})/|\mathbf{A}| \\ (a_{21}a_{32} - a_{31}a_{22})/|\mathbf{A}| & -(a_{11}a_{32} - a_{12}a_{31})/|\mathbf{A}| & -(a_{11}a_{32} - a_{12}a_{31})/|\mathbf{A}| \end{vmatrix} \tag{4A.33b}$$

4-II.7 Rotation Matrices

In two dimensions, the matrix that accomplishes a counterclockwise rotation of the x–y coordinate axes through an angle θ about a line normal to the x–y plane (see Equations 4A.6 in Appendix 4-I.3 and Equation 4A.22 in Appendix 4-II.3.4) is

$$\Omega = \begin{vmatrix} \cos\theta & \sin\theta \\ -\sin\theta & \cos\theta \end{vmatrix} \tag{4A.34a}$$

The three-dimensional form of the equivalent matrix, which rotates the x and y coordinate axes through a counterclockwise angle of θ about the z axis is

$$\Omega_3(\theta) = \begin{vmatrix} \cos\theta & \sin\theta & 0 \\ -\sin\theta & \cos\theta & 0 \\ 0 & 0 & 1 \end{vmatrix} \tag{4A.34b}$$

The subscript "3" in the matrix name refers to the fact that the stationary Cartesian coordinate axis is the z or x_3 axis. The matrix that rotates the x and z axes through a counterclockwise angle of φ about the y or x_2 axis is

$$\Omega_2(\varphi) = \begin{vmatrix} \cos\varphi & 0 & -\sin\varphi \\ 0 & 1 & 0 \\ \sin\varphi & 0 & \cos\varphi \end{vmatrix} \tag{4A.35}$$

and the matrix that rotates the y and z axes through a counterclockwise angle of ψ about the x or x_1 axis is

$$\Omega_1(\psi) = \begin{vmatrix} 1 & 0 & 0 \\ 0 & \cos\psi & \sin\psi \\ 0 & -\sin\psi & \cos\psi \end{vmatrix} \tag{4A.36}$$

Note that the transpose of each of these rotation matrices $\mathbf{\Omega_i}^{\mathrm{T}} = \mathbf{\Omega_i}^{-1}$, the inverse of the original matrix. If the transpose of a matrix is the inverse of that matrix, the matrix is said to be *orthogonal.* The rotation matrices $\mathbf{\Omega_i}$ are orthogonal matrices.

Any general reorientation of coordinate axes can be envisioned as the net result of three successive rotations about different coordinate axes; the specific rotation angles are sometimes called *Euler angles*. One common convention for rotation through Euler angles is: (1) a rotation through an angle θ about the z axis to generate x'–y'–z' axes in which the z and z' axes coincide; (2) a rotation through an angle φ about the y' axis to generate x''–y''–z'' axes in which the y' and y'' axes coincide; and (3) a rotation through an angle χ about the z'' axis to generate x'''–y'''–z''' axes in which the z'' and z''' axes coincide, the total rotation matrix is given by

$$\Omega_{\mathrm{T}} = |\Omega_3(\chi) \cdot \Omega_2(\varphi) \cdot \Omega_3(\theta)| \tag{4A.37}$$

Because matrix multiplication is not commutative, it is imperative that the order of matrices starting at the right side of the product reflect the order of rotations, first rotation of θ about the z axis, next rotation of φ about the y' axis, and finally a rotation of χ the z'' axis. Evaluation of the previous equation yields

$$\begin{vmatrix} \cos\chi\cos\varphi\cos\theta - \sin\chi\sin\theta & \cos\chi\cos\varphi\sin\theta + \sin\chi\cos\theta & -\cos\chi\sin\varphi \\ -\sin\chi\cos\varphi\cos\theta - \cos\chi\sin\theta & -\sin\chi\cos\varphi\sin\theta + \cos\chi\cos\theta & \sin\chi\sin\varphi \\ \sin\varphi\cos\theta & \sin\varphi\sin\theta & \cos\varphi \end{vmatrix} \tag{4A.37a}$$

The components of each of the two-dimensional and three-dimensional rotation matrices in Equations (4A.36) and (4A.37) are the direction cosines defined earlier.

References

Affolter, T. & J.-P. Gratier. 2004. Map view retrodeformation of an arcuate fold-and-thrust belt: the Jura sase. *Journal of Geophysical Research* **109**,381–20. doi:https://doi.org/10.1029/2002JB002270.

Allmendinger, R.W., J. P. Loveless, M. E. Pritchard, & B.Meade. 2009. From decades to epochs: Spanning the gap between geodesy and structural geology of active mountain belts. *Journal of Structural Geology* **31**, 1409–22. doi:https://doi.org/10.1016/j.jsg.2009.08.008.

Árnadóttir, T., Lund, B., Jiang, W. et al. (2009). Glacial rebound and plate spreading: results from the first countrywide GPS survey in Iceland. *Geophysical Journal International* **177**: 691–716.

Cooper, A.F. and Norris, R.J. (1994). Anatomy, structural evolution, and slip rate of a plate-boundary thrust: the Alpine Fault at Gaunt Creek, Westland, New Zealand. *Geological Society of America Bulletin* **106**: 627–633.

Jackson, M. D., & D. D. Pollard. 1988. The laccolith-stock controversy: new results from the southern Henry Mountains, Utah. *Geological Society of America Bulletin* **100**, 117–39. doi:https://doi.org/10.1130/0016-7606(1988)100<0117:TLSCNR>2.3.CO;2.

Lindvall, S.C., Rockwell, T.K., and Hudnut, K.W. (1989). Evidence for prehistoric earthquakes on the Superstition Hills Fault From offset geomorphic features. *Bulletin of the Seismological Society of America* **79**: 342–361.

Sieh, K. and Jahns, R.H. (1984). Holocene activity of the San Andreas fault at Wallace Creek, California. *Geological Society of America Bulletin* **95**: 883–896.

5

Strain

5.1 Overview

Strain is the part of deformation that results in a change in the shape of an object (**distortion**) and/or a change in its volume (**dilation**). In both dilation and distortion, the distances between particles change during deformation causing changes in the lengths of lines. The change in the length of a line is a **longitudinal strain**. Geologists regularly use two measures of line-length change, **stretch** and **elongation**, to quantify longitudinal strains. All directions experience equal longitudinal strain magnitudes in dilation, whereas different directions experience different longitudinal strain magnitudes in distortion. Distortion also causes changes in the angles between most sets of lines. A change in angle between intersecting lines is a **shear strain**. Shear strains are most readily measured by examining the orientations in the deformed state of two lines that were orthogonal in the undeformed state. In every distortion, there are three mutually perpendicular directions, the **principal strain directions**, that experience no shear strain. One principal direction experiences the maximum stretch (or elongation) of the distortion, and another principal direction experiences the minimum stretch (or elongation) of the distortion. The third principal direction necessarily experiences a stretch (or elongation) magnitude between the maximum and minimum values. The longitudinal and shear strain magnitudes for different directions in a distortion relate directly to the values of displacement gradients or position gradients in different directions.

Strain is **homogeneous** when any collection of particles that defined a sphere (circle in two dimensions) prior to deformation is transformed by the deformation into an ellipsoid (ellipse in two dimensions). We call the deformed-state shape of a sphere (circle in two dimensions) the **finite strain ellipsoid** (ellipse in two dimensions). The major, minor, and intermediate axes of the finite strain ellipsoid correspond to the principal directions of the distortion. These three **finite strain axes** exhibit the maximum, intermediate, and minimum stretches of lines. We distinguish between three end-member shapes of the finite strain ellipsoid: (1) cigar-shaped, known as pure constriction; (2) pancake-shaped, known as pure flattening; and (3) surfboard-shaped, known as plane strain. Homogeneous strain occurs in deformations where the values of the displacement or position gradients do not change with position.

Similar to displacements, structural geologists distinguish between **finite strain** (the total distortion of a rock), **strain path** (the succession of strain states during the finite change in shape), and **strain history** (the timing of the succession of strain states). **Pure shearing** and **simple**

An Integrated Framework for Structural Geology: Kinematics, Dynamics, and Rheology of Deformed Rocks,
First Edition. Steven Wojtal, Tom Blenkinsop, and Basil Tikoff.

shearing are two idealized strain paths. During pure shearing, lines that are initially aligned with the strain principal directions remain aligned with those directions throughout deformation. Because these particular lines remain aligned with the finite strain axes throughout deformation, pure shearing is a type of **coaxial deformation**. In simple shearing, lines aligned with the strain principal directions early in the deformation do not remain aligned with the finite strain axes throughout deformation. Simple shearing is an example of **non-coaxial deformation**.

Structural geologists also distinguish between **incremental strain** (which relates two stages in an object's kinematic history separated by a time interval) and **infinitesimal strain** (infinitesimally small deformations that occur instantaneously). Like the finite strain ellipsoid, the **infinitesimal strain ellipsoid** is characterized by three mutually perpendicular axes with specific orientations and magnitudes. These **infinitesimal strain axes** indicate the maximum, intermediate, and minimum rates of stretching of lines. **Strain rate** is the change in strain divided by the time interval of measurement. Typical geological strain rates vary between 10^{-10} s^{-1} (fast) and 10^{-14} s^{-1} (slow).

5.2 Chapter Organization

Like the previous chapter on displacements and displacement fields, the main topics of the chapter are discussed in Conceptual Foundation and Comprehensive Treatment sections. The distinction between the Conceptual Foundation and Comprehensive Treatment sections again depends on the level of mathematical rigor presented. This chapter also contains a section that follows the Conceptual Foundation and precedes the Comprehensive Treatment sections in which we address the practical problem of measuring strain. This discussion of methods for measuring strain in rocks serves as a bridge between the two. Table 5.1 lists the main symbols in this chapter, which also draws on several symbols from Chapter 4 (Table 4.1).

5A Strain: Conceptual Foundation

5A.1 Specifying Strain in Deformed Rocks

In the previous chapter, we noted that the displacement field associated with any deformation can be divided into bulk translation, rigid body rotation, and pure strain components. In this chapter, we focus on the pure strain component of the total displacement field. The bulk translation and rigid rotation components describe the movement of particles with respect to an *external* reference. The pure strain component, on the other hand, describes the movement of particles with respect to a *local* or *internal* reference frame, where particles in an object have different positions *relative to each other* after deformation than they had before deformation.

The movement of particles relative to their neighbors leads to the **dilation** (change in volume) and/or **distortion** (change in shape) of objects. Figure 5.1 is intended to illustrate the distinction between pure strain and the other components of the total displacement field. In Figure 5.1a, the car moves from one location to another without distortion or dilation. The motion may be in a straight line or may include a rotation (such as when a car travels around a curve). As long as the car is neither distorted nor dilated, there is no pure strain. If a car is dilated or distorted only (Figure 5.1b), we see that there is at least one particle (the car's center) that remains stationary while nearby particles move relative to that stationary particle. In the case of distortion, note that the front and rear of the car are closer together after strain, the top and bottom of the car are farther apart after strain, etc. In most geologic

Table 5.1 Symbols used in this chapter in addition to those in Table 4.1.

Scalars		Formulae
l_o, l'	Initial and final lengths of a line between two particles	
T	Stretch	l'/l_o
e	Elongation	$(l'-l_o)/l_o, T-1$
γ	Shear Strain	$\tan\psi$
ψ	Angular strain	
S, s	Major and minor axes of strain ellipse	
$S_1 > S_2 > S_3$	Principal finite strain axes	
r	Radius of the original sphere	
$T_{\max}, T_{\text{int}}, T_{\min}$	Major, intermediate, and minor principal stretches	$S_1/r, S_2/r, S_3/r$
R_s	Strain ellipse axial ratio	$T_{\max}/T_{\min}$
k	Slope of Flinn plot	$(T_{max}/T_{int}-1)/(T_{int}/T_{min}-1)$
R_i, R_f	Axial ratios of undeformed and deformed markers	
θ, ϕ	Orientation of undeformed and deformed strain markers relative to the direction of maximum elongation	
θ', ϕ'	Orientation of undeformed and deformed strain markers relative to an arbitrary direction	
Φ	Vector mean of ϕ measurements	$\frac{1}{2}\arctan[(\Sigma\sin 2\phi)/(\Sigma\cos 2\phi)]$.
H	Harmonic mean of R_f measurements	$N/[(\Sigma R_{fN})^{-1}]$.
I_{Sym}	Symmetry parameter of R_f/ϕ plot	
R_C, ϕ_C	Coordinates of the centre of an R_f/ϕ plot	
A, A'	Initial and deformed area	
W_k	Kinematic vorticity number	
Tensors		
$\mathbf{F}$	Deformation or Position Gradient tensor/matrix	
$\mathbf{\Omega}$	Rotation tensor/matrix	
$\mathbf{G}$	Reciprocal deformation tensor/matrix	$\mathbf{F}^{-1}$
$\mathbf{L}$	Velocity gradient tensor/matrix	
$\mathbf{D}$	Strain rate tensor/matrix	
$\mathbf{W}$	Rotation rate tensor, spin, or vorticity tensor/matrix	

cases, objects are displaced and strained simultaneously (Figure 5.1c). In such cases, all particles move, but particles at different positions have different displacements. In focusing on the pure strain component of the total displacement field, we temporarily disregard the bulk translation and rigid body rotation components in order to understand the relative movements of particles.

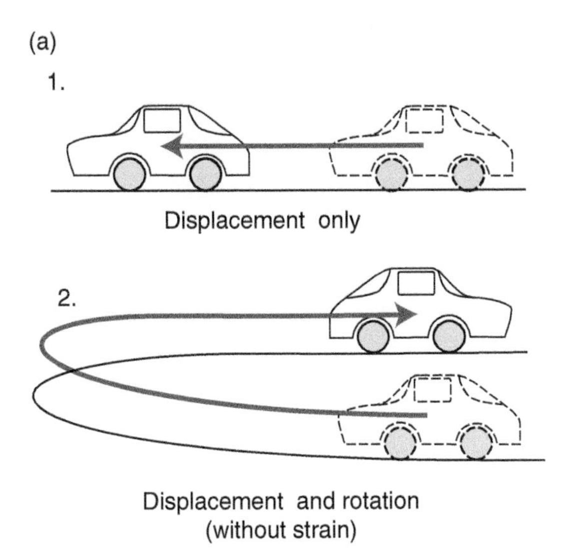

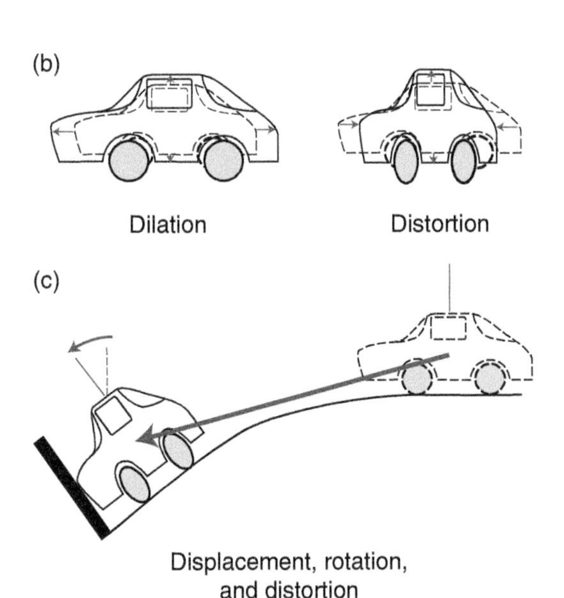

Figure 5.1 (a) The total displacement field associated with a deformation may consist of displacement only (image 1) or displacement and rigid rotation only (image 2). (b) The pure strain component of the total displacement field gives rise to either dilation (volume change in three dimensions or area change in two dimensions), distortion or shape change, or a combination of dilation and distortion. (c) The total displacement field typically consists of a combination of displacement, rotation, and distortion.

The pure strain component is often more readily apparent than bulk translation or rigid body rotation – think of a shrunken clothing item or a crushed can. The change in the size of the clothing or the shape of the can is readily apparent and can be measured more easily than determining the total distance that the objects moved or how much they rotated. We need to only observe particles relative to their near neighbors to see how they moved during deformation.

5A.2 One-dimensional Manifestations of Strain

5A.2.1 Basic Ideas

Longitudinal strain is the change in the distance between two particles along the lines that connects them. Geologists typically calculate longitudinal strain by comparing the ratio of the present (deformed) length of the line between the two particles to its original (undeformed) length. The lines connecting the two particles in the undeformed and deformed states generally will not have the same orientation because the two particles have moved with respect to each other. Geologists have defined many quantities to calculate longitudinal strain, but we focus here on two of those quantities. The first quantity is a simple ratio, called the **stretch** (T), given by the following equation:

$$T = \frac{l'}{l_o}. \tag{5.1}$$

Here l' is the present or final length of a line between two particles and l_o is the original or initial length of the line between the two particles (Note: Some geologists use k to denote the stretch; we have elected to use T rather than k to avoid confusion with another strain-related quantity used in Flinn diagrams described later in the chapter). Because the stretch is a ratio of two lengths measured in the same units, it has no units – it is dimensionless. In fact, all parameters used to quantify strain lack dimensions.

If the length of a line connecting two particles is not changed during deformation ($l' = l_o$), its stretch is unity ($T = 1$). If the line connecting two particles is lengthened, the magnitude of the stretch is greater than one ($T > 1$), and if the line connecting two particles is shortened, the magnitude of the stretch is less than one ($T < 1$). Note that the term "stretch" can refer to the shortening of a line. In cases of extreme lengthening, T can be very large; in cases of extreme shortening, T can be very small. The range of possible values for the stretch is $0 < T < \infty$.

The second quantity used to measure longitudinal strains is a ratio called **elongation** (e); its value is the ratio of the change in the length of the line to the original length of that line:

$$e = \frac{(l' - l_o)}{l_o}, \tag{5.2}$$

where l' is again the present or final length of the line and l_o is again the original or initial length of the line. The magnitude of the elongation is zero ($e = 0$) if a line has the same length before and after deformation ($l' - l_o = 0$). For lines lengthened during deformation ($l' - l_o > 0$), the elongation is greater than zero ($e > 0$). For lines shortened during deformation ($l' - l_o < 0$), the elongation is less than zero ($e < 0$). The range of possible values for the elongation is $-\infty < e < \infty$. Geologists use either decimal values for e (such as an elongation of 1.5 or elongation of 0.5), or they use percentages to denote line length changes (such as a line lengthened 50% or shortened 50%). We use the decimal values in this text. Likewise, some structural geologists use the terms *elongation* and *extension* interchangeably. In this text, we use elongation to refer to the quantity, defined in Eq. (5.2), used to measure one-dimensional strain and extension to refer to a two-dimensional, tectonic deformation that involves vertical shortening and horizontal lengthening. Finally, some structural geologists use different symbols, such as ε or ϵ to denote elongation. We will use e.

The stretch and the elongation are equally valid and common ways to quantify the change in the length of a line. The two quantities are related by the equation:

$$e = \frac{(l' - l_o)}{l_o} = \frac{l'}{l_o} - \frac{l_o}{l_o} = T - 1. \tag{5.3}$$

Because stretch and elongation are inherently one-dimensional measures of strain, geologists often refer to them as "linear strain measures."

5A.2.2 Geological Example

Structural geologists often use distorted fossils to measure strain in deformed rocks. Figure 5.2a is a sketch of a belemnite *guard* or *rostrum* (shell) in deformed Mesozoic limestone from the Valais portion of the western Alps. In undeformed rocks, belemnite guards are elongate cylinders with conical tips; at the opposite end of a complete guard is a hollow region called the *alveolus* where a squid-like organism lived. The belemnite shells in these deformed limestones are broken into a series of segments that have moved relative to each other (i.e. boudinaged). The spaces between segments are filled by secondary mineral growths of calcite and quartz. The fragments of the shell are not extensively internally distorted, and the adjacent segment ends closely match each other. One can conceptually remove the secondary quartz and calcite and piece the segments of the guard back together in order to estimate what was the original configuration of the guard (Figure 5.2b), or alternatively one could measure the cumulative length of the segments (ignoring the secondary minerals) to assess the original length of the guard.

Comparing the pieced-together segments of the deformed belemnite guard with its current configuration, we see that the distance from the tip of the conical end of the guard to the base of the alveolus in the present, deformed configuration is longer than the distance from the tip to the alveolus in the original, undeformed configuration. Thus, a line parallel to the axis of the guard has been lengthened

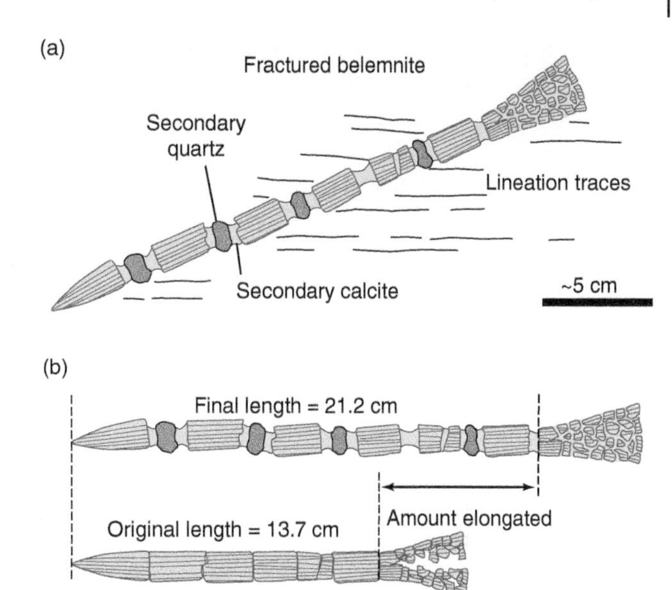

Figure 5.2 (a) Sketch of a distorted belemnite. Source: Badoux (1963)/University of Lausanne. (b) Comparison of the present, fractured and lengthened, belemnite and its reconstructed original form, showing the amount the belemnite was lengthened during deformation

during deformation. By calculating the change in the distance between the particles at the conical tip of the guard and at the base of the alveolus or in the length of the line represented by the belemnite guard, we quantify the longitudinal strain.

The current or final length of the belemnite guard in Figure 5.2, measured from its tip to the base of its alveolus, is ~21.2 cm. The distance between the corresponding particles on the pieced-together belemnite guard is ~13.7 cm. Thus, the stretch magnitude is $T = 21.2\,\text{cm}/13.7\,\text{cm} = 1.55$. Because stretch is a ratio, it has no units. Similarly, the elongation of the belemnite guard in Figure 5.2 is $e = (21.2\text{ cm} - 13.7\text{ cm})/13.7\text{ cm} = 7.5\text{ cm}/13.7\text{ cm} = 0.55$.

Fossils are not the only geological features we can use to measure longitudinal strains. Figure 5.3a is a representation of a section through a metamorphic rock. In this section, a large amphibole porphyroblast, a mineral grain that grew during metamorphism, has been fractured and lengthened in a manner similar to the belemnite in Figure 5.2. Beneath the section showing the current, boudinaged configuration of the porphyroblast is an image of its inferred pre-fracturing-and-lengthening, “original” configuration. The fracturing and lengthening of the porphyroblast capture only the longitudinal strain in this direction experienced by this metamorphic rock after the growth of the porphyroblast. Figure 5.3b is a sketch of a thin section of detrital quartz grains in a deformed quartzite from the central Appalachians of Maryland, USA. Quartz grains in this quartzite often contain thin needles of rutile formed by processes that operated when quartz grain crystallized, long before it was incorporated into the sedimentary rock. In one of the grains from the sketch, a rutile needle was boudinaged and lengthened during deformation (the boudinaged needle is apparent in the photomicrograph in Figure 5.3b). The stretch or elongation value calculated for this needle is likely to capture all of the strain experienced by this detrital grain.

5A.3 Two-dimensional Manifestations of Strain

5A.3.1 Longitudinal Strains in Different Directions

5A.3.1.1 Basic Ideas

Broadening the scope beyond a single, linear dimension, consider a two- or three-dimensional

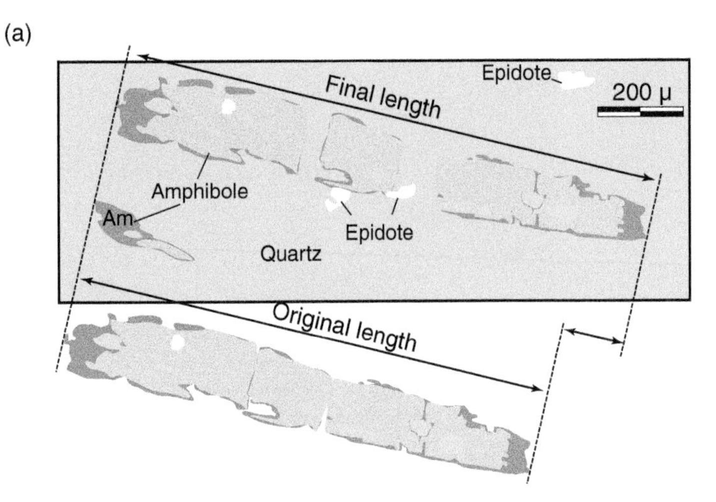

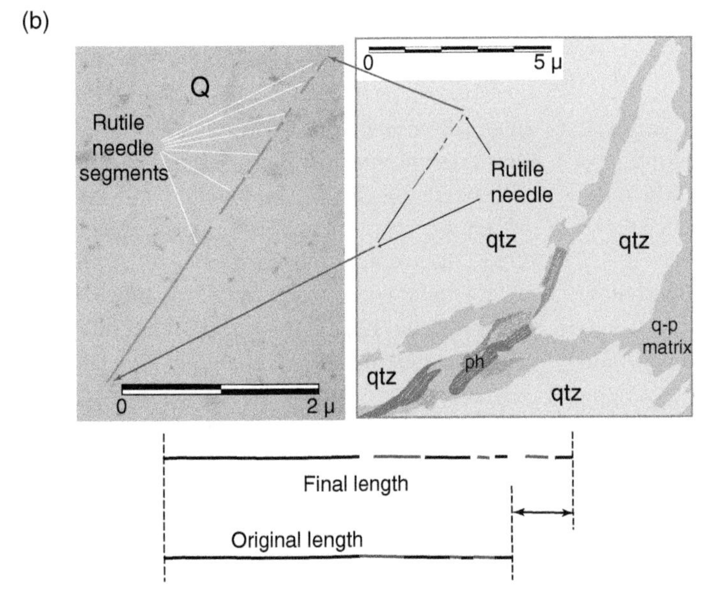

Figure 5.3 (a) Sketch of a portion of a thin section of a metamorphic rock showing a fractured and lengthened amphibole porphyroblast. The reconstructed original shape of the porphyroblast shown here. (b) A fractured and lengthened rutile needle within a quartz grain from a deformed sandstone. Comparison of final length and reconstructed original length shows the amount of lengthening due to deformation.

situation where lines oriented in a single direction are shortened or lengthened but lines perpendicular to this direction do not change their lengths (Figure 5.4). In the two-dimensional section with height *h* and original length l_o, changing the length of lines in one direction with no longitudinal strain perpendicular to this line (not allowing the height to change) must lead to a decrease or increase in area in the plane of the section (Figure 5.4a). Similarly, in the three dimensions, changing only the length of a rock mass while holding its height (*h*) and width (*w*) constant (not allowing longitudinal strains for lines oriented perpendicular to the length) must lead to a decrease or increase in the volume of the rock mass (Figure 5.4b). Geologists do observe changes

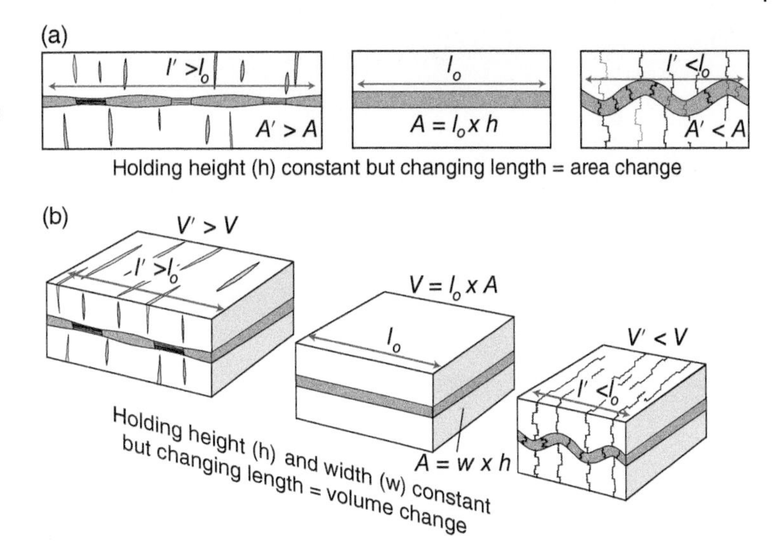

Figure 5.4 (a) In a two-dimensional section, changing the length of a rock mass in one direction only leads to an increase or decrease in the area in the plane of view. (b) In three dimensions, changing the length of a rock mass in one direction only leads to an increase or decrease in the volume of the rock mass.

in area and volume during deformation, but pure strains in which lines are shortened or lengthened in one direction and are neither lengthened nor shortened perpendicular to that direction are not common.

If the pure strain component of a deformation consists only of dilation, that is, a change in volume only, longitudinal strains exist ($T \neq 1$ and $e \neq 0$) in all directions *and* the magnitude of the longitudinal strain in every direction is identical. One can gain some appreciation of this by considering the two-dimensional analog of dilation: a pure change of area. Figure 5.5a is a photomicrograph showing four undeformed ooids; in each ooid two orthogonal lines identify a particular set of particles within the ooid. After a pure area reduction (Figure 5.5b), the particles have moved relative to each other. The change in the area has shortened each of the labeled lines in the ooids, but because all directions exhibit identical stretch or elongation magnitudes (T = a constant <1 and e = a constant <0 for every line), the orientations of the lines do not change. A pure area increase would yield a comparable result with T = a constant >1 and e = a constant >0 for every line.

Figure 5.5c illustrates the more typical two-dimensional situation, where line lengthening occurs in one direction and line shortening in a perpendicular direction. In this example, the line lengthening in one direction balances the line shortening in the perpendicular direction. So, there is no area change – this pure strain consists only of distortion. When lines are shortened in one direction ($T<1$ or $e<0$) and lengthened in a perpendicular direction ($T>1$ or $e>0$), there always exist two directions where lines exhibit no net change in length ($T = 1$ or $e = 0$). The direction that exhibits the most extreme lengthening ($T = T_{max}$ or $e = e_{max}$) is one of the two directions that bisect the directions of no net change in length, and the other bisector of the directions of no net change in length exhibits the most extreme shortening ($T = T_{min}$ or $e = e_{min}$).

When line lengthening is roughly horizontal and line shortening roughly vertical, we call this state of strain *extension*. The opposite situation, that is, roughly horizontal line shortening and roughly vertical line lengthening, is *contraction*. If the magnitudes of lengthening and shortening are equivalent, there will be no area change

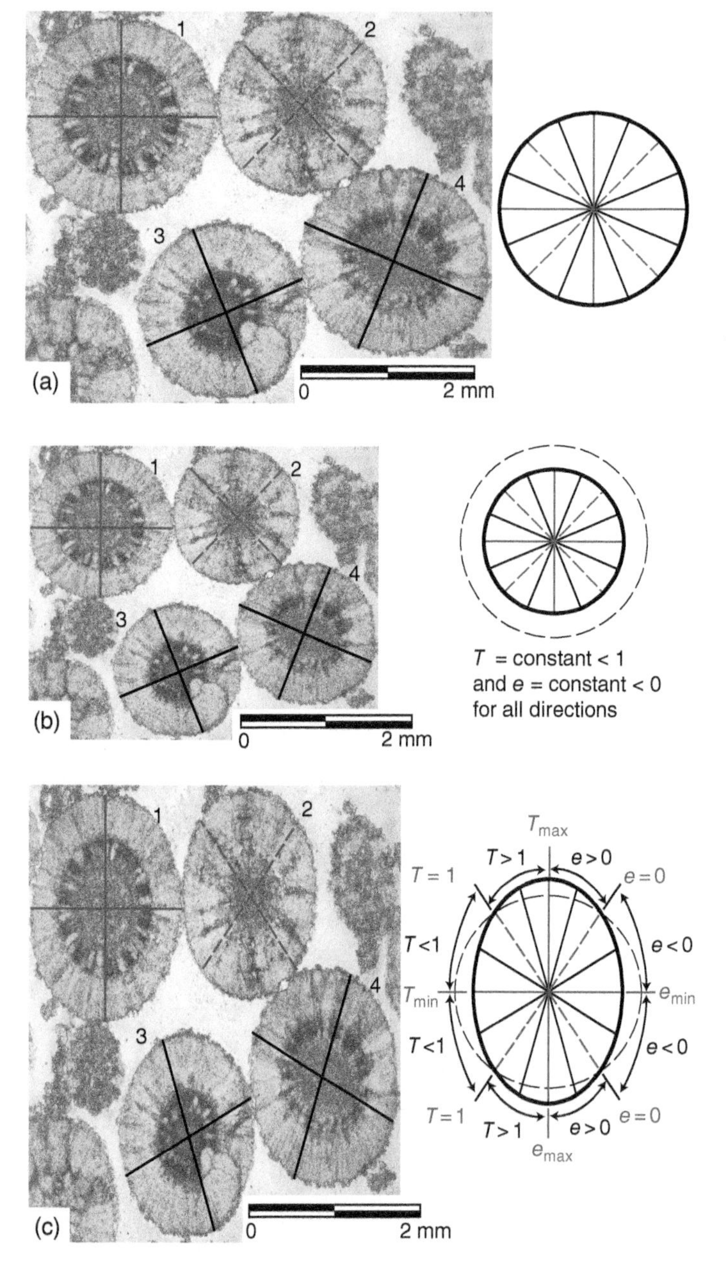

Figure 5.5 (a) Undeformed ooids. Ooids labeled 1, 2, 3, and 4 each have two orthogonal diameters shown. The drawing at right shows that diameters of a circle have the same length regardless of their orientation. (b) Ooids from (a) subjected to dilation, where a reduction in area in the plane of section. In sketch at right, dashed circle gives original shape. Stretches T or elongations e are the same for all directions. (c) Ooids from (a) subjected to a distortion; dashed circle in sketch gives original shape. Diameters with different orientations experience different magnitudes of stretch (T) or elongation (e).

during deformation. Otherwise, the area of the deforming rock will increase or decrease as material is added (for example, in veins) or removed (for example, in stylolites).

5A.3.1.2 Geological Example

We return to the deformed quartzite from the central Appalachians in Maryland, USA. In deformed detrital quartz grains in these rocks, rutile needles with different orientations exhibit different stretch or elongation values. Rutile needles oriented in some directions are boudinaged, like the case illustrated in Figure 5.3c, yielding T magnitudes greater than one (Figure 5.6). Rutile needles oriented at high angles to the elongated rutile needles are bent, buckled, or imbricated. They have current lengths shorter than their original length and T magnitudes less than one. Separating these orientations are two directions of no net change in the length of lines (Figure 5.6).

The examples we have cited so far depict longitudinal strains that are apparent in a hand sample or a thin section. We can also observe longitudinal strains over longer distances by looking at changes in the distance between particles at two ends of a cross section, or using global positioning data at opposite ends of a continental landmass or opposite sides of an ocean basin. Similarly, in each of the examples considered so far, the geologic element we use to document a longitudinal strain has been fractured into numerous segments. Section 5B.3 considers longitudinal strains accommodated by continuous deformation, without the formation of fractures. The critical factor here is the change in the length of the line, not the fracture of the geological element.

5A.3.2 Shear Strain

5A.3.2.1 Basic Ideas

In pure strains, where there are different longitudinal strains in different directions, we recognize another manifestation of strain. In general, the angle at which two intersecting lines meet will be different in the undeformed and deformed states. The change in angle between two intersecting lines or two intersecting families of lines is a

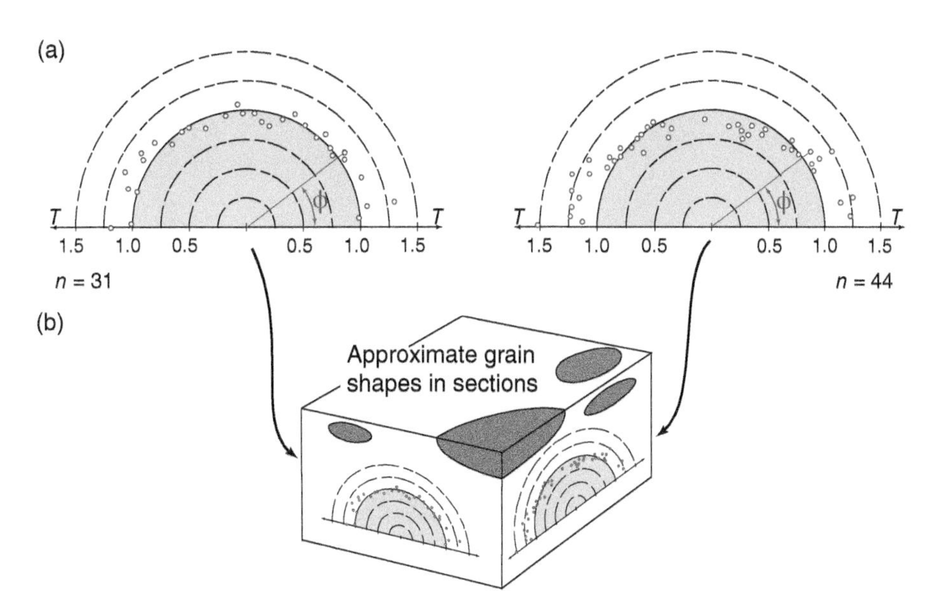

Figure 5.6 (a) Graphs showing how stretch (T) values measured using elongated rutile needles in quartz grains vary as a function of the angle (ϕ) to the trace of foliation in two mutually perpendicular sections of the Weverton Quartzite of Maryland (USA) (data from Mitra (1976, 1977). (b) Block diagram showing inferred shapes of initially spherical quartz grains.

shear strain. Shear strains are apparent when examining distortion in two or three dimensions. Thus, shear strains are apparent in Figure 5.5c, our first encounter with distortion. Note that the pairs of orthogonal lines superposed on ooids #2, #3, and #4 in Figure 5.5a are no longer orthogonal in Figure 5.5c. We needed to add the qualifying phrase "in general" to the second sentence in this paragraph because there are two, mutually perpendicular directions in every two-dimensional pure strain that do not exhibit a change in their *relative* orientations. The relative orientations of the labeled lines in ooid #1 in Figure 5.5a and c are unchanged by the distortion even though their lengths are changed. We measure the maximum stretch or elongation values along one of the two directions exhibiting no shear strain and the minimum stretch or elongation values along the other direction exhibiting no shear strain.

The originally orthogonal lines in ooids #3 and #4 exhibit changes in their length as they are reoriented during deformation. In ooids #3 and #4, one of the labeled lines is shortened and the other is lengthened. More relevant to our present discussion, the lines are reoriented during strain so that they now intersect at an acute angle. Even though the originally orthogonal lines in ooid #2 exhibit no net change in length, they are reoriented by the pure strain. Figure 5.7a shows the pair of orthogonal lines (labeled A and B) in the undeformed ooid #2 from Figure 5.5. Figure 5.7b shows the same pair of lines (now labeled A′ and B′) in the deformed state. Note that line A exhibits a clockwise reorientation relative to line B and that line B exhibits a counterclockwise reorientation relative to line A (Figure 5.7c). We observe the opposite sense of shear strain, clockwise versus counterclockwise or counterclockwise versus clockwise, on opposite sides of the directions of maximum and minimum longitudinal strain (which are also the directions of no shear strain).

5A.3.2.2 Quantifying Shear Strains

We quantify shear strain by envisioning what happens to two lines originally at right angles to each other. We measure the alteration of an original right angle using two related quantities. The first is the **angular shear** ψ, which is the angle measured in radians between the original and final orientations of a line that initially was perpendicular to a particular direction. By positioning lines A′ and B′ in Figure 5.7d over their original

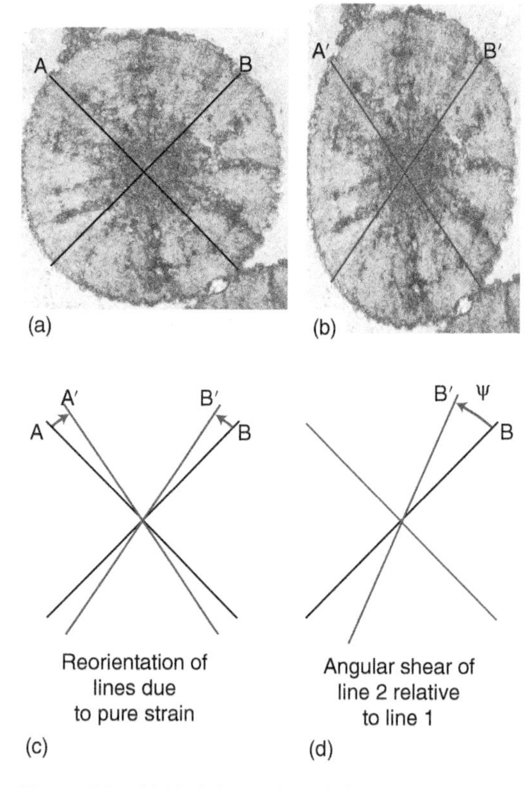

Figure 5.7 (a) Undeformed ooid showing diameters *A* and *B*. (b) Distorted ooid showing deformed state positions, *A′* and *B′*, of diameters *A* and *B*. (c) Diagram comparing the original orientations of *A* and *B* and deformed state positions of *A′* and *B′*. (d) When lines A′ and B′ are reoriented so that line *A′* coincides with line *A*, angular shear of line *B′* relative to line *B* is apparent.

configuration so that line A′ coincides with line A, we can assess the amount of reorientation of the originally perpendicular line B relative to line A. The angle measured from the original orientation of line B to its final orientation B′ is the angular shear ψ. Because radians are a ratio of the length of an arc to the length radius of the arc, both measured in the same units, radians are a dimensionless measure of an angle. In most instances, we consider angular shear to be a positive-valued quantity, whether it is clockwise (= dextral) or counterclockwise (= sinistral). Note that for any pair of material lines, the sense of shear (dextral or sinistral) of the first relative to the second is the opposite of the sense of shear of the second relative to the first. The **shear strain γ** is

$$\gamma = \tan\psi. \tag{5.4}$$

The values of the trigonometric functions are determined by taking the ratios of line lengths, so they are always dimensionless numbers. Thus, like the longitudinal strain measures defined earlier, the angular shear ψ and the shear strain γ are unitless or dimensionless quantities. In the following section, we discuss more fully just what shear strain measures.

5A.3.2.3 Geological Examples

Right angles are not especially common in geological features, but examples of natural right angles do exist. For example, we find natural right angles in sedimentary strata containing the trace fossil *Skolithus*, visible "tubes" or "pipes" oriented 90° to the primary bedding left by a burrowing organism. The deformation of *Skolithus*-bearing strata regularly yields situations where the pipes are inclined at acute angles to bedding (e.g. the layer shaded red in Figure 5.8a). We can measure the shear strain on bedding by measuring the pipes' departure from their original orientation perpendicular to primary bedding. Similarly, the Särv *thrust nappe* (another term for "thrust sheet") in the Caledonides of Sweden consists of arkosic sedimentary strata intruded by basaltic dikes (Figure 5.8b). In the upper portions of the nappe, where the arkoses exhibit weak deformation fabrics, the dikes are approximately perpendicular to bedding, creating right angles between individual sedimentary layers and dike margins. Along the base of the nappe, where the rocks have strong deformation fabrics developed during nappe emplacement, the margins of the dikes are inclined at low angle bedding. We can use the change in the angle between the dike margins and bedding to measure the shear strain across the bedding.

Fossils with bilateral symmetry also provide the right angles needed to measure shear strain. Consider the situation depicted in Figure 5.9a, where a hypothetical stratigraphic package depicted by rectangle *defg* is shortened and thickened to the rectangular shape $d'e'f'g'$. An undistorted brachiopod (in black) is positioned at *c*, the center of the undeformed stratigraphic section *defg* (Figure 5.9a). An undeformed brachiopod possesses a plane of mirror symmetry that is oriented perpendicular to its hinge line, where its two shells attach. As a result of deformation of the stratigraphic section, the particle at *c* moves to a new position c' and the diagonal lines *df* and *eg* meeting at angle ϕ in the original section reorient to $d'f'$ and $e'g'$ meeting at angle ϕ' in the deformed section. During deformation, lines parallel to the hinge of the brachiopod experience shear strains. So, the trace of its plane of mirror symmetry reorients to the acute angle to the hinge line shown in Figure 5.9b. In practice, it would be extremely difficult to determine the absolute movement of an object like a brachiopod, but the orientations of hinge lines and lines of bilateral symmetry are readily apparent in deformed brachiopods. We can, therefore, measure the angular shear exhibited by the brachiopod's hinge line during deformation.

Trilobites are another fossil that exhibits bilateral symmetry and, therefore, have implicit right

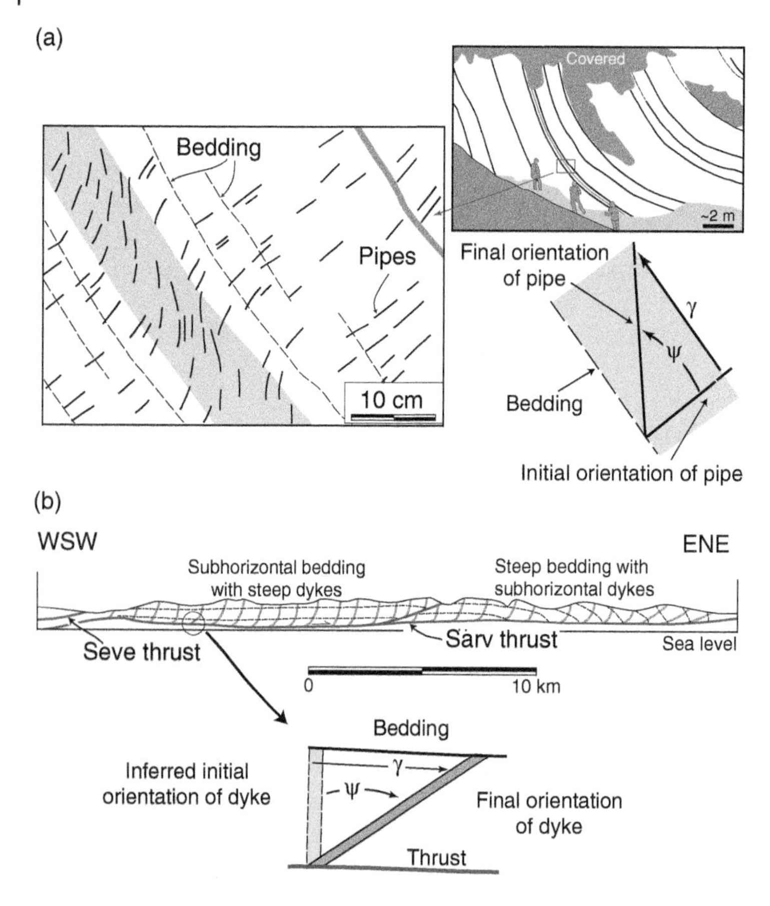

Figure 5.8 (a) In folded Cambrian quartzite with *Skolithus* pipes (see the sketch in the upper right), pipes initially perpendicular to bedding are sheared to acute angles to bedding in layer indicated by red shading. (b) A portion of a section of the Särv nappe from the Scandinavian Caledonides showing the relative orientation of bedding in arkoses and mafic dikes (modified from Gilotti and Kumpulainen, 1986). In the upper portions of the nappe, dikes are normal to bedding. Along the base of the nappe, dikes are sheared to acute angles to bedding and the thrust.

angles between lines connecting corresponding points on opposite sides of their axis of bilateral symmetry. If a trilobite's center line, which is its axis of bilateral symmetry, were aligned parallel to either the direction of the brachiopod's hinge or its normal, the distorted shape of the trilobite would indicate the same shear strain magnitude (provided that both fossils record all the deformation experienced by the surrounding rock). Figure 5.9c shows what would happen to trilobites in either of these two orientations. Note that the two trilobites in Figure 5.9c exhibit opposite senses of shear. From this, we can infer that there is a direction between them that experiences no shear strain.

Returning to the deformed stratigraphic section in Figure 5.9a, note that lines parallel to the top, bottom, and sides of the undeformed stratigraphic section exhibit changes in their length (and were, in some cases, displaced by deformation). They did not, however, change their relative orientations. All lines originally parallel to one of the sides or top/bottom of the section exhibit no angular shear or shear strain from this deformation. In any two-dimensional pure distortion, there are always two directions that exhibit no angular shear or shear strain, whereas lines oblique to those two directions exhibit shear strains. The magnitude of the angular shear or shear strain relative to a line is zero for lines

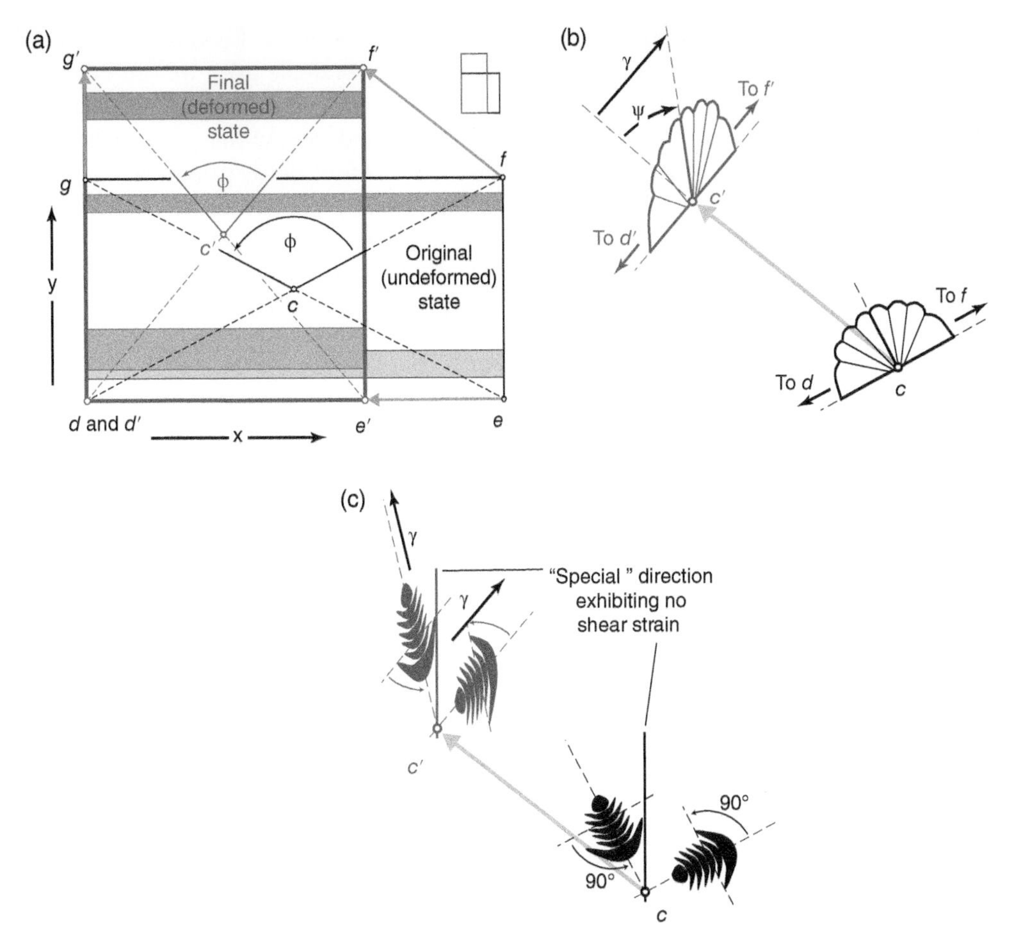

Figure 5.9 (a) Sketch depicting the original shape (in gray) and final shape (in red) of a hypothetical stratigraphic section. Compare the positions of particles at corners and at centers of the undeformed and deformed states. (b) Brachiopod at c displaced to c' and distorted. (c) Equivalent deformation of two trilobites with orientations comparable to brachiopods in (b).

parallel to one of these special directions and increases smoothly to a maximum value for lines presently inclined 45° to the special directions.

5A.4 Relating Strain to Displacements

The discussion in the previous paragraph addresses indirectly the relationship between the movements that particles exhibited during deformation and whether a pure strain component exists in that deformation. We can address this important issue directly by considering the movement of different particles in the examples given earlier. In the previous chapter, we showed that a directed line segment from the original position of a particle to its final position gives the *displacement* of the particle. In the deformed stratigraphic section in Figure 5.9a, we noted that particle at c moved to a new position at c'. The particles that mark the boundaries of the stratigraphic section, d, e, f, and g, also move to new positions d', e', f', and g' as a result of the deformation. The directed line segments giving the displacements of particles at d, e, f, and g are not identical, that is, they have different lengths, different orientations, or both different lengths and orientations. In the previous chapter, we used the position coordinates relative to Cartesian coordinates for the original, undeformed-state position and the final, deformed-state position of any particle to calculate the components of the particle's displacement. Recalling the convention of using (X, Y) to denote the particle's original position and (x, y) to denote the particle's final position, the x and y components of the particle's displacement are:

$$u_X(\text{particle}) = x(\text{particle}) - X(\text{particle})$$
$$u_Y(\text{particle}) = y(\text{particle}) - Y(\text{particle}).$$

Regardless of where and in what orientation one imposes Cartesian coordinates on Figure 5.9a, either the magnitude of the x component, the magnitude of the y component, or the magnitude of both components of the displacement vectors of different particles are different.

In examining any collection of particles that exhibit a pure strain, we find that the magnitude, direction, or both magnitude and direction of the displacement vectors change as the positions of particles change. Drawing on the terminology used when temperatures vary with position (temperature gradient) or when the pressures vary with position (pressure gradient), the variation in displacement vector magnitude and/or direction with position is a *gradient in the displacement field*. We usually focus on the variation of individual components of particles' displacements with changes in an individual coordinate direction. The magnitudes of the resulting four *displacement gradients* ($\Delta u_X/\Delta X$, $\Delta u_X/\Delta Y$, $\Delta u_Y/\Delta X$, $\Delta u_Y/\Delta Y$) characterize the pure strain.

Returning to Figure 5.9a, note that the x axis is parallel to line de and y axis is parallel to line dg. If we consider selected pairs of particles along lines parallel to the coordinate axes, we can examine relationships between the values of the displacement gradients and the resulting strains. Consider the u_X components of the displacements of particles d and e or particles f and g. Each pair of particles lies on a line where Y = constant. $\Delta u_X/\Delta X$ is negative along both lines. Note that u_X goes from 0 to a negative number as the X coordinate value goes from 0 to a positive number. Lines $d'e'$ and $g'f'$ both are shorter than lines de and gf. Next, consider the u_Y components of the displacements of particles d and g or particles e and f. In this instance, each pair of particles lies on a line where X = constant. $\Delta u_Y/\Delta Y$ is positive along each line. Note that u_Y goes from 0 to a positive number as the Y coordinate value goes from 0 to a positive number. Lines $d'g'$ and $e'f'$ both are longer than lines dg and ef. The displacement gradients $\Delta u_X/\Delta X$ and $\Delta u_Y/\Delta Y$ relate to the magnitudes of longitudinal strains, with positive values corresponding to line lengthening, negative values corresponding to line shortening, and zero values corresponding to no change in length.

The displacement gradients $\Delta u_X/\Delta Y$ and $\Delta u_Y/\Delta X$ relate to the magnitudes of shear strains. Figure 5.10 is an enlargement of a portion of the section of the Särv nappe in Figure 5.8b. If the u_X displacement components of particles along the margin of a dike (where all particles have the same X values) were the same, the dike would not change its inclination relative to bedding or the thrust. Particle b, whose Y component is different from the Y component of a, experienced additional movement parallel to the x axis. Then, $\Delta u_X/\Delta Y \neq 0$, and the dike was sheared to a new inclination relative to bedding. If $\Delta u_X/\Delta Y = 0$, particles along lines parallel to the y axis will not exhibit shear strain relative to the x axis. Similarly, if $\Delta u_Y/\Delta X = 0$, particles along lines parallel to the x axis will not exhibit shear strain relative to the y axis. Returning to Figure 5.9a, because $\Delta u_X/\Delta Y = 0$ along line ef and $\Delta u_Y/\Delta X = 0$ along line fg, these directions do not exhibit shear strains.

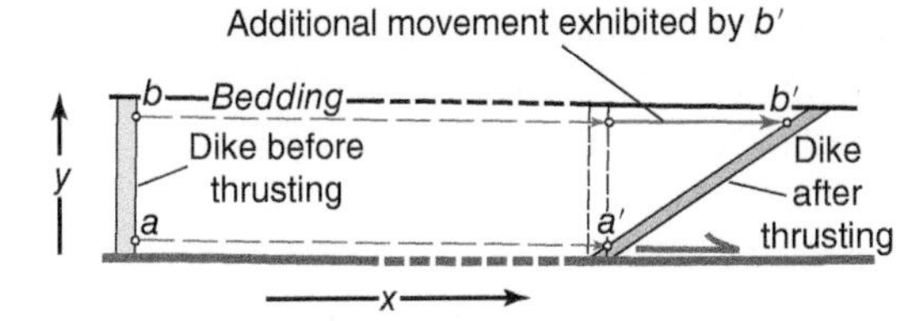

Figure 5.10 Depiction of distortion of the mafic dike at the base of the Särv nappe (see Figure 5.8). Particles a and b displaced to a' and b', respectively. Change in displacement component u_X with position coordinate Y generates shear strain.

The existence of pure strain means that the displacement gradients have nonzero magnitudes, but not all nonzero gradients in a displacement field generate pure strain components in a deformation. Some combinations of displacement gradients give rise to rigid body rotation only. In any and every instance where we recognize pure strain, however, we know that a displacement field with nonzero displacement gradients is responsible for the deformation.

5A.5 Homogeneous and Inhomogeneous Strain

Even though particles at different positions in Figure 5.9a have different displacements, the variations in displacements are such that the longitudinal and shear strains in that deformation conform to a high degree of uniformity. In the final (deformed) state in Figure 5.9a, the magnitudes of the stretch or elongation in a given direction at one position are identical to the magnitudes of stretch or elongation in the same direction at any other position. Similarly, the magnitudes of the shear strain across a line with a given orientation at one position are identical to the magnitudes of shear strain across a line with the same direction at any other position. We use the term **homogeneous** to refer to pure strains that exhibit this level of uniformity.

Homogeneous pure strain is apparent in three effects:

1) Lines that were straight in the undeformed state are straight in the deformed state. In Figure 5.9a, the straight lines defining the boundaries of the original state and the straight lines defining the diagonals of that the rectangular undeformed section are straight lines in the deformed state.
2) Lines that were parallel before deformation are parallel after deformation. In Figure 5.9a, parallel lines *de* and *fg* are transformed into lines $d'e'$ and $f'g'$, which are parallel in the deformed state.
3) The magnitude of the stretch (or elongation) at one point along a line equals the stretch (or elongation) at other points along that line. In Figure 5.9a, the magnitude of the stretch exhibited by line $d'c'(=d'c'/dc)$ is the same as that exhibited by line $c'f'(=c'f'/cf)$.

As a result of these three effects, the homogeneous deformation of a collection of particles that define a square in the undeformed state will transform that collection of particles into a parallelogram. In addition, any two squares with the same orientation, regardless of their size or location within the deformed region, will be transformed into similar (or congruent) parallelograms by homogeneous deformation.

It is useful to compare explicitly homogeneous and heterogenous deformations. The homogeneous character of the deformation arises because the displacement gradients have the same values at different positions within a region, that is, $\Delta u_i/\Delta X_j$ at one point within the region is equal to $\Delta u_i/\Delta X_j$ at all other points in the region. Figure 5.11a illustrates what would happen to the distorted stratigraphic section of Figure 5.9a if the displacement gradients are different at different locations. Along the line *de*, the u_X components of particle's displacements change from 0 at d to $u_X(e)$ at e. Along the line *gf*, the u_X components of particle's displacements change from 0 at g to

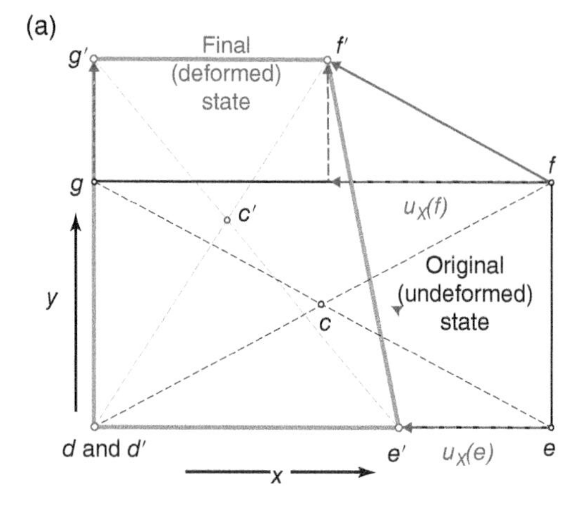

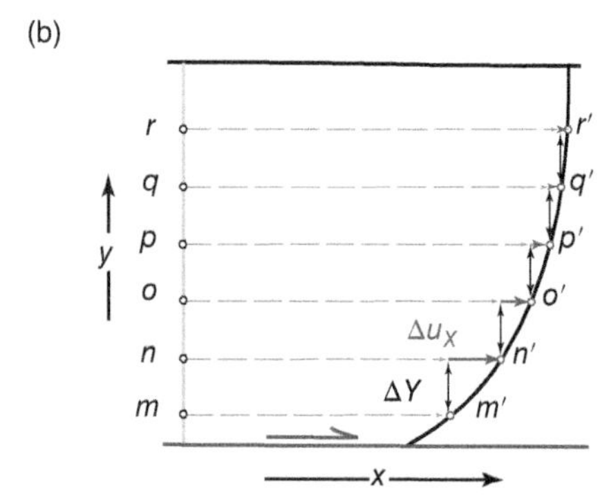

Figure 5.11 Changes in the values of the displacement gradients with position lead to inhomogeneous deformation. In inhomogeneous deformation (a) lines originally parallel, (*dg* and *ef*) are no longer parallel (*d′g′* and *e′f′*); and (b) originally straight lines (*mr*) are no longer straight (*m′r′*).

$u_X(f)$ at *f*. Particles *e* and *f* have equal original *X* coordinates, but $u_X(e) \neq u_X(f)$. As a result, $\Delta u_X/\Delta X$ along *gf* does not equal to $\Delta u_X/\Delta X$ along *de*. Lines *dg* and *ef*, which were parallel in the undeformed state are transformed in the deformed state into lines *d′g′* and *e′f′*, which are not parallel. Further, particle *c* falls on the midpoint of the diagonal line *df* of the original rectangle. Because $u_X(e) \neq u_X(f)$, *c′* does not lie on the midpoint of line *d′f′*. The stretch exhibited by line segment *d′c′* cannot be the same as the stretch exhibited line segment *c′f′*. Figure 5.11b, a variation of the illustration of the distorted dykes of Figure 5.10, illustrates another possible effect if displacement gradients vary, i.e. are different at different locations. The distances ΔY between particles *m*, *n*, *o*, *p*, *q*, and *r* are equal, but the change in the u_X component of displacements of these particles decreases from a relatively large magnitude between particles *m′* and *n′* to a small magnitude between particles *q′* and *r′*. As a result, the originally straight line *mn* is transformed to the curved line *m′n′*.

5A.6 Finite Strain Ellipse and Finite Strain Ellipsoid

5A.6.1 Finite Strain Ellipse

5A.6.1.1 Basic Ideas

In any two-dimensional homogeneous pure strain, there is a single direction that exhibits the maximum stretch or elongation observed within the plane. Lines increasingly oblique to the direction of maximum stretch or elongation exhibit successively smaller stretch or elongation values, culminating with the minimum stretch or elongation for lines perpendicular to the direction of maximum stretch or elongation. Similarly, shear strain magnitudes vary smoothly as a function of orientation relative to the directions of no angular shear (which are the directions of maximum and minimum longitudinal strains). We use the **finite strain ellipse** to illustrate and examine the regular variations in strain magnitude with direction in two-dimensional homogeneous pure strains. Consider the circular collection of particles that define the cross sectional shape of ooids in Figures 5.5 and 5.7. A homogeneous pure strain will transform this circular form into an ellipse in the deformed state. We call the elliptical shape taken by a circle in the undeformed state the finite strain ellipse.

5A.6.1.2 Characterizing the Strain Ellipse

The transformation of a circle into an ellipse provides considerable insight into pure strain. Comparing the circle in the undeformed state to the ellipse in the deformed state is especially useful at illustrating the relative magnitudes of the stretch (or elongation) in different directions. In the undeformed state, all radii or diameters of the circle have equal lengths (r or $2r$, respectively; see Figure 5.12). In the deformed state, radii or diameters in different directions across the ellipse have different lengths, indicating that magnitudes of stretch or elongation are different in different directions. The magnitudes of the stretch or elongation in different directions, like the radii or diameters of the ellipse, vary smoothly with the change in direction (Figure 5.12). An ellipse has a maximum diameter, called its *major diameter*, and a minimum diameter, called its *minor diameter*. The two lines from the center of the ellipse to its periphery along the major diameter are the ellipse's *major axes* S; the lines from the center of the ellipse to its periphery along the minor diameter are the ellipse's *minor axes*.

The major and minor axes (or major and minor diameters) of the ellipse identify the directions of the maximum and the minimum values of the stretch (T) or elongation (e). In many natural deformations and in the situation illustrated in Figure 5.12, the major axis is longer than the radius of the circle and the minor diameter is shorter than the radius of the circle. Further, the longitudinal strain magnitudes along the major and minor axes in Figure 5.12 are such that the area of the ellipse ($= \pi Ss$) equals the area of the circle ($= \pi r^2$). Neither of these conditions is true in all homogeneous pure strains. It is possible that a finite strain ellipse has a major axis longer than the circle's radius and a minor axis shorter than the circle's radius, yet the strain ellipse has an area less than or greater than the area of the circle. If both major and minor axes are shorter than the circle's radius, the deformation must lead to a loss of area in the plane. If both the major and minor axes of the ellipse are longer than the circle's radius, the deformation must lead to an increase of area in the plane.

The major and minor axes of the strain ellipse, like those of any ellipse, are perpendicular. So, the directions of the maximum and minimum stretch or elongation are orthogonal in any homogeneous pure strain. Lines parallel to the major and minor

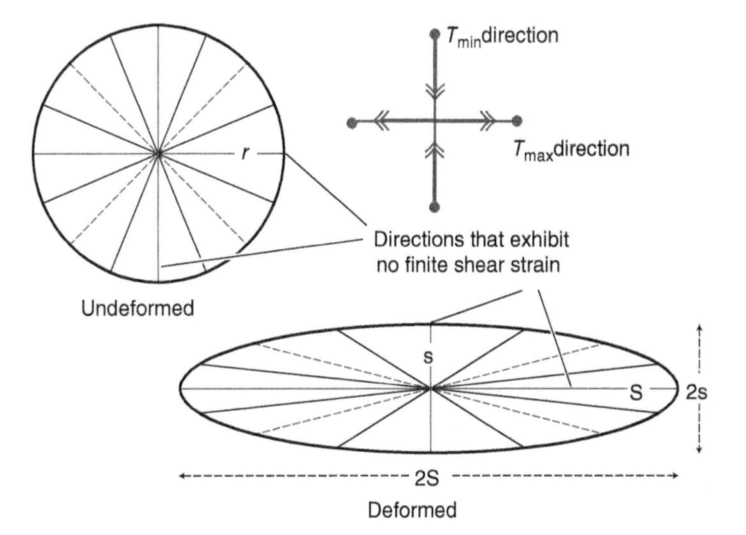

Figure 5.12 In two-dimensional homogeneous pure strain, lines with different original orientations experience different longitudinal strains. The direction of maximum T, which is the direction of greatest lengthening, is perpendicular to the direction of minimum T, the direction of greatest shortening.

axes of the ellipse are perpendicular to each other in the deformed state. Interestingly, those lines were also perpendicular to each other in the undeformed state. Thus, the directions of the maximum and minimum stretch are also directions that exhibit no finite shear strain (see Figure 5.12).

To underscore the special character of the directions of the maximum and minimum stretch, we call them the **principal directions** or the **principal axes of strain**. The values of stretch or elongation parallel to the principal directions are the **principal values** or **principal magnitudes of strain**. If we know the circle's radius, we can calculate its principal stretch magnitudes:

$$T_{max} = \frac{S}{r} \tag{5.5a}$$

$$T_{min} = \frac{s}{r}. \tag{5.5b}$$

We do not need to know the radius of the original circle to determine the *relative* magnitudes of the principal strains because

$$\frac{T_{max}}{T_{min}} = \frac{(S/r)}{(s/r)} = \frac{S}{s} = R_s. \tag{5.6}$$

The ratio of the length of the major axis to the length of the minor axis of this ellipse, called the **axial ratio** or **strain ratio** and denoted R_s, is a convenient measure of the degree to which deformation has changed the shape of an initial circle. Comparing a deformation in which $R_s = 1.5$ to a deformation in which $R_s = 5.0$, it is easy to see that the latter causes more extreme distortion (Figure 5.13a). Note, however, that the axial ratio does not indicate the amount of area change: finite strains where $T_{max} = 0.9$ and $T_{min} = 0.45$ (with a significant reduction of area), where $T_{max} = 1.414$ and $T_{min} = 0.707$ (with no area change), and where $T_{max} = 2.4$ and $T_{min} = 1.2$ (with a significant increase in area) all yield identical axial ratios of $R_s = 2.0$ (Figure 5.13b).

(a)

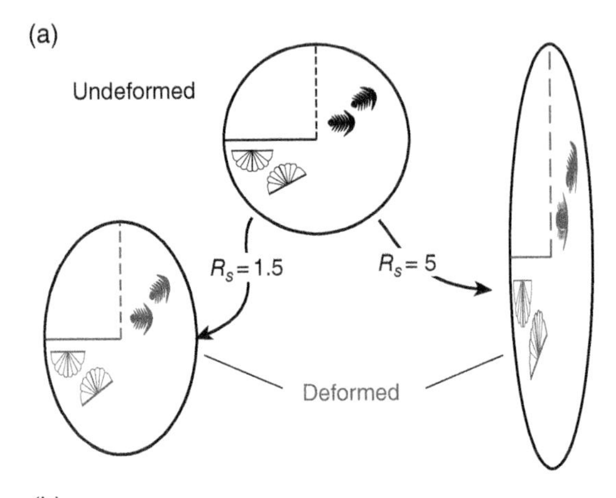

(b)

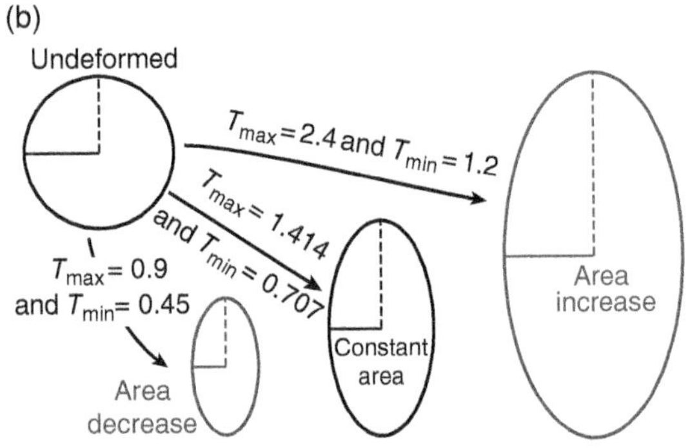

Figure 5.13 (a) In homogeneous pure strain, a circle is transformed into an ellipse. The axial ratio of the ellipse (R_S) measures the intensity of deformation. (b) Three possible deformations with the same axial ratio but different changes in the area.

Reiterating the properties of the two principal directions of strain in homogeneous two-dimensional pure strain, we see that

1) Lines parallel to one strain principal direction exhibit the greatest lengthening (or least shortening) of all lines in the plane; the stretches of lines parallel to this direction are the maximum values observed in this plane.
2) Lines parallel to the second strain principal direction exhibit the greatest shortening (or least lengthening) of all lines in the plane of interest; the stretches of lines parallel to this

direction are the minimum values observed in this plane.

3) The directions of T_{max} and T_{min} are mutually perpendicular.
4) The strain principal directions exhibit no finite shear strain; lines now found parallel to the two, mutually perpendicular strain principal directions were perpendicular to each other in the undeformed state.
5) The ratio of the length of the major axis to the length of the minor axis $S/s = R_s = T_{max}/\mathrm{T}_{min}$ is a convenient measure of the intensity of the deformation.

5A.6.1.3 Geological Example – Using Strain Markers

It is important to keep in mind that the strain ellipse is the shape in the deformed state of an initially circular object. Geologists use a wide variety of geological objects – ooids, detrital grains, cobbles, or pebbles, etc. – as *strain markers* (Figure 5.14), that is, objects whose shape we can use to determine the axial ratio and orientation of the strain ellipse. As noted earlier, an important assumption is that the marker behaves exactly like the surrounding rock and that it has existed for all of the deformation. The preferred strain

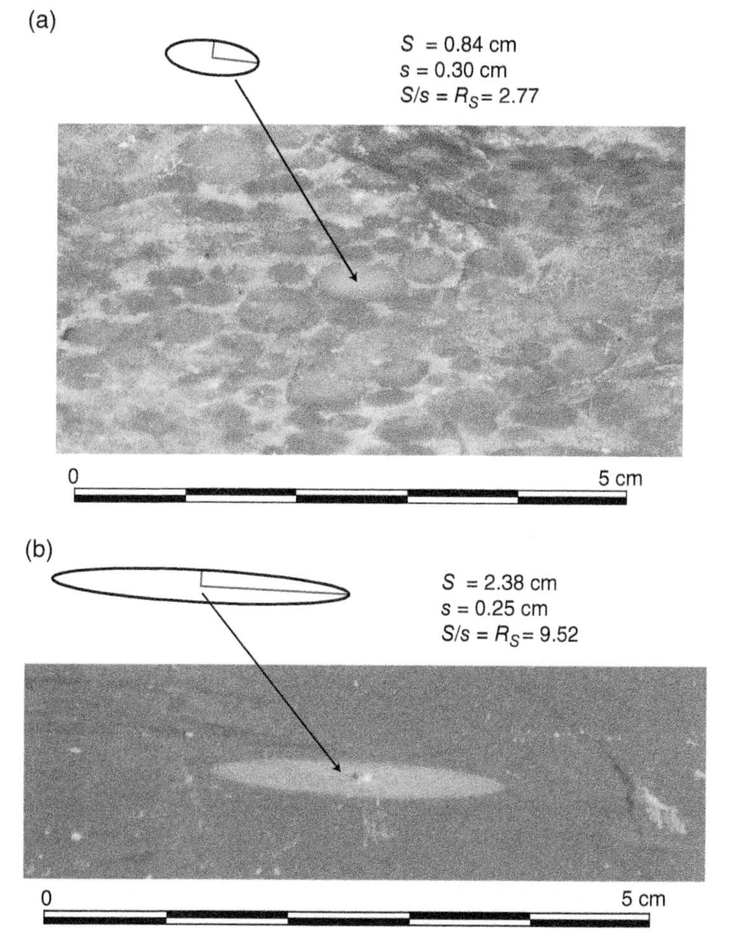

Figure 5.14 Deformed markers yield elliptical shapes that estimate finite strain ellipse. (a) Deformed ooid (indicated by arrow) has an elliptical shape with $R_S = 2.77$. Note other ooids have similar (but not identical) shapes. (b) Deformed reduction spot in slate has an elliptical shape with $R_S = 9.52$.

markers have regular, rounded shapes, but they are rarely truly circular. More typically, their initial shapes are better approximated by ellipses. Figures 5.15 and 5.16 are intended to show how the initial shape of a marker affects its final shape. There are two aspects of a marker's initial elliptical shape that interact with the finite strain ellipse to generate the deformed shape of the marker. One aspect is the initial axial ratio of the

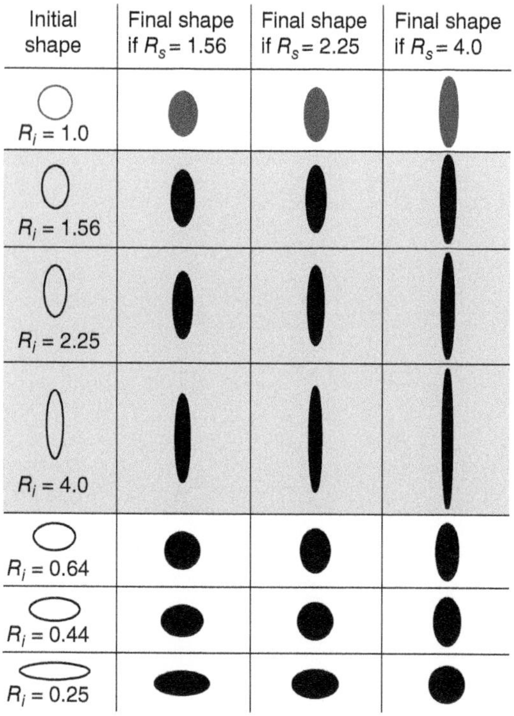

Initial shape	Final shape if R_s = 1.56	Final shape if R_s = 2.25	Final shape if R_s = 4.0
R_i = 1.0			
R_i = 1.56			
R_i = 2.25			
R_i = 4.0			
R_i = 0.64			
R_i = 0.44			
R_i = 0.25			

Figure 5.15 Original axial ratio (R_i) of an elliptical marker (open forms in the column at left) affects the observed shape of the deformed marker. When the major diameter of the elliptical marker is parallel to S_1 (the long axes of the initial shapes of objects are parallel to the long axis of the strain ellipse; shaded regions in the chart), the observed final shape overestimates the axial ratio of strain ellipse. When the major diameter of an elliptical marker is parallel to S_3 (the long axes of the initial shapes of objects are perpendicular to the long axis of the strain ellipse; unshaded regions in the chart), the observed shape underestimates the axial ratio of strain ellipse.

Initial shape	Final shape if R_s = 1.56	Final shape if R_s = 2.25	Final shape if R_s = 4.0
R_i = 1.0			
R_i = 4.0, ϕ_o = 0°			
R_i = 4.0, ϕ_o = 15°			
R_i = 4.0, ϕ_o = 30°			
R_i = 4.0, ϕ_o = 45°			
R_i = 4.0, ϕ_o = 60°			
R_i = 4.0, ϕ_o = 75°			
R_i = 4.0, ϕ_o = 90°			

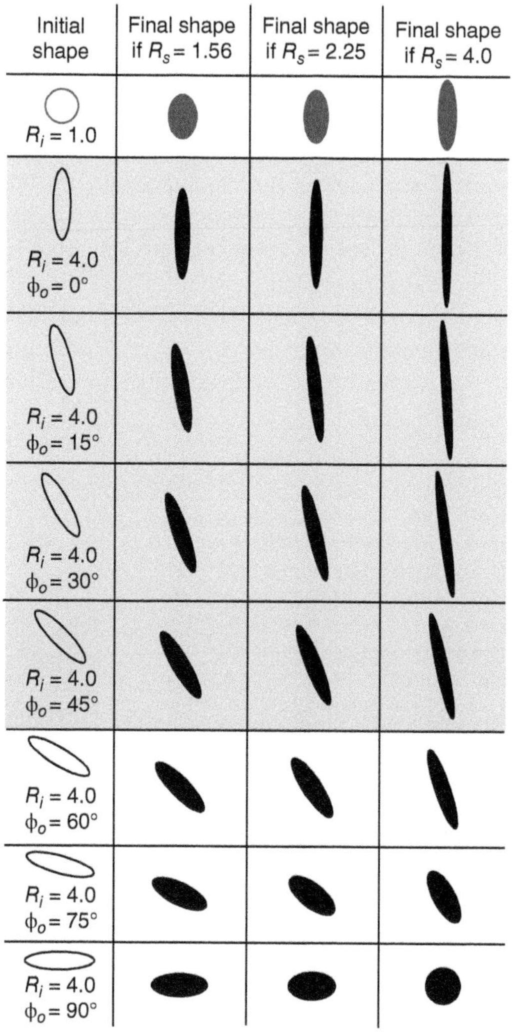

Figure 5.16 Diagram showing how the orientation of an originally elliptical marker relative to the strain principal directions affects the observed shape of the deformed marker. The left side column shows elliptical objects with the same original axial ratio (R_i = 4.0) oriented at different angles to the principal directions of the finite strain ellipse (given by ϕ_o, the angle between the long axis of the undeformed shape and the long axis of the strain ellipse). When $\phi_o \leq 45°$ (shaded region of the chart), the deformed object has an axial ratio greater than the axial ratio of the strain ellipse. When $\phi_o > 45°$ (unshaded region of the chart), the deformed object has an axial ratio less than the axial ratio of the strain ellipse.

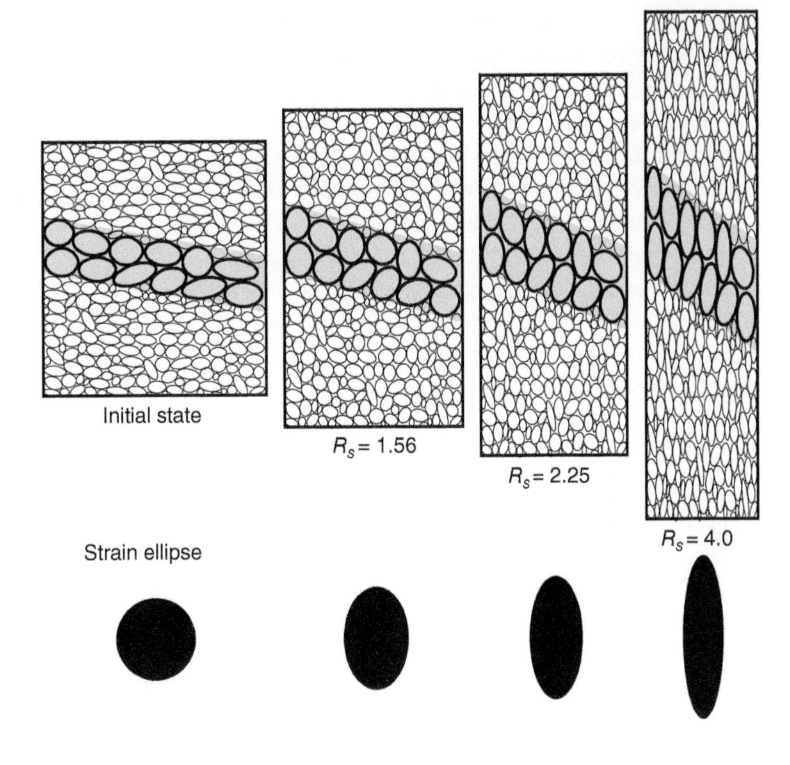

Figure 5.17 Due to the combined effects of variations in the original shapes and original orientations of elliptical markers, no single marker may have the shape of the finite strain ellipse, even though all markers record finite strain.

undeformed elliptical marker (R_i). The second aspect is the orientation of the major and minor axes of an elliptical marker relative to the orientations of the principal axes of the finite strain ellipse. The most significant impact of the initial shape of an initially elliptical marker occurs when the major and minor axes of the marker are oriented parallel to the principal axes of the finite strain ellipse. The deformed shape of the marker will either overestimate or underestimate the axial ratio of finite strain ellipse, with the degree of over- or underestimation greater for markers with more highly elliptical shapes (Figure 5.15). The deformation of originally elliptical markers whose major and minor axes are inclined to the principal axes of the finite strain ellipse will generate deformed shapes whose major and minor axes are inclined to the strain principal directions (Figure 5.16). One can use a collection of deformed markers to calculate the shape of the finite strain ellipse (see Section 5A.11.3 for descriptions of some commonly used methods), but individual deformed markers typically only approximate its shape (Figure 5.17).

5A.6.2 Finite Strain Ellipsoid

5A.6.2.1 Basic Ideas

If an object, such as an ooid, grain of sand, or rounded cobble, has a spherical shape in three dimensions, the smooth variation in the magnitudes of longitudinal strain in different directions within a homogeneous pure strain means that the original sphere will be an ellipsoid after deformation. The **finite strain ellipsoid**, a three-dimensional equivalent of the finite strain ellipse, is the shape in the deformed state of an initially spherical object. Most ellipsoids have a maximum diameter again called the *major diameter*, a minimum diameter again called the *minor diameter*, and a third diameter perpendicular to both the major and minor diameters with a length between the major and minor diameter lengths. We call the diameter perpendicular to both the major and minor diameters the *intermediate diameter*. The length of the intermediate diameter can be close to the length of the minor diameter or close to the length of the major diameter. The line from the center of the ellipsoid to its periphery along the major diameter is the ellipsoid's *major axis* S_1; the line from the center of the ellipsoid to its periphery along the minor diameter is the ellipsoid's *minor axis* S_3. The major and minor axes (or major and minor diameters) of the ellipsoid are parallel to the directions of the maximum and the minimum values of the stretch T or elongation e. The line from the center of the ellipsoid to its periphery along the intermediate diameter defines the ellipsoid's *intermediate axis* S_2. These three directions define the **three principal finite strain axes** ($S_1 > S_2 > S_3$) (Figure 5.18). You may see other symbols used to denote the principal axes of three-dimensional, homogenous finite strains. The major axis of the finite strain ellipsoid (S_1) is alternately given by $1+e_1$ (where e_1 is the greatest elongation), λ_1, or X. Similarly, the intermediate axis of the finite strain ellipsoid (S_2) is sometimes denoted $1+e_2$, λ_2, or Y, and the minor axis of the finite strain ellipsoid (S_3) sometimes given by $1+e_3$ (where e_3 is the smallest elongation), λ_3, or Z.

5A.6.2.2 Characterizing the Strain Ellipsoid

We regularly describe the *shape* of the strain ellipsoid approximately by saying it resembles a pancake, an elongate pancake, a flattened cigar, a cigar, etc. As is the case in examining two-dimensional finite strain, ratios of the principal strain magnitudes characterize more precisely the three-dimensional shape of the finite strain ellipsoid. The values of the stretch measured parallel to the three principal finite strain axes are

$$T_{max} = \frac{S_1}{r} \tag{5.7a}$$

$$T_{int} = \frac{S_2}{r} \tag{5.7b}$$

$$T_{min} = \frac{S_3}{r}, \tag{5.7c}$$

where r is the radius of the original sphere. We can calculate elongation values in the three principal directions by recalling (see Eq. 5.3) that

$$T_i = e_i + 1. \tag{5.8}$$

The shapes of ellipsoids are sufficiently variable to warrant some discussion. Under ideal conditions, measurements of longitudinal strain magnitudes along the principal axes enable us to calculate the ratio of the volume of the finite strain ellipsoid $\left(= \tfrac{4}{3}\pi S_1 S_2 S_3\right)$ to the volume of the original sphere $\left(= \tfrac{4}{3}\pi r^3\right)$:

$$\frac{\left(\tfrac{4}{3}\pi S_1 S_2 S_3\right)}{\left(\tfrac{4}{3}\pi r^3\right)} = \left(\frac{S_1}{r}\right)\cdot\left(\frac{S_2}{r}\right)\cdot\left(\frac{S_3}{r}\right) = T_{max}\cdot T_{int}\cdot T_{min}. \tag{5.9}$$

More typically, such as when we are examining the shape of an ellipsoidal strain marker, we have no independent measure of its volume before deformation. In such a case, we are forced to assume that there was no change in its volume during deformation, that is, $T_{max}\cdot T_{int}\cdot T_{min} = 1$. Even after making this assumption, we can distinguish three different classes of strain ellipsoid shapes, assuming $T_{max} > 1$ and $T_{min} < 1$:

- If $T_{int} = 1$ (this means that $S_2 = r$, $e_{int} = 0$, and $T_{max} = 1/T_{min}$), all particles lying within a particular plane parallel to the XZ plane before deformation lie in that plane after deformation. Thus, the displacement vectors of all particles

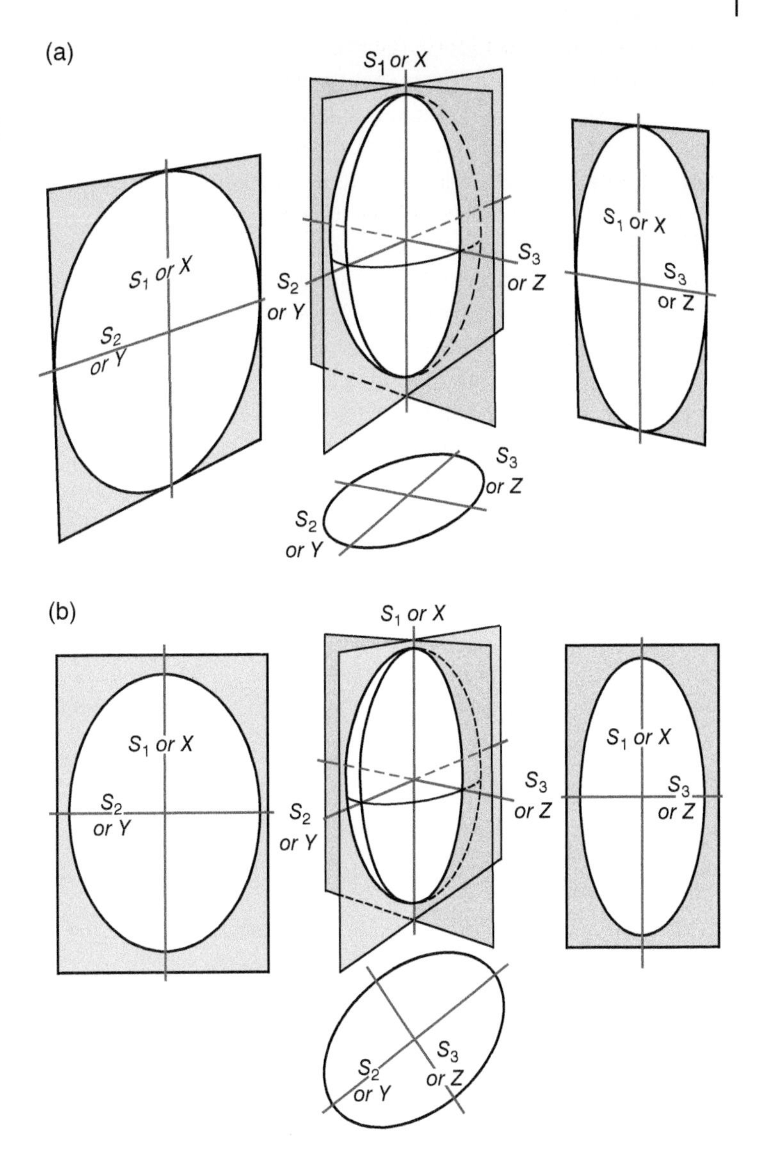

Figure 5.18 A representative finite strain ellipsoid with major axis S_1 or X, intermediate axis S_2 or Y, and minor axis S_3 or Z. Representative XY, XZ, and YZ sections, shown in perspective view in (a) and in orthogonal view in (b), indicate the relative lengths of the three principal axes.

are parallel to the XZ plane. For this reason, we call such deformations **plane strain** strain states.

- If $T_{int} < 1$, we call the deformation **constriction** or say that the rock exhibits a **constrictional** strain state. Constriction generates cigar-shaped strain ellipsoids.
- If $T_{int} > 1$, we call the deformation **flattening** or say that the rock exhibits a **flattening** strain state. Flattening generates pancake-shaped strain ellipsoids.

Truly pancake- or truly cigar-shaped finite strain ellipsoids have no distinct intermediate diameter. They either have: (1) a single minor diameter with all diameters perpendicular to the minor diameter having equal lengths greater than the minor diameter ($S_1 = S_2 > S_3$), creating an ideally circular, pancake-shaped ellipsoid; or (2) a single major diameter with all diameters perpendicular to the major diameter having the same length shorter than the major diameter ($S_1 > S_2 = S_3$), creating a perfectly rounded, cigar-shaped ellipsoid. Both shapes are called ellipsoids of revolution because their shape is what one would generate by spinning an ellipse about its major or minor diameter.

Geologists use a number of different schemes to assess the shape of any finite strain ellipsoid. One scheme uses the values of two axial ratios to characterize the shape of the ellipsoid. The first value is the XY axial ratio of the ellipsoid:

$$R_{XY} = \frac{T_{\max}}{T_{\text{int}}} = \frac{(S_1/r)}{(S_2/r)} = \frac{S_1}{S_2} = \frac{(e_{\max}+1)}{(e_{\text{int}}+1)}. \quad (5.10)$$

The second value is the YZ axial ratio of the ellipsoid:

$$R_{YZ} = \frac{T_{\text{int}}}{T_{\min}} = \frac{(S_2/r)}{(S_3/r)} = \frac{S_2}{S_3} = \frac{(e_{\text{int}}+1)}{(e_{\min}+1)}. \quad (5.11)$$

Since the values of both axial ratios, R_{XY} and R_{YZ}, are needed to characterize the shape of an ellipsoid, geologists have devised a number of ways to depict their relative magnitudes. We compare the magnitudes of R_{XY} and R_{YZ} graphically on a plot called a *Flinn diagram* (Figure 5.19), a Cartesian plot whose abscissa (x axis) is the R_{YZ} value and whose ordinate (y axis) is the R_{XY} value. On such a plot, every distinct ellipsoid shape corresponds to a unique point. Normally, we position the origin of this plot at the point (1,1) rather than (0,0), so the slope of the line from the origin (1,1) to any point is

$$k = \frac{\Delta(\text{ordinate})}{\Delta(\text{abscissa})} = \frac{(R_{XY}-1)}{(R_{YZ}-1)} = \frac{(T_{\max}/T_{\text{int}}-1)}{(T_{\text{int}}/T_{\min}-1)}. \quad (5.12)$$

If $T_{\text{int}} = 1$, Eq. (5.12) becomes

$$k = \frac{(T_{\max}-1)}{(1/T_{\min}-1)}.$$

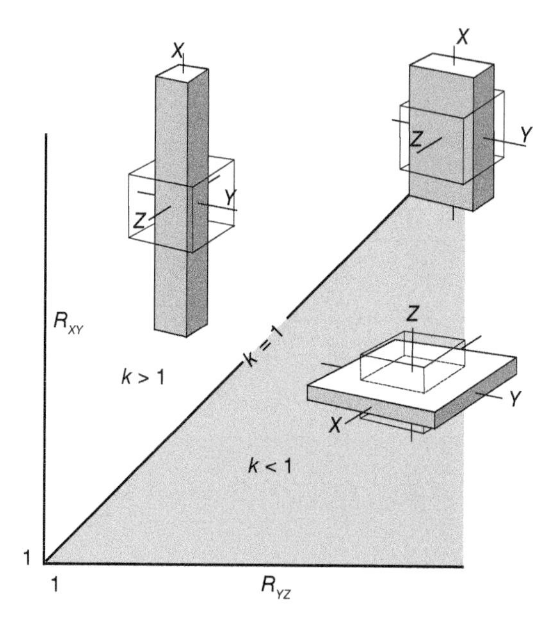

Figure 5.19 Flinn diagram, showing fields where $k>1$, $k=1$, and $k<1$, corresponding to constrictional, plane strain, and flattening finite strain states.

If there has been no change in volume, then $T_{\max} \cdot 1 \cdot T_{\min} = 1$ or $T_{\max} = 1/T_{\min}$ and $k = 1$. Thus, finite strain ellipsoids for plane strain strain states plot along the line with a slope of 1, inclined 45° to the abscissa and ordinate. Ellipsoids that plot above the $k = 1$ line (where the slope is greater than 45°) correspond to constrictional strain states, where $T_{\text{int}} < 1$. Ellipsoids that plot below the $k = 1$ line (where the slope is less than 45°) correspond to flattening strain states, where $T_{\text{int}} > 1$.

It is important to reiterate that when we use ellipsoidal markers, the distinction between constrictional, plane strain, and flattening cases often depends on an *assumption* that there has been no change in the volume of the marker during deformation. To underscore this dependence upon the no volume change assumption, we often describe a finite strain state as one of *apparent constriction* (when $k > 1$), *apparent plane strain* (when $k = 1$), or *apparent flattening* (when $k < 1$).

Another way to describe the shape of the finite strain ellipsoid is to use a triple ratio:

$$\frac{T_{\max}}{T_{\text{int}}} : \frac{T_{\text{int}}}{T_{\text{int}}} : \frac{T_{\min}}{T_{\text{int}}} = \frac{(S_1/r)}{(S_2/r)} : \frac{(S_2/r)}{(S_2/r)} : \frac{(S_3/r)}{(S_2/r)} = \frac{S_1}{S_2} : 1 : \frac{S_3}{S_2}. \tag{5.13}$$

In most instances, $S_1/S_2 > 1$ and $S_3/S_2 < 1$. So, the triple axial ratio defined in Eq. (5.13), which is $R_{XY} : 1 : (1/R_{YZ})$, will consist of (a number greater than one):1:(a number less than one).

To summarize, the finite strain ellipsoid is the shape in the deformed state of an *initially spherical object*. Like analyses of finite strain in two dimensions, we may use any one of several geological objects as **strain markers**. Strain markers are objects whose shape we can use to determine the axial ratios and orientation of the strain ellipsoid. In order to record the finite strain accurately, the marker must behave exactly like the surrounding rock and exist throughout the deformation. Strain markers typically have regular, rounded shapes, but they are rarely truly spherical. Their initial shapes are better approximated as ellipsoids. The initial shape and initial orientation of the marker will, as was the case for deformed markers analyzed in two dimensions (see Section 5A.6.1.3), affect the final shape of the marker and it must be factored into any analysis of the shapes of markers in three-dimensions.

5A.7 States of Strain and Strain Paths

5A.7.1 States of Strain

We have distinguished the *initial, undeformed* configurations of collections of particles or objects from their *final, deformed* configurations. The **state of strain** refers to the totality of the pure strain component, including the magnitudes of longitudinal strains and shear strains in all directions and the manner in which the longitudinal and shear strain magnitudes change with direction. Every state of strain will exhibit specific orientations for the three principal strain axes and specific magnitudes for the principal stretches or elongations. The finite strain ellipsoid is a convenient way to depict the state of strain within a rock.

In the previous chapter, we noted that the positions of a particle or collection of particles define the *instantaneous condition* of a system. In a similar way, the state of strain, which is derived from the relative positions of particles, specifies the components of longitudinal and shear strain in all directions *at a particular instant in time.*

5A.7.2 Strain Paths and Dated Strain Paths

Deformation transforms the *initial* configuration of a collection of particles or of an object to a different *final* configuration. In the previous chapter, we noted that displacements, paths, and dated paths describe *changes in the state* of a system. In the same way, we endeavor to identify and describe how collections of particles or objects transform, that is, how changes in the state of the system take place. **Incremental strains** are the longitudinal and shear strains required to transform the state of strain at one point in time to the state of strain at a subsequent point of time. A particular finite strain ellipsoid will characterize the state of strain at the first point in time. If the first point in time is *prior to* any deformation and the collection of particles or the object is undeformed, the initial finite strain ellipsoid is a sphere. At subsequent points in time, different finite strain ellipsoids will characterize the state of strain.

We can summarize the incremental longitudinal and shear strains responsible for the change in the state of strain with an incremental strain ellipsoid. An incremental strain ellipsoid will transform the finite strain ellipsoid at one point in time to the finite strain ellipsoid at a later point in time (Figure 5.20). As was the situation with finite strains, there are three mutually perpendicular directions of principal incremental strains, one defining the direction of maximum incremental stretch or elongation, one defining the direction of minimum incremental stretch or elongation, and one defining a direction of intermediate incremental stretch or elongation. The three directions of principal incremental strain exhibit no incremental shear strains, whereas all other directions do exhibit incremental shear strains.

The concept of a **strain path** provides another way to examine the manner in which the finite strains accumulated. A strain path refers to the series of finite strain states experienced by a collection of particles as it transforms from its original, undeformed state to its final, deformed state. Each "step" along the path has a subtly but measurably different orientation and magnitude of the finite strain principal axes. It is possible and perhaps even likely that geologists can measure the state of strain in a rock without having firm evidence from which to infer the incremental strains or to define the strain path the collection of particles followed. This problem arises because any individual finite strain state may have been reached via any number of completely different intermediate states. It is imperative to determine the strain path if we wish to understand how structures or fabrics develop in deformed rocks. Further, if we can identify such intermediate states and date them, we have **dated strain path** or **strain history**.

5A.7.3 Coaxial Versus Non-Coaxial Strain Paths

Strain paths fall into one of two general categories (Figure 5.20). In **coaxial** strain paths (shown at the top of Figure 5.20), particles that define a line parallel to one of the principal directions at the

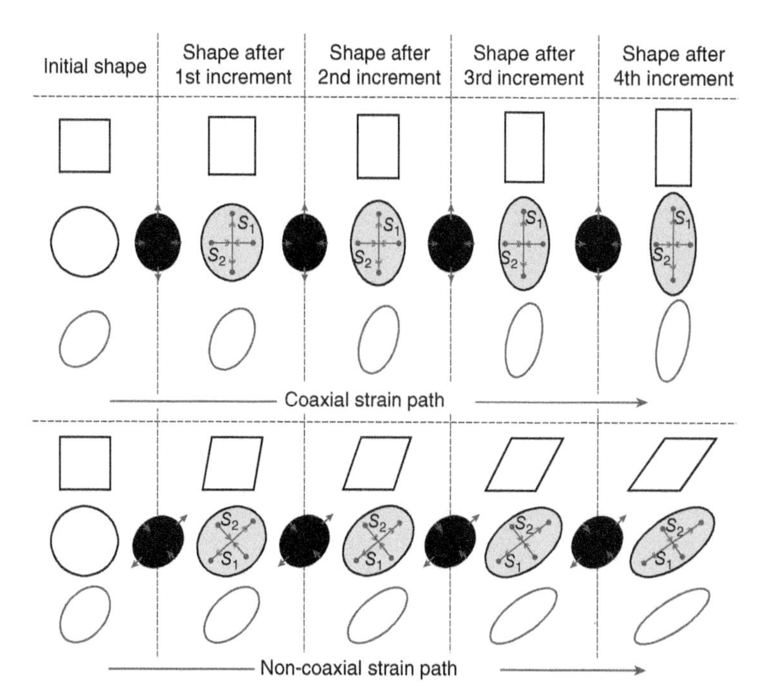

Figure 5.20 Comparison of coaxial strain paths (upper three rows) and non-coaxial strain paths (lower three rows). The column at left shows three different initial shapes (squares, circles, and ellipses) for each strain path. For rows with circular initial shapes, black ellipses give incremental strain ellipses, and gray ellipses show successive finite strain states. The upper three rows show pure shearing, a coaxial strain path. The lower three rows show simple shearing, a non-coaxial strain path.

beginning of deformation will end the deformation parallel to that principal direction. Particles on these lines may converge or diverge, meaning that the lines may shorten or lengthen, but these lines do not experience any shear strains. Stated in a different way, the directions of maximum, minimum, and intermediate shortening during later increments are collinear with the directions of maximum, minimum, and intermediate shortening, respectively, during early increments. In **non-coaxial** strain paths (shown at the base of Figure 5.20), particles that define a line parallel to one of the principal directions at the beginning of deformation need not, and, in general, do not, define a line parallel to that principal direction at the end of deformation. Stated in a different way, the direction of maximum shortening during later increments is generally not collinear with the direction of maximum shortening during early increments, the direction of maximum lengthening during later increments generally is not collinear with the direction of maximum lengthening during early increments, and the intermediate principal direction during later increments often is not collinear with the intermediate principal direction during early increments.

5A.7.3.1 Pure Shearing

In the previous chapter, we described pure shear as a two-dimensional displacement field in which particles converge along one direction and diverge along a second direction perpendicular to the first. **Pure shearing** refers to a strain path in which particle movements during each increment are pure shear displacements and where the directions of particle convergence and divergence are the same in each deformation increment. Pure shearing generates homogeneous, plane-strain finite strain states in which there is no volume change. There is neither shortening nor lengthening parallel to the intermediate principal direction. Moreover, the directions of maximum shortening and maximum lengthening in each increment are parallel to the directions of maximum shortening and maximum lengthening of the finite strain. Pure shearing is, then, a particular type of coaxial strain path.

In pure shearing, particles that define a line parallel to the direction of maximum lengthening during the initial deformation increment will define a line parallel to the direction of maximum lengthening during every increment and will define S_1 for the finite strain. Similarly, particles that define a line parallel to the direction of maximum shortening during the initial deformation increment will define a line parallel to the direction of maximum shortening during every increment and will define S_3 for the finite strain. The particles defining these two directions diverge or converge, but the collections of particles defining the two lines remain orthogonal throughout deformation.

5A.7.3.2 Simple Shearing

Simple shear is a two-dimensional displacement field in which the displacements of particles vary in a single direction; the resulting displacements are similar to those observed in shearing a deck of cards. **Simple shearing** refers to the strain path that results when particle movements during successive increments define simple shear displacement fields. Like pure shearing, simple shearing generates homogeneous, plane-strain finite strain states in which there is no volume change. There is neither shortening nor lengthening parallel to the intermediate principal direction. In simple shearing, all incremental relative displacements are parallel to a *shear direction*, which is the one direction that is both perpendicular to the intermediate principal direction and neither changes its orientation nor experiences any longitudinal strains. The intermediate principal direction and the shear direction together define the orientation of *shear planes*. All particles within any shear plane experience the same incremental displacements, but relative incremental displacement magnitudes vary with distance perpendicular to the shear planes. These relative incremental displacements lead to incremental strain principal directions that are inclined 45° to the shear plane, with the direction of maximum incremental shortening inclined in the shear direction and the direction of maximum lengthening inclined against the shear direction. Lines inclined to the shear plane reorient as a result of these incremental displacements so the particles aligned along either principal incremental strain directions during one increment are no longer aligned with principal incremental strain directions in subsequent deformation increments. The S_1 and S_3 principal directions of the finite strains are initially inclined at 45° to the shear plane, but they change in successive finite strain states. In each successive finite strain state, as deformation proceeds, the inclination of the direction of maximum lengthening (S_1) is more nearly parallel to the shear plane (Figure 5.20). In each successive finite strain state as deformation increments compound, the inclination of the direction of maximum shortening (S_3) is more nearly perpendicular to the shear plane.

In simple shearing, particles that compose a line parallel to either finite strain principal direction at one instant of time will compose a line oblique to the finite strain principal direction at subsequent increments of time. The observations that: (1) the finite strain principal directions reorient relative to a fixed external reference; and (2) different particles compose lines parallel to the finite strain principal directions at different times together confirm that simple shearing is a non-coaxial strain path.

5A.8 Instantaneous Strains and Strain Rates

Recall that incremental strains are the strains required to transform the state of strain at one time to the state of strain at a subsequent time. By envisioning shorter and shorter time intervals during which incremental strains accrue, geologists envision **instantaneous strains**, characterized by an instantaneous strain ellipsoid with three directions of principal instantaneous strains. Instantaneous strains capture the change of the strain state at a particular instant in time. We can approach the idea of the instantaneous change of finite strain in a different way, by calculating **strain rates**, which are changes of strain with time. One way to define strain rates is to calculate the change in longitudinal strains or shear strains per unit time. Alternatively, since the strain values are related to the magnitudes of the displacement gradients, we can assess the time rate of change of the displacement gradients ($\Delta u_X/\Delta X$, $\Delta u_X/\Delta Y$, $\Delta u_Y/\Delta X$, $\Delta u_Y/\Delta Y$, etc.). Note that

$$\frac{\Delta\left(\Delta u_i/\Delta X_i\right)}{\Delta t}=\frac{\Delta\left(\Delta u_i/\Delta t\right)}{\Delta X_i}=\frac{\Delta v_i}{\Delta X_i}. \qquad (5.14)$$

where u_i represents different components of particle displacements, X_i represents different coordinate directions, and v_i represents different components of particle velocities. Stated in words, Eq. (5.14) states that the change with time of the spatial gradient of the displacement field is equivalent to the spatial gradient of the velocity field. We use the symbol $\dot{\varepsilon}$ to denote longitudinal strain rate and the symbol $\dot{\gamma}$ to denote the shear strain rate. The change in strain in the numerator has no units, but the change in time in the denominator has units of time. So, strain rates have units of one over time. In geology, we typically give strain rates as s^{-1} or "per second." Most structural geologists infer that strain rates in natural deformations typically are 10^{-10} to 10^{-14} per second. In deformation experiments, strain rates may be 10^{-4} to 10^{-8} per second.

5A.9 Infinitesimal Strains

In some deformations, the relative displacements of particles remain very small for relatively long time periods. We use the idea of **infinitesimal**

strains to examine the deformation in these settings. In defining infinitesimal strains, geologists are able to disregard certain terms in the mathematical relationships relating displacement gradients to strain values because those terms have such small magnitudes. Infinitesimal strains are also commonly used by engineers to address problems of small deformation of bridges, dams, and other constructed structures.

5A.10 Summary

1) A strain is a change in the length of a line or a change in angle between two lines. The stretch (or elongation) is a measurement of the change in the length of a line, and the shear strain is a measurement of the change in the angle between two lines.
2) The distortion of a rock can be described by considering the change in the shape of a sphere (circle in 2D) to an ellipsoid (ellipse in 2D) with three mutually perpendicular axes that have fixed orientations in space.
3) Structural geologists distinguish between plane strain, constriction and flattening types of 3D strain. A sphere would change to the shape of a surfboard (plane strain), cigar (constriction) and pancake (flattening).
4) Pure shearing and simple shearing are two examples of idealized strain paths. In pure shearing, lines that are initially parallel to the finite strain axes remain parallel to them throughout deformation (coaxial deformation). In simple shearing, lines do not remain parallel to the finite strain axes during deformation (non-coaxial deformation).
5) Structural geologists distinguish between strain (the distortion of a rock), strain path (the progressive change in shape), and strain history (the timing of distortion).
6) The rate of distortion is called the strain rate, which is calculated by the change in strain divided by the time interval of measurement. Typical geological strain rates range from 10^{-10} per second (fast) to 10^{-14} per second (slow).

5A.11 Practical Methods for Measuring Strain

5A.11.1 Using Fabrics to Estimate Strain Ellipsoid Shape

The relative movements of particles responsible for pure strains in rocks also generate fabrics in rocks. For this reason, structural geologists often use fabrics, especially penetrative fabrics, to assess in a general sense the state of strain in the rocks. We infer that a locally uniform fabric indicates a region of homogeneous strain. Foliations regularly form perpendicular to the direction of maximum shortening, and lineations regularly form parallel to the direction of maximum lengthening. The presence of only a foliation in a deformed rock (where the deformed rock is an *S*-tectonite) suggests a flattening strain with the direction of maximum shortening, S_3, oriented perpendicular to the foliation surface. Similarly, the presence of only a lineation in a deformed rock (where the deformed rock is an *L*-tectonite) suggests a constrictional state of strain with the direction of maximum lengthening, S_1, parallel to the lineation. The presence of contemporaneous foliations and lineations in a deformed rock suggests the strain ellipsoid has three unequal principal axes. *SL*-tectonites are consistent with elongate pancake-shaped strain ellipsoids, whereas *LS*-tectonites are consistent with flattened cigar-shaped strain ellipsoids. One advantage of using fabrics to estimate the state of strain is that fabrics are inherently three-dimensional and, therefore, reflect the three-dimensional state of strain. In addition, the smooth variations in the orientations and intensity of fabrics provide an insight into spatial gradients in the orientation and magnitudes of principal strains. Using fabrics to assess the shape of the strain ellipsoid is not quantitative and should be corroborated by measurements whenever possible.

All quantitative determinations of strain are derived from strains measured in one or two dimensions, that is, from measured changes of lengths of lines in different directions, measured shear strains across different directions, or measurement of parameters related to the shapes of elliptical markers in a plane. Structural geologists prefer to make their measurements on sections that contain the strain principal directions whenever possible. If only a single section is used, a section containing the directions of maximum shortening and maximum elongation is preferred. Strain measurements made on two sections parallel to two of the strain principal planes can be manipulated to yield an estimate of the relative magnitudes of the three principal strains. Strain measurements made on any three or more nonparallel sections through a rock will constrain the three-dimensional shape of the strain ellipsoid. We do not describe three-dimensional strain measurement techniques here. Owens (1973, 1974, 1984), Milton (1980), and Wheeler (1984) (and references cited in those papers) describe techniques for measuring strain in three dimensions.

5A.11.2 Types of Methods for Measuring Strain in Two Dimensions

A quantitative description of the strain ellipse specifies the orientations and relative magnitudes of the strain principal directions in a plane. There are three main families of methods for quantifying the orientation and shape of the strain ellipse in a section through a deformed rock. One family of methods uses the shapes of deformed markers within a plane to ascertain the shape and orientation of the strain ellipse in that plane. These methods are conceptually straightforward, based on the inference that shapes of deformed markers relate directly to the strain ellipsoid (see Section 5A.6.1.3). Many deformed rocks lack deformed markers. So, structural geologists developed methods of a second family that utilize measurements of longitudinal or shear strain magnitudes in three or more directions within a plane to determine the orientations and relative magnitudes of the strain principal directions on that plane. The conceptual foundation of these methods is the idea that any change of shape generates distinctive longitudinal or shear strains in different directions across a plane. A third family of methods uses the displacements of particles at different positions within a plane to determine directly values of displacement gradients; from the displacement gradients, one can calculate the orientations and relative magnitudes of the strain principal directions.

In this section, we introduce three different methods from the first family, that is, three techniques that use deformed markers to fix the orientations and relative magnitudes of the strain principal directions. Methods from the second family typically use a geometrical construction called a Mohr circle or another graphical technique to determine the orientations and relative magnitudes of the strain principal directions from the measured longitudinal or shear strain values. Defining the Mohr circle for strain and explaining how to use it are beyond the scope of this chapter. Moreover, by converting those individual strain values into displacement gradients, one can use methods of the third family. Methods from the third family also work with displacement gradient measurements derived directly (direct determination of displacement gradients is increasingly common due to expanding use of global positioning system (GPS) data). Methods from the third family require concepts outlined in the Comprehensive Treatment section that follows. So, we address how to use displacement gradient values to determine the orientation and relative magnitudes of the strain principal values in the following Comprehensive Treatment section of this chapter.

5A.11.3 Measuring Strain in Two Dimensions Using Deformed Markers

A variety of geological elements are potential strain markers, including: (1) reduction spots in sedimentary rocks; (2) ooids in sedimentary rocks; (3) detrital mineral grains (calcite, quartz, or feldspar detrital grains are commonly used); (4) detrital lithic fragments, pebbles, or cobbles; (5) accretionary lapilli, varioles, or amygdules (filled vesicles) in volcanic rocks; (6) mineral grains, grain aggregates, or xenoliths in igneous rocks and gneisses; or (7) cross sections of burrows or pipes. In Section 5A.6.1.3, we noted that in order to record the strain accurately, particles composing the markers must experience displacements and displacement gradients precisely equivalent to the displacements and displacement gradients experienced by particles elsewhere in the rock. With potential markers like reduction spots (Figure 5.14b), where the main difference between the marker and the host rock is the color of the rock, this condition is relatively readily met. In practice, we find that markers composed of different mineral or rock types regularly exhibit systematic differences in their deformed shapes, indicating that this requirement is not met. We might infer that this requirement is met in cases such as detrital quartz grains in quartz-cemented sandstones or calcite ooids in calcite-cemented limestones (cf. Figure 5.14a), but even in those cases, quartz grains with their crystallographic axes oriented in different directions or ooids composed of calcite with different mean grain sizes exhibit systematically different shapes, again indicating that the requirement is violated to some degree. The errors that result from assuming marker deformation replicates exactly the deformation of the rock are reduced by using large numbers of markers. The accuracy of the measured strain increases dramatically as the number of markers measured rises to about 50; the accuracy of strain measurements typically increases more slowly as the number of markers measured rises above 150. Most methods for measuring strain using deformed markers yield acceptable results using measurements of between 50 and 150 markers.

We begin by examining a planar section that contains either the direction of maximum lengthening, S_1, or the trace of a foliation, which defines the direction of greatest lengthening within the section plane. In Figure 5.21a, for example, markers have essentially elliptical shapes in the plane of the section. We characterize each elliptical marker by its longest diameter, $2a$, its shortest diameter, $2b$, and the angle, ϕ, between the marker's longest diameter and the direction of maximum lengthening in the plane. If the rock lacks foliation and lineation, we characterize the orientation of the long dimensions of markers relative to an arbitrary reference direction, ϕ' (e.g. Figure 5.21b). Each marker's axial ratio in the deformed state, its final axial ratio, is $R_f = 2a/2b = a/b$. The following text outlines three techniques for determining the axial ratio of the strain ellipse, R_S, and the orientation of its long axis using deformed markers.

5A.11.3.1 Ellipticity, *R*

The simplest technique to estimate the two-dimensional state of strain from a collection of deformed markers is to calculate the average ellipticity, R, of the deformed markers. The technique is predicated on the assumption that a collection of circular markers was subjected to a homogeneous strain; the final shape of each marker, therefore, estimates the shape of the strain ellipse directly. A plot of the lengths of individual markers' major diameters (on the ordinate) versus the lengths of their minor diameters (on the abscissa) defines a straight line whose slope R defines the average axial ratio of the markers (Figure 5.22a). We take the best-fit slope of that line as the ratio of the strain ellipse. Here $R = R_S = 2.84$. The orientation of the long axis of each deformed marker estimates the

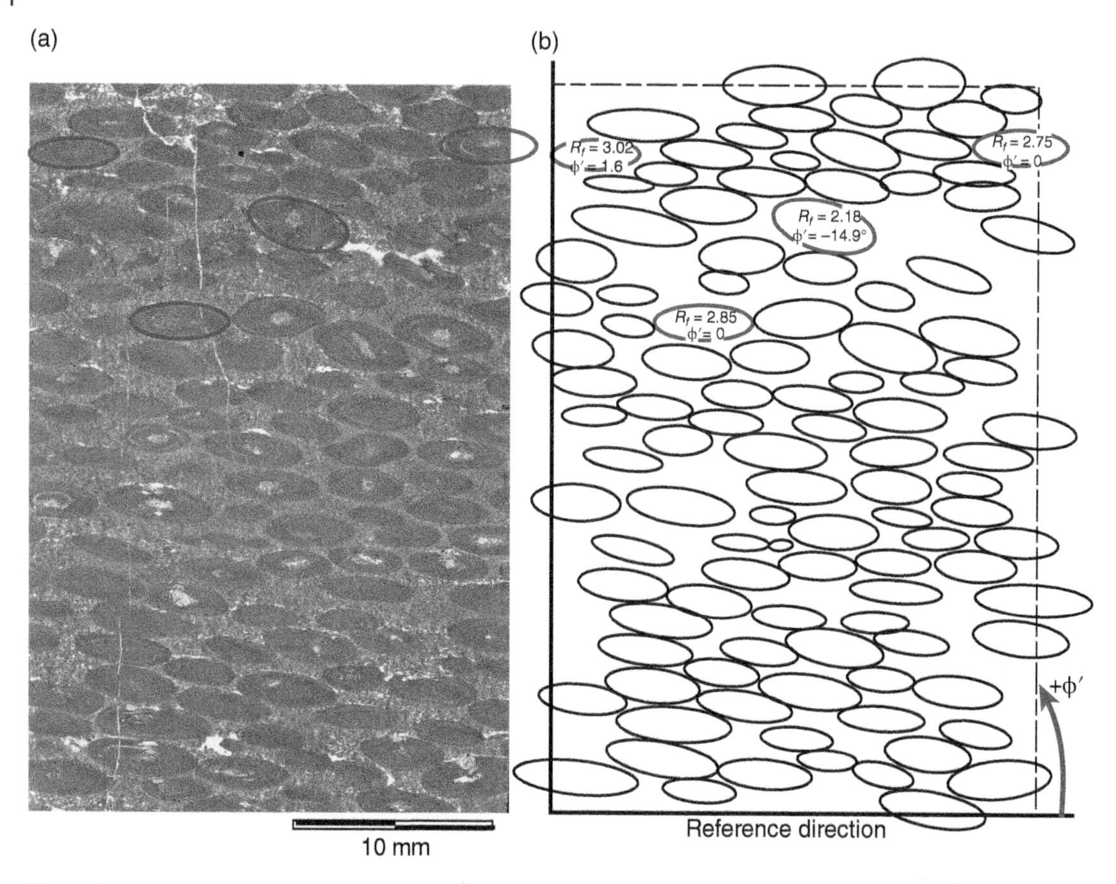

Figure 5.21 (a) Photomicrograph of a deformed oolitic limestone from the central Appalachians (USA). In three dimensions, ooids have elongate pancake shapes; this section plane contains the markers' longest and shortest axes. Red ellipses drawn around four ooids closely approximate the shapes of the deformed ooids. (b) We characterize the elliptical markers by their axial ratio (R_f) and the orientation of their long axis given by an angle (φ') measured relative to an arbitrary reference direction.

orientation of the long axis of the strain ellipse. Thus, the mean of the measured angles φ′ defines the long axis of the strain ellipse. The mean of the angle φ between the direction of maximum lengthening and the long axes of deformed markers should be zero. Measured relative to an arbitrary reference direction, the mean angle to the long axis of the deformed marker shape is −5°. So, the estimated orientation for the long axis of the strain ellipse is 5° clockwise from the reference direction of the image (Figure 5.22b).

The advantages of this technique are that it can be completed rapidly, and the calculations required are easily accomplished using any spreadsheet program. The main disadvantage is that deformed markers rarely have circular initial shapes. Thus, the central premise behind the method is rarely met.

5A.11.3.2 The R_f/φ Method

Many geological elements used as strain markers do have initial shapes that closely approximate

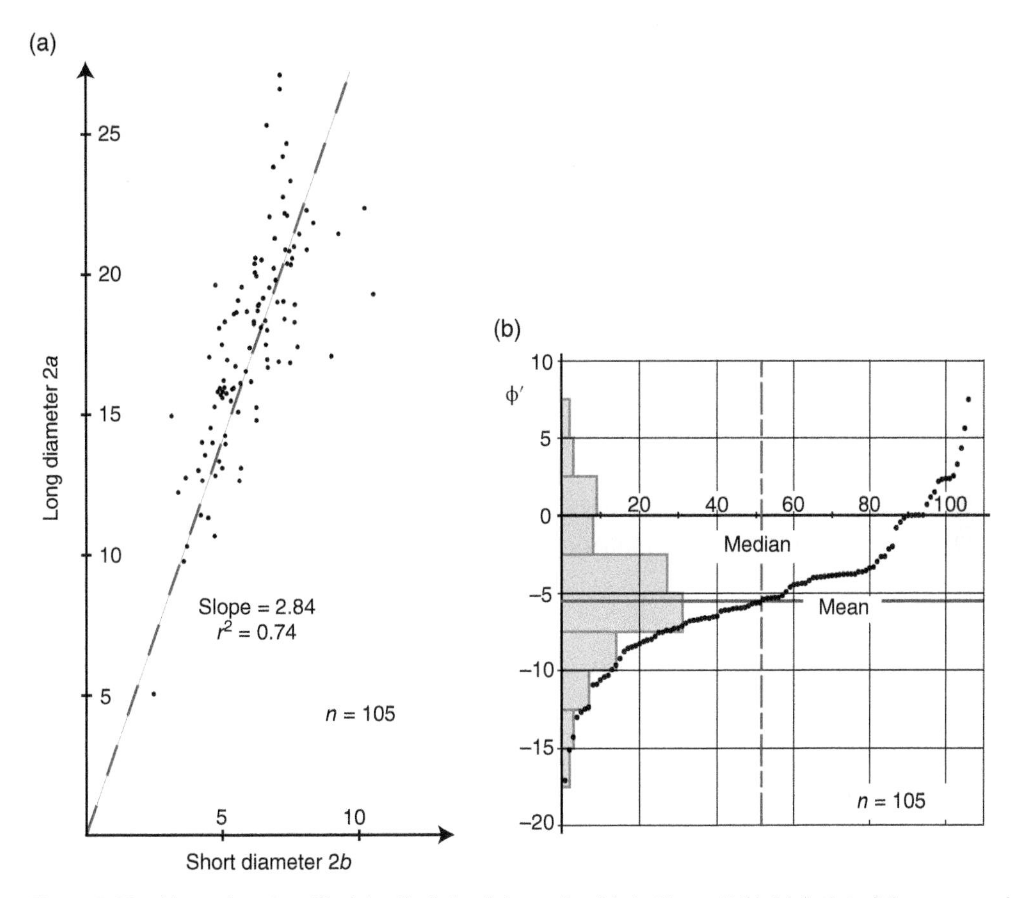

Figure 5.22 Measuring the ellipticity, R, of the deformed ooids in Figure 5.21. (a) A plot of the measured longest diameter of ellipses ($2a$) versus the shortest diameter ($2b$). The best-fit slope = 2.84 estimates the axial ratio of the strain ellipse. (b) Points give the cumulative frequency of measured angles between the longest diameter of the ellipses and an arbitrary reference direction; the bar graph gives the number of ellipses with measured angles within 2.5° bins. The mean angle is ~ −5°.

ellipsoids and, therefore, have elliptical initial shapes in any section. Homogeneous deformation of an initially elliptical marker yields an elliptical deformed marker whose final axial ratio, R_f, and orientation relative to the direction of maximum elongation, ϕ, are a product of the marker's initial axial ratio, R_i, the marker's initial orientation relative to the direction of maximum elongation, θ, and the axial ratio of the strain ellipse, R_S (see Figures 5.15 and 5.16). The R_f/ϕ method is predicated on the idea that statistical analysis of the characteristics of a collection of deformed elliptical markers will define the axial ratio of the strain ellipse, R_S and the orientation of the S_1 principal direction (see Ramsay, 1967; Dunnet, 1969, Elliott, 1970; Lisle, 1985). The analysis is undertaken by examining the characteristics of the deformed markers plotted on orthogonal coordinates whose abscissa is the orientation angle ϕ or ϕ' and whose ordinate is the logarithm of marker's axial ratio.

One can readily understand how to "read" R_f versus ϕ (or ϕ') plots by examining the characteristics of plots generated by deforming collections of elliptical markers. The deformation of an initially circular marker (with $R_i = 1$) generates a deformed marker with $R_f = R_S$ and with $\phi = 0$ (Figure 5.23). A collection of randomly oriented markers with initial axial ratios $R_i = \text{constant} \leq R_S$ subjected to the same deformation defines an onion-shaped curve that is centered on the point $R_f = R_S$ and $\phi = 0$ and is symmetric about the $\phi = 0$ line (e.g. the solid red curve in Figure 5.22 for differently oriented ellipses with $R_i = 2$). The curve intersects the $\phi = 0$ line at two points, $R_{fmax} = R_S \times R_i$ and $R_{fmin} = R_S / R_i$. The maximum value of $\phi = \phi_{max}$, called the *fluctuation*. The fluctuation is defined by

$$\sin 2\phi_{max} = \frac{\left[(R_i - 1)/R_i\right]}{\left[(R_S - 1)/R_S\right]}. \tag{5.15}$$

The value of R_f at ϕ_{max} is

$$R_f(\phi_{max}) = \frac{\left[(R_S - 1)/R_S\right]}{\left[(R_i - 1)/R_i\right]} + \frac{\left[(R_S - 1)/R_S\right]}{\left\{\left[(R_i - 1)/R_i\right]^2 - 1\right\}^{1/2}}. \tag{5.16}$$

One can deduce R_S and R_i from R_{fmax}, R_{fmin}, $R_f(\phi_{max})$, or any combination of these values.

A collection of randomly oriented markers with initial axial ratios $R_i = \text{constant} > R_S$ subjected to the same deformation defines a bell-shaped curve that is symmetric about the $\phi = 0$ line (e.g. the dashed red curve in Figure 5.23 for differently oriented ellipses with $R_i = 3$). The curve intersects the $\phi = 0$ line at one point, $R_{fmax} = R_S \times R_i$. $R_{fmin} = R_S / R_i$ occurs when $\phi = \phi_{max} = 90°$. A collection of markers with different axial ratios but the same initial orientation relative to the long axis of the strain ellipse, where $\theta = \text{constant}$,

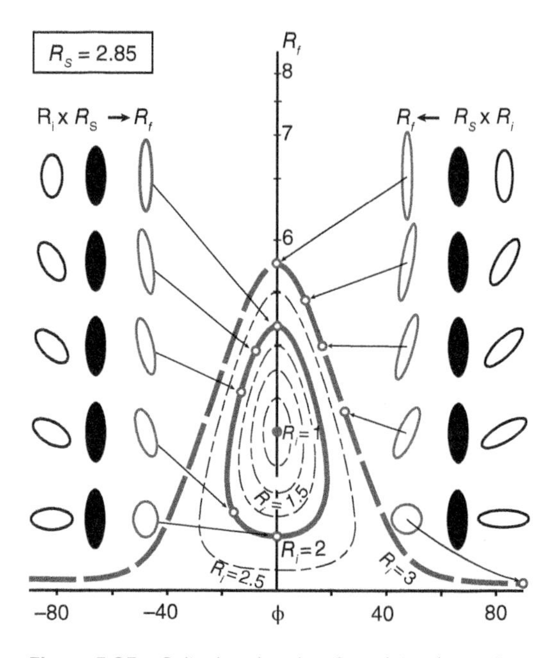

Figure 5.23 R_f/ϕ plot showing R_f and ϕ values of ellipses with identical initial axial ratios but different orientations relative to relative to S_1 for R_S = 2.85. The solid red line gives R_f and ϕ values for ellipses with R_i = 2; dashed black lines give values for other ellipses with $R_i < R_S$. The dashed red line gives R_f and ϕ values for ellipses with Ri_f = 3.

defines half of a W-shaped "θ curve" that intersects the $\phi = 0$ line at R_S and is symmetric about that line (Figures 5.24 and 5.25). The higher values of ϕ at the tails of the θ curve represent deformed markers with the large initial axial ratios, and the maximum value of ϕ is approached asymptotically as R_i increases.

Most collections of deformed markers consist of markers with a range of initial axial ratios and a range of orientations relative to the long axis of the strain ellipse. These deformed markers will define sets of nested onion-shaped curves for markers with $R_i \leq R_S$ and sets of nested bell-shaped curves for markers with $R_i > R_S$.

The data collected for the R_f/ϕ method are identical to those needed to determine the ellipticity,

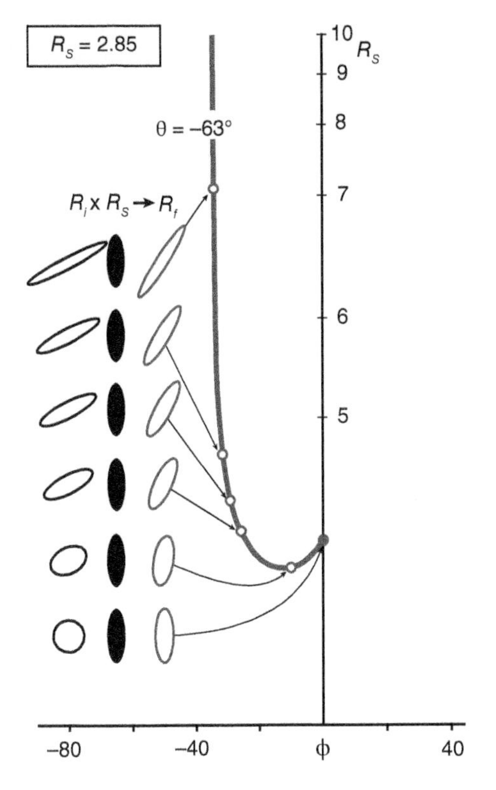

Figure 5.24 R_f/ϕ plot showing R_f and ϕ values of ellipses with identical initial orientations ($\theta = -63°$) relative to S_1 but different axial ratios.

that is, $R_f = 2a/2b = a/b$ and ϕ (or ϕ') for each marker. We plot the measured R_f and ϕ (or ϕ') values for individual markers on orthogonal coordinates whose abscissa is the orientation angle ϕ (or ϕ') and whose ordinate is the logarithm of marker's axial ratio (Figure 5.26). The coordinates (R_C, ϕ_C) of the centroid of the plotted points define the axial ratio $R_S = R_C$ and $\phi' = \phi_C$ for the collection. Visual location of the centroid will work for estimates of the strain, but it is preferable to use the *harmonic mean* and *vector mean* of the cluster of points to define the coordinates of the centroid. The harmonic mean H of a cluster of N points, which defines $R_C = R_S$, is

$$H = \frac{N}{\left[(R_{f1})^{-1} + (R_{f2})^{-1} + (R_{f3})^{-1} + \ldots + (R_{fN})^{-1}\right]}. \tag{5.17}$$

For the data in Figure 5.25, $H = R_C = R_S = 2.83$. The vector mean Φ of that cluster of N points, which defines $\phi' = \phi_C$, is

$$\Phi = \tfrac{1}{2}\arctan\left[\frac{(\Sigma \sin 2\varphi)}{(\Sigma \cos 2\varphi)}\right]. \tag{5.18}$$

For the data in Figure 5.26, $\Phi = \phi' = \phi_C = -5.05°$. If the undeformed markers were randomly oriented, the data plot symmetrically about the centroid and horizontal and vertical lines drawn through the centroid divide the collection into four areas (A, B, C, and D) with roughly equal numbers of points. A quantitative measure of the symmetry of the distribution is the symmetry index

$$I_{\text{Sym}} = 1 - \frac{\left[(n_A - n_B) + (n_A - n_B)\right]}{N}. \tag{5.19}$$

Values near unity indicate a symmetric distribution of data. For a symmetric distribution, the harmonic mean gives R_S and the vector mean gives the orientation of the long axis of the strain ellipse. For the data in Figure 5.26, $I_{\text{Sym}} = 0.90$, and the distribution is symmetric. If the I_{Sym} value is below a critical value that depends upon R_S (see Lisle 1985, p. 15), the markers possessed an initial preferred orientation. By successively excluding markers with high R_i values, one can reduce the collection to an approximately symmetric distribution. Such a symmetric distribution should yield approximately equal numbers of markers on different θ curves. Richard Lisle's *Geological Strain Analysis, A Manual for the R_f/ϕ Technique* contains a series of θ curve plots for different values of R_S that one can use to evaluate test the distribution of the data.

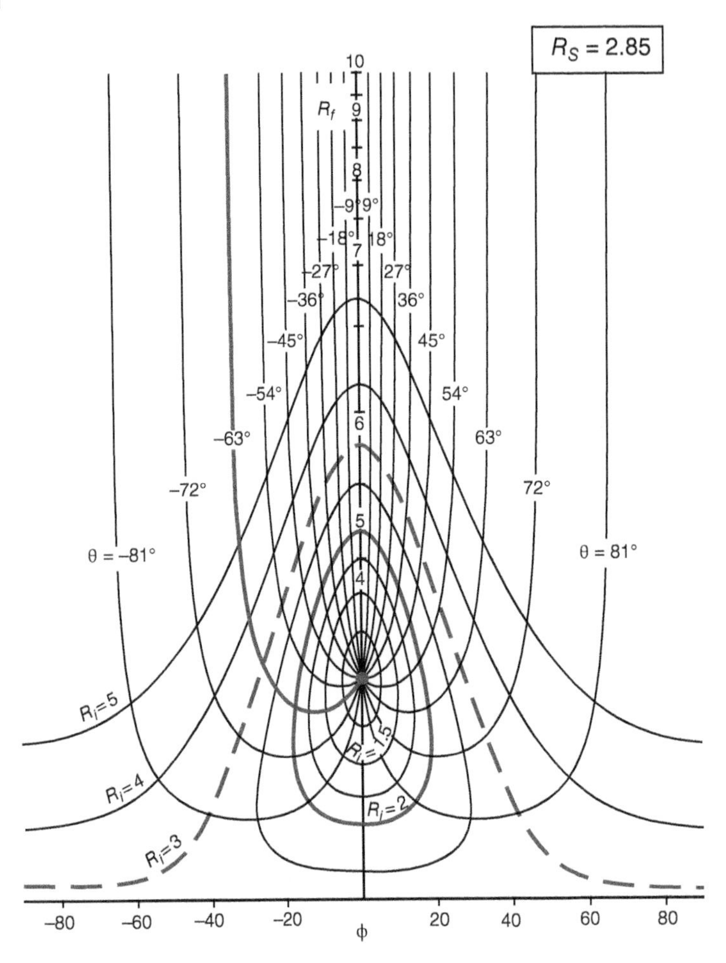

Figure 5.25 R_f/ϕ template for $R_S = 2.85$, with curves showing R_f and ϕ values of ellipses with identical R_i values and for ellipses with identical θ values.

The R_f/ϕ method is widely used by structural geologists because it has a solid theoretical foundation. The technique is most useful when data are evaluated using Lisle's θ curve plots for different values of R_S because this enables one to evaluate the results for internal consistency (using the symmetry test and testing the distribution of data along θ curves for different values of R_S). Such a full analysis requires effort, and results improve with experience. There are a number of software packages available that streamline the analysis and offer the possibility of contouring R_f/ϕ plots (e.g. Vollmer 2017), meaning that this method remains one of the most popular ways to measure strain available.

New methods have been developed for strain analysis that use R_f/ϕ data to explore the strain of rock with initial fabrics in more detail (e.g. Yamaji 2005). A major advance is a recognition that hyperbolic geometry is the appropriate space to analyze ellipse axial ratios and their orientations (Yamaji 2008, 2013). The best-fit ellipse is given by the hyperbolic vector mean within this space, and hyperbolic distances measure differences between ellipses. The unit hyperboloid, and contours of data on it, are readily visualized on two-dimensional projections, for which the equal-area projection is most suitable, and error can be quantified (Yamaji 2008, 2013; Vollmer 2018).

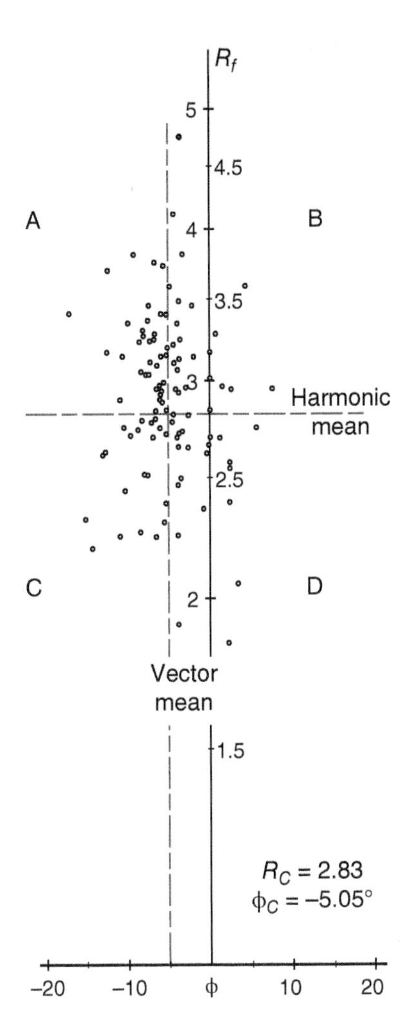

Figure 5.26 R_f/ϕ plot for the deformed ooids in Figure 5.21. Harmonic mean and vector mean calculated using Eqs. (5.17) and (5.18). A, B, C, and D denote quadrants about centroid used in calculating I_{Sym}.

5A.11.3.3 Center-to-Center Methods

A third way to analyze the deformed markers derives strain data from the distances between the centers of markers in different directions within a plane. An initially random distribution of point markers within a plane will be isotropic, that is, distances between grain centers will not vary with direction. Distances between the centers of deformed, elliptical markers are longer in the direction of the maximum elongation S_1 and shorter in the direction of maximum shortening S_3. Plotting the positions of the centers of all grains relative to each marker in the collection yields a plot called a *Fry plot* after its originator Norman Fry, that illustrates trends in the distances between the centers (Fry 1979). To generate a standard Fry plot, one positions the center of a marker over a fixed origin and plots the positions of the centers of all other markers in a collection. Repeating this procedure for each marker in the collection generates a cloud of points that records the relative positions of the centers of the markers (Figure 5.27). In nearly all cases, the cloud of points possesses a central area that lacks points and a dense concentration of points that surrounds the central vacancy. The orientation and axial ratio of the ellipse fit to the dense

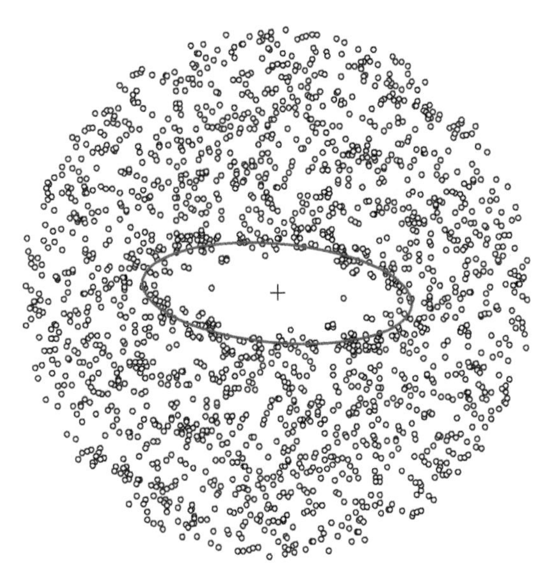

Figure 5.27 Fry plot for deformed ooids in Figure 5.21. An ellipse fit "by eye" through the dense concentrations of points around the open area at the center of the plot has an axial ratio of 2.84 and its long diameter is approximately 5° clockwise from the reference direction.

concentration of points are inferred to define the strain ellipse.

One variation on the Fry method is to plot only the positions of the centers of the nearest neighbors for each marker. This yields an annular cloud of points about the origin that defines the shape and orientation of the strain ellipse. This approach is not appropriate for clustered distributions. A second variation on the Fry method is to normalize the center-to-center distances by the average ellipse diameter to yield *normalized* center-to-center distances. A normalized Fry plot typically has a more distinct central vacancy and stronger concentration of points ringing the central vacancy. The disadvantage of this approach is that it requires additional measurements and, therefore, takes more effort or computation to derive results. There are numerous computer programs available that will construct Fry and normalized Fry plots using the (x, y) positions of centers of the deformed markers. So, these methods are used widely. Objective methods of fitting the Fry plot data and quantifying error are also available through software (e.g. Vollmer 2018).

5B Strain: Comprehensive Treatment

The intent of the Comprehensive Treatment section in this and other chapters is to use mathematics to provide a deeper understanding of the topics addressed in the Conceptual Foundation section. A deeper understanding of the topics addressed in Section 5.A.1 through Section 5.A.3 derives in large measure from the use of vector and tensor mathematics to relate the changes of shape apparent in longitudinal and shear strains to the character of the displacement field responsible for the deformation. For this reason, the Comprehensive Treatment section of this chapter skips over Sections 5.B.1, 5.B.2, and 5.B.3 and begins with Section 5.B.4 Relating Strain to Displacements.

5B.4 Relating Strain to Displacements

In Chapter 4 and in the Conceptual Foundation section of this chapter, we stated that the pure strain component of the total displacement field exists only if the gradients of the displacement field are nonzero. If all particles experience the same displacement, a rock mass is translated but not distorted. Of course, nonzero displacement gradients do not necessarily mean that the pure strain component exists – rigid body rotation also requires nonzero displacement gradients. We can use the example of the elongated belemnite guard to begin exploring the connection between gradients of a displacement field and longitudinal strains. Each frame of Figure 5.28 shows the undistorted and distorted configurations of the belemnite guard from Figure 5.2 superposed over Cartesian coordinate frames. In each frame, the long dimension of the belemnite is aligned parallel to the x axis of the coordinate frame. Orienting both the undeformed and deformed belemnite guards parallel to the x axis is equivalent to envisioning that the deformation is "one-dimensional," or visible along a single linear direction. For this to be true in a real deformation, we would need to show that the belemnite guard did not change its orientation during deformation. We will show later that this need not be true, and it very likely was not true in this case. Nevertheless, treating this deformed belemnite guard as a "one-dimensional" strain problem enables us to address more readily the relationships between strain and displacement gradients. Relative to this coordinate frame, we can determine position vectors $\boldsymbol{X}_i$ for the original or undeformed position of the particles at the center of each segment of the belemnite guard. Similarly, we can determine a position vector $\boldsymbol{x}_i$ for the current or deformed position of each particle. Because we take the x axis to be parallel to the length of the undeformed *and* deformed belemnite guard, the original position

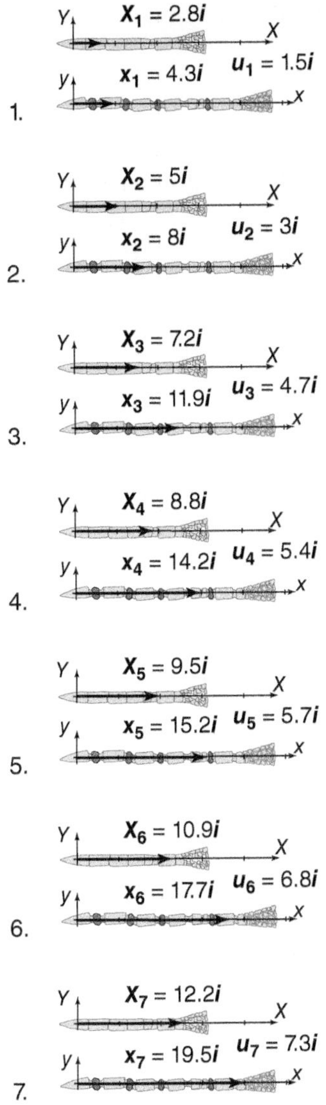

1. $X_1 = 2.8i$; $x_1 = 4.3i$; $u_1 = 1.5i$

2. $X_2 = 5i$; $x_2 = 8i$; $u_2 = 3i$

3. $X_3 = 7.2i$; $x_3 = 11.9i$; $u_3 = 4.7i$

4. $X_4 = 8.8i$; $x_4 = 14.2i$; $u_4 = 5.4i$

5. $X_5 = 9.5i$; $x_5 = 15.2i$; $u_5 = 5.7i$

6. $X_6 = 10.9i$; $x_6 = 17.7i$; $u_6 = 6.8i$

7. $X_7 = 12.2i$; $x_7 = 19.5i$; $u_7 = 7.3i$

Figure 5.28 Panels showing the original position vectors ($\mathbf{X}_i$), final position vectors ($\mathbf{x}_i$), and displacement vectors ($\mathbf{u}_i$) of the different segments of the elongated belemnite from Figure 5.2.

vectors, final position vectors, and displacement vectors of all particles are scalar multiples of the unit vector $\boldsymbol{i}$ parallel to the x axis. This simplifies the mathematics in the discussion that follows, for we can examine the relative magnitudes of vectors simply by comparing the size of the scalar multiples of $\boldsymbol{i}$. From Chapter 4, we know that the displacement of the "ith" segment of the belemnite guard is

$$\boldsymbol{u}(i) = \boldsymbol{x}_i - \boldsymbol{X}_i.$$

Because $\boldsymbol{u}(i) = u_X(i) \cdot \boldsymbol{i}$, $\boldsymbol{x}_i = x_i \cdot \boldsymbol{i}$, and $\boldsymbol{X}_i = X_i \cdot \boldsymbol{i}$, we can divide this equation by $\boldsymbol{i}$ to get

$$u_X(i) = x_i - X_i,$$

which is a scalar equation. In the discussion that follows, then, we need only to determine the length of each vector to specify it completely.

5B.4.1 Longitudinal Strains and Displacement Gradients

We place the origin of our coordinate axis at the center of the segment with the conical tip, which means that we consider that this entire segment is not displaced by the deformation. We also assume that there is no rotation of the segments. By inspecting the drawings of the deformed and restored belemnite guard, you can see that this assumption is not strictly true. For our purposes, however, the errors that arise by envisioning that there is no rotation of segments are small. In Figure 5.28, the numbered image pairs in each frame show the inferred original and current positions of the particles at the centers of seven segments of the belemnite guard. Each segment is numbered, beginning with the segment adjacent to the 0th segment (the one with the conical tip and at the origin of the coordinate axis) and proceeding to the right. Note that the magnitude of the displacement of the second segment is greater than the magnitude of the displacement of the 1st segment, the magnitude of the displacement of the third segment is greater than the magnitude of the displacement of the second segment, etc.

Segments farther from the origin have larger magnitude displacements than segments closer to the origin. The data in Table 5.2 confirm this observation.

The increase in displacement magnitude with distance from the origin arises because the displacement of each successive segment is the sum of the displacement of that segment relative to its neighbor to the left *plus* the displacements of the segments farther to the left relative to the origin. The red line in Figure 5.29 is a plot of the magnitude of the displacement of each segment u_i versus the magnitude of its original position vector X_i. This depicts the displacements of particles as a function of their original position for this one-dimensional example. The slope of this curve, $\Delta u_X/\Delta X$, is the change in displacement with position. This is the only finite gradient of this one-dimensional displacement field, where the total displacement $\boldsymbol{u}$ is a function of X only. There is some variation in the slope from one point on this curve to another, which is given by $\Delta u_X(i)/\Delta X_i$, but these values approximate the average slope $= [u_X(7)-u_X(0)]/(X_7-X_0) = 7.3\text{ cm}/12.2\text{ cm} = 0.6$. Because the displacement vector magnitude is essentially a linear function of the x component of the original position vector, we can determine the magnitude of the displacement of any

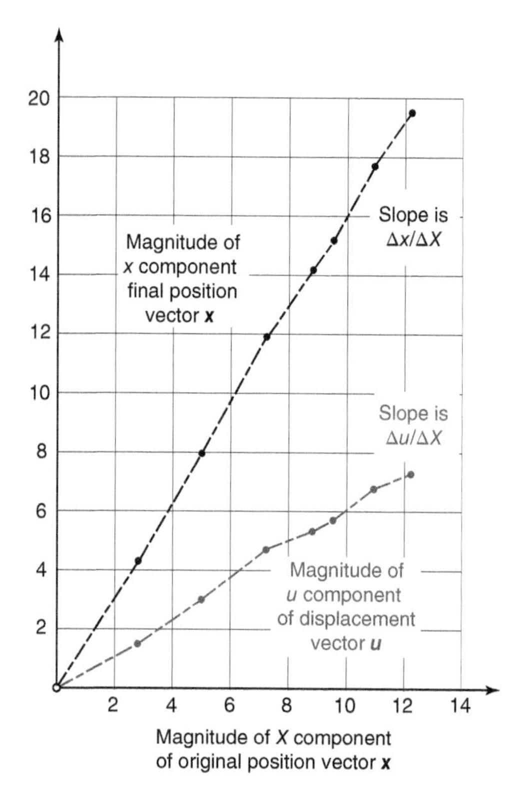

Figure 5.29 Plots of (1) the magnitude of the final position vectors of belemnite sections versus the magnitude of their original position vectors (in black), and (2) the magnitude of the displacement vectors of belemnite sections versus the magnitude of their original position vectors (in red).

Table 5.2 Undeformed and deformed positions of centers of segments of belemnite guard and displacements of segment centers (in cm).

Segment	X_i	x_i	$u_X(i)$	$\Delta u(i) = u(i)-u(i-1)$	$\Delta X_i = X_i - X_{(i-1)}$	$\Delta u(i)/\Delta X_i$	$\Delta x_i = x_i - x_{(i-1)}$	$\Delta x_i/\Delta X_i$	$x_i/X_i \approx k$
0	0	0	**0**	*NA*	*NA*	*NA*	*NA*	*NA*	*NA*
1	2.8	4.3	1.5	1.5	2.8	0.5	4.3	1.5	1.5
2	5.0	8.0	3.0	1.5	2.2	0.7	3.7	1.7	1.6
3	7.2	11.9	4.7	1.7	2.2	0.8	3.9	1.8	1.7
4	8.8	14.2	5.4	0.7	1.6	0.4	2.3	1.4	1.6
5	9.5	15.2	5.7	0.3	0.7	0.4	1.0	1.4	1.6
6	10.9	17.7	6.8	1.1	1.4	0.8	2.5	1.8	1.6
7	12.2	19.5	7.3	0.5	1.3	0.4	2.8	1.4	1.6

particle a if we know the x component of its initial position vector:

$$u_X(a) = \left(\frac{\Delta u_X}{\Delta X}\right) \cdot X_a. \tag{5.20}$$

The displacement of a particle is the vector difference between its current position and its original position. In the one-dimensional example considered here, x_i is the distance from the origin to a particle in the deformed state, that is, the length of a line in the deformed state. Similarly, X_i is the distance from the origin to that particle in the undeformed state, that is, the initial length of that same line. Thus,

$$\frac{\Delta u_X}{\Delta X} = \frac{(l' - l_o)}{l_o} = e. \tag{5.21}$$

For this one-dimensional example, the ratio of the differences between the scalar multiples of the displacement and position vectors, $\Delta u_X(i)/\Delta X_i$, gives the value of the ratio of the differences between the displacement and position vectors, $\Delta \mathbf{u}(i)/\Delta \mathbf{X}_i$. What is relatively easy to see in this one-dimensional case will hold true in later situations: the displacement gradient values correlate directly to elongation values determined for a deformation.

5B.4.2 Longitudinal Strains and Position Gradients

Also listed in Table 1 are the magnitudes of the final position vectors for the particles at the center of the segments, x_i. A plot of x_i versus X_i (the black curve on Figure 5.29) is also essentially a straight line; its slope $\Delta x_i/\Delta X_i (= \Delta \mathbf{x}_i/\Delta \mathbf{X}_i)$ is a **position gradient**. In this one-dimensional example, the ratio $\Delta x_i/\Delta X_i$ corresponds to the final length of a line from the center of the (i−1)th segment to the center of the ith segment divided by the original length of the line from one center to the other. Thus,

$$\frac{\Delta x}{\Delta X} = \frac{l'}{l_o} = T. \tag{5.22}$$

Each value of $\Delta x_i/\Delta X_i$ is a "local" measure of the stretch T. Both Figure 5.29 and Table 1 indicate that $\Delta x_i/\Delta X_i$ (= $\Delta \mathbf{x}_i/\Delta \mathbf{X}_i$) is nearly constant and that the successive values approximate the average slope = $(x_7 - x_0)/(X_7 - X_0)$ = 19.5 cm/12.2 cm = 1.6. Because the final position vector magnitude is a linear function of the x component of the original position vector, we can determine the final position of any particle a if we know the x component of its initial position vector:

$$x_a = \left(\frac{\Delta x}{\Delta X}\right) \cdot X_a. \tag{5.23}$$

5B.4.3 Relating Displacement Gradients and Position Gradients

In Section 5A2.1, we showed that $e = T - 1$. In our one-dimensional example, the displacement gradient, $\Delta u_X/\Delta X$, is the change in the length of a line divided by its original length. This is, of course, the elongation e (see Section 5A2.1). Similarly, in one-dimensional situations, the position gradient, $\Delta x/\Delta X$, is the final length of a line divided by its original length; this is the stretch T (see Section 5A2.1). Combining these observations, we have

$$\frac{\Delta u_X}{\Delta X} = \frac{\Delta x}{\Delta X} - 1. \tag{5.24}$$

You can confirm that this expression holds by examining the values in Table 1. When we move on to consider treating two- and three-dimensional deformations, we will define relationships similar to Eq. (5.24) for longitudinal strains in different directions through a deformation. These relationships underscore the strong connection between the displacement gradients and the position gradients in deformation.

5B.4.4 Longitudinal Strain in Continuous Deformation

In the geological examples described in Sections 5A2 and 5B.4.1–5B.4.3, the lengthening or shortening was accommodated by breaking a distinctive feature into several parts that moved relative to their neighbors. We elected to begin our discussion of strain by focusing on examples of *discontinuous* deformation – where there are breaks or gaps in the feature – because it is easier to determine the original length of a line. We simply needed to sum the lengths of the several segments (or conceptually remove the secondary material that interposes between segments) to assess the original length of the line, which we could then compare to the final length. In many geologic settings, the deformation is *continuous* – lines are elongated or shortened without breaking the feature into segments. In such situations, it is easier to envision gradual changes in displacement as a function of position, but is more difficult to assess the original length.

In some deformed, fine-grained, Paleozoic sedimentary rocks, we find graptolites preserved on bedding planes. Graptolites typically consist of a series of branches, called *stipes*, that emanate from an initial individual organism. In many species, the stipes look like long, thin saw blades, and the serrations, called *thecae*, exhibit highly regular spacings. It is the regularity of spacings that enables geologists to estimate the original length of a stipe for comparison with its length in the deformed rock. Figure 5.30a is a cartoon of an idealized, undeformed graptolite; the total length of the stipe = l_t, is an integer multiple of the species-specific spacing between serrations, l_s, which is the average length of a theca. In Figure 5.30b, the deformed equivalent of this graptolite has an overall length of = l_t', and its thecal spacing is l_s'. Figure 5.30c shows the original and final positions of particles at the centers of the thecae and the displacements experienced by those particles. Table 5.3 presents of measurements taken from Figure 5.30c. In Table 5.3, X_i are the lengths of the initial position vectors to the centers of successive thecae; x_i are the lengths of the final position vectors to the centers of the same

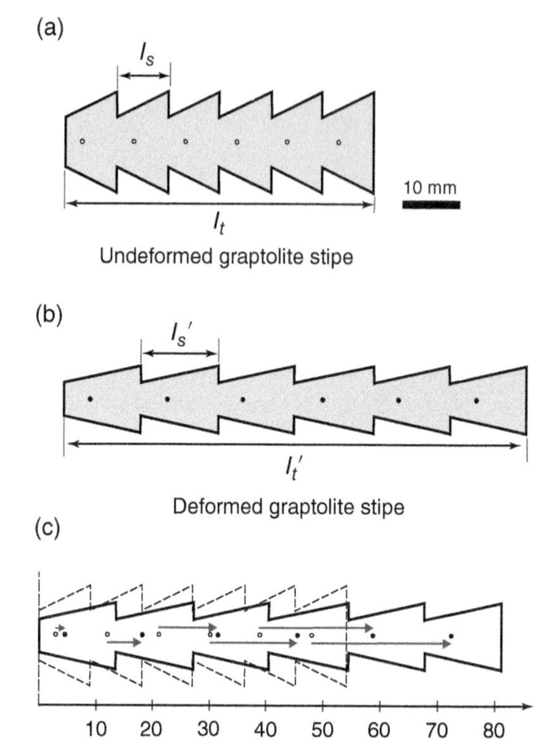

Figure 5.30 (a) An undeformed graptolite stipe with spacing l_s between serrations. (b) A deformed graptolite stipe with spacing l_s' between serrations. (c) Comparison of undeformed and deformed positions of centers of thecae along the graptolite stipe. Red arrows give the displacement of the centers of the thecae.

thecae. u_i are the lengths of the displacement vectors for the centers of the successive thecae. In this idealized case of a continuously elongated line, a plot of x_i versus X_i (the black curve on Figure 5.31) is a straight line. The slope of this line $\Delta x/\Delta X = 1.5 = l'/l_o$. In this continuously deformed example, we can calculate the slope of the line by taking the derivative of the function $x(X)$ and replace $\Delta x/\Delta X$ by dx/dX.

In this continuously elongated line, we see again an increase in displacement magnitude with distance from the origin. The increase again arises because the displacement of each

Table 5.3 Undeformed and deformed positions of centers of segments of graptolite thecae and displacements of thecal centers (in mm).

Theca	X_i	x_i	$u(i)$	$\Delta u(i) = u(i) - u(i-1)$	$\Delta X_i = X_i - X_{(i-1)}$	$\Delta u(i)/\Delta X_i$	$\Delta x_i = x_i - x_{(i-1)}$	$\Delta x_i/\Delta X_i$	$x_i/X_i \approx k$
1	3.0	4.6	1.6	1.6	3.0	0.5	4.6	1.5	1.5
2	12.0	18.2	6.2	4.6	9.0	0.5	13.6	1.5	1.5
3	21.0	31.4	10.4	4.2	9.0	0.5	13.2	1.5	1.5
4	30.5	45.5	15.0	4.6	9.5	0.5	14.1	1.5	1.5
5	39.0	58.6	19.6	4.6	8.5	0.5	13.1	1.5	1.5
6	48.0	72.5	24.5	4.9	9.0	0.5	13.9	1.5	1.5

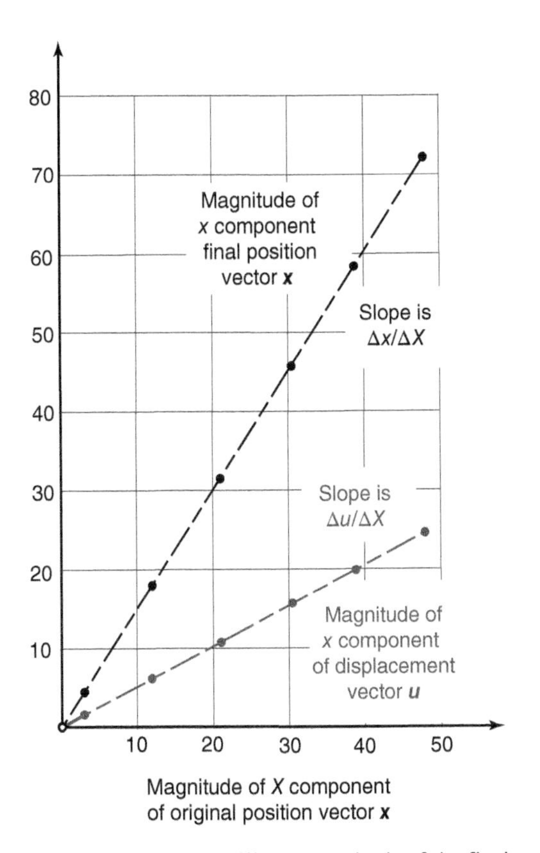

Figure 5.31 Plots of (1) the magnitude of the final position vectors of graptolite thecae versus the magnitude of their original position vectors (in black), and (2) the magnitude of the displacement vectors of graptolite thecae versus the magnitude of their original position vectors (in red).

successive particle along the length of the theca is the sum of its displacement relative to the particle to its left *plus* the displacements of the particles in thecae closer to the origin. The red line on Figure 5.31 is a plot of the magnitude of the displacement of a theca segment u_i versus the magnitude of its original position vector X_i. The slope of this line is $\Delta u_X/\Delta X = 0.5 = (l' - l_o)/l_o$. Here too, we can calculate the slope of this line by taking the derivative of the function $u_X(X)$, that is, replace $\Delta u_X/\Delta X$ by du_X/dX. As was true for the discontinuously deformed belemnite guard,

$$\frac{du_X}{dX} = \frac{\Delta u_X}{\Delta X} = \frac{\Delta x}{\Delta X} - 1 = \frac{dx}{dX} - 1. \tag{5.25}$$

5B.4.5 Consequences of Longitudinal Strains

Our discussion so far has considered longitudinal strains in isolation. We have ignored the context in which longitudinal strains occur, and so we have not examined the consequences of changes in lengths of lines. The shortening or lengthening of a line implies the existence of a gradient in the displacement field in the direction of that line. Shortening indicates convergence in that linear direction, and lengthening indicates divergence in that direction. If deformation is characterized *only* by longitudinal strains in a single direction, the movement of particles *must conform* to a uniaxial convergence displacement field or a uniaxial divergence displacement field. In two dimensions, uniaxial divergence leads to an increase in the area in the plane of section, and uniaxial convergence leads to a decrease in the area in the plane of section (Figure 5.4a). In three dimensions, uniaxial divergence in one direction only requires an increase in the volume occupied by a collection of particles (or a decrease in the density of the material), and uniaxial convergence in a single direction requires a decrease in the volume occupied by a collection of particles (or an increase in the density of the material) (Figure 5.4b). In uniaxial divergence, modest longitudinal strains lead to volume increases far beyond what can be supported by decreases in rock density and/or volume addition due to secondary mineralization. Similarly, in uniaxial convergence, modest longitudinal strains lead to volume decreases that exceed what can, in most situations, be supported by increases in rock density and/or volume removal due to localized solution transfer.

The resolution to this apparent conundrum is not surprising. In most natural deformations, the shortening of lines due to convergence along one direction is compensated by concurrent lengthening of lines due to divergence in one or more directions at high angles to the shortening direction. Or the lengthening of lines due to divergence along one direction is compensated by concurrent shortening of lines due to convergence in one or more directions at high angles to the elongation direction. There are innumerable examples of deformation in which we see concurrent shortening and lengthening in different directions. For example, in the deformed quartzite from the central Appalachians in Maryland, USA (Figure 5.6), detrital grains exhibit a well-defined shape preferred orientation and rutile needles within these grains show a consistent pattern of shortening and lengthening. Figure 5.6 presents plots of T values calculated for individual rutile needles versus the angle ϕ between the trend of the needle and the long dimension of grain-shape fabric. At low angles from the long dimension of the grains, rutile needles have $T > 1$ indicating that they were lengthened. At high angles from the long dimension of the grains, rutile needles have $T < 1$ indicating that they were shortened. Further, this effect is visible in two mutually perpendicular sections through the rock, showing that there are systematic, three-dimensional patterns of longitudinal strain in these deformed rocks.

5B.4.6 Displacement Gradients and Longitudinal Strains in Different Directions

The penetrative deformation that transformed sedimentary strata in North Wales (United Kingdom) to slates (Figure 5.32a), for example, was accompanied by a reduction in volume. It is difficult to document this volume reduction without having a more complete understanding of strain in two and three dimensions. We can, however, ignore the effects of volume change for the time being and use the deformation of sedimentary rocks in slates to begin to gain the insight needed to recognize that slate belt deformation occurred with a concurrent volume loss. The folded and cleaved strata exposed at the surface in Wales are products of extensive shortening of subhorizontal lines and extensive lengthening of subvertical lines. The cartoon images in Figure 5.32b show three stages in an idealization of the development of the slate belt. The filled circles in the undeformed state (at left) experience the same deformation as the external shape of the cartoon image. The horizontal shaded layer in the cartoon cross section and the horizontal diameters of the circles are shortened (for them, $T < 1$), and the vertical dimension of the cartoon cross section and the vertical diameters of the circles are lengthened (for them, $T > 1$). As a result of the deformation, the circles are transformed into

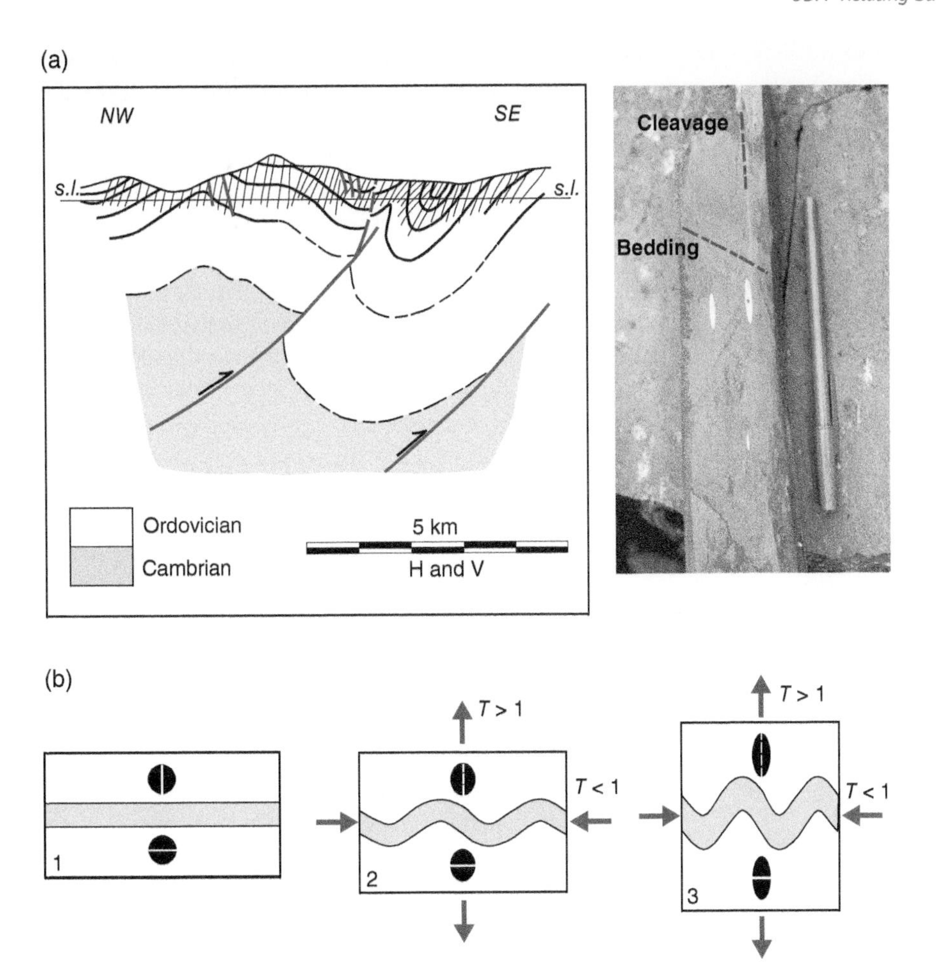

Figure 5.32 (a) At left, a cross section of the Dolwyddelan syncline of the North Wales slate belt. Source: Modified from Campbell et al. (1985). At right, a photograph of slaty cleavage, bedding, and deformed reduction spots in fine-grained Ordovician sedimentary rocks from North Wales. (b) Cartoons depicting the original configuration (1) and two successive stages (2 and 3) in the development of a fold like the Dolwyddelan syncline.

ellipses, mimicking the ellipsoidal reduction spots one sees in samples of slates from Wales (see photograph in Figure 5.32a and in Figure 5.14b). Figure 5.33 compares the positions of selected particles in the undeformed cartoon cross section (whose external shape is indicated by dashed lines) and in the final deformed cartoon cross section (whose external shape is indicated by solid lines). For the purposes of comparison, the particle at the lower left corner of the undeformed and deformed cross sections is assumed to occupy the same position in the undeformed and deformed states. That particle is the origin of a set of Cartesian coordinate axes with a horizontal x axis and vertical y axis. Relative to that origin, we can determine initial and final position vectors for any particle. For example, $\boldsymbol{X}_\mathbf{c} = X_c\boldsymbol{i}+Y_c\boldsymbol{j}$ is the position vector to the particle at the center of the undeformed cross section. $\boldsymbol{x}_\mathbf{c} = x_c\boldsymbol{i}+y_c\boldsymbol{j}$ is the position vector to the corresponding particle in

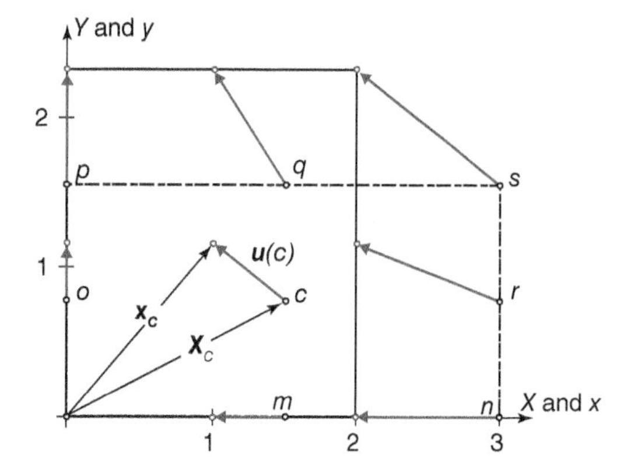

Point	$\mathbf{X}_i$	$\mathbf{x}_i$	$u(i)$
m	$1.5i$	$1i$	$-0.5i$
n	$3i$	$2i$	$-1i$
o	$0.78j$	$1.17j$	$0.39j$
p	$1.56j$	$2.34j$	$0.78j$
c	$1.5i + 0.78j$	$1i + 1.17j$	$-0.5i + 0.39j$
q	$1.5i + 1.56j$	$1i + 2.34j$	$-0.5i + 0.78j$
r	$3i + 0.78j$	$2i + 1.17j$	$-1i + 0.39j$
s	$3i + 1.56j$	$2i + 2.34j$	$-1i + 0.78j$

Figure 5.33 Comparison of the positions of selected particles in stage 1 of Figure 5.32b (shown with open black circles and dashed lines) with their positions in stage 3 of Figure 5.32b (shown with open red circles and solid lines). The red arrows are the displacement vectors of selected particles. The chart gives the components of the original position vectors ($\mathbf{X}_i$), the final position vectors ($\mathbf{x}_i$), and the displacement vectors ($\mathbf{u}_i$).

the deformed cross section. The displacement of that particle $\boldsymbol{u}(c) = u_X(c)\boldsymbol{i} + u_Y(c)\boldsymbol{j}$ is, of course, the vector difference between the final and original position vectors for the particle. The chart below the drawing defines the original position vectors, final position vectors, and displacement vectors for several other particles. For clarity, only the displacement vectors of these particles are shown in the drawing.

Some interesting features are apparent in this image. First, particles that originally fall on the x axis remain on the x axis after deformation. Their displacement vectors are necessarily parallel to the x axis because they have a scalar multiple of 0 for the unit vector $\boldsymbol{j}$. Particles that originally fall on the y axis remain on the y axis after deformation. Their displacement vectors are parallel to the y axis because they have a scalar multiple of 0 for the unit vector $\boldsymbol{i}$. Particles converge along the x axis and diverge along the y axis. Particles off the axis have displacements relative to the origin that are oblique to the coordinate axes. Still, comparing the displacement vectors of all particles (those on the axes and off the axes), you can see that all particles whose original positions fall on a vertical line (have equal X component values, e.g. m, c, and q)) have displacement vectors with equal u_X components (although the u_Y components of the displacement vectors differ). This indicates that the displacement component u_X is a function of X only. Thus, $du_X/dY = 0$. Similarly, all particles whose original positions fall on a horizontal line (have equal Y component values, e.g. p, q, and s) have displacement vectors with equal u_Y components (although the u_X components of the displacement vectors differ). This indicates that the displacement component u_Y is a function of Y only, and $du_Y/dX = 0$. A plot of u_X versus X is a straight line, with a slope $du_X/dX = -0.33$. A plot of u_Y versus Y is a straight line with a slope $du_Y/dY = 0.5$. Combining this information yields

$$u_X = \left(\frac{du_X}{dX}\right)\cdot X + \left(\frac{du_X}{dY}\right)\cdot Y = -0.33X \quad (5.26a)$$

$$u_Y = \left(\frac{du_Y}{dX}\right)\cdot X + \left(\frac{du_Y}{dY}\right)\cdot Y = 0.5Y. \quad (5.26b)$$

5B.4.7 Position Gradients and Longitudinal Strains in Different Directions

Particles whose original position vectors have equal X component values have final position vectors with equal x component values. So, the x component of the final position vector is a function of X only. Thus, $dx/dY = 0$. Particles whose original position vectors have equal Y component values have final position vectors with equal y component values. This indicates that the y component of the final position vector is a function of Y only, and $dy/dX = 0$. Plots of final position components x versus initial position component X and final position components y versus initial position component Y indicate that $dx/dX = 0.67$ and $dy/dY = 1.5$. Combining these observations, we have

$$x = \left(\frac{dx}{dX}\right) \cdot X + \left(\frac{dy}{dY}\right) \cdot Y = 0.67X \quad (5.27a)$$

$$y = \left(\frac{dy}{dX}\right) \cdot X + \left(\frac{dy}{dY}\right) \cdot Y = 1.5Y. \quad (5.27b)$$

5B.4.8 Relating Displacement Gradients and Position Gradients in Two Dimensions

Comparing the original distance between the origin and a particle on the x axis to the distance from the origin to the final position of that particle, we can determine the longitudinal strain along the x axis, either as a stretch T or an elongation e. For particles that begin and end the deformation on the axis, this two-dimensional deformation leads to particle movements that are indistinguishable from those experienced by particles in the one-dimensional cases we examined earlier. Thus, it is easy to see that

$$du_X / dX = e_X, \quad (5.28a)$$

$$dx / dX = T_X, \text{and} \quad (5.28b)$$

$$\frac{du_X}{dX} = \frac{dx}{dX} - 1 \quad \text{or} \quad (5.28c)$$

$$e_X = T_X - 1 \quad (5.28d)$$

for these particles. We noted earlier that particles whose original position vectors have equal X values (and therefore lie along a line perpendicular to the x axis) have displacement vectors with equal u_X components and have final position vectors with equal x components. Thus, for any particle off the x axis, there is a corresponding particle on the x axis that has a displacement vector with the same u_X component and a final position vector with the same X component. Because Eqs. (5.28) describes the variation of u_X and x with X for particles on the x axis, they necessarily hold for all particles in the plane.

Particles that begin and end the deformation on the y axis also experience movements indistinguishable from what occurs in a one-dimensional deformation. Comparing the original distance between the origin and a particle on the y axis to the distance from the origin to the final position of that particle, we can determine the longitudinal strain along the y axis, either as a stretch T_Y or an elongation e_Y. By analogy to particles on the x axis and to the one-dimensional cases examined earlier, we have

$$\frac{du_Y}{dY} = e_Y, \quad (5.29a)$$

$$\frac{dy}{dY} = T_Y, \text{and} \quad (5.29b)$$

$$\frac{du_Y}{dY} = \frac{dy}{dY} - 1 \quad \text{or} \quad (5.29c)$$

$$e_Y = T_Y - 1 \quad (5.29d)$$

for these particles. By again relating any particle whose original position is off the y axis to a corresponding particle on the y axis, we find that the displacement gradients and position gradients calculated along any line parallel to the y axis are equal to those calculated along the y axis. Thus, Eqs. (5.29) holds for all particles in the plane.

5B.4.9 Area Ratios in Two-Dimensional Deformation

The area in the plane of section of the undeformed portion of the cartoon cross section of the slate belt is the product of its length l_X and height l_Y, or $A = l_X \cdot l_Y$. The area in the plane of section of the deformed portion of the cartoon cross section is the product of its length l_x' and height l_y', or $A' = l_x' \cdot l_y'$. In the same way that we used a ratio of deformed lengths to original lengths to compare the state of lines before and after deformation, we use the ratio of the area occupied by a collection of particles after deformation to the area occupied by the same particles before deformation to evaluate how deformation affects an area:

$$\frac{A'}{A} = \frac{\left(l_x' \cdot l_y'\right)}{\left(l_X \cdot l_Y\right)}. \tag{5.30a}$$

If the area occupied by a particular collection of particles is unchanged by the deformation, the area ratio has a value of one. A reduction in area due to deformation leads to an area ratio less than one, whereas an increase in area due to deformation leads to an area ratio greater than one. Rearranging Eq. (5.30a) slightly, we have

$$\frac{A'}{A} = \frac{\left(l_x' \cdot l_y'\right)}{\left(l_X \cdot l_Y\right)} = \left(\frac{l_x'}{l_X}\right) \cdot \left(\frac{l_y'}{l_Y}\right) = T_x \cdot T_y \tag{5.30b}$$

and

$$\frac{A'}{A} = T_x \cdot T_y = \left(\frac{\mathrm{d}x}{\mathrm{d}X}\right) \cdot \left(\frac{\mathrm{d}y}{\mathrm{d}Y}\right) = \frac{\left(\mathrm{d}x \cdot \mathrm{d}y\right)}{\left(\mathrm{d}X \cdot \mathrm{d}Y\right)}. \tag{5.30c}$$

5B.4.10 Discontinuous Deformation in Two Dimensions

In discussing one-dimensional deformation, we observed that the shortening or lengthening of lines can occur continuously or discontinuously. Natural deformations with simultaneous shortening in one direction and lengthening in another often are accommodated by discontinuous deformation processes, such as the formation of stylolitic fractures, extensional fractures (veins), or faults. Whether accommodated by continuous deformation processes or discontinuous processes, we can still evaluate the magnitudes of the displacement gradients or position gradients and from those values, estimate the longitudinal strains in different directions (Figure 5.34).

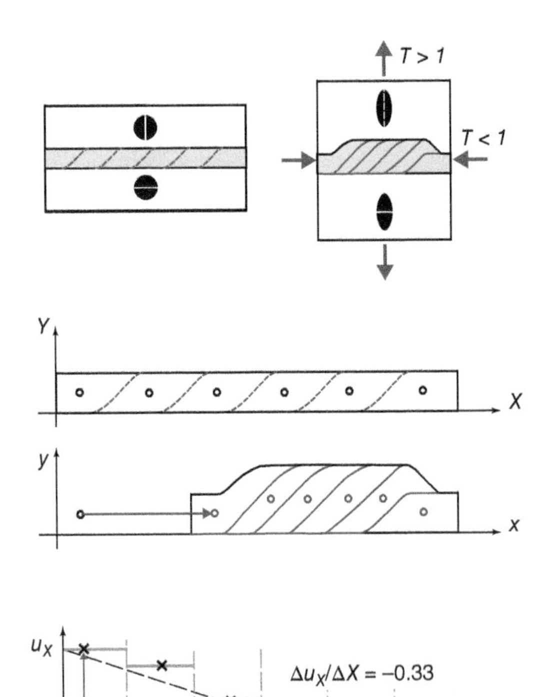

Figure 5.34 The displacement gradients associated with a change in shape equivalent to the transformation of stage 1 to stage 3 in Figure 5.32b can also be accommodated by faults or other discontinuous deformation elements.

5B.4.11 Displacement Gradients and Shear Strains

Longitudinal strains are just one of two indicators of pure strain. Shear strains, the change in the angle between two lines or families of lines, are also a primary indicator of pure strain. In any two-dimensional pure strain, the two principal directions exhibit no finite angular shear or shear strain. Lines oblique to those two directions exhibit shear strains, and the shear strain magnitude increases smoothly from zero for either principal direction to a maximum value for lines presently at 45° to the principal directions. Further, the sense of shear is opposite on either side of any principal direction. As is the case for longitudinal strains, the properties of shear strains relate directly to the magnitudes of the displacement gradients.

Figure 5.35a is an enlargement of the line drawing of the aplite dike displaced by a shear zone shown in Figures 4.9 and 4.31. The dashed black lines in Figure 5.35b give the inferred original positions of the material lines marking the margins of the dike, and the red arrows show the displacements of selected particles within this shear zone. Note that the displacement vectors for these particles are all parallel to the x axis because the u_Y components of the displacements of all particles are zero. The magnitudes of the displacements of particles increase from zero along the x axis, where $Y = 0$, to a maximum when $Y \approx 25$ cm. Displacement magnitudes do not vary parallel to the x axis. $du_X/dX = 0$. Taking together these observations, the displacement component u_X is a function of Y only and the displacement component u_Y is everywhere zero. A plot of u_X versus Y (Figure 5.35c) is essentially a straight line with a slope $\Delta u_X/\Delta Y = du_X/dY = 0.5$. Combining this information yields

$$u_X = \left(\frac{du_X}{dX}\right) \cdot X + \left(\frac{du_X}{dY}\right) \cdot Y$$
$$= 0X + 0.5Y = 0.5Y \qquad (5.31a)$$

$$u_Y = \left(\frac{du_Y}{dX}\right) \cdot X + \left(\frac{du_Y}{dY}\right) \cdot Y$$
$$= 0X + 0Y = 0. \qquad (5.31b)$$

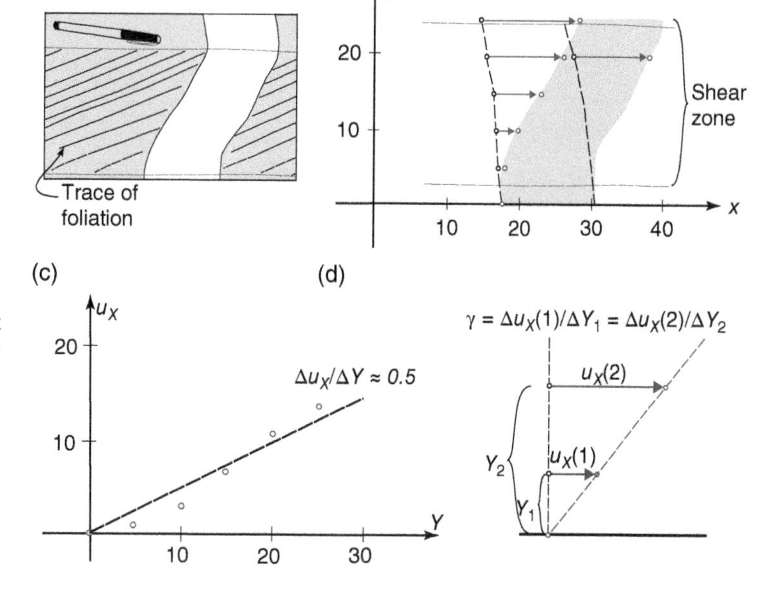

Figure 5.35 (a) Sketch of aplite dike offset by shear zone. Dashed red lines mark edges of the shear zone and black lines show traces of foliation defined by flattened mineral grains. (b) Offset dike in gray, and the inferred initial edges of the dike (dashed lines). Red arrows depict displacement vectors connecting inferred initial positions (open black circles) and observed final positions (open red circles) of particles. (c) The plot of u_X component versus Y component for particles from (b). (d) The shear strain γ determined by plotting measured Δu_X versus ΔY.

How do displacement gradients relate the shear strain γ? The horizontal black line in Figure 5.35d represents a line in a deforming material. The dashed black line is perpendicular to this line, and Y_i measures distances normal to the line. The dashed red line gives the present, deformed position of the dashed black line originally perpendicular to the solid black line. The particle at the intersection of the horizontal and dashed black lines is not displaced during deformation, but particles at increasing distances from the horizontal black line exhibit increasing displacements. The ratio of the difference between the displacement of the intersection point (= 0) and that at a particular distance from the solid black line (such as $u_X(1)$ at Y_1 or $u_X(2)$ at Y_2) is exactly equal to the displacement gradient du_X/dY. Thus, if γ_X is the shear strain parallel to the x axis, then

$$\gamma_X = \frac{du_X}{dY}. \tag{5.32a}$$

We could follow the same line of reasoning to show that the shear strain parallel to the y axis γ_Y is

$$\gamma_Y = \frac{du_Y}{dX}. \tag{5.32b}$$

5B.4.12 Shear Strains and Position Gradients

As we did in previous cases, we can determine position gradients relevant to Figure 5.35. Rewriting Eqs. (5.28c) and (5.29c) in Sections 5.B4.8 and 5.B4.9 yields

$$\frac{dx}{dX} = \frac{du_X}{dX} + 1 \tag{5.28c}$$

$$\frac{dy}{dY} = \frac{du_Y}{dY} + 1. \tag{5.29c}$$

In Figure 5.35, $du_X/dX = du_Y/dY = 0$, so $dx/dX = du_Y/dY = 1$.

The x coordinate of the deformed state position vector is $x = X + u_X$, where u_X is a function of X and Y. For this reason, the change of u_X with Y must capture all of the change of x with Y:

$$\frac{dx}{dY} = \frac{du_X}{dY}. \tag{5.33a}$$

As noted above, in Figure 5.35, u_X is a function of Y only and du_X/dY has a finite value. The y coordinate of the deformed state position vector is $y = Y + u_Y$, where u_Y can be a function of X and Y. Any change of u_Y with X must capture all of the change of x with Y:

$$\frac{dy}{dX} = \frac{du_Y}{dX}. \tag{5.33b}$$

In this instance, $u_Y = 0$ for all points and $dy/dX = 0$. The y coordinate of the deformed state position vector is independent of the magnitude of X.

Collecting these observations,

$$x = \left(\frac{dx}{dX}\right)\cdot X + \left(\frac{dy}{dY}\right)\cdot Y = X + 0.5Y \tag{5.34a}$$

$$y = \left(\frac{dy}{dX}\right)\cdot X + \left(\frac{dy}{dY}\right)\cdot Y = Y \tag{5.34b}$$

5B.4.13 Applying Matrix Algebra to Two-dimensional Deformation

5B.4.13.1 Position Gradient Matrices

In the two-dimensional strain examples illustrated in Figures 5.33–5.35, the coordinates of a particle's deformed state position vector, x and y, are linear functions of X and Y, the coordinates of that particle's position vector in the undeformed state. The components of particle displacement vectors, u_X and u_Y, are also linear functions of X and Y, the coordinates of that particle's position vector in the undeformed state. In some examples, one or more of the position gradients or displacement gradients vanished, that is, were equal to zero, yielding a relatively restricted deformation example that highlighted the relationship between the position gradients or displacement gradients and measurable aspects of the pure strain. Eqs. (5.35) below define a less restricted example where the coordinates of a particle's deformed state position vector, x and y, are linear functions of both X and Y, the coordinates of that particle's position vector in the undeformed state:

$$x = A \cdot X + B \cdot Y \tag{5.35a}$$

$$y = C \cdot X + D \cdot Y. \tag{5.35b}$$

Even when all four coefficients of the linear Eqs. (5.35), A, B, C, and D, are nonzero, the linear equations represent a homogeneous deformation that transforms a square into a general parallelogram (Figure 5.36).

Matrix notation (see Appendix 4-II) is a convenient way to represent a system of linear equations like those in Eqs. (5.35a) and (5.35b). Using matrix notation, the components of the final position vector, x and y, are elements of one 2×1 column matrix, and the components of the original position vector, X and Y, are elements of a second 2×1 column matrix. The coefficients A, B, C, and D are elements of a 2×2 square matrix. A particle's final position column vector is the product of

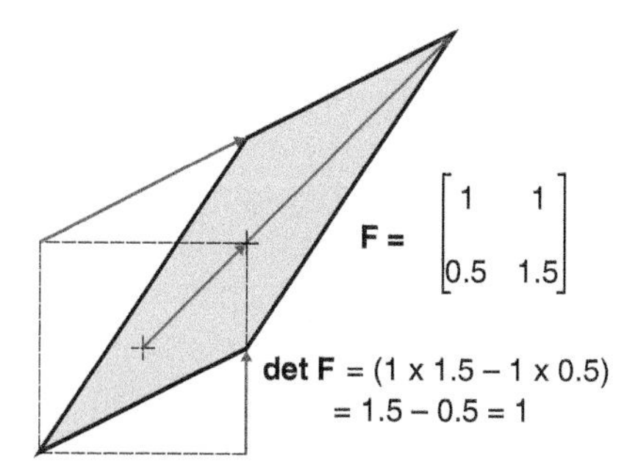

Figure 5.36 Original (in dashed lines) and final (in solid lines) configuration of a figure deformed using the position gradient matrix **F**.

the square matrix and the particle's original position column vector:

$$\begin{vmatrix} x \\ y \end{vmatrix} = \begin{vmatrix} A & B \\ C & D \end{vmatrix} \begin{vmatrix} X \\ Y \end{vmatrix} = \begin{vmatrix} F_{11} & F_{12} \\ F_{21} & F_{22} \end{vmatrix} \begin{vmatrix} X \\ Y \end{vmatrix} \tag{5.35c}$$

Matrix multiplication dictates that the element in the first row of the product matrix (x) is the sum of two products: (1) the first row, first column element in the square matrix (A) times the first-row element of the second column matrix (X); and (2) the second column, first-row element of the square matrix (B) times the second-row element of the second column matrix (Y). The second-row element of the product matrix (y) is also the sum of two products: (1) the second row, first column element in the square matrix (C) times the first-row element of the second-column matrix (X); and (2) the second-row, second-column element of the square matrix (D) times the second-row element of the second column matrix (Y) (See Appendix 4-II). Using matrix shorthand

$$\boldsymbol{x} = \mathbf{F} \cdot \boldsymbol{X} = [\mathrm{F}] \cdot \boldsymbol{X} \tag{5.35d}$$

where $\boldsymbol{x}$ is a particle's position vector in the deformed state, **F** (sometimes denoted [F]) is the 2×2 matrix, and $\boldsymbol{X}$ is the particle's position vector in the undeformed state.

Eqs. (5.35) allow us to take a particle identified by its original position vector and *map* it to its proper position in the deformed state. Stated another way, Eqs. (5.35) transform position vectors in one state (the undeformed state or initial condition) to position vectors in another state (the deformed state or final condition). The idea of mapping one set of positions into another set of positions or transforming one set of vectors into another set of vectors may seem foreign to you, but chances are that you have unknowingly seen or used the simple mathematical concepts outlined here. For example, computer drawing programs (like the drawing programs Adobe Illustrator or Canvas) usually have a set of operations like "Transform," "Scale," or "Shear" that rotate, stretch, shear, or otherwise distort something you have drawn. These programs use position vectors to identify individual points, the sets of points that define curves or grouped sets of curves, or the sets of points that define areas. The drawing operations use matrix multiplication to apply a single transform to the set of original position vectors to generate a set of new position vectors. In this way, the program applies a uniform distortion to the drawing. Returning to our topic because this mathematical operation is comparable to the changes in position that result from deformation, some geologists call the matrix **F** the **deformation matrix**. We use F_{ij} as a shorthand way to refer to the matrix element in the ith row and jth column. In Eq. (5.35c), each of the elements F_{ij} is a position gradient:

$$A = F_{11} = \frac{dx}{dX}, \quad B = F_{12} = \frac{dx}{dY}, \quad C = F_{21} = \frac{dy}{dX}, \quad \text{and} \quad D = F_{22} = \frac{dy}{dY}.$$

For this reason, we also call the matrix **F** the **position gradient matrix**. An alternative version of Eq. (5.35c) is

$$\begin{vmatrix} x \\ y \end{vmatrix} = \begin{vmatrix} \left(\frac{dx}{dX}\right) & \left(\frac{dx}{dY}\right) \\ \left(\frac{dy}{dX}\right) & \left(\frac{dy}{dY}\right) \end{vmatrix} \begin{vmatrix} X \\ Y \end{vmatrix} \tag{5.35e}$$

or using matrix shorthand

$$\boldsymbol{x} = \left[\frac{dx_i}{dX_j}\right] \cdot \boldsymbol{X} \tag{5.35f}$$

where each element of the position gradient matrix dx_i/dX_j is the change in the value of the ith row of the final position column vector with respect to the change in the jth row of the original position column vector.

Consider at the outset position gradient matrices defined with respect to a specific set of Cartesian coordinate axes, where the individual matrix elements specify position gradients parallel to those coordinate axes. Position gradient matrices with different elements refer to deformations with different values for the position gradients and therefore specify different deformations. For example, horizontal shortening and vertical thickening like that depicted in Figure 5.37a results when

$$\begin{vmatrix} F_{11} & F_{12} \\ F_{21} & F_{22} \end{vmatrix} = \begin{vmatrix} T_X & 0 \\ 0 & T_Y \end{vmatrix}$$

if $T_X < 1$ and $T_Y > 1$. If $T_X > 1$ and $T_Y < 1$, the result is horizontal lengthening and vertical shortening. As long as $T_X \cdot T_Y = 1$, the result is a pure shear deformation. If $T_X \cdot T_Y \neq 1$, the deformation must result in a change in area. The matrix elements F_{ij} where $i = j$, such as F_{11} and F_{22}, constitute the *major* or *principal diagonal* of a square matrix. The remaining elements F_{ij} where $i \neq j$, are the *off-diagonal* elements. A matrix with nonzero diagonal elements (F_{11} and F_{22}) and off-diagonal elements equal to zero ($F_{12} = F_{21} = 0$) is a *diagonal* matrix. Diagonal position gradient matrices yield deformations with no net rotation; lines originally parallel to both coordinate axes are parallel to the coordinate axes in the deformed state. In this case, lines parallel to the coordinate directions also experience extreme longitudinal strains. These two properties of lines parallel to the coordinate axes, that they begin and end deformation perpendicular to each other and that they experience the extreme longitudinal strains, define these directions as the principal directions.

Dextral simple shear like that depicted in Figure 5.37b results when

$$\begin{vmatrix} F_{11} & F_{12} \\ F_{21} & F_{22} \end{vmatrix} = \begin{vmatrix} 1 & \gamma_X \\ 0 & 1 \end{vmatrix}$$

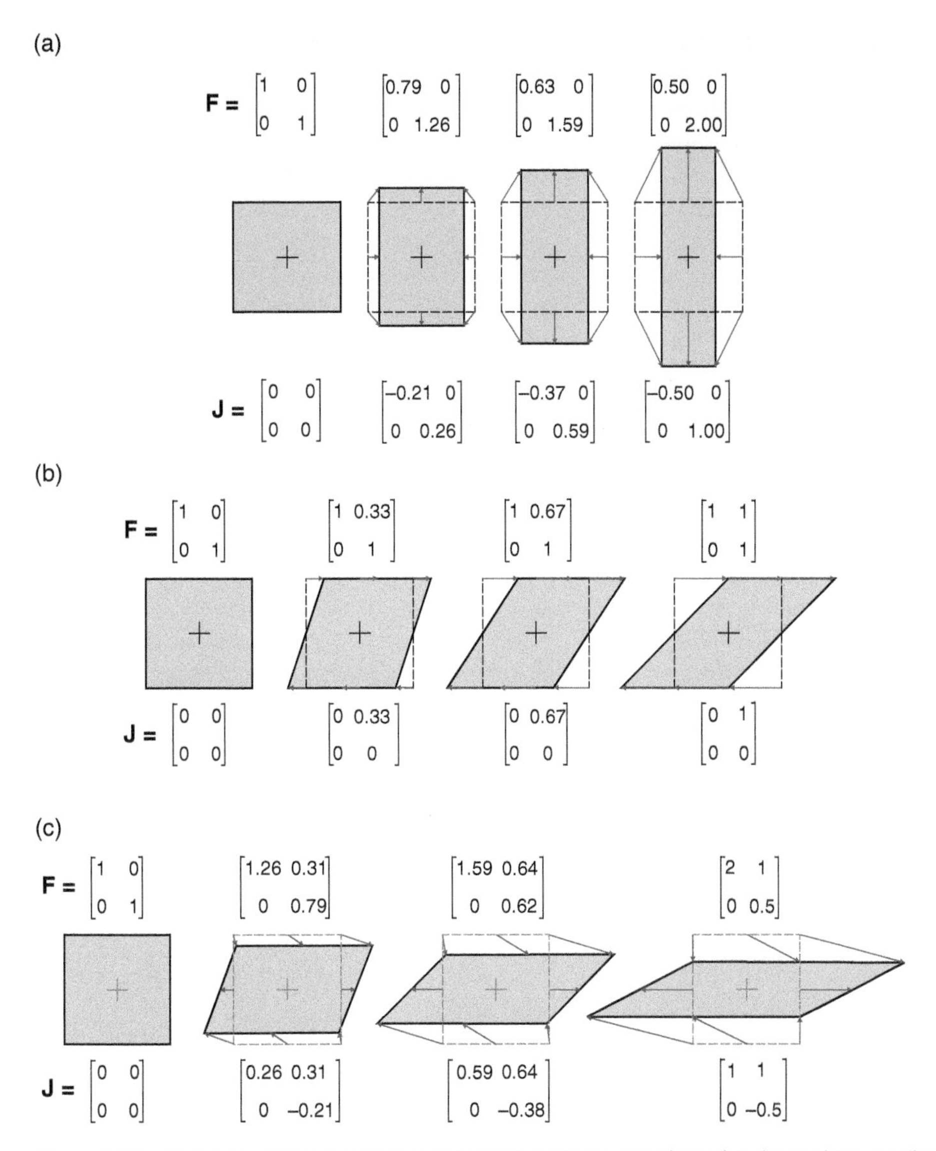

Figure 5.37 Sketches of four stages in a deformation history together with the position gradient matrix and displacement gradient matrix corresponding to each stage. (a) Pure shear deformation. (b) Simple shear deformation. (c) General shear deformation.

In simple shear, lines parallel to the x axis do not shorten, lengthen, or change their orientation. Particle displacements parallel to the x axis cause all other lines to shorten or lengthen at the same time as they change their orientation. When the magnitude of γ_X is small, the principal directions are inclined at ~45° to the coordinate axes. At larger magnitudes of γ_X, the principal direction experiencing the maximum lengthening makes an angle <45° with the x axis and the principal direction experiencing the maximum shortening makes an angle >45° to the x axis. The larger is the magnitude of γ_X, the more nearly parallel to the x axis is the direction of maximum lengthening and the more nearly perpendicular to the x axis is the direction of maximum shortening.

The combination of pure shearing and simple shearing parallel to the x axis we called *general shear* (Figure 5.37c) results when

$$\begin{vmatrix} F_{11} & F_{12} \\ F_{21} & F_{22} \end{vmatrix} = \begin{vmatrix} T_X & \gamma_X \\ 0 & T_Y \end{vmatrix}$$

This matrix with nonzero diagonal elements and elements below and to the left of the principal diagonal equal to zero (that is, $F_{21} = 0$ for a 2×2 matrix) is an *upper-triangular* matrix. Upper-triangular matrices generate deformations with principal directions oblique to the coordinate axes but where lines originally parallel to the x axis are not reoriented. Depending upon the values of the elements along the major diagonal, lines parallel to the x axis shorten or lengthen and move closer together or farther apart due to deformation.

These special cases of the position gradient matrix **F** are useful and interesting, but in nature, deformations develop independently of preconceived orientations for coordinate axes. The values of the elements of the position gradient matrix **F** depend upon the orientation of the coordinate axes selected by the geologists studying that deformation. Unless a geologist is careful in choosing an appropriate coordinate frame orientation, the position gradient matrices of natural deformations will not conform to one of these three special cases. A specific deformation may require that all four elements of **F** take nonzero values. Figure 5.36 depicts a situation where all $F_{ij} \neq 0$. Typically when all $F_{ij} \neq 0$, the position gradient matrix generates a deformation in which all material lines are reoriented by the deformation and the principal directions of the pure strain are oblique to the coordinate axes.

The position gradient matrix (and the corresponding displacement gradient matrix) captures *all* of the changes in the deformed state position vector or displacement vector with the position. In many cases, this includes *both* a pure strain component *and* a rigid rotation component. Depending on specific characteristics of the pure strain component and the rigid rotation component, there may be no lines, one set of lines, or two sets of lines whose orientations after deformation are the same as their orientations before deformation. In cases where one or two sets of lines have the same orientation before and after deformation, those lines typically also shorten or lengthen. Mathematicians use the term *eigenvector* to refer to a vector that "maps" to its original orientation. The length of an eigenvector differs from its original length by a scalar quantity called an *eigenvalue*. The series of particles, each of which had position vectors in the deformed state that are scalar multiples of their position vectors in the undeformed state, define lines emanating from the origin of the coordinate frame that defines the direction(s) that experience no net reorientation as a result of deformation. The eigenvalues are scalar values that reflect the shortening or lengthening along these lines. One can determine the eigenvectors and the eigenvalues by solving the equation

$$\mathbf{F} \cdot \boldsymbol{X} = \lambda \; \boldsymbol{X} \tag{5.36}$$

in which the $\boldsymbol{X}$ are the eigenvectors and λ are the eigenvalues.

If the matrix **F** is *symmetric*, that is, if $F_{ij} = F_{ji}$ (or if $\mathbf{F} = \mathbf{F}^T$) (in a 2×2 matrix, this means that $F_{12} = F_{21}$), then there is no net rotation associated with the deformation. For symmetric deformation matrices, the eigenvectors define the principal directions of the pure strain and the eigenvalues give the stretch along the principal directions. If the matrix **F** is not symmetric, that is, if $F_{ij} \neq F_{ji}$ (in a 2×2 matrix this means that $F_{12} \neq F_{21}$), the eigenvectors do not correspond to the principal directions of the pure strain, even though they indicate the orientations of lines whose orientation is unchanged by the deformation.

Knowing the orientation of the principal directions and the magnitudes of the principal strains is desirable and useful, for the principal directions and principal values control the orientation and intensity of the fabric(s) in a deformed rock. If **F** is symmetric, the eigenvectors of the position gradient matrix are the principal directions. If **F** is not symmetric, one must separate the effect of rigid rotation from the position gradients to assess the symmetric part of the deformation. One approach is to factor **F** into the product of a symmetric matrix and a rotation matrix:

$$\mathbf{F} = \boldsymbol{\Omega} \cdot \mathbf{F_R} \tag{5.37a}$$

or

$$\mathbf{F} = \mathbf{F_L} \cdot \boldsymbol{\Omega}^{-1}. \tag{5.37b}$$

In Eqs. (5.37a) and (5.37b), $\mathbf{F_R}$ and $\mathbf{F_L}$ are the right and left symmetric matrices and $\boldsymbol{\Omega}$ and $\boldsymbol{\Omega}^{-1}$ are rotation matrices. We can overcome the practical problems involved in determining which rotation matrix will yield a symmetric matrix in polar decomposition by invoking a rule from mathematics. The transpose of the product of two matrices is equal to the product of the transpose of the second matrix times the transpose of the first matrix. We have then

$$\begin{aligned}\mathbf{F} \cdot \mathbf{F^T} &= \left(\mathbf{F_L} \cdot \boldsymbol{\Omega}^{-1}\right) \cdot \left(\mathbf{F_L} \cdot \boldsymbol{\Omega}^{-1}\right)^{\mathbf{T}} \\ &= \left(\mathbf{F_L} \cdot \boldsymbol{\Omega}^{-1}\right) \cdot \left(\boldsymbol{\Omega}^{-1}\right)^{\mathbf{T}} \cdot \mathbf{F_L}^{\mathbf{T}}.\end{aligned} \tag{5.38a}$$

The transpose of any rotation matrix is its inverse. So, this equation becomes

$$\begin{aligned}\mathbf{F} \cdot \mathbf{F^T} &= \mathbf{F_L} \cdot \left(\boldsymbol{\Omega}^{-1}\right) \cdot \left(\boldsymbol{\Omega}^{-1}\right)^{\mathbf{T}} \cdot \mathbf{F_L}^{\mathbf{T}} \\ &= \mathbf{F_L} \cdot \left(\boldsymbol{\Omega}^{-1}\right) \cdot \left(\boldsymbol{\Omega}\right) \cdot \mathbf{F_L}^{\mathbf{T}} = \mathbf{F_L} \cdot \mathbf{F_L}^{\mathbf{T}}.\end{aligned} \tag{5.38b}$$

The product of two symmetric matrices (such as $\mathbf{F_L} \cdot \mathbf{F_L}^{\mathbf{T}}$) is always a symmetric matrix. So, the product $\mathbf{F} \cdot \mathbf{F^T}$, known as the *Finger tensor*, must also be symmetric. From Eq. (5.38b), we can infer that the components of the Finger tensor are squares of the pure strain elements of the deformation matrices. The eigenvectors of this tensor define the principal directions of the pure strain, and the square roots of the eigenvalues of the Finger tensor give the principal strain magnitudes.

We can also determine some restrictions on the values of the deformation matrix elements by considering a quantity called the *determinant* of the deformation matrix (see Appendix 4-II). For two-dimensional pure strains, the determinant of **F** is

$$\det \mathbf{F} = |\mathbf{F}| = \left(F_{11}F_{22} - F_{12}F_{21}\right). \tag{5.39}$$

For two-dimensional deformations, the value of det **F** gives the ratio of the area occupied by a collection of particles in the deformed state to the area occupied by that collection in the undeformed state. The deformations illustrated in Figures 5.36 and 5.37 have det $\mathbf{F} = 1$, meaning that the square representing a collection of particles in the undeformed state and the parallelogram representing that same collection of particles in the deformed state occupy equal areas on the plane in each case. As long as det $\mathbf{F} \neq 0$, the deformation matrix could conceivably describe a real deformation. If det $\mathbf{F} = 0$, the deformation matrix maps a collection of particles into a deformed state configuration with no area, thereby requiring that matter be annihilated. Such a deformation matrix cannot describe an actual deformation. In practical terms, position gradient matrices describing two-dimensional deformations are suspect if det **F** is significantly less than or greater than unity.

If particles' position vectors in the deformed state ($\boldsymbol{x}$) are related to their position vectors in the undeformed state ($\boldsymbol{X}$) by Eq. (5.35) and if det $\mathbf{F} \neq 0$, then $\mathbf{F}^{-1}$, the inverse of **F** (see Appendix 4-II) exists. We can explore the physical meaning of $\mathbf{F}^{-1}$ by completing a few straightforward algebraic steps. Begin by premultiplying both sides of Eq. (5.35) by $\mathbf{F}^{-1}$:

$$\mathbf{F}^{-1} \cdot \boldsymbol{x} = \mathbf{F}^{-1} \cdot \mathbf{F} \cdot \boldsymbol{X} = \mathbf{I} \cdot \mathbf{X} = \boldsymbol{X} \tag{5.40a}$$

or taking $\mathbf{F}^{-1} = \mathbf{G}$, then

$$\boldsymbol{X} = \mathbf{G} \cdot \boldsymbol{x}. \quad (5.40b)$$

In component form for a two-dimensional deformation, this is

$$\begin{vmatrix} X \\ Y \end{vmatrix} = \begin{vmatrix} G_{11} & G_{12} \\ G_{21} & G_{22} \end{vmatrix} \begin{vmatrix} x \\ y \end{vmatrix} = \begin{vmatrix} a & b \\ c & d \end{vmatrix} \begin{vmatrix} x \\ y \end{vmatrix} \quad (5.40c)$$

The different forms of Eqs. (5.40) give a particle's position vector in the undeformed state as a function of the particles' position vector in the deformed state. The matrix $\mathbf{G} = \mathbf{F}^{-1}$ is a *reciprocal deformation matrix*, which takes a particle identified by its position vector in the deformed state and maps it to its position in the undeformed state. The matrix $\mathbf{G} = \mathbf{F}^{-1}$ transforms position vectors in the deformed state into position vectors in the undeformed state. For two-dimensional deformations, the components of the matrix **G** are (see Appendix 4-II):

$$\mathbf{G} = \begin{vmatrix} \dfrac{F_{22}}{(F_{11}F_{22} - F_{12}F_{21})} & -\dfrac{F_{12}}{(F_{11}F_{22} - F_{12}F_{21})} \\ -\dfrac{F_{21}}{(F_{11}F_{22} - F_{12}F_{21})} & \dfrac{F_{11}}{(F_{11}F_{22} - F_{12}F_{21})} \end{vmatrix} \quad (5.41)$$

Each element of matrix **G** has the term $F_{11}F_{22} - F_{12}F_{21} = \det \mathbf{F}$ as its denominator. If det **F** = 0, the elements of G cannot be defined. This is another reason that det $\mathbf{F} \neq 0$ for all real deformations.

5B.4.13.2 Displacement Gradient Matrices

The displacement fields that correspond to the strain examples in Figures 5.35–5.37 are described by relatively simple, linear functions of X and Y, the coordinates of that particle's position vector in the undeformed state:

$$u_X = K \cdot X + L \cdot Y \quad (5.42a)$$

$$u_Y = M \cdot X + N \cdot Y. \quad (5.42b)$$

We can again use matrix notation to represent these linear equations, with the displacement components u_X and u_Y defining elements of one 2×1 column matrix, the components of the original position vectors X and Y defining elements of a second 2×1 column matrix, and the coefficients K, L, M, and N defining elements of a 2×2 square matrix. A particle's displacement vector $\boldsymbol{u}$ is the product of the square matrix and the particle's original position vector $\boldsymbol{X}$:

$$\begin{vmatrix} u_X \\ u_Y \end{vmatrix} = \begin{vmatrix} K & L \\ M & N \end{vmatrix} \begin{vmatrix} X \\ Y \end{vmatrix} = \begin{vmatrix} J_{11} & J_{12} \\ J_{21} & J_{22} \end{vmatrix} \begin{vmatrix} X \\ Y \end{vmatrix} \quad (5.42c)$$

or using matrix shorthand

$$\boldsymbol{u} = \mathbf{J} \cdot \boldsymbol{X} = [\mathrm{J}] \cdot \boldsymbol{X}. \quad (5.42d)$$

The elements J_{ij} in this 2×2 matrix are:

$$\mathrm{K} = J_{11} = \frac{\mathrm{d}u_X}{\mathrm{d}X}, \quad \mathrm{L} = J_{12} = \frac{\mathrm{d}u_X}{\mathrm{d}Y},$$
$$\mathrm{M} = J_{21} = \frac{\mathrm{d}u_Y}{\mathrm{d}X}, \quad \text{and} \quad \mathrm{N} = J_{22} = \frac{\mathrm{d}u_Y}{\mathrm{d}Y}.$$

In Eq. (5.42c), the matrix **J** (or **[J]**) is the 2×2 **displacement gradient matrix**. An alternative version of Eq. (5.42c) is

$$\begin{vmatrix} u_X \\ u_Y \end{vmatrix} = \begin{vmatrix} \left(\dfrac{\mathrm{d}u_X}{\mathrm{d}X}\right) & \left(\dfrac{\mathrm{d}u_X}{\mathrm{d}Y}\right) \\ \left(\dfrac{\mathrm{d}u_Y}{\mathrm{d}X}\right) & \left(\dfrac{\mathrm{d}u_Y}{\mathrm{d}Y}\right) \end{vmatrix} \begin{vmatrix} X \\ Y \end{vmatrix} \quad (5.42e)$$

or using matrix shorthand

$$\boldsymbol{u} = \left[\frac{\mathrm{d}u_i}{\mathrm{d}X_j}\right] \cdot \boldsymbol{X} \quad (5.42f)$$

where each element of the displacement gradient matrix $\mathrm{d}u_i/\mathrm{d}X_j$ is the change in the value of the ith row of displacement vector with respect to the change in the *j*th row of the original position column vector.

For any deformation defined by a position gradient matrix, we can define a corresponding displacement gradient matrix. The horizontal shortening and vertical thickening depicted in Figures 5.37a correspond to

$$\begin{vmatrix} J_{11} & J_{12} \\ J_{21} & J_{22} \end{vmatrix} = \begin{vmatrix} e_X & 0 \\ 0 & e_Y \end{vmatrix}$$

if $e_X < 0$ and $e_Y > 0$. If $e_X > 0$ and $e_Y < 0$, the result is horizontal lengthening and vertical thinning. Dextral simple shear like that depicted in Figure 5.37b corresponds to

$$\begin{vmatrix} J_{11} & J_{12} \\ J_{21} & J_{22} \end{vmatrix} = \begin{vmatrix} 0 & \gamma_X \\ 0 & 0 \end{vmatrix}$$

General shear (the combination of pure shearing and simple shearing parallel to the x axis) (Figure 5.37c) results when

$$\begin{vmatrix} J_{11} & J_{12} \\ J_{21} & J_{22} \end{vmatrix} = \begin{vmatrix} e_X & \gamma_X \\ 0 & e_Y \end{vmatrix}$$

As is true of the position gradient matrix, for any particular pure strain, the elements of the displacement gradient matrix **J** depend the orientation of the coordinate axes. For most coordinate frame orientations, the displacement gradient matrices of most actual deformations will not conform to one of these three special cases.

5B.4.13.3 Relating Position Gradient and Displacement Gradient Matrices

In Sections 5B.4.8 and 5B.4.9, we showed that

$$\frac{dx}{dX} = \frac{du_X}{dX} + 1 \tag{5.28c}$$

$$\frac{dy}{dY} = \frac{du_Y}{dY} + 1. \tag{5.29c}$$

In Section 5B.4.12, we showed that

$$\frac{dx}{dY} = \frac{du_X}{dY}. \tag{5.33a}$$

$$\frac{dy}{dX} = \frac{du_Y}{dX}. \tag{5.33b}$$

Compiling these relationships, we have

$$\begin{vmatrix} F_{11} & F_{12} \\ F_{21} & F_{22} \end{vmatrix} = \begin{vmatrix} \frac{dx}{dX} & \frac{dx}{dY} \\ \frac{dy}{dX} & \frac{dy}{dY} \end{vmatrix} = \begin{vmatrix} \frac{du_X}{dX}+1 & \frac{du_X}{dY} \\ \frac{du_Y}{dX} & \frac{du_Y}{dY}+1 \end{vmatrix} = \begin{vmatrix} J_{11}+1 & J_{12} \\ J_{21} & J_{22}+1 \end{vmatrix} \tag{5.43a}$$

Now

$$\begin{vmatrix} J_{11}+1 & J_{12} \\ J_{21} & J_{22}+1 \end{vmatrix} = \begin{vmatrix} J_{11} & J_{12} \\ J_{21} & J_{22} \end{vmatrix} + \begin{vmatrix} 1 & 0 \\ 0 & 1 \end{vmatrix}$$

where the matrix

$$\begin{vmatrix} 1 & 0 \\ 0 & 1 \end{vmatrix} = \mathbf{I}$$

s

is the 2×2 identity matrix (see Appendix 4-II). Thus,

$$\begin{vmatrix} F_{11} & F_{12} \\ F_{21} & F_{22} \end{vmatrix} = \begin{vmatrix} J_{11} & J_{12} \\ J_{21} & J_{22} \end{vmatrix} + \begin{vmatrix} 1 & 0 \\ 0 & 1 \end{vmatrix} \tag{5.43b}$$

or using matrix shorthand

$$\mathbf{F} = \mathbf{J} + \mathbf{I} \text{ or } [\mathrm{F}] = [\mathrm{J}] + [\mathrm{I}] \tag{5.43c}$$

(cf. Figure 5.37).

5B.4.14 Applying Matrix Algebra to Three-dimensional Deformation

A significant advantage of using matrix algebra to describe deformation is the relative ease with which mathematical relationships are extended from two-dimensional cases to three-dimensional cases. In three-dimensional cases, the original position vector for a particle $\boldsymbol{X}$ is a 3×1 matrix with the vector's z axis component Z as its third-row element. Similarly, the particle's position vector in the deformed state is $\boldsymbol{x}$, a 3×1 matrix with the z axis component z as its third-row element. The relationships between position gradient or displacement gradient values and the measures of longitudinal strain or a shear strain derived in the preceding sections still hold. In three dimensions, the position gradient matrix $\mathbf{F}$ is a 3×3 matrix where:

$$F_{11} = \frac{dx}{dX} \quad F_{12} = \frac{dx}{dY} \quad F_{13} = \frac{dx}{dZ}$$
$$F_{21} = \frac{dy}{dX} \quad F_{22} = \frac{dy}{dY} \quad F_{23} = \frac{dy}{dZ}$$
$$F_{31} = \frac{dz}{dX} \quad F_{32} = \frac{dz}{dY} \quad F_{33} = \frac{dz}{dZ}$$

Equation (5.35) still applies, that is

$$\boldsymbol{x} = \mathbf{F} \cdot \mathbf{X} = \left[\frac{dx_i}{dX_j} \right] \cdot \boldsymbol{X}. \tag{5.35}$$

The determinant of the 3×3 matrix $\mathbf{F}$, still denoted det $\mathbf{F}$, is an algebraic expression containing nine terms; it gives the ratio of the volume occupied by a collection of particles in the deformed state to the volume occupied by those particles in the undeformed state.

A particle's displacement vector is $\boldsymbol{u}$, a 3×1 matrix with the z axis component u_Z as its third-row element. In three dimensions, the displacement gradient matrix $\mathbf{J}$ is a 3×3 matrix where:

$$J_{11} = \frac{du_X}{dX} \quad J_{12} = \frac{du_X}{dY} \quad J_{13} = \frac{du_X}{dZ}$$
$$J_{21} = \frac{du_Y}{dX} \quad J_{22} = \frac{du_Y}{dY} \quad J_{23} = \frac{du_Y}{dZ}$$
$$J_{31} = \frac{du_Z}{dX} \quad J_{32} = \frac{du_Z}{dY} \quad J_{33} = \frac{du_Z}{dZ}$$

Equation (5.42) still holds, that is

$$\boldsymbol{u} = \mathbf{J} \cdot \boldsymbol{X} = \left[\frac{du_i}{dX_j} \right] \cdot \boldsymbol{X}. \tag{5.42}$$

Likewise,

$$\mathbf{F} = \mathbf{J} + \mathbf{I} \tag{5.43}$$

where I is the 3 × 3 identity matrix.

Orthogonal shortening and lengthening occur when the position gradient matrix and the displacement gradient matrices are *diagonal* matrices in which the principal diagonal elements are nonzero and the off-diagonal elements are zero:

$$\begin{vmatrix} F_{11} & F_{12} & F_{13} \\ F_{21} & F_{22} & F_{23} \\ F_{31} & F_{32} & F_{33} \end{vmatrix} = \begin{vmatrix} T_X & 0 & 0 \\ 0 & T_Y & 0 \\ 0 & 0 & T_Z \end{vmatrix}$$

and

$$\begin{vmatrix} J_{11} & J_{12} & J_{13} \\ J_{21} & J_{22} & J_{23} \\ J_{31} & J_{32} & J_{33} \end{vmatrix} = \begin{vmatrix} e_X & 0 & 0 \\ 0 & e_Y & 0 \\ 0 & 0 & e_Z \end{vmatrix}$$

Shortening will occur along a coordinate axis if $T_i < 1$ or if $e_i < 0$ and lengthening occurs if $T_i > 1$ if $e_i > 0$. As long as $T_X \cdot T_Y \cdot T_Z = 1$, the result is a pure strain where there is no volume change; the deformation will be a plane strain pure shear if any of the three $T_i \cdot = 1$ and $T_X \cdot T_Y \cdot T_Z = 1$.

Three-dimensional simple shear in the x axis direction with the y axis direction oriented perpendicular to the shear plane occurs when the position gradient matrix and the displacement gradient matrices have the forms:

$$\begin{vmatrix} F_{11} & F_{12} & F_{13} \\ F_{21} & F_{22} & F_{23} \\ F_{31} & F_{32} & F_{33} \end{vmatrix} = \begin{vmatrix} 1 & Y & 0 \\ 0 & 1 & 0 \\ 0 & 0 & 1 \end{vmatrix}$$

and

$$\begin{vmatrix} J_{11} & J_{12} & J_{13} \\ J_{21} & J_{22} & J_{23} \\ J_{31} & J_{32} & J_{33} \end{vmatrix} = \begin{vmatrix} 0 & \gamma & 0 \\ 0 & 0 & 0 \\ 0 & 0 & 0 \end{vmatrix}$$

Simple shear in the x axis direction with the z axis perpendicular to the shear plane occurs when F_{13} or J_{13} is the only nonzero off-diagonal element, simple shear in the y axis direction with the x axis perpendicular to the shear plane occurs when F_{21} or J_{21} is the only nonzero off-diagonal element, etc. If two of the components above and to the right of the principal diagonal of **F** or **J** are nonzero, the shear direction is oblique to the coordinate axes on a shear plane that contains two of the coordinate axes. If all three of the components above and to the right of the principal diagonal are nonzero, the shear plane contains only one of the coordinate axes. Regardless of which of the components above and to the right of the principal diagonal are nonzero, provided that the components below and to the left of the principal diagonal are zero, det **F** = 1 and there is no volume change during deformation.

General shear occurs by combining the pure shear and simple shear, meaning that the general shear matrix is characterized by upper-triangular position gradient and displacement gradient matrices:

$$\begin{vmatrix} F_{11} & F_{12} & F_{13} \\ F_{21} & F_{22} & F_{23} \\ F_{31} & F_{32} & F_{33} \end{vmatrix} = \begin{vmatrix} T_X & \gamma & 0 \\ 0 & T_Y & 0 \\ 0 & 0 & T_Z \end{vmatrix}$$

and

$$\begin{vmatrix} J_{11} & J_{12} & J_{13} \\ J_{21} & J_{22} & J_{23} \\ J_{31} & J_{32} & J_{33} \end{vmatrix} = \begin{vmatrix} e_X & \gamma & 0 \\ 0 & e_Y & 0 \\ 0 & 0 & e_Z \end{vmatrix}$$

The values of the components above and to the right of the principal diagonal control the orientation of the shear plane and the shear direction just as they do in the simple shear matrices. The det **F** $= T_X \cdot T_Y \cdot T_Z$, so the relative magnitudes of the stretches along the coordinate axes control whether this pure strain occurs with or without volume change and whether it is a plane strain or not.

5B.5 Homogeneous and Inhomogeneous Deformation

5B.5.1 Homogeneous Deformation

Homogeneous pure strain is apparent in three effects:

1) Lines that were straight before deformation are straight in the deformed state;
2) Lines that were parallel before deformation are parallel after deformation; and
3) The magnitude of the longitudinal strain measured in a particular direction is the same regardless of where within the area of interest one measures it.

Section 5A5.4 outlined an argument that *homogeneous* pure strains result when the values of the displacement gradients are the same at different positions within an area or region. Using the relationships outlined in Section 5B.4 for two-dimensional pure strains, we can explore the connection between these three effects and the values of the displacement or position gradients.

Begin by considering the equations relating to particles' position vectors in the undeformed and deformed states, that is, Eqs. (5.35) and (5.40):

$$x = AX + BY \tag{5.35a}$$

$$y = CX + DY \tag{5.35b}$$

$$X = ax + by \tag{5.40c}$$

$$Y = cx + dy. \tag{5.40d}$$

Any collection of particles that define a straight line in the undeformed state conform to

$$Y = MX + Y_o, \tag{5.44a}$$

where M is the slope and Y_o is the y-intercept of the line. By substituting for X and Y using Eq. (5.40c), we find the positions in the deformed state occupied by the collection of particles along this line:

$$cx + dy = M(ax + by) + Y_o. \tag{5.44b}$$

Simplifying this equation yields

$$y = \left[\frac{(Ma + c)}{(d - Mb)}\right]x + \frac{Y_o}{(d - Mb)}. \tag{5.44c}$$

Since the elements a, b, c, and d are constants, this is the equation of a straight line in the deformed state,

$$y = mx + y_o \tag{5.44d}$$

with a slope $m = [(Ma + c)/(d - Mb)]$ and an intercept $y_o = Y_o/(d - Mb)$. The slope and intercept of the line in the deformed state typically are not the same as the slope and intercept in the undeformed state. Still, a straight line in the undeformed state becomes a straight line in the deformed state if the position gradients have constant values within an area.

Consider next two parallel lines in the undeformed state. The two lines will have the same slope M but different intercepts Y_1 and Y_2. Equations (5.44) shows that after deformation, the two lines will have the same slope $m = [(Ma + c)/(d - Mb)]$ but have different intercepts $y_1 = Y_1/(d - Mb)$ and $y_2 = Y_2/(d - Mb)$. Thus, the lines in the deformed state are also parallel.

Finally, consider the implications of the situation where the matrix **F** (or **J**) at one point along any line is identical to the matrix **F** (or **J**) at all points along the line. The magnitudes of longitudinal or shear strains at any position are functions of the elements in the position gradient (or displacement gradient) matrix. If the position gradient or displacement gradient matrix has the same value at all locations, then the magnitudes of longitudinal and shear strains in a given direction at one position within the area of interest are identical to the magnitudes of longitudinal and shear strains in the same direction at any other position within the area of interest.

5B.5.2 Inhomogeneous Deformation

Truly homogeneous pure strain, where position gradients and displacement gradients have equal values at all points, is rare. More typically, the value of the position or displacement gradients changes with position. Within limited areas, $\partial x_i/\partial X_j$ (which here denotes the difference in the position vector component x_i at two points separated an infinitesimal distance parallel to the X_j axis) and $\partial u_i/\partial X_j$ (which here denotes the difference in the displacement vector component u_i at two points separated an infinitesimal distance parallel to the X_j axis) are not measurably different from $\Delta x_i/\Delta X_j$ (which here denotes the difference in the position vector component x_i at two points a finite distance apart along the X_j axis) and $\Delta u_i/\Delta X_j$ (which here denotes the difference in the displacement vector component u_i at points a finite distance apart parallel to the X_j axis). Within such an area, then, dx_i/dX_j and du_i/dX_j change very little with position, x_i and u_i are effectively linear functions of the position coordinates X_j, and the pure strain is homogeneous.

In reality, x_i and u_i are rarely linear functions of the position coordinates X_j, meaning that dx_i/dX_j and du_i/dX_j *do* change with position, and the pure strain is **inhomogeneous**. In those cases, the familiar *derivative* of a function of a single variable must be replaced by $\partial x_i/\partial X_j$, which is the *partial derivative* or derivative with respect to a single variable of a function of several variables (this is found by holding the other variables constant; graphically it is akin to measuring the slope of a complex surface along a single direction). At each point in space, we can calculate the values of $\partial x_i/\partial X_j$ or $\partial u_i/\partial X_j$ and determine a unique position gradient matrix or displacement gradient matrix. We can define strain values *at that point* by substituting partial derivatives in the relationships outlined in the previous sections and evaluating the partial derivatives at different locations:

$$\frac{\partial x}{\partial X} = T_X \tag{5.45a}$$

$$\frac{\partial y}{\partial Y} = T_Y \tag{5.45b}$$

$$\frac{\partial u_X}{\partial X} = e_X \tag{5.46a}$$

$$\frac{\partial u_Y}{\partial Y} = e_Y \tag{5.46b}$$

where $\partial u_X/\partial X = \partial x/\partial X - 1$ and $\partial u_Y/\partial Y = \partial Y/\partial Y - 1$.

$$\gamma_X = \frac{\partial u_X}{\partial Y} = \frac{\partial x}{\partial Y} \tag{5.47a}$$

$$\gamma_Y = \frac{\partial u_Y}{\partial X} = \frac{\partial x}{\partial Y}. \tag{5.47b}$$

We can define a position gradient matrix *at a point* from

$$F_{11} = \frac{\partial x}{\partial X} \quad F_{12} = \frac{\partial x}{\partial X}$$
$$F_{21} = \frac{\partial x}{\partial Y} \quad F_{22} = \frac{\partial x}{\partial Y}$$

and, $\boldsymbol{x} = \mathbf{F} \cdot \boldsymbol{X}$ becomes

$$\begin{vmatrix} x \\ y \end{vmatrix} = \begin{vmatrix} \left(\frac{\partial x}{\partial X}\right) & \left(\frac{\partial x}{\partial X}\right) \\ \left(\frac{\partial x}{\partial Y}\right) & \left(\frac{\partial x}{\partial Y}\right) \end{vmatrix} \begin{vmatrix} X \\ Y \end{vmatrix} \tag{5.48}$$

Similarly, the elements of the displacement gradient matrix *at a point* are

$$J_{11} = \frac{\partial u_X}{\partial X} \quad J_{12} = \frac{\partial u_X}{\partial Y}$$
$$J_{21} = \frac{\partial u_Y}{\partial X} \quad J_{22} = \frac{\partial u_Y}{\partial Y}$$

and, $\boldsymbol{u} = \mathbf{J} \cdot \boldsymbol{X}$ becomes

$$\begin{vmatrix} u \\ v \end{vmatrix} = \begin{vmatrix} \left(\frac{\partial u_X}{\partial X}\right) & \left(\frac{\partial u_X}{\partial Y}\right) \\ \left(\frac{\partial u_Y}{\partial X}\right) & \left(\frac{\partial u_Y}{\partial Y}\right) \end{vmatrix} \begin{vmatrix} X \\ Y \end{vmatrix} \tag{5.49}$$

The notion of defining strain values *at a point* may seem counter to intuition – how can one measure the change in the length of a line *at a point*, or measure the changes in the angles between two lines *at a point*, or define the shape and orientation of a finite strain ellipse (see the following section) *at a point*? We use Eqs. (5.45), (5.46), and (5.47) to define the longitudinal and shear strains at individual points, and we use the values of **F** defined by Eqs. (5.48) to determine the shape and orientation of the finite strain ellipse. Because the values of the position or displacement gradients vary from position to position, *inhomogeneous* or **heterogeneous** pure strains differ fundamentally from homogeneous pure strains:

1) Lines that were straight in the undeformed state need not be straight in the deformed state;
2) Lines that were parallel before deformation need not be parallel after deformation; and
3) The magnitude of the longitudinal strain measured in a particular direction at one point along a straight line need not be equal to the magnitude or be the same as the longitudinal strains at other points along that line.

Although most natural deformations are inhomogeneous, geologists use a variety of techniques that enable them to treat portions of real, inhomogeneous deformations as approximately homogeneous. For example, variations in the magnitudes of position gradients or displacement gradients typically may be miniscule within a single thin section or within a microscope field of view at a particular magnification. Within each region, a geologist can quantify the average strain and determine how strains vary from small region to small region. Further, there are mathematical relationships that govern how strain can vary with position, ensuring that variations are smooth and compatible with general principles like the conservation of mass. By combining these two approaches, geologists can cull much information from the variation in strain with the position.

5B.6 Finite Strain Ellipse and Finite Strain Ellipsoid

5B.6.1 Homogeneous Deformations and the Finite Strain Ellipse

Using Eqs. (5.35a) and (5.37c), we can determine precisely what happens to the collection of particles that define a circle in the undeformed state. For simplicity, we consider the collection of particles that define a circle with radius R centered at the origin:

$$X^2 + Y^2 = R^2. \tag{5.50a}$$

Substituting using Eq. (5.37c) yields,

$$(ax + by)^2 + (cx + dy)^2 = R^2. \tag{5.50b}$$

After expanding the square terms on the left side of the equation and simplifying, Eq. (5.50b) becomes:

$$(a^2 + c^2)x^2 + 2(ab + cd)xy + (b^2 + d^2)y^2 = R^2. \tag{5.50c}$$

This is the equation of an ellipse. In most situations, the major and minor axes of this ellipse are inclined to coordinate axes. Consider the special case of pure shear deformation, where the deformation matrix **F** is a diagonal matrix

$$\begin{vmatrix} F_{11} & F_{12} \\ F_{21} & F_{22} \end{vmatrix} = \begin{vmatrix} A & 0 \\ 0 & D \end{vmatrix}$$

with $A \cdot D = 1$. The reciprocal deformation matrix **G** is also a diagonal matrix

$$\begin{vmatrix} G_{11} & G_{12} \\ G_{21} & G_{22} \end{vmatrix} = \begin{vmatrix} a & 0 \\ 0 & d \end{vmatrix}$$

with $a = D$, $d = A$, and $a \cdot d = 1$. Substituting the values of the elements of **G** into Eq. (5.50c) yields the equation giving the deformed state positions of the collection of particles that defined a circle with radius R centered at the origin in the undeformed state:

$$a^2x^2 + d^2y^2 = R^2. \tag{5.51a}$$

Dividing both sides of this equation by a^2d^2 and R^2 yields

$$\frac{x^2}{d^2R^2} + \frac{y^2}{a^2R^2} = \frac{x^2}{A^2R^2} + \frac{y^2}{D^2R^2} = 1, \tag{5.51b}$$

which is the equation of an ellipse with its major and minor axes parallel to the x and y axes. The lengths of the major and minor diameters of this ellipse are $2aR$ ($= 2DR$) and $2dR$ ($= 2AR$). The axial ratio of this ellipse is a/d, underscoring the relationship between the values of the position gradient matrix, the stretch, and the axial ratio of the strain ellipse.

5B.6.2 Working with Strain Markers

Figure 5.38a is a photomicrograph of undeformed ooids from the Carawine Dolomite of the Hamersley Basin in Western Australia. The ooids have nearly circular shapes in section, as you can see by comparing the measured horizontal and

(a)

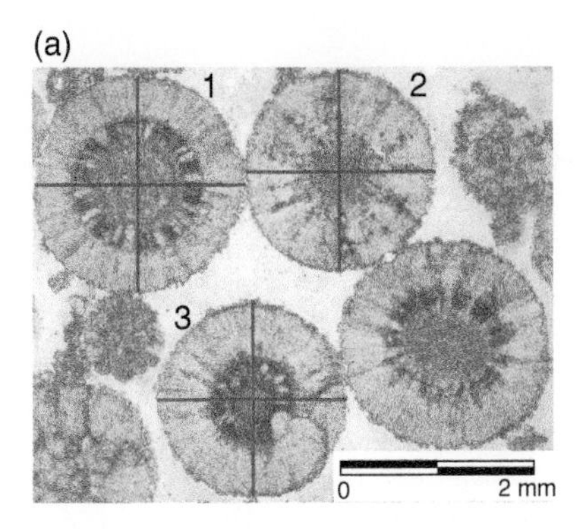

1: d_h = 2.19 mm; d_v = 2.19 mm
d_{aver} = 2.19 mm
2: d_h = 1.99 mm; d_v = 2.03 mm
d_{aver} = 2.01 mm
3: d_h = 1.98 mm; d_v = 1.98 mm
d_{aver} = 1.98 mm

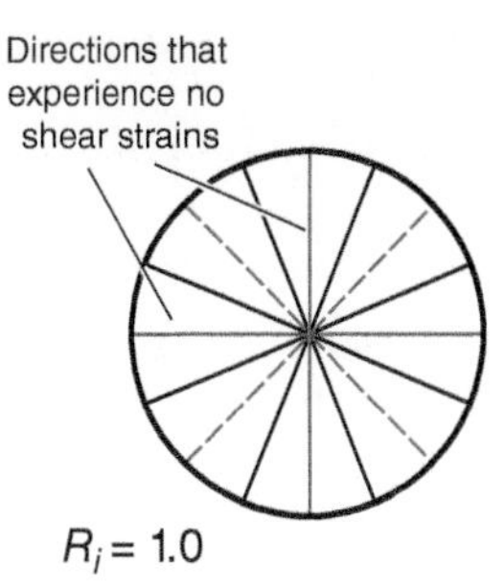

R_i = 1.0

(b)

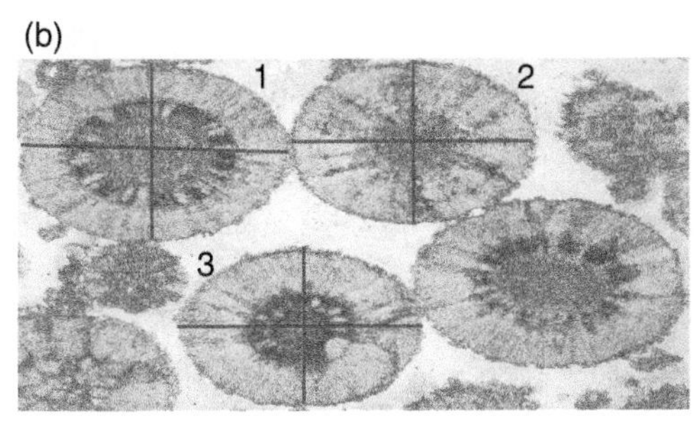

1: d'_{max} = 2.73 mm;
d'_{min} = 1.75 mm
2: d'_{max} = 2.53 mm;
d'_{min} = 1.63 mm
3: d'_{max} = 2.5 mm;
d'_{min} = 1.56 mm

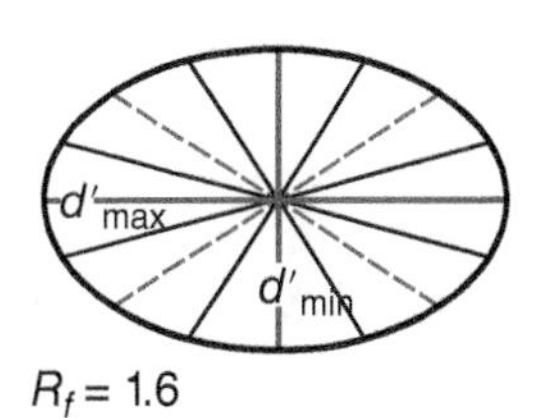

R_f = 1.6

1: d'_{max} = 3.49 mm; d'_{min} = 1.39 mm
2: d'_{max} = 3.19 mm; d'_{min} = 1.29 mm
3: d'_{max} = 3.15 mm; d'_{min} = 1.24 mm

(c)

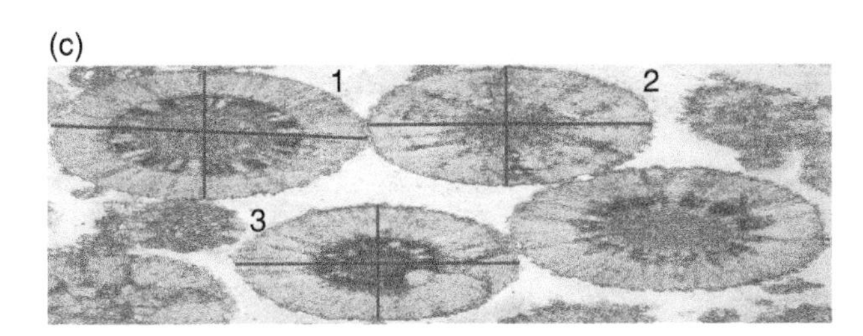

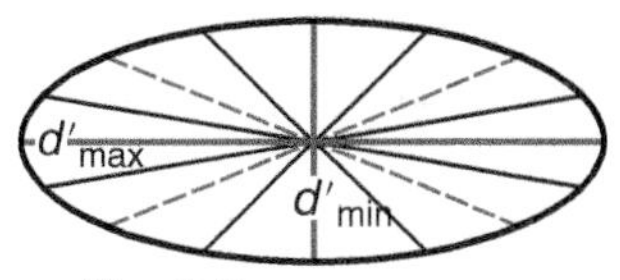

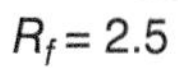

R_f = 2.5

1: d'_{max} = 4.34 mm;
d'_{min} = 1.08 mm
2: d'_{max} = 3.99 mm;
d'_{min} = 1.02 mm
3: d'_{max} = 3.96 mm;
d'_{min} = 0.97 mm

(d)

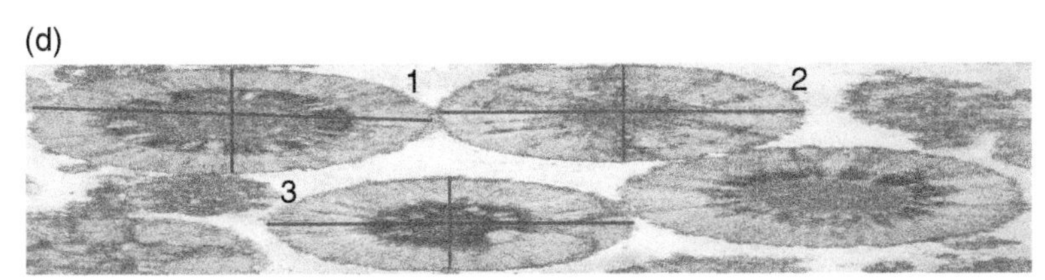

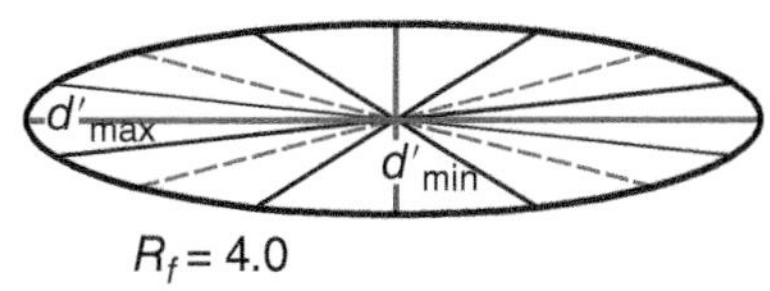

R_f = 4.0

Figure 5.38 (a) Photomicrograph at left shows undeformed ooids. In the middle are measurements of horizontal and vertical diameters of ooids 1, 2, and 3. At right is a depiction of the pure shear imposed on the image. With no strain, R_S = 1.0 (b) Photomicrograph distorted in pure shear to R_S = 1.6 (depicted at right). Center column gives measurements of the longest and shortest diameters of ooids. (c) Same as (b) but with R_S = 2.5. (d) Same as (b) but with R_S = 4.0.

vertical diameters of ooids 1, 2, and 3. Using a drawing program, we subjected this image to three different pure shear deformations (Figure 5.38b–d). In each frame, the external dimensions of the image show that the direction of maximum shortening is "vertical," from the top to the bottom of the image, and that the direction of maximum elongation is "horizontal," from right to left across the image. The right angles between the top or bottom and sides of the image are unchanged. So, the principal directions are "horizontal" and "vertical." The nearly circular ooids are distorted into shapes that closely approximate ellipses in each case, and all the ooids in the field of view take on similar shapes. The longest and shortest diameters of the distorted ooid images are parallel to the principal directions of the deformation. Ratios of the major and minor diameters to the original diameters of the ooids give the magnitudes of the principal stretches. In Figure 5.38b, the average of the maximum stretches for ellipses 1, 2, and 3 is $T_{max} \approx 1.26$, and the average of the minimum stretches is $T_{min} \approx 0.79$. $R_s = T_{max}/T_{min} \approx 1.26/0.79 = 1.6$. In Figure 5.38c, the comparable values are $T_{max} \approx 1.59$, $T_{min} \approx 0.63$, and $R_s = T_{max}/T_{min} \approx 2.5$, and in Figure 5.38d, they are $T_{max} \approx 1.99$, $T_{min} \approx 0.49$, and $R_s = T_{max}/T_{min} \approx 4.0$.

Figure 5.39 shows the same image of detrital ooids deformed in simple shear using the drawing program. The external dimensions of the distorted images show that lines parallel to the top edge of the images are displaced to the right but otherwise are not distorted. Once again, the nearly circular outlines of the ooids are distorted into shapes that closely approximate ellipses. Furthermore, the several ooids in the field of view take on similar shapes in each case. The longest and shortest diameters of the distorted ooids, which define the principal directions of the deformation, are inclined at different angles to the "horizontal" base in the different images. The average of the maximum stretch calculated for ellipses 1, 2, and 3 in Figure 5.39b is $T_{max} \approx 1.27$, and the average of the minimum stretch for those ellipses is $T_{min} \approx 0.81$, so $R_s \approx 1.6$. For Figure 5.39c, $T_{max} \approx 1.59$, $T_{min} \approx 0.63$, and $R_s \approx 2.5$. For Figure 5.39d, $T_{max} \approx 2.0$, $T_{min} \approx 0.5$, and $R_s \approx 4.0$.

Figure 5.40 shows the image distorted in general shear. The external dimensions of the distorted images indicate that lines parallel to the top edges of the images are displaced to the right, lengthened, and they are closer to the base in successive frames. Still, the nearly circular outlines of the ooids are distorted into shapes that closely approximate ellipses. The longest and shortest diameters of the distorted ooids, which define the principal directions of the deformation, are inclined at different angles to the "horizontal" base in each image. The average of the maximum stretch calculated for ellipses 1, 2, and 3 in Figure 5.40b is $T_{max} \approx 1.27$, and the average of the minimum stretch is $T_{min} \approx 0.79$; so $R_s \approx 1.6$. For Figure 5.40c, $T_{max} \approx 1.59$, $T_{min} \approx 0.63$, and $R_s \approx 2.5$. For Figure 5.40d, $T_{max} \approx 2.0$, $T_{min} \approx 0.5$, and $R_s \approx 4.0$.

In these three examples, the original image was subjected to homogeneous pure strains. Because each ooid experienced exactly the same pure strain as the entire image, the ooid shapes captured the entire pure strain and the ooid axial ratios faithfully indicated that strain. Because the external shape of the distorted images reflects the character of the pure strain (pure shear versus simple shear versus general shear), the three (b), (c), or (d) frames of Figures 5.38, 5.39, and 5.40 can be distinguished on the basis of the orientation of the principal directions relative to the edges of the image. Note, however, that if the edges of the images were not visible and an observer saw only the shapes of individual distorted ooids, she or he would not be able to distinguish whether a particular pure strain accrued as pure shear or simple shear or general shear. Thus, although measuring finite strain is an important step toward understanding a deformation, determining the orientation and axial ratio of the strain ellipse does not provide a complete picture of the deformation.

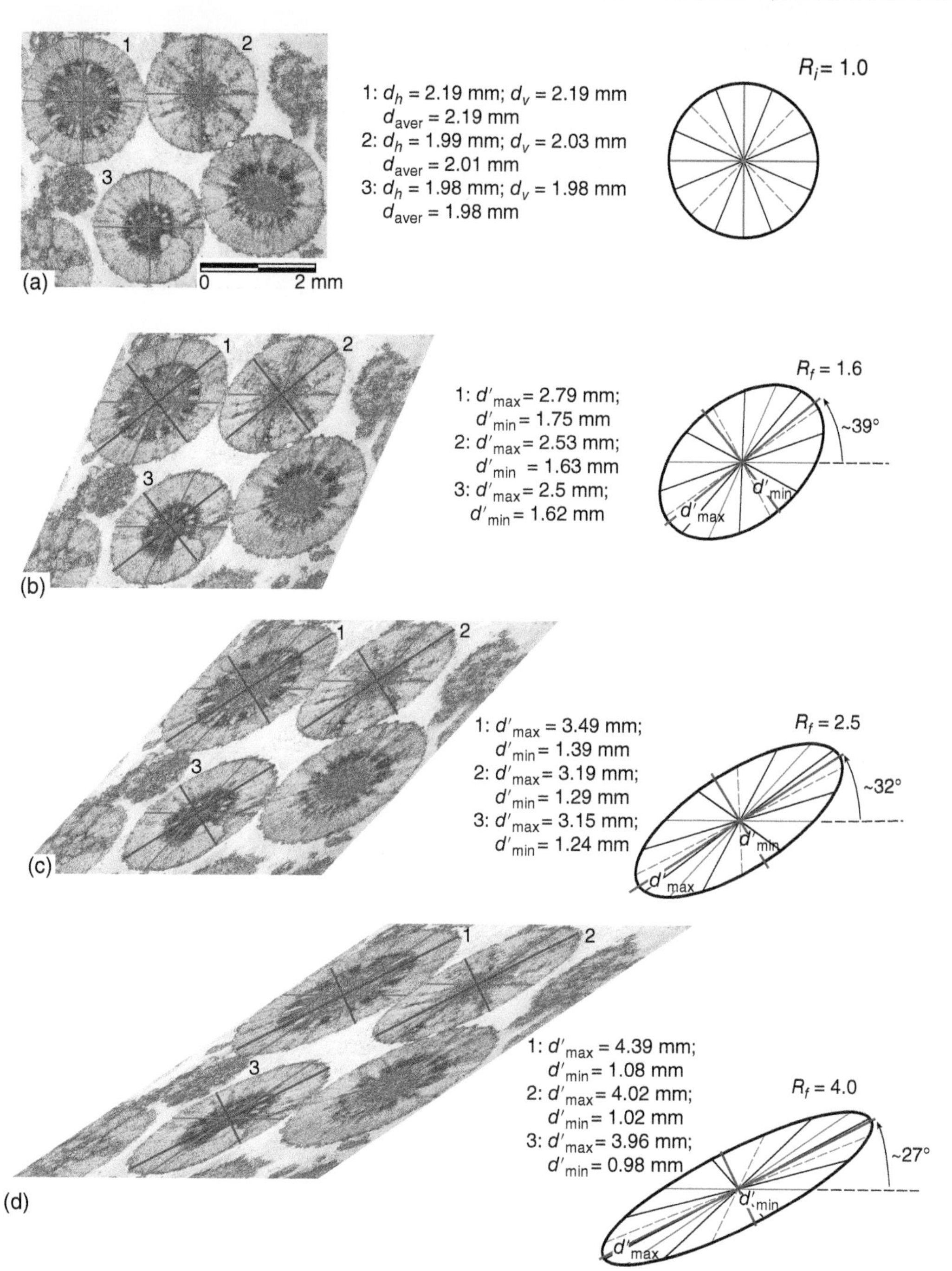

Figure 5.39 (a) Photomicrograph at left shows undeformed ooids. In the middle are measurements of horizontal and vertical diameters of ooids 1, 2, and 3. At right is a depiction of the simple shear imposed on the image. With no strain, R_S = 1.0 (b) Photomicrograph distorted in simple shear to R_S = 1.6 (depicted at right). In the middle are measurements of the longest and shortest diameters of ooids. (c) Same as (b) but with R_S = 2.5. (d) Same as (b) but with R_S = 4.0.

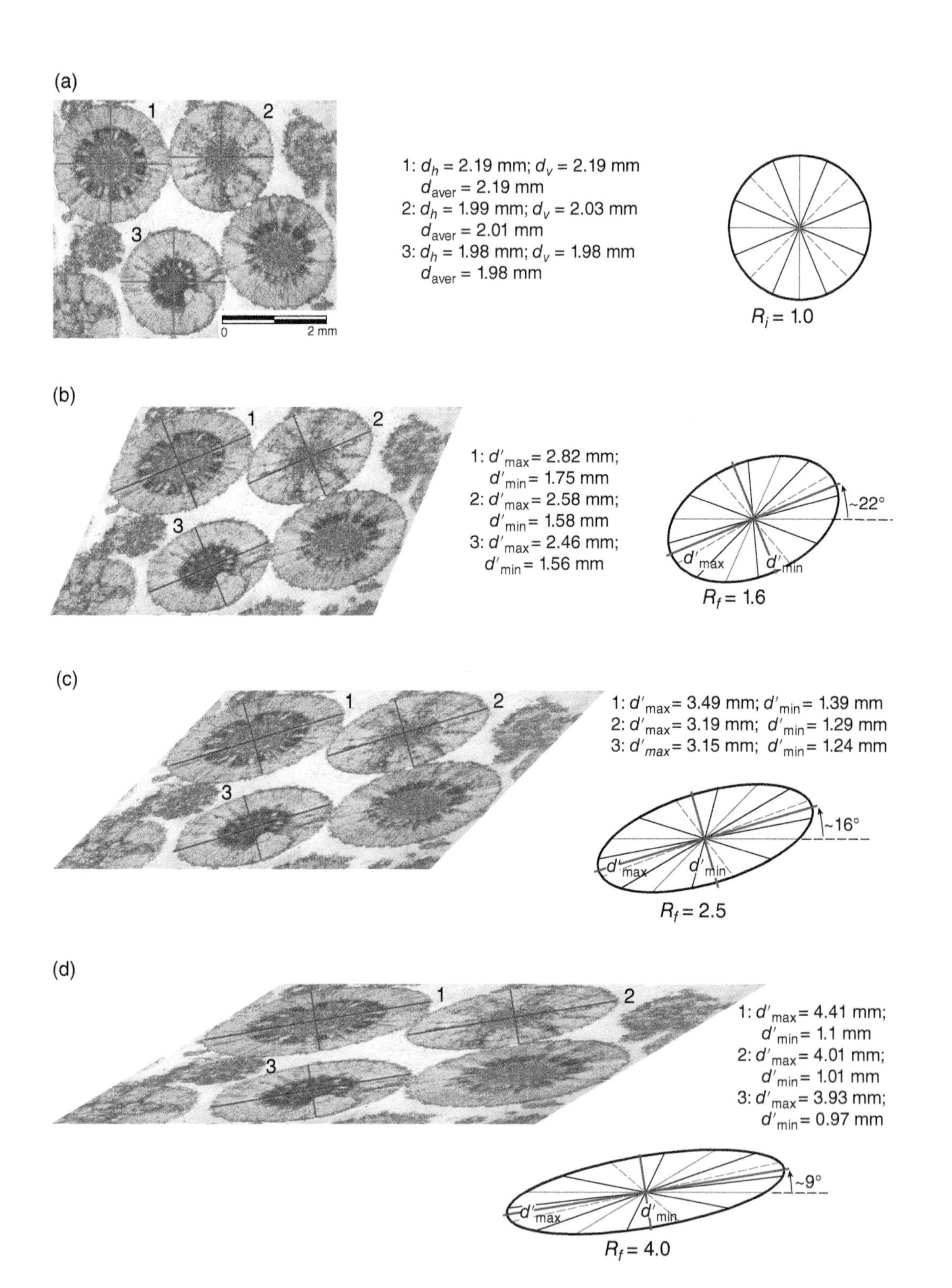

Figure 5.40 (a) Photomicrograph at left shows undeformed ooids. In the middle are measurements of horizontal and vertical diameters of ooids 1, 2, and 3. At right is a depiction of the general shear imposed on the image. With no strain, R_S = 1.0 (b) Photomicrograph distorted in general shear to R_S = 1.6 (depicted at right). In the middle are measurements of the longest and shortest diameters of ooids. (c) Same as (b) but with R_S = 2.5. (d) Same as (b) but with R_S = 4.0.

5B.6.3 Finite Strain Ellipsoid

By following an analysis similar to that outlined in Section 5B.6.1, it is possible to demonstrate that homogeneous, three-dimensional pure strain transforms an initial sphere into an ellipsoid whose shape and orientation reflect the pure strain, that is, the finite strain ellipsoid. Because the algebra required is more involved, we do not show it here.

5B.7 States of Strain and Strain Paths

5B.7.1 States of Strain

In Section 5A.6.1, we defined the state of strain as the totality of the pure strain component, including the magnitudes of longitudinal strains and shear strains in all directions and the manner in which the longitudinal and shear strain magnitudes change with direction. We used the finite strain ellipsoid as a convenient way to depict the state of strain. We can derive local values of the longitudinal strain or shear strain, or the shape and orientation of strain ellipse, from the values of the position gradient matrix **F** or the displacement gradient matrix **J**. In a real sense, then, either of these matrices fully encompasses the state of strain.

In Section 5B.4.13.1, we noted that changes in the magnitudes of the position gradient matrix for a particular Cartesian coordinate system lead to different states of strain. For any individual state of strain, particle position vectors in the undeformed state, particle position vectors in the deformed state, and particle displacement vectors exist independent of the orientation of the coordinate frame in which we specify them. Using the *vector transformation law* (see Appendix 4-I), we can use the magnitudes of a vector's two or three components in one coordinate frame to calculate the magnitudes of its two or three components in other, differently oriented coordinate frames. In fact, because the vectors are independent of the frame of reference (or coordinate frame) in which we specify them, no set of vector components in one coordinate frame is more fundamental than the set of components in any other coordinate frame. In a similar way, we can use the elements of the position gradient matrix or the displacement gradient matrix specified relative to a particular orientation of coordinate axes to calculate the elements of the position gradient or displacement gradient matrix defined relative to differently oriented coordinate axes. Determining or specifying the elements of either matrix with respect to a single orientation of coordinate axes is equivalent to determining or specifying the elements of the matrix relative to all orientations of coordinate axes. This is what is implied by the term "state of strain" – specifying the magnitudes of the matrix elements in one coordinate frame necessarily determines what will be their magnitudes in all other coordinate frames.

Two components specify a vector in two dimensions, and three components specify a vector in three dimensions. Four matrix elements specify position gradients or displacement gradients in two dimensions, and nine matrix elements specify position gradients or displacement gradients in three dimensions. The greater number of elements needed to specify position or displacement gradients is an indication that they exhibit greater complexity than vectors. We call entities like position gradients or displacement gradients, which exhibit this level of complexity beyond that of vectors, *tensor quantities* or *tensors*. The formal way to show that a quantity is a tensor is to show that the mathematical relationship between the elements in one coordinate frame and the elements in other, differently oriented coordinate frames is:

$$\mathbf{F}' = \mathbf{\Omega}^{-1}\mathbf{F}\mathbf{\Omega}. \tag{5.52}$$

This relationship is known as the *tensor transformation rule* and a quantity is a tensor if it follows the tensor transformation law. In Eq. (5.52), **F** denotes the position gradient in the current coordinate frame orientation (the coordinate frame in which one has specified the matrix elements), $\mathbf{F}'$ denotes the position gradient in the desired coordinate frame orientation (a new coordinate frame with a different orientation for which one wishes to determine the matrix elements), $\boldsymbol{\Omega}$ is the matrix that will rotate the current coordinate frame orientation into parallelism with the desired, new coordinate frame orientation, and $\boldsymbol{\Omega}^{-1}$ is the inverse of $\boldsymbol{\Omega}$. Because values of the position gradient or displacement gradient matrices in coordinate frames with different orientations conform to Eq. (5.52), position gradients and displacement gradients are tensors.

Vector quantities relate two scalar quantities (position coordinates at two points in space such as the origin of a coordinate frame and an identifiable particle) or "map" one scalar quantity (position coordinates of an individual particle in the undeformed state) into another scalar quantity (position coordinates of that particle in the deformed state). In the same way, tensor quantities relate two vector quantities (displacement vectors to position vectors) or "map" one vector quantity into another vector quantity (position vectors in the undeformed state to position vectors in the deformed state). This hierarchy of complexity, from scalar to vector to tensor, is apparent in a second nomenclature scheme for the different types of quantities. In that scheme, scalar quantities are *tensors of rank zero*, vector quantities are *tensors of rank one*, and tensor quantities are *tensors of rank two*. Thus, the position gradient and displacement gradients are *second-rank tensors*.

Knowing the components of the position gradient tensor or the displacement gradient tensor relative to any one coordinate frame, we can use Eq. (5.52) to determine the tensor components relative to a coordinate frame with a different orientation. The overall character of the tensor, the gestalt of the state of strain, remains the same regardless of which coordinate frame we use to specify it. Thus, although the magnitude of an individual tensor component will change, that is, $J_{12} \neq J'_{12}$, the overall character of the tensor does not. One manifestation of the independence of the state of strain from the coordinate system in which we specify it is the existence of quantities related to the tensor that do not change – are *invariant* – regardless of the coordinate system we choose to specify the tensor.

For second-rank tensors, there are three invariant quantities. The first invariant is

$$J_{11} + J_{22} = J'_{11} + J'_{22} \tag{5.53}$$

in two dimensions and

$$J_{11} + J_{22} + J_{33} = J'_{11} + J'_{22} + J'_{33} \tag{5.54}$$

in three dimensions. The second invariant is

$$\frac{\left(F_{ij}F_{ij} - F_{ii}F_{jj}\right)}{2} = \frac{\left(F_{12}^{\ 2} + F_{21}^{\ 2} - 2F_{11}F_{22}\right)}{2} \tag{5.55}$$

in two dimensions and

$$\frac{\left(F_{ij}F_{ij} - F_{ii}F_{jj}\right)}{2} = \frac{\left(F_{12}^{\ 2} + F_{13}^{\ 2} + F_{21}^{\ 2} + F_{23}^{\ 2} + F_{31}^{\ 2} + F_{32}^{\ 2} - 2F_{11}F_{22} - 2F_{11}F_{33} - 2F_{11}F_{22}\right)}{2} \tag{5.56}$$

in three dimensions. The third invariant of a second-rank tensor is

$$\det \mathbf{F} \tag{5.57}$$

in two dimensions or three dimensions.

5B.7.2 Strain Paths

We can use the matrix algebra outlined in the preceding sections to examine the positions of particles in successive increments of a single protracted deformation or at different stages in the evolution of rocks subjected to multiple deformations. In the equations that follow, $\boldsymbol{X}$ is the original position vector of a particle, $\boldsymbol{x}_1$ is the particle's position vector after the first deformation increment (or after the first of multiple deformations), $\boldsymbol{x}_2$ is the particle's position vector after the second deformation increment (or after the second of multiple deformations), $\boldsymbol{x}_3$ is the particle's position vector after the third deformation increment (or after the third of several deformations), and so on. Similarly, $\mathbf{F}_1$ is the position gradient matrix for the first deformation increment (or the first of the multiple deformations), $\mathbf{F}_2$ is the position gradient matrix for the second deformation increment (or the second of the multiple deformations), $\mathbf{F}_3$ is the position gradient matrix for the third deformation increment (or the third of the multiple deformations), and so on. From Eq. (5.35), we know that

$$\boldsymbol{x}_1 = \mathbf{F}_1 \cdot \boldsymbol{X}.$$

Similarly,

$$\boldsymbol{x}_2 = \mathbf{F}_2 \cdot \boldsymbol{x}_1$$

and

$$\boldsymbol{x}_3 = \mathbf{F}_3 \cdot \boldsymbol{x}_2.$$

Combining these equations, we relate the position vector at the end of the third increment to the original position vector by

$$\boldsymbol{x}_3 = \mathbf{F}_3 \cdot \mathbf{F}_2 \cdot \mathbf{F}_1 \cdot \boldsymbol{X}.$$

We relate the position vector at the end of the nth increment to the original position vector by

$$\boldsymbol{x}_n = \mathbf{F}_n \cdot \mathbf{F}_{n-1} \cdot \cdots \cdot \mathbf{F}_3 \cdot \mathbf{F}_2 \cdot \mathbf{F}_1 \cdot \boldsymbol{X}. \tag{5.58}$$

Note that the position gradient matrices for successively later increments or deformation episodes premultiply the position gradients for earlier increments or deformation episodes. The position gradient matrix for the total deformation is given by

$$\boldsymbol{F}_T = \mathbf{F}_n \cdot \mathbf{F}_{n-1} \cdot \cdots \cdot \mathbf{F}_3 \cdot \mathbf{F}_2 \cdot \mathbf{F}_1. \tag{5.59}$$

Matrix multiplication is not, in general, commutative; thus, $\mathbf{A} \cdot \mathbf{B} \neq \mathbf{B} \cdot \mathbf{A}$. In general, a change in the order in which deformation increments occur will result in a different total deformation.

From the magnitudes of the incremental position gradient matrices, we can define incremental principal strain directions, incremental principal strain magnitudes, and an incremental strain ellipse. Similarly, from the magnitudes of the total deformation position gradient matrix, we can define finite strain principal directions, finite principal strain magnitudes, and the finite strain ellipse. Only under special circumstances, when the incremental position gradient matrices $\mathbf{F}_i$ have a particular form, will the principal directions of the total deformation parallel the principal directions of the deformation increments.

5B.7.3 Velocity Gradient Tensor and Decomposition

As deformations proceed, the position vectors and displacement vectors of particles vary with time. Thus, the position in the deformed state, $\boldsymbol{x} = \boldsymbol{x}(\boldsymbol{X}, t)$, where

$$\boldsymbol{x}(\boldsymbol{X}, t) = x(\boldsymbol{X}, t)\boldsymbol{i} + y(\boldsymbol{X}, t)\boldsymbol{j} + z(\boldsymbol{X}, t)\boldsymbol{k}. \tag{5.60}$$

The displacement of the particle, $\boldsymbol{u}(\boldsymbol{X}) = \boldsymbol{u}(\boldsymbol{X}, t)$ is

$$\boldsymbol{u}(\boldsymbol{X}, t) = u_X(\boldsymbol{X}, t)\boldsymbol{i} + u_Y(\boldsymbol{X}, t)\boldsymbol{j} + u_Z(\boldsymbol{X}, t)\boldsymbol{k}. \tag{5.61}$$

The derivative of the vector $\boldsymbol{x}$ taken with respect to the scalar argument t gives the change in position per unit time or a velocity of the point:

$$\frac{\mathrm{d}\boldsymbol{x}}{\mathrm{d}t} = \left[\frac{\mathrm{d}x(\boldsymbol{X})}{\mathrm{d}t}\right]\boldsymbol{i} + \left[\frac{\mathrm{d}y(\boldsymbol{X})}{\mathrm{d}t}\right]\boldsymbol{j} + \left[\frac{\mathrm{d}z(\boldsymbol{X})}{\mathrm{d}t}\right]\boldsymbol{k}. \tag{5.62}$$

In examining the kinematics of deformation, it is convenient to change the perspective adopted elsewhere in the chapter and view parameters as functions of the spatial coordinates rather than the initial coordinates of particles. Thus, we consider particles identified by their position vectors $\boldsymbol{x}$, and the velocity of the particle referred to spatial coordinates is

$$\boldsymbol{v} = v_x(\boldsymbol{x})\boldsymbol{i} + v_y(\boldsymbol{x})\boldsymbol{j} + v_z(\boldsymbol{x})\boldsymbol{k}. \tag{5.63}$$

Equation (5.63) gives the velocities of particles in a region at one instant in time. This is a vector field, the *velocity field*. If the region is undergoing deformation, then the velocity at one position in space differs from the velocities of the nearby positions. The **velocity gradient tensor** is the spatial derivative of the velocity field. In two dimensions, the velocity gradient tensor $\mathbf{L}$ is a 2×2 matrix where:

$$L_{11} = \frac{dv_x}{dx} \quad L_{12} = \frac{dv_x}{dy}$$
$$L_{21} = \frac{dv_y}{dx} \quad L_{22} = \frac{dv_y}{dy}$$

In three dimensions, the velocity gradient tensor $\mathbf{L}$ is a 3×3 matrix where:

$$L_{11} = \frac{dv_x}{dx} \quad L_{12} = \frac{dv_x}{dy} \quad L_{13} = \frac{dv_X}{dz}$$
$$L_{21} = \frac{dv_y}{dx} \quad L_{22} = \frac{dv_y}{dy} \quad L_{23} = \frac{dv_Y}{dz}$$
$$L_{31} = \frac{dv_z}{dx} \quad L_{32} = \frac{dv_z}{dy} \quad L_{33} = \frac{dv_z}{dz}$$

Within a region undergoing homogeneous deformation, where the displacement gradients have equal values at different positions, the velocity gradients will also have equal magnitudes at different locations. The velocity at a point is a product of the velocity gradient with respect to position times the position. For a two-dimensional deformation:

$$\begin{vmatrix} v_x \\ v_y \end{vmatrix} = \begin{vmatrix} L_{11} & L_{12} \\ L_{21} & L_{22} \end{vmatrix} \begin{vmatrix} x \\ y \end{vmatrix} = \begin{vmatrix} \frac{dv_x}{dx} & \frac{dv_x}{dy} \\ \frac{dv_y}{dx} & \frac{dv_y}{dy} \end{vmatrix} \begin{vmatrix} x \\ y \end{vmatrix} \tag{5.64a}$$

Using matrix shorthand yields a form appropriate for two- or three-dimensional flows:

$$\boldsymbol{v} = \mathbf{L} \cdot \boldsymbol{x} = \left[\frac{dv_i}{dx_j}\right] \cdot \boldsymbol{x}. \tag{5.64b}$$

As is true of the position gradient tensor and the displacement gradient tensor, the velocity gradient tensor captures both the pure strain and the rigid rotation components of the total deformation. We postpone outlining explicit expressions for the **strain rate**, **deformation rate tensor**, or **rate-of-deformation tensor**, which captures the pure strain component of the deformation, and the **rotation rate**, **spin tensor**, or **vorticity tensor**, which captures the rigid rotation component of the deformation, in order to address some characteristics of velocity gradient tensors in relatively straightforward flows.

The eigenvectors of $\mathbf{L}$ define the orientations of *instantaneously stable lines* in a flow, that is, lines that do not, at that particular instant in time, tend to reorient relative to each other. Depending upon the overall kinematics of the flow, the eigenvectors may rotate relative to a fixed external reference. Such a rotation relative to an external reference does not affect the state of strain, so we ignore its effects at the outset. Consider a material line, consisting of a collection of particles, aligned parallel to an eigenvector direction. That material line will not change its orientation at a particular instant, although the collection of particles defining it may tend to converge (that is, exhibit a finite shortening rate parallel to the eigenvector), tend to diverge (that is, exhibit a finite lengthening rate parallel to the eigenvector), or tend to retain their relative positions (that is, exhibit a zero rate of change in length parallel to the eigenvector). In coaxial deformations, the two eigenvectors in two-dimensional deformations or three eigenvectors define three unique, mutually perpendicular directions. In non-coaxial deformations, the two eigenvectors in two-dimensional deformations or three eigenvectors in three-dimensional deformations are no longer mutually perpendicular. In some non-coaxial deformations, two eigenvectors can collapse into a single stable orientation.

A flow is *steady* if the values of the velocity gradient matrix **L** do not change with time. In steady flows, the orientations of instantaneously stable lines do not change from one instant in time to the next. The eigenvectors of **L**, therefore, define the orientation of lines whose orientations remain fixed as the strain accrues. In Chapter 4, we called the equivalent directions in a displacement field "eigendirections." The name applied to these directions in steady velocity fields is *flow apophyses*. Lines parallel to a flow apophysis may shorten, lengthen, or remain the same length as deformation proceeds depending upon the values of **L**. The important point is that any movement of particles along a flow apophysis is parallel to it. Lines oblique to flow apophyses are not stable and reorient during deformation. The apparent movement of lines is away from flow apophyses that experience shortening and toward flow apophyses that experience no change in length or experience lengthening.

We can see the effects of the reorientation of lines by examining coaxial and non-coaxial two-dimensional flows. In Chapter 4, we showed coaxial and non-coaxial, two-dimensional displacement fields where particle displacements gave rise to the finite reorientation of lines oblique to the eigendirections (cf. Figures 4.37 and 4.38). At that time, we noted that the movement of the particles that define lines oblique to the eigendirections is fundamentally different from that involved in the solid body or rigid rotation because the distance between them changed as they reoriented. The same is true for the movements of the particles along lines oblique to the flow apophyses – they do not undergo rigid rotation because the distance between them changes as they move. Some geologists use the term "internal rotation" to refer to the change in the orientation of lines during deformation. In order to underscore the unique character of this movement, we prefer to say that the lines *reorient* rather than undergo any type of rotation.

Lines oblique to flow apophyses reorient so that they become more nearly parallel to flow apophyses along which there is either zero rate of change in length (such as the shear direction in simple shearing flows) or a measurable rate of lengthening (such as the direction experiencing the maximum rate of elongation in pure shearing flows). It is now common to call such a flow apophysis a *fabric attractor*. In two-dimensional coaxial deformations, the principal directions of the deformation rate matrix (D_i) are flow apophyses. In a steady coaxial deformation, lines oblique to the flow apophyses become more nearly parallel to the D_1 ($= S_1$) direction, which experiences a measurable rate of lengthening. The D_1 direction is a fabric attractor in these deformations. In simple shearing, the shear direction is a flow apophysis along which there is zero rate of change in length. In steady simple shearing, lines oblique to the shear direction become more nearly parallel to it over time. The shear direction is, then, a fabric attractor. In general shearing, the flow apophysis experiencing a measurable rate of line lengthening is the fabric attractor.

Coaxial three-dimensional deformations are similar to coaxial two-dimensional deformations in that the principal directions of the deformation rate matrix (D_i) are flow apophyses and fix the orientations of the strain principal directions (S_i) for steady flows. The $D_1 = S_1$ direction is a fabric attractor. Three-dimensional non-coaxial deformations can exhibit a wide range of flow geometries. When the velocity gradient tensor, **L**, is an upper-triangular matrix, the range of flow geometries is restricted sufficiently that one can recognize patterns resembling those seen in two-dimensional deformations. In flows where $\mathbf{L}$ is an upper-triangular matrix, one of the three eigenvectors is perpendicular to a plane that contains the other two eigenvectors. Within the plane, the two eigenvectors take orientations comparable to the orientations of eigenvectors in two-dimensional, non-coaxial flows, that is, they can: (1) define two directions oblique to each other, where the one with a measurable rate of lengthening is the fabric attractor; or (2) define a single direction, the shearing direction in three-dimensional simple shearing, that is, the fabric attractor.

In order to understand how the values of velocity gradient tensor components relate to the coaxial versus non-coaxial character of deformation or to whether the flow apophyses are orthogonal or not, we factor the velocity gradient tensor $\mathbf{L}$ into two components:

$$\mathbf{L} = \frac{\left[\mathbf{L}+\mathbf{L}^{\mathrm{T}}\right]}{2} + \frac{\left[\mathbf{L}-\mathbf{L}^{\mathrm{T}}\right]}{2} = \mathbf{D} + \mathbf{W}. \quad (5.65)$$

The component $\mathbf{D}$, called the *rate-of-deformation tensor*, *deformation rate tensor*, or *strain rate tensor*, is symmetric. Thus, $D_{ij} = D_{ji}$ or $\mathbf{D} = \mathbf{D}^{\mathrm{T}}$. The component $\mathbf{W}$, called the *rotation rate tensor*, the *spin tensor*, or the *vorticity tensor*, is antisymmetric or skew-symmetric. Thus, $W_{ij} = -W_{ji}$ or $\mathbf{W} = -\mathbf{W}^{\mathrm{T}}$. The eigenvectors of $\mathbf{D}$ are mutually perpendicular, and they define the orientations of lines that experience no instantaneous shear straining. One eigenvector is the direction that experiences the maximum instantaneous rate of lengthening, and another is the direction that experiences the maximum instantaneous rate of shortening. For this reason, eigenvectors of $\mathbf{D}$ are known as *instantaneous stretching axes*. $\mathbf{W}$ gives the rate of spin of the particles that fall along the principal directions of $\mathbf{D}$ at a particular instant in time.

If $\mathbf{W} = 0$ in a steady flow, the $\mathbf{L}$ must equal $\mathbf{L}^{\mathrm{T}}$ and must itself be symmetric. The rate of spin is zero, and the eigenvectors of $\mathbf{D}$ maintain the same orientations as deformation proceeds. If $\mathbf{D} = 0$ in a steady flow, the $\mathbf{L}$ must equal $-\mathbf{L}^{\mathrm{T}}$ and must itself be antisymmetric. The deformation in this case devolves into rigid rotation with $\mathbf{W}$ giving the rate of spin.

5B.8 Vorticity

The vorticity of a velocity field in a particular direction refers to the tendency for particles to spin in that direction. To begin exploring vorticity, consider the differences between coaxial and non-coaxial two-dimensional deformations. The velocity gradient tensor $\mathbf{L}$ is symmetric for a coaxial deformation, that is, $\mathbf{L} = \mathbf{L}^{\mathrm{T}}$. Thus, the rotation rate tensor $\mathbf{W} = 0$ and $\mathbf{L} = \mathbf{D}$. In a steady flow, the strain rate principal directions (D_1 and D_2) do not change their orientations and they coincide with the strain principal directions (S_1 and S_2) at all times. Lines oblique to the principal directions will reorient during deformation, tending to align with the direction of maximum lengthening (S_1). Lines at the same angle from, but on opposite sides of the S_1 principal direction, experience equal tendencies to reorient, but their reorientation has the opposite sense and they "cancel each other." Such deformation has no net vorticity along a direction perpendicular to the plane of deformation.

The velocity gradient tensor $\mathbf{L}$ is not symmetric for non-coaxial deformation, that is, $\mathbf{L} \neq \mathbf{L}^{\mathrm{T}}$. The rotation rate tensor $\mathbf{W}$ is, therefore, nonzero. Particles falling along the strain rate principal directions (D_1 and D_2) at any instant experience a tendency to spin about an axis perpendicular to the plane. As a result, the collections of particles experiencing the maximum rate of shortening or lengthening tend to move off the strain rate principal directions. As deformation progresses, the directions of greatest finite lengthening and greatest finite shortening, the strain principal directions S_1 and S_2, diverge from the instantaneous shortening axes D_1 and D_2. Further, the particles along either set of principal directions at one instant are different from those along the principal directions at the next instant. Such deformation has net vorticity along a direction perpendicular to the plane of deformation.

Coaxial three-dimensional deformations are similar to coaxial two-dimensional deformations in that the principal directions of the deformation rate matrix (D_i) parallel the strain principal directions (S_i) in steady flows. Lines at equal angles from, but on opposite sides any of the strain principal direction, experience equal tendencies to reorient, but their reorientation has the opposite sense. Since the two lines with opposite tendencies to reorient "cancel each other," these deformations have no net vorticity about lines perpendicular to any principal plane. When the velocity gradient tensor **L** for a three-dimensional non-coaxial deformation is an upper-triangular matrix, one eigenvector will be perpendicular to a plane containing the other two eigenvectors. That unique eigenvector also parallels one of the strain rate and strain principal directions. In any plane that contains that unique eigenvector of **L**, two lines at equal angles but opposite sides of the eigenvector experience an equal but opposite tendency to reorient. So, there is no net vorticity about the normal to the plane containing the unique eigenvector. However, eigenvectors lying in planes perpendicular to the unique eigenvector are not parallel to the instantaneous shortening axes D_i and D_j or the strain principal directions $_i$ and S_j. As deformation progresses, the strain principal directions within the plane diverge from the instantaneous shortening axes in the plane, the particles along either set of principal directions in the plane at one instant are different from those along the principal directions at the next instant, and the deformation has net vorticity along the normal to the plane.

5B.8.1 Vorticity Vector

We calculate the vorticity of a velocity field about a direction using

$$\text{Vorticity} = \nabla \times \boldsymbol{v} = \left(\frac{dv_z}{dy} - \frac{dv_y}{dz}\right)\boldsymbol{i} + \left(\frac{dv_x}{dz} - \frac{dv_z}{dx}\right)\boldsymbol{j} + \left(\frac{dv_y}{dx} - \frac{dv_x}{dy}\right)\boldsymbol{k}. \tag{5.66}$$

In two-dimensional deformations where there are no velocity components parallel to the z axis and no change in the other velocity components in the z direction, the only term in Eq. (5.66) that can have a finite value is the coefficient of $\boldsymbol{k}$. The vorticity vector in two-dimensional flows must be perpendicular to the deformation plane of the deformation if it exists. The magnitude of the vorticity vector measures the rate of spin of particles about the vorticity vector. If **L** is symmetric and $dv_y/dx = dv_x/dy$, then even the component of $\boldsymbol{k}$ vanishes, and the deformation possesses no vorticity.

A three-dimensional deformation where **L** has an upper triangular form and only dv_y/dx or dv_x/dy is nonzero defines a flow with a component of simple shear whose shear direction lies within the x-y plane. In such a case, the vorticity vector has only a component parallel to the $\boldsymbol{k}$ unit vector, and it is parallel to the z axis. The vorticity vector is perpendicular to the shear direction and with it defines a shear plane that is perpendicular to the x-y plane. The magnitude of the vorticity vector again measures the rate of spin within the x–y plane, that is, about the z axis. A three-dimensional deformation where **L** has an upper triangular form and only dv_x/dz is nonzero defines a flow with a component of simple shear whose shear direction is parallel to the x axis. In this case, only the coefficient of the $\boldsymbol{j}$ unit vector of the vorticity vector is nonzero, that is, the vorticity vector is parallel to the y axis. This fixes the shear plane parallel to the x-y plane, and the magnitude of the vorticity vector measures the rate of spin perpendicular to the y axis, that is, the rate of spin in the x–z plane. When all three off-diagonal terms are nonzero, the deformation has a simple shear component oblique to the three coordinate axes. Regardless of the orientation of that shear direction, the vorticity vector is oriented perpendicular to it and therefore defines a shear plane. The magnitude of the vorticity vector measures the rate of spin in the plane that is normal to the vorticity vector and contains the shear direction. One can determine the orientation and magnitude of the vorticity vector for a deformation where **L** has a more general form. The orientation of the vorticity vector defines the direction about which particles have the greatest rate of spin, and the magnitude of the vorticity vector gives the rate of spin.

In many tectonic settings, geologists identify a shear direction that does not reorient during the deformation and is a fabric attractor. These observations are consistent with steady simple shearing or general shearing deformations. By tradition, geologists define the shear direction as the *tectonic a axis*. The direction that lies in the shear plane and is normal to the shear direction is the *tectonic b axis*. The normal to the shear plane defines the *tectonic c axis*. In such a frame of kinematic axes, the tectonic b axis will parallel the vorticity vector. Looking along the b axis or vorticity vector direction affords the best chance to see shear sense indicators in these tectonic settings.

In simple shearing velocity fields, the orientations of the instantaneous stretching axes and the vorticity vector do not change with time. Note, however, that the directions of maximum finite lengthening (S_1) and maximum finite shortening (S_3) rotate like the hands of a clock about the vorticity vector as the deformation proceeds (the motion may, however, be either clockwise or counterclockwise, depending upon the components of **L**). The vorticity vector will parallel the intermediate principal direction of the finite strain (S_2) at all stages during simple shearing deformation. Here too, the best perspective to view simple shear deformation is along the vorticity vector direction, which yields the displacement field depicted in Figures 4.17 and 4.35.

General shearing deformations are combinations of simple shearing and orthogonal convergence and divergence that produce particle movements consistent with upper-triangular position gradient matrices. In steady general shearing deformations too, the orientations of the instantaneous stretching axes and the vorticity vector do not change with time. Different relative magnitudes of the components of **L** yield variations in the geometry of the flow. In some cases, the vorticity vector is perpendicular to the plane that contains the direction of greatest instantaneous lengthening and the direction of greatest instantaneous shortening. In this case, the directions of maximum finite lengthening (S_1) and maximum finite shortening (S_3) rotate like the hands of a clock about the vorticity vector direction as the deformation proceeds, and the vorticity vector parallels the intermediate principal direction of the finite strain state (S_2). In the second class of steady general shearing flows, the vorticity vector is parallel to the direction of maximum instantaneous rate of lengthening and is perpendicular to a plane that contains the direction of greatest instantaneous rate of shortening and the intermediate instantaneous stretching axis. The vorticity vector direction itself is a fabric attractor in this case. This flow yields directions of maximum finite lengthening (S_1) that parallel the vorticity vector, and intermediate strain principal direction (S_2) and maximum shortening principal strain direction (S_3) rotate like the hands of a clock about the vorticity vector direction as the deformation proceeds. In the third class of steady general shearing flows, the vorticity vector is parallel to the direction of maximum instantaneous rate of shortening and is perpendicular to a plane that contains the directions of the greatest and intermediate instantaneous rate of lengthening. This flow yields directions of maximum finite shortening (S_3) parallel to the vorticity vector, and intermediate strain principal direction (S_2) and maximum lengthening principal strain direction (S_3) rotate like the hands of a clock about the vorticity vector as the deformation proceeds.

The movement of particles during deformation need not conform to orthogonal convergence and divergence, simple shearing, or general shearing. In these most complex flows, the vorticity vector will be oblique to all three principal planes of the strain tensor, and the finite strains will not have readily predictable orientations relative to the vorticity vector. Fortunately, many geological deformations conform to at least approximately one of the more restrictive types of flow, such as orthogonal convergence and divergence, simple shearing, or general shearing flows. These flows are easier to decipher in the field, and we can test our understanding of their development using relatively simple numerical models.

5B.8.2 Kinematic Vorticity Number

In Chapter 4, we described general shearing displacement fields as combinations of pure shearing and simple shearing displacement fields. In Section 5B.8.1, we again referred to general shearing flows as combinations of pure shearing and simple shearing. We assess the relative contributions of the coaxial and non-coaxial components of the deformation by calculating the magnitude of a dimensionless number called the *kinematic vorticity number*, W_k. For two-dimensional deformations,

$$W_k = \frac{(W_1 - W_2)}{(D_1 - D_2)}, \tag{5.67}$$

where W_1 and W_2 are the principal values of the rotation rate tensor **W** and D_1 and D_2 are the principal values of the strain rate tensor **D**. The kinematic vorticity number W_k is 0 for coaxial deformations, such as in cases of pure shear or orthogonal convergence and divergence. The kinematic vorticity number W_k is one in simple shearing, the archetype of non-coaxial deformation. For general shearing deformations, which can be envisioned as mixtures of simple shearing and orthogonal convergence and divergence, $0 \leq W_k \leq 1$. The kinematic vorticity number does not increase linearly with an increase in the amount of simple shearing.

One can use finite strain values to evaluate the kinematic vorticity number of the steady flow responsible for the flow. Using an upper-triangular form for the position gradient tensor where

$$\begin{vmatrix} F_{11} & F_{12} \\ F_{21} & F_{22} \end{vmatrix} = \begin{vmatrix} T_x & \gamma \\ 0 & T_y \end{vmatrix}$$

the kinematic vorticity number is given by

$$W_k = \frac{\gamma}{\sqrt{\left[2\ln^2(T_x) + 2\ln^2(T_y) + \gamma^2\right]}}. \tag{5.68a}$$

If $T_x \cdot T_y = 1$,

$$W_k = \cos\left\{\tan^{-1}\left[2\ln(T_x)/\gamma\right]\right\}. \tag{5.68b}$$

5B.9 Summary

Deformations relate the original positions of particles, denoted by position vectors $\boldsymbol{X}$, to the final positions of those particles, denoted by position vectors $\boldsymbol{x}(\boldsymbol{X})$ through a displacement field, $\boldsymbol{u}(\boldsymbol{X})$. When particles at different initial positions experience different displacements, the deformation produces solid-body rotation, pure strain, or pure strain with solid body rotation depending upon the values of the gradients of particle displacements. Pure strains are manifest as changes in the lengths of lines (longitudinal strains) and changes in the angles between lines (shear strains). Longitudinal and shear strains are explicit functions of the gradient of the final position vectors,

$d\boldsymbol{x}/d\boldsymbol{X}$, or gradients of the displacement vectors, $d\boldsymbol{u}/d\boldsymbol{X}$.

The position gradient tensor $\mathbf{F} = [dx_i/dX_j]$ and the displacement gradient tensor $\mathbf{J} = [du_i/dX_j]$ define explicitly the state of strain. From the values of the components of either tensor, one can determine the longitudinal strain magnitudes and shear strain magnitudes in different directions through a deformed material. In homogeneous deformations, the components of the position gradient tensor and the components of the displacement gradient tensor, do not change from point to point. In those cases, the final position $\boldsymbol{x}$ of a particle originally at $\boldsymbol{X}$ is given by $\boldsymbol{x} = \boldsymbol{F} \cdot \boldsymbol{X}$ and the displacement $\boldsymbol{u}$ of that particle is given by $\boldsymbol{u} = \boldsymbol{J} \cdot \boldsymbol{X}$. The values of the components of these tensors determine explicitly the axial ratio of the strain ellipsoid, the orientation of the strain principal directions, and the volume or area ratio of the deformation.

The tensor formulation provides particular insight into deformation histories and strain paths. The matrix representing the total deformation position gradient tensor is the product of the several matrices representing incremental position gradient tensors. The order in which incremental deformations occur control the character of the final deformation, and this is reflected by the fact that because matrix multiplication is not commutative, different orders of incremental deformations yield different total deformations.

A tensor formulation can also be used to analyze instantaneous deformation by using an analogue of the position gradient tensor – the velocity gradient tensor, **L**. The velocity gradient tensor differs significantly from the position gradient tensor and displacement gradient tensors described here in that velocity gradients are measured with respect to current positions, $\boldsymbol{x}$. Eigenvectors of **L** define instantaneously stable lines in a flow. By decomposing the velocity gradient tensor into the symmetric rate of deformation tensor **D** and the skew-symmetric vorticity tensor **W**, one can determine the orientations of the instantaneous shortening axes (the eigenvectors of **D**) and whether particles along them tend to remain on them (making the deformation coaxial) or tend to spin relative to them (making the deformation non-coaxial).

Coaxial flows have no net vorticity. The net vorticity of a non-coaxial flow is captured by the existence of a vorticity vector, which defines the direction about which particles tend to spin. The magnitude of the vorticity vector measures the rate of spin. The kinematic vorticity number, W_k, is a measure of the relative magnitudes of coaxial and non-coaxial components of the flow. If a velocity field is steady, that is, if the velocity gradient values do not change with time, the velocity gradient, the eigenvectors of **L** define the orientations of stable lines in a flow, called flow apophyses. Lines oblique to flow apophyses steadily reorient during steady flows. With time, lines tend toward parallelism with particular flow apophyses, known as fabric attractors.

Appendix 5-I

Geologists typically measure strain by comparing the lengths of lines in the undeformed and deformed states, by comparing the angles between pairs of lines in the undeformed and deformed states, or by evaluating the shapes of deformed markers. Longitudinal strain measures, shear strain measures, and the shape and orientation of the finite strain ellipse are explicitly, or functionally, related to the spatial gradients of the displacement field. Thus, one can measure strain by measuring the magnitudes of the spatial gradients of the displacement field. This approach is particularly useful when treating situations where it is easier to measure the displacements of particles' direction, whether this is done using GPS data or using the relative movements of rock across deformation elements like stylolites, mineral-filled veins, or faults. The most straightforward way to convert measured displacements into displacement gradients from which one can calculate strain requires a slight change in perspective from that adopted elsewhere in the chapter. Instead of identifying particles by their positions in the undeformed state $\boldsymbol{X}$, we here use their positions in the deformed state $\boldsymbol{x}$. We can calculate the original position of a particle using

$$\boldsymbol{X} = \mathbf{G} \cdot \boldsymbol{x}, \tag{5.40b}$$

where $\mathbf{G}$ is the inverse of the position gradient matrix $\mathbf{F}$. $\mathbf{G}$ is also a position gradient matrix referred to the deformed state; we here call it the *reciprocal position gradient* matrix. Following reasoning parallel to that outlined in Eqs. (5.36) through (5.38), one can show that the product $\mathbf{G}^{\mathrm{T}} \cdot \mathbf{G}$ is a symmetric matrix $[\boldsymbol{\lambda}']$, the *reciprocal quadratic elongation* matrix. The components of $[\boldsymbol{\lambda}']$ are squares of the pure strain elements of the reciprocal position gradient matrix. The eigenvectors of $[\boldsymbol{\lambda}']$ define the principal directions of the pure strain (referred to the coordinate axes in the deformed state), and the reciprocals of the square roots of the eigenvalues of $[\boldsymbol{\lambda}']$ give the principal strain magnitudes.

The reciprocal position gradient matrix is related to the *reciprocal displacement* of a particle

$$\boldsymbol{U} = \boldsymbol{X} - \boldsymbol{x}, \tag{5A.1}$$

$\boldsymbol{U}$ is the displacement vector required to move a particle from its position in the deformed state to its position in the undeformed state. Combining Eqs. (5.40b) and (5A.1) yields

$$\boldsymbol{U} = \mathbf{G} \cdot \boldsymbol{x} - \boldsymbol{x} = \left[\mathbf{G} - \mathbf{I}\right] \cdot \boldsymbol{x} = \mathbf{J}' \cdot \boldsymbol{x} = \left[\frac{\mathrm{d}U_i}{\mathrm{d}x_j}\right] \cdot \boldsymbol{x}. \tag{5A.2}$$

$\mathbf{J}' = [\mathbf{G} - \mathbf{I}] = [\mathrm{d}U_i/\mathrm{d}x_j]$ is the *reciprocal displacement gradient* matrix. $\mathbf{G}$ and $\mathbf{J}'$ are 2×2 matrices for two-dimensional deformations and 3×3 matrices for three-dimensional deformations from which one can determine the orientations and magnitudes of the principal strain directions. We here consider the case of a homogeneous two-dimensional deformation.

By measuring the reciprocal displacements of a series of particles, we can constrain the values of the reciprocal displacement gradient matrix $\mathbf{J}'$. In component form, the reciprocal displacement of particle a is

$$\boldsymbol{U}(\boldsymbol{a}) = U_x(a)\boldsymbol{i}' + U_y(a)\boldsymbol{j}', \tag{5A.3}$$

where: (1) $\boldsymbol{U}_x(a)$ is the component of the reciprocal displacement vector parallel to the x axis of a coordinate frame superposed on the deformed state and $\boldsymbol{i}'$ is the unit vector parallel to that x axis; and (2) $\boldsymbol{U}_y(a)$ is the component of the reciprocal displacement vector parallel to the y axis of a coordinate frame superposed on the deformed state and $\boldsymbol{j}'$ is the unit vector parallel to that y axis. Written in terms of the reciprocal displacement gradient matrix, Eq. (5A.3) is

$$\begin{aligned} U_x(a) = &\left[\mathrm{d}U_x(a)/\mathrm{d}x\right] x_a \\ &+ \left[\mathrm{d}U_x(a)/\mathrm{d}y\right] y_a \end{aligned} \tag{5A.4a}$$

$$\begin{aligned} U_y(a) = &\left[\mathrm{d}U_y(a)/\mathrm{d}x\right] x_a \\ &+ \left[\mathrm{d}U_x(a)/\mathrm{d}y\right] y_a \end{aligned} \tag{5A.4b}$$

To measure the reciprocal displacement gradients, determine the values of the selected particles along the coordinate axes x and y. For i particles that fall on the x axis, the measured $U_x(i)$ and $U_y(i)$ values are functions of x only. Plots of $U_x(i)$ versus $x(i)$ and $U_y(i)$ versus $x(i)$ should yield linear curves whose slopes are $\Delta U_x/\Delta x \approx \mathrm{d}U_x/\mathrm{d}x = \mathrm{G}_{11} = \mathrm{J}'_{11} + 1$ and $\Delta U_x/\Delta y \approx \mathrm{d}U_x/\mathrm{d}y = \mathrm{G}_{12} = \mathrm{J}'_{12}$, respectively. Similarly, for j particles that fall on the y axis, the measured $U_x(j)$ and $U_y(j)$ values are functions of y only. Plots of $U_x(j)$ versus $y(j)$ and $U_y(j)$

versus $y(j)$ should yield linear curves whose slopes are $\Delta U_y/\Delta x \approx dU_y/dx = G_{21} = J'_{21}$ and $\Delta U_y/\Delta y \approx dU_y/dy = G_{22} = J'_{22} + 1$, respectively. The reciprocal quadratic elongation matrix $[\lambda']$ is then

$$[\lambda'] = \mathbf{G}^{\mathrm{T}} \cdot \mathbf{G} = [\mathbf{J}' + \mathbf{I}]^{\mathrm{T}} \cdot [\mathbf{J}' + \mathbf{I}] = \left[\frac{\Delta U_x}{\Delta x} + \mathbf{I}\right]^{\mathrm{T}} \cdot \left[\frac{\Delta U_x}{\Delta x} + \mathbf{I}\right]. \tag{5A.5}$$

References

Badoux, H. (1963). Les bélemnites tronçonnees de Laytron (Valais). *Bulletin des Laboratoires de Géologie, Minéralogie, Géophysique et du Musée géologique de l'Université de Lausanne* **138**: 1–7.

Campbell, S.D.G., Reedman, A.J., and Howells, M.F. (1985). Regional variations in cleavage and fold development in North Wales. *Geological Journal* **20**: 43–52.

Dunnet, D. (1969). A technique of finite strain analysis using elliptical particles. *Tectonophyics* **7**: 117–136.

Elliott, D. (1970). Determination of finite strain and initial shape from deformed elliptical objects. *Geological Society of America Bulletin* **81**: 2221–2236.

Fry, N. (1979). Random point distributions and strain measurement in rocks. *Tectonophysics* 60: 89–105.

Gilotti, J.A. and Kumpulainen, R. (1986). Strain softening induced ductile flow in the Särv thrust sheet, Scandinavian Caledonides. *Journal of Structural Geology* **8**: 441–455.

Lisle, R.J. (1985). *Geological Strain Analysis: A Manual for the R_f/ϕ Method*, 95. Oxford: Pergamon Press.

Milton, N.J. (1980). Determination of the strain ellipsoid from measurements on any three sections. *Tectonophysics* **64**: T19–T27.

Mitra, S. (1976). A quantitative study of the deformation mechanisms and finite strain in quartzites. *Contributions to Mineralogy and Petrology* **59**: 203–226.

Mitra, S. (1977). Studies on deformation mechanisms and finite strain in quartzites and their relation to structures of various scales within the South Mountain Anticline. Unpublished Ph. D. dissertation. Johns Hopkins University, p. 292.

Owens, W.H. (1973). Strain modification of angular density distributions. *Tectonophysics* **16**: 249–261.

Owens, W.H. (1974). Representation of finite strain by three-axis planar diagrams. *Geological Society of America Bulletin* **85**: 307–310.

Owens, W.H. (1984). The calculation of a best-fit ellipsoid from elliptical sections on aribtrarily orientated planes. *Journal of Structural Geology* **6**: 571–578.

Ramsay, J.G. (1967). *Folding and Fracturing of Rocks*, 568. New York: McGraw-Hill.

Vollmer, F.W. (2017). Ellipsefit 3.4.0: strain and fabric analysis software. http://www.frederickvollmer.com/ellipsefit/index.html.

Vollmer, F.W. (2018). Automatic contouring of geologic fabric and finite strain data on the unit hyperboloid. *Computers and Geosciences* ***115***: 134–142. https://doi.org/10.1016/j.cageo.2018.03.006.

Wheeler, J. (1984). A new plot to display the strain of elliptical markers. *Journal of Structural Geology* **6**: 417–423.

Yamaji, A. (2005). Finite tectonic strain and its error, as estimated from elliptical objects with a class of initial preferred orientations. *Journal of Structural Geology* ***27***: 2030–2042.

Yamaji, A. (2008). Theories of strain analysis from shape fabrics: a perspective using hyperbolic geometry. *Journal of Structural Geology* ***30***: 1451–1465.

Yamaji, A. (2013). Comparison of methods of algebraic strain estimation from R_f/φ data: a unified theory of 2D strain analysis. *Journal of Structural Geology* **49**: 4–12.

6

Stress

6.1 Overview

Geologists draw upon ideas from continuum mechanics and approach the relative movements in rocks by examining a sort of average of the effects of forces acting on and through rock masses. This average of force is **stress**, which is a local measure of the force per unit area acting across a surface within a substance. A focus on stress enables us to assess the internal effects of the forces that act on and through rock masses.

Stress at a point is given by a matrix (tensor), with two (in two dimensions) or three (in three dimensions) **principal stress axes**. The principal stress axes are always mutually perpendicular. Those stresses, when resolved on any plane result in **traction** or **stress vectors**, which have both magnitudes and directions. We resolve the traction or stress vector acting on a plane into a **normal** stress component, which acts perpendicular to the plane, and a **tangential** stress component, which acts parallel to the plane. Normal stress components directed toward the plane are **compressive**, and those directed away from the plane are **tensile**. Because stresses are generally compressive within the Earth, most geologists adopt the convention that compressive normal stresses are positive. Likewise, tangential stress components may impart either a counter-clockwise (left-lateral) or clockwise (right-lateral) sense of shear of the surface on which they act. We use the convention of left-lateral as positive and right-lateral as negative. There are two planes (in two dimensions) and three planes (in three dimensions) – each oriented perpendicular to one of the principal stress axes – that have zero tangential stress; we call these **principal stress planes**. In cases in which all the principal stresses are positive, the stress matrix can be visualized as a **stress ellipse** (in two dimensions) or **stress ellipsoid** (in three dimensions).

There are two ways to determine the traction vectors on any plane in two dimensions. One is to use the **fundamental stress equations**. The second is to use a **Mohr Circle** convention. They are completely equivalent.

The stress matrix can be mathematically separated (decomposed) into a **hydrostatic stress component** and **deviatoric stress component**. Both hydrostatic and deviatoric components are given by matrices. The hydrostatic component describes an average or **mean stress** applied equally in all directions. The deviatoric component measures departures from the mean stress. It is characterized by an effective compression axis (parallel to the maximum principal stress) and an effective tension axis (parallel to the minimal principal stress). Importantly, deviatoric stress is related to the distributed deformation of rock volumes. For comparison, the stress matrix itself is thought to

An Integrated Framework for Structural Geology: Kinematics, Dynamics, and Rheology of Deformed Rocks,
First Edition. Steven Wojtal, Tom Blenkinsop, and Basil Tikoff.

control deformation related to discrete deformation along surfaces.

We use the term **stress state** to refer to the combined orientations and magnitudes of stresses acting within a volume of rock. A wide range of stress states occurs in the Earth. The simplest stress state is that from an overlying column of water (**hydrostatic**) or rock (**lithostatic**). Associated with tectonics, we consider the role of both **gravitational stresses** (body forces) and a variety of **tectonic stresses** (surface forces).

In this chapter, we focus on two-dimensional treatments of stress for three reasons. First, considering stress in two dimensions reduces the complexity of the algebra and trigonometry required, thereby enabling readers to focus on physical principles rather than mathematics. Second, it is easier to depict concepts on figures in two dimensions than in three dimensions. The reader will regularly encounter situations where an illustration depicts only lines, yet the accompanying text or figure caption refers to "planes." In each such instance, the line used to depict a plane on an illustration is the trace of that planar surface, which we take to be oriented perpendicular to the plane of the illustration (see Figure 6.1). An important corollary is that the area of any "plane" depicted on illustrations is determined by the length of its trace on the illustration times a unit depth below or height above the plane of the illustration. Third, once a reader understands the concepts and principles of stress in two dimensions, it is not difficult to extend those concepts and principles to three dimensions.

Finally, this chapter utilizes a host of symbols, many of them Greek letters, to denote parameters encountered in addressing stress. These symbols are likely to be unfamiliar, so we include Table 6.1 to introduce and define the symbols used in this chapter.

6A Stress: Conceptual Foundation

In order to understand rock deformation, one must understand how forces act on and through rocks. In physics courses, students learn to relate forces acting on objects to the accelerations experienced by those objects. Analyses of this sort help one understand the movement of rock detritus suspended in a stream or bounced along the ground surface by the wind. Using a similar approach to track the forces that act within and through rocks to cause deformation is not practical, however. One would need to determine the

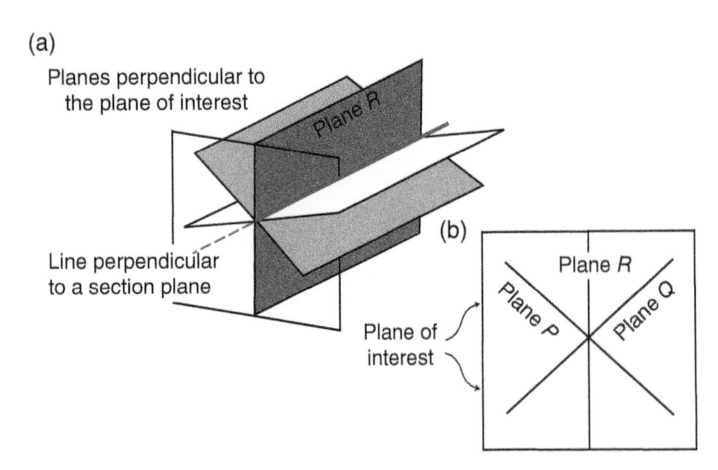

Figure 6.1 (a) A perspective image of three planes P, Q, and R, all of which are perpendicular to a plane of the section. In this image, the planes intersect along a line that is also perpendicular to the plane of the section. (b) On the plane of section, the planes P, Q, and R are lines, and the line of their intersection is a point.

Table 6.1 List of symbols.

Scalars		**Formulae**
A	Area	
$\Delta\sigma$	Differential stress	$\sigma_1 - \sigma_2$ (2D) or $\sigma_1 - \sigma_3$ (3D)
σ_m	Mean stress	$(\sigma_1 + \sigma_2 + \sigma_3)/3$
σ_l	Lithostatic stress or lithostatic pressure	ρgz
P_f	Pore fluid pressure	P_f
λ	Pore fluid factor	$P_f/\rho gz$
R	Stress ratio	$(\sigma_2 - \sigma_3\}/(\sigma_1 - \sigma_3)$
k	Ratio between vertical and average horizontal stress	$\sigma_v/1/2(\sigma_H + \sigma_h)$
Vectors		
$\boldsymbol{F}$	Force: $\boldsymbol{F}_s$ parallel to a plane; $\boldsymbol{F}_n$ normal to a plane	
$\boldsymbol{s}$	Traction or stress vector	
σ	Stress	
σ_N	Normal stress on plane P	
σ_T or τ	Tangential stress on plane P	
$\sigma_1\ \sigma_2\ \sigma_3$	Principal stresses $\sigma_1 \geq \sigma_2 \geq \sigma_3$	
σ_v	Vertical stress	
σ_H, σ_h	Maximum and minimum horizontal stresses	
Tensors		**Tensor components**
$\mathbf{T}$	Stress tensor	σ_{ij}
$\mathbf{T}_{\mathbf{dev}}$	Deviatoric part of the stress tensor	σ_{dij} or σdev_{ij}
$\mathbf{T}_{\mathbf{hyd}}$	Hydrostatic part of the stress tensor	σhyd_{ij}.
$\mathbf{T}_{\mathbf{eff}}$	Effective stress tensor	σeff_{ij}
$\mathbf{T}_{\mathbf{fluid}}$	Pore fluid tensor	P_f
$\mathbf{T}_{\mathbf{red}}$	Reduced stress tensor	

N.B. Magnitudes of vectors, which are scalar quantities, are given by the same symbols are the vectors but not in bold. Tensor components are also not in bold.

forces acting on each of the myriads of particles within any rock, determine whether any net force exists to cause acceleration, and then plot the accelerations of the individual particles. Instead of undertaking such a daunting task, we use **stress**, a local measure of the force per unit area acting across surfaces within a substance, to examine the effects of forces acting on and through rock masses.

For many students, this task is made more difficult by the fact that we cannot observe stress in rocks. This is true for rocks that are under stress today in the near-surface settings that we can examine *in situ* as well as for rocks that experienced stress at depth inside the Earth at some time in the distant past and have been exhumed by erosion. When we look at a deformed rock, we see the *effects* of stress, not the stress itself.

Determining stress is then fundamentally different from recording displacements and strains. We can *measure directly* displacements and strains in rocks with a ruler, tape measure, compass, or GPS unit. We must *deduce* the existence of stress. The distinction between displacements and strain on the one hand and stress, on the other hand, is so fundamental that it warrants repeating. We come back to this point at the end of Part A of this chapter: Section 6A.9, which clarifies these points, is the single most important part of this chapter. It alone should be read even if the rest of the chapter is ignored.

There is a sizable body of work to support deductions on stresses. So, we do not mean to imply that stresses are imaginary just because we cannot "see" them or measure them directly. We simply want to underscore why we take a different approach in this chapter from the previous two chapters, where empirical observations were the basis for developing theory. In this chapter, we follow an alternative to the empirical approach that we outlined in Chapter 1: Reasoning from first principles. In this chapter, we develop a theoretical understanding of stress, and then ground that understanding in physical reality through the use of examples.

If stress cannot be observed, why is it important? Consider a few simple practical examples to answer this question. We need to know the strength of rock beneath buildings, or the rock in which an underground mine is developed, in order to make these constructions safe. Rock "strength" is the stress that the rock can bear without failing. In the same way, for the geological record, we also want to understand the conditions that led to the formation of structures such as faults or shear zones, which are types of failure. These conditions are determined partly by the geological stress acting in the rock. A complete understanding of the mechanics of rock deformation cannot be achieved without an understanding of stress.

6A.1 Forces, Tractions, and Stress

6A.1.1 Accelerations and the Forces that Act on Objects

We think of forces as pushes or pulls, but going beyond that definition or conceptualization is difficult. Much of the difficulty arises because we can never measure forces directly; we measure them indirectly by the *accelerations* they produce. Acceleration is a change in an object's velocity per unit time, that is, (final velocity – initial velocity) divided by the length of time over which the velocity changed (final time [t_f] – initial time [t_i]). Written as an equation, we have:

$$\text{acceleration} = \text{velocity change} / \text{elapsed time}. \tag{6.1a}$$

The standard units of accelerations are (m/s)/s = m/s^2 = ms^{-2}. Velocities are vectors, which means that they have both magnitude and direction. Accelerations must necessarily also be vectors and have both magnitude and direction. In two dimensions, an acceleration will have two components.

$$\text{acceleration} = \boldsymbol{a} = \left(a_x, a_y\right), \tag{6.1b}$$

where

$$a_x = \frac{\left[v_x\left(t_f\right) - v_x\left(t_i\right)\right]}{\left(t_f - t_i\right)} \tag{6.2a}$$

$$a_y = \frac{\left[v_y\left(t_f\right) - v_y\left(t_i\right)\right]}{\left(t_f - t_i\right)}. \tag{6.2b}$$

The magnitude of the acceleration a is:

$$a = \sqrt{\left(a_x^2 + a_y^2\right)} = \left(a_x^2 + a_y^2\right)^{1/2}. \tag{6.2c}$$

The rule for calculating the magnitude of a force from the acceleration observed is:

$$\boldsymbol{a} = \frac{\boldsymbol{F}}{\text{m}}. \tag{6.3}$$

Stated in words, any acceleration ($\boldsymbol{a}$) observed is directly proportional to the *force* ($\boldsymbol{F}$) applied and inversely proportional to the *mass* (m) of the accelerating object. Rearranging Eq. (6.3) yields the more familiar equation $\boldsymbol{F} = \mathrm{m}\cdot\boldsymbol{a}$ that enables us to calculate the force that causes an observed acceleration. Because the acceleration has both magnitude and direction, the force causing that acceleration must also have magnitude and direction, that is:

$$F_x = \mathrm{m}\cdot a_x \tag{6.4a}$$

$$F_y = \mathrm{m}\cdot a_y. \tag{6.4b}$$

Where the magnitude of the force F is:

$$F = \sqrt{\left(F_x^{\,2} + F_y^{\,2}\right)} = \left(F_x^{\,2} + F_y^{\,2}\right)^{1/2}. \tag{6.4c}$$

The mass of an object has only magnitude. So, the ratio of the two components of the force F_y/F_x will equal the ratio of the two components of the acceleration a_y/a_x. This means that a force always has the same direction as the acceleration it produces. Note also that forces must have units of mass × acceleration. The SI units of force are Newtons (N), where $1\,\mathrm{N} = 1\,\mathrm{kg} \times 1\,\mathrm{m/s^2} = 1\,\mathrm{kg{\cdot}m/s^2}$.

When n forces act upon an object, we must add together the corresponding components of the forces together to determine the *resultant* force that determines an object's acceleration.

$$F_x(\text{total}) = F_x(1) + F_x(2) + F_x(3) + \cdots + F_x(n) \tag{6.5a}$$

$$F_y(\text{total}) = F_y(1) + F_y(2) + F_y(3) + \cdots + F_y(n) \tag{6.5b}$$

If the forces sum to zero, there is no *net force* acting on an object, and it will experience no net acceleration.

6A.1.2 Forces Transmitted Through Objects

The discussion in the previous section is adequate for examining the movement of an object from an *external* perspective. To consider the state of forces in an object, a different viewpoint is needed. In Figure 6.2a, no forces act on the object. In Figure 6.2b, $\boldsymbol{F}$ and $-\boldsymbol{F}$, two forces with equal magnitudes but opposing directions, act on the object (recall that we use the bold italic typeface to denote vectors). The net force acting on the object is zero in both cases. So, we observe no acceleration in either case. From an external perspective, the two cases are equivalent. There are, however, significant differences in the interiors of the two objects in the two situations.

In the situation depicted in Figure 6.2a, atoms or ions in the object occupy positions dictated by the character of the chemical bonds holding them together. The bonds between any two atoms (or ions) determine an *equilibrium distance* for the atoms, the distance at which attractive and repulsive forces between the two are balanced and at which the atoms have their lowest energy configuration. Atoms are not rigidly fixed at those relative positions. Due to thermal vibrations, they move about their equilibrium positions, but averaged over time, their position is determined by the equilibrium distances to their neighboring atoms. In the situation depicted in Figure 6.2b, the forces acting at the two sides of the object displace the atoms away from their equilibrium spacing, either moving them closer together if the forces acting on the sides of the object are directed toward each other or moving them farther apart if the forces acting on the sides of the object are directed away from each other. The forces acting on the sides of the object are *transmitted* along the network of bonds within the object. So, atoms are either closer together or farther apart than in the situation depicted in Figure 6.2a.

6A.1.3 Traction – A Measure of "Force Intensity" within Objects

Planes *ab* and *cd* in Figure 6.2a have different areas. The number of bonds that extend across

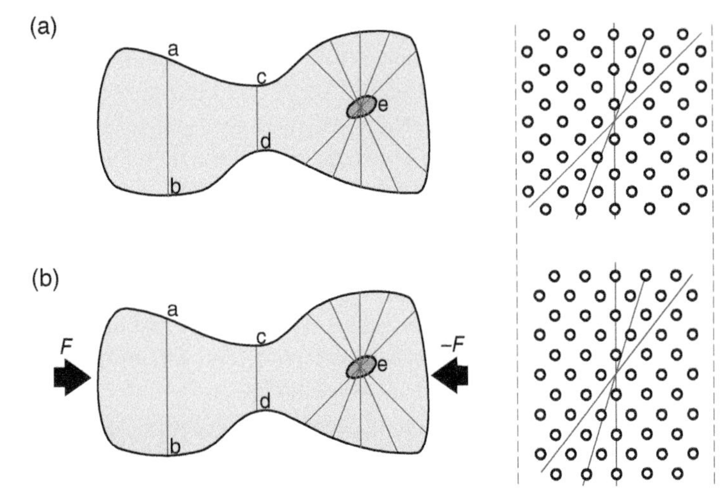

Figure 6.2 (a) An irregularly shaped object with no external forces. At right is a schematic depiction of the object's atomic-scale microstructure. Circles represent atoms or ions that occupy positions at an equilibrium spacing, where attractive and repulsive forces are balanced. The horizontal and vertical spacing of atoms/ions are equal. (b) The same object subjected to horizontal forces ***F*** and −***F***. The net force acting on the object is zero because forces ***F*** and −***F*** are transmitted through the object along the network of bonds. This yields different horizontal and vertical spacing for atoms/ions.

plane *ab*, whose area is larger, is larger than the number of bonds that extend across plane *cd*, whose area is smaller. Applying forces ***F*** and −***F*** to the sides of the object necessarily yields different conditions for atoms or ions in the vicinity of plane *ab* than for those in the vicinity of plane *cd*. Because the area of plane *cd* is smaller than the area of plane *ab*, a smaller number of bonds must transmit the forces meaning that the "intensity" of the effect of the applied forces is greater for individual bonds near plane *cd*. Because the area of plane *ab* is larger than the area of plane *cd*, a larger number of bonds transmit the forces so the "intensity" of the effect of the applied forces is lower for individual bonds near plane *ab*. Even in this relatively straightforward situation, determining the resultant of all forces acting on an object does not provide insight into the *internal* effects of that net force. A better measure of the magnitude of local "force intensity" within this object is a quantity called the **traction**, ***s***, defined as the force acting on one side of a plane, ***F***, divided by the area *A* of the plane on which the force acts:

$$\mathbf{s} = \frac{\mathbf{F}}{A}. \tag{6.6}$$

Because all forces have direction and magnitude, the tractions derived from those forces must have direction and magnitude. Tractions are vectors, sometimes called **traction vectors**, to emphasize that point. We use ***s*** to denote the traction vector defined by the two components s_x and s_y:

$$s_x = \frac{F_x}{A} \tag{6.6a}$$

$$s_y = \frac{F_y}{A} \tag{6.6b}$$

In this instance and in all instances we will consider in this text, both the external forces acting on this object and the net force intensity at any point within the object sum to zero. Thus, tractions of equal magnitude and opposite directions act on the opposing sides of any plane on which we focus.

In Figure 6.2b, the area of plane *ab* is two times the area of plane *cd*. As a result, the magnitude of the traction acting on either side of plane *cd* is twice that acting on either side of plane *ab*. Tractions, therefore, provide direct insight into the intensity of the forces to which bonds within an object are subjected. Because tractions include both force and the area on which that force acts,

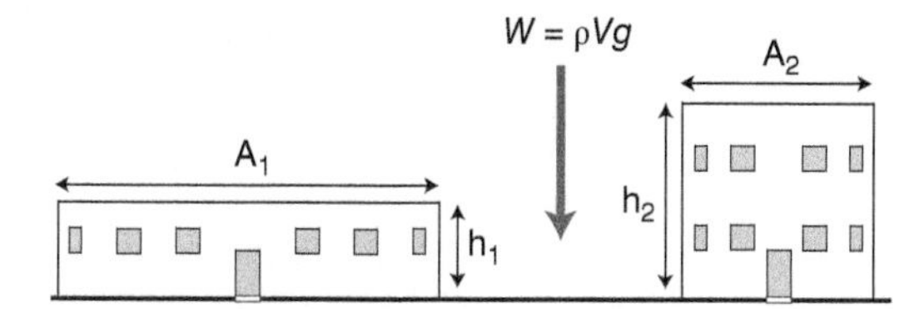

Figure 6.3 The lower and wider building at left and the taller and narrower building at right have equal weights. Because the force of their weight is resolved across planes with different areas the two buildings exert different stress = force/unit area on the ground beneath them.

they are generally a more useful quantity in the context of rock deformation than force alone.

Consider a building on a horizontal ground surface. Figure 6.3 shows two possible designs for the building. The building on the right is twice as high but half the width of the one on the left. The two buildings have the same volume and density, and therefore they exert the same total vertical force due to gravity on the bedrock below their bases. Is the bedrock strong enough to support either design? There is clearly an important difference between the two cases. The narrow base of the tall building concentrates the force on a smaller area, leading to greater tractions on the bedrock than those generated by the wide building. To assess the effect of the buildings on the bedrock beneath them, we need to determine the "intensity" of forces transmitted across bonds within the bedrock that result from each design. Calculating the magnitudes of the tractions beneath the buildings provides the means to make the relevant assessment.

6A.1.4 Stress

In solving deformation problems in geology, we make an important but reasonable assumption: that the rocks (or buildings) in the problem are at equilibrium. This assumption holds in many regions, and it will lead to only small errors for most non-seismic deformation because the actual accelerations are finite but very small. This assumption does not hold in settings where rock masses experience sizable accelerations, such as during earthquakes. The nonequilibrium physics of earthquakes is a topic too advanced to be addressed at this point. At this point, however, we use the assumption of equilibrium to place a powerful constraint on problems: it means, according to Newton's third law, that any force acting on an object is balanced by an equal and opposite force. Therefore, although we show traction acting in a single direction (for example in Figures 6.3 and 6.4), the assumption of equilibrium specifies that this traction is counterbalanced by one acting in the opposite direction. We call the combination

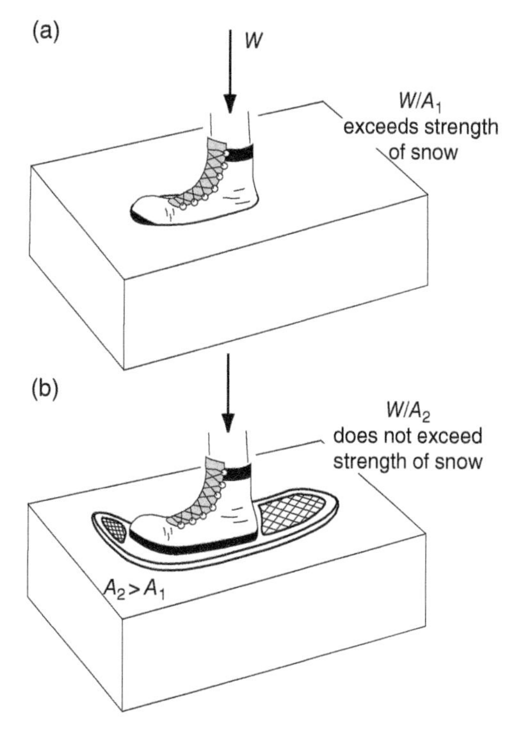

Figure 6.4 (a) Because their boot has a relatively small area, the hiker's weight $\boldsymbol{W}$ is transmitted across that area may generate a stress (= force/unit area) that exceeds the strength of the snow. (b) The same hiker's weight $\boldsymbol{W}$ transmitted across the larger area of the snowshoe generates lower stress that does not exceed the strength of the snow.

of the tractions of equal magnitude and opposing directions acting on the opposite sides of a plane the **stress**, regularly denoted by the symbol σ (a lower-case Greek sigma). Typically, a geologist depicting the stress acting across a plane will show only the traction vector acting on one side of that plane. For this reason, geologists use the terms "traction vector" or "traction" and "stress vector" or "stress" interchangeably. In this chapter, we will use the terms traction vector or traction when we show the force per unit area acting on both sides of a plane and will use stress vector or stress when we depict the force per unit area acting on one side of a plane that is at equilibrium. Clearly, the magnitude of the stress vector acting on a plane equals the magnitude of the traction vector because we assume that the material is at equilibrium.

Also, at the outset, we focus on regions in which stresses are *uniform*. In these regions, planes that are parallel experience identical stresses. In the following sections, we will analyze stresses beneath a building or within a laboratory sample, and we assume that the stresses are uniform within the region beneath that building or within that laboratory sample. You will see that it is the orientation of a plane, not its position that is the critical factor in determining the type and magnitude of stresses that act on the plane. Even with this simplification, you will discover that a complete description of stresses requires significant attention to detail.

Note that each of the buildings Figure 6.3 exert a downward directed traction and the bedrock below exerts an upward-directed traction with equal magnitude. If the buildings exerted a traction with greater magnitude than the bedrock, the buildings would accelerate downward, that is, sink. If the bedrock exerted upward-directed traction with greater magnitude, the buildings would accelerate upward, that is rise off the ground. Since the buildings are at equilibrium – the upward and downward directed tractions on them are equal, we can refer to the vertical stress σ_v, with magnitude σ_v, across planes beneath them. The mass m of a building is equal to its density $\rho \times$ its volume V. The volume of each building in Figure 6.3 is equal to its height $h \times$ the area A of its base (recall that we assume that the two buildings extend the same distance "into" the plane of the drawing). We stated earlier that the two buildings have equal mass. The weight of each building is the downward force $\boldsymbol{W}$ exerted due to Earth's gravitational attraction. The magnitude of $\boldsymbol{W}$ is $W = mg = V\rho g = hA\rho g$ (where g is the acceleration due to gravity). We can use Eq. (6.6) to calculate the magnitude of the vertical stress:

$$\sigma_v = \frac{F_v}{A} = \frac{W}{A} = \frac{hA\rho g}{A} = h\rho g. \tag{6.7}$$

Since the tall and thin building is twice as high as the short and wide one, the stress beneath it is twice as great. We previously suggested that the rock beneath the tall building was affected differently than the rock under the wide building. This calculation shows why this is so, and by how much. Notice that the width of the building does not appear in the final formula of Eq. (6.7). This is true because tractions and stress are related to the force per unit area, and the area in this instance is directly related to the width. It is only the height of the building that is important for the vertical stress because the force per unit area increases with the height.

Another analogy may help explain stress. Snowshoes are sheets of plastic or metal fixed to ordinary shoes or boots to allow people to walk on snow (Figure 6.4). Without snowshoes, the downward traction exerted by a person is regularly greater than the upward traction the snow can exert. The net downward force causes the person's feet to sink into the snow. The snowshoe distributes the weight of the person over a larger area, producing lower stress that can be supported by the snow.

Equations (6.6) and (6.7) show that tractions and stress have units of force per area. The SI unit

Table 6.2 Conversion factors between stress units.

	Pascal, Pa	Bar, b	Kilobar, Kb	Pound/square inch, PSI
1 bar	10^5	1	0.001	14.5
1 MPa	10^6	10	0.01	145
1 kb	10^8	1000	1	14,500
1 GPa	10^9	10,000	10	145,000

for traction and stress is the *Pascal*, where 1 Pa = 1 N/m^2 = 1 (kg·m/s^2)/m^2 = 1 kg/ms^2. Stresses in typical geological situations regularly exceed several hundred thousand Pascals. So, we typically use *Megapascals*, where 1 MPa = 10^6 Pa, to measure geological stresses. To give you a physical sense of stress magnitudes, a column of air in the atmosphere exerts vertical stress of ~1 bar (for *barometric* pressure) = 10^5 Pa = 0.1 MPa on the ground surface at sea level. A 100 kg person balanced on a single pencil (with a cross sectional area of 1 cm^2) would exert vertical stress of about 1 MPa on the ground. A column of granite 100 m tall would exert a stress of approximately 2.5 MPa across its base. Table 6.2 shows some conversions between other units of stress.

6A.2 Characteristics of Stress in Two Dimensions

6A.2.1 Normal and Tangential Stress Components

Returning to the building example, suppose it is necessary to build on a slope, for example, in a city like Hong Kong. Now one must consider not only whether the ground can support the building but also whether the building will slide downhill (Figure 6.5). The weight of the building is a vertical force that can be resolved into two components: one component acting perpendicular to the inclined ground surface, and the other acting parallel to it. Each of the two components of the force gives rise to traction on the ground surface, one perpendicular to the ground surface and the other parallel to it. These two types of traction are called **normal** and **tangential tractions**, respectively. Because the ground exerts equal and opposite tractions on the ground surface, we refer in this situation to **normal stress** and **tangential stress** vectors acting across the ground surface. Siting the building on ground surfaces with progressively greater inclinations yields situations in which the normal stress component decreases in magnitude as the tangential component increases in magnitude.

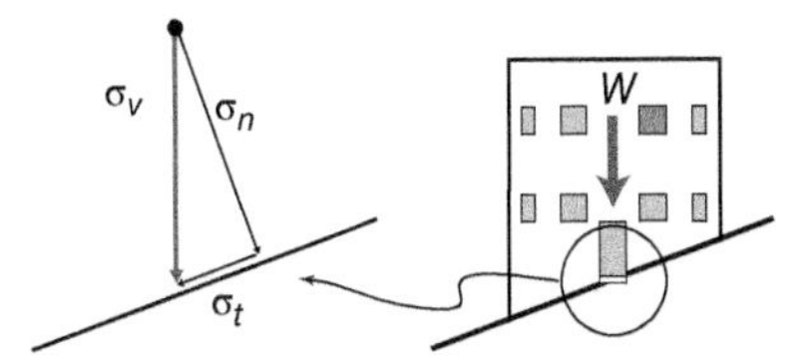

Figure 6.5 We can resolve the vertically directed stress vector σ_v into two components. The component σ_n is perpendicular to the inclined ground surface, and the component σ_t is parallel to the inclined ground surface.

Normal stresses, sometimes called *direct stresses*, are perpendicular or normal to the surfaces on which they act. Geologists often use σ or σ_n to denote normal stresses. *Tangential* or *shear stresses* are parallel to the surfaces on which they act. Geologists often use τ (lower case Greek letter tau) or σ_t to denote tangential stresses. In the case of our building, the tangential stress is what tends to cause the building to slide downhill, although whether it does so or not depends on the magnitude of the normal stress.

The normal stress under the building pushes the building and the ground surface together, a situation clearly different from the stress in a cable supporting a mass hanging from a crane, where the cable is being pulled apart. We distinguish these two types of normal stress by different names: The stress like that beneath a mass supported from below (Figures 6.3–6.5) is a **compression**, compared to the **tension** like that in the cable carrying the mass from above (Figure 6.6a). In any case where the traction vectors on opposite sides of a plane are directed toward each other, we say that the normal stresses are *compressive*. In any case where the traction vectors on opposite sides of a plane are directed away from each other, we say that the normal stresses are *tensile*. We also distinguish these two cases by the sign of the stress component. In this book and in most of the geological literature, we assign positive values to compressive normal stresses, whereas tensile normal stresses have negative values. However, you will need to exercise some care about signs because some geology textbooks, and nearly all of the engineering literature, adopt the opposite sign conventions. We also distinguish two types of tangential or shear stresses according to whether they tend to cause anticlockwise (counterclockwise; left-lateral) or clockwise (right-lateral) sense of shear of the surface on which they act (Figure 6.6b). We use the convention of counterclockwise (left-lateral) as positive and clockwise (right-lateral) as negative.

In our example of the building, the only stress was the vertical stress due to gravity, which is inevitably present at all places on and under

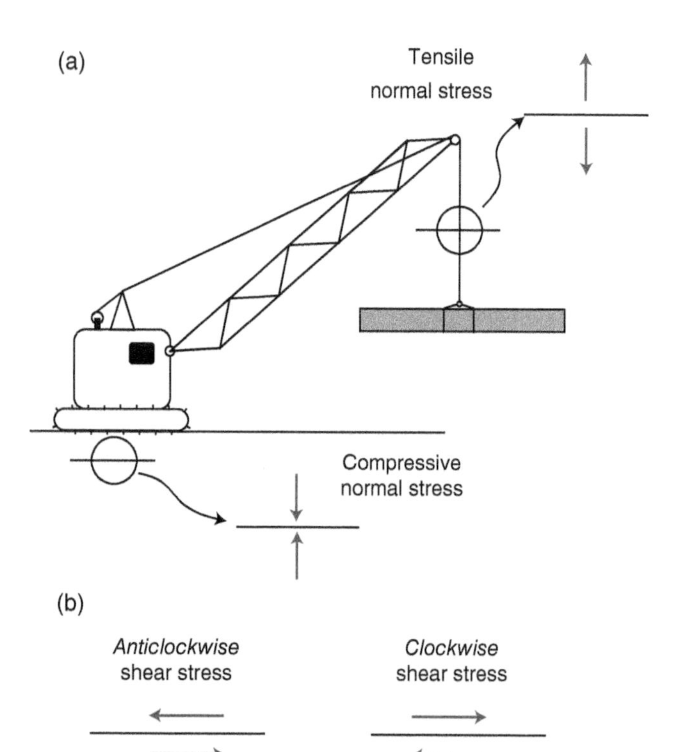

Figure 6.6 (a) Normal traction vector components are compressive when both are directed toward a surface and tensile when both are directed away from a surface. (b) Tangential traction vector components are clockwise or anticlockwise.

Earth's surface. However, the existence of plate motions in Earth and the active uplift of mountainous terrains indicate that there must also be other stresses. These stresses act in a rock in addition to stress due to gravity, and they can give rise to both normal and tangential stresses on planes with a variety of orientations.

6A.2.2 Stresses on Planes with Different Orientations

Another way to envision the situation of the building on an inclined surface is to consider that the vertical stress $\boldsymbol{\sigma}_v$ acting on an inclined plane gives rise to normal and tangential stress components acting on that plane. The original situation of a building sited on a horizontal ground surface is then just a special case in which the normal stress is equal to the vertical stress and the tangential stress is zero. A dipping plane within the bedrock beneath that building will experience finite normal and tangential stress (Figure 6.7a). Similarly, a horizontal plane within the bedrock beneath a building sitting on an inclined surface will experience a normal stress component whose magnitude is equal to the magnitude of the vertical stress due to the building and a tangential or shear stress component with a magnitude of zero (Figure 6.7b). Adding the load of the building affects all planes within the bedrock beneath the building. Regardless of whether that load is applied across a horizontal or inclined ground surface, planes of all orientations experience an effect of the load. Planes with different orientations experience normal and tangential stresses with different magnitudes. Note that (1) the magnitudes of the normal and tangential stresses acting on any one plane are not independent of the magnitudes of the normal and tangential stresses acting on all other planes, and (2) these normal and tangential stresses with different magnitudes acting across on planes with different orientations exist at the same time.

6A.2.3 Principal Stresses and Differential Stress

We call any plane with no shear stresses acting on it a principal plane of stress and the direction perpendicular to that plane a **principal stress axis.** The magnitude of the normal stresses along the principal stress axis is the **magnitude** of the **principal stress**.

6A.2.3.1 The Case with Vertical Loads Only

Horizontal planes beneath buildings on both horizontal and inclined surfaces are special in the following respects: the normal stress component acting across horizontal planes is the full magnitude of vertical stress due to the force of gravity acting on the buildings, and these planes have no shear stress component acting on them (Figures 6.3 and 6.7a). In contrast, planes inclined to the horizontal have smaller magnitude normal stress components and finite shear stress components acting on them (Figure 6.7b).

We can, using trigonometry, demonstrate just how special are horizontal planes in this situation. The vertical stress due to gravity, $\boldsymbol{\sigma}_V$, results from the vertical force of the weight of the building, $\boldsymbol{W}$, acting across the horizontal area of the building's base, here labeled A_H to underscore its horizontal orientation. That is, $\boldsymbol{\sigma}_V = \boldsymbol{W}/A_H$ where $\boldsymbol{W}$ is perpendicular to A_H. Figure 6.7c shows the weight of the building also contributes to the stress acting on a plane A_I dipping at an angle δ relative to the horizontal. The area of any inclined plane within the column of bedrock beneath the building is $A_I = A_H/\cos\delta$. Since the plane A_I makes an angle δ with A_H, a line perpendicular to A_I will make an angle δ with a line perpendicular to A_H. Thus, W_N, the component of the weight acting normal to plane A_I, lies at an angle δ from $\boldsymbol{W}$ (Figure 6.7c). By trigonometry, we have

$$W_N = W\cos\delta, \tag{6.8a}$$

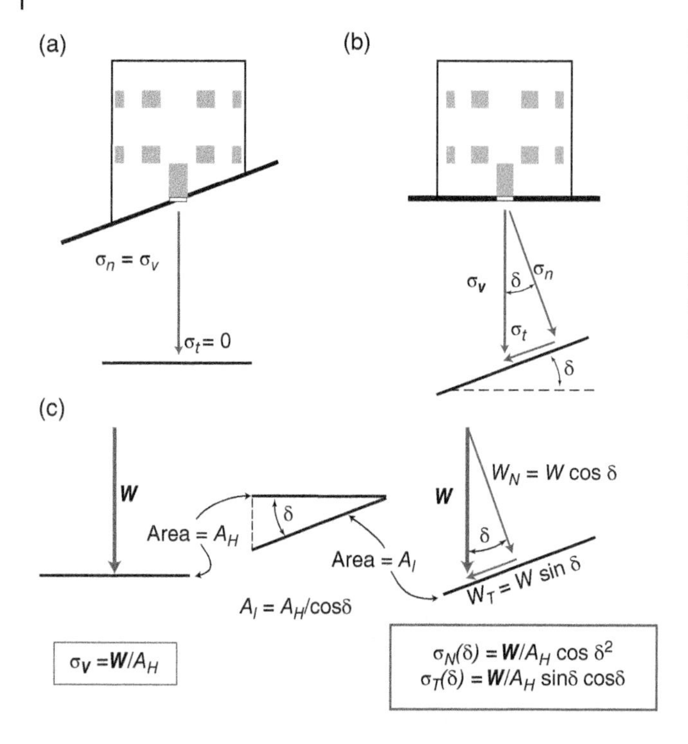

Figure 6.7 (a) A horizontal plane beneath a building on an inclined ground surface experiences a normal stress component equal to the vertically directed stress σ_v and a tangential stress component of zero. (b) An inclined plane beneath a building on a horizontal ground surface experiences finite normal and tangential stress components. (c) Diagram relating the normal and tangential stress components acting across an inclined plane to the magnitude of the vertically directed stress.

where W is the magnitude of $\mathbf{W}$. W_T, the component of the weight acting parallel to plane A_I, is

$$W_T = W\sin\delta. \tag{6.8b}$$

Thus, the building's contribution to the normal stress acting across a dipping plane is

$$\sigma_{VN}(\delta) = \frac{(W\cos\delta)}{(A_H/\cos\delta)} = \left(\frac{W}{A_H}\right)\cos^2\delta = \sigma_V\cos^2\delta, \tag{6.9a}$$

and its contribution to the tangential stress acting across that plane is

$$\sigma_{VT}(\delta) = \frac{(W\sin\delta)}{(A_H/\cos\delta)} = \left(\frac{W}{A_H}\right)\sin\delta\cos\delta = \sigma_V\sin\delta\cos\delta. \tag{6.9b}$$

where σ_V is the magnitude of $\boldsymbol{\sigma}_V$. Note that this tangential stress is anticlockwise or positive-valued because the vertical stress is positive (compressive) and both sinδ and cosδ are positive for $0 < \theta < 90°$ (recall that anticlockwise angles are considered positive).

We can check our analysis by determining the predicted values of normal and tangential stress acting on planes with critical dip values. When $\delta = 0°$, $\sigma_{VN}(\delta) = \sigma_V$ and $\sigma_{VT}(\delta) = 0$. This conforms to the conditions we set at the outset. When $\delta = 90°$, $\sigma_{VN}(\delta) = 0$ and $\sigma_{VT}(\delta) = 0$. These values seem reasonable because the vertically directed weight of the building has no component acting across a vertical plane. For $0 < \delta < 90°$, that is, when planes dip to the left, $0 < \sigma_{VN}(\delta) < \sigma_v$ and $\sigma_T(\delta) > 0$. The load of the building generates compressive normal stresses less than σ_v and anticlockwise shear stress on left-dipping planes in Figure 6.7. The dip angle δ enters Eq. (6.9a) through the "$\cos^2\delta$" term. $\cos^2\delta > 0$ even if $\cos\delta < 0$, so the building generates compressive normal stresses across all inclined planes. The dip angle δ enters Eq. (9b) through the "sinδ cosδ" term. For right-dipping planes where $-90 < \delta < 0°$, $\cos\delta > 0$ and $\sin\delta < 0$, so the "sinδ cosδ" term is negative-valued. Thus, the building generates clockwise (negative-valued) shear stresses on planes dipping to the right. The tangential stresses on right-dipping planes have magnitudes similar to those on left-dipping planes.

Equation (6.9b) indicates that only horizontal and vertical planes have zero-valued tangential stress components, that is, have no tangential stress components. This analysis also underscores the unique qualities of these two planes – they are principal planes. Note also that the normal stress magnitudes acting across these two planes, $\sigma_{VN}(\delta) = \sigma_V$ and $\sigma_{VN}(\delta) = 0$, are the extremes, that is, the greatest and least magnitudes, for normal stresses.

6A.2.3.2 The Case with Vertical and Horizontal Loads

If the setting for the building is a tectonically active area, there might be significant horizontal stresses. How would additional horizontal stress with magnitude σ_H contribute to the state of stress beneath the building? We again use trigonometry to calculate the components generated by this compressive horizontal stress. The horizontal stress vector $\boldsymbol{\sigma_H}$ is the product of a horizontal force $\boldsymbol{F_H}$ acting across a vertical plane A_V (Figure 6.8). That is, $\boldsymbol{\sigma_H} = \boldsymbol{F_H}/A_V$. The horizontal force $\boldsymbol{F_H}$ has components acting perpendicular and parallel to the inclined plane A_I, which dips at an angle δ. Plane A_I makes an angle (90°–δ) with the vertical plane A_V. The area of the inclined plane affected by σ_H is $A_I = A_V/[\cos(90-\delta)] = A_V/\sin\delta$. F_N, the component of the horizontal force acting perpendicular to the inclined plane, makes an angle (90°–δ) with $\boldsymbol{F_H}$. The angle measured from the horizontal force $\boldsymbol{F_H}$ to F_N is a clockwise angle of magnitude (90°–δ). So, the magnitude of that angle is –(90°–δ). Because $\cos(-\phi) = \cos\phi$ for any angle φ,

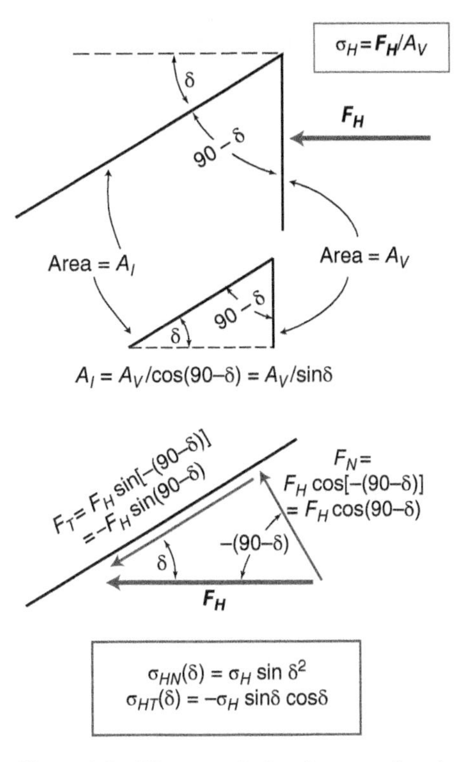

Figure 6.8 Diagram relating the normal and tangential stress components acting across an inclined plane to the magnitude of a horizontally directed stress.

$$F_N = F_H \cos\left[-\left(90° - \delta\right)\right] = F_H \cos\left(90° - \delta\right) = F_H \sin\delta, \tag{6.10a}$$

where F_H is the magnitude of $\boldsymbol{F_H}$. Because the $\sin(-\phi) = -\sin\phi$ for any angle φ, the component of the horizontal force acting parallel to plane A_I is

$$\begin{aligned} F_T &= F_H \sin\left[-\left(90° - \delta\right)\right] = -F_H \sin\left(90° - \delta\right) \\ &= -F_H \cos\delta. \end{aligned} \tag{6.10b}$$

Thus, the horizontal force acting across the vertical plane A_V generates normal stress acting across an inclined plane of the amount

$$\sigma_{HN}(\delta)=\frac{(F_H\sin\delta)}{(A_V/\sin\delta)}=\left(\frac{F_H}{A_V}\right)\sin^2\delta$$
$$=\sigma_H\sin^2\delta, \quad (6.11a)$$

and generates tangential stress acting across that plane of the amount

$$\sigma_{HT}(\delta)=-\frac{(F_H\cos\delta)}{(A_V/\sin\delta)}=-\left(\frac{F_H}{A_V}\right)\sin\delta\cos\delta$$
$$=-\sigma_H\sin\delta\cos\delta. \quad (6.11b)$$

The negative sign in Eq. (6.11b) indicates that a positive or compressive horizontal normal stress acting across a plane with a positive or anticlockwise dip yields tangential component stress that is negative-valued or clockwise.

We again check our analysis by determining the predicted values of normal and tangential stress acting on planes with critical dip values. When $\delta = 90°$, $\sigma_{HN}(\delta) = \sigma_H$ and $\sigma_{HT}(\delta) = 0$. This conforms to the conditions we set at the outset. When $\delta = 0°$, $\sigma_{HN}(\delta) = 0$ and $\sigma_{HT}(\delta) = 0$, reasonable values. For $0 < \delta < 90°$, $0 < \sigma_{HN}(\delta) < \sigma_H$ and $\sigma_{HT}(\delta) < 0$. That is, the horizontal stress generates compressive normal stresses and clockwise shear stress on planes that dip to the left in Figure 6.8. When $-90° < \delta < 0°$, that is, for planes dipping to the right, the angle from the horizontal force $\boldsymbol{F}_H$ to the normal to the inclined plane A_I is anticlockwise, that is, positive valued. Substituting a positive-valued angle $[90°-\delta]$ in Eq. (6.11a) still yields compressive normal stresses across inclined planes. Substituting a positive-valued angle $[90°-\delta]$ in Eqs. (6.10b) or (6.11b) yields positive-valued results, indicating that compressive horizontal stresses yield tangential stresses on right-dipping planes that are anticlockwise.

Combining Eqs. (6.9) and (6.11) gives the magnitudes of the normal and tangential stresses that result from both the vertical load due to the building and a horizontal compressive stress. The total magnitude of the normal stress acting across a plane dipping at angle δ is

$$\sigma_N(\delta)=\sigma_{VN}(\delta)+\sigma_{HN}(\delta)=\sigma_V\cos^2\delta+\sigma_H\sin^2\delta. \quad (6.12a)$$

For reasons that will become apparent in the following section, geologists often use the double-angle formulae to rewrite Eq. (6.12a). The relevant double angle formulae are $\cos^2\delta = (1 + \cos 2\delta)/2$ and $\sin^2\delta = (1-\cos 2\delta)/2$. Substituting for $\cos^2\delta$ and $\sin^2\delta$, Eq. (6.12a) becomes

$$\sigma_N(\delta)=\frac{\sigma_V(1+\cos 2\delta)}{2}+\frac{\sigma_H(1-\cos 2\delta)}{2}$$
$$=\frac{1/2(\sigma_V+\sigma_H)}{2}+\frac{1}{2}(\sigma_V-\sigma_H)\cos 2\delta. \quad (6.13a)$$

The total magnitude of the tangential stress acting across a plane dipping at angle δ is

$$\sigma_T(\delta)=\sigma_{VT}(\delta)+\sigma_{HT}(\delta)$$
$$=\sigma_V\sin\delta\cos\delta-\sigma_H\sin\delta\cos\delta$$
$$=(\sigma_V-\sigma_H)\sin\delta\cos\delta. \quad (6.12b)$$

We can make a similar substitution using another double angle formula (again the reasons for doing so will become apparent later). The relevant double angle formula is

$$\sin 2\delta = 2\sin\delta\cos\delta.$$

Substituting for $\sin\delta\cos\delta$, Eq. (6.12b) becomes

$$\sigma_T(\delta)=(\sigma_V-\sigma_H)\sin\delta\cos\delta=\frac{1}{2}(\sigma_V-\sigma_H)\sin 2\delta. \quad (6.13b)$$

Equations (6.12) and (6.13) define the normal and tangential components on a plane with dip δ relative to the horizontal in a region subjected to normal stresses on its horizontal and vertical boundaries. As long as $\sigma_V \neq \sigma_H$, the tangential stress component magnitudes are zero only when $\delta = 0°$ and $90°$. Thus, horizontal and vertical planes are unique principal planes in this setting. When $\delta = 0°$, $\cos 2\delta = 1$, and when $\delta = 90°$, $\cos 2\delta = -1$. These values yield the largest and smallest magnitudes for the normal stress. In this example, and in every case, the principal planes are characterized by zero-valued shear stress components and by normal stress components that are the largest and smallest attainable for a given loading situation. In order to emphasize the distinctive character of the normal stresses acting parallel to the two principal stress axes, we denote one principal stress value σ_1 and the other σ_2, with $\sigma_1 \geq \sigma_2$. Using the sign convention that compressive stresses are positive, σ_1 is the more compressive principal stress and σ_2 is the less compressive principal stress. Defining the magnitudes and orientations of the principal stresses is the most straightforward way to describe the orientations of stress. In almost all stress problems, structural geologists prefer to work with the principal stresses.

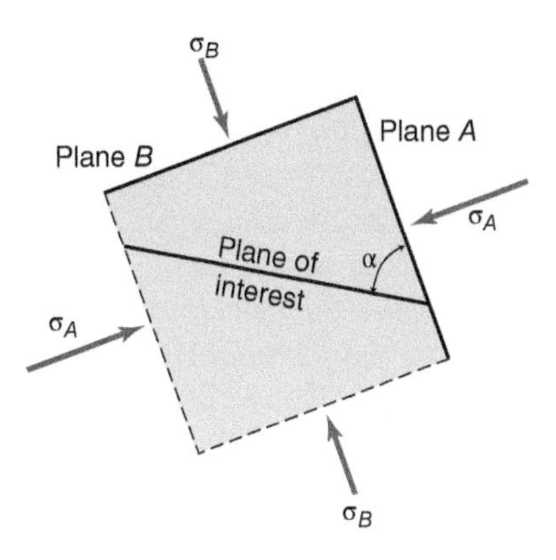

Figure 6.9 The orientation of a plane of interest within an object subjected to stresses σ_A acting on plane *A* and σ_B acting on plane *B*, which is perpendicular to plane *A*.

6A.2.4 The Fundamental Stress Equations

In the analysis outlined earlier, the angle δ denotes the dip of a plane with respect to the horizontal, which is, in this instance, a principal plane. Consider the region depicted in Figure 6.9 where planes *A* and *B*, the principal planes subjected to normal stresses σ_A and σ_B, respectively, are both inclined to the horizontal. Using α to denote the angle between plane *A* and a plane of interest within the region, an analysis along the lines of that completed in two sections earlier would show that the normal and tangential stress components acting on the plane of interest are:

$$\sigma_N(\alpha) = \frac{1}{2}(\sigma_A + \sigma_B) + \frac{1}{2}(\sigma_A - \sigma_B)\cos 2\alpha \tag{6.14a}$$

$$\sigma_T(\alpha) = \frac{1}{2}(\sigma_A - \sigma_B)\sin 2\alpha \tag{6.14b}$$

Note that the principal stress σ_A acting on plane *A* can be either the most compressive (σ_1) or the least compressive (σ_2) normal stress in this region. The relevant issue here is that the orientation of the plane of interest is determined relative to plane *A*, which can have any orientation in space. By convention, we usually orient planes by measuring the angle α between the plane of interest and the plane subjected to σ_1, the most compressive normal stress. In Eq. (6.15), we have substituted σ_1 for σ_A and σ_2 for σ_B.

$$\sigma_N(\alpha) = \frac{1}{2}(\sigma_1 + \sigma_2) + \frac{1}{2}(\sigma_1 - \sigma_2)\cos 2\alpha \tag{6.15a}$$

$$\sigma_T(\alpha) = \frac{1}{2}(\sigma_1 - \sigma_2)\sin 2\alpha \tag{6.15b}$$

Equations (6.15a) and (6.15b) are called the **fundamental stress equations**. They determine the normal and tangential stresses on any plane, provided that the normal to that plane lies in the plane that contains both σ_1 and σ_2. They form the basis for most calculations of stress in two dimensions, but they require knowledge of both the magnitudes and orientations of σ_1 and σ_2.

Because the principal plane subjected to the most compressive stress, σ_1, can have any orientation in space, the variable α in Eq. (6.14) denotes the angle between the plane of interest and the principal plane and has no implication regarding the inclination of a plane relative to the horizontal.

The difference between the magnitudes of the principal stresses is the **differential stress**, $\Delta\sigma = \sigma_1 - \sigma_2$. This scalar quantity turns out to be very useful in determining how likely a rock is to fail. It also controls the overall character of stresses acting on planes with different orientations. If, for instance, $\Delta\sigma = \sigma_1 - \sigma_2 = 0$ or $\sigma_1 = \sigma_2 = \sigma_m$, Eq. (6.14a) indicates that the normal stress component acting on any plane, regardless of the value of α, is $(\sigma_m + \sigma_m)/2 = \sigma_m$. Equation (6.14b) indicates that the tangential stress component acting on any and all planes, regardless of the value of α, is $(\sigma_m - \sigma_m)/2 = 0$. One way to describe such a situation is to say that all planes are principal planes. Another way to describe this situation is to say that there are no unique principal planes.

If the differential stress, $\Delta\sigma = \sigma_1 - \sigma_2$ is finite, nearly every pair of planes at right angles will experience different magnitudes of the normal stress. If one plane makes an angle of $\alpha = 45°$ to the principal plane experiencing a normal stress magnitude of σ_A, then the plane perpendicular to it makes an angle of $\alpha = -45°$ to the principal plane experiencing a normal stress magnitude of σ_A. Because $\cos(90°) = \cos(-90°) = 0$, Eq. (6.14a) indicates that $\sigma_N(45°) = \sigma_N(-45°) = ½(\sigma_A + \sigma_B)$. Because $\sin(90°) = 1$ and $\sin(-90°) = -1$, Eq. (6.14b) indicates that $\sigma_T(45°) = -\sigma_N(-45°) = ½(\sigma_A - \sigma_B)$. Regardless of the magnitude of the differential stress, planes at ±45° from the two principal planes experience equal magnitudes of normal stress. The shear stresses acting on those two planes have equal magnitudes and opposite senses - one is clockwise and the other is anticlockwise.

6A.2.4.1 Stresses Acting on Perpendicular Planes

Through the use of a different trigonometric substitution, we can demonstrate an interesting property of the magnitudes of the shear stresses acting on *any* two planes at right angles to each other. We first use Eq. (6.14b) to calculate the tangential stress magnitude acting on a plane at an angle $(\alpha + 90°)$ from the principal plane experiencing a normal stress magnitude of σ_A:

$$\sigma_T(\alpha+90°) = (\sigma_A - \sigma_B)\sin(\alpha+90°)\cos(\alpha+90°).$$

Using identities defining the sines and cosines for sums of angles, that is sin (P + Q) = (sinP cosQ + cosP sinQ) and cos(P + Q) = (cosP cosQ − sinP sinQ), we have

$$\sigma_T(\alpha+90°) = (\sigma_A - \sigma_B)\big[(\sin\alpha\cos 90° + \sin 90°\cos\alpha)(\cos\alpha\cos 90° - \sin 90°\sin\alpha)\big].$$

$\sin 90° = 1$ and $\cos 90° = 0$, so

$$\sigma_T(\alpha+90°) = (\sigma_A - \sigma_B)\big[(1\cdot\cos\alpha)(-1\cdot\sin\alpha)\big] = -(\sigma_A - \sigma_B)\sin\alpha\cos\alpha.$$

The plane at an angle $(\alpha + 90°)$ from the principal plane experiencing a normal stress magnitude of σ_A has a tangential stress component with an equal magnitude but an opposite sense of that acting on the plane at an angle α from the principal plane experiencing a normal stress magnitude of σ_A. For a plane at an angle $(\alpha - 90°)$ from the principal plane experiencing a normal stress magnitude of σ_A,

$$\sigma_T(\alpha-90°) = (\sigma_A - \sigma_B)\sin(\alpha-90°)\cos(\alpha-90°).$$

Using the same trigonometric identities with P + (−Q),

$$\sigma_T(\alpha-90°) = (\sigma_A - \sigma_B)\big[(\sin\alpha\cos(-90°) + \sin(-90°)\cos\alpha)(\cos\alpha\cos(-90°) - \sin(-90°)\sin\alpha)\big].$$

$\sin(-90°) = -1$ and $\cos(-90°) = 0$, so

$$\sigma_T(\alpha - 90°) = (\sigma_A - \sigma_B)[(-1 \cdot \cos\alpha)(-(-1 \cdot \sin\alpha)]$$
$$= -(\sigma_A - \sigma_B)\sin\alpha\cos\alpha.$$

The plane at an angle (α −90°) from the principal plane experiencing a normal stress magnitude of σ_A also has a tangential stress component with an equal magnitude but an opposite sense to the plane an angle α from the principal plane experiencing a normal stress magnitude of σ_A. This must be the case, of course, since the plane −90° from a given plane is identical to the plane 90° from the given plane. The main point of this analysis is, however, that two planes at right angles to each other will typically have different magnitudes of normal stress, but they *always* have tangential stress magnitudes with equal magnitudes and opposite senses.

6A.3 State of Stress in Two Dimensions

6A.3.1 The Stress Matrix

One of the most important points to take away from the previous section is that within a two-dimensional region at equilibrium (where the forces sum to zero), the magnitudes and orientations of the traction vectors acting on two mutually perpendicular planes characterize fully the nature of forces in that region. Since we view each traction vector as the resultant of a normal and a tangential component, one needs to define the magnitudes of these four components to specify the loads transmitted through the area. One common scheme that geologists use to depict the magnitudes of these four traction vector components that define the nature of forces is to define a matrix:

$$\mathbf{T} = [\sigma] = \begin{vmatrix} \sigma_{11} & \sigma_{12} \\ \sigma_{21} & \sigma_{22} \end{vmatrix}$$

Here, $[\sigma]$ (lower case Greek letter sigma within square brackets) denotes the array of four values that together constitute the **stress matrix**. Every two-dimensional stress matrix has two rows and two columns. We use a lower-case sigma with subscripts, σ_{ij}, to denote different positions within the array. In the nomenclature convention used in this text, the first subscript (i) gives the row in which the component appears, and the second subscript (j) gives the column in which the component appears. Thus, σ_{11} denotes the component in the first row, first column position, σ_{12} denotes the component in the first row, second column position, σ_{21} denotes the component in the second row, first column position, and σ_{22} denotes the component in the second row, second column position. Further, the two subscripts identify the traction vector component to which that value refers, with one of the two subscripts defining the plane on which the component acts and the other defining the direction in which the component points. There is general agreement on which subscript identifies the plane on which the component acts and which identifies the orientation of the component (Figure 6.10a):

- The first subscript i identifies the orientation of the plane on which the traction vector acts by defining the coordinate direction that is perpendicular (or normal) to the plane on which the traction vector acts. Here, 1 indicates the x_1 or x coordinate axis, and 2 indicates the x_2 or y coordinate axis.
- The second subscript j identifies the orientation of the component by defining the coordinate direction that is parallel to the traction vector component, that is, gives the direction in which the traction vector acts. Again, 1 indicates the x_1 or x coordinate axis, and 2 indicates the x_2 or y coordinate axis.

Thus, σ_{11} in the upper left corner of the array and σ_{22} in the lower right corner of the array denote normal stress components. Similarly, σ_{12}

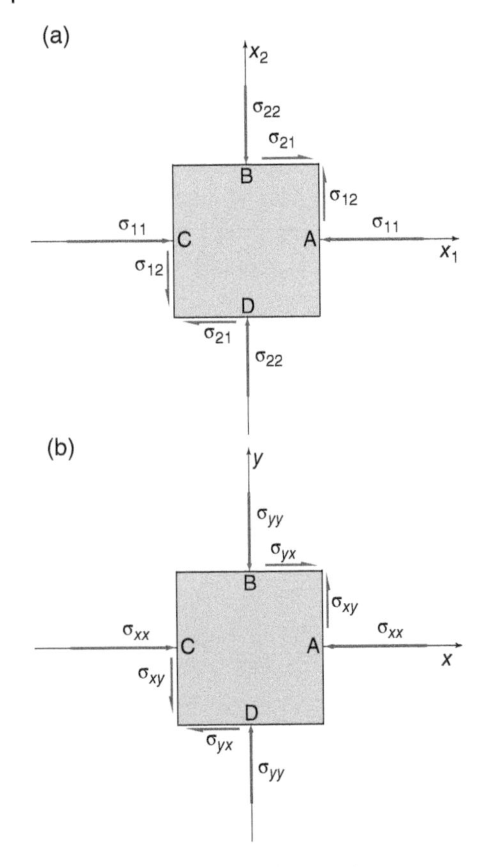

Figure 6.10 Convention for naming stress vector components σ_{ij}. For planes A and C, which are perpendicular to the x_1 or x axis, $i = 1$ or x. The normal stress component on A or C is parallel to the x_1 or x axis, so $j = 1$ or x. This yields σ_{11} or σ_{xx}. The tangential or shear stress component on A or C is parallel to the x_2 or y axis, so $j = 2$ or y. This yields σ_{12} or σ_{xy}. For planes B and D, which are perpendicular to the x_2 or y axis, $i = 2$ or y. The normal stress component on B or D is parallel to the x_2 or y axis, so $j = 2$ or y. This yields σ_{22} or σ_{yy}. The tangential or shear stress component on B or D is parallel to the x_1 or x axis, so $j = 1$ or x. This yields σ_{21} or σ_{yx}.

in the upper right corner of the array and σ_{21} in the lower-left corner of the array denote shear stress components. Each row of the array consists of two components that act together on a single plane. σ_{11} and σ_{12} are, respectively, the normal and tangential stress components acting on a plane perpendicular to the x_1 or x axis. σ_{21} and σ_{22} are, respectively, the tangential and normal stress components acting on a plane perpendicular to the x_2 or y axis. Some geologists prefer a nomenclature scheme that uses x and y for the subscripts, yielding traction vector components such as σ_{xx} or σ_{yx} (Figure 6.10b). Note that

$$\mathbf{T} = \left[\sigma\right] = \begin{vmatrix} \sigma_{11} & \sigma_{12} \\ \sigma_{21} & \sigma_{22} \end{vmatrix} = \begin{vmatrix} \sigma_{xx} & \sigma_{xy} \\ \sigma_{yx} & \sigma_{yy} \end{vmatrix}$$

Sections 6A.2.3.2 and 6A.2.3.3 showed that, as long as there are no net forces acting on a region, the tangential or shear stress components acting on any two planes oriented at right angles to each other within that region must have equal magnitudes. Thus, only three stress matrix components have independent magnitudes: either $\sigma_{12} = \sigma_{21}$ or $\sigma_{xy} = \sigma_{yx}$. The physical condition that the shear stresses on orthogonal planes have equal magnitudes constrains the stress matrix to be *symmetric*, that is, where $\sigma_{ij} = \sigma_{ji}$.

6A.3.2 The Stress Ellipse

As we have seen earlier, the stress at some location can be the product of more than one traction. The result of all the tractions acting at any point in Earth is called the **state of stress**. The terminology here is analogous to that used in the previous chapter to describe strain. The state of stress encompasses the normal and tangential or shear stress components acting on *all* planes passing through a particular point as opposed to the specific stress components acting on any individual plane. One way to visualize the state of stress is as an ellipse defined by the ends of the stress vectors acting across the myriad of planes that pass through a single point (Figure 6.11a). This ellipse, called the **stress ellipse**, is similar in some respects to the finite strain ellipse. In the same way that a radius of the strain ellipse depicts the relative magnitude of the elongation in a particular direction, each radius

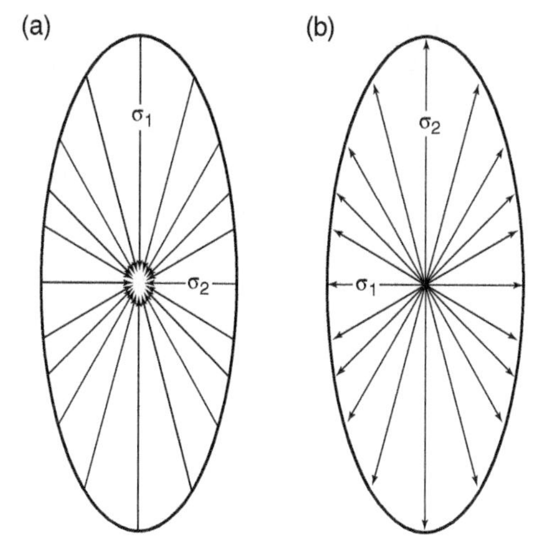

Figure 6.11 Stress ellipses for (a) a stress state where all normal stresses are compressive, and (b) a stress state where all normal stresses are tensile.

of the stress ellipse relates to the magnitude and direction of the stress vector acting on a plane passing through a given point. This relationship is simple and straightforward for the major and minor axes of the stress ellipse: those radii are parallel to the principal stress axes acting on planes perpendicular to the radii, and their lengths are proportional to the magnitudes of the principal stresses. Radii oblique to the principal axes of the stress ellipse do not depict directly the stress vectors acting on planes perpendicular to them, but the smooth variation in the lengths of the radii of the stress ellipse does show in a general way how the stress vectors acting on planes with different orientations vary smoothly from a maximum magnitude on the plane perpendicular to σ_1 to a minimum magnitude on the plane perpendicular to σ_2.

For geological applications, where all planes experience compressive normal stresses, we show the vectors relating to the stresses acting on planes with different orientations pointing toward the center of the ellipse. In engineering applications, where all planes with positive stresses experience tensile normal stresses, those vectors point from the center of the ellipse toward the periphery (Figure 6.11b). We cannot draw a stress ellipse for situations where some planes experience compressive normal stresses and others experience tensile normal stresses. Still, for most geological settings, the stress ellipse is an effective way to depict the local orientation and relative magnitudes of the principal stresses σ_1 and σ_2.

6A.3.3 The Mohr circle

The **Mohr circle** (or **Mohr diagram**) for stress is another way to represent a uniform state of stress in two dimensions. Karl Culmann (1821–1881), a 19th-century German engineer, recognized that one could use a circle centered on one of two Cartesian coordinate axes to represent graphically the solutions to Eqs. (6.15a) and (6.15b), which give the normal and shear stress components acting on different planes as a function of the angle α between the planes and one of the principal stress planes. Stated in a different way, the coordinates of each point on such a circle correspond to the magnitudes of normal and shear stresses acting on a plane in a particular stress state. The German engineer Otto Mohr (1835–1918) explored more fully how to use this simple graphical construction to analyze stresses, so we associate his name with the construction. The construction's simple, circular form underscores the interrelationships between the magnitudes of the principal stresses and the normal and shear stress components on planes with different orientations – in order to change the position or size of a circle centered on the abscissa all points on the circle must move. Moreover, that a single, simple plot depicts simultaneously the magnitudes of the normal and shear stresses on *all* planes within an area of interest, regardless of their orientation, emphasizes that it is the *state of stress*, not the magnitudes of normal and shear stress components on any individual plane, that is, fundamental. Mohr circles provide both qualitative and quantitative insight into the state of stress, so it is well worth learning to use the Mohr circle to solve many problems related to stress in rocks.

We plot Mohr circles in a special graphical space called the "Mohr plane" or "Mohr space." We use a special name for this graphical space to differentiate it from the "physical plane" or "real space," in which we draw sketches of the situations being analyzed. A unique set of Cartesian coordinates define the Mohr plane (Figure 6.12a): Distances parallel to the abscissa ("x axis") give the magnitudes of normal stress components acting on different planes, so we call the horizontal coordinate axis the "σ_N axis." Distances parallel to the ordinate ("y axis") give the magnitudes of tangential stress components acting on different planes. Thus, the vertical coordinate axis is the "σ_T axis." Each point in Mohr space represents a paired set of normal and shear stress components, which we can envision as the normal and tangential shear stresses acting on a plane in some physical plane situation. In order to emphasize the correspondence between points lying on the Mohr plane and planes in the physical plane, we use a lower case letter (e.g. p) to denote point on the Mohr plane and the upper case of the same letter (e.g. P) to denote the corresponding plane in real space. The Mohr plane point p (a, b) represents the physical plane P with $\sigma_N = a$ and $\sigma_T = b$ (Figure 6.12a, b). The Mohr plane origin (0, 0) corresponds to a stress-free plane, with $\sigma_N = 0$ and $\sigma_T = 0$. A principal plane in real space, which is a plane with no shear stress component, will have a Mohr plane correlative that lies somewhere on the abscissa; its Mohr space components will be (σ_N, 0). A point on the Mohr plane with coordinates (0, σ_T) represents an actual plane subjected only to a shear stress component.

In this text, we adhere to the following conventions for Mohr circles:

- Points to the right of the σ_T axis correspond to planes subjected to compressive normal stresses, whereas those to the left of the σ_T axis correspond to planes subjected to tensile normal stresses (Figure 6.12c).
- Points above the σ_N axis correspond to planes subjected to anticlockwise shear stresses; those below the σ_N axis correspond to planes subjected to clockwise shear stresses (Figure 6.12c).

The assumption that there are no net forces or no unbalanced stresses in regions we consider dictates that the center of every stress Mohr circle lies on the σ_N axis.

Figure 6.13 is a Mohr circle for stress for a typical, two-dimensional geological example. This particular Mohr circle defines the state of stress for a specific region of real space. For every point p on the circumference of the circle with coordinates (σ_{Np}, σ_{Tp}), there is a corresponding plane P in that region of real space whose normal stress component has magnitude σ_{Np} and whose shear stress component has magnitude σ_{Tp}. Conversely, for every plane Q in that region of real space, there is a corresponding point q on the circumference of the Mohr circle whose coordinates (σ_{Nq}, σ_{Tq}) give the magnitude of the normal stress component (σ_{Nq}) and shear stress component (σ_{Tq}) acting on that plane. There are no planes in that region of real space with pairs of stress components that correspond to the coordinates of points that lie inside or outside the circle. In the example in Figure 6.13, the entire circle plots to the right of the σ_T axis because the normal stress components acting on all planes, like the situations described in Sections 6A.2.2 and 6A.2.3, are compressive. The point on the circle farthest from the origin corresponds to the plane that experiences the greatest magnitude of compressive stress and zero shear stress. This plane has normal and shear stress components of (σ_1, 0). The point on the circle closest to the origin corresponds to the plane that experiences the least compressive stress and zero shear stress. This plane has normal and shear stress components of (σ_2, 0).

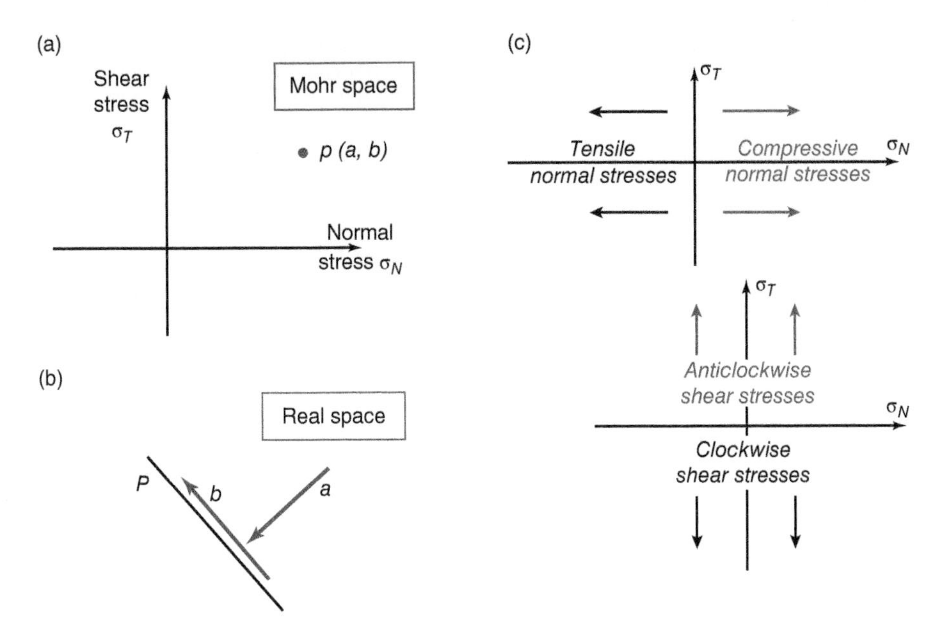

Figure 6.12 (a) Mohr space, defined by Cartesian coordinates in which the abscissa gives the magnitudes of normal stress components acting on planes and the ordinate gives the magnitudes of the tangential stress components acting on planes. The point p with coordinates (a, b) represents a plane P whose normal stress component has magnitude a and whose shear stress component has magnitude b. (b) The real space depiction of plane P. (c) The conventions we follow for Mohr space: Compressive normal stress components and anticlockwise shear stress components are positive-valued; tensile normal stress components and clockwise shear stress components are negative-valued.

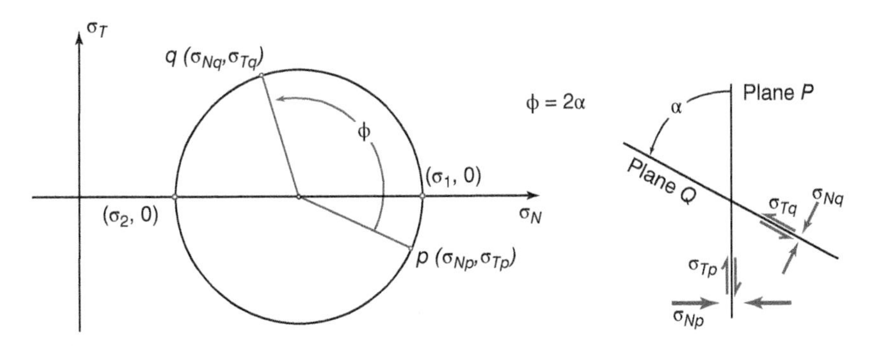

Figure 6.13 At left is Mohr space, with a Mohr circle depicting a two-dimensional stress state. At right is a corresponding real space depiction of the stress state. The polar angle between points p and q on the Mohr circle is ϕ. The angle between the corresponding planes P and Q in real space is $\alpha = \phi/2$.

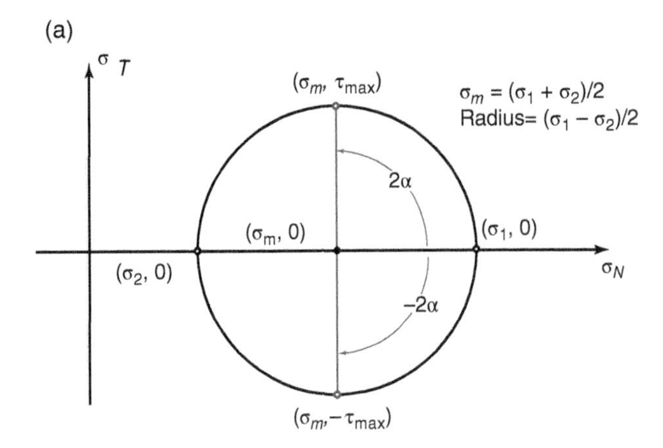

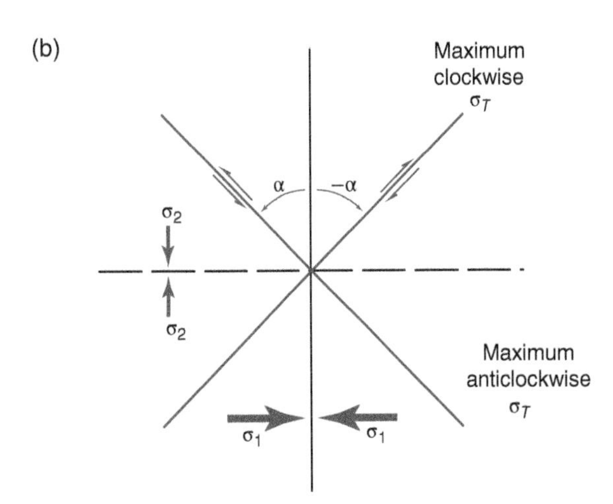

Figure 6.14 (a) A Mohr circle depicting a two-dimensional stress state. The center of the Mohr circle is $(\sigma_m, 0)$, where $\sigma_m = (\sigma_1 + \sigma_2)/2$, and the radius of the circle = $(\sigma_1 - \sigma_2)/2$. The points corresponding to planes experiencing the maximum shear stress have polar angles of ±90° from the points representing the planes subjected to the most and least compressive normal stresses. (b) The real space depiction corresponding to the Mohr space depiction of the stress state. The planes experiencing maximum shear stress make ±45° angles with the planes subjected to the most and least compressive normal stresses.

The center of the circle lies midway between these two extremes, at $(\sigma_m, 0)$ where $\sigma_m = \frac{1}{2}(\sigma_1 + \sigma_2)$, the *average* or *mean* of the two extreme normal stress magnitudes (Figure 6.14a). Since the center of the circle *does not lie on the circle*, in the region of real space experiencing this stress state there is no plane whose total stress components are $(\sigma_m, 0)$. Only if the circle had a radius with zero length would the center of the circle lie on the circle and therefore indicate the stress components acting on a plane in real space. There are, however, two points on the circle with normal stress magnitudes equal to the **mean stress** σ_m. They are the points farthest above and below the σ_N axis. One of those points corresponds to the plane with stress components = (σ_m, τ_{max}), where τ_{max} is the magnitude of the maximum anticlockwise shear stress. This point on the Mohr circle is separated from the point representing the plane that experiences the most compressive stress $(\sigma_1, 0)$ by a polar angle of $2\alpha = +90°$ (an anticlockwise rotation of 90°). In real space, the plane that experiences the greatest positive shear stress magnitude is $\alpha = +45°$ from the plane that experiences the most compressive normal stress (Figure 6.14b). In Section 6A.2.3.3, we showed that the magnitude of the shear stress acting on a plane inclined +45° from a principal plane is

½(σ_1−σ_2). Since the diameter of the circle is (σ_1−σ_2) and the center of the Mohr circle lies on the σ_N axis, τ_{max} must equal ½(σ_1−σ_2). The other point on the circle with normal stress magnitudes equal to σ_m is (σ_m, $-\tau_{max}$), which corresponds to the plane in real space with normal stress σ_m and shear stress $-\tau_{max}$. In real space, the plane that experiences the greatest shear clockwise stress lies $\alpha = -45°$ from the plane that experiences the most compressive normal stress (Figure 6.14b). The angle between the radii to the points (σ_1, 0) and the point (σ_m, $-\tau_{max}$) is $2\alpha = -90°$.

Consider again the pair of points *p* and *q* on the Mohr circle in Figure 6.13, which correspond to the planes *P* and *Q* in real space. The polar angle from the point *p* to the point *q*, which is the angle between the radii to those two points, is twice the angle from plane *P* to plane *Q* measured in a physical plane sketch. This statement holds for any two points on the Mohr circle and the planes they represent in the physical plane. The polar angle between the points (σ_1, 0) and (σ_2, 0) is 180°, so the angle between the corresponding principal planes in real space is 90° (Figure 6.15a). Points on the circle at opposite ends of any diameter, such as points *e* and *f* in Figure 6.15b, have a polar angle of 180° and so correspond to planes in real space oriented at right angles to each other. In Section 6A.2.3.3, we showed that two planes oriented at right angles to each other generally have different normal stress magnitudes whereas the absolute magnitudes of their shear stress components must be

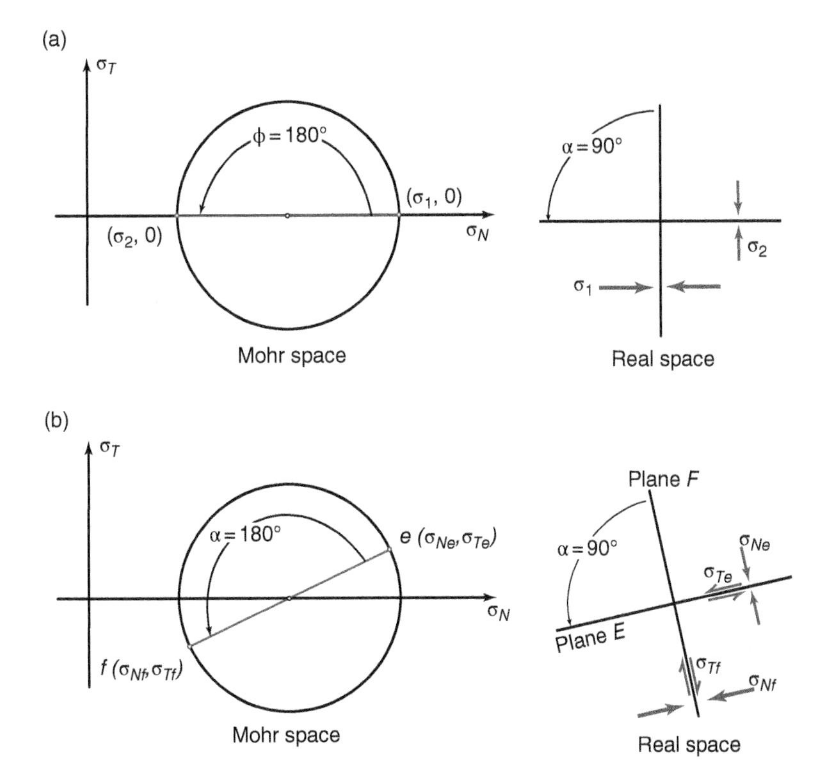

Figure 6.15 (a) On a Mohr circle, the points representing the planes subjected to the most and least compressive stresses lie at opposite ends of a diameter, meaning that the corresponding planes in real space are perpendicular to each other. (b) On a Mohr circle, the points on opposite ends of a diameter represent two perpendicular planes. Two planes oriented at right angles to each other experience shear stresses with an equal magnitude but opposite shear sense.

equal. When shear stresses are nonzero, one plane (plane *E*) has an algebraically positive shear stress component indicating anticlockwise shear stress, whereas the other plane (plane *F*) has an algebraically negative shear stress component indicating clockwise shear stress. On the Mohr circle, the point representing the first of those two planes (point *e*) lies at one end of diameter and the point representing the second plane (point *f*) lies at the opposite end of that diameter.

The Mohr space coordinates of points on the opposite ends of any diameter of a Mohr circle define the magnitudes of the traction vector components acting on two mutually perpendicular planes. Taken together, the four coordinate values define the components in a stress matrix referred to a particular pair of orthogonal planes. Clearly, each Mohr circle has an infinite number of unique diameters, meaning that any given stress state can be represented by any one of an infinite number of different stress matrices. Of the infinite number of different stress matrices, a few are particularly straightforward to visualize and use. The stress matrix giving the stress components acting on the two principal planes is:

$$\mathbf{T}' = [\sigma'] = \begin{vmatrix} \sigma_1 & 0 \\ 0 & \sigma_2 \end{vmatrix}$$

The stress matrix giving the magnitudes of the traction vector components acting on planes at 45° to the principal planes is:

$$\mathbf{T} = [\sigma] = \begin{vmatrix} \sigma_{11} & \sigma_{12} \\ \sigma_{21} & \sigma_{22} \end{vmatrix} = \begin{vmatrix} \dfrac{(\sigma_1 + \sigma_2)}{2} & \dfrac{(\sigma_1 - \sigma_2)}{2} \\ \dfrac{(\sigma_1 - \sigma_2)}{2} & \dfrac{(\sigma_1 + \sigma_2)}{2} \end{vmatrix}$$

where $(\sigma_1+\sigma_2)/2 = \sigma_m$ and $(\sigma_1-\sigma_2)/2 = \tau_{max}$. The many different stress matrices are not, of course, independent of each other. The following section outlines one way to relate one stress matrix representing the state of stress in a region to another stress matrix representing the same state of stress. The Comprehensive Treatment section of this chapter addresses more fully the relationships between the different stress matrices representing a particular state of stress.

6A.3.3.1 Relating Mohr Circles to the Equations Defining Stress Components

Equations (6.15a) and (6.15b), reprinted here, define the values of the normal and shear stress components acting on a plane oriented at an angle α from the plane subjected to σ_1, the most compressive normal stress.

$$\sigma_N(\alpha) = \frac{(\sigma_1 + \sigma_2)}{2} + (\sigma_1 - \sigma_2)\cos 2\alpha \quad (6.15a)$$

$$\sigma_T(\alpha) = \frac{1}{2}(\sigma_1 - \sigma_2)\sin 2\alpha \quad (6.15b)$$

By substituting a value for the variable α, one can calculate the magnitude of the normal and shear stress components acting on a plane of any orientation. The basis of the Mohr circle is the recognition that Eqs. (6.15a) and (6.15b) correspond to the pair of parametric equations

$$x = c + r\cos\phi \quad (6.16a)$$

$$y = r\sin\phi \quad (6.16b)$$

that define the set of all points that lie on a circle whose center plots at the point $(c, 0)$ where $c = ½(\sigma_1 + \sigma_2)$ and whose radius $= r$ where $r = ½(\sigma_1 - \sigma_2)$. By substituting a value for the variable α and thereby determining the value of ϕ, one can calculate the position of the point on a circle on the Mohr plane where the x coordinate gives σ_N and the y coordinate gives σ_T. The analogy between Eqs. (6.15) and (6.16) requires that the polar angle between two points on the Mohr circle is twice the angle between the corresponding planes in real space, that is, $\phi = 2\alpha$. Thus, two points on opposite ends of any diameter of a Mohr diagram give the normal and shear stress magnitudes on two planes at right angles to each other in real space. Points falling inside or outside the circle have $x = \sigma_N$ and $y = \sigma_T$ values that do not conform with solutions to Eq. (6.15) and therefore do not correspond to the normal stress-shear stress combinations observed on any planes in the particular situation described by this Mohr circle.

If we know the magnitudes of the stress components acting on two planes oriented at right angles to each other in a two-dimensional analysis, we have sufficient information to construct the Mohr circle that describes this two-dimensional situation and thereby determine the stresses acting on planes at any angle from those planes. If the original planes are not principal planes, we can use the Mohr circle to determine the relative orientation and magnitudes of the principal planes in that region. If the original two planes are principal planes, we can use the Mohr circle to calculate the magnitudes of the stress components that act on *any* of the infinite numbers of other planes in that region.

6A.3.3.2 Recapping the Main Characteristics of Mohr Circles

It is worthwhile to emphasize several important points associated with stress Mohr circles:

1) Mohr space is a special Cartesian coordinate space where distances parallel to the x axis give the magnitudes of σ_N, and distances parallel to the y axis give the magnitudes of σ_T. Thus, every point in Mohr space represents a paired set of normal and shear stress values, which could conceivably give the normal and tangential shear stresses acting on a plane in some physical plane situation.
2) For any uniform stress state there exists a collection of points in Mohr space, which fall on a circle with its center at $(c, 0)$ where $c = ½(\sigma_1 + \sigma_2) = \sigma_m$ (the mean stress) and with a radius $r = ½(\sigma_1 - \sigma_2)$, that define the different combinations of normal and shear stress components on planes within that stress state. We call that circle a *stress Mohr circle.*
3) Every point p on the circumference of a stress Mohr circle corresponds to a plane P in the corresponding two-dimensional region of real space that experiences that state of stress.
4) If the coordinates of p are $(\sigma_{Np}, \sigma_{Tp})$, the normal stress component acting on P is σ_{Np} and the shear stress component acting on P is σ_{Tp}. Since Mohr circles apply to regions of uniform stress, the plane P in fact represents any one of an infinite number of parallel planes within the region.
5) The Mohr plane coordinates of points inside or outside of the circle give combinations of normal stress and shear stress that do not exist in the corresponding region of real space.
6) If the polar angle between two points p and q on a Mohr circle, which is the angle between the radii to each of the points, is an angle ϕ, then the angle between the corresponding planes P and Q is $\alpha = \phi/2$.
7) Knowing the magnitudes of the normal and shear stress components acting on the two planes at right angles to each other enables one to construct a Mohr circle for that region.
8) The absolute magnitude of the maximum tangential stress experienced by any plane in the region, $|\tau_{max}|$, is equal to the radius of the circle (Figure 6.14a). This value is precisely one half the magnitude of the differential stress $\Delta\sigma = \sigma_1 - \sigma_2$. The normal stress acting on each of the two planes subjected to the maximum tangential stress is the mean stress σ_m.

6A.3.3.3 Examples of Mohr Circles

Figure 6.16a depicts a hypothetical laboratory experiment in which a piston applies normal stress to one end of a cylindrical rock sample. The sample sits on a rigid base so that it does not accelerate. The piston generates normal stress, called the *end load*, and no shear stresses across all horizontal planes in

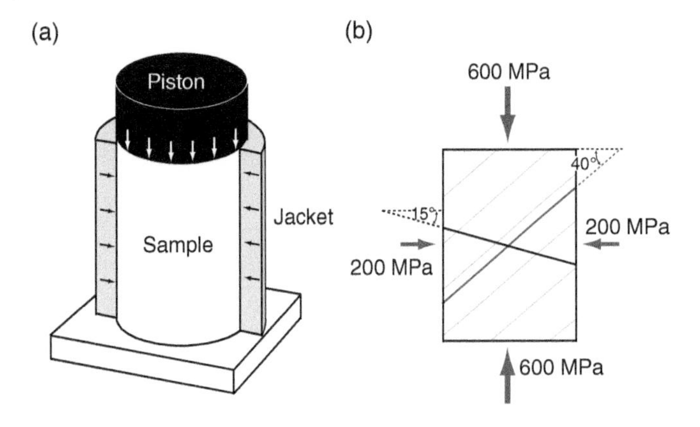

Figure 6.16 (a) Schematic representation of a cylindrical sample of rock whose sides are subjected to compressive normal stress by a pressure jacket and whose end is subjected to greater compressive stress by a piston. (b) Sketch of a section through the sample. The red line inclined 40° to the sample end is aligned parallel to compositional layering planes within the sample. The black line indicates a plane inclined −15° from the sample end. (c) Mohr circle depicting stresses in the sample.

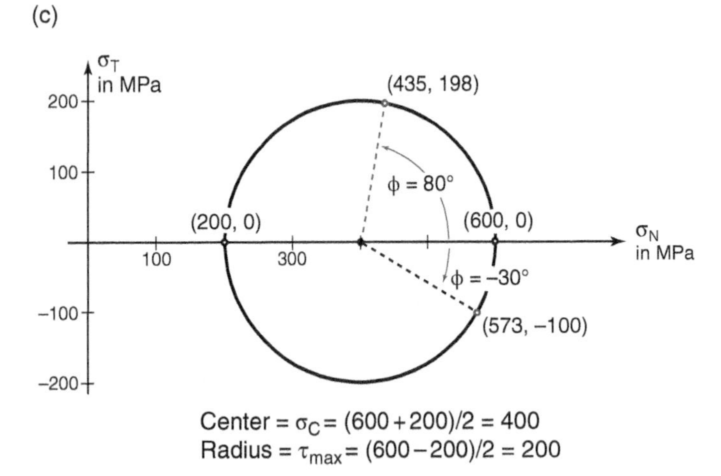

the sample. This cylindrical sample is jacketed by an apparatus that applies uniform normal stresses, called the *confining pressure*, and no shear stresses to the sides of the sample. The magnitude of the end load in this experiment is 600 MPa, and the magnitude of the confining pressure is 200 MPa. Figure 6.16b is a planar cross section of this laboratory experiment; it is a physical plane or real space sketch of the stress state in the sample. We can readily construct a Mohr circle describing the stresses within that cross section of the sample. Because the sample ends and sides are subjected only to normal stresses, they are necessarily the principal planes of the stress state. The most compressive stress is the end load of 600MPa, and the least compressive stress is the confining pressure of 200MPa. The Mohr circle describing the stresses in this section has its center at $(c, 0)$ where $c = ½(\sigma_1 + \sigma_2) = 400$ MPa and its radius $r = ½(\sigma_1 - \sigma_2) = 200$ MPa (Figure 6.16c).

Imagine that the rock sample has a well-defined compositional banding inclined at an angle $\alpha = 40°$ from the end of the sample. We show the orientation of the surfaces separating those bands as gray lines on the physical plane sketch of the sample. What are the stress components acting on those surfaces? The radius of the Mohr circle at an angle of $\phi = 2\alpha = 80°$ intersects the circle at the point (435, 198), indicating that the surfaces separating the bands experience compressive normal

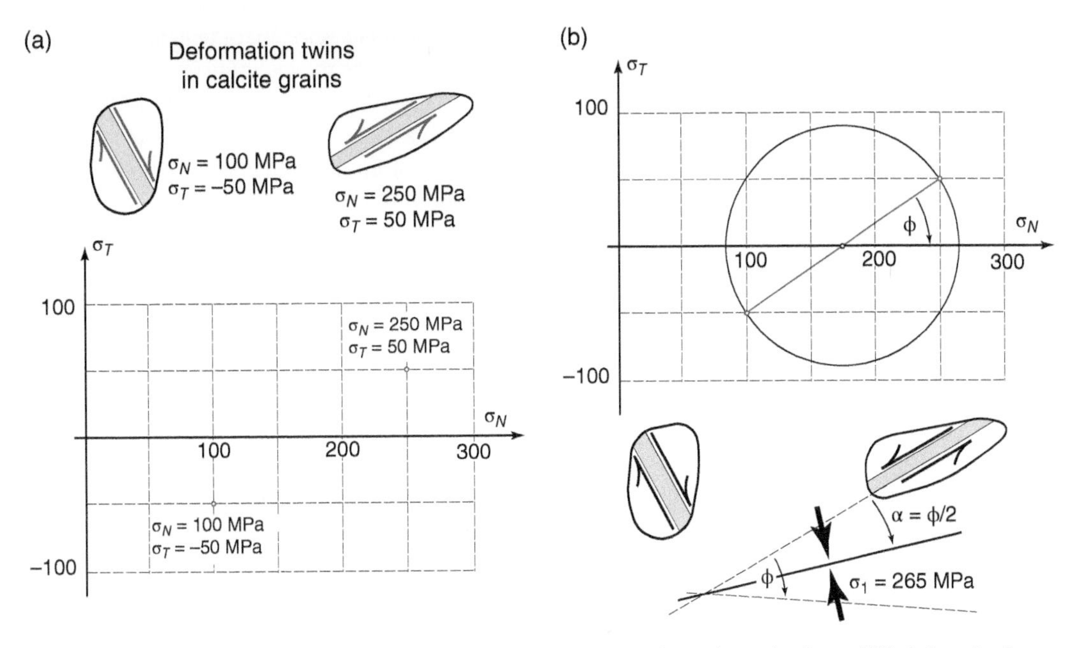

Figure 6.17 (a) Schematic representation of calcite grains in which deformation twins formed. We infer twins form when the shear stress magnitude reaches 50 MPa; the normal stresses acting on the twin planes are 100 MPa and 250 MPa. Points representing the two twin planes are plotted on the Mohr space plot. (b) Mohr circle constructed using the observations summarized in (a). The principal plane subjected to the most compressive normal stress makes a clockwise angle of $\alpha = \phi/2$ from the twin plane exhibiting anticlockwise shear sense, and the magnitude of σ_1 = 265 MPa.

stress of 435 MPa and anticlockwise shear stress of 198 MPa. What are the stress components acting on plane oriented at an angle of α = -15° from the end of the piston (shown as a single black line on the physical plane sketch)? The radius of the Mohr circle at an angle of $\phi = 2\alpha$ = -30° intersects the circle at the point (573, −100), indicating that the plane experiences compressive normal stress of 573 MPa and clockwise shear stress of 100 MPa. This example illustrates many features of the use of Mohr circles, but since the data available at the outset included the orientations of the principal directions and the magnitude of the stress difference, it is apparent that one could equally well have substituted the values of the angles α into Eq. (6.15) to determine the stress components acting on these surfaces.

A second example better shows the utility of Mohr circles. Figure 6.17a is a real space sketch depicting schematically two calcite grains in a deformed limestone in which there are deformation twins. Deformation twinning occurs when shear stresses resolved across a particular crystallographic plane in calcite crystals reaches a critical value. For simplicity, we show two only differently oriented grains with deformation twins and assume that the critical shear stress for twinning is 50 MPa. In the grain at left, where twinning has a clockwise shear sense, the normal stress acting perpendicular to the deformation twin is 100 MPa. Thus, the stresses acting on the crystallographic planes conducive to the formation of deformation twins in that grain are σ_N = 100 MPa and σ_T = −50 MPa. In the grain at right, where twinning has an anticlockwise shear sense, the normal stress acting perpendicular to the deformation twin is 250 MPa. Thus, the stresses acting on the crystallographic planes conducive to the formation of deformation twin in that grain are $\sigma_N = 250$ MPa and $\sigma_T = 50$ MPa. From the data available, the precise orientations of the principal directions and the magnitude of the stress difference are not immediately apparent. We have sufficient information to construct a Mohr circle, however, and from that, we can readily orient the principal directions and determine the principal stress magnitudes.

On the Cartesian σ_N–σ_T axes that define Mohr space, we plot a point representing the crystallographic planes of the deformation twin in the left grain, (100, −50), and a point representing the crystallographic planes of the deformation twin in the right grain, (250, 50) (Figure 6.17a). Since these two planes have equal magnitudes and opposite senses of shear stress, they must lie on opposite sides of a diameter of the Mohr circle. Thus, we can connect the two points with a straight line, find the center of the Mohr circle where that straight line intersects the σ_N axis, and draw the circle that passes through the two points. Inspection of that circle indicates that the most compressive normal stress in the stress state responsible for generating the deformation twins was approximately 265 MPa, and the least compressive stress was approximately 85 MPa. The stress difference was approximately 180 MPa, and the maximum shear stress acting on any plane in the rock at the time the deformation twins formed was about 90 MPa. On the Mohr circle, the point representing the σ_1 plane lies a clockwise angle ϕ from the point representing the deformation twin in the right grain. In the corresponding physical plane sketch of the stress state, we can draw a line representing the plane a clockwise angle ϕ from the crystallographic planes of the deformation twin in the right grain. Bisecting that angle, we find the orientation of the plane that experiences the principal normal stress $\sigma_1 = 265$ MPa and no shear stresses. The plane that experiences the least compressive principal normal stress $\sigma_2 = 85$ MPa is perpendicular to the σ_1 plane.

6A.3.3.4 Relating Mohr Space to Real Space – The Pole to the Mohr Circle

Once a geologist has constructed the Mohr circle that pertains to a particular stress state, she or he can use the Mohr circle to analyze characteristics of that stress state. To do so, she or he must relate one plane in the real space depiction of stresses (plane P) to its corresponding point on the Mohr circle (point p). Having completed that step, the geologist can determine the magnitudes of stress components acting on a plane at an angle α from plane P by finding a point on the Mohr circle at a polar angle $\phi = 2\alpha$ from point p. Similarly, to determine the orientation of a plane with particular stress components, she or he needs to identify the point on the circle with those components, measure the polar angle ϕ between that point and point p, and find the plane at an angle $\alpha = \phi/2$ from the plane P in a physical plane sketch.

There is a neat, graphical construction that one can use to avoid the need to measure angles between planes or polar angles between points on a Mohr circle. On any stress Mohr circle, a geologist can identify a specific point called the **pole to the Mohr circle** that connects that Mohr circle to a particular physical plane sketch. By finding the position of the pole, she or he relates all planes in real space to their corresponding points on the Mohr circle. To find the pole to the Mohr circle, the geologist identifies a point p on the Mohr circle that he or she can correlate to a particular plane in real space. On the physical plane sketch of the region, draw a *normal* to the plane P (Figure 6.18a). Then, on the Mohr circle in Mohr space, draw a line parallel to the normal through the point p, which corresponds to the plane P. Typically, that line, extended if necessary, will intersect the circle at two points: the point p and the pole to the Mohr circle. In rare cases, a line normal to a plane X drawn through the corresponding point x, the point corresponding to plane X, will pass only through point x because it is tangent to the Mohr circle. In such a case, the

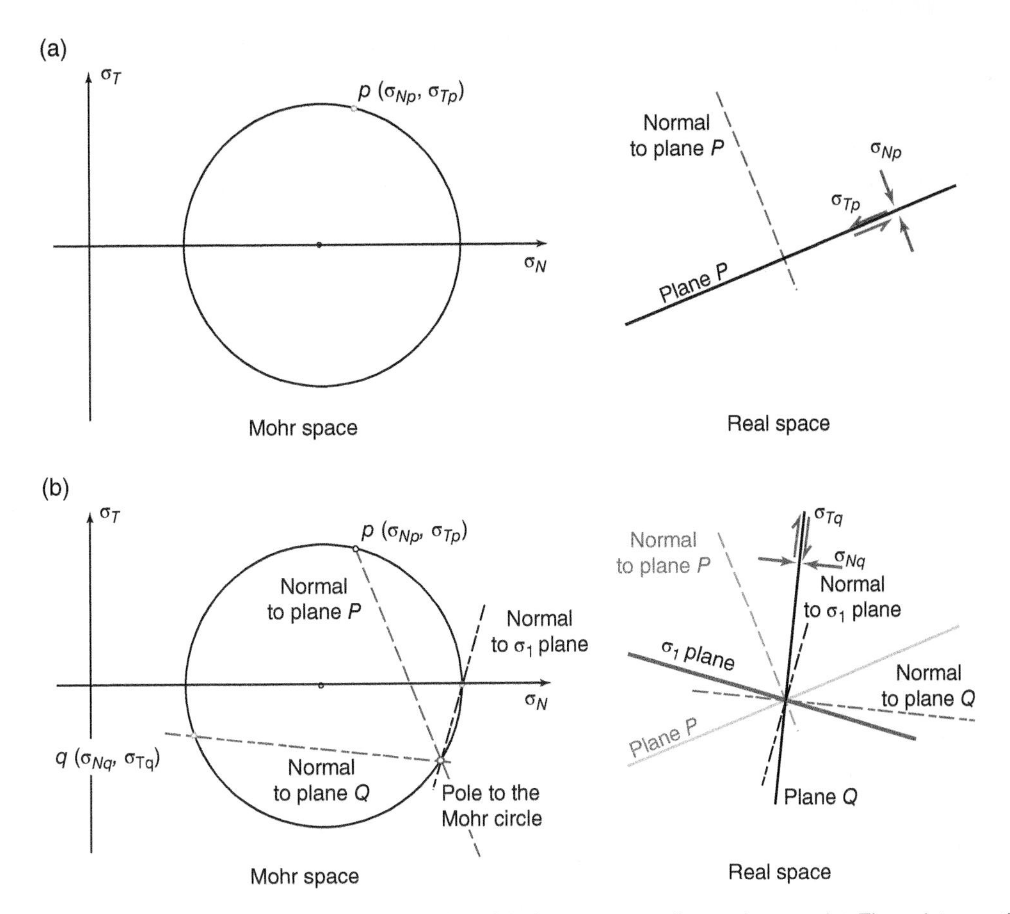

Figure 6.18 (a) At left is a Mohr circle, and at right is a corresponding real space plot. The point p on the Mohr circle corresponds to plane P in the real space drawing. (b) A line that is perpendicular to the plane P in real space drawn through point p intersects the Mohr circle at the pole to the Mohr circle. Lines from a pole to the Mohr circle defined in this way to any point on the Mohr circle (e.g. q) are perpendicular to the corresponding plane in real space (e.g. Q).

point x is the pole to the Mohr circle. Once the pole is identified, a line drawn from the pole to the Mohr circle to any point q on the circle will be parallel to the normal to the corresponding plane Q in the real space depiction (Figure 6.18b). Using the pole to the Mohr circle, it is easy to identify either of the principal planes of the stress state (Figure 6.18b).

Consider the example introduced in Figure 6.17. We label the crystallographic plane with the deformation twin in the right grain plane D and the crystallographic plane with the deformation twin in the left grain plane E. Points d and e on the Mohr circle correspond, respectively, to those planes. We draw a normal to one of those planes, plane E for example, and then draw a line parallel to the normal to plane E through point e on the Mohr circle. That line also intersects the circle at the pole to the Mohr circle (Figure 6.19). Having found the pole to the Mohr circle, we can draw a

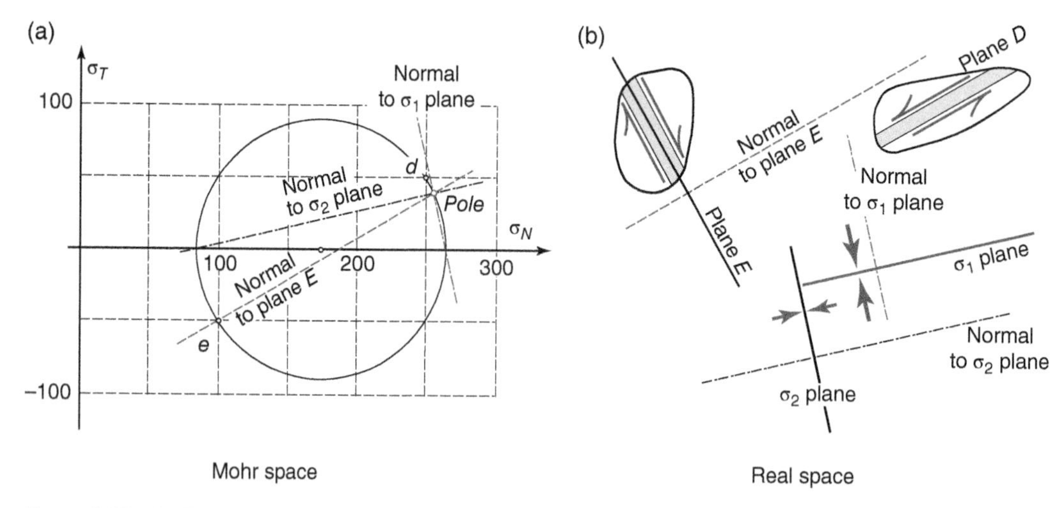

Figure 6.19 (a) The Mohr circle corresponding to the twinned calcite situation described in Figure 6.16. (b) A real space depiction of the twinned calcite situation described in Figure 6.16. The pole to the Mohr circle is identified by drawing a line parallel to the normal to plane E. Lines subsequently drawn emanating from the pole to the Mohr circle to the points corresponding to the two principal planes indicate the orientations of the principal planes.

line from the pole to any other point on the Mohr circle, such as the point representing the σ_1 principal plane. That line gives the attitude of the normal to the σ_1 principal plane, and we can use it to orient the σ_1 principal plane in our real space sketch of the stress state (Figure 6.19). Similarly, we can draw a line from the pole to the point representing the σ_2 principal plane. That line gives the attitude of the normal to the σ_2 principal plane, and we can use it to orient the σ_2 principal plane in our real space sketch of the stress state.

6A.3.4 Hydrostatic vs. Non-hydrostatic Stress

The simplest possible state of stress is **hydrostatic stress**, in which all planes experience equal normal stress components and the stress ellipsoid is a circle (Figure 6.20). In theory, the normal stresses can be either compressional or tensional, although all familiar examples have compressional normal stresses. In hydrostatic stress, there are no unique principal directions because all pairs of mutually perpendicular planes have normal stresses with equal magnitudes and no planes have tangential stresses acting on them. Note that the magnitude of the normal stress acting on any and all planes is necessarily the mean stress σ_m. We call this stress state *hydrostatic* because it is the stress state that prevails within liquids and gases, neither of which can sustain shear stresses. Because the stress difference is zero in hydrostatic stress, a Mohr circle for hydrostatic stress has a radius of zero length, and the Mohr circle collapses to a single point on the σ_N axis. In this situation, the stress matrix for any two mutually perpendicular planes is:

$$\mathbf{T_m} = \left[\sigma_m\right] = \begin{vmatrix} \sigma_m & 0 \\ 0 & \sigma_m \end{vmatrix}.$$

When the magnitude of the normal stress component acting on one of two mutually perpendicular planes in a region is greater than the magnitude of the normal stress component acting on the other plane, Eqs. (6.15) indicate that finite tangential stress components act on all but one pair of mutually perpendicular planes *and* that the two planes with the zero tangential stress components experience the greatest and least magnitudes of normal

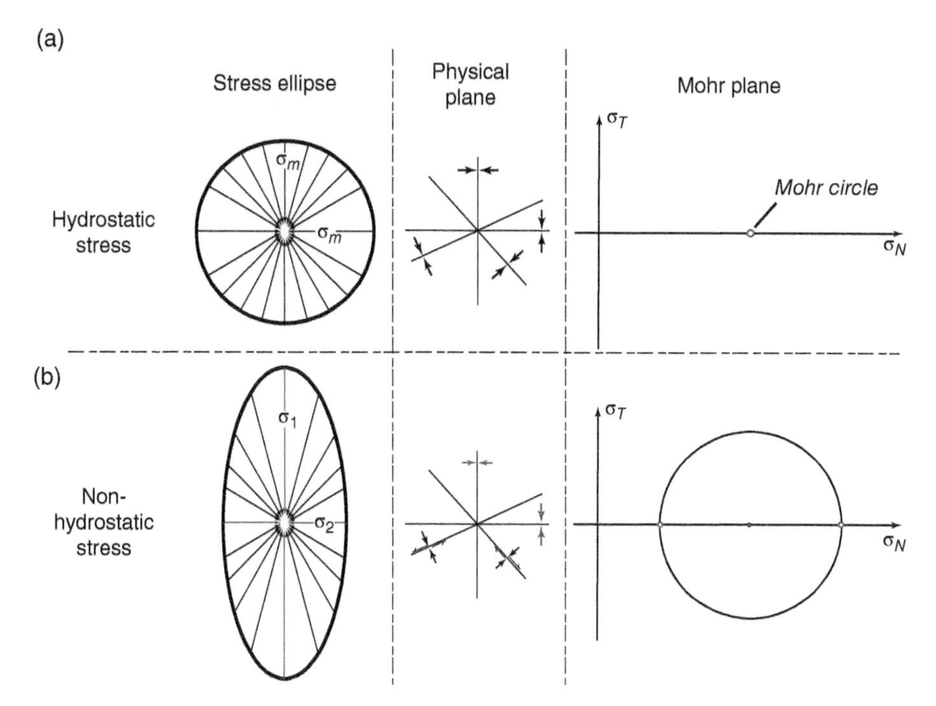

Figure 6.20 (a) From left to right, stress ellipse, physical plane or real space depiction, and Mohr plane depiction of a hydrostatic stress state. (b) From left to right, stress ellipse, physical plane or real space depiction, and Mohr plane depiction of a non-hydrostatic stress state.

stress components. Such a stress state, called **non-hydrostatic stress**, prevails at most places in Earth (Figure 6.20). It is only within solids, which can sustain finite differential stress or finite tangential or shear stresses, that non-hydrostatic stress can be attained. The stress ellipse for non-hydrostatic stress is an ellipse, and the stress Mohr circle has a finite radius. The stress matrix for a typical pair of mutually perpendicular planes in such a stress state is:

$$\mathbf{T} = [\sigma] = \begin{vmatrix} \sigma_{11} & \sigma_{12} \\ \sigma_{21} & \sigma_{22} \end{vmatrix}.$$

Structural geologists often conceptualize such a non-hydrostatic stress state as the sum of two components: (1) a **hydrostatic stress component** in which each plane experiences normal stress with a magnitude given by $\sigma_m = ½(\sigma_1 + \sigma_2)$ and no tangential stress, and (2) a **deviatoric stress component** which is the normal stress *and* tangential stress magnitude that must be added to the hydrostatic stress component to attain the total stress state. Recall that the mean stress is the arithmetic average of the two extreme normal stress magnitudes, that is, $\sigma_m = ½(\sigma_1 + \sigma_2)$. Note that the deviatoric stress component measures the departure, or deviation from, a hydrostatic stress state.

When using a stress matrix to represent the state of stress, it is possible to factor the total stress matrix into a hydrostatic stress component matrix and a deviatoric stress component matrix:

$$\mathbf{T} = [\sigma] = \begin{vmatrix} \sigma_{11} & \sigma_{12} \\ \sigma_{21} & \sigma_{22} \end{vmatrix}$$

$$= \begin{vmatrix} \sigma_m & 0 \\ 0 & \sigma_m \end{vmatrix} + \begin{vmatrix} \sigma_{11} - \sigma_m & \sigma_{12} \\ \sigma_{21} & \sigma_{22} - \sigma_m \end{vmatrix}.$$

6A.3.5 Homogeneous vs. Inhomogeneous Stress

Up to this point, we have focused on states of stress within limited regions and have argued that the stresses acting on different planes vary only with the orientation of the planes. In actuality, a state of stress refers to the magnitudes of the stress components acting on planes with different orientations *at a particular point in space*. The states of stress can and do vary from place to place in Earth. The state of stress at a mid-ocean ridge is not like that in a collisional mountain belt, or even like that in the middle of an oceanic plate. We use the notion of a **stress field**, analogous to the displacement or strain fields outlined in the last two chapters, to describe whether and how states of stress vary from point to point within Earth.

If parallel planes have equal normal and tangential stress magnitudes at all points within a region, the stress field is **homogeneous** within that region. Some geologists refer to homogeneous stress fields as *regions of uniform stress*. If parallel planes have unequal normal and/or tangential stress magnitudes at different points within a region, the stress field in that region is **inhomogeneous** or **heterogeneous**. The orientations of the stress principal axes and magnitudes of the principal stresses are the same at all points in a region of homogeneous stress. Either the orientations of the stress principal axes, the magnitudes of the principal stresses, or both the orientations of the stress principal axes and the magnitudes of the principal stresses vary with position in a region of heterogeneous stress. Figure 6.21 contrasts a vertical cross section across a homogeneous stress field with a vertical cross section across an inhomogeneous stress field. The principal stress magnitudes and orientations do not vary across the homogeneous field but they vary smoothly across the inhomogeneous stress field.

Stresses are rarely uniform in Earth. At the scale of plates – 1000's of km – stresses vary significantly: planes with the same orientation might experience the most compressive normal stress in the vicinity of a deep ocean trench along one end of a plate and experience the least compressive normal stress in the vicinity of a mid-ocean ridge at the other end of that plate. At the scale of geologic maps – tens of km – comparable variations occur across map-scale faults or folds. Even at the scale of thin sections – tens of mm – the state of stress within a quartz clast in sandstone may differ from the state of stress in an adjacent calcite clast in that sandstone. Still, there are settings, ranging in size from tens of m^2 to thousands of km^2 depending on the tectonic environment, in which stresses vary sufficiently little from one point to another that we can treat stress fields as homogeneous. We can recognize such regions of uniform stress by insignificant variations in the orientation of stress principal axes and insignificant variations in the magnitudes of the principal stresses across the region.

There are important constraints on stress fields in Earth. Since liquids and gases are unable to support shear stresses, the surface of contact between rocks and the atmosphere must be free of shear stresses. The ground surface is, then, a principal plane. Any evidence that stress principal planes at depth are inclined confirms that the stresses in that region are nonuniform. Similarly, the contact between rocks and any enclosed liquid (a magma body, a fluid-filled pore, etc.) must also be free of shear stresses. These contacts too are necessarily principal planes. Any evidence of changes in the orientations of stress principal directions away from such contacts confirms that stresses are inhomogeneous.

6A.4 Stress in Three Dimensions

In three dimensions, there are (1) three mutually perpendicular principal planes of stress on

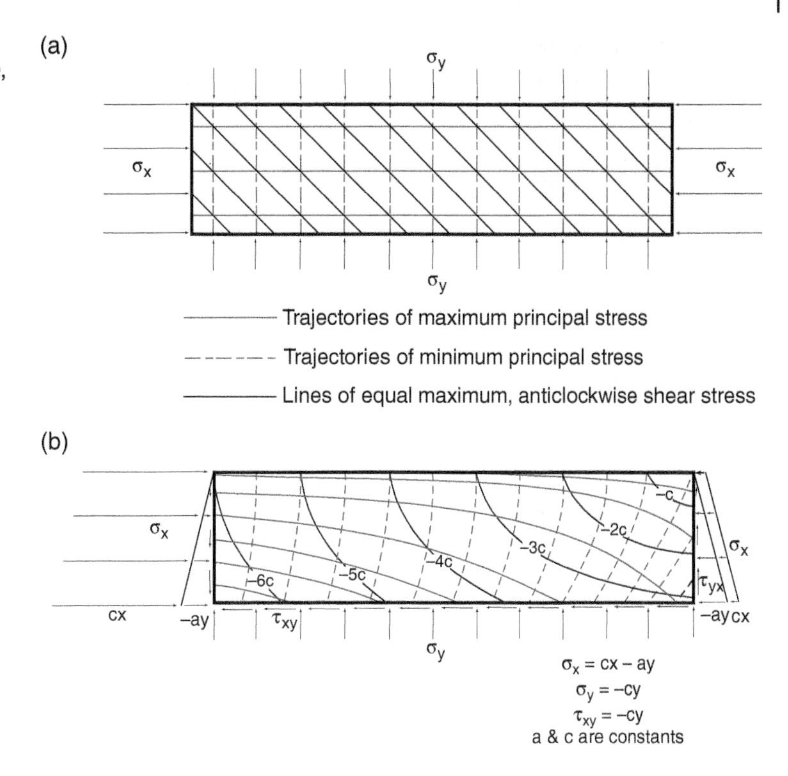

Figure 6.21 (a) Diagram illustrating a uniform stress state, where orientations and magnitudes of principal stresses are everywhere the same. Thus, the orientations of planes experiencing the maximum anticlockwise shear stress are therefore also everywhere parallel, and the magnitude of the maximum shear stress does not change with position. (b) Diagram, after Hafner (1951) illustrating a nonuniform stress state. Orientations and magnitudes of principal stresses vary with position, as do the orientations of planes experiencing the maximum anticlockwise shear stress and the magnitude of the maximum shear stress on those planes.

which there are no tangential stresses, (2) three principal stress axes, each one perpendicular to one of the three principal planes of stress, and (3) three principal stresses, with magnitudes $\sigma_1 \geq \sigma_2 \geq \sigma_3$ (Figure 6.22). By analogy with the treatment of two-dimensional stress states, defining the stress vectors that act on three mutually perpendicular faces of a cubic volume at equilibrium necessarily specifies the stress vectors acting on the three mutually perpendicular faces on the opposite sides of the cube. We use a 3×3 matrix to represent the stress vector components acting on three mutually perpendicular planes. The row-column position of an element in the matrix is again fixed by its subscripts, and the subscripts again also denote the plane on which the different stress vector components act and the direction in which the different stress vector components point. In the 3×3 stress matrix

$$\mathbf{T} = \left[\sigma\right] = \begin{vmatrix} \sigma_{11} & \sigma_{12} & \sigma_{13} \\ \sigma_{21} & \sigma_{22} & \sigma_{23} \\ \sigma_{31} & \sigma_{32} & \sigma_{33} \end{vmatrix} = \begin{vmatrix} \sigma_{xx} & \sigma_{xy} & \sigma_{xz} \\ \sigma_{yx} & \sigma_{xy} & \sigma_{yz} \\ \sigma_{zx} & \sigma_{zy} & \sigma_{zz} \end{vmatrix}$$

the first subscript gives the coordinate direction that is perpendicular to a plane and the second subscript gives the coordinate direction in which a component points (Figure 6.23). For example, for the plane perpendicular to the x_1 or x coordinate axis, σ_{11} is a normal stress component parallel to the x_1 or x axis, σ_{12} is a shear stress component parallel to the x_2 or y axis, and σ_{13} is a shear stress component parallel to the x_3 or z axis. Because that there are no net moments in a region at equilibrium, σ_{ij} must equal σ_{ji} and the stress matrix $[\sigma]$

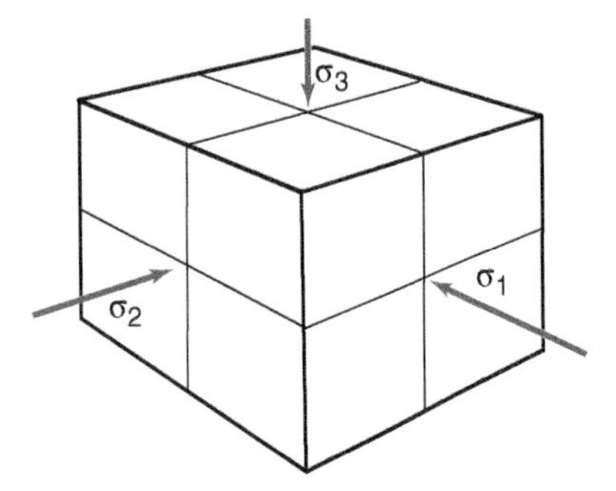

Figure 6.22 In three dimensions, there are three mutually perpendicular principal planes on which the shear stress component vanishes. One principal plane has the most compressive normal stress (σ_1), a second principal plane has the least compressive normal stress (σ_3), and the third principal plane has a normal stress magnitude σ_2 where $\sigma_3 \leqslant \sigma_2 \leqslant \sigma_1$.

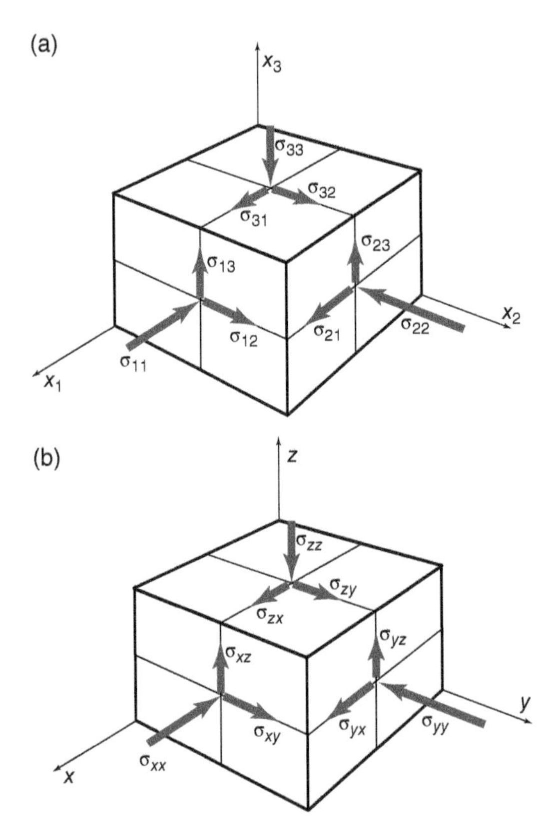

Figure 6.23 Illustration showing two common schemes for naming the nine components that define stresses in three dimensions. (a) With coordinate axes labeled x_1, x_2, and x_3. (b) With coordinate axes x, y, and z. On the plane perpendicular to the x_1 or x axis, the normal stress component is σ_{11} or σ_{xx} and the shear stress components are σ_{12} or σ_{xy} and σ_{13} or σ_{xz}. On the plane perpendicular to the x_2 or y axis, the normal stress component is σ_{22} or σ_{yy} and the shear stress components are σ_{21} or σ_{yx} and σ_{23} or σ_{yz}. On the plane perpendicular to the x_3 or z axis, the normal stress component is σ_{33} or σ_{zz} and the shear stress components are σ_{31} or σ_{zx} and σ_{32} or σ_{zy}.

must be symmetric. Only six of the nine stress vector components are free to vary independently. We call the line running from the upper left corner to the lower right corner of the array of σ_{ij} values (the line runs through σ_{xx}, σ_{yy}, and σ_{zz}) the *principal diagonal* of the array. The other six σ_{ij} values are off-diagonal components. Another way to recognize that the stress matrix is symmetric is to note that the array remains unchanged if we "reflect" values across the principal diagonal. You can understand why we call the line from the upper left to lower right line across the array the principal diagonal by noting that the 3x3 matrix referred to the three orthogonal principal planes is:

$$\mathbf{T}' = \left[\sigma'\right] = \begin{vmatrix} \sigma_1 & 0 & 0 \\ 0 & \sigma_2 & 0 \\ 0 & 0 & \sigma_3 \end{vmatrix}$$

Only along the principal diagonal of the array referred to the principal planes do the stress vector components have nonzero values. The orientations and magnitudes of the three principal stresses define the state of stress in three dimensions completely, so determining the principal stresses is the simplest way to deal with stress. In some settings in Earth, a three-dimensional state of stress might have one of its principal stress axes vertical and two of its principal stress axes horizontal. In such a setting, the vertical principal stress magnitude will be the stress due to gravity.

The principal axes of a three-dimensional stress state need not have vertical or horizontal principal stress axes, and commonly no principal stress axis is horizontal or vertical.

6A.4.1 The Stress Ellipsoid

Just as one can use an ellipse to represent the state of stress in two dimensions, an ellipsoid known as the **stress ellipsoid** depicts three-dimensional stress states. As is the case for the stress ellipse, radii of the stress ellipsoid represent the stress vectors acting across planes with different orientations that pass through a single point (Figure 6.24). Each of the three principal radii of the stress ellipsoid parallels one of the three principal stress axes, represents the stress vector that acts on the plane perpendicular to that radius, and has a length proportional to the magnitude of one of the principal stresses. Radii oblique to the principal radii of the stress ellipse again depict the smooth variation in the magnitude of the stress vectors acting on planes with different orientations. For geological situations, where all planes experience compressive normal stresses, the stress vectors point toward the center of the ellipsoid. In some geological settings, stress ellipsoids have three unequal principal radii, so the stress ellipsoid is a triaxial ellipsoid. In other settings, one principal radius is longer or shorter than all other radii, so the stress ellipsoid is an ellipsoid of revolution. Finally, in some geologic settings, all radii of the stress ellipsoid are equal in length, and the stress ellipsoid is a sphere.

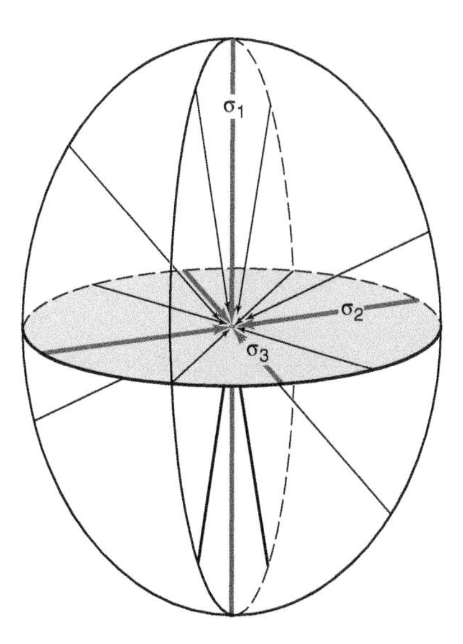

Figure 6.24 A stress ellipsoid for a three-dimensional stress state where all normal stresses are compressive.

6A.4.2 Hydrostatic, Lithostatic, and Deviatoric Stresses

A three-dimensional hydrostatic stress state is one in which the normal stresses acting across planes with any orientation have the same magnitudes and where no planes experience any shear stress. The stress ellipsoid for a three-dimensional hydrostatic stress state is a sphere. The stress matrix for a three-dimensional hydrostatic stress state is:

$$\mathbf{T_m} = \left[\sigma_m\right] = \begin{vmatrix} \sigma_m & 0 & 0 \\ 0 & \sigma_m & 0 \\ 0 & 0 & \sigma_m \end{vmatrix}$$

As is true of a two-dimensional hydrostatic stress state, there are no unique principal axes in a three-dimensional hydrostatic stress state because all planes passing through a point experience the same stress. One example of a hydrostatic stress state with geologic relevance is a **lithostatic stress state** in which the magnitude of normal stresses acting on any plane is the stress due to the weight of the overlying material. We calculate the value of the vertical lithostatic stress σ_L at a depth z from the same formula used to calculate the vertical stress under a building, with z (depth below ground surface) replacing h (compare with Eq. (6.7)):

$$\sigma_L = \rho g z. \tag{6.17}$$

Table 6.3 Special and general states of stress (modified from Means 1976). The Mohr circles are drawn for the σ_1–σ_3 plane; σ_2 is not shown on the Mohr circles.

Stress State	Principal stresses	Mohr circle	Geologic occurrence
1. Hydrostatic tension	$\sigma_1 = \sigma_2 = \sigma_3 < 0$	σ_T, σ_N; Mohr circle is a point	Not likely
2. General tension	$\sigma_1 \neq \sigma_3$, both < 0	σ_T, σ_N	Possible at shallow depths
3. Uniaxial tension	$\sigma_1 = \sigma_2 = 0, \sigma_3 < 0$	σ_T, σ_N	Possible at shallow depths
4. Tension and compression	$\sigma_1 > 0, \sigma_3 < 0$	σ_T, σ_N	Possible at shallow depths
5. Pure shear stress	$\sigma_1 = -\sigma_3, \sigma_2 = 0$	σ_T, σ_N	Possible
6. Uniaxial compression	$\sigma_2 = \sigma_3 = 0, \sigma_1 > 0$	σ_T, σ_N	Possible
7. General compression	$\sigma_1 \neq \sigma_3$, both > 0	σ_T, σ_N	Common
8. Hydrostatic compression	$\sigma_1 = \sigma_2 = \sigma_3 > 0$	σ_T, σ_N; Mohr circle is a point	Possible at great depths; a pure lithostatic stress

In the upper crust, a typical value for ρ is $2.7\ g/cm^3 = 2700\ kg/m^3$. Taking $g = 9.8\ ms^{-2}$, Equation (6.17) indicates that σ_L increases at a rate of 26.5 MPa/km $\approx$ 27 MPa/km. This is a useful figure to remember. In Earth, the hydrostatic component of stress dominates the state of stress at intermediate to deep crustal levels, and hydrostatic stress is the value that is recorded by pressure estimates from metamorphic petrology. States of stress in the Earth are shown in Table 6.3.

In most geologic settings that interest structural geologists, the stresses acting on planes with different orientations are not equal. In those cases, the three-dimensional **differential stress** is $\Delta\sigma = \sigma_1 - \sigma_3$. When the differential stress is finite, the stress ellipsoid is no longer spherical and planes with some orientations have shear stresses acting on them. For three-dimensional stress states too, it is useful to divide the total state of stress into *hydrostatic stress* and *deviatoric stress components*. For three-dimensional stress states too, the **mean stress** σ_m is equal to the average of three principal stresses or the average of the normal stresses acting on any three, mutually perpendicular planes: $\sigma_m = (\sigma_1 + \sigma_2 + \sigma_3)/3 = (\sigma_{11} + \sigma_{22} + \sigma_{33})/3$. The hydrostatic component of the state of stress again consists of normal stress with magnitude σ_m acting equally in planes with any orientation. When the differential stress $\Delta\sigma$ is finite, we again envision that the total stress acting across any plane is the sum of mean stress and a contribution that causes a deviation from a hydrostatic stress state. In three dimensions as in two dimensions, the *deviatoric stress* encompasses how the total stress departs from a hydrostatic reference state. Stated in a different way, the total stress is the sum of the hydrostatic component and deviatoric component. In matrix form, we have

$$\mathbf{T} = [\sigma] = \begin{vmatrix} \sigma_m & 0 & 0 \\ 0 & \sigma_m & 0 \\ 0 & 0 & \sigma_m \end{vmatrix} + \begin{vmatrix} \sigma_{11} - \sigma_m & \sigma_{12} & \sigma_{13} \\ \sigma_{21} & \sigma_{22} - \sigma_m & \sigma_{23} \\ \sigma_{31} & \sigma_{32} & \sigma_{23} - \sigma_m \end{vmatrix}$$

The deviatoric stress component is that part of the total stress that causes rocks to change shape, whereas the hydrostatic stress component is that part of the total stress that causes a change in rock volume.

6A.5 Pore-fluid Pressure and Effective Stress

Fluids – such as water, hydrocarbons, CO_2, or melt – regularly occur in the pores of a rock. Because these fluids resist compaction, the fluids trapped in pores exert a stress on the surrounding rock that is called the **pore fluid pressure**, P_f. Since fluids cannot support shear stresses, the stress across the fluid-rock contact can only be hydrostatic stress. Terzaghi (1923) proposed that the total state of stress in a rock containing fluids could be divided into two components, the pore fluid pressure and an **effective stress**. He defined effective stress as the total stress minus the pore fluid pressure. Because the pore fluids are unable to exert shear stresses on rocks, the effective stress acting across any plane consists of (1) a normal stress component equal to the total stress normal component minus the pore fluid pressure, and (2) a tangential stress component equal to the total tangential shear stress component.

The stress state within the pore fluid is always **hydrostatic**, that is, it consists of normal stresses with equal magnitude acting across all planes passing through a point and no shear stresses acting on any planes. If the pores in a rock are interconnected, so that fluid can move from one pore to another, the pressure at any point in the fluid is due to the weight of the overlying column of fluid (Figure 6.25). In that case, we can rewrite Eq. (6.17) as

$$P_f = \rho_f g z, \tag{6.18}$$

where ρ_f is the density of fluid and z is the height of the column of fluid. The nature of the pore fluid pressure is critically important in analyzing the stresses within layers of saturated, unconsolidated sediment or saturated sedimentary rock sealed under an impermeable layer. Because the density of water is approximately 1/3 the density of the surrounding rock, the pore fluid pressure in a water-saturated rock with interconnected pores

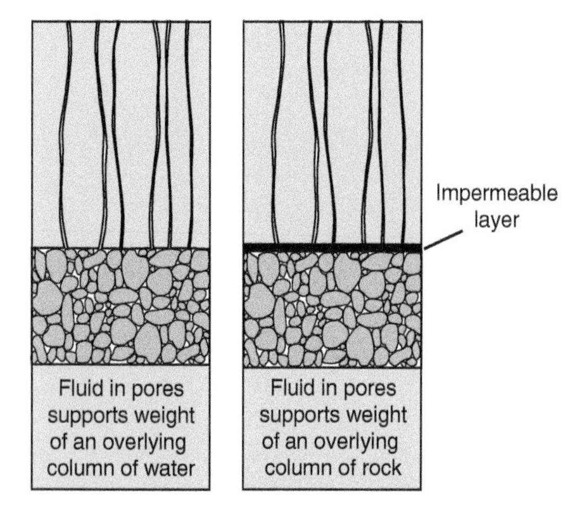

Figure 6.25 If open conduits of some sort connect fluids filling the pores between minerals with the surface (drawing at left), the pressure supported by the fluids is that due to the weight of the overlying column of fluid. If an impermeable layer interposes, isolating the fluids filling the pores between minerals (drawing at right), the pressure supported by the fluids is that due to the weight of the overlying column of rock.

is $P_f \approx (\sigma_L)/3$. If the pores in a rock are not interconnected, however, and the fluid cannot move from one pore to another, the fluid within a pore must support the weight of the overlying column of rock (Figure 6.25). In such a situation, P_f can approach and even equal σ_L, the lithostatic stress at that depth. Any additional stress placed on fluids confined to isolated pores can raise the pore fluid pressure to magnitudes that exceed the lithostatic stress. Pore fluid pressures are important factors in the formation of mineral deposits and in seismic activity, and high pore fluid pressures can be very dangerous when drilling for oil.

6A.6 Three-dimensional States of Stress

We distinguish three general types of three-dimensional stress states (See Table 6.4). In *uniaxial* stress states, only one of the three principal stress vectors is nonzero. For such a stress state, the stress "ellipsoid" is a straight line. In *biaxial* stress states,

Table 6.4 Nomenclature for three-dimensional stress states, with comments on their applicability to Earth and rock mechanics experiments.

Stress State Name	Definition	Comments
Uniaxial	Two principal stresses = 0; third principal stress can be compressive or tensile	Stress state approximated by some rock mechanics experiments
Biaxial	One principal stress = 0; other principal stresses can be compressive or tensile	Stress state approximated by some rock mechanics experiments
Triaxial	No principal stresses = 0	Typical state of stress in Earth and in rock deformation experiments
Hydrostatic	$\sigma_1 = \sigma_2 = \sigma_3 = \sigma_m$	*Stress state in the liquid outer core, approximated in the inner core and lower mantle; occurs locally in upper mantle and lower crust*
Axial	*Two principal stresses are equal*	*Stress state in most rock deformation experiments*
Axial Compression	$\sigma_1 > \sigma_2 = \sigma_3$	*Stress state in many rock deformation experiments*
Axial tension	$\sigma_1 = \sigma_2 > \sigma_3$	*Stress state in some rock deformation experiments*
Polyaxial	$\sigma_1, \sigma_2, \sigma_3$ *all different:* $\sigma_1 \neq \sigma_2 \neq \sigma_3$	*Typical state of stress in Earth*

two of the three principal stress vectors have nonzero magnitudes and the third principal stress vector has a magnitude of zero. For biaxial stress states, the stress "ellipsoid" is an ellipse. Some engineering texts refer to biaxial stress states as "plane stress" stress states. In *triaxial* stress states, all three of the principal stress vectors have nonzero magnitudes, and the stress ellipsoid is the three-dimensional figure described earlier.

In all geological settings and in many laboratory settings, stress states are triaxial. As noted earlier, the distinction between hydrostatic and non-hydrostatic stress states is relevant to triaxial stress states. Three-dimensional hydrostatic stress states have equal normal stress magnitudes acting on all planes and no shear stresses acting on any planes passing through a point. We distinguish several different classes of three-dimensional, non-hydrostatic stress states. In the first class, *axial stress states*, there is one principal stress vector with a magnitude different from the other two principal stress magnitudes. The stress ellipsoids for this class of stress states are ellipsoids of revolution, so the stress vectors along any direction perpendicular to the single distinct principal stress vector are equivalent. Any and all of those directions are principal directions, and the stress state has axial symmetry about the one distinct principal stress vector. Many modern rock deformation experiments are *axial compression* experiments, where the single distinct stress vector is more compressive than the innumerable, equivalent compressive stress vectors that generate lateral confinement for a sample. Some rock deformation experiments and many experimental deformations of engineering materials are axial tension experiments, where the single distinct stress vector is less compressive than the innumerable, equivalent compressive stress vectors confining the sample. Triaxial stress states in which the three principal stress vectors with unequal magnitudes are *polyaxial* stress states. Stresses at many locations in Earth are polyaxial. One polyaxial stress state warrants additional comment. If the magnitude of the intermediate principal stress vector equals the mean stress, there will be no deviatoric stress in the σ_2 direction and no tendency for rock to shorten or lengthen in that direction. In that case, $\sigma_1 + \sigma_2 + \sigma_3 = \sigma_m$ becomes $\sigma_1 + \sigma_m + \sigma_3 = \sigma_m$ or $(\sigma_1 + \sigma_3)/2 = \sigma_2 = \sigma_m$. Stress states with this characteristic may prevail along relatively straight plate boundaries, where along-strike changes in structures or fabrics are minimal.

Geologists often refer to a **stress-free state**. This is an idealization, for materials can never be truly stress-free. If the stresses acting on planes within a geological material are small (especially when compared to typical geological stresses), a material is approximately **stress-free**. We sometimes use a "stress-free" state as a reference state against which some loading conditions may be compared. Similarly, because the normal stress exerted on a rock surface by the atmosphere, $\sigma_n \approx 10^5$ Pa = 0.1 MPa, is so small relative to the magnitude of the stresses operating within rocks, geologists often refer to rock-atmosphere contacts as *stress-free surfaces*. In most examinations of variations in stress within Earth's crust, geologists envision the ground surface as a stress-free surface. Similarly, in rock deformation experiments an unjacketed cylindrical rock sample subjected to stresses only at its ends is said to experience uniaxial compression or uniaxial tension because the stresses acting on the sides of the cylinder are essentially zero. An experiment in which a rectangular rock sample is loaded across four of its six bounding surfaces might be said to be subjected to a biaxial stress state because the stresses on the remaining two surfaces are essentially zero. The stresses on the atmosphere-rock contact surfaces do not have zero magnitude, but they are essentially zero relative to the typical stress magnitudes within rocks.

6A.7 The State of Stress in Earth

The forces that act on rocks in Earth are one of two general types: (1) **Body forces**, which arise from properties that are internal to a volume of

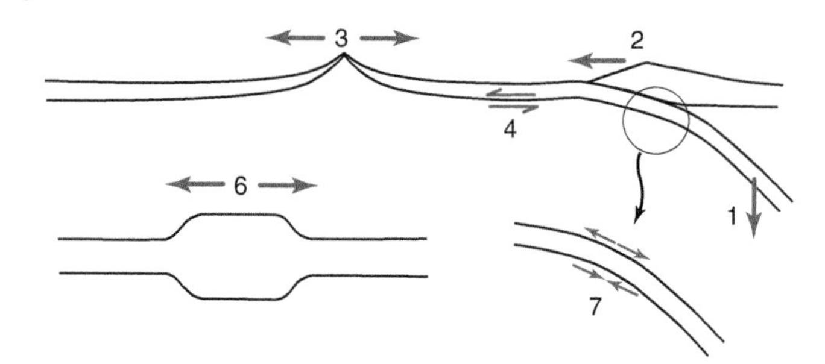

Figure 6.26 Sources of tectonic stresses in the lithosphere: slab pull (1), subduction suction (2), ridge push (3), mantle drag (4), plateau uplift (6), and lithospheric bending (7). See text for explanation.

rock and act on every particle in the rock; and (2) **Surface forces** (also called *contact forces*), which act on the boundaries of a rock body and are transmitted to the interiors of rock volumes along networks of bonds within the volume. Body forces that are geologically important are those due to gravity, which are always present in any rock. Other body forces, such as those due to magnetic and electric fields, rarely are sufficient to be a factor in geological settings. Body forces give rise to surface forces – for example, our weight creates a force that is transmitted through the soles of our shoes to the floor or snow beneath us.

The distinction between body and surface forces is a useful way to analyze forces and stresses in Earth. Natural stresses in Earth can be divided into these two types: body forces due to gravity give rise to **gravitational stresses**, and surface forces, mostly applied at plate boundaries, give rise to **tectonic stresses**, also called *applied stresses*. The state of stress in Earth is given by the combination of gravitational and tectonic stresses.

Among the possible sources of tectonic stress in the lithosphere are:

1) Slab pull – caused by the density contrast between cold, dense subducted oceanic crust and mantle, pulling the subducted plate into the mantle.
2) Subduction suction – a stress induced by the subducting slab that pulls the overriding plate toward the subduction zone.
3) Ridge push – stress caused by the buoyancy of hot mid-ocean ridges.
4) Mantle drag – caused by convection currents in the mantle exerting shear on the overlying plates.
5) Resistance forces – various types that balance other sources of stress.
6) Plateau uplift – uplifted areas have higher potential energy than adjacent lower areas, causing them to be in a state of tension and exert compression on surrounding lithosphere.
7) Lithosphere bending – bending or unbending of plates during subduction can induce outer arc tension and inner arc compression.
8) Membrane stress – the variation of the radius of the Earth from a maximum at the equator to the minimum at the poles necessitates that plates change shape as they move across lines of latitude.
9) Thermal stress – contraction due to cooling, expansion due to heating.

Plate motion is considered to be driven mainly by stresses caused by 1–4 (Figure 6.26), which are capable of inducing large strains in the Earth and are referred to as *renewable*. Stresses due to 6–9 can be relieved by very small strains, and are thus referred to as *nonrenewable*. They typically do not cause large deformation, and are unlikely to be observed in the geological record.

6A.8 Change of Stress: Paleostress, Path, and History

The geological record contains many examples of structures that have formed in one stress regime and been reactivated in another. Such changes of stress regime may be expected from the notion of the Wilson cycle, by which constructive plate boundaries are superseded by destructive plate boundaries and vice versa. The different types of plate boundary are associated with different stress regimes, so rock masses near current or former plate boundaries are likely to have experienced changes in their state of stress over time. In some situations, geologists can determine the orientations and/or the relative or absolute magnitudes of the principal stresses at some time in the past. The orientations *and* the relative magnitudes of the principal stresses are important aspects of the stress state, but ideally one would also like to know the absolute stress magnitudes. Structural geologists use the term **paleostress** to refer to a characterization of the orientations, relative magnitudes, or absolute magnitudes of the stresses that pertained at some point in the past. If a geologist can determine the paleostress states that prevailed at different relative times in the past, she or he can compile the record of the changes from earlier stress states to later stress states into a **stress path**. In the Comprehensive Treatment section of this chapter, we show a number of ways to depict stress paths.

In rare cases it is possible to determine the absolute time at which a stress state prevailed. The simplest way in which a stress state might be attributed to an absolute time interval is by observing that structures or fabrics associated with a particular stress state occur only in rocks of a given age or older. The age of the youngest unit in which these structures or fabrics occur is therefore a maximum age for the stress field. When absolute ages or age ranges can be attributed to different stress states along a stress path, the change of stress with time is called the **stress history**. Age ranges can be given for stress states by using the age of the host rocks as a maximum age, and the age of an overprinting stress state as a minimum age.

6A.9 Comparison of Displacements, Strain and Stress

Chapter 4 (Displacements), Chapter 5 (Strain), and this chapter on stress set out the basis of understanding deformation in rocks in the framework of continuum mechanics. Some of the most common and serious misunderstandings in structural geology occur because displacements, strain and stress are conflated or confused. For example, using the term compression to describe a structure or fabric instead of the term shortening is *not* simply an inappropriate word choice – it indicates a fundamental misunderstanding of these related but distinct concepts. We therefore return, as promised in the introduction to this chapter, to address the distinctions between displacements, strain, and stress.

Table 6.5 and Figure 6.27 provide a framework for this discussion. The first row of Table 6.5 emphasizes the strong similarities between the displacement and strain. The **displacement** at a point is defined by a vector $\boldsymbol{u}$ that connects the original position of a particle to its deformed position. It has the properties of magnitude, with dimensions of length, and direction. The **longitudinal strain** at the same point is a dimensionless quantity that compares initial and final lengths in a given direction. The **stress** at that point is a vector σ with the properties of magnitude, with dimensions of Force/Length^2, and direction, but its magnitude and direction are derived in part from the orientation of the surface on which the stress acts. The inclusion of the plane on which the stress vector acts means that stresses differ fundamentally from strain or displacement. There are fundamental characteristics of displacements, strain, and stress that require at least a plan view to define (row 2 of Table 6.5 and Figure 6.27b). Specifying transcurrent or oblique movement depends on a comparison of the movements of two particles relative to the line that connects them. Angular shear and shear strain are defined

Table 6.5 Vocabulary for displacements, strain and stress. Mutually exclusive pairs of terms are joined by /.

	Displacement	**Strain**	**Stress**
1D units	Displacement vector $\boldsymbol{u}$ Divergent/Convergent	Longitudinal strains: Stretch S Elongation e	Traction vector $\boldsymbol{s}$ or stress vector $\boldsymbol{\sigma}$ Compression/Tension
2D units	Transcurrent/Oblique	Shear strains: Angular shear ψ Shear strain γ	Stress vector $\boldsymbol{\sigma}$ Normal stress component σ_N/ tangential stress component σ_T or τ
2D states	Pure shear/ Simple shear/ General shear displacement fields	Strain Tensor **E** Extension/Contraction/ Wrench/Wrench contraction/ Wrench extension	Stress Tensor **T** Hydrostatic/ Non-hydrostatic
3D states	Constriction/Flattening Oblique divergence/Oblique convergence displacement fields	Strain Tensor **E** Constriction/Flattening Wrench Extension (transtension)/ Wrench Contraction (transpression)	Stress Tensor **T** Hydrostatic/Deviatoric components

by the change in angle between two lines. The two still have properties of magnitude and direction and dimensionless measurement of change, respectively, whereas resolving a stress vector into its normal and tangential components again requires both the stress vector and the plane on which it acts. These two rows of Table 6.5 underscore the unique character of stress, and we advocate maintaining a rigorous separation of the vocabulary as perhaps the single most important step to clarifying the differences between displacement, stress and strain.

The third and fourth rows of Table 6.5 illustrate how strain and stress are sometimes conflated and why they must not be. In terms of similarities, displacements, strain and stress all define **fields** in two or three dimensions, where every point is associated with a value of the quantity. Those fields can be **homogeneous**, in which the value of the quantities do not change with position, or **heterogeneous**, where there is spatial variation. We use similar geometric forms, the ellipse and ellipsoid, to represent graphically the state of strain (using a two-dimensional strain ellipse or a three-dimensional strain ellipsoid) or the state of stress (using a two-dimensional stress ellipse or a three-dimensional stress ellipsoid). It is tempting to assume that the strain and stress ellipses (ellipsoids) have parallel and proportional principal radii. This need not be and generally is not the case, for in addition to the differences in the dimensions of strain and stress, there is no generally applicable relationship between the orientation and relative magnitudes the principal radii of these two ellipses or two ellipsoids. In any geologic setting, pure shearing, simple shearing, or general shearing displacement fields are likely to result from different stress states. Yet the particular orientation and shape for the strain ellipse observed at a point could be the product of *any* of

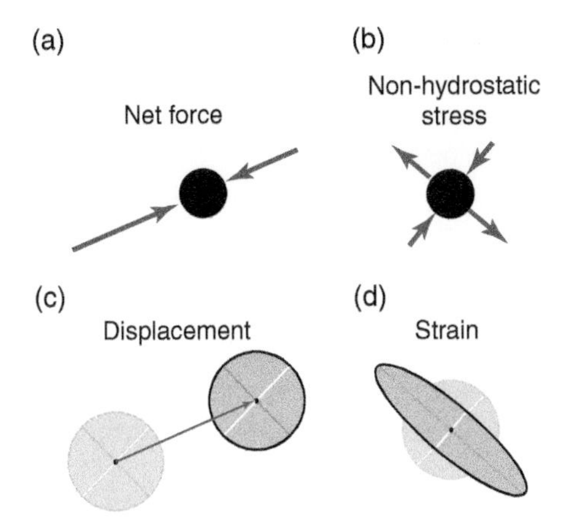

Figure 6.27 (a) An object subjected to a net force will experience an acceleration in a direction parallel to the net force. (b) An object may be subjected to forces that are balanced yet give rise to non-hydrostatic stresses within the object. (c) The unbalanced forces in (a) cause an object to accelerate and displace the object's center of mass. (d) Non-hydrostatic stresses drive changes in shape, which can occur with no net movement of the object if the forces responsible for the stresses are balanced.

those three types of displacement fields. It is imperative, therefore, that geologists are scrupulous in distinguishing, conceptually and in terminology, these aspects of deformation.

The only one of these quantities that can be measured directly from an outcrop is displacement, and even the measured displacement is in many cases only a component of the true displacement, that is, an apparent displacement. Strains can be inferred from outcrops, but only quantitatively after linear measurements are manipulated mathematically to define some aspect or all of the strain ellipse (ellipsoid). We cannot measure stress directly; we *infer* it from displacements or strain using a number of assumptions, such as the rheology of the rock (the topic of the following chapter). These considerations have important implications for field measurements and descriptions of rocks. All measurements recorded on the outcrop should be taken in dimensions of length, and any methods of strain calculation should be specified explicitly. In general there is no need for any reference to stress in outcrop description.

Finally, we consider cause and effect in deformation. Geologists sometimes state that "Stress X caused deformation event Y" or "The fold/fault/shear zone was caused by stress Z." The attribution of cause and effect in these cases implies a relative importance of deformation factors that operates at a larger scale than the observed feature. This relative importance is encompassed in the "boundary conditions" inferred for the deformation – whether displacements, strains, or stresses were specified or fixed along the boundaries of the deforming rock by external conditions. On a very large scale, much deformation in the crust can be said to be driven or caused by plate motions. Thus, it is as valid to refer to displacements as the cause of much strain and stress as it is to refer to stress as the cause of displacements and strain. Commenting on causes of deformation has philosophical overtones that are best avoided. For this reason, we advocate using language such as: "During deformation event Y the differential stress was X" or "Stress Z is associated with the fold/fault/shear zone."

6A.10 Summary

1) We quantify the effects of the forces that act on and through rock masses by determining the magnitudes and directions of the **traction vectors** or **stress vectors(*s*)** that act on all planes passing through a point in a rock mass. Traction vectors have both a direction and a magnitude, measured in units of force per unit area, and are defined in part by the plane on which they act.
2) The SI unit for stress or traction vectors is the *Pascal*, where 1 Pa = 1 kg/ms^2. In geology, stresses are much more than 10,000s of Pascals, so the standard unit of stress magnitude is the MPa = 10^6 Pa.

3) In most geological settings, rock masses are at equilibrium. Traction vectors acting on opposite sides of a plane have equal magnitudes and opposite directions. We use a single vector to depict the balanced traction vectors acting on a plane and refer to the **stress** (σ) that acts on that plane.
4) We resolve the stress acting on a plane into two components, a **normal stress component** (σ_N) that is oriented perpendicular to the plane and a **tangential** or **shear stress component** (σ_T) that acts parallel to the plane.
5) Normal stress is **compressive** when two traction vectors are directed toward the plane on which they act and **tensile** when two traction vectors are directed away from the plane on which they act. Compressive normal stresses are positive-valued, and tensile normal stresses are negative-valued.
6) Shear stress is positive-valued if it tends to cause an anticlockwise rotation and is negative-valued if it tends to cause a clockwise rotation.
7) The magnitudes of the stress vectors acting on different planes passing through a point are related and vary smoothly with changes in the orientations of the planes. The manner in which the magnitudes change reflects the general character of the **state of stress** at that point.
8) A geometric figure known as the **stress ellipse** (for two-dimensional situations) or **stress ellipsoid** (for three-dimensional situations) represents the magnitudes of the stress vectors acting planes with different orientations if all principal stresses are compressional. The shape and orientation of the stress ellipse or stress ellipsoid capture the overall state of stress at that point.
9) Any normal stress component acting parallel to a principal axis is a **principal stress component** or **principal stress**. One principal stress is the most compressive normal stress acting on any plane that passes through the point, and another principal stress is the least compressive normal stress acting on any plane passing through the point. In three dimensions, there is a third, intermediate principal stress. We denote the principal stresses by $\sigma_1 \geq \sigma_2 \geq \sigma_3$.
10) We define the **differential stress** as $\Delta\sigma = (\sigma_1 - \sigma_2)$ (in two dimensions) and $\Delta\sigma = (\sigma_1 - \sigma_3)$ (in three dimensions). This is an important quantity in determining whether a rock will deform.
11) Stress states are **hydrostatic** if all planes passing through a point experience the same stress, characterized by finite normal stress components and no shear stress components.
12) Stress states are **non-hydrostatic** if different planes passing through a point experience the different stresses, characterized by finite normal stress components and finite shear stress components on nearly all planes.
13) Non-hydrostatic stress states have two (in two dimensions) or three (in three dimensions) **principal axes**. Each principal axis is perpendicular to a plane whose tangential or shear stress component has zero magnitude.
14) In non-hydrostatic stress states, we envision that the total stress state consists of a **hydrostatic component**, with equal normal stresses and no shear stresses acting on all planes, and a **deviatoric component**, with the normal stresses and shear stresses required to bring the hydrostatic stress component to the total stress. The hydrostatic component of the stress state tends to cause changes in rock volume, whereas the deviatoric component of the total stress state tends to cause changes in the shapes of rock masses.
15) In rocks containing pore fluids, we calculate **effective stress** = the total stress minus the pore fluid pressure.

16) At most locations in Earth's crust and upper mantle, the total stress is the sum of **gravitational stresses**, due to the action of gravity on rock masses, and **tectonic stresses**, related to the motion of plates.

6A.11 Practical Methods for Measuring Stress

As with displacements and strain, structural geologists are interested in both the present-day values of stress in Earth and in states of stress that have existed in the past. For reasons not entirely clear to us, it has become the current jargon to refer to determinations of stress in the crust today as *in situ* (literally in place) measurements, although we do not commonly refer to measurements of modern displacements or strain in this way. Determinations of past stress states also have a particular jargon associated with them: they are known as *paleostress* measurements.

Both *in situ* and paleostress measurements would ideally determine the complete stress state, that is, specify both the orientation and absolute magnitudes of the principal radii of the stress ellipsoid. Such a full determination of the state of stress is rarely possible, however. Several types of measurement constrain the orientations of the principal stresses and their relative magnitudes, that is, which principal stress is most and least compressive. Earth scientists regularly use a parameter $\phi = (\sigma_2 - \sigma_3)/(\sigma_1 - \sigma_3)$ to define the relative values of the principal stresses. It is considerably more difficult to determine the absolute magnitudes of the principal stresses.

6A.11.1 *In situ* Stress Measurements

Unlike displacement or strain, there is no method that can determine the complete state of stress directly. Aspects of the *in situ* stress are inferred indirectly from displacements by assuming that strains are proportional to the stress, as in a homogeneous, elastic rock.

6A.11.1.1 The Flatjack Technique

The flatjack technique evaluates *in situ* stress by determining the normal stresses that act across several planes with different orientations. The technique requires a locally planar rock face, like the ground surface or the wall of a tunnel or mine, and each of the planes on which normal stresses are measured is perpendicular to that rock face (like the planes illustrated in Figure 6.1). Because the rock face is a free surface and the only stresses acting on it are those exerted by the atmosphere, the rock face is a principal plane and the magnitude of the normal stress acting on it is zero. By determining the range of normal stress magnitudes resolved across several planes perpendicular to the rock face, geologists can determine the orientations and magnitudes of the other two principal values of the *in situ* stress.

The flatjack itself is a flexible metal membrane, either rectangular or in the shape of a thin semicircular cylinder with a diameter of ~ 0.4 m that is filled by a hydraulic fluid (Figure 6.28). In order to determine stress, a geologist cements two lines of pins or marker points onto the rock surface, each parallel to a line along which a saw cut will be made (Figure 6.28, 1). Geologists measure the precise distance between the pins, and then cut a slot perpendicular to the rock face between the pins. The slot breaks the network of bonds transmitting tractions through the rock, thereby removing the tractions that rock on one side of the slot exerted on the other side of the slot. As a result, the pins move to new positions (Figure 6.28, 2). If the rock removed in the slot supported compressive normal stress, cutting the slot causes the pins to converge. Convergence occurs because tractions directed toward the slot by the surrounding rock are no longer balanced by tractions directed away from the slot. By cementing the flatjack into the slot and pumping hydraulic fluid into the flatjack,

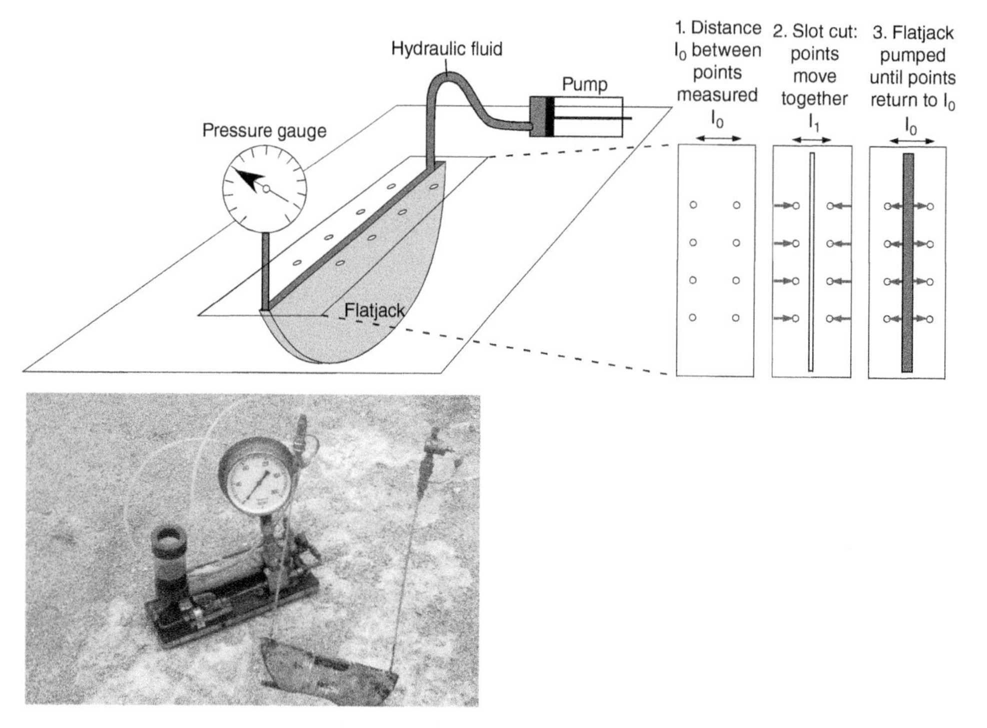

Figure 6.28 Principle of *in situ* stress measurement by the flatjack method. The flatjack is a flexible metal membrane, shown at left in a slot under the ground surface, connected to a pump and a pressure gauge. Before installation, pairs of marker points are cemented onto the ground surface and their initial separation l_0 is measured. The flatjack is inserted into a slot cut between the marker points. Fluid is pumped into the flatjack to restore the markers to their original positions, and the pressure is recorded. The procedure is repeated for several different azimuths. The photograph shows the flatjack partially entered into its slot, a pump and the pressure gauge.

geologists regenerate tractions directed away from the slot and displace the pins back toward their original positions. With sufficient pressure, the marker points return to their original spacing (Figure 6.28, 3). We take this pressure exerted by the flatjack on the surrounding rock, sometimes called the *cancellation pressure*, to be equal to the compressive normal stress that originally acted across the slot. If the slot supported tensile normal stress, then cutting the slot removes tractions directed toward the slot from the rock beneath the array of marker points so they diverge. Because the flatjack can only exert compressional normal stresses on the walls of the slot, there is no way the flatjack can mimic the stress originally acting across the slot. Geologists can use the flatjack to estimate the magnitude of tensile stress, however. By cementing the flatjack into the slot and increasing the pressure until the marker points diverge farther by an amount equal to their initial separation, we assume that the magnitude of the pressure across the slot equals the magnitude of the tensile stress originally acting across the slot.

By repeating the procedure for slots in a variety of orientations, geologists can determine the magnitudes of normal stresses acting in different directions parallel to the rock face. Table 6.6 gives flatjack measurements of normal stress at different values of the azimuth of the line perpendicular to the flatjack. These data are shown on the

Table 6.6 Normal stress as a function of azimuth, determined by flatjack measurements from the floor of a quarry.

Azimuth of Normal Stress	Normal Stress MPa (compression positive)
001	−0.971
008	−0.228
021	−0.376
032	−0.191
040	−0.347
051	−0.175
058	−0.202
089	−0.407
090	0.404
118	0.178
124	0.492
136	0.351
138	−0.060
139	−0.638
141	−0.411
145	−0.67
159	−0.138
170	−0.646
170	−1.120

radial plot of Figure 6.29. Equation (6.15a) gives the normal stress σ_N acting on a plane as a function of its angle θ from the plane that experiences the most compressive principal stress (here denoted θ to underscore that it is a variable whose value is to be determined):

$$\sigma_N(\theta) = \frac{1}{2}(\sigma_1 + \sigma_2) + \frac{1}{2}(\sigma_1 - \sigma_2)\cos 2\theta. \quad (6.15a)$$

The flatjack results can be fitted to this equation to solve for three variables: the two principal stresses σ_1 and σ_2 and the orientation of σ_1. The results suggest that $\sigma_1 = 0.17$ MPa, $\sigma_2 = -0.69$ MPa and the azimuth of σ_1 is 102°.

An important advantage of the flatjack technique compared to the method outlined in the next section is that it samples a larger volume of rock, so the measured stresses may be more representative of the *in situ* state of stress. To ensure that the measurements made at Earth's surface are representative of the *in situ* stresses, geologists avoid weathered or extensively jointed exposures. Similarly, because the excavation of a mine or tunnel may alter the stress field, geologists must exercise caution in interpreting measurements made underground.

6A.11.1.2 Overcoring

The overcoring technique is similar to the flatjack technique in that it uses the displacements that result when the stresses that act rock masses are removed to evaluate *in situ* stress. Overcoring is undertaken in boreholes drilled into rock (Figure 6.30), usually to a depth of at least several meters in order to sample unweathered rock. At the bottom of the borehole, a smaller diameter hole is extended deeper into the rock. Then, very sensitive *strain gauges*, which measure changes in length along a line, are cemented into the smaller hole. A cylinder of rock containing the strain gauges is "overcored" by drilling around it at the original diameter of the larger hole. Overcoring isolates the cylinder of rock from the stresses that were acting on it (Figure 6.30). If the *in situ* stresses were compressional (which is true in nearly all geological settings), the cylinder will expand and its change in shape is measured very accurately by the strain gauges. In order to determine the stresses that were acting on the rock, geologists assume that the rock exhibits *linearly elastic behavior*, where the magnitudes of stresses are linear multiples of observed strains. We describe elastic behavior and the conditions under which rock exhibits this type of behavior in greater detail in the following chapter. In order to convert the observed strains to stress magnitudes, the geologist must know the rock's *elastic*

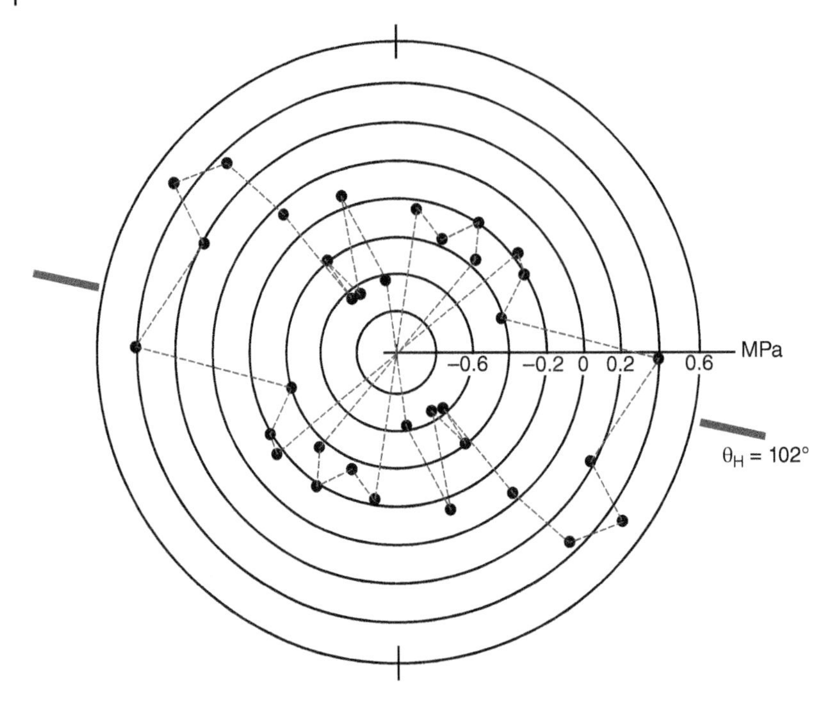

Figure 6.29 Results from flatjack determinations plotted as normal stress (compressive positive) by azimuth. The best-fit orientation for the maximum horizontal stress is 102°.

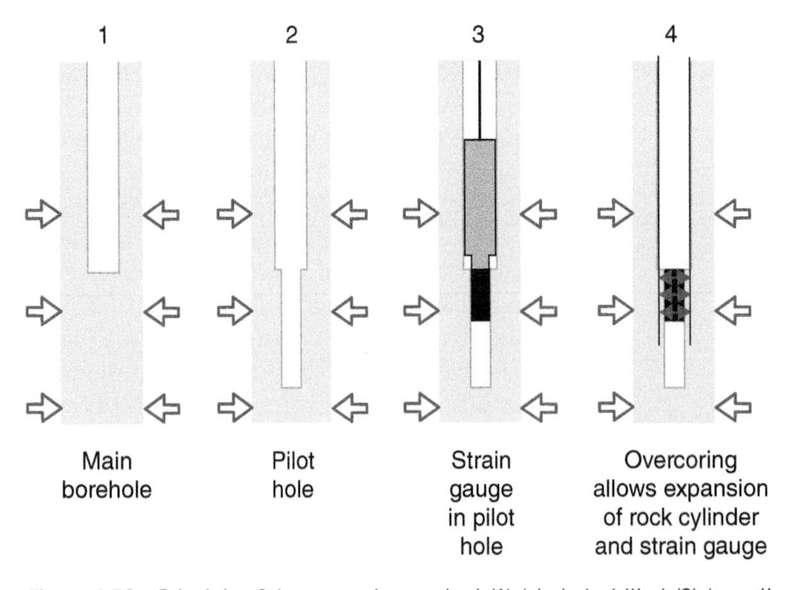

Figure 6.30 Principle of the overcoring method. (1) A hole is drilled. (2) A smaller diameter hole is extended from the bottom of the hole in (1). (3) An array of strain gauges is inserted into to smaller hole. (4) The cylinder of rock containing the strainmeters is isolated by drilling around it at the original diameter of the hole – "overcoring." The strains due to the elastic expansion of the cylinder are recorded.

constants, the values of the coefficients that relate the strain and stress magnitudes. With an assumption of linearly elastic behavior and laboratory-derived values of the elastic constants, one can calculate the stresses that were acting on the cylinder of rock to maintain its original shape in the borehole from the observed expansion. The lengthening of the overcored cylinder is measured in at least three horizontal directions: these are combined to find the direction of maximum lengthening, which is inferred to be the azimuth of the maximum horizontal compression.

6A.11.1.3 Hydrofrac (Hydrofracture)

Unlike the two previous methods described, the hydrofrac technique derives from a different property of rocks subjected to differential stresses – that the orientations of fractures induced in rock relate to the orientations of stresses to which the rock is subjected. To generate hydraulically induced fractures (*hydrofractures* or *hydrofracs*), an interval of a borehole is sealed and connected to a pump and fluid reservoir. The pressure on fluid pumped into the sealed interval of the borehole is increased until the rock fractures (Figure 6.31). In order to induce the formation of a fracture, the fluid pressure must first equal and then exceed the magnitude of the least compressive principal stress. At that point, fractures initiate at the periphery of the borehole and propagate into isotropic rock perpendicular to σ_3. Thus, the orientation of hydrofractures shows the orientation of σ_3: horizontal fractures indicate a vertical least principal stress, and vertical fractures indicate a horizontal least principal stress. After the fracturing operation is completed, an instrument is lowered into the drillhole to make oriented image of the drillhole wall, from which the orientation of the hydrofractures is measured. The magnitude of σ_3 can also be estimated by this method: it is the pressure required to hold the hydrofracture open during continued pumping.

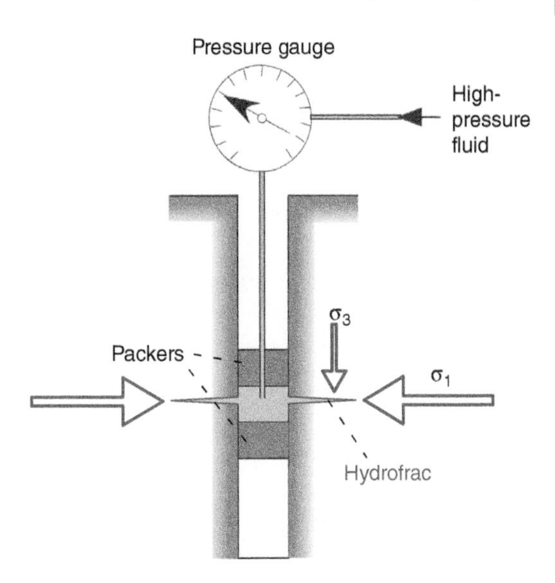

Figure 6.31 Principle of the hydrofrac technique. Two packers in a drill hole isolate a chamber that is pumped with high-pressure fluid until the tensile strength is exceeded. A fracture is created perpendicular to σ_3. The pressure required to hold the fracture open is equal to the magnitude of σ_3. If σ_3 is horizontal, the azimuth of the fracture will indicate the direction of σ_1.

6A.11.1.4 Borehole Breakouts and Drilling-Induced Fractures

When a hole is drilled into a stressed rock, removal of rock by the drill bit alters the stresses in the immediate vicinity of the borehole (Figure 6.32). The side walls of the borehole that are perpendicular to the least compressive *in situ* stress experience increases in compressional stresses tangential to the borehole. The localized increase in differential stress (zero stress perpendicular to the borehole wall and greater compressive tangential to the borehole wall) can generate fractures in that part of the borehole wall, which lead, in turn, to localized excavations called *borehole breakouts*. We must postpone a full explanation of the formation of these fractures to the later chapters. At this point, however, we note that excavations tend to enlarge the borehole in the direction

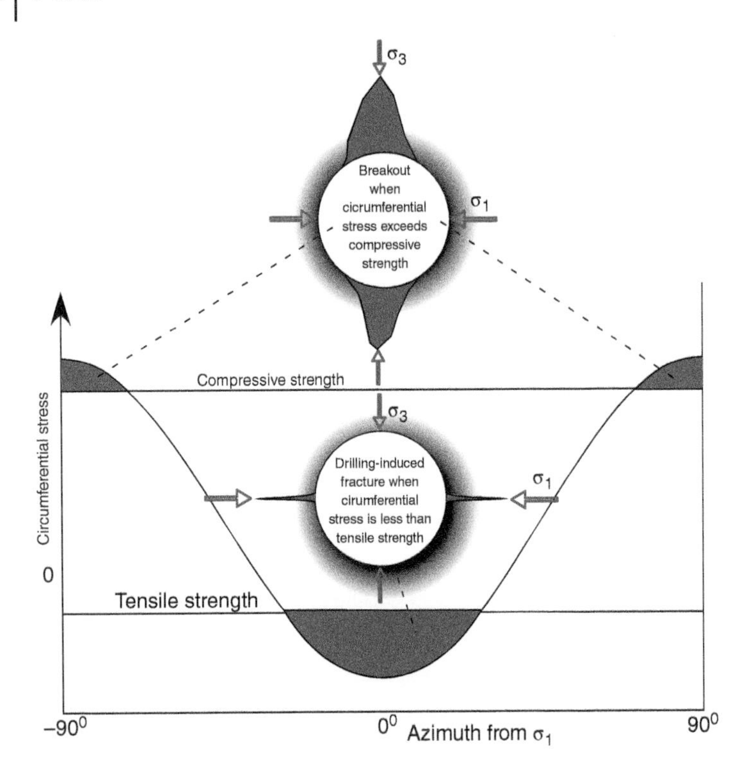

Figure 6.32 Borehole breakouts and drilling-induced fractures. When a drill hole is created in a stressed rock, compressive and tensile stresses are created around the circumference of the drill hole. Compressive stresses may result in shear fracture perpendicular to σ_1 if they exceed the compressive strength, creating a borehole breakout. Tensile stress may result in drilling-induced fracture in the direction of σ_1 if they are less than the tensile strength of the rock.

of the least compressive stress, altering the initially circular hole into an approximately elliptical shape. The elongated shape of the borehole can be measured by a device called a four-arm caliper, consisting of two pairs of opposed arms that are expanded so that they maintain contact with the walls of the borehole under pressure. The calliper is lowered into the borehole to measure the maximum and minimum diameters of the hole, and their orientation, as a function of depth. These instruments also measure resistivity, and borehole breakouts can be interpreted from the resistivity logs because they are typically more conductive.

The borehole radius can be measured more sensitively by instruments such as the borehole televiewer, which measures borehole diameter at any depth in many radial directions. Details of the shape of the breakout can be used to solve for the magnitudes of the horizontal stresses if the compressive strength of the rock is known. A reasonable agreement has been found between these calculations and the hydrofrac methods.

As a result of the altered stresses in the vicinity of the borehole, large tensile stresses occur on the portion of the borehole walls that are perpendicular to the direction of the greatest horizontal stress. In some instances, extension fractures form perpendicular to the borehole wall (Figure 6.32). These extension fractures, known as *drilling-induced fractures*, can also be detected by the televiewer.

6A.11.1.5 Seismology

Analysis of spatial variations of characteristics of the seismic waves (specifically the characteristics of *primary* or *P waves*) associated with an earthquake can be used to obtain a **focal mechanism** or **fault plane solution**. A fault plane solution identifies the orientations of two planes, called *nodal planes*, at the location of the earthquake. One of the nodal planes parallels the fault surface and the other, called the *auxiliary plane*, is a plane perpendicular to the fault. A fault plane solution cannot determine which nodal plane is the fault and which is the auxiliary plane, but it does indicate: (1) what sense of slip on either nodal plane is compatible with the seismic wave data; and (2) the direction of slip on either plane because the two nodal planes intersect along a line that must be perpendicular to the fault slip vector (Figure 6.33). Even though every fault plane solution is inherently ambiguous because it cannot distinguish which of two possible fault plane-fault slip vector pairs characterize a particular earthquake, each solution does indicate local directions of shortening and elongation associated with the seismic event (Figure 6.33). The displacements during an earthquake do not indicate directly the stress state, but some geologists infer that, to a first order, the intersection of the nodal planes parallels the direction of the intermediate principal stress and the local directions of shortening and elongation derived from a fault plane solution correspond, respectively, to the directions of the most and least compressive stress.

An alternative approach combines numerous fault plane solutions within a region to calculate three-dimensional strain ellipsoids associated with the slips on many faults and then infers principal stress directions from the strain principal directions. Here too, additional steps are required to use these kinematic data to infer stresses. Usually, these steps involve the assumptions that faults slipped in the direction of maximum shear stress on the fault plane, and the different focal mechanisms within a region pertain to the same stress field. If these assumptions are true, relative stress magnitudes and orientations can be inferred by the technique of *inversion* from earthquake data, as described here for fault slip data. Earthquake focal mechanisms are perhaps best used to distinguish the three tectonic regimes, normal, strike-slip, and reverse, corresponding to the situations in which σ_1, σ_2, and σ_3 are vertical, respectively.

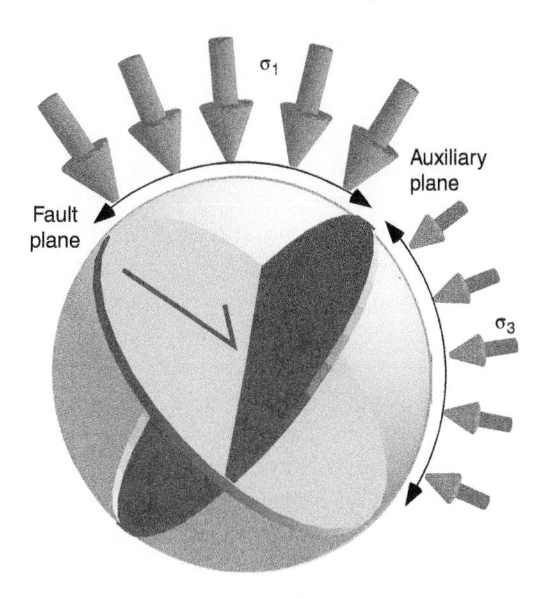

Figure 6.33 Relationship of principal stresses to fault plane solutions. Seismological data provide the orientation of the two nodal planes: additional data is required to select the fault plane unambiguously. The maximum and minimum principal stresses will lie in opposite quadrants, separated by the nodal planes, but their orientation is not more accurately prescribed by the fault plane solution itself.

6A.11.2 Paleostress

The aim of paleostress investigations is to find the stress state in past. As with *in situ* stress measurements, there are no methods by which one can directly determine paleostresses. To be able to do that, we would need both a "stress meter" of some kind and a sample site in which old stresses were still somehow maintained in the rock, an uncommon circumstance. Instead, we use visible manifestations of stresses preserved in rocks deformed during a particular past event to assess paleostresses. As was the case with *in situ* stress measurement, what we usually observe is a deformation record of past displacements, which can only be related to stress indirectly by assuming the properties of the rocks at the time of deformation and calculating the stress that would have caused such displacements. Paleostresses, then, are always inferred, in contrast to displacements in the geological record that can be directly measured. This is a most important distinction, which means that we should generally only refer to displacements and strains when we are describing rocks in the field.

A common problem for paleostress techniques is to determine the timing of the inferred stress. Standard principles of stratigraphy can be applied: if a feature used to infer paleostress is present in rocks up to a certain age but no younger, then the paleostress can be ascribed to the age of the most recent rocks that contain the feature. Cross-cutting relationships can also be used to date features in a relative sense.

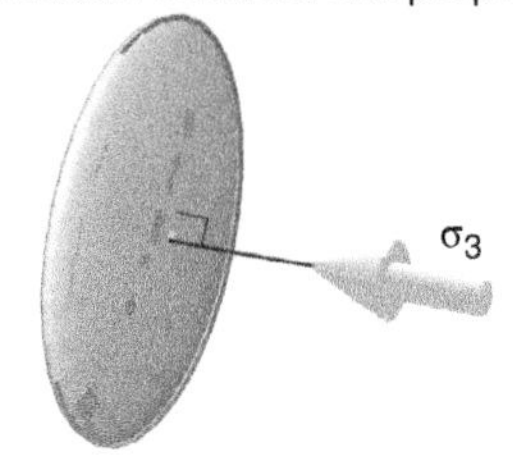

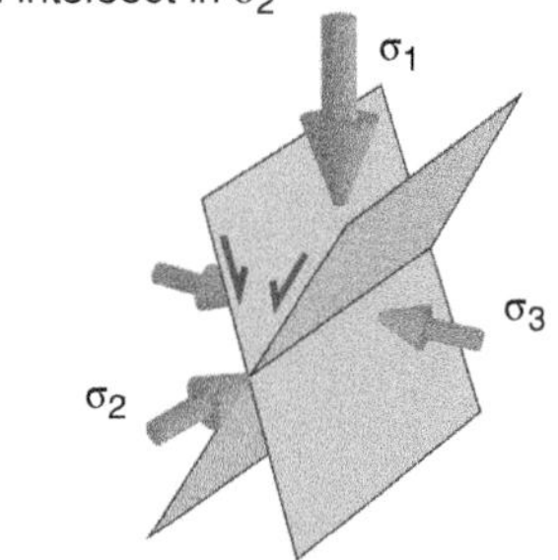

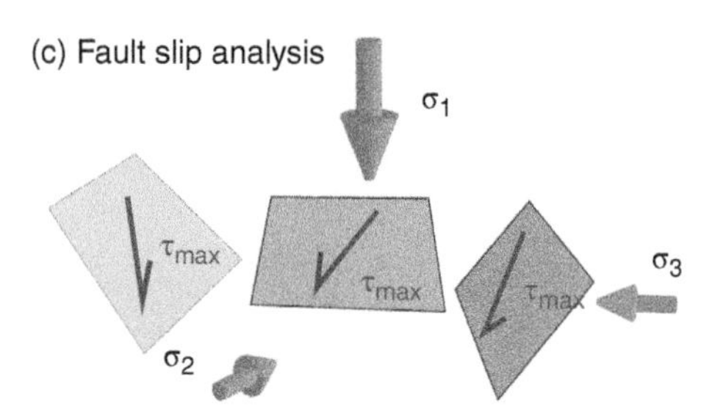

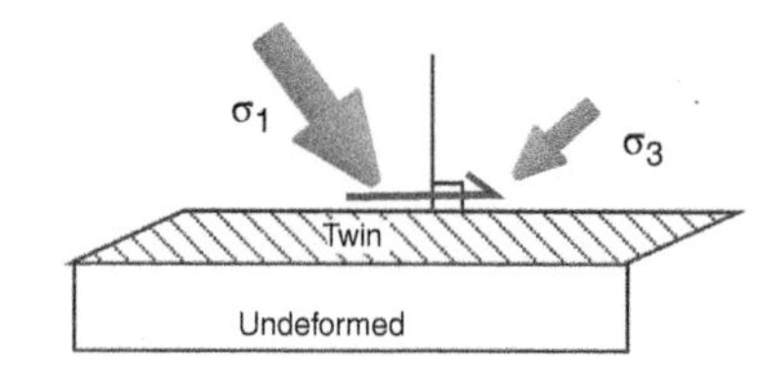

Figure 6.34 Use of geological features for paleostress analysis. (a) Extension fractures are perpendicular to σ_3. (b) Conjugate faults contain σ_2 in their acute bisector. (c) A variety of fault slip data can be related to the optimum stress state to generate the observed displacements, assuming that slip occurred in the direction of maximum resolved shear stress, τ_{max}. (d) Calcite twins are assumed to have a geometrical relationship to the principal stresses: the plane formed by the slip direction and the normal to the twin plane contains σ_1 and σ_3: the principal stresses lie at 45° to the slip direction.

6A.11.2.1 Extension fractures

Extension fractures in isotopic materials form parallel to the plane of least normal stress, that is, perpendicular to σ_3. In appropriate settings, geologists can infer that the normal to an extension fracture is parallel to σ_3 (Figure 6.34a). This approach has been used on a variety of scales from microfractures to the intrusion of continental-scale dykes, which may occur along extension fractures. The method can only generally give the orientation of the least principal stress because extension fractures do not necessarily form at a constant stress magnitude, even in the same rock.

6A.11.2.2 Conjugate Faults

Conjugate faults consist of a pair of faults with complementary senses of displacement that formed in the same deformation event. In natural examples, conjugate faults intersect, or would intersect if extended, in an angle of 40–60° (Figure 6.34b). Laboratory analyses and deformation theory outlined in the following chapters suggest that the symmetry of conjugate faults developed in isotropic, homogenous rock relates closely to the stress state during faulting, with the maximum principal stress σ_1 parallel to the acute bisector of conjugate faults, σ_2 parallel to the line of intersection of the faults, and σ_3 parallel to the obtuse bisector of the faults (Figure 6.34b). Geologists, therefore, use conjugate faults to deduce the orientations but not the magnitudes of the principal stresses. It is imperative to demonstrate that the faults analyzed in this way formed in the same deformation event, that their orientations are restricted to two directions, and that they are not part of more complex fault geometries such as multiple fault sets.

6A.11.2.3 Fluid Inclusions

Fluid inclusions are microscopic pockets of fluid trapped in minerals. In favorable circumstances, the compositions of the fluids in the inclusions do not change after entrapment, and the pressure and temperature of inclusion formation can be inferred from the properties of the fluids in the inclusions. Since the pressure in the inclusion is in a fluid, it is a hydrostatic state of stress. In Section 6A.5, we compared the fluid pressure (P_f) to the lithostatic stress (σ_L); here we compare the fluid pressure to the vertical normal stress component (σ_v) as defined in Eq. (6.7).

6A.11.2.3.1 Fluid Pressure is Hydrostatic In a rock that is permeable and open to the surface, the fluid pressure will be *hydrostatic* and given by

$$P_f = \rho_f g z, \tag{6.18}$$

where ρ_f is the density of fluid, g is the gravitational constant 9.8 ms^{-2}, and z is the height of the column of fluid. Geologists use the word "hydrostatic" in two ways. We refer to *hydrostatic fluid pressure* to indicate specifically the magnitude of the fluid pressure is given by the previous equation. The same word is also used to describe a general state of stress (*hydrostatic stress*) in which there are no shear stresses and normal stresses are equal in all directions – such as is always the case on a fluid at any pressure. The lithostatic stress σ_v is given by a slightly modified version of Eq. (6.7):

$$\sigma_v = \rho g z.$$

In the upper crust, a typical value for ρ is 2.7 g/cm^3 = 2700 kg/m^3. If the pore fluid is water water, the density of water ρ_w is 1.0 g/cm^3 = 1000 kg/m^3. Therefore, the ratio of fluid pressure to vertical normal stress, often represented by λ, is:

$$\lambda = \frac{P_f}{\rho g z} = \frac{\rho_w g z}{r g z} \approx \frac{1000}{2700} = 0.37$$

in a hydrostatic fluid pressure state.

6A.11.2.3.2 Fluid Pressure is Suprahydrostatic, but < σ_v Fluid pressure can exceed hydrostatic values when fluid from a source such as an intrusion or regional metamorphism of wet sedimentary rocks is sealed by some sort of trap in a rock. In these cases,

$$0.37 < \lambda < 1.$$

Values of λ > 0.37 are called *suprahydrostatic.*

6A.11.2.3.3 Fluid Pressure is > σ_v If λ > 1, the fluid pressure is sometimes said to be *supralithostatic.* It is possible, however, that vertical stress in a rock due to gravity (σ_v) is not equal to the lithostatic pressure. so $\lambda > 1$ just describes situations in which $P_f > \sigma_v$.

Fluid pressures measured in rocks from a single generation of fluid inclusions commonly have a considerable range. Such variability can be interpreted as indicating fluid inclusion entrapment under different values of $\boldsymbol{\sigma}_v$ due to loading by, for example, thrusting, or unloading by normal faulting or erosion. Alternatively, the range in fluid pressures may reflect pressures of entrapment varying between hydrostatic and suprahydrostatic, perhaps related to dynamic changes in stress during earthquake cycles.

6A.11.2.4 Fault Slip Analysis

If a state of stress is known at some point, it is possible to calculate the direction across a plane with an arbitrary orientation in which the shear stress attains its maximum magnitude. If the rock mass subjected to known stresses is cut by a pre-exiting surface like a fracture or joint, slip direction on that surface will parallel the direction in which shear stress attains its maximum magnitude. Fault slip analyses reverse this principle: from measurements of slip directions on faults in outcrop, geologists calculate the state of stress that generates the maximum magnitude of shear stress in directions parallel to the observed slip directions (Figure 6.34c).

There are several different methods devised to calculate the stress state from fault plane-slip vector data, of which the *right dihedra* or *trihedra* methods and the *inversion* method are the most common. The right dihedra method utilizes reasoning like that behind the focal mechanism solutions outlined earlier. In this method, each fault plane-slip vector pair defines the orientations in which σ_1 and σ_3 can lie. By superimposing the σ_1 and σ_3 orientations determined from several different fault plane-slip vector pairs, the range of possibilities for the principal stress orientations can be reduced to almost unique values. The right trihedral method is similar but even more efficient in its solution. Inversion methods are iterative computational routines that seek to find the optimum stress state that would generate maximum shear stress magnitudes parallel to observed slip vectors on planes with the corresponding orientations. These methods are computationally intensive but can obtain the relative magnitudes of stresses as well as their orientation.

Paleostress analyses require measurements of fault orientation (dip and dip azimuth), slip direction (trend and plunge), and, for most methods, sense of slip on the fault. Geologists must take the following precautions in collecting fault slip data for paleostresses: (1) all the fault slips must to belong to a single episode of deformation; (2) slip on one fault must not interfere with slip on others; and (3) a minimum of eight fault plane-slip vector pairs with a variety of orientations must be collected within a limited region; results are more reliable with more than this minimum number.

6A.11.2.5 Paleopiezometry

The word "paleopiezometry" is literally the measurement of past stress states (*paleo* – from the Greek *palaios* = ancient; *piezo* – from the Greek *piezein* = to squeeze; *metry* – from the Greek *metron* = measure). Laboratory data and deformation theory indicate that the development of or characteristics of some microscopic structures depend only or dominantly on the applied stress. Such microstructures are known as *paleopiezometers*, and their specific relation to stress can be calibrated by laboratory experiments.

6A.11.2.5.1 Twins and deformation lamellae

Several minerals form deformation twins, calcite being one of the most common. Experiments suggest that the proportion of calcite grains that are twinned, the number of twins/grain, and the volume of twinned grain are all parameters that depend on the differential stress magnitude. One can readily measure these parameters from a thin section and then use equations derived from deformation experiments to relate the parameters to the differential stress magnitude.

Figure 6.34d illustrates how deformation twins can be used to infer the orientations of the principal stresses. The method derives from an assumption that the twin plane shearing indicates the direction of the maximum shear stress, thus the σ_1 and σ_3 principal directions are inclined 45° from any twin plane and lie within an imaginary plane defined by a normal to the twin plane and direction of the twin-related shearing. Therefore, measuring the orientation of a twin and its host calcite crystal gives the orientation of the principal stresses. Measuring a suitable number of twins for a sample can define a representative stress state. The principles of determining stress orientations from calcite twins are similar to those used in inversion from fault slip data: that is, the orientation of the stresses are inferred from the observed directions of slip and slip planes. Because twinning occurs only when a threshold magnitude of shear stress is resolved across an appropriately crystallographic plane in a crystal, it is possible to use twin data to estimate the principal stress magnitudes.

The spacing of deformation lamellae in quartz has been proposed as a paleopiezometer. In particular, the magnitude of differential stress $\Delta\sigma$ is inferred to be proportional to the inverse of the lamellae spacing s raised to a power k:

$$\Delta\sigma \propto s^{-k}.$$

k is a constant inferred from laboratory experiments. Quartz deformation lamellae have also been used to deduce principal stress orientations, using principles similar to those used for calcite twins and faults.

6A.11.2.5.2 Dislocation Density Dislocations are linear imperfections in a crystal lattice whose movement accomplishes changes in the shapes of crystals without destroying the crystal's lattice (see Chapter 8). Only under extreme conditions are dislocations created by applying stress to a crystal lattice. Typically, dislocations form as accidents during crystal nucleation and growth, but interactions between dislocations or between dislocations and other types of imperfections in crystal lattices during deformation can increase the length of dislocation per unit volume of crystal. We refer to the length of dislocations per unit volume of crystal as the dislocation density ρ_D. ρ_D has units of length (L) divided by volume (L^3), typically cm^{-2}. One can think of dislocation density as the number of dislocations per unit area of a section plane, though techniques to measure dislocation density by optical or transmission electron microscopy are more involved than simply counting numbers of dislocation per unit area. Theoretical considerations and experimental evidence suggest that the dislocation density ρ_D is related to the magnitude of differential stress $\Delta\sigma$ by an equation of the form

$$\Delta\sigma \propto D{\rho_D}^u,$$

where D and u are constants that have been calibrated for quartz, calcite and olivine.

6A.11.2.5.3 Subgrain Size Early experimental work on the deformation of crystalline materials demonstrated that individual crystals deformed by the movement of dislocations regularly develop distinct regions in which the crystal lattice is misoriented slightly relative to neighboring regions of the crystal. The boundaries separating these different regions are called *low-angle grain boundaries* (or sometimes *low-angle tilt boundaries*) because the lattice misorientation (tilting) across the boundaries is less than ~5°. We call these regions *subgrains* because they are visible regions of what was once a single grain, now slightly misoriented relative to their surroundings and separated from other subgrains or from the host grain by low-angle grain boundaries. Subsequent studies have demonstrated that subgrains are volumes with relatively low dislocation densities and that the boundaries that separate subgrains from other subgrains or the host grain are three-dimensional arrays of dislocations. The size of subgrains (d) is inversely dependent on differential stress $\Delta\sigma$ according to a relationship such as:

$$\Delta\sigma \propto cd^{-\nu},$$

where c and ν are constants, known for quartz, olivine and halite.

There are potential problems with the use of dislocation-density and subgrain-size paleopiezometers. First, the measurement of both quantities is not straightforward. Different methods of measuring either dislocation density or the mean size of subgrains may yield varying results. Second, the relationship between these either dislocation density and subgrain size and differential stress may also vary with temperature or water content of minerals and these effects are not yet fully quantified.

6B Stress: Comprehensive Treatment

6B.1 Force, Traction, and Stress Vectors

6B.1.1 Accelerations and Forces

Vector quantities or *vectors* have both *magnitude* (or length) and *orientation* (or attitude) (see the Comprehensive Treatment section of Chapter 4). A particle's displacement vector, $\boldsymbol{u}$, is the difference between the particle's position vectors at two instants in time. With respect to Cartesian coordinate axes x_1, x_2, and x_3 (in this section we use subscripts to denote different coordinate axes rather than naming axes x, y, and z), a particle's displacement vector is

$$\boldsymbol{u} = u_1\boldsymbol{i} + u_2\,\boldsymbol{j} + u_3\,\boldsymbol{k}.$$

The displacement vector component parallel to the x_1 axis is the product of u_1 (a scalar) and $\boldsymbol{i}$ (a unit vector parallel to the x_1 axis), the component parallel to the x_2 axis is the product of u_2 (a scalar) and $\boldsymbol{j}$ (a unit vector parallel to the x_2 axis), and the component parallel to the x_3 axis is the product of u_3 (a scalar) and $\boldsymbol{k}$ (a unit vector parallel to the x_3 axis). The particle's velocity $\boldsymbol{v}$, the time rate of change of its displacement, and its acceleration $\boldsymbol{a}$, the time rate of change of its velocity, are both also vectors:

$$\boldsymbol{v} = v_1\boldsymbol{i} + v_2\,\boldsymbol{j} + v_3\,\boldsymbol{k}$$

$$\boldsymbol{a} = a_1\boldsymbol{i} + a_2\,\boldsymbol{j} + a_3\boldsymbol{k}$$

where $v_1 = \mathrm{d}u_1/\mathrm{d}t$, $v_2 = \mathrm{d}u_2/\mathrm{d}t$, $v_3 = \mathrm{d}u_3/\mathrm{d}t$, $a_1 = \mathrm{d}v_1/\mathrm{d}t = \mathrm{d}^2u_1/\mathrm{d}t^2$, $a_2 = \mathrm{d}v_2/\mathrm{d}t = \mathrm{d}^2u_2/\mathrm{d}t^2$ and $a_3 = \mathrm{d}v_3/\mathrm{d}t = \mathrm{d}^2u_3/\mathrm{d}t^2$.

The force $\boldsymbol{F}$ vector responsible for an acceleration $\boldsymbol{a}$ is a scalar multiple of the acceleration vector, that is, $\boldsymbol{F} = \mathrm{m}\boldsymbol{a}$ where m is the *mass* of the particle (see the *Conceptual Foundations* section of this chapter). The vector $\boldsymbol{F}$ can be resolved into components parallel to the coordinate axes.

$$\boldsymbol{F} = F_1\boldsymbol{i} + F_2\,\boldsymbol{j} + F_3\boldsymbol{k},$$

where $F_1 = \mathrm{m}\,a_1$, $F_2 = \mathrm{m}\,a_2$, and $F_3 = \mathrm{m}\,a_3$. Because the mass of an object, m, is a scalar, the vector $\boldsymbol{F}$ is a scalar multiple of the acceleration vector $\boldsymbol{a}$, that is, the acceleration experienced by an object is in the same direction as a force acting on that object.

When M forces act on an object, the total or net force $\boldsymbol{F_t}$ is the *resultant* of the M force vectors acting on the particle, that is:

$$\boldsymbol{F}_t = \boldsymbol{F}_A + \boldsymbol{F}_B + \boldsymbol{F}_C + \ldots\ \boldsymbol{F}_M = \Sigma \boldsymbol{F}_i. \tag{6.19}$$

This holds for all components of the total force,

$$F_{1t} = F_{1A} + F_{1B} + F_{1C} + \ldots\ F_{1M} = \Sigma F_{1i}. \tag{6.19a}$$

$$F_{2t} = F_{2A} + F_{2B} + F_{2C} + \ldots\ F_{2M} = \Sigma F_{2i}. \tag{6.19b}$$

$$F_{3t} = F_{3A} + F_{3B} + F_{3C} + \ldots\ F_{3M} = \Sigma F_{3i}. \tag{6.19c}$$

The net acceleration $\boldsymbol{a}_t$ is parallel to the *resultant* of the *M* force vectors that act on an object, that is,

$$\boldsymbol{a}_t = \frac{\boldsymbol{F}_t}{\mathrm{m}} \text{ or } \boldsymbol{F}_t = \mathrm{m}\,\boldsymbol{a}_t. \tag{6.20}$$

When the total or net force $\boldsymbol{F}_t$ is zero, the total acceleration $\boldsymbol{a}_t$ is also zero – the object does not accelerate. When $\boldsymbol{F}_t$ is zero, the forces acting on the object are "balanced" and the object is "at equilibrium."

Force vectors, like position, displacement, velocity, and acceleration vectors, exist independently of the Cartesian coordinate system in which we originally specify them. Therefore, the magnitudes of the components of a force vector in different Cartesian coordinate systems are functionally related by the *vector transformation law.* Written in matrix form (see Sections 4-I.3 and 4-II.3), the transformation law is

$$\boldsymbol{F}' = \boldsymbol{\Omega} \cdot \boldsymbol{F}. \tag{6.21a}$$

In Eq. (6.21a), $\boldsymbol{F}$ is a column matrix (or vector) whose components are F_1, F_2, and F_3, the scalar coefficients of the unit vectors $\boldsymbol{i}$, $\boldsymbol{j}$ and $\boldsymbol{k}$ in the original x_1–x_2–x_3 coordinate frame. $\boldsymbol{F}'$ is a column matrix (or vector) whose components are F_1', F_2', and F_3', the scalar coefficients of the unit vectors $\boldsymbol{i}'$, $\boldsymbol{j}'$ and $\boldsymbol{k}'$ in the new x_1'–x_2'–x_3' coordinate frame. $\boldsymbol{\Omega}$ is a matrix whose components are direction cosines relating to the two coordinate frames. For two coordinate frames related by a rotation through an angle θ about the x_3 axis, the component forms of Eq. (6.21a) are:

$$F_1' = F_1 \cos\theta + F_2 \sin\theta + 0 \tag{6.21b}$$

$$F_2' = -F_1 \sin\theta + F_2 \cos\theta + 0 \tag{6.21c}$$

$$F_3' = 0 + 0 + F_3 \tag{6.21d}$$

6B.1.2 Traction or Stress Vectors

The Conceptual Foundations section of this chapter addressed how forces are transmitted through objects along networks of bonds. Further, that section addressed how different combinations of balanced forces, each of which yields no net acceleration of the object, give rise to different "local loading conditions" within the object. A *force per unit area vector* or *traction vector* describes the local loading conditions experienced within an object. Even in a region of uniform loading, where parallel planes at different positions experience identical traction vectors, the magnitude of traction vectors varies with the orientation of the plane on which the forces acted.

In order to define traction vectors, consider the interactions across a surface at some point within the object. The mass on one side of a portion of the surface (δA) near the point exerts a force ($\boldsymbol{\delta F}$) on the mass on the opposite side. As the area of the surface element (δA) becomes smaller and smaller, the force per unit area approaches a constant value in that small area (Figure 6.35), that is:

$$\operatorname*{Limit}_{\delta A \to 0} \frac{\boldsymbol{\delta F}}{\delta A} = \boldsymbol{s}. \tag{6.22}$$

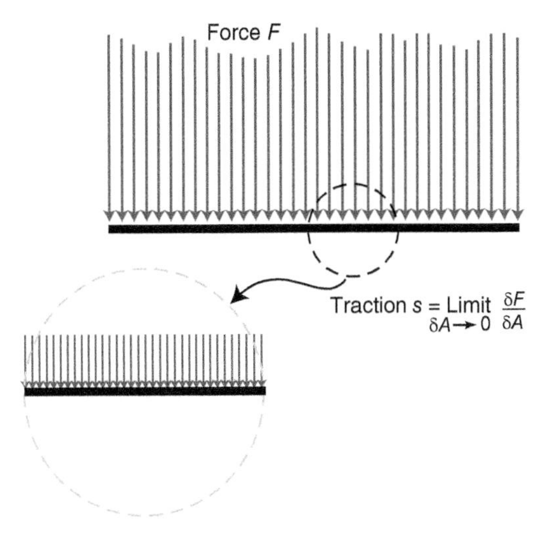

Figure 6.35 The force vectors $\boldsymbol{F}$ acting on an arbitrary plane within an object are unlikely to have a single magnitude, but as the area of consideration shrinks the magnitude of the force per unit area approaches a constant value.

The vector $\boldsymbol{s}$, with units of force/area, is the *traction* or *stress* that mass on one side of the surface exerts on mass on the other side of the surface. Because the object is at equilibrium and the local forces sum to zero, mass on the second side of the surface exerts a force ($-\boldsymbol{\delta F}$) with equal magnitude and the opposite direction on mass on the first side of the surface. For smaller and smaller areas of consideration, the force per unit area again approaches a limit:

$$\operatorname*{Limit}_{\delta A \to 0} \frac{-\delta \boldsymbol{F}}{\delta A} = -\boldsymbol{s}.$$

In addition to the requirement that the two sides of a surface have equal and opposing tractions acting on them, an object at equilibrium must have no net ***moments***, which are force vectors not directed at the center of an object body and which therefore cause the object to rotate or spin. The condition of no net moments is met if, as the area considered (δA) gets smaller and smaller, the moments $\boldsymbol{\delta M}$ conform to the condition that

$$\operatorname*{Limit}_{\delta A \to 0} \frac{\delta \boldsymbol{M}}{\delta A} = 0. \quad (6.23)$$

Any traction vector $\boldsymbol{s}$ has a specific direction and magnitude, defined by the magnitudes of its coefficients $s_1 = \delta F_1/\delta A$, $s_2 = \delta F_2/\delta A$, and $s_3 = \delta F_3/\delta A$:

$$\boldsymbol{s} = s_1 \boldsymbol{i} + s_2 \boldsymbol{j} + s_3 \boldsymbol{k}.$$

The coefficients s_1, s_2 and s_3, however, provide less insight into the character of a traction vector than is true of other vectors such as acceleration or force vectors because traction vectors "carry" more information than typical vectors. One must specify *both* a force vector and the plane on which it acts to define a stress or traction vector. The plane on which the traction vector acts typically is inclined to the coordinate axes x_1, x_2, and x_3, so there is no regular or predictable relationship between the components of the traction vector and the plane. For that reason, as outlined in Conceptual Foundations section of this chapter, we typically resolve stress or traction vectors into components parallel and perpendicular to the plane on which they act. This underscores the tie between the stress vector and the plane on which it acts. Section 6A.3 outlined how to use trigonometric relationships or the Mohr circle to relate the magnitudes of normal and tangential stresses acting on one set of planes in a two-dimensional setting to the normal and tangential stresses acting on planes with different orientations in that setting. The following section demonstrates that the mathematical expressions that relate traction or stress vector components in one coordinate frame to traction or stress vector components in a different coordinate frame are fundamentally different from the expression defined in Eq. (6.21). The "extra" information carried by stress vectors means that they are fundamentally different from other vector quantities.

6B.1.3 Relating Traction or Stress Vector Components in Different Coordinate Frames

Recall that, in specifying the stresses acting on a surface in a region at equilibrium, one need only depict the traction or stress vector, denoted by $\boldsymbol{\sigma}$, directed at one side of the surface. The existence of an opposing traction vector on the opposite side of the surface is implied. Further, recall that determining the magnitudes of the stress vectors acting on three mutually perpendicular planes at a location within a region specifies fully the state of stress in that region. The statement that the state of stress is "specified fully" indicates that one can calculate the magnitudes of the stress vector components acting on any other plane within the region as a function of the known magnitudes of the stress vector components on those three mutually perpendicular planes.

In order to derive an expression that gives the stress components referred to any "new" coordinate axes as a function of the stress components referred to three known planes ("old" stress components), consider the situation shown in Figure 6.36a. x_1, x_2, and x_3 are coordinate axes in which the stress components are known. The known or "old" stress components σ_{ij} act on planes perpendicular to the "old" coordinate axes, that is, the left, right, and bottom faces of the form in Figure 6.36a. For example, σ_{11}, σ_{12}, and σ_{13} are the normal stress component and two shear stress components acting on the left-side face, which is perpendicular to x_1 coordinate axis. In Figure 6.36a, the plane ABC is perpendicular to an arbitrary "new" x_1' coordinate axis. The x_1', x_2', and x_3'

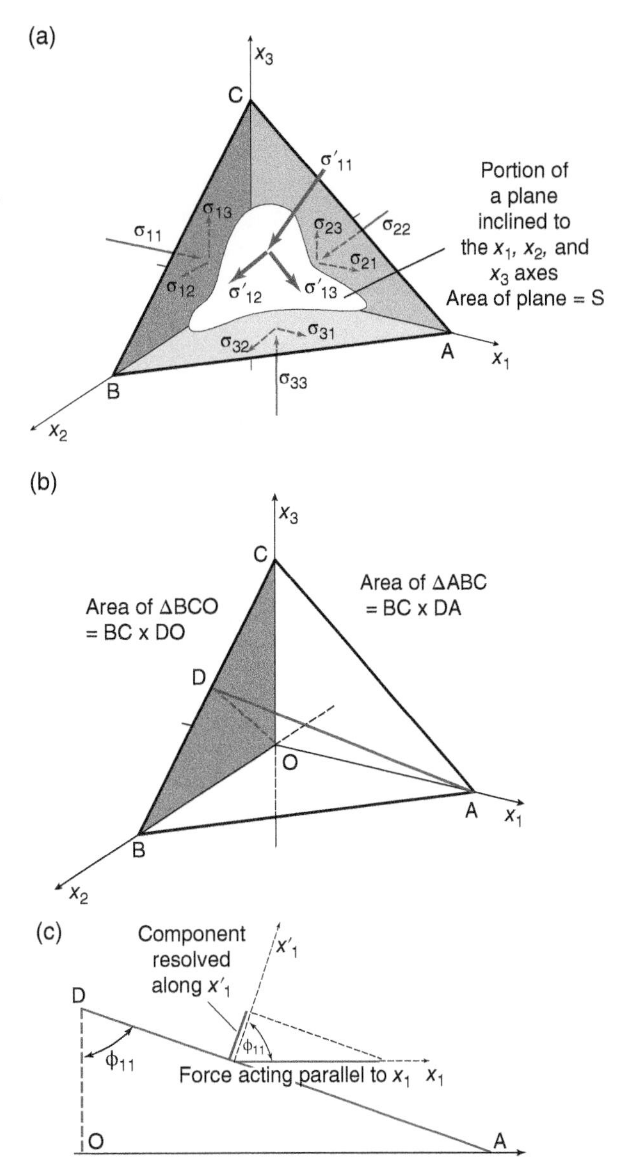

Figure 6.36 (a) The plane S inclined to the Cartesian coordinate axes x_1, x_2, and x_3 has stress vector components $\sigma'_{11}, \sigma'_{12}$ and σ'_{13} acting on it. The magnitudes of those stress components are such that the net force on the object is zero. See text for explanations. (b) If the area of plane $ABC = S$, then the area of the face $BCO = S \cos \phi_{11}$, where ϕ_{11} is the angle between the normal to face S and the normal to face BCO. (c) A force vector parallel to the x_1 axis has a component parallel to the x'_1 axis proportional to $\cos \phi_{11}$.

coordinate axes are not shown, but their orientations can be inferred from the orientations of the "new" stress components σ'_{11}, σ'_{12}, and σ'_{13}, the normal stress component and two shear stress components that act on plane ABC (stress components referred to the new coordinate axes are σ'_{ij}).

In order to derive an expression defining the new stress components σ'_{ij} as a function of the old stress components σ_{ij}, we take advantage of the fact that the region illustrated in Figure 6.36a is in equilibrium and so the net force acting in any direction must sum to zero. The area of plane ABC is S, and the normal stress acting on plane ABC is σ'_{11}. The total force on face ABC in the new x_1' coordinate direction, $\sigma'_{11}S$, must equal the sum of the components of forces acting on the left side, right side and bottom faces resolved in the same direction. The stress component σ_{11} on plane BCO results in a force component that is parallel to the x_1 coordinate axis. Its magnitude is the product of σ_{11} × Area of BCO. To determine the area of BCO, consider Figure 6.36b. The area of the face ABC is equal to ½ length BC × DA, the perpendicular distance across face ABC. The area of BCO is equal to ½ length BC × DO, the perpendicular distance across face BCO. The x_1' coordinate axis is perpendicular to face ABC, and so is perpendicular to all lines that lie in that face. Thus, DA is perpendicular to the x_1' coordinate axis. Similarly, the x_1 coordinate axis is perpendicular to face BCO and so is perpendicular to DO. From these statements, we deduce that the angle between DA and DO is equal to the angle between x_1' and x_1. Further, the length of DO is (from the right triangle ADO) is DA × cos (ϕ_{11}), where ϕ_{ij} denotes the angle between the new x_i' axis and the old x_j axis. Thus, the area of BCO = (area of ABC) × cos (ϕ_{11}). Inspection of Figure 6.36c shows that a force vector parallel to the x_1 axis will have a component parallel to the x_1' axis proportional to the cosine of the angle between the two axes. Using similar reasoning, one can calculate the force components parallel to the x_1' axis due to the σ_{12} and σ_{13} components acting on face BCO and then the contributions due to the stress components acting on faces ACO and ABO.

In order to simplify the resulting expression, we again make use of the *direction cosine* λ_{ij}, which is the cosine of the angle between a new x_i' axis and an old x_j axis (see Section 4-II.3.4 in Appendix II to Chapter 4). The cosine of the angle between the x_1' (new) and x_1 (old) axes, cos (ϕ_{11}) = λ_{11}, and the cosine of the angle between the x_1' and x_3 axes is λ_{13}. Using direction cosines, the area of the face BCO (perpendicular to the x_1 axis) is $\lambda_{11}S$, the area of the face ACO (perpendicular to the x_2 axis) is $\lambda_{12}S$, and the area of the face ABO (perpendicular to the x_3 axis) is $\lambda_{13}S$. Note also that a force of magnitude F acting parallel to x_1 will have a component parallel to the x_1' axis = $\lambda_{11} F$ (Figure 6.36c). Compiling all the forces resolved along the new x_1' direction, we have:

$$\sigma'_{11}S = \sigma_{11}(\lambda_{11}S)\lambda_{11} + \sigma_{12}(\lambda_{11}S)\lambda_{12} + \ldots\ldots$$

Force on face ABC↲ stress × (area) = force ↲ × ↓ (direction cosine)

or

$$\begin{aligned}\sigma'_{11}S = &(\lambda_{11}\lambda_{11}\sigma_{11})S + (\lambda_{11}\lambda_{12}\sigma_{21})S + (\lambda_{11}\lambda_{13}\sigma_{31})S\\ + &(\lambda_{12}\lambda_{11}\sigma_{12})S + (\lambda_{12}\lambda_{12}\sigma_{22})S + (\lambda_{12}\lambda_{13}\sigma_{32})S\\ + &(\lambda_{13}\lambda_{11}\sigma_{13})S + (\lambda_{13}\lambda_{13}\sigma_{23})S + (\lambda_{13}\lambda_{13}\sigma_{33})S\end{aligned}$$

After dividing both sides of the expression by S and substituting subscripts, we have

$$\sigma'_{11} = \sum_{k,l=1-3} \lambda_{1k}\lambda_{1l}\sigma_{kl}. \tag{6.24}$$

Following similar reasoning yields similar expressions for each of the remaining eight stress vector components referred to the new coordinate frame. Using general subscripts i and j to represent the nine equations defining the stress vector components referred to the new coordinate frame (σ'_{ij}) as functions of the angles between the new and old coordinate axes (λ_{ik} and λ_{jl}) and the magnitudes of the stress components referred to the original coordinate frame (σ_{kl}) results in a ***general transformation law*** for stress vector components:

$$\sigma'_{ij} = \sum_{k,l=1-3} \lambda_{ik}\lambda_{jl}\sigma_{kl}. \tag{6.25a}$$

A further simplification of the notation for the general transformation law for stress components results from invoking the Einstein repeated suffix convention, which prescribes that a repeated subscript (or suffix) on the right side of an equation indicates automatically that one should sum over the subscripts' range of values. Thus, the summation symbol $\sum$ is unnecessary, and the transformation law for stress components becomes

$$\sigma'_{ij} = \lambda_{ik}\lambda_{jl}\sigma_{kl}. \tag{6.25b}$$

Rewriting Eq. (6.25b) in matrix form, the components σ'_{ij} are the components of a stress matrix $\mathbf{T}'$ referred to the new coordinate axes, the direction cosine terms are a rotation matrix $\mathbf{\Omega}$ and its transpose $\mathbf{\Omega}^{\mathrm{T}}$, and the components σ_{kl} are the components of a stress matrix $\mathbf{T}$ referred to the old coordinate axes:

$$\mathbf{T}' = \mathbf{\Omega}\cdot\mathbf{T}\cdot\mathbf{\Omega}^{\mathrm{T}}. \tag{6.25c}$$

In Appendix 4-II we showed that direction cosine arrays are rotation matrices, and that the transpose of any rotation matrix is the inverse of that rotation matrix. Some workers, therefore, write Eq. (6.25c) as

$$\mathbf{T}' = \mathbf{\Omega}\cdot\mathbf{T}\cdot\mathbf{\Omega}^{-1}. \tag{6.25d}$$

Equations (6.25) indicate that $\mathbf{T}$, the matrix composed of stress or traction vector components, follows the tensor transformation law outlined in the Comprehensive Treatment section of Chapter 5 (see Eq. 5.52 in Section 5B.7.1). Thus, stress is not a vector quantity but a tensor quantity.

6B.1.4 Stress Transformation Law in Two Dimensions and the Mohr Circle

Applying the stress transformation law (Eq. 6.25c) to two-dimensional situations provides additional insight into the Mohr circle construction introduced in Section 6A.3.2. Consider a situation in which one knows the orientations of the principal planes and the magnitudes of the principal stresses and wishes to determine the stress components acting on planes other than the principal planes. In such a situation, the original or "old" coordinate axes x_1 and x_2 are the principal axes, and the known or "old" stress components are

$$\mathbf{T} = \begin{vmatrix} \sigma_1 & 0 \\ 0 & \sigma_2 \end{vmatrix}$$

Stress components referred to different, "new" coordinate axes x_1' and x_2' are

$$\mathbf{T}' = \begin{vmatrix} \sigma'_{11} & \sigma'_{12} \\ \sigma'_{21} & \sigma'_{22} \end{vmatrix}$$

The two descriptions of this one stress state are related by the stress transformation law, Eq. (6.25c). For this two-dimensional case, the direction cosine array is

$$\mathbf{\Omega} = \begin{vmatrix} \cos\alpha & \sin\alpha \\ -\sin\alpha & \cos\alpha \end{vmatrix}$$

(Figure 6.37). In component form, the transformation law is:

$$\begin{vmatrix} \sigma'_{11} & \sigma'_{12} \\ \sigma'_{21} & \sigma'_{22} \end{vmatrix} = \begin{vmatrix} \cos\alpha & \sin\alpha \\ -\sin\alpha & \cos\alpha \end{vmatrix} \begin{vmatrix} \sigma_1 & 0 \\ 0 & \sigma_2 \end{vmatrix} \begin{vmatrix} \cos\alpha & -\sin\alpha \\ \sin\alpha & \cos\alpha \end{vmatrix} \quad (6.26a)$$

After matrix multiplication this becomes

$$\begin{vmatrix} \sigma'_{11} & \sigma'_{12} \\ \sigma'_{21} & \sigma'_{22} \end{vmatrix} = \begin{vmatrix} \sigma_1\cos^2\alpha + \sigma_2\sin^2\alpha & -(\sigma_1-\sigma_2)\sin\alpha\cos\alpha \\ -(\sigma_1-\sigma_2)\sin\alpha\cos\alpha & \sigma_1\sin^2\alpha + \sigma_2\cos^2\alpha \end{vmatrix} \quad (6.26b)$$

Using the double-angle trigonometric identities $\cos^2\alpha = (1 + \cos2\alpha)/2$, $\sin^2\alpha = (1-\cos2\alpha)/2$, and $\sin2\alpha = 2\sin\alpha\cos\alpha$ to simplify the expression, Equation (6.26b) becomes

$$\begin{vmatrix} \sigma'_{11} & \sigma'_{12} \\ \sigma'_{21} & \sigma'_{22} \end{vmatrix} = \begin{vmatrix} \frac{1}{2}(\sigma_1+\sigma_2)+\frac{1}{2}(\sigma_1-\sigma_2)\cos2\alpha & -\frac{1}{2}(\sigma_1-\sigma_2)\sin2\alpha \\ -\frac{1}{2}(\sigma_1-\sigma_2)\sin2\alpha & \frac{1}{2}(\sigma_1+\sigma_2)-\frac{1}{2}(\sigma_1-\sigma_2)\cos2\alpha \end{vmatrix} \quad (6.26c)$$

Letting $\frac{1}{2}(\sigma_1 + \sigma_2) = c$ and $\frac{1}{2}(\sigma_1-\sigma_2) = r$ yields a matrix form of Eq. (6.16) from Section 6A.3.3.1

$$\begin{vmatrix} \sigma'_{11} & \sigma'_{12} \\ \sigma'_{21} & \sigma'_{22} \end{vmatrix} = \begin{vmatrix} c + r\cos2\alpha & -r\sin2\alpha \\ -r\sin2\alpha & c - r\cos2\alpha \end{vmatrix}$$

As noted in Section 6A.3.3.1, using Cartesian σ_N–σ_T coordinate axes, with different magnitudes of the angle α the components of this matrix outline a circle with center c on the abscissa (σ_N axis) and with radius = r. Since the algebraic equations defining the circle follow directly from the transformation law for stress components, the collection of points that fall on the circle, points with coordinates ($\sigma_N = c \pm r\cos2\alpha$, $\sigma_T = -r\sin2\alpha$) for different magnitudes of α, are the collection of stress component pairs related by the stress component transformation law. Note that the angle between the two corresponding coordinate axes in real space, such as x_1 and x_1', is α, and the angle between the corresponding pairs of points on the Mohr circle giving the stress components is 2α.

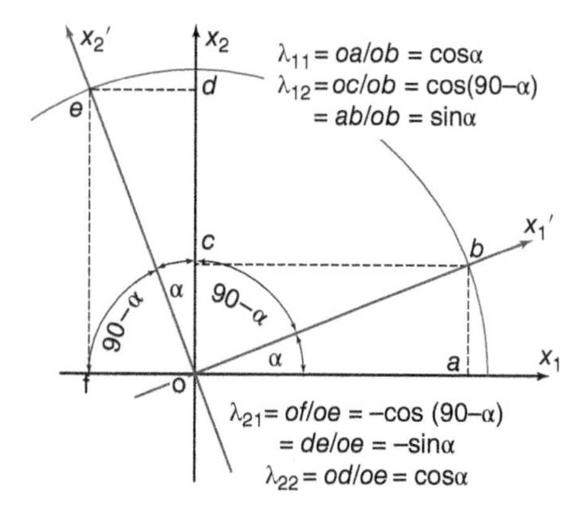

Figure 6.37 Diagram showing the origin of the direction cosines for an anticlockwise rotation of rotation axes. See text for explanation.

The Mohr circle for stress is, then, a graphical version of the transformation law for the two-dimensional stress tensors. Material scientists, engineers, and geologists regularly use Mohr circles to depict graphically the transformation law for other tensor quantities, including the position gradient tensor, the displacement gradient tensor, or a variety of strain tensors derived from the position gradient or displacement gradient tensors. In addition, you might encounter Mohr circles for velocity gradient or strain rate tensors, which are quantities derived from the time derivatives of displacement gradients.

6B.1.5 Stress Transformation Law in Three Dimensions and the Mohr Diagram

One can use a plot in Mohr space to calculate the stress components acting on differently oriented planes in a three-dimensional stress state, though most workers now rely on a computer algorithm of Eq. (6.25) to analyze in detail stresses in three dimensions. Still, there are some advantages to using Mohr space representations to depict three-dimensional stress states; one can quickly distinguish visually the general character of stress states, such as whether stress state is uniaxial, biaxial, triaxial, etc. (see Table 6.5), with a glance at its Mohr space representation. The plot in Mohr space (on Cartesian $\sigma_N - \sigma_T$ coordinate axes) that represents any three-dimensional stress state consists of three related Mohr circles and a series of circular arcs within a confined region of Mohr space. Because the plot is more than a single circle, we use the term *Mohr diagram* to refer to the Mohr-space representation of any three-dimensional stress state.

The character of any three-dimensional stress state is fully specified by defining the orientation and magnitudes of the three principal stresses. Three-dimensional Mohr diagrams highlight the relative magnitudes of the principal stresses σ_1, σ_2, and σ_3. Consider first a planar section through a three-dimensional stress state that contains the most and least compressive principal stresses, σ_1 and σ_3. A Mohr circle centered at $[\frac{1}{2}(\sigma_1 + \sigma_3), 0]$ and with radius $= \frac{1}{2}(\sigma_1 - \sigma_3)$ defines magnitudes of traction vectors that lie within the $\sigma_1 - \sigma_3$ plane and act on planes perpendicular to the $\sigma_1 - \sigma_3$ plane. Similarly, a Mohr circle centered at $[\frac{1}{2}(\sigma_1 + \sigma_2), 0]$ with radius $= \frac{1}{2}(\sigma_1 - \sigma_2)$ defines magnitudes of traction vectors that lie within the $\sigma_1 - \sigma_2$ plane and act on planes perpendicular to the $\sigma_1 - \sigma_2$ plane). Finally, a Mohr circle centered at $[\frac{1}{2}(\sigma_2 + \sigma_3), 0]$ with radius $= \frac{1}{2}(\sigma_2 - \sigma_3)$ defines magnitudes of traction vectors that lie within the $\sigma_2 - \sigma_3$ plane and act on planes perpendicular to the $\sigma_2 - \sigma_3$ plane. Plotting the three Mohr circles on a single set of Cartesian $\sigma_N - \sigma_T$ coordinate axes defines the outlines of a Mohr diagram that depicts the three-dimensional state of stress (Figure 6.38a).

In Section 6A.3.3 we took pains to make clear that Mohr space points inside or outside of a Mohr circle do not and cannot correspond to planes within a particular stress state – only points in Mohr space with (σ_N, σ_T) coordinates that fall on a Mohr circle correspond to normal and shear stress component pairs observed within the particular stress state. This restriction must be relaxed when considering three-dimensional Mohr diagrams. Points in Mohr space that lie outside the $\sigma_1 - \sigma_3$ Mohr circle *or* inside the $\sigma_1 - \sigma_2$ and $\sigma_2 - \sigma_3$ circles do not and cannot correspond to planes within the three-dimensional stress state. Mohr space points falling inside the $\sigma_1 - \sigma_3$ Mohr circle *and* outside the $\sigma_1 - \sigma_2$ and $\sigma_2 - \sigma_3$ circles (within the shaded region in Figure 6.38a) can and do correspond to planes within the three-dimensional stress state. Figure 6.38b illustrates that one can draw a series of circular arcs within the region of Mohr space inside the $\sigma_1 - \sigma_3$ Mohr circle *and* outside the $\sigma_1 - \sigma_2$ and $\sigma_2 - \sigma_3$ circles and use them to define the magnitudes and of normal and tangential stress components acting planes oblique to the principal stress axes. The arcs are concentric about the centers of the $\sigma_1 - \sigma_2$ and $\sigma_2 - \sigma_3$ circles, and they define the magnitudes of the normal and shear stress components acting on planes at specific inclinations to the stress principal directions (Figure 6.38c).

A more common use of Mohr diagrams for three-dimensional stress states is to represent graphically the relative magnitudes of the principal stresses. Figure 6.39 shows three three-dimensional stress states with the same differential stress $\Delta\sigma = (\sigma_1 - \sigma_3)$ but different characteristics. If $\sigma_1 = \sigma_2$ (in a state of uniaxial tension) or if $\sigma_2 = \sigma_3$ (in a state of uniaxial compression), we can use a single Mohr circle to determine the stresses acting on any plane

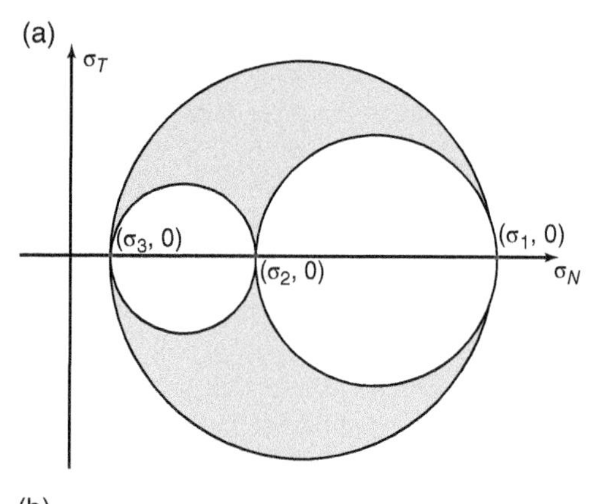

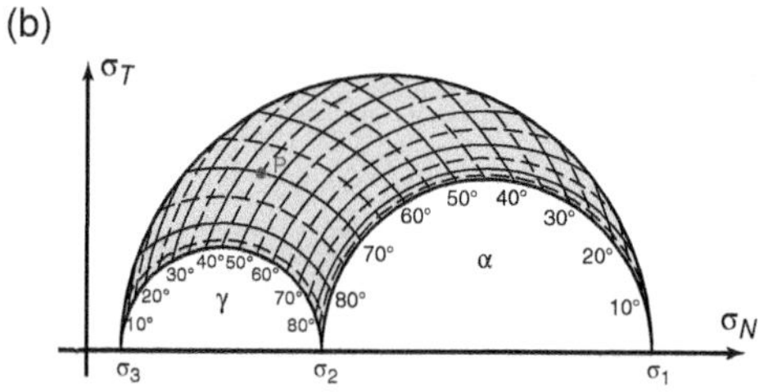

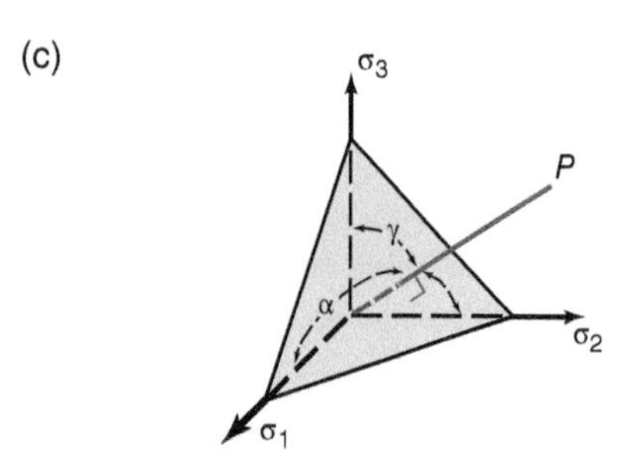

Figure 6.38 (a) Mohr circles for the three principal planes define the portion of Mohr space (the shaded area) where points in Mohr space with components (σ_N, σ_T) denote the magnitudes of normal and shear stress, respectively, on planes within a region with principal stresses σ_1, σ_2, and σ_3. (b) The shaded region of Mohr space identifies possible magnitudes of (σ_N, σ_T) attainable within this three-dimensional stress state. The specific points fall on particular arcs with specific angles from the principal directions σ_1, σ_2, and σ_3. (c) Diagram illustrating the angles α and γ between the normal to a plane (*P*) and the principal stress vectors σ_1 and σ_3, respectively.

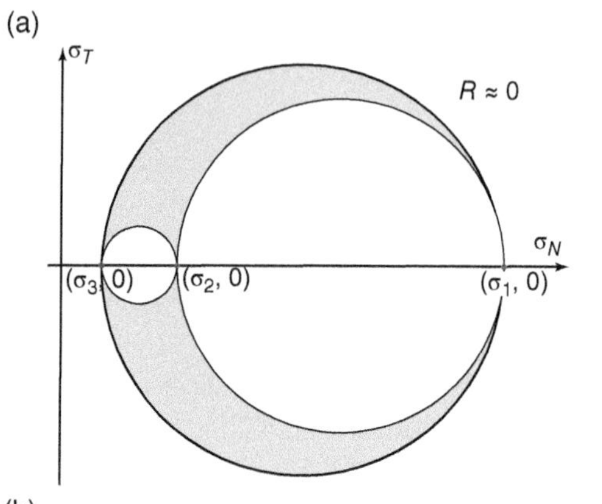

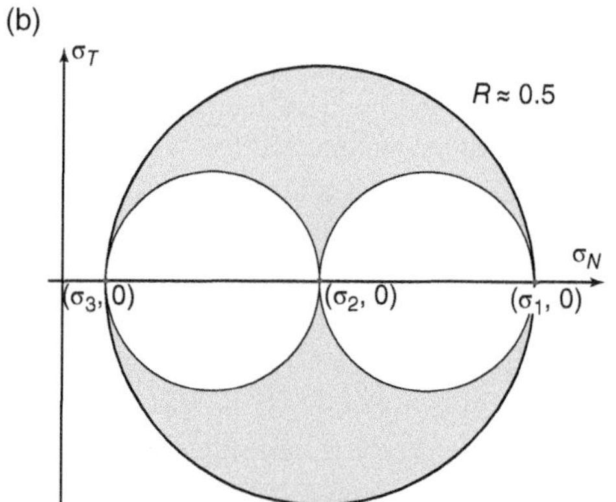

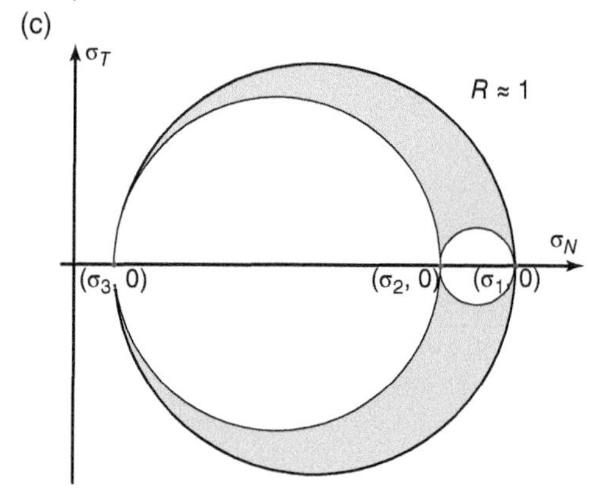

Figure 6.39 The relative sizes of the two-dimensional Mohr circles within a three-dimensional Mohr diagram indicate the relative size of the three principal stress magnitudes. (a) When $\sigma_2 \approx \sigma_3, R \approx 0$. (b) When $\sigma_2 \approx (\sigma_2 + \sigma_3)/2, R \approx 0.5$. (c) When $\sigma_1 \approx \sigma_2, R \approx 0$.

inclined to the unique σ_1 or σ_3 axis. Figure 6.39a illustrates a triaxial stress state where σ_2 is less than but not greatly different from σ_1. In Figure 6.39b, the stress state is again triaxial stress but here σ_2 is greater than but not greatly different from σ_3.

In considering the shape of the three-dimensional strain ellipsoid (see Section 5A.6.2.2), we introduced Flinn's parameter k, whose magnitude defines the relative sizes of the principal stretches. Many workers rely upon a similar parameter, R, to communicate information on the relative magnitudes of the principal stresses.

$$R = \frac{(\sigma_2 - \sigma_3)}{(\sigma_1 - \sigma_3)}. \tag{6.27}$$

Values of the parameter R range from 0 to 1. When $R \approx 0$, $\sigma_2 \approx \sigma_3$, and the stress state approximates uniaxial compression. When $R \approx 1$, $\sigma_2 \approx \sigma_1$, and the stress state approximates uniaxial tension. Intermediate values. such as $R \approx 0.5$, correspond to more nearly plane stress states.

6B.1.6 An Alternative Way to Define Traction or Stress Vectors

In the course of examining homogeneous deformation in the Comprehensive Treatment section of Chapter 5, we considered the relationship between a particle's position vectors in the deformed and undeformed states and the relationship between a particle's displacement vector and its position vector in the undeformed state. In the first instance, a deformation matrix or position gradient matrix relates a particle's final position vector to its original position vector. From Eq. (5.35) in Chapter 5,

$$\boldsymbol{x} = \mathbf{F} \cdot \boldsymbol{X},$$

where $\boldsymbol{x}$ is a particle's final position vector, $\boldsymbol{X}$ is the particle's original position vector, and $\mathbf{F}$ is the position gradient or deformation matrix. Equation (5.52) in Chapter 5 indicates that $\mathbf{F}$ is tensor quantity. From Eq. (5.42) in Chapter 5,

$$\boldsymbol{u} = \mathbf{J} \cdot \boldsymbol{X},$$

where $\boldsymbol{u}$ is a particle's displacement vector, $\boldsymbol{X}$ is the particle's original position vector, and $\mathbf{J}$ is the displacement gradient matrix or Jacobian. The displacement gradient matrix is related to the position gradient matrix by $\mathbf{F} - \mathbf{I} = \mathbf{J}$, where $\mathbf{I}$ is the identity matrix. Thus, one can easily determine the components of one of the two matrices from the components of the other matrix. $\mathbf{J}$ is also a tensor quantity. Position gradient and displacement gradient tensors characterize the nature of the general deformation component of the total displacement field. Moreover, specifying the components of either one of these tensors defines the degree to which solid-body rotation and pure strain contribute the general deformation component of the total displacement field.

In a region of uniform stress, an analogous relationship holds between the stress vector acting on a plane and a unit vector normal to the plane. That is,

$$\boldsymbol{s} = \mathbf{T} \cdot \boldsymbol{n}, \tag{6.28}$$

where $\boldsymbol{s}$ is the traction vector acting on a plane N, $\boldsymbol{n}$ is a vector with a unit length that is normal to the plane N and is directed outward, and $\mathbf{T}$ is a matrix whose components characterize the state of stresses within the region. We can describe fully the orientation of any line by giving its direction cosines relative to the coordinate axes. Let l be the direction cosine relative to the x_1 axis, m be the direction cosine relative to the x_2 axis, and n be the direction cosine relative to the x_3 axis. The components of the traction vector acting on the plane N are

$$\begin{vmatrix} s_1 \\ s_2 \\ s_3 \end{vmatrix} = \begin{vmatrix} \sigma_{11} & \sigma_{12} & \sigma_{13} \\ \sigma_{21} & \sigma_{22} & \sigma_{23} \\ \sigma_{31} & \sigma_{32} & \sigma_{33} \end{vmatrix} \cdot \begin{vmatrix} l \\ m \\ n \end{vmatrix}$$

$$= \begin{vmatrix} l\,\sigma_{11} + m\,\sigma_{12} + n\,\sigma_{13} \\ l\,\sigma_{21} + m\,\sigma_{22} + n\,\sigma_{23} \\ l\,\sigma_{31} + m\,\sigma_{32} + n\,\sigma_{33} \end{vmatrix} \tag{6.29a}$$

The traction vector components acting parallel to the x_1, x_2, and x_3 axes, respectively, are

$$s_1 = l\,\sigma_{11} + m\,\sigma_{12} + n\,\sigma_{13} \tag{6.29b}$$

$$s_2 = l\,\sigma_{21} + m\,\sigma_{22} + n\,\sigma_{23} \tag{6.29c}$$

$$s_3 = l\,\sigma_{31} + m\,\sigma_{32} + n\,\sigma_{33} \tag{6.29d}$$

The traction vector component acting perpendicular to the face N is the sum of the projections of s_1, s_2, and s_3 on a line normal to N, whose direction cosines are l, m, and n:

$$\sigma_N = l\,s_1 + m\,s_2 + n\,s_3 \tag{6.30a}$$

$$\begin{aligned}\sigma_N = {} & l\left(l\sigma_{11} + m\sigma_{12} + n\sigma_{13}\right) \\ & + m\left(l\sigma_{21} + m\sigma_{22} + n\sigma_{23}\right) \\ & + n\left(l\sigma_{31} + m\sigma_{32} + n\sigma_{33}\right)\end{aligned} \tag{6.30b}$$

$$\begin{aligned}\sigma_N = {} & l^2\sigma_{11} + m^2\sigma_{22} + n^2\sigma_{33} + 2\,l\,m\,\sigma_{12} \\ & + 2\,l\,n\,\sigma_{13} + 2\,m\,n\,\sigma_{23}\end{aligned} \tag{6.30c}$$

Similarly, the shear stress components acting on plane N are given by products of the traction vector components s_1, s_2, and s_3 on lines parallel to N.

6B.1.7 Determining Stress Principal Directions and Magnitudes

Equation (6.30c) gives the magnitude of the normal stress acting on a plane whose orientation is defined by the direction cosines of its normal $\boldsymbol{n}$. The three principal planes in any three-dimensional stress state are planes for which the normal stress component is finite and the shear or tangential component vanishes. Further, one of the three principal planes experiences the greatest magnitude of normal stress (the most compressive normal stress), and another experiences the minimum magnitude of the normal stress (the least compressive normal stress). One way to identify two of the principal directions is to find the values of the direction cosines l, m, and n that yield the maximum and minimum values of σ_N (because the third principal direction is perpendicular to those two directions). To find a maximum or minimum of a function, we find the values of the argument of the function where the function's derivatives are zero. Beginning with this general principle, we follow the analysis outlined by J. C. Jaeger in his classic text *Elasticity, Fracture, and Flow* to identify the stress principal directions and magnitudes (see Jaeger 1969, pp. 10–18).

6B.1.7.1 Jaeger's Method for Defining Stress Principal Directions and Magnitudes

Only two of three direction cosines defining the orientation of the plane N, let us say l and m, are independent of each other. The third direction cosine, n in this case, is related to l and m by

$$l^2 + m^2 + n^2 = 1. \tag{6.31}$$

The maximum or minimum values of σ_N occur when its derivative with respect to the two independent direction cosines l and m are zero:

$$\frac{\partial \sigma_N}{\partial l} = 0 \text{ and } \frac{\partial \sigma_N}{\partial m} = 0. \tag{6.32}$$

Because n is a function of l, evaluating the derivative of Eq. (6.30c) with respect to l yields

$$\begin{aligned}\frac{\partial \sigma_N}{\partial l} = {} & 2l\,\sigma_{11} + 2\,n\,\sigma_{33}\frac{\partial n}{\partial l} + 2\,m\,\sigma_{12} + 2\,n\,\sigma_{13} \\ & + 2\,l\,\sigma_{13}\frac{\partial n}{\partial l} + 2\,m\,n\,\sigma_{23}\frac{\partial n}{\partial l} \\ & + 2\,m\,\sigma_{23}\frac{\partial n}{\partial l} = 0\end{aligned}$$

or

$$\begin{aligned} & 2l\,\sigma_{11} + 2\,m\,\sigma_{12} + 2\,n\,\sigma_{13} \\ & + \left(2\,l\,\sigma_{13} + 2\,m\,\sigma_{23} + 2\,n\,\sigma_{33}\right)\frac{\partial n}{\partial l} = 0\end{aligned}$$

Using Eqs. (6.29b) and (6.29d) to substitute for the first three terms and the terms inside the parentheses, respectively, yields

$$s_1 + s_3\frac{\partial n}{\partial l} = 0. \tag{6.33}$$

Because n is a function of m, evaluating the derivative of Eq. (6.30c) with respect to m yields

$$\begin{aligned}\frac{\partial \sigma_N}{\partial m} = {} & 2m\,\sigma_{22} + 2\,n\,\sigma_{33}\frac{\partial n}{\partial m} + 2\,l\,\sigma_{12} \\ & + 2\,l\,\sigma_{13}\frac{\partial n}{\partial m} + 2\,n\,\sigma_{23} + 2\,m\,\sigma_{23}\frac{\partial n}{\partial m} = 0\end{aligned}$$

or

$$\begin{aligned} & 2\,l\,\sigma_{12} + 2\,m\,\sigma_{22} + 2\,n\,\sigma_{23} \\ & + \left(2\,l\,\sigma_{13} + 2\,m\,\sigma_{23} + 2\,n\,\sigma_{33}\right)\frac{\partial n}{\partial m} = 0\end{aligned}$$

Using Eqs. (6.29c) and (6.29d) to substitute for the first three terms and the terms inside the parentheses, respectively, yields

$$s_2 + s_3 \frac{\partial n}{\partial m} = 0. \quad (6.34)$$

Further simplification results by differentiating Eq. (6.31) with respect to l and m to find

$$l + n\frac{\partial n}{\partial l} = 0 \quad (6.35a)$$

or

$$\frac{\partial n}{\partial l} = \frac{-l}{n} \quad (6.35b)$$

and

$$m + n\frac{\partial n}{\partial m} = 0 \quad (6.36a)$$

or

$$\frac{\partial n}{\partial m} = -\frac{m}{n}. \quad (6.36b)$$

Combining first Eqs. (6.33) and (6.35b) yields

$$s_1 + s_3\left(\frac{-l}{n}\right) = 0$$

or

$$\frac{s_1}{l} = \frac{s_3}{n}.$$

Combining first Eqs. (6.34) and (6.36b) yields

$$s_2 + s_3\left(-\frac{m}{n}\right) = 0$$

or

$$\frac{s_2}{m} = \frac{s_3}{n}.$$

Taken together, the maximum and minimum normal stresses have components that adhere to the condition

$$\frac{s_1}{l} = \frac{s_2}{m} = \frac{s_3}{n}. \quad (6.37)$$

For normal stress with magnitude Λ, this condition holds if $s_1 = l\,\Lambda$, $s_2 = m\,\Lambda$, and $s_3 = n\,\Lambda$ (the magnitude is $\sqrt{[(\Lambda l)^2 + (\Lambda m)^2 + (\Lambda n)^2]} = \sqrt{[\Lambda^2 l^2 + \Lambda^2 m^2 + \Lambda^2 n^2]} = \sqrt{[\Lambda^2(l^2 + m^2 + n^2)]} = \sqrt{[\Lambda^2]} = \Lambda$).

Inserting these values for s_1, s_2, and s_3 into Eq. (6.29) yields

$$l\,\Lambda = l\,\sigma_{11} + m\,\sigma_{12} + n\,\sigma_{13}$$

$$m\,\Lambda = l\,\sigma_{21} + m\,\sigma_{22} + n\,\sigma_{23}$$

$$n\,\Lambda = l\,\sigma_{31} + m\,\sigma_{32} + n\,\sigma_{33}$$

Rearranging, these equations become

$$l(\sigma_{11} - \Lambda) + m\,\sigma_{12} + n\,\sigma_{13} = 0 \quad (6.38a)$$

$$l\,\sigma_{21} + m(\sigma_{22} - \Lambda) + n\,\sigma_{23} = 0 \quad (6.38b)$$

$$l\,\sigma_{31} + m\,\sigma_{32} + n(\sigma_{33} - \Lambda) = 0 \quad (6.38c)$$

A collection of k linear equations in k variables, like Eqs. (6.38), is a *system of linear equations*. The equations are *homogeneous* if the value of the constant term in each equation is 0. Thus, Eqs. (6.38) constitute a *homogeneous system* of linear equations. A homogeneous system of the linear equation has nontrivial roots only if the determinant of the matrix defined by the coefficients of variables is zero. For Eqs. (6.38) the matrix of coefficients is

$$\begin{vmatrix} \sigma_{11} - \Lambda & \sigma_{12} & \sigma_{13} \\ \sigma_{21} & \sigma_{22} - \Lambda & \sigma_{23} \\ \sigma_{31} & \sigma_{32} & \sigma_{33} - \Lambda \end{vmatrix}$$

Calculating the determinant of this yields a cubic equation in Λ:

$$\Lambda^3 - \Lambda^2(\sigma_{11} + \sigma_{22} + \sigma_{33}) - \Lambda\left(\begin{vmatrix} \sigma_{22} & \sigma_{32} \\ \sigma_{23} & \sigma_{33} \end{vmatrix} + \begin{vmatrix} \sigma_{11} & \sigma_{21} \\ \sigma_{12} & \sigma_{22} \end{vmatrix} + \begin{vmatrix} \sigma_{11} & \sigma_{31} \\ \sigma_{13} & \sigma_{33} \end{vmatrix}\right) - \begin{vmatrix} \sigma_{11} & \sigma_{12} & \sigma_{13} \\ \sigma_{21} & \sigma_{22} & \sigma_{23} \\ \sigma_{31} & \sigma_{32} & \sigma_{33} \end{vmatrix} = 0 \quad (6.39)$$

The three roots of Eq. (6.39), Λ_1, Λ_2, and Λ_3, correspond to values of Λ for which the condition outlined in Eq. (6.37) holds, that is, values for which the traction vector acting on a plane is parallel to the normal to the plane. Substituting the roots into Eq. (6.38) and solving for three sets of values for l, m, and n defines the orientations of the three principal planes of the stress state. The three roots of Eq. (6.39), Λ_1, Λ_2, and Λ_3, define the magnitudes of the principal stresses.

6B.1.7.2 An Alternative Method for Defining Stress Principal Directions and Magnitudes

The traction vector acting on any plane in a three-dimensional stress state is a product of the stress tensor $\mathbf{T}$ and the unit normal to the plane $\boldsymbol{n}$ (see Eq. (6.28) in Section 6B.1.6). That is,

$$\boldsymbol{s} = \mathbf{T} \cdot \boldsymbol{n}. \tag{6.28}$$

For each of the principal planes, the traction vector is parallel to the normal to the plane. Two vectors are parallel if they are scalar multiples of each other, that is:

$$\boldsymbol{s} = \Lambda \boldsymbol{n}, \tag{6.40}$$

where $\boldsymbol{s}$ is the traction vector acting on the plane N, $\boldsymbol{n}$ is outward directed normal to the plane N with unit length, and Λ is a scalar. Combining Eqs. (6.28) and (6.40) yields

$$\mathbf{T} \cdot \boldsymbol{n} = \Lambda \boldsymbol{n} \tag{6.41a}$$

or

$$\mathbf{T} \cdot \boldsymbol{n} - \Lambda \boldsymbol{n} = 0. \tag{6.41b}$$

The scalar multiple of any matrix is equivalent to multiplying the matrix by a scalar matrix, and any scalar matrix is equivalent to the scalar multiple of the identity matrix $\mathbf{I}$ (see Section 4-II.3.5). Thus, Eq. (6.41b) becomes

$$\mathbf{T} \cdot \boldsymbol{n} - \Lambda \mathbf{I} \cdot \boldsymbol{n} = \left(\mathbf{T} - \Lambda \mathbf{I}\right) \cdot \boldsymbol{n} = 0. \tag{6.41c}$$

In component form, Eq. (6.41c) is

$$\begin{vmatrix} \sigma_{11} - \Lambda & \sigma_{12} & \sigma_{13} \\ \sigma_{21} & \sigma_{22} - \Lambda & \sigma_{23} \\ \sigma_{31} & \sigma_{32} & \sigma_{33} - \Lambda \end{vmatrix} \cdot \begin{vmatrix} l \\ m \\ n \end{vmatrix} = 0 \tag{6.41d}$$

where l, m, and n are the direction cosines that define the orientation of the normal to the plane N. Completing the matrix multiplication yields the homogeneous system of linear equations. (6.38a), (6.38b), and (6.38c). The homogeneous system of linear equations will have non-trivial roots only if the determinant of the matrix $(\mathbf{T} - \Lambda \mathbf{I})$ is zero. Calculating this determinant again yields Eq. (6.39), the cubic equation in Λ. The three roots to this equation, Λ_1, Λ_2, and Λ_3, are known as the *eigenvalues* (from the German *eigen* meaning unique or idiosyncratic) of the stress state. Each eigenvalue defines the magnitude of one of the principal stresses σ_1, σ_2, and σ_3. Corresponding to each eigenvalue is an eigenvector $\boldsymbol{n_i}$ which defines one of the three principal directions. Eigenvectors of any tensor quantity have the following unique characteristic: the product of the tensor times an eigenvector yields a vector parallel to the eigenvector. In the case of the stress tensor, multiplication of an eigenvector (normal to a principal plane) by the tensor yields a traction vector that is a scalar multiple of the normal to the plane.

6B.1.8 Stress Invariants

The components of the stress tensor $\mathbf{T}$ have different values when referred to different coordinate frames, that is, $\sigma_{12} \neq \sigma'_{12}$. The overall character of the stress tensor is the same, however, regardless of the coordinate frame in which it is specified. The different forms of the stress transformation equation (6.25), in which the components in any one coordinate frame are used to calculate another coordinate frame, insure the uniformity of character of the stress tensor. There is another way to demonstrate that the intrinsic character of the stress state does not change as we change the coordinate frame in which we specify it. The cubic Eq. (6.39) must yield the same roots Λ_1, Λ_2, and Λ_3, regardless of the orientation of the coordinate frame in which one defines the stress state. Thus, the coefficients of this cubic equation cannot vary with the coordinate frame in which the tensor coordinates are specified. This is the origin of the statement made in the Appendix to Chapter 5, that a tensor's independence of the coordinate system in which we specify it is apparent in the existence of tensor *invariants* – quantities that have the same magnitude regardless of the coordinate system in which we choose to specify the tensor. For second rank tensors like stress, the three invariant quantities are the coefficients of

Eq. (6.39). The first invariant I_1 is the coefficient of the $\Lambda 2$ term in Eq. (6.39):

$$I_1 = \sigma_1 + \sigma_2 + \sigma_3 = \sigma_{11} + \sigma_{22} + \sigma_{33} = \sigma'_{11} + \sigma'_{22} + \sigma'_{33}. \quad (6.42)$$

The first invariant is the sum of the normal stresses acting on any three mutually perpendicular planes. As was the case in the two-dimensional treatment of stress, geologists often define a three-dimensional *mean stress* $\sigma_m = (\sigma_{11} + \sigma_{22} + \sigma_{33})/3$. The three-dimensional mean stress $\sigma_m = (I_1/3)$. The three-dimensional mean stress defines the magnitude of the three-dimensional hydrostatic stress and corresponds to the pressure as defined by metamorphic petrologists, geophysicists, and geochemists. Wherever possible, a structural geologist should use the three-dimensional mean stress when determining if a stress state is lithostatic, calculating deviatoric stresses, or evaluating pore fluid ratios, etc.

The second invariant I_2 is the coefficient of the Λ term in Eq. (6.39):

$$\begin{aligned} I_2 &= -\left(\sigma_1\sigma_2 + \sigma_2\sigma_3 + \sigma_3\sigma_1\right) \\ &= -\left(\sigma_{11}\sigma_{22} + \sigma_{22}\sigma_{33} + \sigma_{33}\sigma_{11} - \sigma_{12}\sigma_{21} - \sigma_{23}\sigma_{32} - \sigma_{13}\sigma_{31}\right) \\ &= -\left(\sigma'_{11}\sigma'_{22} + \sigma'_{22}\sigma'_{33} + \sigma'_{33}\sigma'_{11} - \sigma'_{12}\sigma'_{21} - \sigma'_{23}\sigma'_{32} - \sigma'_{13}\sigma'_{31}\right). \end{aligned} \quad (6.43)$$

Because the stress tensor is symmetric and $\sigma_{ij} = \sigma_{ji}$, the second invariant is sometimes written as

$$I_2 = -\left(\sigma_{11}\sigma_{22} + \sigma_{22}\sigma_{33} + \sigma_{33}\sigma_{11} - \sigma_{12}^2 - \sigma_{23}^2 - \sigma_{13}^2\right)$$

or

$$\frac{\left(\sigma_{ij}\sigma_{ij} - \sigma_{ii}\sigma_{jj}\right)}{2}.$$

The second invariant is related to the degree of ellipticity of the stress ellipsoid, and as such, measures a three-dimensional stress state's departure from the hydrostatic state. In a sense, I_2 is analogous to the radius of a two-dimensional Mohr circle; the analogy is not precise, but I_2 captures the departure from the hydrostatic in the same way that the radius of a two-dimensional Mohr circle correlates to how much a two-dimensional stress state departs from the center of the circle.

The third invariant I_3 is the final term in Eq. (6.39)

$$I_3 = \det \mathbf{T} = \det \mathbf{T}'. \quad (6.44)$$

In component form, the third invariant is

$$\begin{aligned} I_3 &= \sigma_1\sigma_2\sigma_3 = \sigma_{11}\sigma_{22}\sigma_{33} + \sigma_{12}\sigma_{23}\sigma_{31} + \sigma_{13}\sigma_{21}\sigma_{32} - \sigma_{11}\sigma_{23}\sigma_{32} - \sigma_{22}\sigma_{13}\sigma_{31} - \sigma_{33}\sigma_{12}\sigma_{21} \\ &= \sigma'_{11}\sigma'_{22}\sigma'_{33} + \sigma'_{12}\sigma'_{23}\sigma'_{31} + \sigma'_{13}\sigma'_{21}\sigma'_{32} - \sigma'_{11}\sigma'_{23}\sigma'_{32} - \sigma'_{22}\sigma'_{13}\sigma'_{31} - \sigma'_{33}\sigma'_{12}\sigma'_{21}. \end{aligned}$$

Invoking the symmetric character of the stress tensor, where $\sigma_{ij} = \sigma_{ji}$, the third invariant is sometimes written as

$$I_3 = \sigma_{11}\sigma_{22}\sigma_{33} + 2\sigma_{12}\sigma_{23}\sigma_{31} - \sigma_{11}\sigma_{23}\sigma_{32} - \sigma_{22}\sigma_{13}\sigma_{31} - \sigma_{33}\sigma_{12}\sigma_{21}.$$

There is no straightforward physical concept to correlate to the third invariant, but the magnitude of the third invariant does figure in some formulations of rock deformation subjected to truly triaxial stress states.

6B.1.9 Spatial Variation in Stress

Other than the brief introduction of inhomogeneous stress in Section 6A.3.5, this chapter has focused on characterizing and understanding homogeneous stress states. Homogeneous stress states are, of course, stress states in which parallel planes at different locations within the region experience identical stress or traction vectors. Figure 6.21 compared the uniform orientations of the principal directions and constant magnitude of the stress difference in a region of homogeneous stress with the curved principal directions and the changing magnitude of the stress difference in a region of inhomogeneous stress. Since stresses in Earth are almost always inhomogeneous, it is appropriate to consider what constraints there are on how stresses vary within regions of inhomogeneous stress.

A primary constraint on stresses in Earth arises from the fact that liquids and gases are unable to support any stress differences. As a result, there are no shear stresses within liquids or gases, and there can be no shear stresses acting on a surface where solid rock contacts a volume of gas or liquid. Earth's surface must always be one of the stress three principal surfaces because the air or water above it cannot support any shear stress (e.g. the ground surface in Figure 6.21b). In two-dimensional analyses of stress states, the second principal must be perpendicular to Earth's surface immediately below the surface. In three-dimensional analyses, the other two principal planes are perpendicular to the ground surface immediately below the surface. With increasing distance from the ground surface, principal planes curve gradually to new orientations due to changes in the magnitudes of the stress components within any uniformly oriented coordinate frame.

Any changes in the magnitudes of the stress components must not disturb the conditions of equilibrium in which stresses are defined; that is since the net acceleration of a region is zero the forces acting on it must sum to zero. Figure 6.40 shows a cubic element in a region of varying stresses. Because the element is at equilibrium, the forces acting on the element must sum to zero along any linear direction through the element. Three of the nine stress components are aligned parallel to the x_1 coordinate axis (Figure 6.40b): σ_{11} acting on planes perpendicular to x_1, σ_{21} acting on planes perpendicular to x_2, and σ_{31} acting on planes perpendicular to x_3. If there is a difference in the magnitude of σ_{11} between the two planes perpendicular to x_1 at the "front" and "back" of the cubic element, then

$$\sigma_{11}(\text{back}) = \sigma_{11}(\text{front}) - \left(\frac{\partial\sigma_{11}}{\partial x_1}\right) \times \delta x_1,$$

where $\partial\sigma_{11}/\partial x_1$ measures the rate of change of σ_{11} in a direction parallel to x_1 and δx_1 is the length of the cubic element parallel to the x_1 axis. If there is a difference in the magnitude of σ_{21} between the two planes perpendicular to x_2 at the "left" and "right" of the cubic element, then

$$\sigma_{21}(\text{left}) = \sigma_{21}(\text{right}) - \left(\frac{\partial\sigma_{21}}{\partial x_2}\right) \times \delta x_2,$$

where $\partial\sigma_{21}/\partial x_2$ measures the rate of change of σ_{21} in a direction parallel to x_2 and δx_2 is the length of the cubic element parallel to the x_2 axis. Finally, if there is a difference in the magnitude of σ_{31} between the two planes perpendicular to x_3 at the "top" and "bottom" of the cubic element, then

$$\sigma_{31}(\text{bottom}) = \sigma_{31}(\text{top}) - \left(\frac{\partial\sigma_{31}}{\partial x_3}\right) \times \delta x_3,$$

where $\partial\sigma_{31}/\partial x_3$ measures the rate of change of σ_{31} in a direction parallel to x_3 and δx_3 is the length of the cubic element parallel to the x_1 axis.

The net force acting in the x_1 direction on the cubic element due to variation in σ_{11} is the difference in the stresses at the front and back of the cubic element times the area of a section perpendicular to x_1:

$$\left[\sigma_{11}(\text{front}) - \sigma_{11}(\text{back})\right] \times \text{Area} = \left[\left(\frac{\partial\sigma_{11}}{\partial x_1}\right) \times \delta x_1\right] \times (\delta x_2 \delta x_3).$$

Similarly, the net forces acting in the x_1 direction on the cubic element due to variations in σ_{21} and σ_{31} are the difference in the stresses at the right and left of the cubic element times the area of a section perpendicular to x_2 and the difference in the stresses at the top and bottom of the cubic element times the area of section perpendicular to x_3:

$$\left[\left(\frac{\partial\sigma_{21}}{\partial x_2}\right) \times \delta x_2\right] \times (\delta x_1 \delta x_3) \text{ and } \left[\frac{\partial\sigma_{31}}{\partial x_3} \times \delta x_3\right] \times (\delta x_1 \delta x_2).$$

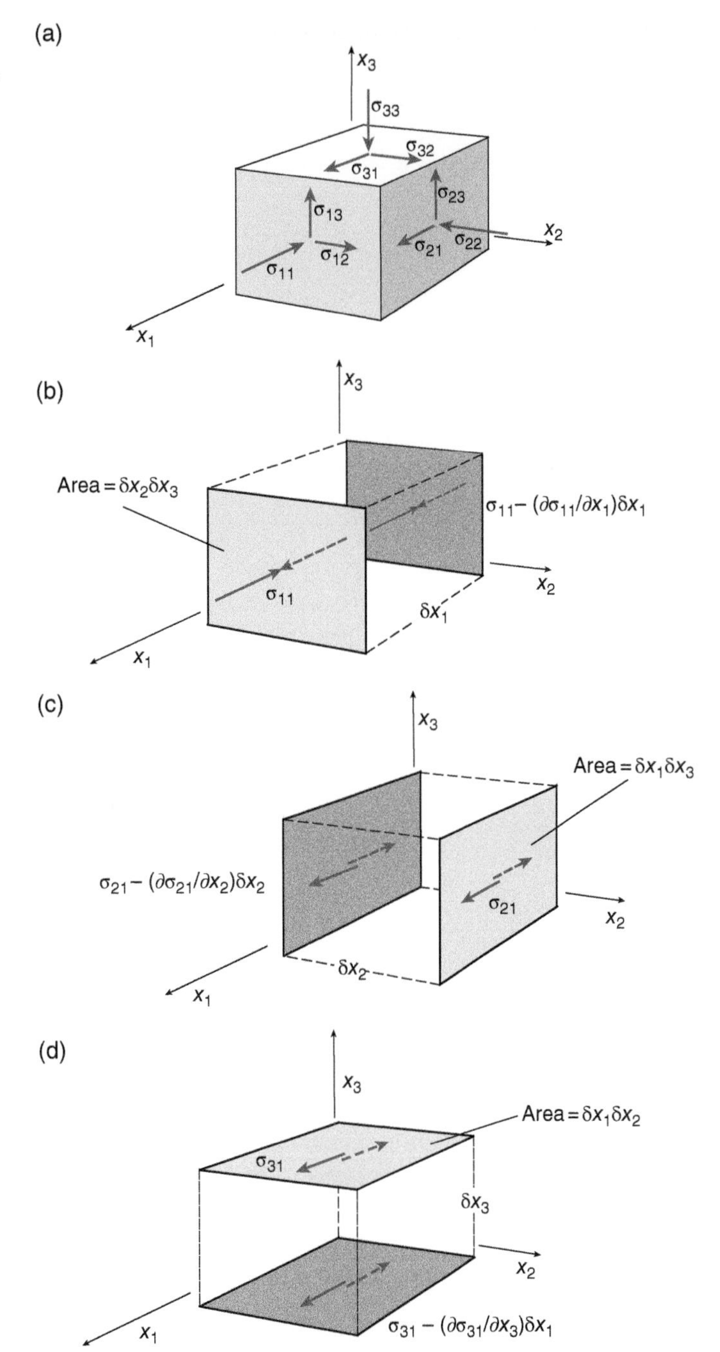

Figure 6.40 (a) Nine individual stress components acting on three mutually perpendicular planes define the state of stress in three dimensions. (b) Variations in three of the nine stress components, σ_{11} in (b), σ_{21} in (c), and σ_{31} in (d) can contribute to the balance of forces in the x_1 coordinate direction. See text for explanation.

The sum of these three contributions is the net force acting in the direction parallel to x_1, which must be zero if the element experiences no net acceleration:

$$\left[\left(\frac{\partial\sigma_{11}}{\partial x_1}\right)\times\delta x_1\right]\times(\delta x_2\delta x_3)+\left[\left(\frac{\partial\sigma_{21}}{\partial x_2}\right)\times\delta x_2\right]\times(\delta x_1\delta x_3)+\left[\left(\frac{\partial\sigma_{31}}{\partial x_3}\right)\times\delta x_3\right]\times(\delta x_1\delta x_2)=0$$

$$\left(\frac{\partial\sigma_{11}}{\partial x_1}+\frac{\partial\sigma_{21}}{\partial x_2}+\frac{\partial\sigma_{31}}{\partial x_3}\right)\times(\delta x_1\delta x_1\delta x_2)=0$$

$$\frac{\partial\sigma_{11}}{\partial x_1}+\frac{\partial\sigma_{21}}{\partial x_2}+\frac{\partial\sigma_{31}}{\partial x_3}=0 \tag{6.45a}$$

Conducting an analogous balance of forces in the x_2 coordinate direction (Figure 6.40c) yields

$$\frac{\partial\sigma_{12}}{\partial x_1}+\frac{\partial\sigma_{22}}{\partial x_2}+\frac{\partial\sigma_{32}}{\partial x_3}=0 \tag{6.45b}$$

The analysis has ignored the effects of body forces such as gravity on the element. Given the orientation of the coordinate axes in Figure 6.40 (with x_1 and x_2 horizontal), gravity exerts a force on the element in the x_3 direction only. The force due to gravity is, of course, the product of the mass of the element and the gravitational acceleration g. The mass of the element is the product of its volume ($\delta x_1\delta x_1\delta x_2$) and its density ρ. Undertaking the balance of forces due to variations in stress (Figure 6.40c) and the effect of gravity yields

$$\left[\left(\frac{\partial\sigma_{13}}{\partial x_1}\right)\times\delta x_1\right]\times(\delta x_2\delta x_3)+\left[\left(\frac{\partial\sigma_{23}}{\partial x_2}\right)\times\delta x_2\right]\times(\delta x_1\delta x_3)+\left[\left(\frac{\partial\sigma_{33}}{\partial x_3}\right)\times\delta x_3\right]\times(\delta x_1\delta x_2)+\rho g(\delta x_1\delta x_1\delta x_2)=0$$

$$\left(\frac{\partial\sigma_{13}}{\partial x_1}+\frac{\partial\sigma_{23}}{\partial x_2}+\frac{\partial\sigma_{33}}{\partial x_3}\right)\times(\delta x_1\delta x_1\delta x_2)+\rho g(\delta x_1\delta x_1\delta x_2)=0$$

$$\frac{\partial\sigma_{13}}{\partial x_1}+\frac{\partial\sigma_{23}}{\partial x_2}+\frac{\partial\sigma_{33}}{\partial x_3}+\rho g=0 \tag{6.45c}$$

Some geologists refer to the three Eqs. (6.45a), (6.45b), and (6.45c) collectively as the *equilibrium equations*. There is no reason a priori that any one or two of the coordinate axes x_1, x_2, and x_3 are horizontal. It is relatively easy to adjust Eq. (6.45) to accommodate coordinate axes inclined to the horizontal. With inclined axes, the force due to gravity might have a finite component parallel to each of the three coordinate axes, in which case the equilibrium equations become

$$\frac{\partial\sigma_{11}}{\partial x_1}+\frac{\partial\sigma_{21}}{\partial x_2}+\frac{\partial\sigma_{31}}{\partial x_3}+\rho g_1=0 \tag{6.46a}$$

$$\frac{\partial\sigma_{12}}{\partial x_1}+\frac{\partial\sigma_{22}}{\partial x_2}+\frac{\partial\sigma_{32}}{\partial x_3}+\rho g_2=0 \tag{6.46b}$$

$$\frac{\partial\sigma_{13}}{\partial x_1}+\frac{\partial\sigma_{23}}{\partial x_2}+\frac{\partial\sigma_{33}}{\partial x_3}+\rho g_3=0 \tag{6.46c}$$

where the ρg_i term refers to the component of the gravitational acceleration in the x_i coordinate direction.

As an example, consider again the situation with spatially varying stress depicted in Figure 6.21. The formulation for varying stress in Figure 6.21b, $\sigma_x = \sigma_{11} = cx_1 - ax_2$, $\sigma_y = \sigma_{22} = -x_2$, and $\tau_{xy} = \sigma_{12} = cx_1 = -\sigma_{21}$, ignored the effects of gravity. Thus, the two-dimensional form of Eq. (6.45) become

$$\frac{\partial\sigma_{11}}{\partial x_1}+\frac{\partial\sigma_{21}}{\partial x_2}=c-0+(-c)=0 \tag{6.47a}$$

$$\frac{\partial \sigma_{12}}{\partial x_1} + \frac{\partial \sigma_{22}}{\partial x_2} = c + (-c) = 0 \tag{6.47b}$$

The equilibrium equations enumerate Newton's second "law," and therefore they constitute a fundamental constraint that must pertain in any geological deformation – that the forces acting through a region of deforming rock are balanced. The equilibrium equations can be modified to include unbalanced forces if one can determine the acceleration experienced by a rock mass during deformation – that sort of modification is widely used in analyses of catastrophic fault slip events. By combining the equilibrium equations with the equations that enumerate the condition that no mass is created or destroyed – the *continuity equations* of Chapter 5 – one can begin to analyze or model deformation mathematically.

Appendix 6-I

As noted in Section 6A.2.1, throughout the rest of Section 6A, and throughout Section 6B, we have adhered to the common geological sign convention that compressional normal stresses are positive and tensional normal stresses are negative. This convention has been followed by a significant majority of geologists for some time. One obvious practical reason for this sign choice for normal stresses is that most stresses in Earth's crust are compressional, even in extensional tectonic settings such as rifts. Many geologists also follow the convention enumerated in Section 6A.2.1 that shear stresses are positive if they are counterclockwise and negative if they are clockwise. This convention for shear stresses is not as widely followed as the convention for normal stresses.

These two conventions, taking compressive normal stresses and counterclockwise shear stresses positive and tensile normal stresses and clockwise shear stresses negative, are collectively called the *geological Mohr circle sign convention*. We illustrated these conventions in Figure 6.12, utilized those conventions in Figures 6.13 to 6.20, and adhered to them throughout the chapter. As indicated above, some workers follow different conventions. In particular, engineers have exactly the opposite convention, the *engineering Mohr circle convention*, in which compressive normal stresses and counterclockwise shear stresses are negative and tensile normal stresses and clockwise shear stresses are positive. The two Mohr circle conventions are compared in the top three rows of Table 6A.1.

Not only do engineers conceive of tensional normal stress as negative and have different Mohr

Table 6A.1 Note on geological conventions (right hand column) and engineering conventions (left column) for the sign of stress components for Mohr circles (first row) and stress tensors (second row).

Compression positive Mohr Circle convention: the geological convention	**Tension positive Mohr Circle convention: the engineering convention**
Conventional in structural geology thinking and used in the majority of geology textbooks	Typical in the widespread and useful engineering literature
Most stresses in the crust are compressional	
With a North-East-Down coordinate framework, vertical (compressional) stresses due to gravity are positive	With an East-North-Up coordinate framework, vertical (compressional) stresses due to gravity are negative
Compression positive Tensor Convention	**Tension positive tensor convention**
The geological tensor convention is consistent with the geological Mohr circle convention (but it is a unusual in terms of conventional tensor practice, because it uses the inward pointing normal)	The sign of an outward pointing normal to a face is opposite to the direction of a compressional traction. The engineering tensor values of compressional stress are negative, consistent with the engineering Mohr circle convention and with general tensor conventions

circle sign conventions, they use different conventions to represent stress components in tensor form. A simple way to understand the differences between these two conventions is as follows: in a tensor representation of stress at equilibrium, shear stresses on adjacent sides of a cube (e.g., σ_{31} and σ_{13}) must have equal magnitudes but exert opposite torques, one clockwise and one anticlockwise (Figure 6A.1a). Because the stress tensor is symmetrical, these two tensor shear stress components must have the same algebraic sign, whereas in the Mohr circle sign convention, they have opposite signs. We therefore need to define a system for determining the signs of the tensor components of stress – a tensor sign convention. There are is a *geological tensor sign convention* and an *engineering tensor sign convention* (Twiss & Moores 1992, pp. 146–147).

For both tensor sign conventions, one must consider 1) the direction of the stress vector; and 2) the direction of the normal to the surface on which the stress vector acts. One then can determine the sign of a tensor component by multiplying the signs of the stress vector and the normal (i.e., both positive or negative for a positive tensor sign versus one positive and one negative for a negative tensor sign). In diagrams, the direction of the stress vector is determined by comparing the direction that the stress vector arrow points to the positive direction of the coordinate axis parallel to the stress vector (Figure 6I.1b). Although normals to surfaces are not usually shown with an arrowhead, we assign them a direction by comparing them to the coordinate axis that is perpendicular to the surface. Thus, in Figure 6A.1b, the outward directed segments of the normals drawn on two faces are assigned "+" signs because they "point" in the positive direction on the corresponding coordinate axis. Similarly, the inward directed segments

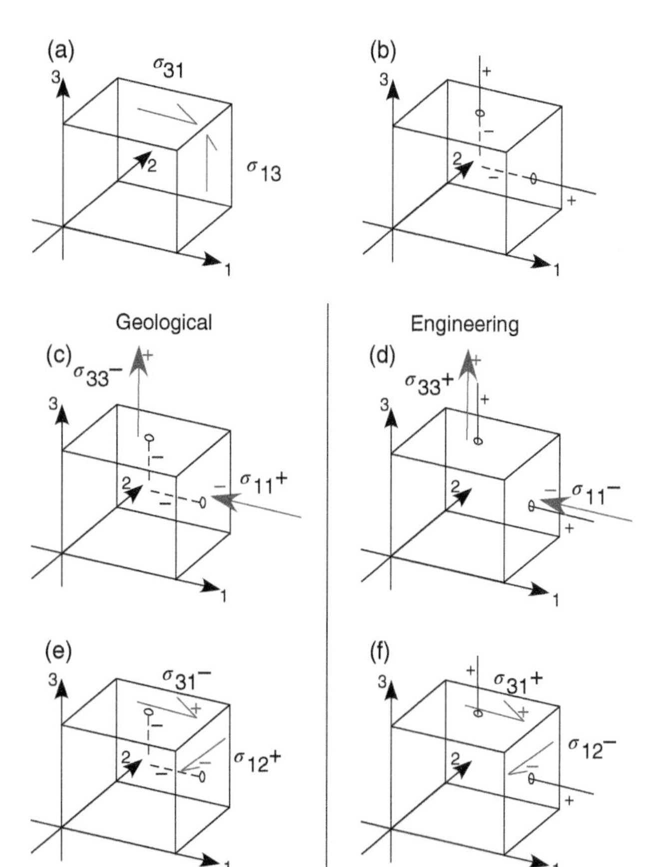

Figure 6A.1 Tensor component sign conventions. a) At equilibrium, shear stresses σ_{31} and σ_{13} must be equal but opposite in sense. b) Normals to faces of a cube have positive signs if the point outwards, and negative signs if they point inwards. Geological tensor component sign conventions in (c) and (e) are compared to engineering tensor sign conventions in (d) and (f). c) in the geological tensor sign convention for normal stresses, a tensional s_{33} on the top of the cube will have a negative sign because of the negative inward pointing normal. A compressional σ_{11} will have a positive sign because the direction of the stress is negative, but so is the inward pointing normal. d) The same principles are used for the engineering tensor sign conventions, but the outward pointing normals are used. e) Geological tensor sign conventions for shear stresses. σ_{31} is negative, because of the inward pointing normal; σ_{12} is positive, because both the stress and the inward pointing normal are negative. f) The same principles are applied to shear stresses in the engineering tensor sign convention, but the use of outward pointing normals leads to the opposite signs.

of the normals drawn on the two faces in Figure 6A.1b are assigned "–" signs because they "point" in the negative direction on the corresponding coordinate axis.

In the geological tensor sign convention, we compare the relative directions the stress vector arrow points to that of the *inward* pointing normal to the surface. For example in Figure 6A.1c, the sign of σ_{11} is positive, because the direction of the (compressional) stress is negative (red arrow pointing in the negative direction of the x_1 axis) and the direction of the inward pointing normal is also negative (dashed red end of the normal has – sign). Similarly, the sign of σ_{33} is negative, because the direction of the (tensile) stress is positive (red arrow pointing in the positive direction of the x_3 axis) and the direction of the inward pointing normal is also negative (dashed red end of the normal has – sign). The same convention can be applied to shear stresses. Figure 6A.1e shows that σ_{31} is negative because the stress vector points in the positive direction along x_1 axis and the inward pointing normal is negative. By similar reasoning, σ_{12} is positive because the stress vector points in the negative direction along x_2 axis and the inward pointing normal is negative.

The engineering tensor convention is similar to the geological sign convention, with the exception that one compares the direction of the *outward* pointing normal with the stress vector arrow. Figures 6A.1d and f on the right of the figure illustrate that by using the outward pointing normals, the signs of the engineering stress tensor component examples are the opposite of the signs of the geological stress tensor components. The outcome of both sign conventions can be summarized as follows:

- If the direction of the normal (inward for geology or outward for engineering) is the same as the direction of the stress vector, the sign of the tensor component is positive.
- If the two directions are opposite, the sign is negative.

The bottom row of Table 6A.1 compares the two tensor sign conventions. This comparison reveals that the engineering sign convention has one advantage over the geological one: the use of an outward pointing normal is standard in geometry. The geological sign convention is slightly unconventional in this regard. However, because of the geological tensor sign convention matches the well-accepted geological Mohr diagram convention, we have elected to use the geological sign conventions throughout the book. The existence of the two conventions have the potential to give rise to "an unending source of confusion" (Twiss & Moores 2007), but clarity can be achieved if the sign conventions used are stated.

References

Hafner, W. 1951. Stress distributions and faulting. *Geological Society of America Bulletin* **62**, 373–398.

Jaeger, J. C. 1969. *Elasticity, Fracture, and Flow with Engineering and Geological Applications.* Chapman & Hall, London.

Means, W. D. 1976. *Stress and Strain – Basic Concepts of Continuum Mechanics for Geologists.* New York. Springer-Verlag. 339pp.

Terzhagi, K., Von, (1923) Die Berechnungder Durchlässigkeitsziffer des Tones au dem Verlauf der hydrodynamishen Spannungserscheinungen. *Akademie der Wissenschaften in Wein, Mathematisch-naturwissenschaftliche Klasse, Part IIa* **132** (3/4), 125–138.

Twiss, R.J. & E. M. Moores. 1992. *Structural Geology*. New York: W. H. Freeman and Company.

Twiss, R.J. & E. M. Moores. 2007. *Structural Geology, 2nd Edition*. New. York: W.H. Freeman and Company.

7

Rheology

7.1 Overview

In order to understand more fully the movements that generate rock fabrics and rock structures or accommodate the motion of lithospheric plates, we need to relate those movements to forces and stresses. As outlined in Chapter 1, the term **rheology** was coined to refer to the study of the relations between displacements or velocities and forces or stresses. Thus, the field of rheology links together the contents of the three previous chapters; it is essential for a full understanding of tectonics.

Geologists who study the rheology of Earth materials commonly pursue two related lines of inquiry. One approach uses conceptual models of how materials behave, where each conceptual model is represented by a mathematical equation that either relates the orientations and magnitudes of stresses to the orientations and magnitudes of strains or relates the orientations and magnitudes of stresses to the orientations and magnitudes of strain rates. We call those equations **constitutive relationships** or **flow laws** for deforming materials. Table 7.1 presents a list of the symbols we use in flow laws and related equations in this chapter. These conceptual models and constitutive relationships are the subjects of this chapter. Using constitutive relationships, geologists can predict local strains or strain rates if they know the magnitudes and orientations of the stresses at different positions in a deforming rock mass. Geologists can use flow laws to infer the orientations and magnitudes of stresses at different positions in a deforming rock mass if they can measure the local strains or strain rates.

The second approach for geologists studying the rheology of Earth materials is to associate constitutive relationships or flow laws with the different physical processes that together constitute the mechanisms by which the rock masses deform. This approach uses laboratory analyses of deformation experiments and theoretical investigations of the structure and character of aggregates of grains as the basis for analyzing how rocks have deformed. This latter approach is the topic of the following chapter.

An Integrated Framework for Structural Geology: Kinematics, Dynamics, and Rheology of Deformed Rocks,
First Edition. Steven Wojtal, Tom Blenkinsop, and Basil Tikoff.

Table 7.1 List of symbols.

Elasticity			
E	Young's modulus	c_{ijkl}	Elastic stiffness constants or *stiffnesses*
ν	Poisson's ratio	s_{ijkl}	Elastic *compliances*
K	Bulk modulus	k	Spring constant
G	Shear modulus		
V_p, V_s	Velocity of seismic P waves & S waves, respectively	A, B, N, Q	Coefficients relating stress and strain in anisotropic elastic materials isochoric deformation
Fracture/fault formation			
C	Half-length of an elliptical crack	R_c	Radius of curvature at tip of an elliptical crack
T_0	Tensile strength of a material	P_f	Pore fluid pressure
σ_D	Fracture driving stress	$\sigma_n{}^*$	Effective normal stress or effective stress
$\mathbf{K}$	Stress intensity factor in fracture mechanics	$\mathbf{K_c}$	Critical stress intensity factor
Fault slip			
μ	Coefficient of friction	μ_s	Static coefficient of friction
μ_o	Steady-state coefficient of friction	D_c	Characteristic slip magnitude
μ_d	Dynamic coefficient of friction	$V(t)$	Slip velocity as a function of time
V_o	Slip velocity at which coefficient of friction $= \mu_o$	$\theta(t)$	State variable that measures changes in fault character
a	Coefficient defining the magnitude of rate-dependent effect $a \ln(V(t)/V_o)$	b	Coefficient defining magnitude of the state-dependent effect $b \ln(V_o\, \theta(t)/D_c)$
Viscous flow			
η	Material viscosity	μ_{FL}	Material fluidity
E_a	Activation energy in Arrhenius-type equation where $r = r_o exp\{-Ea/kT\}$	A, B, N, Q	Coefficients relating stress & strain rate in anisotropic viscous materials in isochoric flows
k	Boltzmann's constant	η_N, η_S	Viscosities for normal and shear components, respectively, in anisotropic viscous flows

7A Rheology: Conceptual Foundation

7A.1 Moving Beyond Equilibrium

In earlier chapters, we noted that particles in rock masses change their positions over time during deformation. We then used the notion of changing positions with time to distinguish between finite strains, incremental strains, and strain rates. When analyzing the forces that act on rock masses or the stresses prevailing within rock masses, however, we focused on **equilibrium situations**, where the stresses are balanced and there are no net accelerations. A world in which true equilibrium settings prevailed would be a static one – and that is not true of our world. In order to understand the rheology of Earth materials, we need to examine the manner in which rocks *do* respond to the stresses, or loads, applied to them. Two simple demonstrations using familiar materials can provide some insight into what can happen when stresses applied to a material exceed those the material can support at equilibrium.

Consider first a wooden pencil or a thin, dry twig. Applying a bending moment to the pencil or twig will cause it to bend (Figure 7.1a). If the magnitude of the bending moment is small, the bending will be slight. The pencil or twig will remain bent only as long as the bending moment is applied, and it will return to its original configuration when the moment is removed (Figure 7.1a, *ii.*). We say that this impermanent deformation is **recoverable** – it can be "undone" or "recovered" by removing the applied stresses. If the bending moment is too large, the pencil or twig will bend and then break (Figure 7.1a, iii.). Once the pencil or twig breaks, it can no longer support any bending moment. Notice that the transition from supporting a bending moment to failing to support a bending moment is, in this instance, abrupt or catastrophic.

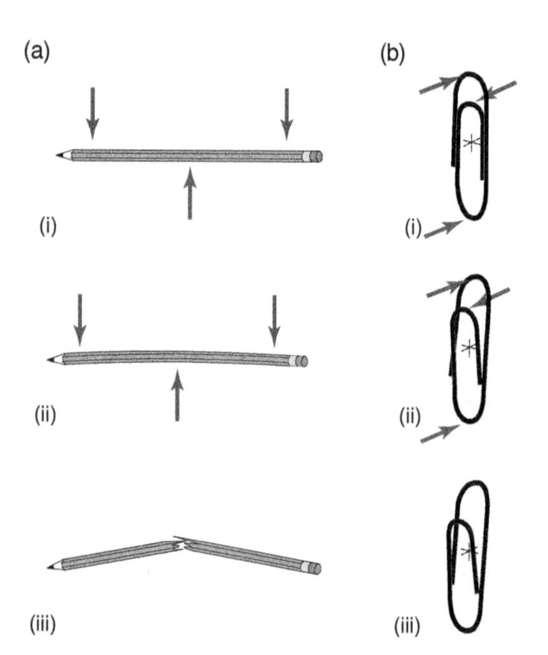

Figure 7.1 (a) A pencil subjected to a bending moment initially distorts elastically (*i* and *ii*). If the bending moment is too great, the pencil fails abruptly by fracture (*iii*). (b) A paper clip subjected to a bending moment initially distorts elastically (*i* and *ii*). If the bending moment is too great, the paper clip fails gradually by flow and remains permanently distorted (*iii*).

Next, consider a situation in which a bending moment is applied to a paper clip (Figure 7.1b). Once again, if the magnitude of the bending moment is small, the bending will be slight, the paper clip will again remain bent only as long as the bending moment is applied, and the paper clip will return to its original shape when the moment is removed. If the bending moment is too large, the paper clip will give way and distort. Distortion will continue as long as the bending moment is sufficiently large, but if its magnitude is reduced enough the paper clip will support it without further distortion. Further, the paper clip will remain bent even when the bending moment is removed (Figure 7.1b, *iii.*). Thus, the paper clip is permanently distorted but has not lost its ability to support a load. A material scientist would say that the paper clip yielded to the applied stresses, although its response in this instance was gradual. The concepts of recoverable versus permanent deformation shown by Figure 7.1 are very important for rock deformation, as are the differences between fracture and flow (Figure 7.1a, *iii*, b, *iii.*).

7A.1.1 Conducting and Interpreting Deformation Experiments

Geologists investigating the deformation of minerals and rocks in the laboratory have shown that under appropriate conditions minerals and rocks exhibit behaviors that are broadly analogous to those described earlier. One apparatus commonly used to study the deformation of minerals and rocks has a cylindrical sample sitting on a stationary platform, held inside a pressure jacket, and with a piston at one end (Figure 7.2a). At the beginning of an experiment, the platform and the piston will exert the same stress on the ends of the

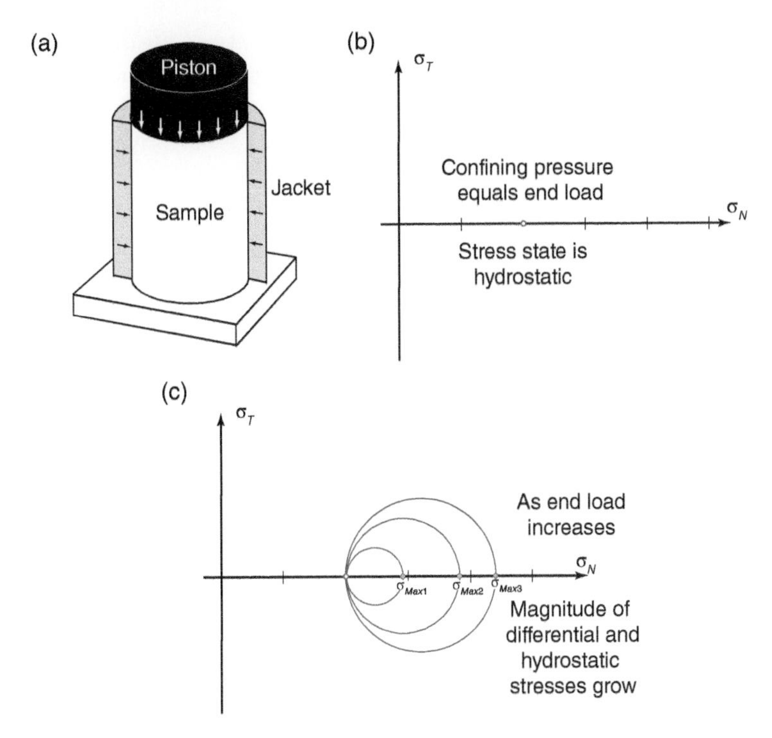

Figure 7.2 (a) A common experimental rock deformation configuration: a cylindrical sample subjected to confining pressure and end loads. (b) Mohr space representation of stresses on the sample prior to shortening. (c) Mohr space plots showing stress states as end loads increase, shown by three increasing values of σ_1. Note that both differential stress and mean stress increase as the end load is raised.

cylinder as the pressure jacket exerts on the sides of the cylinder. Thus, stresses in the sample are hydrostatic. As the motor drives the piston toward the platform, the magnitude of the compressive stresses across the ends of the cylinder increase, and the stress state becomes non-hydrostatic.

Figures 7.2b and c use Mohr circles to depict the changes in loading during the experiment. Recall that Mohr space is a graphical space where we can represent the state of stress diagrammatically. The abscissa ("*x* axis") of Mohr space gives the magnitude of normal stress components, and the ordinate ("*y* axis") gives the magnitude of tangential or shear components. In Mohr space, each point denotes the normal stress and tangential stress components that act on a particular plane. For any stress state, there is a circle whose points collectively define the normal and shear stress components on planes with any different orientation for a particular stress state.

In the case of hydrostatic stresses, where the principal stresses have equal magnitudes, the Mohr circle devolves to a point (Figure 7.2b). The three Mohr circles in Figure 7.2c show the impact of increasing the end load on the cylinder. Note that both the mean stress (the location of the center of each successive Mohr circle) and the differential stress (the magnitude of the diameter of each successive Mohr circle) vary as the end load increases. Increasing the difference between the magnitude of the compressive stress acting on the ends of the cylinder and the stresses acting on the sidewalls of the cylinder will cause the sample to shorten, with greater end loads leading to greater shortening.

Geologists regularly use another kind of diagram, called a **stress-strain plot**, to depict the stress and strain magnitudes at different stages of a deformation experiment; the succession of stress and strain values define a **stress-strain curve** (Figures 7.3 and 7.4). We typically plot the magnitudes of the normal stresses across the sample ends on the vertical axis and the corresponding

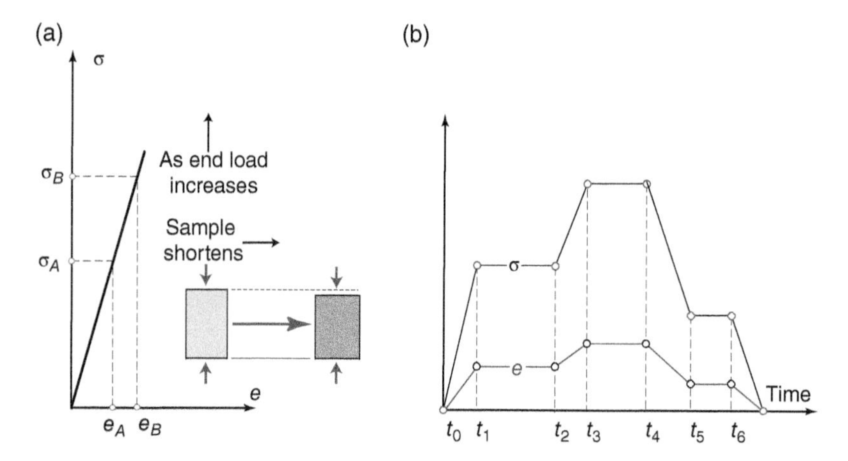

Figure 7.3 (a) A stress (σ)-strain (*e*) plot for elastic deformation. (b) A plot showing stress and strain versus time for the deformation history described in the text.

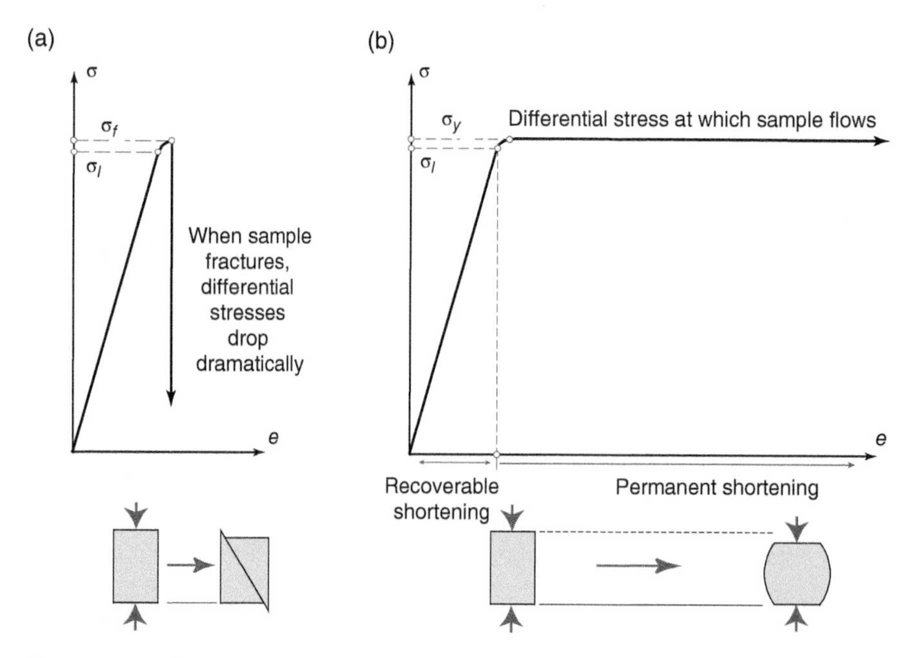

Figure 7.4 (a) Stress-strain plot for a material that behaves elastically unit it fractures. σ_l – limiting stress at which deformation of the sample is recoverable (elastic limit), σ_f – stress at failure. (b) Stress-strain plot for material that behaves elastically and then yields by flow; σ_y – stress at failure.

shortening strain magnitudes on the horizontal axis. This may seem contrary to expectations since, according to the earlier description, you might think of strain as a response to applied stress, where the differential stress is the independent variable and strain is the dependent variable. In experiments it is the displacement of the piston that results in increased differential stress *and* strain within the sample; neither is an independent variable. A similar situation pertains to natural deformations, where it is helpful to think of the displacement of a mass of rock (a lithospheric plate or the convecting mantle) relative to its neighbors (a neighboring lithospheric plate or the overlying lithosphere) leading to increased differential stresses and strains within both rock masses.

Stress-strain plots help us visualize what has occurred during deformation experiments. In addition, the plots provide important constraints for discerning the mathematical formulae, that is the constitutive equations, which relate stress and strain. As we will discuss later in the chapter, in many experimental and natural settings, there is another parameter critical to how deformation proceeds – time. In those cases, stress-strain curves cannot characterize fully the deformation. Instead, plotting differential stress versus the strain rate to construct a **stress-strain rate plot** provides important insight into the deformation and constrains a different, more appropriate, constitutive relationship between stress and strain rate.

7A.1.2 Recoverable Deformation versus Material Failure

In deformation experiments conducted across a broad range of boundary conditions, the initial shortening of rock samples is not permanent. Figure 7.3a shows a stress-strain curve for a deformation experiment focused on this stage of deformation. As illustrated in Figure 7.3, an end load magnitude of σ_A leads to a shortening magnitude of e_A. If the end load is held at σ_A for moderate lengths of time (from time t_1 to time t_2 on Figure 7.3b), the shortening magnitude remains at e_A. Increasing the end load to σ_B leads to greater shortening, that is a greater magnitude of e_B. If the end load is held at σ_B for moderate lengths of time (from time t_3 to time t_4 on Figure 7.3b), the shortening magnitude remains at e_B. The magnitude of shortening reverts to a lower value if the end load is reduced (for example from time t_3 to time t_4 on Figure 7.3b). Further, the sample regains its original length if the piston is backed out and the compressive stress across the end of the cylinder is reduced so that the differential stress is zero. As you might imagine, this sort of material behavior is necessarily "nondestructive" – reducing the differential stress magnitude to zero brings the rock sample to its original, unstrained configuration. For this recoverable deformation, the stress-strain curve is identical for loading and unloading. There is, then, no unique correlation between any stress and strain values on the curve in Figure 7.3a and at any point during the experiment. This stress-strain curve does not provide any insight into the progression or history of the deformation.

If the persistent movement of the piston raises the magnitude of the normal stress across the ends of the sample to a value σ_l, where the subscript l indicates that it is the limiting stress at which deformation of the sample is recoverable, any further increase in end load will lead to permanent deformation (Figure 7.4). In engineering applications, permanent deformation of a component indicates the failure of a design because a distorted component in a structure alters the overall shape and utility of the design. Engineers will therefore say that a component has **failed** if it becomes permanently distorted. J. C. Jaeger, in his classic book *Elasticity, Fracture, and Flow* (1971; Metheun and Company), uses the term **failure** in s similar way, that is to refer to situations where rock is permanently deformed in some way. It is worth noting that some geologists use the term "failure" only when the onset of permanent deformation causes an abrupt drop in stresses, as one sees in some rock deformation experiments. In this book, we use failure to refer to the onset of permanent deformation of rock, whether or not there is an associated drop in stresses (note also that the permanent deformation *succeeds* in producing the rock structures and microstructures that we geologists seek to study). Depending on conditions, failure can be abrupt (in some cases catastrophic) or gradual (see Figure 7.1).

Abrupt failure is manifest as a dramatic decrease in a sample's ability to support its end load. This abrupt drop in stress typically is associated with the formation of one or several *extensional fractures* or *faults* in a sample, depending on the relative magnitude of the confining pressure. **Extension fractures**, which form in samples deformed with little confining pressure, are discrete discontinuities across which a rock has lost cohesion and where the rock masses on opposite sides of the surface have moved relative to each other perpendicular to the surface, opening the fracture. **Faults** are thin zones of localized shearing that eventually form a discrete surface across which a rock loses cohesion and where the rock masses on opposite sides of the surface have moved parallel to the surface, past each other. They form in samples deformed at moderate confining pressures. Figure 7.4a illustrates a situation in which a sample fails abruptly and in which extension fracture or fault formation occurs at an end load magnitude σ_f, which is only slightly larger than σ_l. In such a situation, the sample would have accrued a negligible amount of permanent, or non-recoverable, deformation prior to the formation of the extensional fractures or faults, which are, of course, permanent features. In Figure 7.4a, the formation of extensional fractures or faults is accompanied by an abrupt, dramatic drop in the sample's ability to support an end load. Abrupt failure with little to no precursory deformation predominates in samples shortened at low temperatures and, as noted earlier, when the pressure jacket exerts low to moderate confining pressures.

In gradual failure, samples **yield** or begin to deform permanently without a dramatic drop in stress. Yielding may be accommodated by permanent deformation that is localized in one or more portions of a sample, or deformation may be distributed throughout much or all of the sample. In either case, we say the sample **flows**. Figure 7.4b illustrates a situation in which a sample yields (deforms permanently) at an end load slightly above σ_l. In this case, however, the sample accrues significant permanent deformation and still supports its end load. In experiments like that depicted in Figure 7.4b, sample deformation would cease if the magnitude of the end load dropped below σ_y. As long as the compressive stresses across the sample ends equal σ_y, the sample will continue to deform (Figure 7.4b). This sort of gradual failure without a dramatic stress drop predominates in samples shortened at higher temperatures and/or when the pressure jacket exerts high confining pressure.

Geologists often correlate particular modes of failure to specific structures and expected laboratory behavior (Figure 7.5). It is important, however, to recognize that failure by the formation of an extension fracture or fault with negligible precursory deformation and the failure by localized or distributed flow at constant stress σ_y are end-member behaviors for rock failure. By changing the conditions under which samples of a particular rock type deform, or by changing the rock type of the samples tested under particular conditions, experimentalists see a range of failure types in addition to either of these two end members. In some cases, gradual failure follows a course in which each successive increment of permanent deformation requires a greater magnitude of normal stress across the sample ends (Figure 7.6a). In such an instance, the sample gets stronger as a result of each successive deformation increment

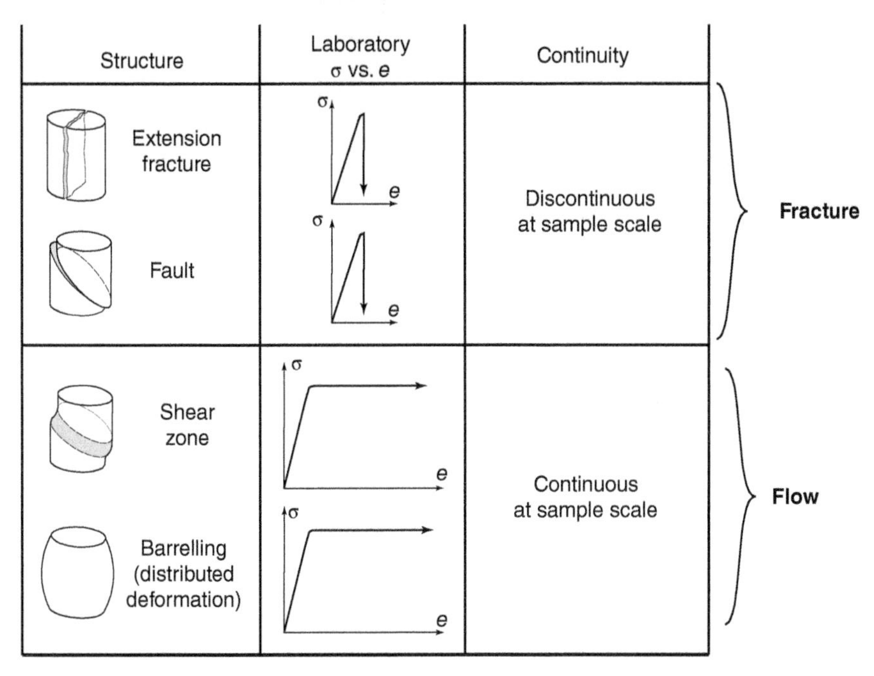

Figure 7.5 Chart illustrating fracture and flow as two end-members of a continuum of modes of failure. At one end member, samples experience negligible permanent deformation prior to the formation of a fault or extension fracture. At the other end member, samples flow at constant differential stress forming shear zones or distributed deformation.

and is able to support a greater magnitude of differential stress or exert greater resistance to deformation. Because the deformed sample resists deformation more effectively, material scientists say that it "work strengthens," "work hardens," or "strain hardens" during deformation. As shown in Figure 7.6a, an end load of σ_A is required for the sample to be shortened to the strain magnitude e_{Ax} (the subscript "x" is added to indicate the strain magnitude during the experiment). Should the experimenter decide to reduce the magnitude of the normal stress from σ_A to zero, the sample would follow an unloading path (dashed line) in which a recoverable portion of the shortening is relieved, meaning that the net shortening of the sample would be e_{Af} at the end of the experiment (the subscript "f" is added to indicate the final strain magnitude).

The opposite effect can occur, such that each successive increment of the deformation requires lower differential stress than the previous deformation increment. In such an instance, the material becomes weaker as a result of each deformation increment, and can only support a lower magnitude of differential stress. Material scientists say that the sample has experienced "strain weakening" or "strain softening" (Figure 7.6b).

Under appropriate conditions, a sample can undergo recoverable shortening, then undergo permanent deformation with associated strain hardening as illustrated in Figure 7.6c, with the eventual result that an extensional fracture or slip

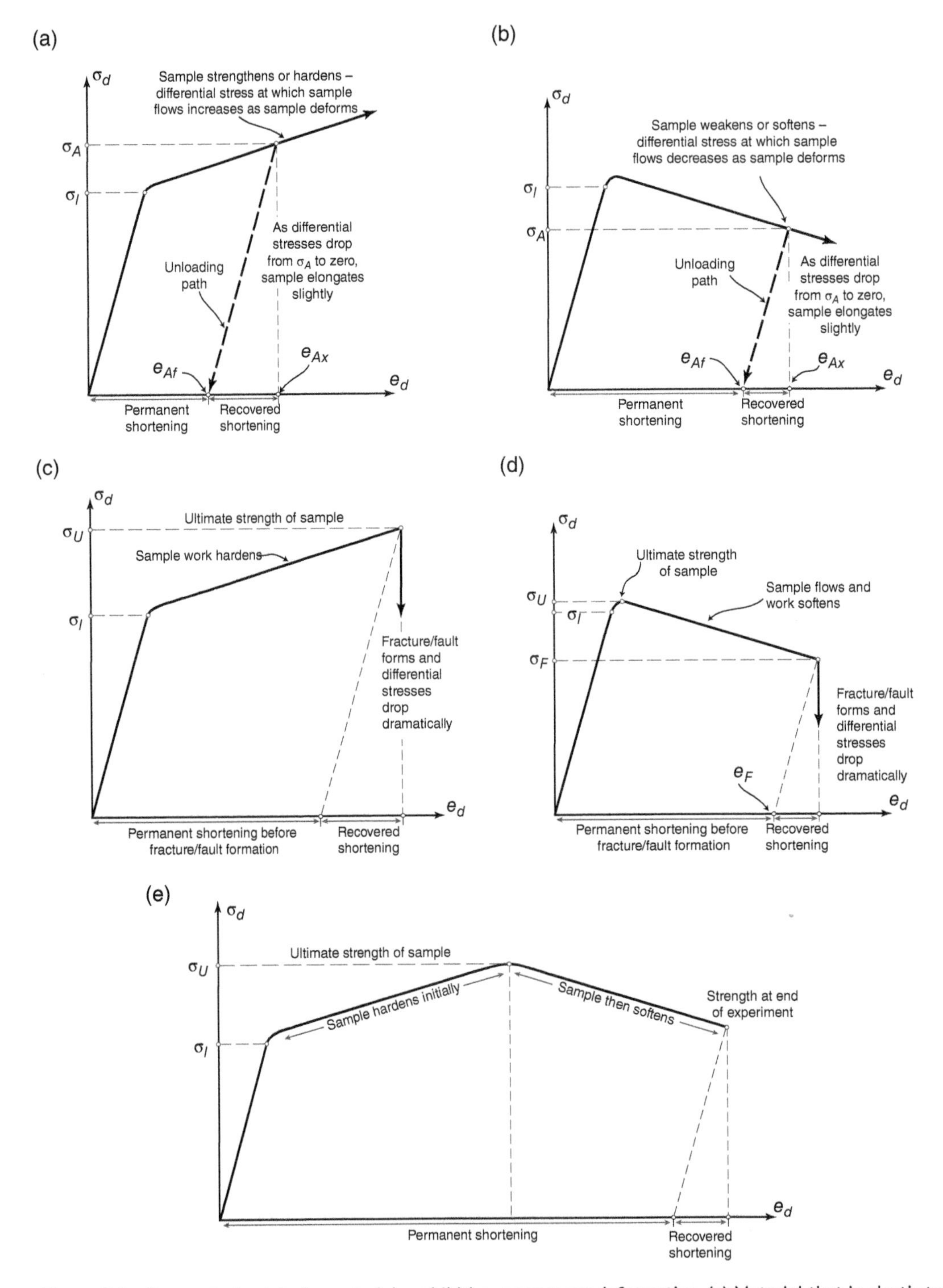

Figure 7.6 Stress-strain plots for materials exhibiting permanent deformation. (a) Material that is elastic to yield and then work hardens. σ_A, e_{Ax} – Stress and strain when the experiment is stopped, e_{Af} – Final permanent strain. (b) Material that is elastic to yield and then work softens. (c) Material that is elastic to yield, work hardens, and eventually fractures. (d) Material that is elastic to yield, work softens, and eventually fractures. σ_u – Ultimate strength, σ_F – stress at fracture. (e) Material that is elastic to yield, work hardens and eventually work softens.

surface develops in the sample. The formation of the fracture or fault in the case illustrated in Figure 7.6c causes the applied stresses to drop precipitously. As the stresses drop, the portions of the sample separated by the fracture or fault will unload in such a way that each gives up a recoverable portion of shortening. Still, the portions of the sample separated by the fracture or fault will be measurably, permanently deformed. Figure 7.6d illustrates a situation in which a sample begins to deform by flow, experiences strain softening as deformation accrues, and ultimately forms a fracture or fault, at which point the sample can no longer support the applied stresses.

In situations like those illustrated in Figure 7.6e, a sample could shorten by flow, accompanied by strain hardening or work strengthening, to a particular amount of shortening at a particular magnitude of applied stress. Instead of developing a fracture or fault at that point, the sample could continue to flow but under conditions where it then work softens. In the particular instance illustrated, the sample never develops a fracture or fault before the end of the experiment. Under some circumstances (not illustrated here), samples may first strain harden, then strain soften, and eventually develop fractures or faults.

7A.1.3 Moving from Deformation Experiments to Mathematical Relations

In translating the behavior of materials in laboratory experiments into conceptual models of deformation, and in conceiving the corresponding mathematical formulae (constitutive equations or flow laws) that describe the deformation models, we make the following assumptions: (1) that rocks constitute a *continuum* – they fill space completely and there are no "holes" in the material; (2) that rocks are *uniform* – they behave the same way at all locations; and (3) that rocks are *isotropic* – they respond in the same way regardless of the orientation of the loads.

The assumption of continuity means that we assume there is "material" at all points in the area of interest. Although rocks often have open pores or voids, we will assume at the outset that all of a deforming volume is filled by rock. Because most rocks are composed of different minerals with different physical properties, you might think that deforming volumes would never conform to the assumption of uniformity. In most of this chapter, we follow the thinking of many geologists studying rock deformation and focus on volumes where rock properties are uniform. This might mean focusing on a sufficiently small area that the area of interest is a single mineral grain. Alternatively, it might mean focusing on a sufficiently large area, composed of a myriad of mineral grains, that any two portions of the area are statistically equivalent. The assumption of isotropy implies that the instantaneous direction of maximum *shortening* in a material (referring to its response, that is strains) is parallel to the direction of the *most compressive stress* (referring to the applied loads, that is stresses). Likewise, the instantaneous direction of maximum *elongation* is parallel to the direction of the *most tensile stress* (the least compressive stress in geological situations). This parallelism between the principal directions of the applied stresses and the observed instantaneous strains often does not occur in natural settings. Most minerals are *anisotropic*, meaning that they exhibit differing resistances to deformation along different directions through the material. Individual quartz crystals, for example, exhibit greater resistance to deformation (are stronger) when compressed parallel to their *c* axes than if they are compressed along lines oblique to or perpendicular to their *c* axes. Since rocks are composed of aggregates of numerous small crystals that have a wide range of orientations, geologists studying deformation regularly assume that their average (bulk) response is essentially the same along any direction through the rock. A preferred orientation of crystals will render this assumption invalid.

Even if we assume that rocks are isotropic, there is another aspect of deformation that we must keep in mind. If the strain path is non-coaxial, the instantaneous directions of shortening and elongation (which we presume to indicate the directions of the most compressive and least compressive stresses, respectively) are not parallel to the directions of maximum finite shortening and maximum finite elongation. Stated in another way, the principal directions of the finite strains need not be parallel to the principal directions of the stresses, even though the incremental directions of shortening and elongation are parallel to the directions of most compressive and least compressive stress respectively. This is just one reason to differentiate between stresses, displacements and strains within materials.

In describing experimental data on recoverable deformation, we noted that neither the magnitude of the stress nor the magnitude of the strain were necessarily functions of time and that the stress-strain curve could not be related to the deformation history in a unique way. Once experimental samples fail, however, physical changes to the samples that occur during one stage of an experiment have an effect on the sample's behavior later in the experiment. Thus, stress-strain curves reflect the deformation history. Carefully conceived and constrained deformation experiments that lead to sample failure provide significant insights into deformation in nature. For all the insights into deformation geologists have gained from experimental or laboratory deformation, it is nevertheless important to remain cognizant of some important differences between rock deformation in the laboratory and in nature.

- Experimental rock samples typically have dimensions on the order of cm, and the materials are chosen for their uniformity. Both the size and uniformity mean that samples rarely include heterogeneities regularly found in nature, such as cm to dm scale features like fossils, bedding planes, joints, or some fabrics.
- Extension fractures or faults formed in laboratory experiments often extend completely across a sample, leading to a dramatic drop in the sample's ability to support the applied stresses. It is not clear that the formation of extension fractures or slip surfaces in nature, where these features may not extend across a rock mass, always leads to a comparable drop in stresses.
- Rock mechanics experimental designs, such as the axially symmetric loading shown in Figure 7.2a, often have two principal stresses with equal magnitude (the normal stresses exerted by the jacket). This sort of loading may be uncommon in nature, and it leads to constitutive relationships that overlook the effects of the magnitude of the intermediate principal stress. Experimentalists increasingly use more sophisticated apparatuses that provide more realistic loading configurations, but much of our understanding of rock rheology is derived from data collected using axially symmetric experiments.
- Deformation experiments are typically conducted at a fixed deformation rate in order to constrain the constitutive relationship for a particular deformation process or mechanism. In nature, deformation rates typically vary with time, and individual deformation processes or mechanisms may "switch on" and "switch off" as deformation accrues.
- Deformation experiments are more likely to last days than to last months. Experiments on seismic slip can mimic the temporal character of the natural process, but the flow of rock in nature may accrue over decades, centuries, or millennia. The rates of deformation in experiments can be many orders of magnitude faster than the deformation rates in many natural settings. These higher deformation rates may require differential stresses that are higher than those in natural settings, ambient temperatures that are higher than those in natural settings, or

both higher differential stresses and higher temperatures.

Earth scientists increasingly draw upon experimental deformation of artificial and natural two (or more) phase aggregates, experiments conducted at varying deformation rates, analog modeling, and computational modeling to augment inferences from traditional deformation experiments and to explore the rheology of rocks. The main thrust of this introductory section is to introduce different model rock rheologies and their associated constitutive equations.

7A.2 Models of Rock Deformation

7A.2.1 Elastic Behavior

The idealized rock deformation experiments depicted in Figures 7.2, 7.4, and 7.6 all exhibit similar behavior at *low strain values*:

- Samples deform as soon as stresses are applied. Thus, sample deformation is *instantaneous*. It is not apparent in the stress-strain curves, but at low strains, the rock samples respond instantaneously to both hydrostatic or differential stresses.
- There is a unique relationship between stress and strain – each strain magnitude is associated with a particular magnitude of applied stress.
- Samples remain deformed as long as the stresses are applied.
- If the stresses are removed, samples return to their original configurations as soon as the stresses are removed. The deformation is recoverable, and recovery is also instantaneous.

We use the word **elastic** to describe this recoverable and time-independent deformation, and we say that samples exhibit **elasticity**. To visualize elastic behavior, think of a coiled spring of a certain length. As soon as one pulls it, it stretches to a new length. The spring remains stretched as long as one is willing to maintain the pull upon it, but it returns to its original length as soon as one stops pulling on it. Similarly, a coiled spring will shorten immediately when compressed and return to its original length when the compressive load is removed.

- For many elastic materials, doubling the stress magnitude produces twice the magnitude of elastic strain, and halving the magnitude of the stress yields half the magnitude of elastic strain. Materials conforming to a *linear* relationship between applied stress and the resulting strain exhibit **linear elasticity**.

Rubber bands (elastics) are familiar objects that readily exhibit elastic behavior. It is easy to demonstrate, however, that there is no linear relationship between the pull and the elongation of rubber bands. Rubber does not exhibit linear elasticity. The elasticity of rubber is not restricted to small strains; rubber bands can experience sizable elongations and return to their original shape. Still, it is possible to stretch a rubber band to the point where it snaps. If one stretches the rubber band just a bit less than the stretch required to break it, segments of the rubber band will be permanently thinned and may exhibit a surface texture different from the band's original texture: the rubber band exhibits permanent distortion prior to breaking.

The strain magnitude below which strain is recoverable and above which strain is permanent is known as the elastic limit e_l. In Figures 7.4a, 7.4b, and 7.6, the elastic limit is reached when the normal stress applied across the sample ends is σ_l. In those figures, as long as strain magnitudes are below e_l and normal stress magnitudes are below σ_l, samples behave elastically.

We can gain some insight into elastic behavior by considering what occurs microscopically during elastic deformation (Figure 7.7). Solids are composed of innumerable atoms or ions held together by chemical bonds. Most rock-forming minerals have combinations of ionic bonds (such as those between alkali cations and oxygen), bonds with mixed ionic-covalent character (such

(a)

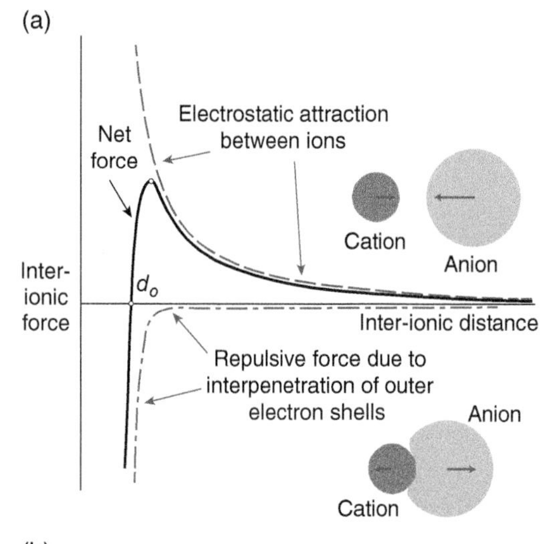

(b)

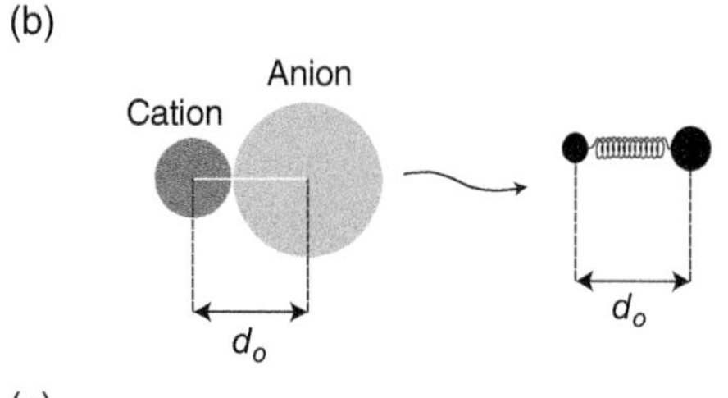

(c)

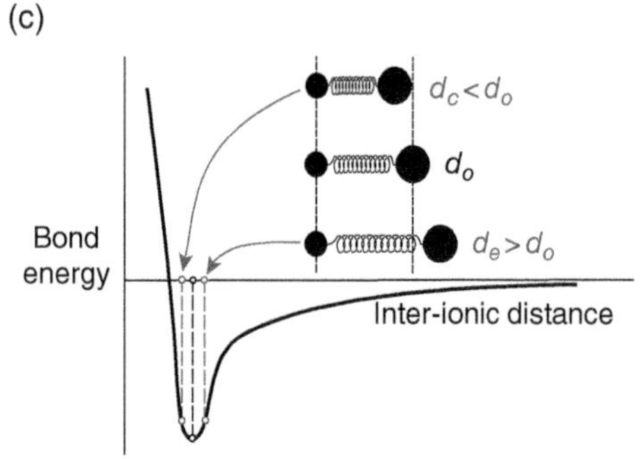

Figure 7.7 (a) Plot showing attractive, repulsive, and net forces in an ionic bond as a function of the interion distance. d_o – Equilibrium bond length. (b) The bonded ions envisioned as two spheres connected by a spring. (c) Plot of bond energy versus interionic distance. Energy minimum defines equilibrium spacing for bonded ions.

as those between silicon or aluminum and oxygen), and covalent bonds (such as those between carbon or sulfur and oxygen). Regardless of bond type, there is a preferred distance d_o between the centers of the atoms or ions bound together at which the forces that attract and repel two atoms or ions are balanced and there is no net force operating on the atom or ion. The bonds draw the atoms or ions together with a force whose magnitude decreases nonlinearly with increasing distance. In all types of bonds, if atoms or ions are forced too close to each other, repulsive forces generated by resistance to the interpenetration of outer electron shells (resulting from the so-called Pauli exclusion principle) force them apart. Figure 7.7a shows the attractive and repulsive forces for an idealized ionic bond. Note that the attractive and repulsive forces have equal magnitudes, and of course opposing directions, at the distance d_o that defines an equilibrium bond length. The two atoms or ions bound together exhibit behavior akin to two objects connected by spring with length d_o (Figure 7.7b).

In this idealized, isolated bond, increasing the interionic distance requires a force to pull the atoms apart. When atoms (or ions) are separated by a distance d_e greater than the equilibrium bond length, the bond has slightly higher energy (Figure 7.7c), and the two atoms experience a net attractive force that tends to return them to their equilibrium spacing. Similarly, decreasing the interionic distance requires a force to push the atoms together. When atoms or ions are separated by a distance d_c less than the equilibrium bond length, the bond again has slightly higher energy (Figure 7.7c), and the two atoms experience a net repulsive force. As a result, any perturbation in the position of atoms or ions is corrected. In real materials, atoms experience thermal vibrations, and the actual distance between the centers of atoms involved in bonds is thought to vary over time. Averaging over time, the atoms are found at an equilibrium bond length d_o from their nearest neighbors. In a similar fashion, the several bonds between one atom or ion and its neighbors take equilibrium positions, so that the bonds assume equilibrium bond angles.

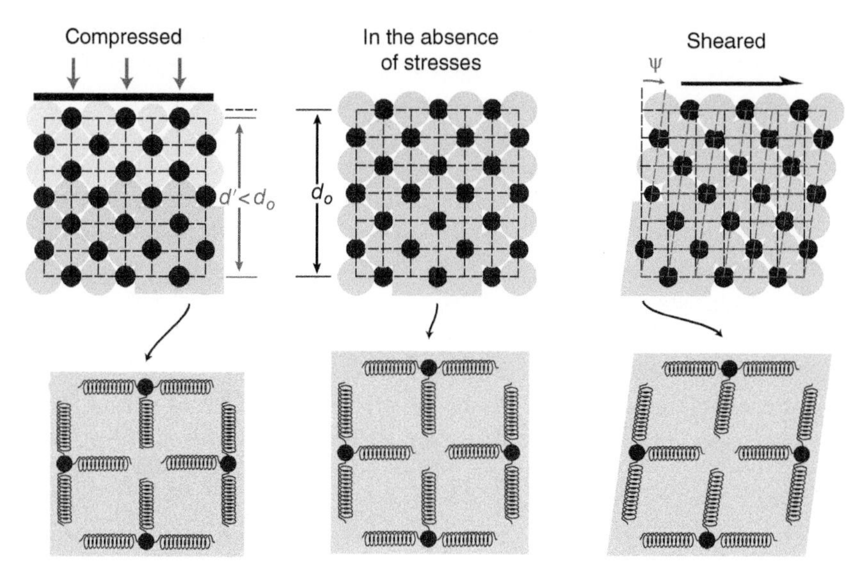

Figure 7.8 In an ionically bonded crystal, ions have characteristic spacing defined by energy minima for bonds. When subjected to compressive or shear stresses, bonds shorten, elongate, or shear (ψ = angular shear) as long as stresses are applied.

During elastic deformation, the applied loads shorten or elongate equilibrium bond lengths or change equilibrium bond angles, but the bonds *do not break* (Figure 7.8). Thus, on removing the applied loads, the bonds return to their equilibrium configuration. In the silicates and oxides that compose most rocks, the bonds between atoms are strong. As a result, the stresses needed to stretch or shorten bond lengths or to change bond angles are very large, typically on the order of hundreds of MPa. Further, in the silicates and oxides that compose most rocks, elastic deformation is normally restricted to extremely small strains. In silicates, elastic deformation typically is limited to stretches between 0.99 and 1.01 or elongations less than $\pm 1\%$.

Figure 7.7c indicates that holding an individual atom at a position other than its equilibrium spacing from its neighbors requires adding energy to the atom or ion. The product of the force applied to the atom times the resulting displacement of the atom gives the energy required to hold it at a distance other than d_o from its neighbors. Because elastic deformation entails the application of forces to displace atoms from their equilibrium positions, elastic deformation adds energy to the deformed material. Note that the product

$$\sigma \times e = \frac{\text{Force}}{\text{Unit Area}} \times \frac{\text{Displacement}}{\text{Unit Length}} = \frac{\text{Force} \times \text{Displacement}}{\text{Unit Area} \times \text{Unit Length}} = \frac{\text{Energy}}{\text{Unit Volume}}. \tag{7.1}$$

An elastically deformed substance stores an amount of energy per unit volume proportional to the area under the elastic stress-strain curve, and the energy added to the material will be released if atoms or ions return to their equilibrium positions. That is to say that the elastic strain energy is released when the strains are *recovered*.

Mechanical clocks and kitchen timers have a coiled mainspring that must be wound to run the clock or timer. The process of winding the mainspring distorts the spring elastically, thereby adding energy to it. That stored energy is released slowly by the mechanism in the clock or timer and enables the mechanism to run. In Earth, rock masses adjacent to locked faults are distorted elastically, and that stored elastic energy is released when the fault slips. Rapid slip can lead to an abrupt release of stored elastic strain energy. The rapidly released energy will propagate away from the slipped portion of a fault as seismic waves, with large magnitudes of slip over large portions of faults leading to large releases of stored elastic strain energy that travel as large-amplitude seismic waves.

7A.2.1.1 Constitutive Equations for Elastic Deformation

Because plots of elastic strain versus the applied stress for most geological materials are *linear*, that is plot as straight lines, the equations describing stress-strain "curves" for elastic behavior have a familiar format: $y = m\,x + b$, where σ is the ordinate (y axis), e is the abscissa (x axis), and the intercept $b = 0$:

$$\sigma = C \cdot e. \tag{7.2}$$

The slope of the elastic stress-strain curve is the *elastic constant* C, which will have a fixed value for any particular material. Note that in order for this equation to hold, C must have units of stress (Pascals = Pa = N/m^2 = $kg/m\,s^2$). Because elastic deformation is recoverable, elastic stress-strain curves hold for both loading and unloading.

The precise value of a material's elastic constant depends upon the configuration of the experiment in which it is measured. The following sections outline two different schemes that geologists use to describe the elastic behavior of rocks. As you will see later, it takes *two independent* elastic constants to describe completely the elastic distortion of an isotropic material. For anisotropic materials, an array of values, known as *stiffnesses*, are required to relate stresses to the strains in a manner analogous to Eq. (7.2).

7A.2.1.2 Elastic Constants Derived from Uniaxial Tests

When a sample is tested in a uniaxial loading apparatus, the resulting stress-strain curve compares longitudinal stresses with longitudinal strains. The slope of that elastic stress-strain curve is denoted as E = **Young's modulus** (where the word *modulus* is derived from Latin for *measure*).

$$\sigma_{\text{long}} = E \cdot e_{\text{long}}. \tag{7.3}$$

Given the strengths of bonds in silicates and oxides, for most minerals, E is 10–300 GPa. For a typical granite, $E \approx$ 50 GPa, and for a typical diabase, $E \approx$ 100 GPa.

The uniaxial test setting, where one measures the elongation of a material in response to the stress applied to the end of a sample, appears to be a simple setting that isolates and compares the normal stress acting in one direction and the longitudinal strain in the same direction. In reality, axial testing is a complex loading situation in which lateral expansion (or contraction) counteracts longitudinal shortening (or elongation). Young's modulus only measures a part of the sample response.

A sample's tendency to resist lateral expansion contributes to its resistance to shortening. For that reason, knowing only a material's Young's modulus will not provide complete insight into the material's behavior in other settings. By measuring at the same time the lateral expansion associated with axial shortening, we find another fundamental elastic constant. Most investigators use the radial expansion to calculate a second material constant, **Poisson's ratio**, which is the ratio of lateral expansion to longitudinal shortening. We use the Greek letter ν (nu) to denote Poisson's ratio. Because it is a ratio of two strains, ν has no units. If a material were completely incompressible, $\nu = 0.5$, but because materials are compressible, typically $0 \leq \nu \leq 0.5$. For many geological materials, $\nu \approx 0.25$.

7A.2.1.3 Elastic Constants Derived from Three-Dimensional Stresses and Strains

In a three-dimensional setting, the mean stress $\sigma_m = [(\sigma_{11}+\sigma_{22}+\sigma_{33})/3] = [(\sigma_{xx}+\sigma_{yy}+\sigma_{zz})/3]$ measures the hydrostatic component of the stress state, which tends to cause a change in the volume of a material. The strain equivalent to mean stress is $e_m = [(e_{11}+e_{22}+e_{33})/3] = [(e_{xx}+e_{yy}+e_{zz})/3]$. The change in volume of the sample, the volumetric strain $e_v = (e_{11}+e_{22}+e_{33}) = (e_{xx}+e_{yy}+e_{zz}) = 3e_m$.

A plot of the magnitudes of mean stress and the resulting magnitudes of elastic volumetric strain yields a stress-strain curve whose slope is called the **bulk modulus** K:

$$\sigma_m = K \cdot e_v = 3K \cdot e_m. \tag{7.4}$$

The bulk moduli of most geological materials are ~100 GPa. The bulk modulus is the reciprocal of the compressibility defined in many chemistry courses and texts. The bulk modulus is also directly related to the curvature at the base of the bond energy versus the interionic spacing plot (Figure 7.7c).

In the previous chapter, we noted that it is deviatoric stresses that cause changes in shape. In the same way that the components of the deviatoric stress tensor are defined by $[\sigma'] = [\sigma]-[\sigma_m]$, the shape change component of the strain, the deviatoric strain, is $[e'] = [e]-[e_m]$.

A plot of a range of magnitudes of the components of deviatoric stress and the resulting magnitudes of elastic deviatoric strain yields a stress-strain curve whose slope is a material constant called the **shear modulus** G:

$$\left[\sigma'\right] = G \cdot \left[e'\right]. \tag{7.5}$$

Some geologists and material scientists use the Greek letter **μ** (mu) to denote the shear modulus instead of G. The Greek letter μ is used for other parameters, so we use G to denote the shear modulus throughout this book. Empirically, the relative magnitudes of G and K for most solids are $G \approx 3K/5-2K/3$. Rewriting Eq. (7.5) in component form, we have

$$\sigma'_{ij} = G \cdot e'_{ij}. \tag{7.6}$$

for elastic strains. Because $e' = \sigma'/G$, as G approaches zero the deviatoric strain generated by particular differential stress grows. That is, the material has less and less resistance to changes in its shape. If $G = 0$, the material is unable to resist changes in its shape – it is a fluid. As G approaches ∞, the stresses required to change the material's shape becomes immensely large – the material approaches perfect rigidity.

Using the constants K and G, we can relate any elastic strain to the stresses responsible for it:

$$\sigma_{ij} = \sigma_m + \sigma'_{ij} = 3K \cdot e_m + G \cdot e'_{ij}. \tag{7.7}$$

We noted earlier that the propagation of seismic waves entails elastic deformation of a mass of rock. For longitudinal or P waves, the elastic deformation is localized compression and dilation of rocks. For transverse or S waves, the elastic deformation is localized shearing deformation in one direction or another. Theoretical analyses, confirmed by empirical studies, indicate that the velocities of elastic waves traveling through a rock are functions of the rock's elastic constants.

$$V_P = \sqrt{\left[\frac{(K + 4G/3)}{\rho}\right]} \tag{7.8}$$

$$\text{and } V_S = \sqrt{\left(\frac{G}{\rho}\right)}, \tag{7.9}$$

where V_P is the velocity of a longitudinal wave, V_S is the velocity of a transverse wave, and ρ is the density of the rock. Using Eqs. (7.8) and (7.9), geophysicists use measured seismic velocities to infer K, G, and ρ for rocks in Earth's interior, where we are unable to acquire samples. Note also that when $G = 0$, and the material is a liquid, V_S must be zero.

7A.2.1.4 Limits to Elastic Behavior

Because elastic deformation occurs without breaking any bonds, it is not permanent. If all geological deformation were elastic, there would

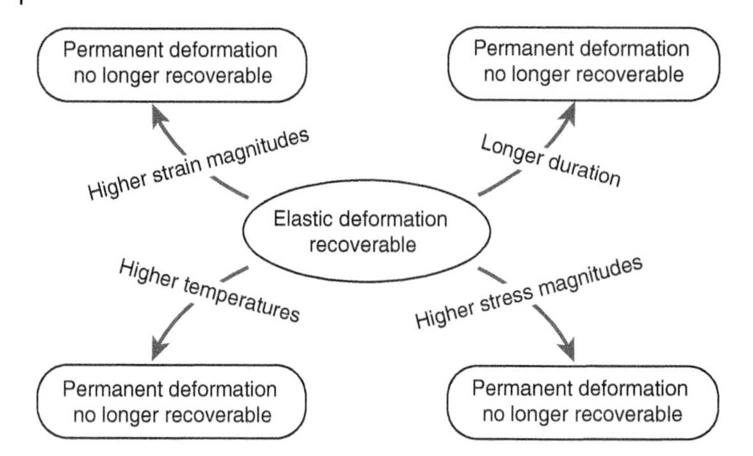

Figure 7.9 A transition from recoverable, elastic deformation to permanent deformation can result from high stresses or strains, high temperature, or the long duration of loading.

be no deformation structures or microstructures. All geological structures are products of conditions where either: (1) the applied stresses were large enough ($>\sigma_l$) to exceed the range of strains that can be accommodated elastically; (2) where conditions were such (e.g. temperatures so high) that materials' shear moduli were reduced and the materials distorted at low differential stresses; or (3) as outlined later, the materials were subjected to stresses for long periods of time. In those situations, *permanent* deformation results because bonds are broken and reformed in new configurations. Some geologists use the term *inelastic* to describe deformation that is permanent.

The length of time over which stresses are applied is an important factor affecting whether deformation is or is not recoverable. When stresses are applied to materials over extended time periods, the idealization that elastic deformation is independent of time breaks down. You may have noticed that a paper clip that holds a thick sheaf of papers together for a short time can be removed, and it will return to its original shape. On the other hand, a paper clip holding together the same sheaf of papers for a longer time period, such as several months or more, often remains distorted permanently after being taken off the sheaf of papers. In this instance, permanent deformation accrued in response to differential stresses $\leq \sigma_l$ applied over a long time period. By including the length of time one applies stresses or observes a material, the situation becomes significantly more complicated. In the previous chapter, we described overcoring as a method for determining crustal stresses. In fact, cores taken of crustal rocks subjected to stress for long time periods do not only exhibit an initial recovery of stored elastic strain, they also typically exhibit a slower, time-dependent recovery of *anelastic strain*. Figure 7.9 is intended to show schematically that any geologic material may exhibit elastic behavior under certain conditions but deform permanently under different conditions. The conditions also influence whether the permanent deformation occurs by fracture, flow, or some combination of the two.

7A.2.2 Criteria for Fracture or Fault Formation

The first example of material failure mentioned in this chapter was the breakage – fracture – of a wooden pencil. Bending a pencil until it breaks generates extension fractures. In most of the "every day" fracture-formation situations with which we have experience, the fractures are extension fractures initiated in response to tensile stresses (the outer arc of the pencil in Figure 7.1

experiences tensile stresses). In contrast, in nearly all natural deformation situations and in the deformation apparatus illustrated in Figure 7.2, the extension fractures or faults formed, provided the applied stresses were sufficiently high and conditions were appropriate, in rocks subjected to compressive stresses.

In laboratory experiments that exhibit a dramatic drop in the magnitude of the differential stresses (e.g. Figures 7.4a, and 7.6c or d), samples typically are cut by extension fractures or faults, depending on the magnitude of the confining pressure (Figure 7.10). The drop in differential stress occurs because the extension fractures or faults cut completely across or through the sample, and stresses cannot be transmitted across either type of surface. In naturally deformed rocks, extension fractures and faults need not cut completely across a rock mass, so fractures and faults in nature can form with or without a dramatic drop in the local differential stress levels.

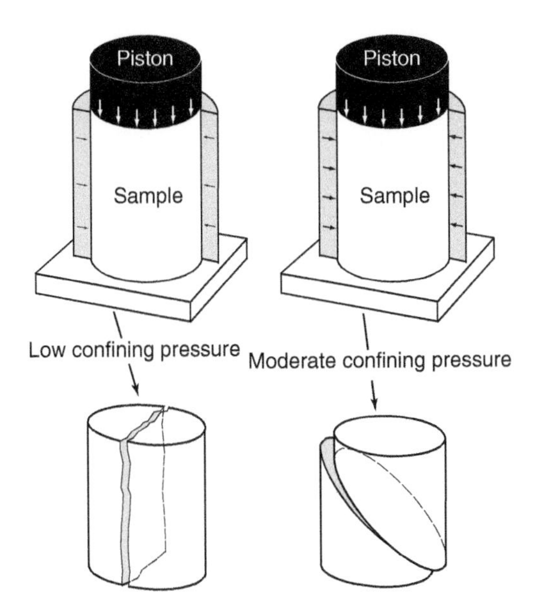

Figure 7.10 At left, a sample subjected to end load at low confining pressures forms an extension fracture. At right, a sample subjected to end load at moderate confining pressures forms a fault.

Studies of the characteristics of features in naturally deformed rocks *and* the results of computational modeling confirm that extension fracture and fault formation and subsequent growth can, but need not, be accompanied by a dramatic drop in the differential stress magnitude.

Experimental studies indicate that extension fractures form at very low confining pressures and/or when the rate of deformation is relatively high. In those laboratory experiments (e.g. Figure 7.4a), materials accrue negligible permanent deformation prior to the formation of the extension fracture or fault. Fracture formation like this is similar to what occurs when one drops a favorite coffee mug or teacup and the handle breaks off; it occurs with little to no appreciable change of shape prior to fracturing. For that reason, the mended mug or cup, with the handle glued back on, often looks as good as new.

Faults result when confining pressure is increased, and they can form after small amounts of precursory permanent deformation. Geologists and material scientists say that the solids deforming in this way, by forming fractures or faults with little precursory permanent deformation, exhibit **brittle** behavior. The first definition for the term brittle, enumerated earth scientists more than 50 years ago, is:

- Deformation is brittle if a sample fails by forming extension fractures or faults after less than 3% permanent strain (stretches $T = l'/l_o$ have magnitudes of $0.97 < T < 1.03$ or elongations $e = (l' - l_o)/l_o$ have magnitudes of $-0.03 < e < 0.03$).

In other laboratory experiments (e.g. Figure 7.6c or d), samples experience significant permanent deformation by flow prior to the formation of an extension fracture or fault. Using a definition also enumerated more than 50 years ago, those samples exhibit **ductile** behavior:

- Deformation is **ductile** if a sample accrues at least 5% permanent strain (stretches T have magnitudes of $0.95 \leq T$ or $T \geq 1.05$ or elongations e have magnitudes of $e \leq -0.05$ or $e \geq 0.05$) before extension fractures or faults form.

A brief look at an index of structure geological literature indicates that the terms brittle and ductile are widely used. Unfortunately, the terms carry different meanings for different geologists. Some geologists use the term brittle to indicate that deformation generated extension fractures, faults, or fault zones (narrow, tabular zones with a finite thickness that record relative movement of rock masses), regardless of the amount of precursory strain, and they use the term ductile to indicate a lack of fracturing or faulting, where deformation is characterized by folds, shear zones, and deformation fabrics. Other geologists argue that this usage does not conform with the original definitions and that it is better to envision that faulting is "semi-brittle" or "semi-ductile" deformation. For still others, deformation is brittle only if mineral grains are fractured but otherwise not visibly distorted, and deformation is ductile only if mineral grains exhibit evidence that their shapes have changed, without fracturing, during deformation. In many deformed rocks, some mineral grains change their shapes by flow at the same time as other mineral grains fracture with little additional distortion, suggesting that the terms brittle and ductile are better applied to individual mineral types in the deformed rock. We, therefore, suggest that any use of either term be accompanied by a description of the criteria (including a scale) for assigning the label, that is whether the rock possesses fractures or not, what is their extent and orientation, whether mineral grains are measurably deformed or not, etc.

7A.2.2.1 Fault Formation

In this section, we consider first the formation of faults, using an empirical approach to derive mathematical relationships predicting the applied stresses at which faults form and the orientation of the faults relative to the principal stresses. We then turn to analyses of the formation of extension fractures, for which there is a firmer theoretical foundation relating the processes by which fracture forms to the mathematical relationships predicting the stresses at which the fracture will form. We then briefly consider how theories for extension fracture formation can be adjusted to encompass the formation of faults.

In naturally deformed rocks, we regularly find faults in conjugate sets that meet at 60 and 120° angles (cf. Figure 4.26 in Chapter 4 – Displacements). Offsets on these faults indicate that the rock masses are shortened along the acute bisector and elongated along the obtuse bisector of the conjugates. Taking the direction of maximum shortening to be parallel to the most compressive principal stress, this geometry suggests that faults form along surfaces that experience shear stresses of relatively high magnitude and normal stresses less than the mean stress. Laboratory experiments yield similar insight: individual faults form at ~30° angles to the direction of the most compressive stress and principal shortening (Figure 7.10). Comparing laboratory experiments undertaken at different confining pressures, experiments run at higher confining pressures require greater differential stresses to form faults. Both of these observations suggest that there is a critical ratio between the magnitude of the shear stress and the normal stress on surfaces that become faults. The notion of a critical ratio of shear stress magnitude relative to normal stress magnitude is similar to that observed in sliding friction. Most geologists see analyses of frictional sliding as a starting point for their understanding of fault formation.

To understand frictional sliding, envision a model consisting of a block of rock on an inclined

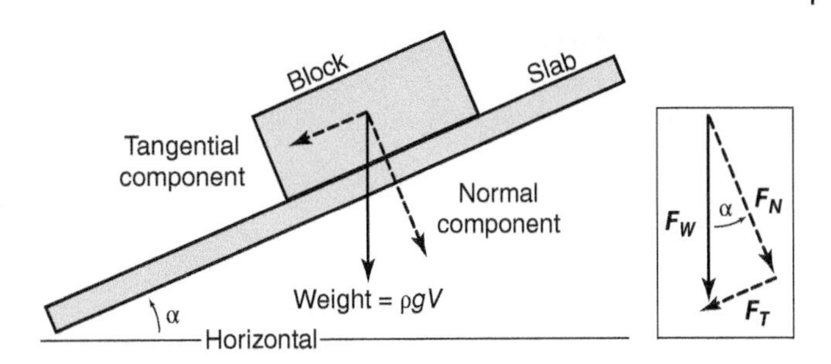

Figure 7.11 A block of rock with density ρ and volume V, on a slab with inclination α. g is the acceleration due to gravity. The downward force of the weight of the block can be resolved into a component F_N perpendicular to the base of the block and a component F_T parallel to the base of the block. Trigonometry indicates $F_N = F_W \cos \alpha$ and $F_T = F_W \sin \alpha$.

slab of the same rock type (Figure 7.11). The block has density ρ and volume V. Its weight, the force $\boldsymbol{F_W} = \rho g V$, acts vertically downward. The component of $\boldsymbol{F_W}$ acting perpendicular or *normal* to the base of the block is $\boldsymbol{F_N} = \boldsymbol{F_W} \cos \alpha$. The component acting parallel or *tangential* to the base of the block is $\boldsymbol{F_T} = \boldsymbol{F_W} \sin \alpha$ (see inset in Figure 7.11). The slab is not fixed, so one can change its inclination α. As the slab inclination gradually increases from zero (horizontal), the block will begin to slide down the slab at some value of α. Repeated runs of this simple experiment demonstrate that the block will begin to slide at a particular value of α we call the critical angle for sliding. We use the Greek letter ϕ to denote this critical angle and call it the "angle of friction."

At the critical angle, the ratio of the tangential component $\boldsymbol{F_T}$ to the normal component $\boldsymbol{F_N}$ attains a critical magnitude μ:

$$\frac{\boldsymbol{F_T}}{\boldsymbol{F_N}} = \frac{\sin \phi}{\cos \phi} = \tan \phi = \mu. \quad (7.10)$$

We call μ the **coefficient of friction** of the material from which the slab and block are made. Different rock types will have different magnitudes of μ, but they are often close to 0.6. Note that there is *no* dependence upon the area of the sliding surface in this type of behavior.

Materials that exhibit this type of sliding behavior, where there is a critical value of the material's coefficient of friction, are said to follow *Amonton's Laws*, named after the seventeenth century French scientist who first enumerated them. Amonton's Laws are:

1) Frictional resistance (a force) is proportional to the normal load across a surface, and
2) Frictional resistance is independent of the area of surface contact as long as the normal force remains constant.

We can deduce the origin of these two "laws" by noting that the nature of the surface of contact is ultimately a function of the normal force across the base. At a microscopic level, the contact between any two solids is rough, with both sides of the contact covered with small protuberances called *asperities*. The roughness of the surfaces means that some portions of the surfaces of the two solids are not in contact. If A is the area of the plane separating the two rock masses, the area of contact A_C is always less than A. As the two surfaces are forced together by the forces acting normal to the surface, the asperities deform, so the true contact area A_C is a function of the normal force across the base.

$$A_C = \frac{\boldsymbol{F_N}}{\sigma_Y}, \quad (7.11a)$$

where $\boldsymbol{F_N}$ is the normal force and σ_Y is the stress at which the material deforms readily or yields. Rearranging, Equation (7.11a) becomes

$$\boldsymbol{F_N} = \sigma_Y A_C. \quad (7.11b)$$

Slip occurs when the shearing force is sufficient to distort the material along the contact area, given by

$$F_T = \sigma_S A_C, \tag{7.12}$$

where σ_S is the stress at which the material deforms in shear. Note that

$$\frac{F_T}{F_N} = \frac{\sigma_S A_C}{\sigma_Y A_C} = \frac{\sigma_S}{\sigma_Y}. \tag{7.13}$$

Inasmuch as σ_S and σ_Y have characteristic values for a given material, the coefficient of friction will have a characteristic value for that material and will be independent of the contact area.

By recasting this analysis in terms of stresses rather than forces, we can apply the results to geological situations. Instead of considering F_T, the tangential force acting across the contact area, we consider σ_T, the tangential or shear stress acting on a plane within the rock. Instead of considering F_N, the normal force acting across the contact area, we consider σ_N, the normal stress acting across a plane within the rock. We envision that a fault can form when the ratio of the tangential stress to the normal stress acting on a plane reaches a critical value. We hypothesize that the fault will form when the ratio of the shear stress to the normal stress attains a magnitude equal to the coefficient of friction. Thus, if $\sigma_T < \mu\sigma_N$, the material will resist fault formation. If

$$\sigma_T = \mu\sigma_N, \tag{7.14}$$

a fault forms. Another way to express this equation is that the ratio σ_T/σ_N must exceed μ. The ratio σ_T/σ_N is therefore critical to evaluating stability: it is called the **slip tendency.**

Geologists often turn to Mohr diagrams to examine this hypothesis for fault formation. Replacing σ_T by y, σ_N by x, and μ by m, we see that Eq. (7.14) plotted in Mohr space corresponds to $y = mx$, a straight line emanating from the origin with slope $m = \mu$ (Figure 7.12a). Since the only distinction between algebraically positive and negative shear stresses in Mohr space is the sense of shear across the surface (anticlockwise for positive values and clockwise for negative values), the hypothesis also predicts that faults should form if $\sigma_T = -\mu\sigma_N$.

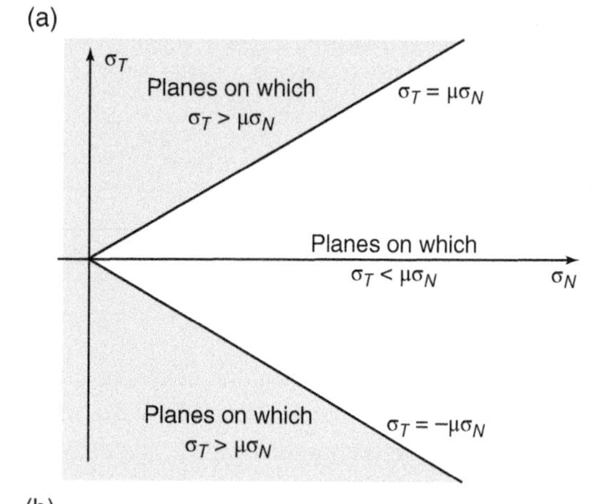

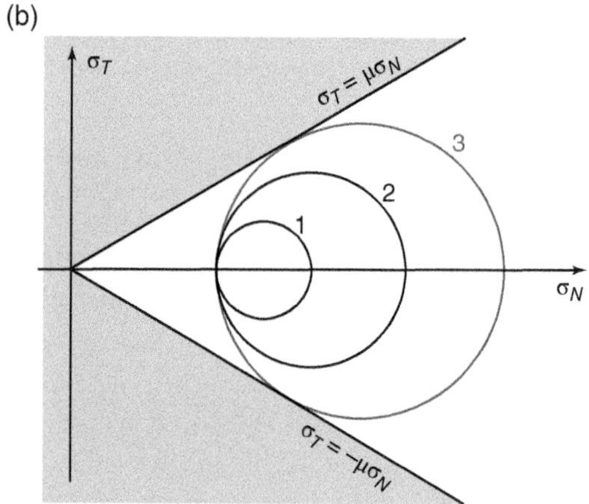

Figure 7.12 (a) Mohr space plots showing a failure envelope defined by $\sigma_T = \mu\sigma_N$. The shaded portions of the graph denote normal stress and shear stress components unattainable since $\sigma_T > \mu\sigma_N$ or $\sigma_T < -\mu\sigma_N$. (b) Mohr circles showing successive stress states in which end load increases till the Mohr circle intersects the failure envelope, leading to the formation of faults.

Mohr circles 1, 2 and 3 in Figure 7.12b depict successive stress states in a mass subjected to fixed

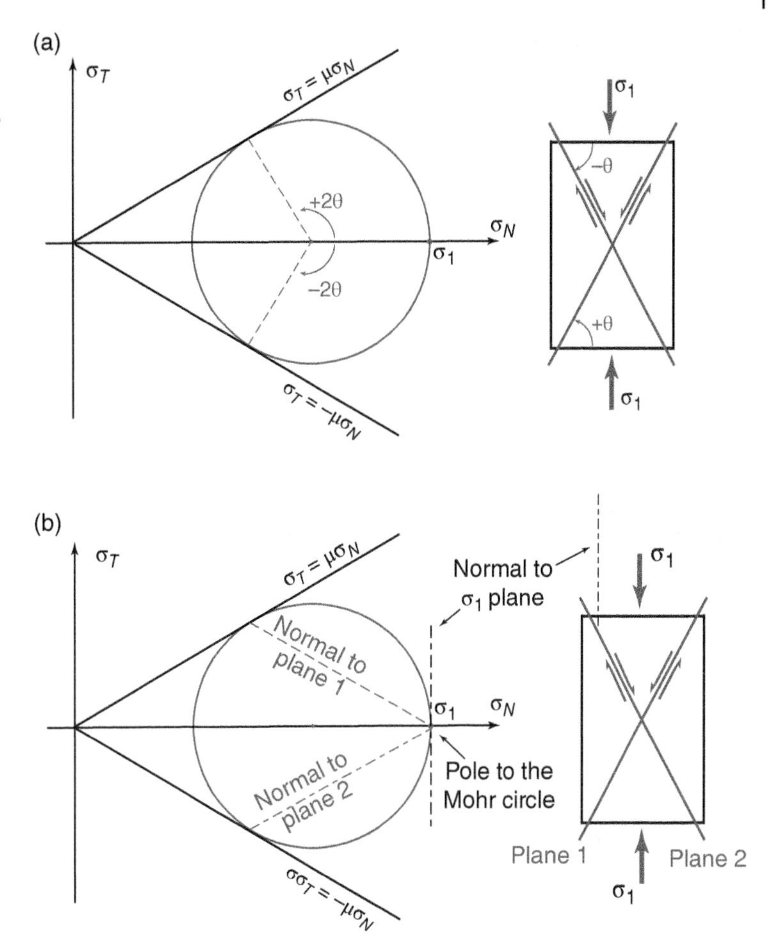

Figure 7.13 (a) Mohr circle at failure. Angle between plane experiencing most compressive stress and faults are ± 2θ. (b) Same Mohr circle with a pole to Mohr circle. Lines from pole to points of intersection denote normals to fault surfaces.

least compressive stress and increasing most compressive stress. All points on Mohr circles 1 and 2 have stress values such that $|\sigma_T| < \mu\sigma_N$, meaning that no planes have the appropriate ratio of shear stress to normal stress to form a fault. The stress state depicted by Mohr circle 3 (in red) has sufficiently large differential stresses that two points on the circle intersect the lines $\sigma_T = \pm\ \mu\sigma_N$. For the two planes represented by those two points, the ratio of tangential stress to normal stress has attained the appropriate value to initiate faults. The polar angles between the point on the Mohr circle representing the plane that experiences the most compressive stress and the points representing either fault are ± 2θ (Figure 7.13a). Thus, the faults should form at angles ± θ from the plane experiencing most compressive stress. Alternatively, one can locate the pole to the Mohr circle and thereby find the relative orientations of the potential faults (Figure 7.13b). Either method shows that faults would form conjugates at the ~60 and 120° angles regularly observed.

The two lines $\sigma_T = \pm\ \mu\sigma_N$ bound or envelop a region of Mohr space in which stress states are stable, that is where no faults form. Because faults form when a Mohr circle representing a stress state intersects the lines, we often refer to the lines $\sigma_T = \pm\ \mu\sigma_N$ as a *failure envelope*. There is one problem with this particular failure envelope. If the normal stress acting on a plane $\sigma_N = 0$, the shear stress required to form a fault is also zero. This is clearly unrealistic. Adding a term to account for the cohesion across planes in materials is a straightforward way to "correct" this problem. If τ_o is the **cohesion**, the shear stress required to form a fault on a plane in the absence of normal stress, then we can hypothesize that faults should form if

$$\sigma_T = \tau_o + \mu\sigma_N. \tag{7.15}$$

Again, we replace σ_T with y, σ_N with x, and μ with m, we see that the equation $\sigma_T = \tau_o + \mu\sigma_N$ corresponds to $y = mx + b$, a straight line with slope $m = \mu$ and y intercept $b = \tau_o$. Noting again that the positive and negative values of shear stress simply refer to anticlockwise or clockwise shear senses, we see that the two lines $\sigma_T = \tau_o + \mu\sigma_N$ and $\sigma_T = -\tau_o - \mu\sigma_N$ together compose a failure envelope defining the limiting stress states at which faults form (Figure 7.14a). Many geologists call this adjusted criterion for the formation of faults the ***Coulomb criterion***, refer to the μ as the **coefficient of internal friction**, and define $\phi = \tan^{-1}\mu$ as the **angle of internal friction**. Measured values of the coefficient of internal friction μ in intact rocks are 0.5–0.8, yielding angles of internal friction $\phi \approx$ 30–43°. Commonly measured values of the cohesion τ_o in rocks are 10 to 75 MPa. Values of μ and τ_o, especially in sedimentary rocks composed of clay minerals with well-defined bedding planes, may be significantly lower than those values.

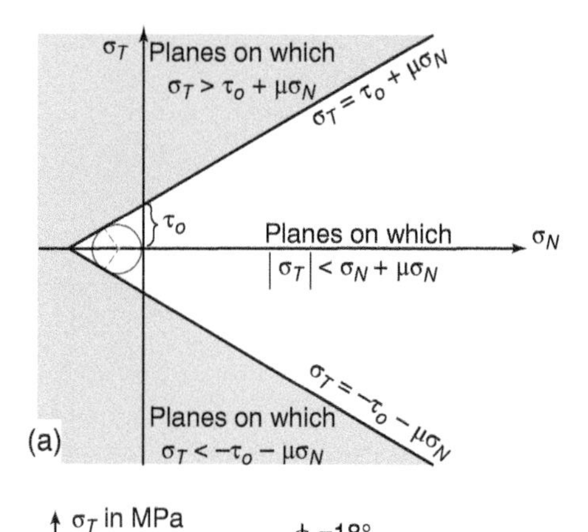

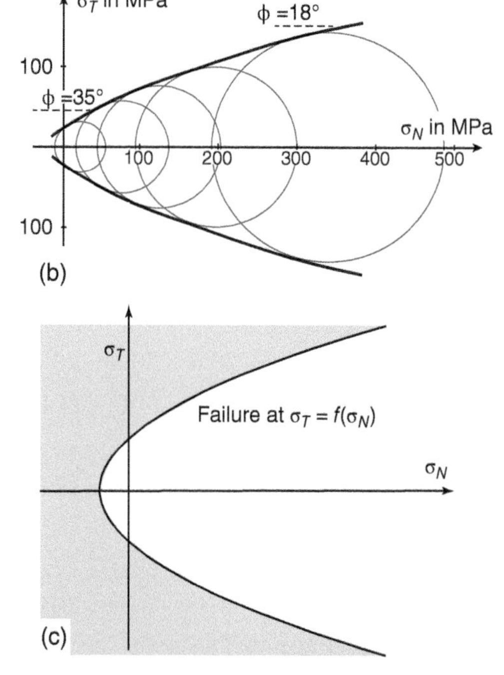

Figure 7.14 (a) Mohr space plot showing a failure envelope defined by $\sigma_T = \tau_o + \mu\sigma_N$. (b) Experimentally defined failure envelope for Green River shale, redrawn from Handin and Hager (1957). ϕ – angle of internal friction. (c) Mohr space plot showing a failure envelope predicting realistic geometry for extension fractures.

The Coulomb criterion for the formation of faults generally conforms with two important characteristics of experimental data on the formation of faults. First, the most compressive stress direction makes an acute angle, typically ~30°, to the plane of the fault. In many instances, the most compressive stress direction is the acute bisector of conjugate faults meeting at ~60° angles. Second, experiments show that the differential stresses required to form faults increase with increasing mean stress (Figure 7.14b). Examining the experimental data on fault formation reveals, however, disparities between the behavior predicted by the

Coulomb criterion and the actual behavior of rocks. The Coulomb criterion does not, for example, predict realistic geometries for extension fractures. The failure envelope defined by Eq. (7.15) predicts conjugate fractures bisected by the most compressive and most tensile stresses even in cases where rocks are subjected to tensile stresses (Figure 7.14a). Extension fractures, however, tend to form perpendicular to the most tensile stress and parallel to the most compressive stress. Further, experimental data typically show that the slope of the failure envelope also varies with the magnitude of the hydrostatic pressure. On Figure 7.14b, for example, $\phi = 35°$ and $\mu \approx 0.7$ at low hydrostatic stresses and $\phi = 18°$ and $\mu \approx 0.3$ at greater hydrostatic stresses. Thus, failure envelopes derived from laboratory tests at different mean stresses depart from the Coulomb criterion in that

$$\sigma_T = f(\sigma_N) \tag{7.16}$$

are nonlinear curves (Figure 7.14c). Before we can consider a more generally applicable theory that provides insight into both faults and extensional fractures we need to examine the formation of extension fractures under tension, for which there is a robust theoretical framework that effectively ties our understanding of the nature of bonds within materials to the mechanisms by which we break those bonds along a fracture surface.

7A.2.2.2 Extension Fracture Formation

Earth scientists, ceramicists, material scientists, and engineers alike have been interested in developing theoretical models describing failure by fracturing that conform to the definition earlier for brittle deformation, where the material exhibits its elastic behavior nearly to the point at which the fracture forms and there is little deformation other than the formation of the fracture. We can gain some insight into the formation and propagation of fractures in tension by re-examining the forces on atoms or ions in crystals and the bond energy versus interionic spacing plot (Figure 7.7), which is the basis of much of our understanding of crystal structure and behavior. We follow the analysis of A. A. Griffith (1893–1963), originally outlined in Griffith (1921), in: (1) taking an energy-balance approach to calculate theoretical values of the stresses required to fracture a material; and (2) using an understanding of stress distributions within real substances to explain why most materials fracture when applied differential stress magnitudes are far below the theoretical fracture strengths determined for those materials.

In order to increase the spacing between planes of atoms or ions from their equilibrium value d_o to a new value d', one must apply a tensile force to counteract the bonds between atom or ion planes. Figure 7.15 is a modified version of Figure 7.7a in which the ordinate is not interatomic or interionic force but the average force per unit area, or stress,

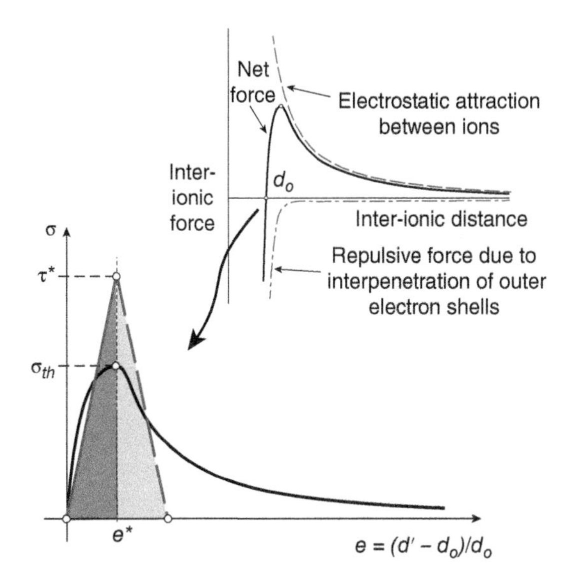

Figure 7.15 Stress-strain curve derived from interionic force versus interionic distance plot. The red line illustrates stress-strain relation assuming material exhibits elastic behavior to fracture. Area of dark and light red triangles is used to the approximate area under stress-strain curve. τ^*, e^* – stress and strain at failure. d – bond length.

acting across planes within a solid. Instead of the spacing d', the abscissa is a relative measure of the spacing between planes of atoms or ions, the strain magnitude $e = (d'-d_o)/d_o$. Note that $e = 0$ when $d' = d_o$. Figure 7.15 shows only the portion of this stress-strain curve where spacing values $d' > d_o$. The tensile stresses required to stretch the bonds connecting planes of atoms or ions increase until $d' = d^*$ and $e = e^*$. Because increasing the spacing between planes of atoms or ions beyond d^* requires lower stress magnitudes, the tensile stress at spacing d^* is the theoretical maximum tensile stress the material can support, σ_{th}. σ_{th} is the magnitude of the tensile stress required to initiate a fracture.

In order to estimate the magnitude of σ_{th}, we need to look closely at what happens during fracture formation. Fracture formation creates two new external surfaces in the material (Figure 7.16). The atoms or ions along the outer surface of a substance are drawn toward the interior of the material but no longer have "outboard" atoms or ions holding them at distance d_o from their "inboard" neighbors. For this reason, bonds holding atoms or ions along the external surface must, according to the bond energy versus interatomic spacing curve in Figure 7.7c, have higher energy than those within the material. There is then a characteristic energy per unit area associated with any bounding surface of a substance. In this context, it is clear that the formation of the fracture requires an *input* of energy to account for the formation of the new bounding surfaces in the material where there were none before.

The amount of energy *expended* (U_E) to create those two new bounding surfaces is the product of the energy per unit area of the surface, γ_S, times the total area the new surface created. Since that new area created is twice the area of fracture A,

$$U_E = 2A \times \gamma_S. \tag{7.17}$$

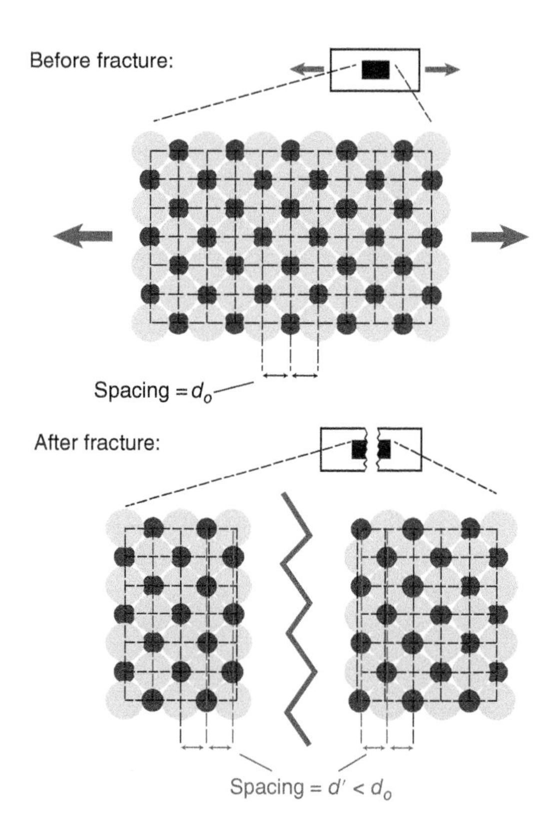

Figure 7.16 Fracture of material creates two new surfaces. Along those surfaces, ions sit at distance $d' < d_o$ from their inboard neighbors. Figure 7.8c indicates that bonds holding the ions along surfaces must have higher energy than bonds within the crystal.

An important insight to fracture formation is the recognition that the energy to form the fracture surfaces is supplied by the release of elastic energy stored in displacing atoms or ions from their equilibrium positions. In the vicinity of what will become the fracture surface, the stored energy is the elastic strain energy. Recall that the product $\sigma \times \varepsilon$, or the area under an elastic stress-strain curve, gives the strain energy per unit volume of an elastically deformed material (Eq. 7.1). The area under the stress-strain curve in Figure 7.15 defines the total energy of elastic distortion (U_S) available to drive fracture formation. By deriving a formula for U_S and equating it to the formula for U_E in

Eq. (7.17), we can use an understanding of the character of bonds in the material to estimate the stress magnitudes required to fracture the material.

Instead of integrating the stress-strain curve in Figure 7.15, we derive a rough estimate for U_S by assuming that the stress-strain behavior of the material is linear right up to the point of fracture. This hypothetical situation is depicted by the red line on the stress-strain plot in Figure 7.15, with fracture formation occurring when the stress magnitude is $\tau^{\dagger}$ and the strain magnitude is $e^{\dagger}$ (the asterisks indicating they are crude estimates of the true magnitudes of stress and strain at fracture formation). For simplicity, we envision that the stresses fall during fracture formation along the course of the dashed red line. In effect, this analysis uses the dark and light red-shaded, triangular areas to approximate the area under the actual stress-strain curve.

The area of the dark-red shaded triangle is ½($\tau^{\dagger} \times e^{\dagger}$), thus the strain energy available drive fracture formation is

$$U_S \approx 2 \times \frac{1}{2}\left(\tau^{\dagger} \times e^{\dagger}\right) \times \left(\text{volume of deformed solid}\right). \quad (7.18a)$$

Assuming that the strain energy used to form the fracture is derived "locally," the volume of deformed solid is the product of the area of the fracture surface times the stretched bond length d^*. In addition, assuming that the material exhibits linear elastic behavior right to the point of a fracture means that $e^{\dagger} = \tau^{\dagger}/E$, where E is the Young's modulus of the material. Thus,

$$U_S \approx \frac{\left(\tau^{\dagger}\right)^2}{E} \times \left(A \times d^{\dagger}\right). \quad (7.18b)$$

Equating U_E from Eq. (7.17) with U_S from Eq. (7.18b) yields,

$$2A \times \gamma_S = \frac{\left(\tau^{\dagger}\right)^2}{E} \times \left(A \times d^{\dagger}\right). \quad (7.19)$$

Solving for $\tau^{\dagger}$ yields

$$\tau^{\dagger} = \sqrt{\left[\frac{\left(2E\gamma_S\right)}{d^{\dagger}}\right]}. \quad (7.20)$$

We have laboratory measurements of Young's moduli and average surface energies for rock-forming minerals. By assuming that $d^* \approx$ the length of bonds in crystals, derived from X-ray crystallography measurements, we can use Eq. (7.20) to calculate $\tau^{\dagger}$ values, thereby estimating σ_{th}, for different minerals. Those calculations yield $\tau^{\dagger}$ values of 20.5×10^{10} Pa = 205 GPa $\approx E/6$ for diamond, 1.4×10^{9} Pa = 1.4 GPa $\approx E/7$ for graphite, 4.3×10^{9} Pa = 4.3 GPa $\approx E/10$ for halite, and 4.6×10^{10} Pa = 46 GPa $\approx E/10$ for corundum. This analysis estimates the stress required to fracture minerals at between $E/5$ and $E/10$.

The observed macroscopic fracture strength for diamond, that is the differential stress required to fracture a macroscopic crystal, is 6×10^{10} Pa = 60 GPa $\approx E/20$, approximately one third the theoretical estimate. For other minerals, the observed fracture strengths are between $^1/_{10}$ and $^1/_{100}$ of the theoretical estimates. Improving the estimate by making a more accurate evaluation of the area under the stress-strain curve or taking into account the three-dimensional character of crystals leads to improved estimates of σ_{th}, but even these improved estimates yield calculated fracture strengths larger than measured fracture strengths by factors of 10–100.

What is the source of error in this analysis? The evaluation of bonding and bond strength behind Figures 7.7 and 7.15 generally holds too well in too many other situations for us to infer that this is incorrect. Likewise, most workers believe there are sound theoretical reasons to infer that the energy-balance approach taken here has value. Some other factor must affect materials to reduce their fracture strengths.

Griffith was aware that the fracture strengths of thin glass fibers increased as one made the fibers thinner, approaching the calculated theoretical fracture strengths for the thinnest fibers. He inferred these thin fibers were less likely to possess flaws than larger masses and envisioned that the flaws affected their macroscopic strength. Another scientist, C. E. Inglis (1875–1952), had determined how stresses vary in the vicinity of an elliptical hole in an elastic solid (Inglis, 1913). For an elliptical crack oriented perpendicular to the direction of the most tensile stress acting across an object, Inglis' analysis indicated that the macroscopic, applied tensile stresses, σ_A, were magnified near the ends of the elliptical hole by a factor proportional to the half-length of the hole C and inversely proportional to the radius of curvature at the end of the hole R_c (Figure 7.17). The concentrated stress σ_C is:

$$\sigma_C = \sigma_A \sqrt{\left(\frac{C}{R_c}\right)}. \tag{7.21}$$

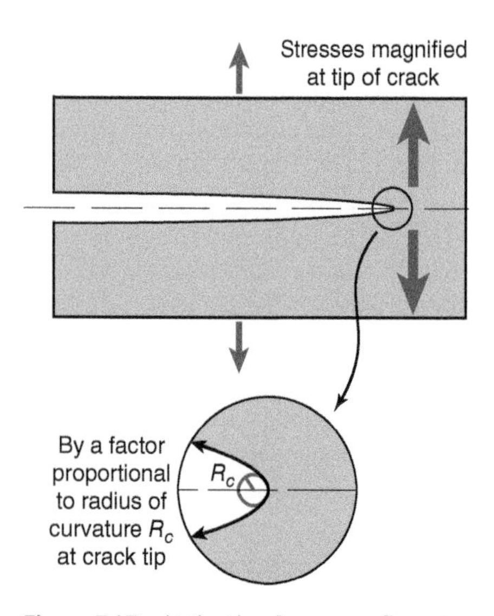

Figure 7.17 At the tip of a narrow flaw, stress is magnified by a factor that is inversely proportional to the radius of curvature, R_c, of the flaw. At sharp flaws, R_c is very small, leading to significant stress magnification.

For highly elliptical holes, those with very sharp tips, the inverse proportionality to the radius of curvature means that the degree of stress concentration is considerable. Griffith inferred that appropriately oriented flaws – flaws with their long dimensions in the plane experiencing the most tensile stress – would magnify the applied stresses sufficiently at the tip of the flaw to approach a substance's theoretical fracture strength, σ_{th}. The flaws thereby provide sufficient elastic strain energy to account for the increase in surface area of the flaw.

Griffith's insights are the foundation of all subsequent work in a vast field of physics and engineering known as fracture mechanics. Early analyses assumed that there were actual, open cracks (called Griffith cracks or Griffith flaws) within materials and that statistically there would exist a tiny, open crack with the correct orientation and sharpness to grow under the applied stress. Such flaws have not been observed, but a modified theory, citing the existence of closed cracks with cohesion, produces sufficiently great stress concentrations to drive fracture propagation. The assumption that materials in the vicinity of crack tips deform elastically up to the point of fracture has been shown to be incorrect, so modern fracture mechanics include adjustments to account for small regions of inelastic deformation in the vicinity of the tip of the flaw. Still, fracture mechanics relies upon the idea that stresses are concentrated in the vicinity of flaws and that one can determine stress concentration factors for different types of fractures subjected to different stresses. Similarly, the validity of the energy-balance calculation has been verified using more complex analyses that capture the nature of the energy expenditure during fracture formation more accurately.

7A.2.2.3 Griffith Criterion for Fracture

Griffith (1924) extended the analysis of crack growth to settings other than the formation of fractures in tension. In his two-dimensional analysis, the stress distributions about randomly oriented cracks are again determined by assuming that the material behaves elastically until the fracture forms and that the critical measure of the material behavior is T_0, its tensile strength. When the most compressive stress $\sigma_1 < 3|\sigma_2|$, where σ_2 is the least compressive stress, fracture formation is governed by

$$\sigma_2 = T_0. \tag{7.22}$$

In "strongly compressive settings," those in which the most compressive stress $\sigma_1 > 3|\sigma_2|$, fault formation is governed by

$$\left(\sigma_1 - \sigma_2\right)^2 - 8T_0\left(\sigma_1 + \sigma_2\right) = 0. \tag{7.23}$$

This condition is typically recast in terms of the critical shear stress and normal stress acting on the plane that will become the fault:

$$\left(\sigma_T\right)^2 - 4T_o\sigma_N = 4T_0^2. \tag{7.24}$$

Equation (7.24) defines a curved failure envelope similar to those derived from experimental data (cf. Figure 7.14c).

Current thinking on the formation of faults envisions that macroscopic faults result, in fact, from the linking of a series of closed cracks inclined at acute angles to the most compressive stress principal axis (Figure 7.18a). The presence of the crack alters the uniform distribution of stresses, leading to stress concentrations that cause the crack to grow as an extensional crack inclined to the original, closed crack and perpendicular to the most tensile principal stress. With sufficiently large magnitudes of applied stresses, these extension cracks link together to form the macroscopic fault. The failure criterion expressed in Eq. (7.24) and depicted in Figure 7.14c is the macroscopic manifestation of those microscopic processes.

Curved failure envelopes derived using this modified Griffith approach conform reasonably well with the results of laboratory experiments on fault formation under compressive loads at moderate pressures. Likewise, the extensional fractures that result from tensile loading form with orientations – perpendicular to the most tensile stress – that are consistent with the curved failure envelope. When one principal stress is tensile and the other compressive, the modified Griffith analysis predicts that fractures/faults should form oblique to the

(a)

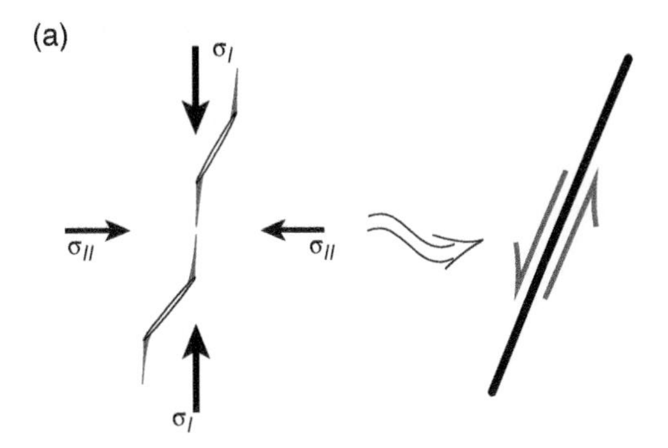

(b)

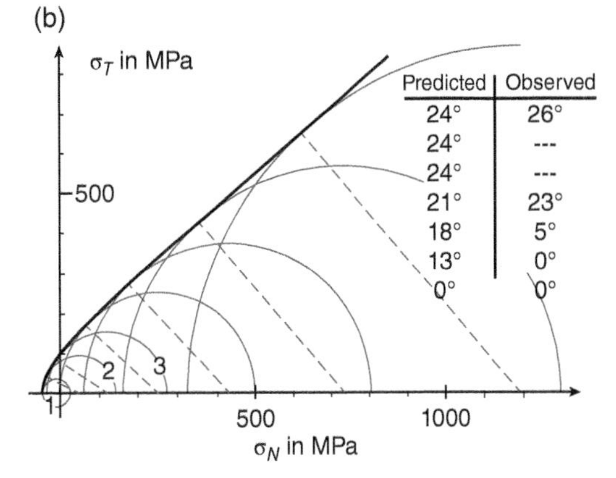

Figure 7.18 (a) As fractures inclined to the stress principal directions grow, the fracture surface curves into an orientation normal to the least compressive stress. The linkage of numerous curved fractures can eventually form a macroscopic fault. (b) Mohr envelope after Engelder (1999) using data from Brace (1964), drawn to account for observed fractures normal to tensile stresses, predicts that extensional shear fractures should form. Table giving predicted versus observed orientations of faults suggests there is a problem with the theoretical understanding of fracture formation.

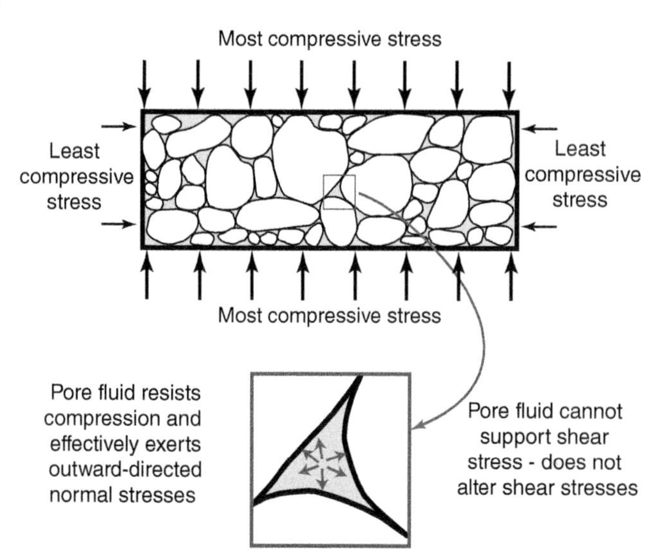

Figure 7.19 Diagram showing idealized rock composed of rounded grains. A fluid phase, shaded, fills the pores. Enlargement shows that fluid resists compression, ultimately negating some of the impacts of the normal stresses.

principal stresses and that the relative movement on the faults should be a combination of slip parallel to and opening perpendicular to the fault surface. There are neither experimental (Figure 7.18b) nor field data that support the existence of such extensional-shear fractures, indicating that the modified Griffith analysis does not fully capture the details of fault formation under different loading configurations.

7A.2.2.4 Effect of Pore Pressure on Fracture Formation

Soil scientists and geologists have found that there is another parameter that can have a significant impact on fracture formation - the pore fluid pressure P_f. In near-surface settings, it is pore water that impacts the formation of shear fractures. At most points deeper within Earth, pressures are sufficiently great that the most common pore-filling constituents, water and CO_2, exist as fluids regardless of the temperatures. These fluids are unable to support shear stresses but resist compression. One can envision the fluids' resistance to compression within pores exert as an effective outward pressure on the walls of pores, thereby reducing the impact of normal stresses in the surrounding rock (Figure 7.19). Recall the quantity known as effective stress. The normal component of the effective stress here denoted $\sigma_N{}^*$, is

$$\sigma_N{}^* = \sigma_N - P_f. \tag{7.25a}$$

Because fluids cannot support shear stresses, pore fluids have no effect on shear stresses. Thus, the shear component of the effective stress is precisely equal to the shear stress.

$$\sigma_T{}^* = \sigma_T. \tag{7.25b}$$

The impact of pore fluid pressure is readily visualized by comparing the Mohr circle depicting the true stress state and that depicting the effective stress state for a particular magnitude of pore pressure (Figure 7.20a). Planes 1 and 2 have finite shear stress components. Because the pore fluid pressure does not affect the magnitude of shear stresses, the points depicting the effective stresses acting on these planes are the same distance from the σ_N axis in the Mohr circles depicting the actual stress state and the effective stress state. This is true for all planes with finite shear stresses. Similarly, the planes that experience the most and

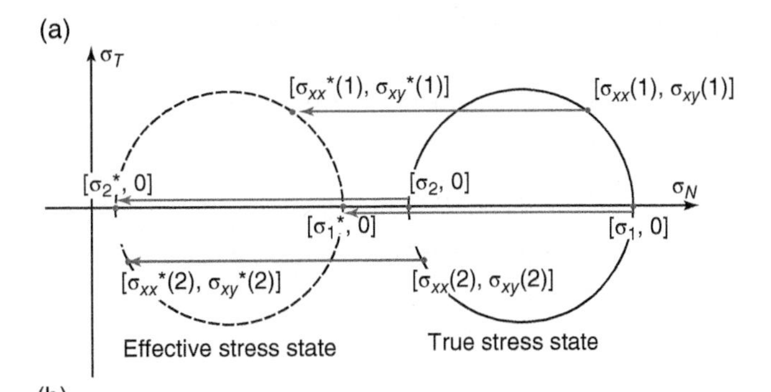

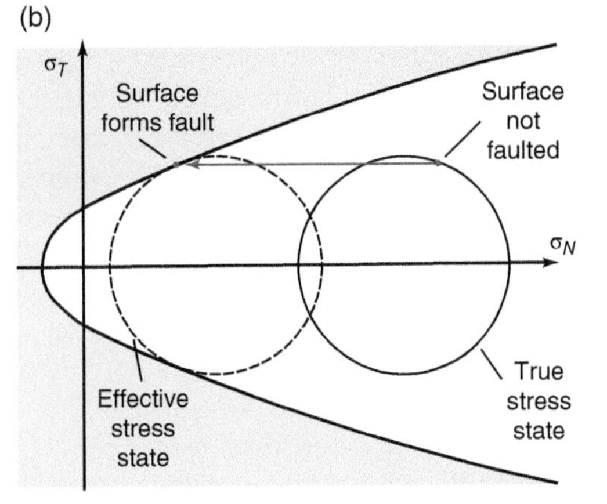

Figure 7.20 (a) Mohr diagrams representing true stress state (solid) and effective stress state (dashed). (b) Increasing pore fluid pressure can bring rock resisting applied stresses (solid circle) to a condition where shear fractures form (dashed circle).

least compressive stresses in the actual stress state experience the most and least compressive stresses in the effective stress state. The radius of the effective stress Mohr circle is the same as that of the true stress Mohr circle because the differential stress is the same in both:

$$\sigma_1{*} - \sigma_2{*} = (\sigma_1 - P_f) - (\sigma_2 - P_f)$$
$$= (\sigma_1 - \sigma_2) - P_f + P_f = (\sigma_1 - \sigma_2).$$

In essence, the Mohr circle for the effective stress state is translated to the left by a distance corresponding to the magnitude of the pore fluid pressure. In instances where fracture formation is a function of the normal stresses acting across potential fracture surfaces, as described by Eqs. (7.15), (7.16), or (7.24), increasing the pore fluid pressure alone, provided that the differential stresses remain unchanged, can bring planes from a condition of stability, where the stress magnitudes are insufficient to generate a shear or extension fracture, to a condition of failure. This effect occurs in near-surface settings when heavy rainfall raises the pore pressures sufficiently to trigger landslides in regions that were stable with lower pore fluid pressure. This effect is exploited by petroleum companies to generate fractures in source rocks, where is it is known as "hydrofracturing" or "fracking." The effect is also responsible for triggering earthquakes when the disposal of liquid by pumping it into wells raises pore fluid pressure sufficiently.

7A.2.3 Yield and Creep

Under appropriate conditions, rock samples yield and are able to deform at constant differential stress magnitudes or even as the magnitudes of differential stresses decrease (e.g. Figures 7.4b, 7.6b). When materials deform without an increase in differential stress magnitudes, we say that they undergo **creep** – the material "creeps" away from the applied load. The flow of glacial ice, apparent in the distortion of moraines or arrays of markers on the surfaces of glaciers or in the slow deflection of originally straight and vertical boreholes to successively more curved shapes (Figure 7.21), is one natural example of creep of rock. The convection of the Earth's mantle is accomplished by creep.

In the two cases cited earlier, creep in response to applied loads typically is flow distributed throughout the material. It is possible, however, that offsets on arrays of intersecting shear zones or arrays of faults can accrue slowly and give rise to macroscopic behavior that we would identify as a creep.

As illustrated in Figures 7.4b and 7.6d, once a material begins to creep there no longer is a unique relationship between the magnitude of differential stress and the strain magnitude. In Figure 7.4b the strain magnitude increases over time even though the differential stress remains constant at σ_y, and in Figure 7.6d, the strain magnitude increases over time even as differential stresses fall. In creep, then, relationships between the magnitude of differential stress and strain, like the straightforward stress-strain relationship of linear elasticity, no longer hold. In these situations, stress-strain rate plots and constitutive equations relating stress to strain rate are a more useful way to characterize deformation.

7A.2.4 Viscous Behavior

We observe that for many materials there is a unique relationship between the differential stress and the rate at which creep occurs. Very slow creep begins at vanishing small magnitudes of deviatoric stress, and the rate at which creep increases with the increasing magnitude of the deviatoric stress. Material scientists refer to this kind of creep as **viscous** creep. Two familiar materials, water and oil, flow by viscous creep. In water, for example, a hand or foot can be moved through the water slowly with very little effort – the flow of water around your hand or foot requires very little stress at all. However, should one dive into the water from a height and enter the water at an incorrect angle, one "asks" the water to flow more rapidly. That higher rate of creep requires higher stresses. Those stresses are sufficiently high that your body feels a "slap." Note that there can be no viscous creep in response to hydrostatic stress, otherwise a mass would shrink at a small but finite rate in response to the hydrostatic stresses in Earth until it disappeared. For this reason, we envision that viscous creep results only from differential stresses and relates only to the deviatoric component of the stress tensor.

Recall from Chapter 5 that longitudinal strain rates are

$$\dot{e} = \frac{de}{dt} \tag{7.26a}$$

and shear strain rates are

$$\dot{\gamma} = \frac{d\gamma}{dt}. \tag{7.26b}$$

Thus, a rock is viscous if

$$\sigma' \propto \dot{e} \tag{7.27a}$$

and

$$\sigma' \propto \dot{\gamma}. \tag{7.27b}$$

For viscous materials, a plot of stress versus strain rate is a smooth curve emanating from the origin (Figure 7.22a). The curve shows that there is a unique magnitude of strain rate for each magnitude of differential stress (e.g. $\dot{e}_A$ for σ'_A and $\dot{e}_B$ for σ'_B).

7A.2.5 Plastic Behavior

There are other materials in which creep occurs at negligibly small rates at differential stress magnitudes below a threshold value, but creep rates increase dramatically when differential stresses

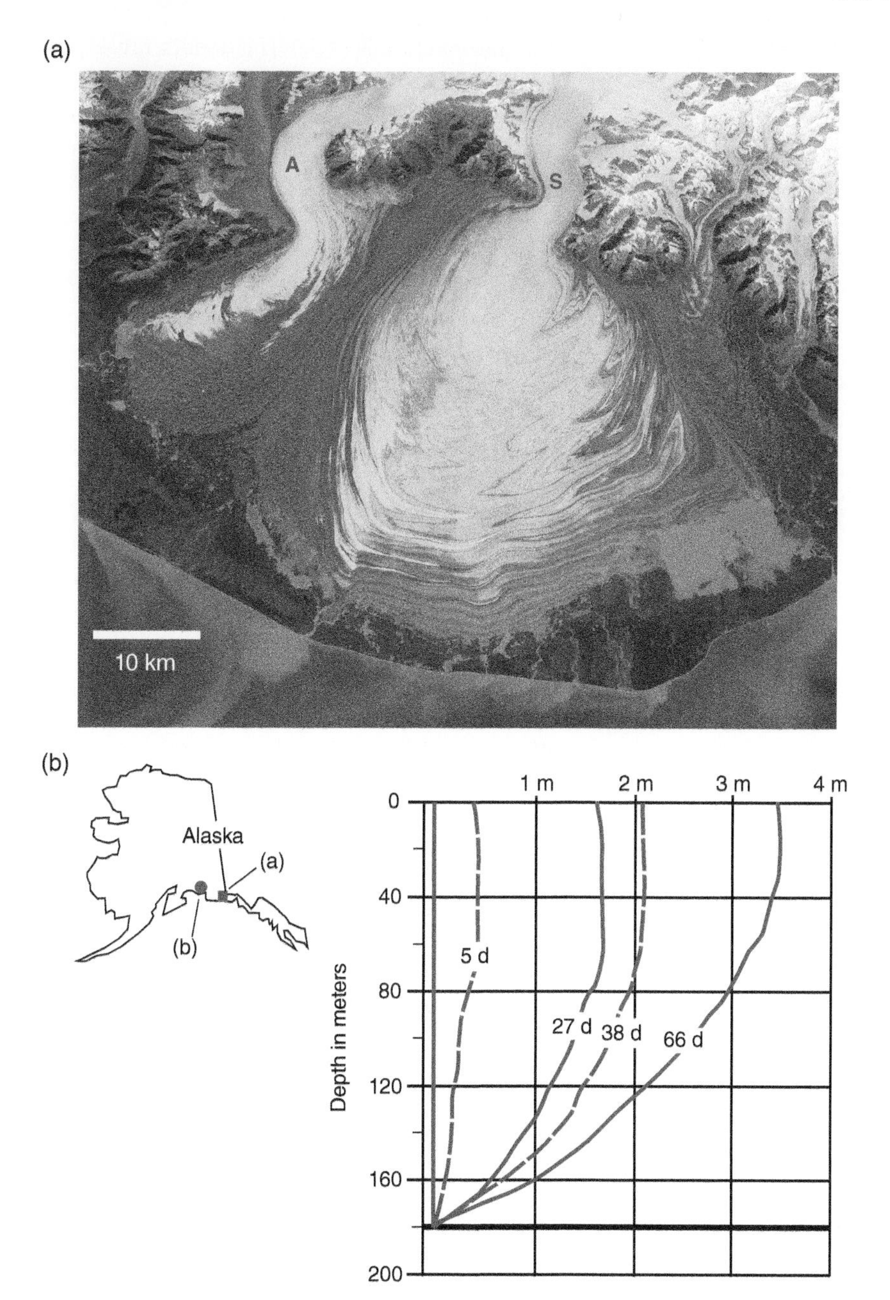

Figure 7.21 (a) A NASA Landsat image of dramatic folded surface moraines in the Malaspina glacier piedmont lobe at the termination of the Seward glacier (S). The adjacent Agassiz glacier (A) also feeds a separate piedmont lobe with distorted surface moraines. (b) Diagram showing the distortion of a vertical borehole in the Worthington glacier, after Harper et al. (1999). Inset gives the locations of the Malaspina and Worthington glaciers.

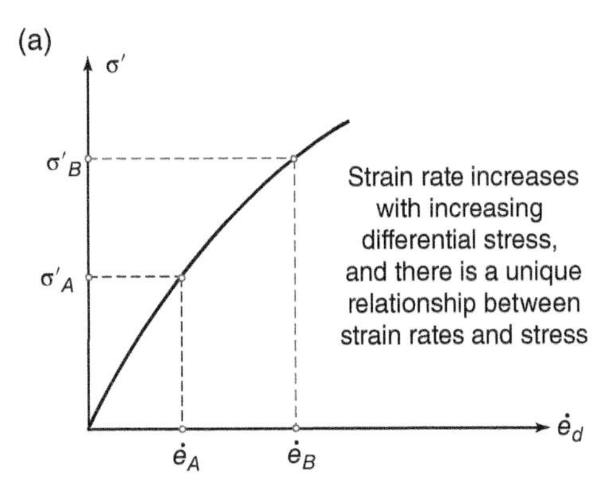

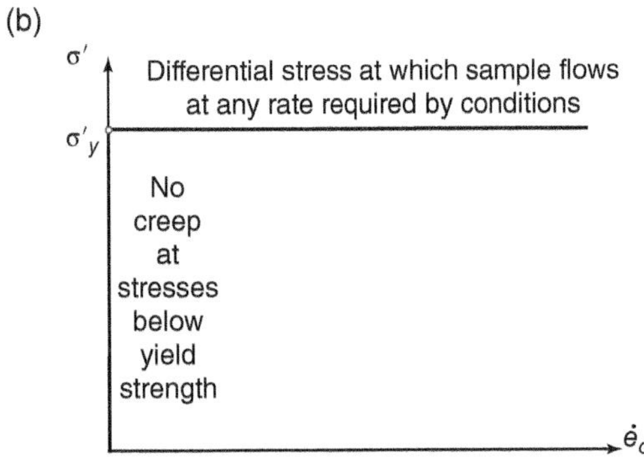

Figure 7.22 (a) Stress-strain rate plot for a viscous material. Strain rate increases with increasing differential stress, and there is a unique strain rate for every differential stress value. (b) Stress-strain rate plot representing perfectly-plastic flow. σ'_y – Yield strength.

reach the threshold value. We say that these materials are **plastic** and that they undergo plastic creep. Modeling clay is a familiar material that exhibits plastic behavior. A block of modeling clay will retain its shape, but it can be formed into a new shape when your hand exerts sufficient stress on it.

We approximate this sort of behavior with the idealization of a **rigid-perfectly plastic material**. Such a material resists all deformation at differential stresses below a threshold value. Should differential stresses reach the threshold value, the material deforms at a rate determined by the external conditions of deformation. In this idealization, differential stresses in the deforming material can never exceed the limiting differential stress value known as the material's **yield stress** or **yield strength**. If σ'_y is the perfectly plastic material's yield strength,

$$\dot{e} = 0 \text{ if } \sigma' < \sigma'_y, \tag{7.28a}$$

and

$$e = \text{some value if } \sigma' = \sigma'_y. \tag{7.28b}$$

The stress-strain rate plot for a perfectly plastic material consists of two straight-line segments: a vertical line from the origin to the point (σ'_y, 0) and a horizontal line extending from (σ'_y, 0) to whatever strain rate a deformation setting requires (Figure 7.22b). Note that in perfectly plastic material, there is no longer a unique relationship between differential stress and the rate of creep.

Viscous behavior and plastic behavior are used here in the context of rheological behavior. Neither viscous nor plastic rheologies require that the material maintains continuity or that the deformation is distributed throughout the material. Deformation that is localized within limited portions of a material or even is accommodated by slip along discrete surfaces across which material has no cohesion can still exhibit a bulk behavior that is viscous or plastic.

7A.2.6 Constitutive Equations for Viscous Creep and Plastic Yield

7A.2.6.1 Linear Viscosity

Many materials, including many geological materials, exhibit a linear relationship between the rate of deformation and the applied differential stress. For those materials, plots of differential stress against the deformation rate are straight lines (Figure 7.23a). The equation, known as a constitutive equation, describing stress-strain rate relationship takes the form $y = mx$, where the ordinate (y) is the differential stress σ' and the abscissa (x) is the strain rate $\dot{e}$:

$$\sigma' = \eta\dot{e}. \tag{7.29}$$

The slope of the stress-strain rate plot is the constant η (the Greek letter eta). We call η the **viscosity** of the deforming material. Materials that exhibit a constant viscosity are said to be *linearly viscous* or to exhibit *linear viscosity*. A linearly viscous material may be said to exhibit *Newtonian viscosity*. Note that in order for Eq. (7.29) to hold, the constant η must have units of stress × time. The standard unit for viscosity is the *Pascal-second* or *poiseuille* (*Pl*) = $N/m^2 \times s = (kg/m\,s^2) \times s = kg/m\,s$. Viscosities are also commonly given in the CGS unit *Poise* (*P*) = $dyne/cm^2 \times s = 10^{-5}\ N/m^2 \times s = 10^{-5}$ *Pl*. If a material has a low viscosity, its stress-strain rate curve has a gentle slope, and an increase of the magnitude of differential stress from σ'_A to σ'_B yields a relatively large increase in the strain rate magnitude from $\dot{e}_{Al}$ or $\dot{e}_{Bl}$ (Figure 7.23a). If a material has a high viscosity, its stress-strain rate curve has a steep slope, and an increase of the magnitude of differential stress from σ'_A to σ'_B yields a comparatively small increase in the strain rate magnitude from $\dot{e}_{Ah}$ or $\dot{e}_{Bh}$ (cf. Figure 7.23a). The viscosity of water is low, on the order of 10^{-2} *P*, meaning that the slope of its stress-strain rate curve is very shallow. The viscosities of oils are higher, typically ~1–10 *P*. Basaltic magmas have viscosities of 10^3–10^4 *P*, and granitic magmas have viscosities of 10^{10}–10^{12} *P*. Analyses of postglacial rebound yield estimates of viscosity for mantle rock at ~10^{23} *P*, indicating a very steep stress-strain rate curve.

The statement that a material is linearly viscous and has constant viscosity is understood to imply that its viscosity is not dependent upon the differential stress magnitude. The viscosities of linearly viscous materials do vary with temperature and other physical conditions. Cooking oil will pour more slowly when cool than when warmed in a pan.

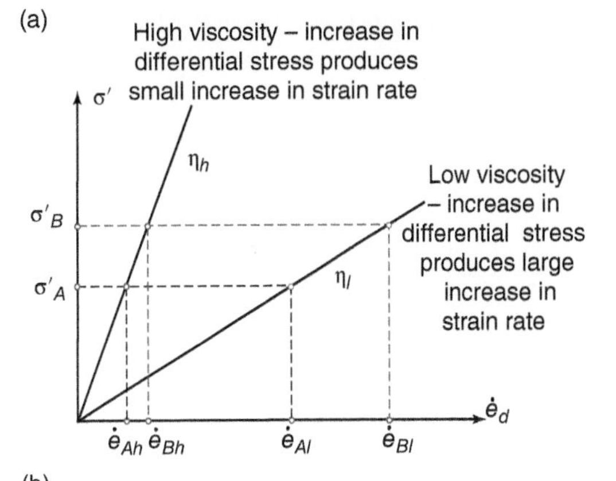

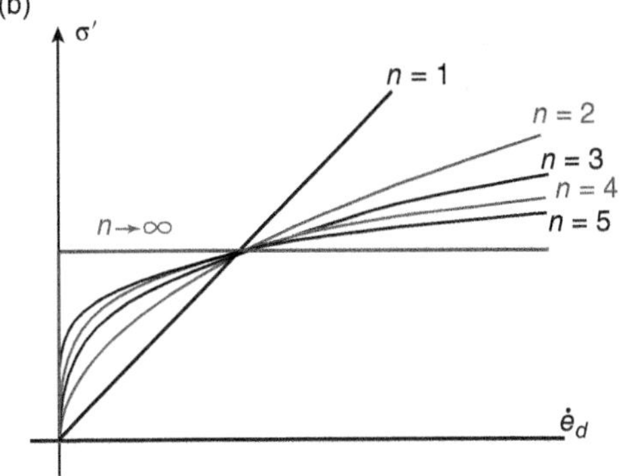

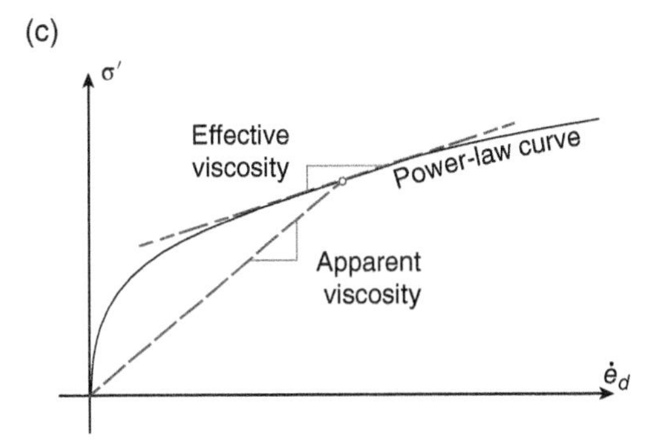

Figure 7.23 (a) Stress-strain rate plot for linearly viscous materials with high viscosity (η_h) and low viscosity (η_l). (b) Stress-strain rate plots for nonlinear viscous materials approximated by power-law relations with different power-law exponents n. (c) Diagram showing the difference between the apparent viscosity and the effective viscosity.

In a three-dimensional linearly viscous flow, the magnitudes of the different components of the strain rate are proportional to the magnitudes of the respective deviatoric stress components. Thus,

$$\sigma'_{ij} = \eta \dot{e}_{ij}. \tag{7.30a}$$

Written in matrix form, we have

$$[\sigma'] = \eta[\dot{e}]. \tag{7.30b}$$

7A.2.6.2 Nonlinear Viscosity

Many geological materials exhibit viscous behavior, with a unique relationship between differential stress and rate of creep, but are not linearly viscous. Their viscosity varies with the magnitude of the differential stresses. For example, the magnitudes of differential stresses in glaciers increase linearly with depth but the strain rates increase nonlinearly with depth (see Figure 7.21b), indicating that the viscosity of ice varies with the magnitude of differential stress. In ice, then, the slope of a stress-strain rate curve is steep at low differential stresses (viscosity is high) and shallow at high differential stresses (viscosity is low). We typically represent nonlinear viscosity using a ***power-law*** relationship (Figure 7.23b). In a material that undergoes viscous flow according to a power law, the constitutive equation has a form

$$\dot{e} = f(\sigma') = B\left(\frac{\sigma}{S_o}\right)^n. \tag{7.31}$$

In Eq. (7.31), B is a material constant with the units of strain rate and S_o is a material constant with units of stress, and n is a dimensionless material constant. Laboratory deformation experiments indicate that $n \approx 3$–4 for many geological materials. In materials exhibiting power-law behavior, we distinguish between the **apparent viscosity** and the **effective viscosity** (Figure 7.23c). The apparent viscosity is the ratio of the differential stress divided by the observed strain rate; one can envision this as the slope of a line from the origin of the stress versus strain rate axes to a point along the power-law curve. The effective viscosity is the local slope of the power-law curve, which measures the increase in differential stress required to generate a characteristic increase in the strain rate. Note that power-law materials have higher effective viscosities at low differential stresses and lower effective viscosities at higher differential stresses.

When $n = 1$,

$$\sigma' = \left(\frac{S_o}{B}\right)\dot{e}. \tag{7.32}$$

Equation (7.32) has the same form as Eq. (7.29), with $S_o/B = \eta$ (the viscosity) (Figure 7.23b).

As the exponent n gets larger, the flow described by Eq. (7.31) increasingly resembles plastic flow, where viscosity is high for low differential stress magnitudes and low for higher differential stress values (Figure 7.23b). As $n \rightarrow \infty$, the power-law depicts perfectly plastic flow.

$$\text{If } \sigma' < S_o, \text{ then} \left(\frac{\sigma'}{S_o}\right)^n \rightarrow 0. \tag{7.33a}$$

Whatever the value of B, the strain rate will be zero. On the other hand,

$$\text{if } \sigma' = S_o, \text{ then} \left(\frac{\sigma'}{S_o}\right)^n = 1. \tag{7.33b}$$

The strain rate will have a finite value (B). Finally,

$$\text{if } \sigma' > S_o, \text{ then} \left(\frac{\sigma'}{S_o}\right)^n \rightarrow \infty. \tag{7.33c}$$

An attempt to increase the differential stress above S_o will yield a dramatic increase in the strain rate. This is essentially the behavior described by Eq. (7.28), that is, a perfect plastic material with a yield stress $\sigma_y = S_o$.

7A.3 Summary

Rock-forming materials exhibit different responses to applied loads depending upon the conditions and the length of time over which the loads are applied. When subjected to relatively low stresses or when subjected to moderate stresses for short periods of time, rocks deform but the deformation is not permanent. We say that this deformation is recoverable and refer to it as elastic deformation. Elastic deformation requires an input of energy, called strain energy, given by the product of the magnitudes of the applied stress and the observed strain. Strain energy is released on unloading. In elastic deformation, the magnitude of the strain has a unique relationship to the magnitude of stress. In many settings of elastic deformation stress and strain are linearly related. That is, the slope of a stress versus elastic strain plot is a constant with units of stress. Geologists regularly use one of two sets of measured material parameters to characterize elastic deformation. The first, derived from tests of axial shortening or elongation, is Young's modulus (E). Young's modulus is the ratio of axial compression or tension to axial shortening or elongation. A related parameter, Poisson's ratio (ν), measures the associated lateral expansion or contraction, respectively. The second set of elastic constants are the bulk modulus K, the ratio of mean stress to volumetric strain, and the shear modulus G, the ratio of deviatoric stress to deviatoric strain.

When differential stresses are sufficiently great they cause permanent deformation. Geologists distinguish two end-member types of permanent deformation, fracture and flow. It is important to recognize that fracture and flow are not mutually exclusive. In nature and in experimental studies, fractures may form with or without significant flow of the surrounding rock. Materials that deform by forming fractures with little associated flow are said to exhibit brittle behavior. Materials that accrue significant deformation by flow, even if they eventually form fractures, are said to exhibit ductile behavior.

In this chapter, however, we examine fracture formation without significant flow, what is known as brittle fracture. An analysis of shear fracture formation addresses the empirical data showing that the magnitude of differential stress required to form shear fractures increases with the magnitude of the mean stress. A relatively straightforward analysis of shear fracture formation gives rise to a failure criterion in which the magnitude of the shear stress (σ_T) acting across the surface that will become the shear fracture is linearly related to the magnitude of the normal stress acting on that plane:

$$\sigma_T = \tau_o + \mu\sigma_N, \tag{7.15}$$

where τ_o is the cohesion, the shear stress required to form a shear fracture on a plane in the absence of any normal stress, and μ is the coefficient of friction.

The measured fracture strengths of Earth materials in tension are significantly lower than what one would anticipate from the strengths of the bonds within those materials. Following the analysis of Griffith, we understand that the presence of microscopic flaws leads to local stress concentrations, where stresses are sufficiently great first to distort materials elastically and eventually to break strong bonds. The released strain energy is consumed in the formation of the fracture surfaces.

Earth materials may also accrue permanent deformation by creep, which is flow in response to stresses applied over long time periods. Because materials can accrue deformation at constant or even decreasing magnitudes of differential stress, we cannot relate stress and strain magnitudes, instead, we relate differential stress to the rate of creep. There are two end-member types of creep. In viscous deformation, there is a unique relationship between the deviatoric stress and the rate of creep. In linearly viscous materials, flow occurs at vanishingly small magnitudes of differential stress and occurs at a rate proportional to the differential stress magnitude. The proportionality constant relating deviatoric stress to strain rate or the slope of a stress versus strain rate plot is the viscosity (η), which has units of stress×time. In perfectly plastic deformation, there is no longer a unique relationship between the deviatoric stress and the rate of creep. In perfectly plastic materials, the materials resist flow at small magnitudes of differential stress. Should the differential stress reach a threshold value, known as the yield strength or yield stress (σ_y), flow will occur at any rate dictated by external conditions. In fact, most geological materials exhibit creep behavior intermediate between linear viscous and perfectly plastic creep; there is a unique relationship between differential stress and creep rate but a plot of differential stress versus strain rate is nonlinear. A power-law relationship, where strain rate is proportional to differential stress raised to a power, represents the flow of such materials. Experiments show that many geologic materials exhibit power-law constitutive relationships with exponents n ~3–4.

7B Rheology: Comprehensive Treatment

As in earlier chapters, the Comprehensive Treatment section provides a deeper exploration of the topics addressed in the Conceptual Foundation section. This section begins by combining models of rock deformation introduced in Section 7A to gain greater insight into the behavior of real materials over geological time scales. As in earlier chapters, we use tensor and/or three-dimensional analyses to extend concepts introduced in Section 7A. Those approaches make it possible to address material anisotropy in elastic and viscous deformation and to modify failure criteria to accommodate three-dimensional deformation in a greater range of deformation conditions. The Comprehensive Treatment section closes with a discussion of how geologists have attempted to address how vertical gradients of temperature and pressure impact the rheology of different portions of the lithosphere.

7B.1 Combining Deformation Models to Describe Rock Properties

At different points throughout Section 7A Rheology – Conceptual Foundation but mainly in Section 7A.2.4, we noted the significance of time as a factor in the deformation of rock. Having learned that deformation by viscous flow can occur at low-stress magnitudes, you will have a clearer understanding of the need for qualifying statements such as, "If the end load is held at σ_B (a stress value below that required to reach the elastic limit) for moderate lengths of time. . .." in a discussion of recoverable deformation (for example Section 7A.1.2). Over short periods of time, rocks often exhibit nearly ideal elastic behavior. Seismic waves travel regularly through rock masses via elastic deformation with no permanent effects. Further, seismic waves regularly travel through the lithosphere with little energy loss along any individual segment of their path, meaning that nearly all of the imposed strain energy associated with that elastic deformation is recovered. Over longer time periods, however, rock masses experiencing the small strains and moderate stress levels characteristic of elastic deformation begin to accrue permanent deformation.

This effect can be seen in stress versus time plots derived from a particular type of deformation experiment known as a "relaxation test." In such a test, a sample is held at stresses and strains below its elastic limit for an extended period of time. Figure 7.24 is a plot showing the changes in stresses in the sample versus time. Segment O to A in Figure 7.24 shows the initial effect of the

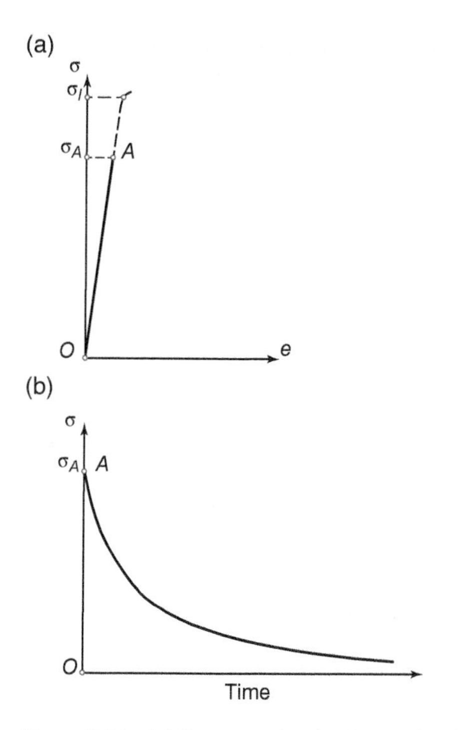

Figure 7.24 (a) Stress-strain plot for an elastic material. Applied stress σ_A causes elastic strain $e = A$. (b) A plot of stress versus time for a typical relaxation test.

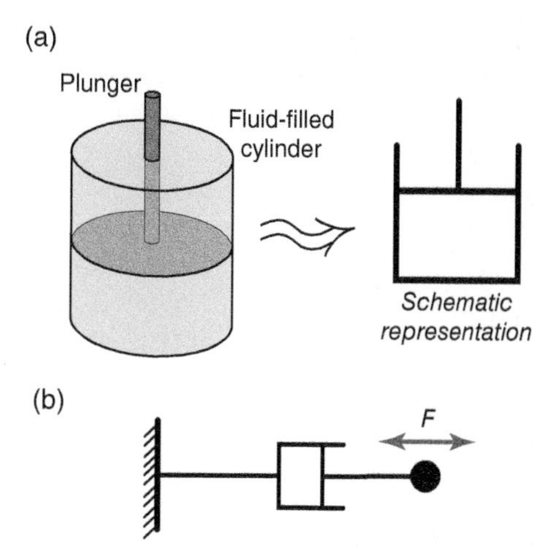

Figure 7.25 (a) At left is a rendering of dashpot, a fluid-filled container with a plunger. At right is the typical schematic representation of a dashpot. (b) Schematic representation of a viscous material. The force ***F*** required to move the component (solid ball) relative to the wall will vary directly with the rate of movement of the component.

loading of the sample – displacement of the piston at a constant rate causes the end load to rise at a constant rate to the value σ_A, which is below the sample's elastic limit σ_l. If the piston is held at the position initially corresponding to the end load σ_A for an extended period of time, we observe that stresses fall and permanent strains accrue slowly. The sample that initially exhibited elastic behavior begins to creep at stresses below the elastic limit. The creep rates are slow, and the accrued permanent strains are small in experiments. In nature, however, stresses and strains below the experimentally determined elastic limit may prevail for time periods sufficiently long that creep leads to significant permanent deformation.

One approach used to model both the time-independent and time-dependent character of real materials is to envision that their properties are combinations of the properties of elastic and viscous materials. Materials that exhibit properties consistent with both elastic and viscous behavior are known as **visco-elastic materials**. In Section 7A.2.1, we used the analogy of spring to describe elastic behavior. Like a spring, elastic deformation is directly proportional to the stresses applied and recoverable. Material scientists regularly use a *dashpot* to portray viscous behavior. A dashpot (Figure 7.25) is a fluid-filled container with a plunger. Because the plunger fits loosely in the container or has one or more holes in it, its movement within the container is constrained by the rate that the fluid can flow around the plunger or through the hole(s) in it. The force required to move the plunger depends on the rate of movement, with a greater force needed for more rapid movement. Dashpots are used to damp, that is resist and/or slow, the movement of whatever is attached to the plunger. You may have encountered a dashpot used to slow the rate of movement of hinged features like doors, desktops, or lids on large containers. Dashpots are also components of the shock absorbers found on cars, motorcycles, and some bicycles. Material scientists portray the combination of elastic and viscous behavior described earlier, and those outlined later, by envisioning that rocks' properties are those exhibited by one of two combinations of a spring (elastic behavior) and dashpot (viscous behavior) (see Jaeger 1971, pp. 99–106; Twiss and Moores 2007, pp. 460–464).

Perhaps the simplest way to combine springs and dashpots to portray visco-elastic (that is combined elastic and viscous) behavior is to envision a movable component connected to a fixed feature via a spring attached to the end of the plunger of a dashpot (Figure 7.26a). A material that exhibits behavior consistent with this combination is said to be a *Maxwell material* or to exhibit **elastico-viscous behavior**. A load applied to an elastico-viscous material generates an immediate response – an elastic strain. If the load is transient, applied briefly and then removed, that elastic strain is fully recovered and there is no permanent deformation (Figure 7.26b). Drawing upon the schematic representation of elastico-viscous behavior in Figure 7.26a, this is equivalent to seeing the spring compress or stretch and then return to its original configuration before the dashpot is able to respond. If the load is applied over a longer period of time, the dashpot begins to take up the stretching or compression initially accommodated by the spring. If stress magnitudes are held constant, permanent deformation due to viscous flow would accrue at fixed rates (Figure 7.26b). In a

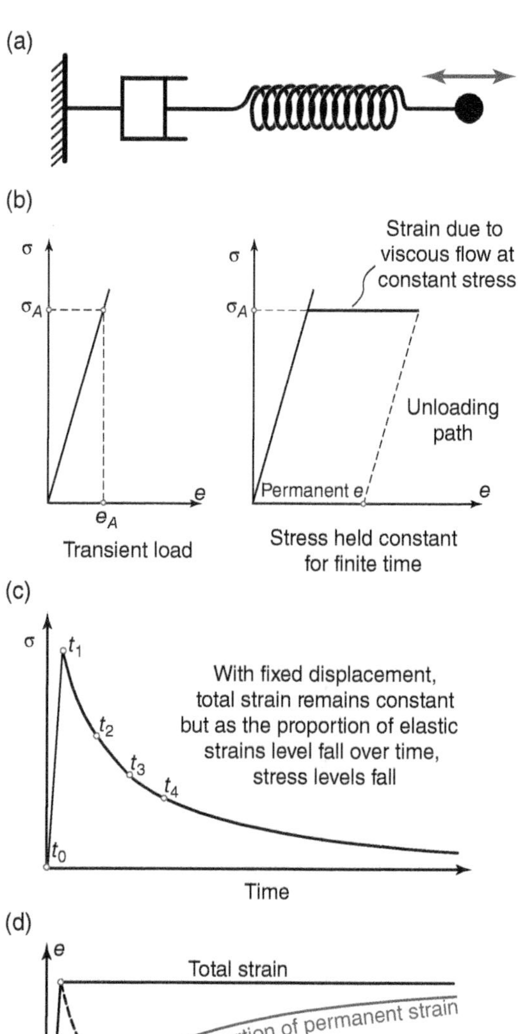

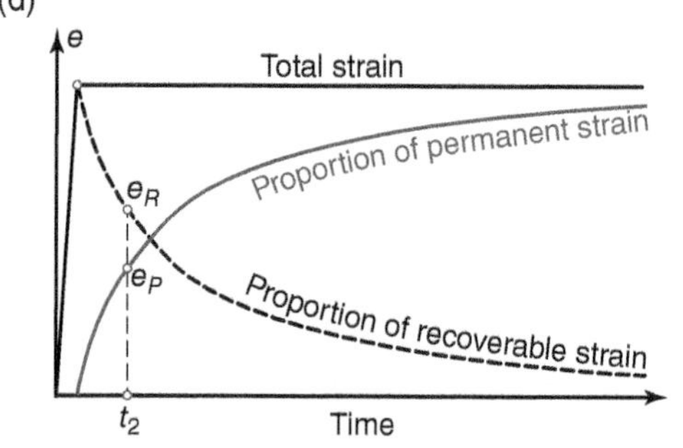

Figure 7.26 (a) Schematic (spring and dashpot) representation of an elastico-viscous material. (b) Stress-strain plot at the left shows an elastico-viscous material's response to transient load. Stress-strain plot at right shows an elastico-viscous material's response to sustained load. (c) Stress versus time plot showing the response of an elastico-viscous material to a fixed displacement. (d) Strain versus time plot corresponding to (c). Total strain remains constant. Initially, the strain is entirely elastic (and recoverable); over time an increasing proportion of the strain is permanent (viscous).

deformation experiment where the load is due to a moving piston (cf. Figure 7.2), continued movement of the piston would be required to maintain constant stress levels (and a constant magnitude of elastic strain).

Figure 7.26c and d illustrates the situation where the piston in a laboratory experiment (or of a neighboring rock mass in nature) moves and then halts at a position sufficient to generate strains and stress levels below the material's elastic limit. As time passes, an elastico-viscous laboratory sample or rock mass accrues permanent strain by viscous flow at a rate proportional to the instantaneous stress magnitude. The total strain is determined by how far the piston or neighboring rock mass moved, but over time permanent strain accounts for a larger proportion of that total strain. As the magnitude of recoverable, elastic strain falls, deviatoric stresses also drop. This, in turn, causes the creep rate to slow. The net result is an asymptotic decay in stress magnitudes (Figures 7.24 and 7.26c) as the magnitude of permanent strains increase. If the load is removed at some point (e.g. time t_2, see Figure 7.26d), the elastic component of the total strain (e_R) can be recovered but the material will also exhibit permanent strain (e_P).

A different type of visco-elastic behavior, known as **firmo-viscous behavior**, is consistent with the movable component directly connected to the fixed feature via both a spring and a dashpot (Figure 7.27a). Materials that exhibit behavior consistent with this combination are said to be Kelvin-Voigt materials. A Kelvin-Voigt material's response to an applied load is a time-dependent (viscous) increase in strain, eventually attaining a strain magnitude proportional to the applied load (an elastic strain). On unloading, deformation is recovered, but this too is time-dependent (Figure 7.27b). Automobile shock absorbers rely on this time-dependent damping on loading and unloading over short periods of time. The isostatic subsidence due to glacial ice

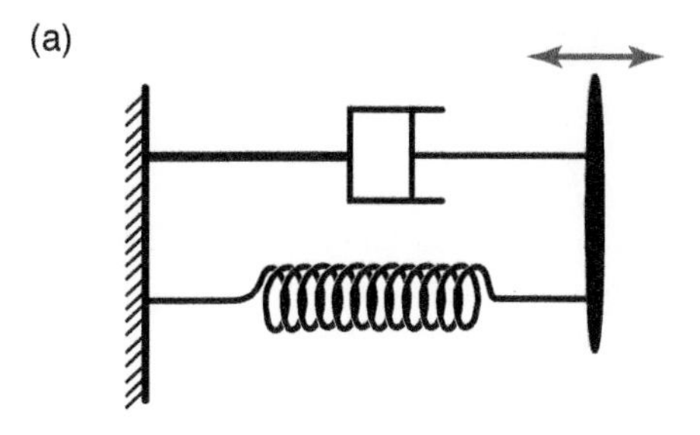

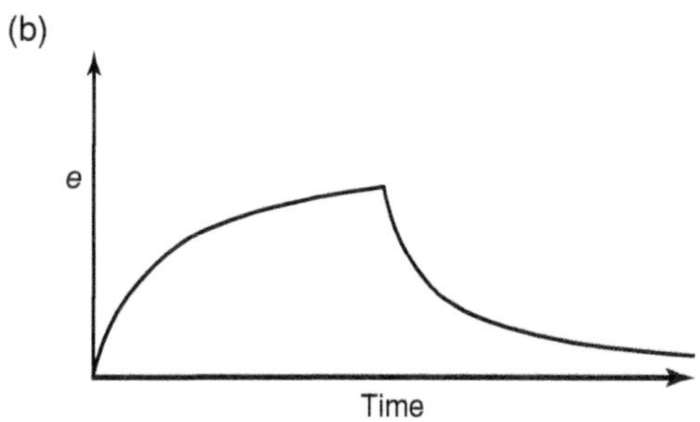

Figure 7.27 (a) Schematic (spring and dashpot) representation of a firmo-viscous response. (b) A strain versus time plot showing the firmo-viscous response to loading and unloading.

sheets and the subsequent postglacial isostatic rebound can be modeled as a firmo-viscous response.

In some settings, rocks exhibit **elasto-plastic behavior**. In this model, applied loads generate an immediate elastic strain. Should the applied load reach the material's yield strength, the material will accrue time-dependent permanent (plastic) deformation (Figure 7.28a). On unloading, there will be a component of the deformation, the elastic portion, that is recovered. Stable, frictional sliding is one physical setting where elasto-plastic deformation occurs (Figure 7.28b). Less commonly used are models of **visco-plastic behavior**, in which materials resist deformation as long as the stresses remain below yield stress and then deform at a rate proportional to the amount that stresses exceed that yield stress.

(a)

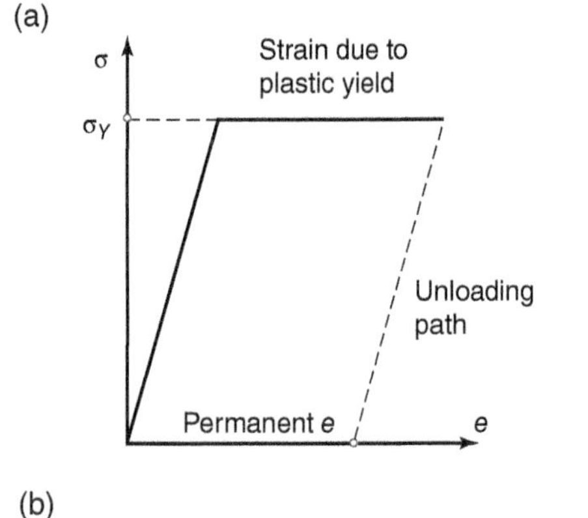

(b)

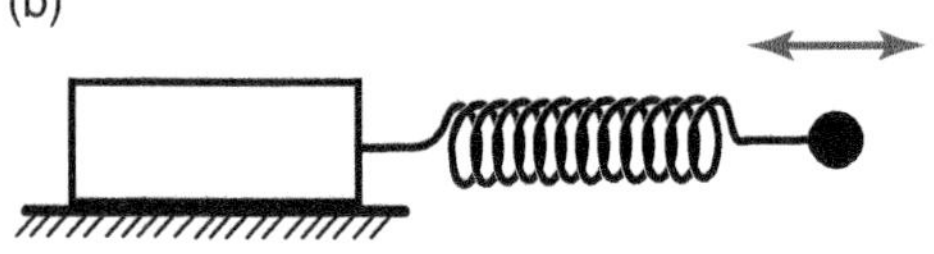

Figure 7.28 (a) Stress-strain plot for an elastico-plastic material. (b) Schematic (spring and slider block) representation of elastico-plastic response.

7B.2 Rock Deformation Modes

7B.2.1 Elasticity

7B.2.1.1 Equivalence of Elastic Constants

Section 7A.2.1.2 showed that two elastic constants are required to characterize fully elastic strains. Thus, one can use Young's modulus E and Poisson's ratio ν, the former relating the longitudinal stress and strain of a sample subjected to uniaxial loading and the latter determined from the radial expansion or contraction of that sample, to relate corresponding components of the stress and strain tensor in elastically deformed samples. Section 7A.2.1.3 showed that two different elastic constants, bulk modulus K and shear modulus G, relate the hydrostatic and differential stresses to the elastic volume change and shape change respectively under different loading situations. By examining how these two loading situations relate to each other, we can see that one pair of elastic constants can be derived from the other. The approach outlined in the following paragraphs is based on the analysis presented in Gundmundsson (2011, pp. 92–100).

In uniaxial loading, the longitudinal strains perpendicular to the shortening direction are

$$e_{22} = e_{33} = -\nu e_{11}. \tag{7.34}$$

Thus, normal stress acting along the axis of the sample, σ_{11}, generates radial strains perpendicular to the axis:

$$e_{22} = e_{33} = -\frac{(\nu\sigma_{11})}{E}. \tag{7.35}$$

The same is true for the mutually perpendicular, radial normal stresses σ_{22} and σ_{33}. Thus, the longitudinal strains parallel to the three mutually perpendicular axes are:

$$e_{11} = \frac{\sigma_{11}}{E} - \frac{(\nu\sigma_{22})}{E} - \frac{(\nu\sigma_{33})}{E} = \frac{[\sigma_{11} - \nu(\sigma_{22} + \sigma_{33})]}{E} \tag{7.36a}$$

$$e_{22} = \frac{\sigma_{22}}{E} - \frac{(\nu\sigma_{33})}{E} - \frac{(\nu\sigma_{11})}{E} = \frac{[\sigma_{22} - \nu(\sigma_{33} + \sigma_{11})]}{E} \tag{7.36b}$$

$$e_{33} = \frac{\sigma_{33}}{E} - \frac{(\nu\sigma_{11})}{E} - \frac{(\nu\sigma_{22})}{E} = \frac{[\sigma_{33} - \nu(\sigma_{11} + \sigma_{22})]}{E} \tag{7.36c}$$

One can show that comparable relations hold for the principal strains:

$$e_1 = \frac{\sigma_1}{E} - \frac{(\nu\sigma_2)}{E} - \frac{(\nu\sigma_3)}{E} = \frac{[\sigma_1 - \nu(\sigma_2 + \sigma_3)]}{E} \tag{7.37a}$$

$$e_2 = \frac{\sigma_2}{E} - \frac{(\nu\sigma_3)}{E} - \frac{(\nu\sigma_1)}{E} = \frac{[\sigma_2 - \nu(\sigma_3 + \sigma_1)]}{E} \tag{7.37b}$$

$$e_3 = \frac{\sigma_3}{E} - \frac{(\nu\sigma_1)}{E} - \frac{(\nu\sigma_2)}{E} = \frac{[\sigma_3 - \nu(\sigma_1 + \sigma_2)]}{E} \quad (7.37c)$$

The volumetric strain $e_V = (e_1 + e_2 + e_3)$ is related to the mean stress by the bulk modulus K. In Section 7A.2.1.3, we noted that stresses and strains in elastically distorted isotropic substances are related by

$$\sigma_m = K \cdot e_v. \quad (7.4)$$

Rearranging yields

$$e_v = \sigma_m / K$$

Using Eqs. (7.37), one can relate e_V to stresses using E and ν:

$$e_v = \frac{[\sigma_1 - \nu(\sigma_2 + \sigma_3)]}{E} + \frac{[\sigma_2 - \nu(\sigma_3 + \sigma_1)]}{E} + \frac{[\sigma_3 - \nu(\sigma_1 + \sigma_2)]}{E}, \quad (7.38a)$$

which simplifies to

$$e_v = \frac{[\sigma_1 - 2\nu\sigma_1 + \sigma_2 - 2\nu\sigma_2 + \sigma_3 - 2\nu\sigma_3]}{E} = \frac{[\sigma_1 + \sigma_2 + \sigma_3](1 - 2\nu)}{E}. \quad (7.38b)$$

Now $[\sigma_1 + \sigma_2 + \sigma_3] = 3\sigma_m$, so Eq. (7.38b) becomes

$$e_v = \frac{3\sigma_m(1 - 2\nu)}{E}. \quad (7.39a)$$

Substituting from Eq. (7.4b) yields

$$\frac{\sigma_m}{K} = \frac{3\sigma_m(1 - 2\nu)}{E} \quad (7.39b)$$

or

$$K = \frac{E}{3(1 - 2\nu)}. \quad (7.40)$$

Using similar reasoning, we can relate the shear modulus G to the Young's modulus E and Poisson's ratio ν. The maximum shear strain is $e_{13} = (e_1 - e_3)$, and the maximum shear stress is $\sigma_{13} = (\sigma_1 - \sigma_3)/2$. Shear stress and shear strain magnitudes are related by Eq. (7.6)

$$\sigma'_{ij} = G \cdot e'_{ij}. \quad (7.6)$$

In terms of the maximum shear stress and shear strain, we have

$$\sigma_{13} = \frac{(\sigma_1 - \sigma_3)}{2} = G \cdot e_{13} = G(e_1 - e_3) \quad (7.41a)$$

or

$$\frac{(\sigma_1 - \sigma_3)}{2G} = (e_1 - e_3). \quad (7.41b)$$

Substituting from Eqs. (7.37) yields

$$\frac{(\sigma_1 - \sigma_3)}{2G} = \frac{[\sigma_1 - \nu(\sigma_2 + \sigma_3)]}{E} - \frac{[\sigma_3 - \nu(\sigma_1 + \sigma_2)]}{E}, \quad (7.42a)$$

which simplifies to

$$\frac{(\sigma_1 - \sigma_3)}{2G} = \frac{(\sigma_1 - \sigma_3)(1 + \nu)}{E} \quad (7.42b)$$

or

$$G = \frac{E}{2(1 + \nu)}. \quad (7.43)$$

Equations (7.40) and (7.43) demonstrate the equivalence of the two approaches to evaluating elastic deformation and confirm that two distinct elastic constants are required to characterize fully the relationship between stresses and elastic strains in isotropic materials.

7B.2.1.2 Elasticity of Anisotropic Materials

In Section 7A.2.1.3, we noted that stresses and strains in elastically distorted isotropic substances are related by

$$\sigma_m = K \cdot e_v \quad (7.4)$$

and

$$[\sigma'] = G[e']. \tag{7.5}$$

We can rewrite Eq. (7.4) in matrix form by noting that the matrix form of the mean stress is the scalar matrix $[\sigma_m]$ and the matrix form of the volumetric strain is $[e_v]$. Thus, we have

$$[\sigma_m] = K[e_v], \tag{7.4a}$$

Since the total stress matrix $[\sigma_M]$ and volumetric strain matrix $[e_V]$ are both scalar matrices. Since the total stress $[\sigma] = [\sigma_m] + [\sigma']$, the relationship between the stress and strain in for a linearly-elastic isotropic material is

$$[\sigma] = K[e_v] + G[e'] = Ke_v\mathbf{I} + G[e'], \tag{7.7a}$$

where **I** is the identity matrix. This is the matrix form of Eq. (7.7).

Most rock-forming minerals are anisotropic, that is their three-dimensional arrangements of atoms and ions are such that the characteristic distances between atoms or ions and strengths of bonds vary with orientation relative to the mineral's crystallographic axes. For linearly elastic anisotropic materials, even hydrostatic stresses result in different elastic strain magnitudes along different directions relative to the mineral's crystallographic axes. Thus, the relationships between elastic strains and stresses in anisotropic minerals cannot be captured by either of the pairs of independent elastic constants (E and ν or G and K) described earlier. We follow the approach outlined by Nye (1976, pp. 131–149) in examining the elastic behavior of anisotropic materials.

In describing linearly elastic anisotropic materials, each component of the stress tensor is, in the most general case, related to the elastic strain components by nine linear equations of the form

$$\begin{aligned}\sigma_{11} = {} & c_{1111}e_{11} + c_{1112}e_{12} + c_{1113}e_{13} + c_{1121}e_{21} \\ & + c_{1122}e_{22} + c_{1123}e_{23} + c_{1131}e_{31} \\ & + c_{1132}e_{32} + c_{1133}e_{33}\end{aligned} \tag{7.44a}$$

$$\begin{aligned}\sigma_{12} = {} & c_{1211}e_{11} + c_{1212}e_{12} + c_{1213}e_{13} \\ & + c_{1221}e_{21} + c_{1222}e_{22} + c_{1223}e_{23} \\ & + c_{1231}e_{31} + c_{1232}e_{32} + c_{1233}e_{33}\end{aligned} \tag{7.44b}$$

etc. Using suffix notation, one can represent these nine equations with

$$\sigma_{ij} = c_{ijkl} \cdot e_{kl}. \tag{7.44}$$

The 81 elements c_{ijkl} are elastic **stiffness constants** or stiffnesses. Alternatively, each elastic strain component is related to the stress components by nine linear equations of the form

$$\begin{aligned}e_{11} = {} & s_{1111}\sigma_{11} + s_{1112}\sigma_{12} + s_{1113}\sigma_{13} \\ & + s_{1121}\sigma_{21} + s_{1122}\sigma_{22} + s_{1123}\sigma_{23} \\ & + s_{1131}\sigma_{31} + s_{1132}\sigma_{32} + s_{1133}\sigma_{33}\end{aligned} \tag{7.45a}$$

$$\begin{aligned}e_{12} = {} & s_{1211}\sigma_{11} + s_{1212}\sigma_{12} + s_{1213}\sigma_{13} \\ & + s_{1221}\sigma_{21} + s_{1222}\sigma_{22} + s_{1223}\sigma_{23} \\ & + s_{1231}\sigma_{31} + s_{1232}\sigma_{32} + s_{1233}\sigma_{33}\end{aligned} \tag{7.45b}$$

etc., or in suffix notation

$$e_{ij} = s_{ijkl} \cdot \sigma_{kl}. \tag{7.45}$$

Here the 81 elements s_{ijkl} are known as **compliances**.

In Eqs. (7.44) and (7.45), the elastic stiffness coefficients and the compliances relate the second rank stress tensor to the second rank strain tensor. The c_{ijkl} and s_{ijkl} arrays are, then, fourth rank tensors.

One can gain insight into the tensor components c_{ijkl} and s_{ijkl} by examining the characteristics of the elastic strains associated with idealized stress distributions. If the only nonzero stress component is σ_{11}, then $e_{11} = s_{1111}\,\sigma_{11}$, $e_{12} = s_{1211}\,\sigma_{11}$, $e_{13} = s_{1311}\,\sigma_{11}$, $e_{21} = s_{2111}\,\sigma_{11}$, etc. Now $e_{12} = e_{21}$ and $e_{13} = e_{31}$ so s_{1211} must equal s_{2111} and s_{1311} must equal s_{3111}, etc. This line of reasoning leads to the general conclusion that

$$s_{ijkl} = s_{jikl}. \tag{7.46}$$

Since there are no net torques in stress states, a nonzero shear stress σ_{12} is necessarily accompanied by shear stress $\sigma_{21} = \sigma_{12}$. If all other stress components are zero, Equation (7.45a) yields $e_{11} = s_{1112}\,\sigma_{12} + s_{1121}\,\sigma_{21}$. Because $\sigma_{21} = \sigma_{12}$, $e_{11} = s_{1112}\,\sigma_{12} + s_{1121}\,\sigma_{12} = (s_{1112} + s_{1121})\,\sigma_{12}$. It is not possible to determine individual values for s_{1112} and s_{1121}. To avoid the need to introduce in the analysis an arbitrary constant to relate s_{1112} and s_{1121}, we take the two coefficients to be equal. Extending this line of reasoning, we conclude that

$$s_{ijkl} = s_{ijlk}. \tag{7.47}$$

The conditions (7.46) and (7.47) constrain the number of *independent* compliances to 36.

Similarly, envisioning a situation where all strain components except for e_{11} vanish, we see that $\sigma_{11} = c_{1111}\,e_{11}$, $\sigma_{12} = c_{1211}\,e_{11}$, $\sigma_{13} = c_{1311}\,e_{11}$, $\sigma_{21} = c_{2111}\,e_{11}$, etc. Now $\sigma_{12} = \sigma_{21}$ and $\sigma_{13} = \sigma_{31}$, so c_{1211} must equal c_{2111} and c_{1311} must equal c_{3111}, etc., again leading to the conclusion that

$$c_{ijkl} = c_{jikl}. \tag{7.48}$$

When the only nonzero strain components are $e_{12} = e_{21}$, $\sigma_{11} = c_{1112}\,e_{12} + c_{1121}\,e_{21}$. Because $e_{21} = e_{12}$, $\sigma_{11} = c_{1112}\,e_{12} + c_{1121}\,e_{12} = (c_{1112} + c_{1121})\,e_{12}$. Following reasoning similar to that outlined earlier, we see that $c_{1112} = c_{1121}$ and in general

$$c_{ijkl} = c_{ijlk}. \tag{7.49}$$

Conditions (7.48) and (7.49) again indicate that the number of *independent* stiffnesses is 36.

The fourth-order tensor relations expressed in Eqs. (7.44) and (7.45) are readily handled computationally, but even with the restrictions outlined in Eqs. (7.47) and (7.49) it is difficult to characterize the relationships between elastic stresses and strains in anisotropic materials. A three-step method that portrays the stress and strain tensors as column matrices and the fourth-order stiffness coefficient and compliance tensors as 6×6 matrices (see Nye 1976, pp. 138–149) provides insight into the relationships between stress and strain in anisotropic materials.

The first step in this method replaces nine elements of the 3×3 stress and strain tensors by a series of elements identified by a single suffix. Both stress and strain tensors are symmetric, with six independent components. Thus, the stress or strain components can be portrayed by elements with a single suffix whose values range from 1 to 6. Elements along the principal diagonal of either the stress or strain tensors are assigned a single suffix from 1 to 3, starting in the upper left. Components in the third column of the stress or strain tensors are assigned consecutive indices of 3–5 moving upward from the lower right to the upper right of the array. The final independent stress or strain component, the element in the middle of the first row, is given the single suffix 6. The remaining elements in the array have suffix values that preserve the symmetric character of the stress tensor. In this way, either tensor can be sequentially rewritten in the form of a column matrix. Beginning with the stress tensor:

$$\begin{vmatrix} \sigma_{11} & \sigma_{12} & \sigma_{13} \\ \sigma_{21} & \sigma_{22} & \sigma_{23} \\ \sigma_{31} & \sigma_{32} & \sigma_{33} \end{vmatrix} \rightarrow \begin{vmatrix} \sigma_1 & \sigma_6 & \sigma_5 \\ \sigma_6 & \sigma_2 & \sigma_4 \\ \sigma_5 & \sigma_4 & \sigma_3 \end{vmatrix} \rightarrow \begin{vmatrix} \sigma_1 \\ \sigma_2 \\ \sigma_3 \\ \sigma_4 \\ \sigma_5 \\ \sigma_6 \end{vmatrix} \tag{7.50}$$

A similar scheme applied to the strain tensor has off-diagonal elements multiplied by ½:

$$\begin{vmatrix} e_{11} & e_{12} & e_{13} \\ e_{21} & e_{22} & e_{23} \\ e_{31} & e_{32} & e_{33} \end{vmatrix} \rightarrow \begin{vmatrix} e_1 & \frac{e_6}{2} & \frac{e_5}{2} \\ \frac{e_6}{2} & e_2 & \frac{e_4}{2} \\ \frac{e_5}{2} & \frac{e_4}{2} & e_3 \end{vmatrix} \rightarrow \begin{vmatrix} e_1 \\ e_2 \\ e_3 \\ \frac{e_4}{2} \\ \frac{e_5}{2} \\ \frac{e_6}{2} \end{vmatrix} \tag{7.51}$$

It is tempting to consider these column matrices as vectors. However, under coordinate frame reorientations, they do not adhere to the vector transformation law. They are not, then, vectors.

Next, the components of the c_{ijkl} and s_{ijkl} tensors are depicted as 6×6 arrays of elements with two suffices. The values of the suffixes of each element are derived from the values of the first two or last two tensor suffixes. A "11" pair of tensor suffixes is replaced by a single suffix "1," a "22" pair of tensor suffixes is replaced by a single suffix "2," and a "33" pair of tensor suffixes is replaced by a single suffix "3." A "23" or "32" pair of tensor suffixes is replaced by a single suffix "4," a "31" or "13" or pair of tensor suffixes is replaced by a single suffix "5," and a "12" or "21" pair of tensor suffixes is replaced by a single suffix "6." In addition, the following substitutions are made to simplify the equations that substitute for Eq. (7.45):

s_{ijkl} are replaced by s_{mn} when m and n are 1, 2, or 3

$2s_{ijkl}$ are replaced by s_{mn} when either m and n are 4, 5, or 6

$4s_{ijkl}$ are replaced by s_{mn} when both m and n are 4,5, or 6

The fourth rank tensors c_{ijkl} and s_{ijkl} are then depicted by square arrays:

$$\begin{vmatrix} c_{11} & c_{12} & c_{13} & c_{14} & c_{15} & c_{16} \\ c_{21} & c_{22} & c_{23} & c_{24} & c_{25} & c_{26} \\ c_{31} & c_{32} & c_{33} & c_{34} & c_{35} & c_{36} \\ c_{41} & c_{42} & c_{43} & c_{44} & c_{45} & c_{46} \\ c_{51} & c_{52} & c_{53} & c_{54} & c_{55} & c_{56} \\ c_{61} & c_{62} & c_{63} & c_{64} & c_{65} & c_{66} \end{vmatrix}$$

and

$$\begin{vmatrix} s_{11} & s_{12} & s_{13} & s_{14} & s_{15} & s_{16} \\ s_{21} & s_{22} & s_{23} & s_{24} & s_{25} & s_{26} \\ s_{31} & s_{32} & s_{33} & s_{34} & s_{35} & s_{36} \\ s_{41} & s_{42} & s_{43} & s_{44} & s_{45} & s_{46} \\ s_{51} & s_{52} & s_{53} & s_{54} & s_{55} & s_{56} \\ s_{61} & s_{62} & s_{63} & s_{64} & s_{65} & s_{66} \end{vmatrix}$$

Equations (7.44) and (7.45) can then be written as matrix multiplication using one of the two arrays earlier and the column matrices defined by (7.50) or (7.51). Writing this matrix multiplication in suffix notation, we have simplified representations of the fourth-order tensor equations:

$$\sigma_i = c_{ij} \cdot e_j \text{ with } i \text{ and } j = 1 \text{ to } 6 \qquad (7.44^*)$$

and

$$e_i = s_{ij} \cdot \sigma_j \text{ with } i \text{ and } j = 1 \text{ to } 6, \qquad (7.45^*)$$

where the asterisk denotes the simplified version of the equation.

In Section 7A.2.1, we equated the product of the stress × strain with the energy per unit volume. Following the derivation outlined in Nye (1976), if an elastically distorted crystal with initial stresses σ_i and strains e_i experiences an additional increment of strain, the work done or energy added to the crystal will be

$$dW = \sigma_i \, de_i \text{ where } i = 1 \text{ to } 6, \qquad (7.52a)$$

or in tensor form

$$dW = \sigma_{ij} \, de_{ij} \text{ where } i = 1 \text{ to } 3. \qquad (7.52b)$$

For isothermal and reversible deformation, the work done equals the change in internal energy (Ψ) of the crystal. For linear elasticity (Eqs. 7.44 hold),

$$d\Psi = \sigma_i \, de_i = c_{ij} \cdot e_j \, de_i, \qquad (7.53)$$

or

$$\frac{\partial \Psi}{\partial e_i} = c_{ij} e_j. \qquad (7.54)$$

Differentiating Eq. (7.54) with respect to e_j yields

$$\frac{\partial}{\partial e_j}\left(\frac{\partial \Psi}{\partial e_i}\right) = c_{ij}. \qquad (7.55)$$

The internal energy Ψ is defined by the strain components, so the order of differentiation of the left-hand side of Eq. (7.55) does not matter. Thus,

$$\frac{\partial}{\partial e_j}\left(\frac{\partial \Psi}{\partial e_i}\right)=\frac{\partial}{\partial e_i}\left(\frac{\partial \Psi}{\partial e_j}\right), \quad (7.56)$$

but $\partial/\partial e_i(\partial\Psi/\partial e_j) = c_{ji}$, so

$$c_{ij} = c_{ji}. \quad (7.57)$$

Equation (7.57) indicates that in its most general form, the 6×6 c_{ij} array is symmetric, with only 21 independent elements. Similarly, the most general form of the 6×6 s_{ij} array also can be shown to be symmetric with 21 independent elements.

The most general forms of c_{ij} or s_{ij} arrays correspond to minerals with triclinic symmetry. Crystals with greater symmetry have c_{ij} or s_{ij} arrays with fewer independent elements. The numbers of independent elements for the different crystal classes are: (1) monoclinic – 13 independent elements; (2) orthorhombic – 9 independent elements; (3) tetragonal and trigonal – 7 or 6 independent elements; (4) hexagonal – 5 independent elements, and (5) cubic (isometric) – 3 independent elements. Isotropic substances have but two independent elements, s_{11} and s_{12}. These two elements are readily related to Young's modulus and Poisson's ratio, e.g. $s_{11} = 1/E$ and $s_{12} = -\nu/E$.

7B.2.2 Fracture or Fault Formation

7B.2.2.1 A Closer Examination of Fault Slip Criteria

In Section 7A.2.2.1, the approach taken to understand fault formation combined empirical observations on the geometry of conjugate faults with a hypothesis concerning the roughness of fault surfaces and the yield strengths of rock to derive criteria for fault slip (see Eqs. 7.14 and 7.15) in which

$$\sigma_T \propto \mu\sigma_N,$$

where σ_T is the shear stress acting on a surface, σ_N is the normal stress acting across the surface, and μ is the coefficient of friction of the rock. The hypothesis that μ is explicitly related to the relative magnitudes of the normal and shear stresses at which materials yield (cf. Eqs. 7.11, 7.12, and 7.13) leads one to infer that: 1) coefficients of friction should be constant under fixed physical conditions (such as temperature, pressure, rock composition); and 2) the magnitudes of coefficients of friction should vary systematically with rock type and/or composition.

The first of these inferences is the basis for citing, as we did in Section 7B.1, that stable frictional sliding is an example of elasto-plastic behavior, where deformation occurs at any imposed rate once differential stresses reach a threshold level. However, in experimental analyses conducted in a different type of testing apparatus (Figure 7.29a), fault sliding often departs from the anticipated idealized elasto-plastic behavior. Figure 7.29b, after Byerlee (1978), is an idealized plot of shearing force versus displacement often observed in response to shearing at a constant displacement rate. Instead of steady slip at the imposed rate, fault movement often is characterized by distinct slip increments, each associated with a drop in shearing force, separated by intervals of no slip in which the shearing force magnitude increases. This behavior, known as **stick-slip behavior**, is thought to mimic the behavior of some natural faults, where slip occurs in discrete events – earthquakes – separated by periods of no movement on the fault. In other cases (corresponding to the dotted segment of the plot in Figure 7.29b), fault slip may occur continuously, but typically the coefficient of friction drops as slip accrues. The observation that the shearing force required for movement is not constant contradicts the inference that coefficients of friction are constant. Byerlee (1978) also compiled and analyzed laboratory data on fault slips that call into question the inference that rock strength is the most important factor in determining coefficients of friction. Byerlee's compilation suggested that under conditions comparable to those in Earth's lithosphere, a variety of rock types exhibit similar coefficients of friction. It seems, then, that Eqs. (7.14) and (7.15)

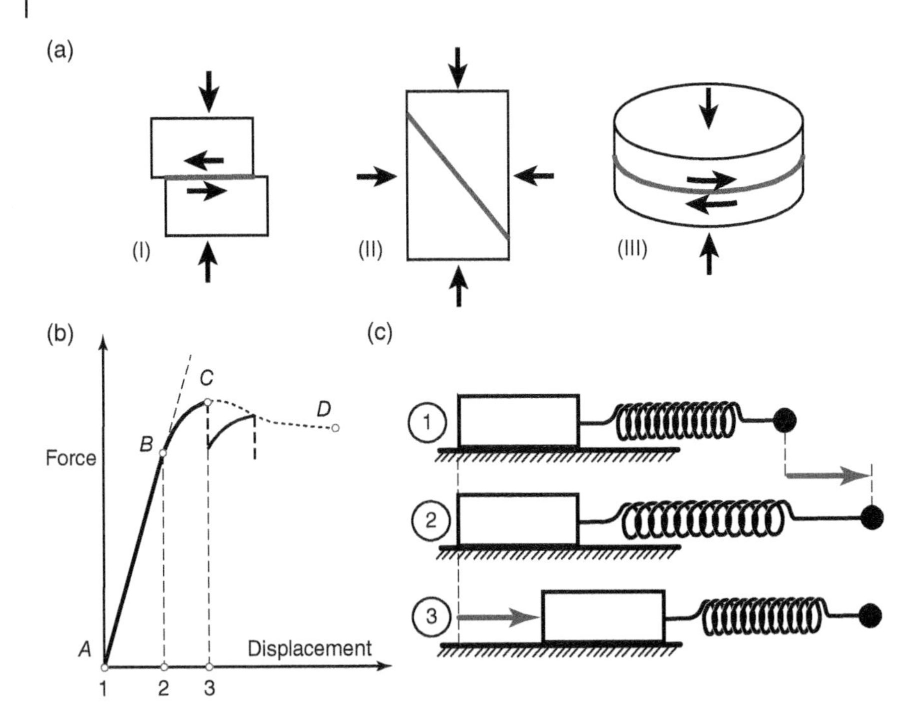

Figure 7.29 (a) Three common experimental set-ups for examining fault sliding. I. Direct shear test. II. Biaxial loading test. III. Rotary shear test. (b) A force versus displacement plot (after Byerlee 1978) showing schematically curves for stick-slip and stable sliding (dotted). (c) Schematic representations of three stages in stick-slip fault movement, corresponding to three displacement values labeled in (b).

are at best a significant simplification of fault formation and fault slip. Thus, rock friction warrants a closer look.

7B.2.2.1.1 Interpreting Rock Friction Data Figure 7.29c is a schematic representation showing how geologists conceptualize and interpret fault sliding experiments: a rider sitting on a rigid flat, connected by a spring to a component that applies a tangential force as it moves at constant velocity. In actuality there is no spring – the spring is intended to represent the elastic response of the system, consisting of the two samples separated by the fault and the surrounding apparatus, to the applied shearing. Prior to any movement, at stage 1 in Figure 7.29c, the force is zero (point *A* on the graph in Figure 7.29b). As shearing force is applied, the samples and the apparatus distort elastically. In the schematic representation, the tangential force initially extends the spring but the rider remains stationary (stage 2). The force rises linearly with displacement from *A* to *B* on the force-displacement graph. In actuality, of course, it is the elastic distortion of the samples and the surrounding apparatus that is measured by the linear displacement in response to the applied shearing. Segment *BC* on the force-displacement graph either records the beginning of slip on the fault or nonelastic deformation of the samples along the fault. In either case, the force departs from a linear force-displacement relationship. When the shear stress reaches a critical value, the samples will slip past each other. This is represented by point *C* on the force-displacement curve, where tangential force reaches its maximum value, and in the schematic representation, it corresponds to the rider moving to a new position (3 in Figure 7.29c), lowering the force. Continued application of the force may lead to a series of slip increments, each preceded by increases in force along nonlinear curves and then drops in force associated with the slip increment. Alternatively, the system may transition to stable sliding at a force magnitude D, as depicted by the dotted line on the force-displacement graph.

On its discovery, stick-slip behavior in frictional sliding experiments in the laboratory was immediately considered to provide an analogue to the behavior of faults in nature. Earthquakes are commonly considered as manifestations of slip events, and the interseismic period is considered the stick interval. Moreover, stable sliding has a natural analogue in the form of faults or segments of faults, such as the San Andreas fault and the Northern Anatolian fault, which move aseismically or "creep."

In the schematic representation of sliding experiments, the weight and area of the base of the rider are constant, so the normal stress across the base of the rider is constant. In the actual experiments, the normal stress acting across the fault contact is held constant. Thus, plots of tangential stress across the contact versus displacement will take a form similar to the curve in Figure 7.29. The ratio of shear stress to normal stress at a point equivalent to B on the force-displacement curve in Figure 7.29b is known as the *initial friction*. At a point equivalent to C on the force-displacement curve, the ratio of shear stress to normal stress is known as the *maximum friction* or the *static friction*. At a point equivalent to D on the force-displacement curve, the ratio of shear stress to normal stress is known as *residual friction* or *dynamic friction*.

In geologic settings, static friction (μ_s) is nearly always greater than dynamic friction (μ_d). Experimental data indicate that the magnitude of dynamic friction itself depends upon the sliding velocity. In some circumstances, samples exhibit **velocity strengthening** behavior, where dynamic friction increases with sliding velocity. An increase in the coefficient of friction with velocity will counteract any tendency to accelerate slip, thereby preventing the abrupt slip events characteristic of stick-slip behavior. In other circumstances, samples exhibit **velocity weakening** behavior, where dynamic friction decreases with sliding velocity. Velocity weakening behavior is a prerequisite for the accelerating slip that occurs during stick-slip behavior, but it alone is not sufficient to cause stick-slip behavior.

7B.2.2.1.2 Byerlee's Law Byerlee reported the shear stress required for slip compiled from a variety of sources. He acknowledged the potential errors in comparing reported values that could correspond to the initial, maximum, or residual friction. Nevertheless, he was able to draw several conclusions from the compiled data (Figure 7.30) that are still widely accepted:

- There is significant variability in the shear stress required for slip, especially at low values of normal stress ($\sigma_N < 5$ MPa).
- There is a weak correlation between rock type and maximum friction at low magnitudes of normal stress and no discernible correlations between rock type and maximum friction at intermediate or high magnitudes of normal stress. That is, at intermediate and high magnitudes of normal stress neither sedimentary, igneous, nor metamorphic rocks required consistently higher or lower shear stress magnitudes for slip.
- At low to intermediate magnitudes of normal stress, where $5\,\text{MPa} < \sigma_N \leq 200\,\text{MPa}$, the data conform to a relationship where the maximum tangential stress required for slip

$$\sigma_T = 0.85\sigma_N. \tag{7.58a}$$

- At higher magnitudes of normal stress, where $200\,\text{MPa} < \sigma_N \leq 2000\,\text{MPa}$, the data conform to a

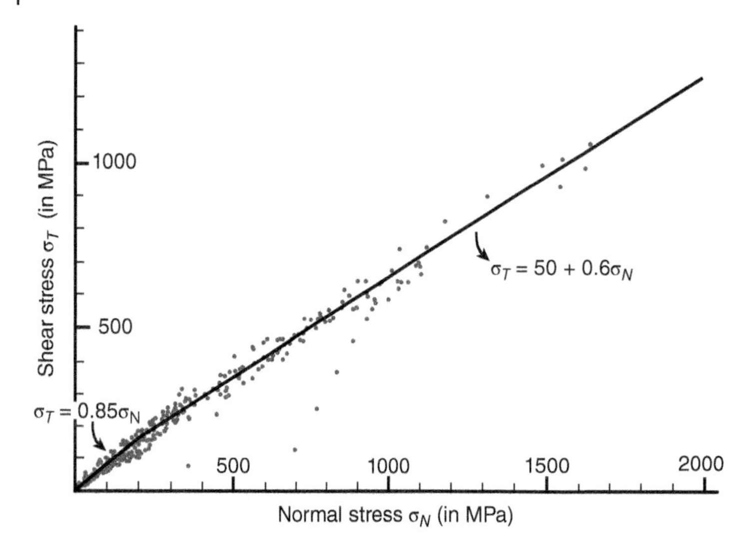

Figure 7.30 Plot, from Byerlee (1978), of laboratory measurements of the normal and shear stresses at which faults slip. The best fit lines for $\sigma_N < 200$ MPa and $\sigma_N > 200$ MPa define Byerlee's law.

relationship where the maximum tangential stress required for slip

$$\sigma_T = 50 + 0.6\sigma_N. \tag{7.58b}$$

The empirical relationships in Eqs. (7.58a) and (7.58b) together define a criterion for lithospheric faulting known as **Byerlee's law.**

The low normal stress relation in Byerlee's law has the form of Eq. (7.14) derived in Section 7A.2.2:

$$\sigma_T = \mu\sigma_N. \tag{7.14}$$

The higher normal stress relation in Byerlee's law has the form of the Coulomb criterion outlined in Section 7A.2.2:

$$\sigma_T = \tau_o + \mu\sigma_N. \tag{7.15}$$

Comparing Eqs. (7.14) and (7.15) to Eqs. (7.58a) and (7.58b), we see that under Byerlee's law the coefficient of friction $\mu = 0.6$–0.85 and the cohesion $\tau_o = 50$ MPa. Recall that $\mu = \tan\phi$, where ϕ is the angle of internal friction and where conjugate faults tend to form at an angle of 2ϕ. With $\phi = \tan^{-1}\mu = \tan^{-1}(0.6$–$0.85)$, Byerlee's law predicts that conjugate faults should form at 30 to 40° angles from the most compressive stress direction, with angles of 60–80° between the faults. The cohesion value in Byerlee's law is consistent with many laboratory measurements of cohesion.

Byerlee's law is regularly used to estimate the strength of rocks in the lithosphere. One can estimate where within the lithosphere Byerlee's law applies by assuming that lithospheric stresses are approximately lithostatic and that the lithostatic pressure is a reasonable estimate for the magnitude of the normal stress acting across a fault. Recall that the lithostatic pressure $P \approx \rho g z$, where z is the thickness of lithosphere above a mass of rock, ρ is the mean density of the lithosphere above the rock mass, and g is the gravitational acceleration. Taking the mean density of rocks as 2.5–$3.0 \times 10^3\,\text{kg/m}^3$, Equation (7.58a) holds to a depth of 7–8 km and Eq. (7.58b) holds at depths from 7–8 to 70–80 km. These equations predict that the strength of the lithosphere increases regularly with depth. In the continental lithosphere, there is considerable evidence for increasing rock strength to depths of ~15 km, but there is no consensus on how rock strength varies with greater depth. We consider other factors involved in estimating the strength of different segments of the lithosphere in Section 7B.2.7.

7B.2.2.1.3 Limits to Byerlee's Law As outlined in Eqs. (7.14), (7.15), and (7.58), the Coulomb criterion and Byerlee's law are predicated on the notion that a single parameter, the coefficient of static friction μ_S, relates the applied stresses to fault formation or slip on an existing fault: knowing the magnitude of μ_S, one can predict whether specific stress magnitudes are stable or lead to fault slip. These criteria are useful, first-order guides to understanding fault slip, but laboratory data on fault sliding, like those presented in a general form in Figure 7.29, indicate that rock friction entails more than what is captured in this single parameter.

In Section 7A.2.2.1, we argued that: (1) fault surfaces are rough, possessing numerous asperities; (2) that the true area of contact between two rock masses is a function of the normal stress acting across the surface because asperities deform by plastic yield; (3) that sliding requires shearing through the asperities; and (4) the coefficient of friction is related to the relative magnitudes of the strength of the material in shear to its strength in compression. This view of frictional sliding applies to "mature" surfaces. In laboratory studies of the initiation of slip, surfaces typically exhibit low coefficients of friction ($\mu \approx 0.3$) during the first small increments of slip. The low coefficients of friction during the first increments of slip are consistent with models in which elastic distortion of asperities figures prominently. As the displacement increases, surfaces typically "harden," that is coefficients of friction increase. Once the displacement reaches a particular threshold value, the coefficient of friction attains a steady-state value. The slip required to reach the steady-state coefficient of friction is related to the size of the asperities. It is on the order of a millimeter or less in experiments but may be larger for natural faults. Hardened surfaces exhibit steady-state coefficients of friction of 0.5–0.7, values that are consistent with ratios of shearing strengths to strengths in compression. This suggests that plastic deformation of asperities contributes to shearing resistance once the fault slip exceeds the characteristic displacement. Since frictional sliding often generates angular wear particles – *fault gouge* – asperities probably also deform by fracture.

Changes in the nature of asperity deformation and the development of gouge along fault surfaces are two factors that contribute to changes in the resistance to fault slip over time. These physical changes to faults are significant reasons that the Coulomb criterion or Byerlee's law, both of which assume that the coefficient of friction has a single value, are at best first-order descriptions of fault behavior. The following section addresses some of the factors that are thought to control frictional slip on mature faults. It is appropriate to note here, however, that structures and microstructures in some natural fault zones are consistent with the inference that physical changes wrought during slip, particularly reductions in the grain size of mineral components of rock along the fault zone, facilitate a switch from frictional sliding to other mechanisms of deformation. We examine those other deformation mechanisms in Chapter 8 and return to our discussion of fault movement at that point.

7B.2.2.1.4 Dynamic Friction and Stick-Slip Behavior We noted earlier that stick-slip behavior is associated with velocity weakening, where a fault's dynamic coefficient of friction μ_d is less than its static coefficient of friction μ_S. Our discussion of velocity weakening during fault slip follows the approaches outlined in Scholz (2002, pp. 81–94) and Twiss and Moores (2007, pp. 466–475); see Aharonov and Scholz (2017) for a more recent review of this subject. Figures 7.29

and 7.31 depict schematically the conceptual model, derived from fault slip experiments, used to understand velocity weakening: a rider that slides across a "fault" surface is connected by a spring to a component that moves a constant velocity. In these experiments, the fault surfaces are "mature," that is they have experienced slip magnitudes sufficiently large that they exhibit steady-state static friction coefficients. In the actual experiments, elastic and nonelastic distortion in response to shearing have brought the force to the peak of the tangential force (F) versus displacement (u) graph, corresponding to point C on Figure 7.29b.

Figures 7.31a and b present idealized views of the force "run-up" to the point of slip and the system response to slip. In the schematic rider-and-spring representation, at the outset the rider is stationary, and the shearing displacement is accommodated by the stretching of the spring. As the spring stretches, the force or "pull" it exerts on the rider increases linearly proportional to its

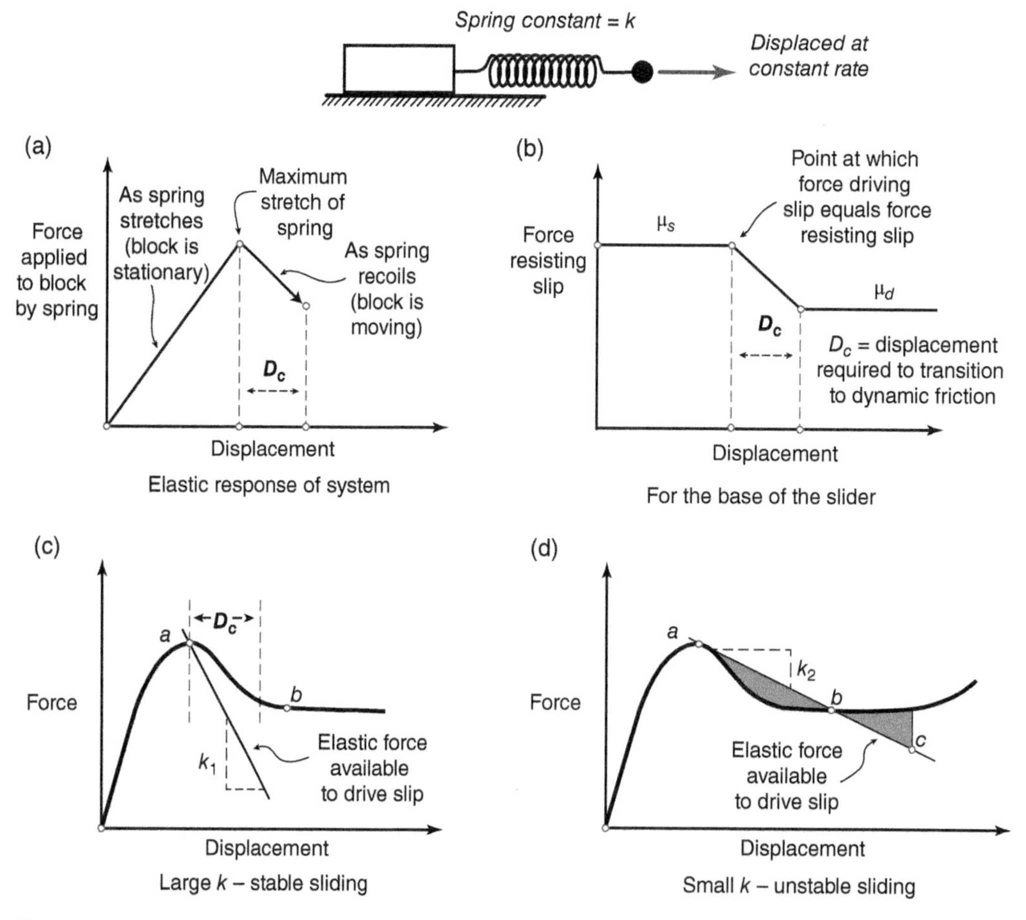

Figure 7.31 (a) Idealized plot showing the variation in elastic driving force with displacement during stick-slip event. (b) Idealized plot showing the variation in frictional resistance force with displacement during stick-slip event. D_c – Characteristic slip, μ_s, μ_d – Coefficients of static and dynamic friction. (c) Force versus displacement plot for the system that slides stably. (d) Force versus displacement plot for system exhibiting stick-slip behavior.

spring constant k (Figure 7.31a). (We follow the widely accepted convention in Physics of using k to denote the spring constant. Note that k is *not* the bulk modulus K. In these shearing settings, k will scale with the bulk modulus K and the shear modulus G, or with Young's modulus E, but it is not equivalent to K.) During this time, the resistance to slip has a constant magnitude given by the static friction (Figure 7.31b). In this idealized depiction, the "pull" exerted by the spring increases linearly until it equals the product of the normal force acting across the base of the rider times the static coefficient of friction, that is the pull $\boldsymbol{F} = \boldsymbol{F}_N\, \mu_s$. When that happens, the rider begins to move. As the rider begins sliding, the elastic energy stored in the system contributes to its slip. One can visualize the effect of the stored elastic energy as a force exerted on the rider by the stretched spring. As slip magnitude increases and the spring gradually returns to its "unstretched" configuration, the force exerted on the rider due to the release of elastic strain energy decreases as the magnitude of the slip increment increases. The force decreases at a rate that is linearly proportional to its spring constant k (Figure 7.31a). At the same time, the frictional resistance drops from the static friction value to a dynamic friction value. This occurs over a slip magnitude with a characteristic value, D_c.

The system behavior depends on the relative magnitudes of the decrease in frictional resistance per unit of slip ($\partial \boldsymbol{F}/\partial u$) and the decrease in elastic stress per unit of relieved elastic strain energy. The decrease in frictional resistance with slip is (see Figure 7.31b):

$$\frac{\partial \boldsymbol{F}}{\partial u} = \frac{\left(\boldsymbol{F}_N \mu_s - \boldsymbol{F}_N \mu_d\right)}{D_c} = \frac{\boldsymbol{F}_N\left(\mu_s - \mu_d\right)}{D_c}. \tag{7.59a}$$

Recasting this in terms of stresses yields

$$\frac{\partial \sigma}{\partial u} = \frac{\sigma_N\left(\mu_s - \mu_d\right)}{D_c}. \tag{7.59b}$$

The decrease in elastic stress available to drive slip ($\partial\sigma$) per unit of relieved elastic strain energy (∂u) depends on the stiffness or spring constant k of the system.

Figures 7.31c and d show two different situations with different relative magnitudes of these two parameters. The point labeled a on Figure 7.31c and d corresponds to point C on the force versus displacement curve in Figure 7.29. Figure 7.31c depicts a system where $k > \sigma_N (\mu_s - \mu_d)/D_c$), meaning the system is relatively stiff. With the onset of slip at point a, each unit decrease in elastic strain is associated with a significant drop in elastic stress (given by the straight line with a slope $= -k_1$ on Figure 7.31c). The elastic stress driving slip is always lower than the frictional resistance, and the rider's movement, therefore, remains constrained by the system's frictional resistance. As long as the shearing force is sufficient, the rider will slip at the same velocity as the moving component. If not, a slip will cease. This type of slip is "stable." (It is important to distinguish "stable slip" from what we earlier called a "stable stress state," which is a stress state where no slip occurs.) Figure 7.31d depicts a system where $k < \sigma_N (\mu_s - \mu_d)/D_c$), meaning the system is less stiff. With the onset of slip, each unit decrease in elastic strain is associated with a less significant drop in elastic stress (the straight line with a slope $= -k_2$ on Figure 7.31d). Because the system's decrease in frictional resistance per unit of slip ($\partial \boldsymbol{F}/\partial u$) is greater than the decrease in elastic stress per unit of slip, elastic energy stored in the system exerts a force on the rider in excess of the frictional resistance. Where the elastic stress exceeds the frictional resistance (between points a and b), the rider accelerates and slip is "unstable." Eventually, the elastic stress driving slip falls to a value equal to the frictional resistance to slip (at point b). Even though these two forces acting on the rider are equal, the inertia of the rider causes it to keep moving. The energy added to the rider is proportional to the area between the

$\boldsymbol{F}$ versus u curve and the line with slope $-k_2$ between points a and b. The energy will be dissipated by driving the rider's displacement to c, the point at which the shaded area between the $\boldsymbol{F}$ versus u curve and the line with slope $-k_2$ between points b and c is equal to the shaded area between a and b. The situation illustrated in Figure 7.31d, where the frictional resistance falls with slip more rapidly than the system releases elastic energy, is said to exhibit **slip weakening**.

Integrating dynamic effects into the analysis of stick-slip behavior allows for a deeper understanding of the relevance of stick-slip to earthquakes. It is the combined effects of velocity-weakening and slip-weakening behaviors that lead to unstable slip comparable to an earthquake, provided the difference between the static and dynamic coefficients of friction and the elastic properties of the fault zone have appropriate magnitudes. Further, in order for this process to recur, some process or combination of processes must return the fault zone to conditions comparable to its state prior to the unstable slip event.

7B.2.2.1.5 Rate- and State-Dependent Models for Friction The magnitudes of the static friction and dynamic friction are significant factors determining whether a fault will exhibit stable or unstable slip. The hardening of fault surfaces suggests that static friction is a function of the true contact area and the manner in which asperities deform. Scientists studying fault slip have devised two types of fault slip experiments intended to provide insight into the parameters responsible for the onset of and recovery from unstable slip events.

Experiments known as slide-hold-slide tests (Figure 7.32) illustrate how geologists measure static and dynamic friction and confirm that the physical state of the fault impacts the static friction value. A slide-hold-slide test begins with two samples sliding steadily past each other at constant velocity and a steady-state, dynamic coefficient of friction μ_d. Slip is halted for a fixed time interval. When the slip is reinitiated and brought back to the original sliding velocity, we see a change in the friction coefficient of $\Delta\mu_s$, bringing the coefficient to a higher value of μ_s, the static friction. Figure 7.32a shows that the magnitude of $\Delta\mu_s$ is larger with longer hold times. Data from numerous experiments indicate that $\Delta\mu_s$ varies with the logarithm of the hold time, a relationship indicating that changes in the physical state of the fault (or the fault gouge) affect the value of the coefficient of friction.

The second type of experiment again begins with a fault slipping steadily. Instead of a hold, slip velocity is abruptly stepped up to a new value until a new steady state is attained and then stepped down to the original velocity until the original steady state is attained. The abrupt increase in slip velocity causes an immediate jump in the coefficient of friction value followed by a gradual drop to a lower steady-state coefficient of friction at the higher velocity (Figure 7.32b). With the abrupt return to the original slip velocity, the coefficient falls and then gradually returns to the original steady-state coefficient of friction (Figure 7.33a).

The temporal variation of the coefficient of friction is described by the following equation (see Marone 1998, p. 650; Scholz 1998; Scholz 2002, pp. 85–97), which is the most commonly used form of a "rate- and state-dependent friction" law:

$$\mu(t) = \mu_o + a \ln\left(\frac{V(t)}{V_o}\right) + b \ln\left(\frac{V_o\theta(t)}{D_c}\right). \tag{7.60}$$

In Eq. (7.60), μ_o is the steady-state coefficient of friction, V_o is the velocity at which the coefficient of friction is μ_o, $V(t)$ is the new slip velocity, and D_c is the magnitude of slip required for the slip behavior to change. D_c is a characteristic length for the system. D_c may be as small as a 10's of microns in laboratory experiments but has been

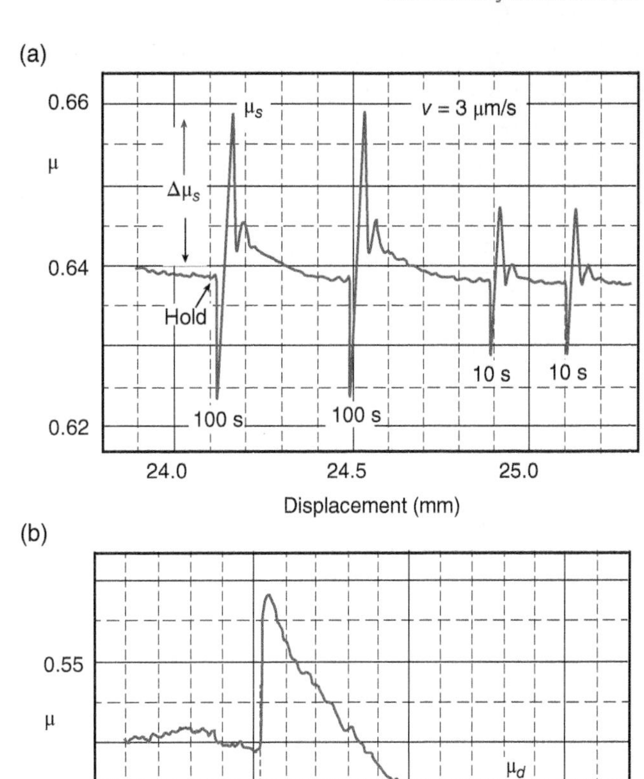

Figure 7.32 (a) μ versus displacement plot (from Marone 1998) from a slide-hold-slide test of quartz gouge along the fault. (b) Variation of μ of a thin layer of quartz gouge in response to a step-up in slip velocity (Marone 1998 / Annual Reviews).

estimated to be between 10 cm and 5 m for natural faults. $\theta(t)$ is a "state variable" that measures the physical changes of the fault zone, and a and b are parameters that describe the magnitudes of the rate-dependent and state-dependent effects.

The term $a \ln (V(t)/V_o)$ describes the rate-dependent effect. If $V(t) = eV_o$, where e is the base of the natural logarithm, then the rate-dependent term becomes $a \ln (e\, V_o/V_o) = a \ln e = a$. The rate-dependent effect alone leads to an increase of the coefficient of friction to $(\mu_o + a)$ as long as $V(t) = eV_o$ (Figure 7.33b).

The term $b \ln (V_o\theta(t)/D_c)$ describes *the state-dependent effect*. Empirically, the variation of θ with time is

$$\frac{d\theta}{dt} = 1 - \frac{\theta V}{D_c}. \tag{7.61}$$

The system is at a steady state when $d\theta/dt = 0$ and, from Eq. (7.61), $\theta = D_c/V$. The state-dependent term reduces the coefficient of friction when $d\theta/dt < 0$ and $\theta < D_c/V$. The state-dependent term increases the coefficient of friction when $d\theta/dt > 0$ and $\theta > D_c/V$. If V is constant, the more

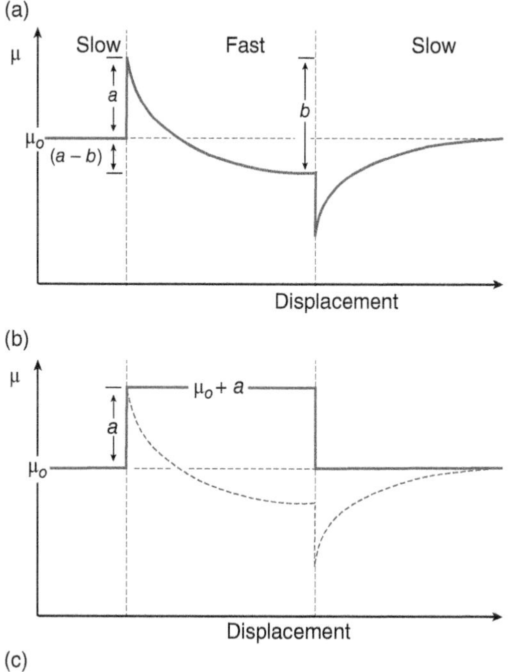

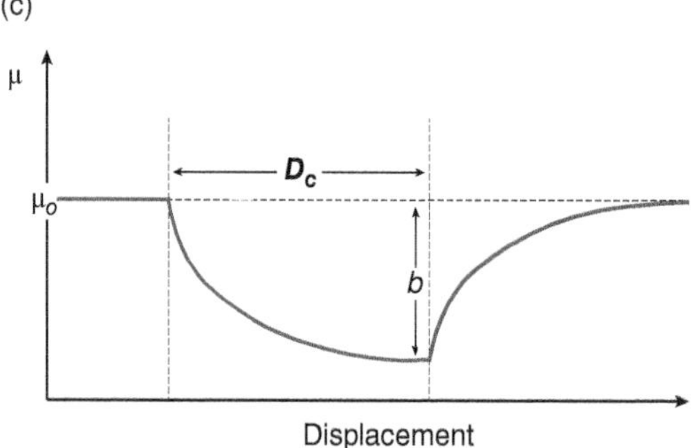

Figure 7.33 (a) Idealized variation of μ in response to a step up then step down in slip velocity. The overall response is the sum of rate-dependent and state-dependent effects. (b) Hypothesized variation of μ due to rate-dependent effect. (c) Hypothesized variation of μ due to state-dependent effect.

nearly θ is equal to D_c/V, the smaller is $d\theta/dt$ and the more slowly θ changes. Stated in different terms, when slip velocity is constant the system trends toward a steady state over time (D_c/V) and the state variable θ describes how the coefficient of friction changes over a displacement of D_c (Figure 7.33c).

The parameter a determines the size of the instantaneous increase of the coefficient of friction to an increase in sliding velocity, and the parameter b describes the rate at which a fault evolves from its perturbed condition to a new steady state while slipping over a characteristic distance D_c. These parameters are envisioned to depend upon the true area of contact between the rocks separated by the fault and the manner in which areas of contact, the asperities, deform, which is in turn dependent upon presence or absence of fault gouge, physical characteristics (composition, grain size, etc.) of any gouge, wall rock characteristics, ambient temperature, and the presence of fluid phases.

Equation (7.60) implies that the change in the coefficient of friction with the natural logarithm of velocity, ln(V) is given by the difference between the rate- and state-dependent parameters (a–b):

$$\frac{d\mu}{d\left(\ln V\right)} = a - b. \qquad (7.62)$$

The magnitude of (a–b) is therefore the critical parameter that distinguishes velocity weakening from velocity strengthening. When (a–b) is negative, slip causes a reduction in the coefficient of friction. When (a–b) is positive, slip causes an increase in the coefficient of friction. Simplistically, velocity weakening, that is negative (a–b), can be interpreted as one condition for slip to become seismic. Experiments indicate that negative values of (a–b) are favored by increased shear displacement, rougher sheared surfaces and more localized deformation within fault zones. On the other hand, the production of thicker gouge layers leads to positive (a–b).

More generally, the rate-and state-dependent friction laws suggest a spectrum of possible fault slip rates, which is consistent with recent observations on natural faults. These show a range of behaviors that include slow slip, tremor, and low-frequency earthquakes. The rate- and state-dependent laws are successful in explaining laboratory friction results and how they relate to the behavior of natural faults.

7B.2.2.1.6 Failure Criterion for Polyaxial Stress States The Coulomb failure criterion

$$\sigma_T = \tau_o + \mu\sigma_N \quad (7.15)$$

can be rewritten as:

$$\sigma_T - \mu\sigma_N = \tau_o. \quad (7.15a)$$

In any three-dimensional stress state, the greatest magnitude of shear stress is $(\sigma_1-\sigma_3)/2$. The planes that experience this magnitude of shear stress are inclined 45° to the σ_1 and σ_3 principal directions and intersect along lines parallel to the σ_2 principal direction. These shear traction vectors with maximum magnitude are perpendicular to the σ_2 principal direction. Similarly, the maximum values of $\sigma_T-\mu\sigma_N$ occur on planes inclined to the σ_1 and σ_3 principal directions and intersecting along lines parallel to the intermediate principal stress σ_2 (Figure 7.34a). For $\sigma_1-\sigma_3$ sections, the Coulomb criteria can be rewritten using the equations for tangential and normal stress in terms of σ_1 and σ_3:

$$\sigma_1\left[\left(\mu^2+1\right)^{1/2}-\mu\right]-\sigma_3\left[\left(\mu^2+1\right)^{1/2}+\mu\right]=2\tau_o \quad (7.63)$$

(see Jaeger and Cook 1971, pp. 95–97). Reorganizing this equation to

$$\sigma_1=\sigma_3\frac{\left\{\left[\left(\mu^2+1\right)^{1/2}+\mu\right]\Big/\left[\left(\mu^2+1\right)^{1/2}-\mu\right]\right\}+2t_o}{\left[\left(\mu^2+1\right)^{1/2}-\mu\right]} \quad (7.63a1)$$

or, after multiplying the σ_3 term by the fraction $[(\mu^2+1)^{1/2}+\mu]/[(\mu^2+1)^{1/2}+\mu]$ which has the value of 1,

$$\sigma_1=\frac{\sigma_3\left[2\mu^2+\mu\left(\mu^2+1\right)^{1/2}+1\right]+2\tau_o}{\left[\left(\mu^2+1\right)^{1/2}-\mu\right]} \quad (7.63a2)$$

makes clear that it defines a straight line on Cartesian coordinate axes σ_1 and σ_3, analogous to the linear appearance of the Coulomb criterion on the Mohr diagram (Figure 7.34b). Although the slope of the line depends only on the coefficient of friction, the value of μ is not readily apparent from the inclination of the line. Also, the intercept of the line is a function of both μ and τ_o, and the plot on $\sigma_1-\sigma_3$ axes does not provide a straightforward relationship to the fault orientation in the way that Mohr diagrams do.

This expression does not depend on σ_2, thus the Coulomb criteria predict failure along planes whose relative orientations do not vary with the magnitude of σ_2. This condition requires that the failure planes have familiar conjugate orientations and intersect along a line parallel to the σ_2 principal direction. Moreover, the lack of influence of σ_2 implies that slip directions as well as fault planes lie in the $\sigma_1-\sigma_3$ plane, limiting the resulting displacements and strain to plane strain conditions (no stretching perpendicular to the $\sigma_1-\sigma_3$ plane) (Figure 7.34c).

Data from natural faults and the results of some rock mechanics experiments indicate that simple conjugate faults with slip vectors perpendicular to the fault intersections (and therefore perpendicular to σ_2) are a special case. This is not surprising, for faults regularly occur in tectonic settings where plane strain deformation is not expected. In general, then, one should expect that σ_2 has an effect on the onset and geometry of failure and should therefore be included in a failure criterion.

The most common type of rock mechanics experiments are conducted on cylindrical samples using a confining medium for which $\sigma_2 = \sigma_3$ (an *axial* stress state). Despite the experimental convenience of this set-up, it has the drawback that σ_2 cannot be changed relative to σ_3 and therefore the effects of σ_2 cannot generally be investigated. Only by using prismatic samples in a special apparatus in which all three principal stresses can be varied independently can this limitation be overcome. Such a stress state configuration is *triaxial* (Figure 7.35). Since common axial stress

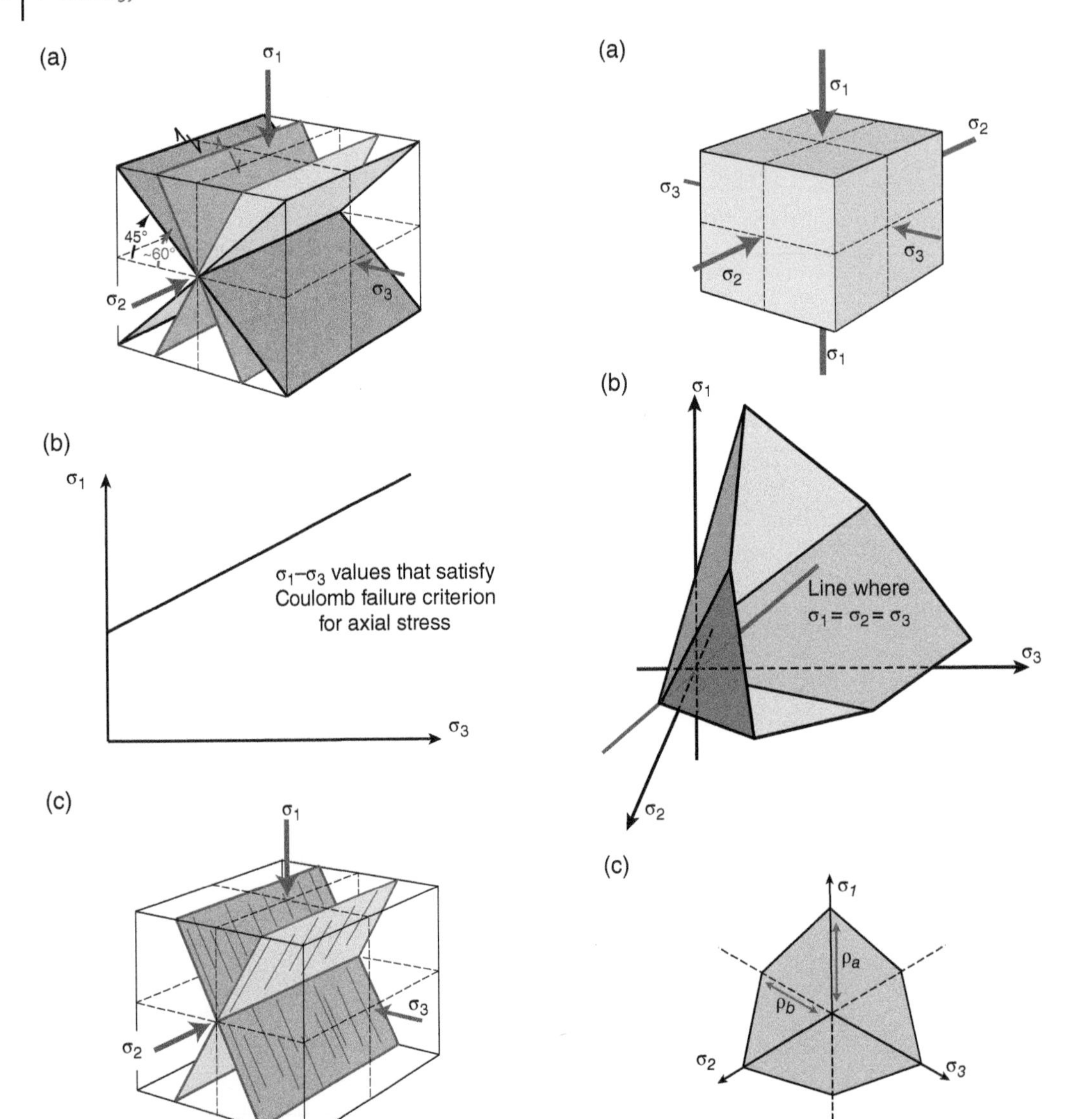

Figure 7.34 (a) The orientation of planes experiencing maximum shear stress σ_T in gray and planes with the maximum ratio of σ_T/σ_N in red. (b) The Coulomb criterion plotted on Cartesian σ_1–σ_3 axes. (c) Schematic representation of faults, with striations, accommodating a plane strain deformation.

Figure 7.35 (a) Illustration of rock subjected to triaxial loading. (b) Coulomb failure envelope for triaxial loading plotted on Cartesian coordinate axes corresponding to the principal stresses (modified from Lee et al. 2012). Failure envelope is symmetric about line $\sigma_1 = \sigma_2 = \sigma_3$. (c) View of the shape of the failure envelope on the octahedral plane, which is perpendicular to the line $\sigma_1 = \sigma_2 = \sigma_3$.

experiments and testing devices are regularly, and incorrectly, called triaxial tests, the experiments on prisms are sometimes distinguished as *true triaxial* tests.

Geologists have long known from true triaxial experiments that σ_2 has an important influence on both the geometry of faults and the stresses at failure (cf. Nádai 1950; Mogi 1971a, b). The angle between σ_1 and the fault plane normal increases as σ_2 increases, and the value of σ_1 required for failure increases over the axial stress failure condition as σ_2 increases. The differential stress $\sigma_1-\sigma_3$ is, therefore, higher as well.

Experiments as well as theoretical concepts support the idea that triaxial failure criteria are best expressed in terms of the octahedral shear (τ_{oct}) and normal (σ_{oct}) stresses. These are defined as:

$$\tau_{oct} = \frac{1}{3}\left[\left(\sigma_1-\sigma_2\right)^2+\left(\sigma_2-\sigma_3\right)^2 +\left(\sigma_3-\sigma_1\right)^2\right]^{1/2} \tag{7.64}$$

$$\sigma_{oct} = \frac{1}{3}\left(\sigma_1+\sigma_2+\sigma_3\right) \tag{7.65}$$

Figures 7.35b and c depicts the form of a modified Mohr-Coulomb failure envelope plotted on Cartesian coordinate axes corresponding to the three principal stresses σ_1, σ_2, and σ_3 (Lee et al. 2012). The envelope is symmetric about the line $\sigma_1 = \sigma_2 = \sigma_3$. More sophisticated formulations of failure criteria have the form of

$$\tau_{oct} = a\sigma_{oct}^{b}, \tag{7.66}$$

where a and b are experimentally or theoretically derived constants. For example, in the Murrell extended criterion, a is given by 8 times the uniaxial tensile strength T_0, and $b = ½$. On a graph of the three principal stresses along three axes, this equation is represented by segments of a paraboloid of revolution about the line $\sigma_1 = \sigma_2 = \sigma_3$. Values of b reported from experiments range from 0.5 to 0.8.

Alternative forms of true triaxial failure criteria that have been used include:

$$\tau_{oct} = a\left(\sigma_1-\sigma_3\right)^b \tag{7.67}$$

and

$$\tau_{oct} = c + a\sigma_{oct} \tag{7.68}$$

(cf. Al-Ajmi and Zimmerman 2005). There are some experimental data about the effect of σ_2 on shear failure criteria, but almost nothing is known about the effect of σ_2 on extensional failure. There is, then, no truly triaxial failure criterion of comparable stature or utility to Griffiths criteria.

7B.2.2.2 Linear Elastic Fracture Mechanics

The discussion of fracture formation in Sections 7A.2.2.2 and 7A.2.2.3 abstracted the work of Griffith and focused on the energetics of the process – examining how elastic strain in the neighborhood of existing microscopic cracks provides the energy needed for cracks to extend and connect to form macroscopic fractures. We briefly described analytical work published by Inglis at the beginning of the last century, showing how cracks in elastic bodies alter the stress distributions in those solids. Of particular importance was the recognition that the cracks amplify the **remote stresses** σ_r. "Remote" in this context implies that the source of stress is at a distance that is large compared to the fracture dimensions. Section 6B.1.9 introduced the *equilibrium equations*, the differential equations that govern spatial variations of stresses. In many circumstances, the combination of the equilibrium equations, the linearly elastic constitutive equation that defines the functional relationship between stresses and strains, and known values of stresses or displacements along the boundary of a region result in "well-defined" problems whose solutions are analytical expressions for stresses and strains which conform simultaneously to all three constraints. Inglis solved just such a *boundary value problem*

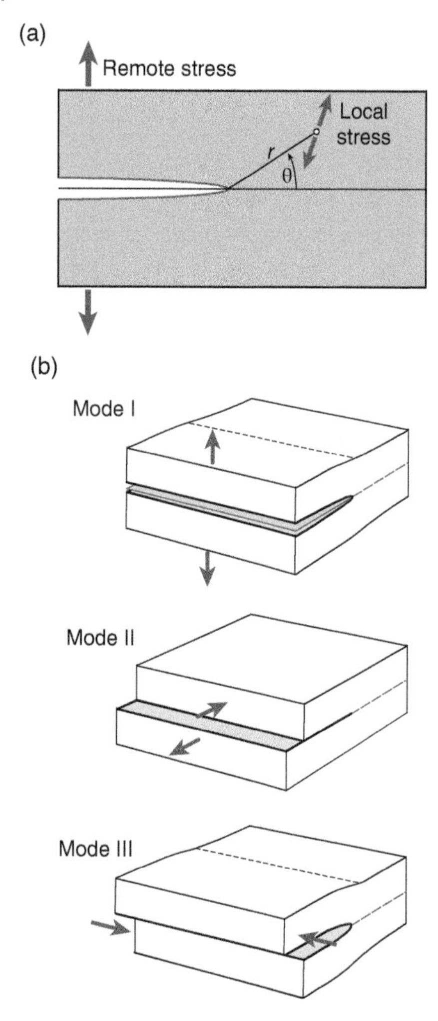

Figure 7.36 (a) Diagram illustrating the polar coordinates used to identify positions relative to the tip of a crack. (b) Different fracture modes. Mode I is the opening mode. Mode II is the in-plane shearing mode. Mode III in anti-plane or tearing mode.

and was able to calculate stresses at different positions around an elliptical hole in an isotropic elastic solid in relation to the remote σ_r field acting on the boundaries of the solid (Figure 7.36a). His analysis showed that stresses adjacent to the hole or flaw have magnitudes many times that of the applied stress when the radius of curvature of the fracture tip is small. This realization was, and is today, fundamental to the scientific study of the formation of fractures.

With advanced techniques, often drawing on the results of numerical (that is computational) analyses, scientists are now able to characterize stress distributions around cracks with other than ellipsoidal shapes, such as cracks with infinitely sharp terminations. All of these approaches share the foundational assumption that materials in the vicinity of a crack exhibit linearly elastic behavior prior to fracture, thus workers often use the term **linear elastic fracture mechanics** when referring to this approach. Lawn (1993, pp. 16–50) presents a cogent, concise, and accessible introduction to the linear elastic fracture mechanics approach, and this section follows the approach outlined there.

Although exact analytical expressions for the stresses as a function of position relative to the crack exist for a range of crack shapes, in many instances the expressions defining stresses are functions of parameters that obscure the spatial variation of stresses. For that reason, it is now common practice to use equations of the form

$$\sigma_{ij} = \frac{K}{\sqrt{2\pi r}} f(\theta) \tag{7.69}$$

to define *local stresses* at different positions. In Eq. (7.69), r and θ are the polar coordinates of a point relative to an origin at the crack tip. The function $f(\theta)$ takes different forms depending on the **mode** of the fracture (Figure 7.36b). Mode I fractures are "opening mode" fractures, where the two sides of the fracture move apart from each other. Previously in this book, we have been called such fractures "extension fractures." Mode II fractures are "in-plane shear" fractures, where two sides of the fracture shear past each other in a direction perpendicular to the fracture edge. Mode III fractures are "anti-plane shear" or "tearing" fractures, where two sides of the fractures shear past each other parallel to the fracture edge.

The fracture modes are defined relative to the fracture plane and the fracture edge, and not relative to geographic coordinates, The two sliding modes (II and III) cannot, therefore, be equated to the displacements of specific types of fault, e.g. dip slip, strike slip.

In Eq. (7.69), K is the **stress intensity factor**, a mathematical function that is formulated in terms of parameters having physical meaning and that closely approximates the exact analytical expressions mentioned earlier. K takes the form

$$K = Y \sigma_r \sqrt{\pi C}, \tag{7.70}$$

where C is the crack half-length and Y is a geometric parameter, which depends on the shape and loading of the fracture. Lawn (1993, 25–26) gives expressions for Y, K, and $f(\theta)$ for different fracture modes.

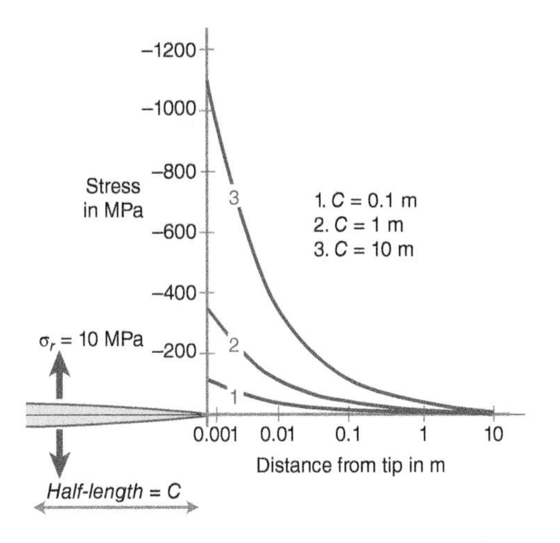

Figure 7.37 Plot of stress magnitudes at different distances from a Mode I fracture crack tip for cracks of different lengths.

In linear elastic fracture mechanics, the conditions for extension fracture formation are given by a simple criterion. A fracture will grow when K reaches a critical value, called the **critical stress intensity factor, K_c**. Eq. (7.70) shows that K depends on both the remote stress and the crack length, which must combine to achieve the critical value. The value also depends on the geometry of the fracture, through the parameter Y, which varies with the mode of the fracture. Acknowledging that there are a number of factors affecting the magnitude of K, there are a few general conclusions that one can draw about the stress distributions about cracks. The most important aspect of the earlier equations is that the stresses increase towards the fracture tip proportional to $1/\sqrt{r}$, emphasizing that stresses can be very large in the vicinity of fracture tips as is shown in Figure 7.37 for a mode I fracture. In fact, Equation (7.69) predicts infinite stress at the fracture tip itself, but this is an artifact of the assumption that the fracture is infinitely sharp, and it is dealt with by suitable modifications of the equation for regions very close to the fracture tip. The implication of the increase in stress toward the fracture tip is that even moderate remote stresses can cause stress concentrations at the fracture tip sufficient to exceed the solid's strength and propagate the fracture. Equation (7.70) indicates that stresses near the crack are proportional to $\sqrt{C}$. Thus, longer cracks have higher stress concentrations at their tips (Figure 7.37). In a constant remote stress field, the propagation of longer fractures is favored. This may be why in rocks cut by sets of systematically oriented fractures or *joints*, we often find individual fractures, called *master joints*, which have considerably greater lengths than others of the same set. In Section 7.A.2.2.2, we noted that Griffith used the radius of curvature of an elliptical crack tip, R_c, in defining the magnitude of the stress concentration at the end of an elliptical flaw. The radius of curvature at the tip of an ellipse is an indirect measure of the "flatness" of a crack. The amplifying effects of a crack grow with the ratio of its long to short dimensions, but even equant flaws, such as pores, amplify stresses dramatically relative to the remote stresses. Solutions to Eqs. (7.69) and (7.70) indicate that tensile stresses can form around cracks subjected to compressional stresses, and the tensile stress magnitudes can be much greater than the applied stresses.

In summary, both experiments and independent geological observations of stress orientations confirm the theoretical prediction that extension fractures form perpendicular to σ_3 in isotropic rocks. Joints and other mode I fractures can be, then, reliable indicators of the orientation of σ_3, often providing evidence of a remarkably regular pattern of stresses on a regional scale.

7B.2.2.3 Growth of Shear Fractures

A fracture will propagate when its critical stress intensity is exceeded. The path followed by a growing fracture depends, however, on the orientation of the fracture relative to the remote stress. We find evidence for this in nature and in the laboratory. Natural fractures oriented obliquely to principal directions of the remote stress regularly have subsidiary joints or mineral-filled extension fractures that branch off the main fracture (Figure 7.38a). In laboratory experiments, pre-existing fractures oblique to the principal directions of the remote stress respond consistently to increasing remote stress. As stresses increase, the fracture starts to slide but does not lengthen in a plane. Rather, two new extension fractures initiate at the ends of the original fracture and grow with characteristic curved shapes until they are oriented perpendicular to the most tensile remote stress. These branching fractures, sometimes called **wing cracks**, closely resemble those in nature (Figure 7.38b). These experiments suggest that growing cracks tend toward orientations perpendicular to the most tensile principal stress (Figure 7.38c).

A fracture mechanics analysis provides insight into what controls the development of branching fractures. Because the original crack is inclined to the principal directions of the remote stress, it is a mixed-mode fracture. The stress distribution around this crack will have contributions from a mode I (opening mode) stress field and a mode II (in-plane shearing mode) stress field. The maximum tensile stress around a fracture at a distance r from the fracture tip and angle θ from the plane of the original fracture (Figure 7.39) is given by:

$$\sigma_t = \frac{\cos(\theta/2)}{\sqrt{2\pi r}}\left[K_{\mathrm{I}}\cos\left(\frac{\theta}{2}\right)^2 - \left(\frac{3}{2}\right)K_{\mathrm{II}}\sin\theta\right]. \tag{7.71}$$

Figure 7.39 plots calculated values of σ_t as a function of orientation relative to the crack tip plotted for $K_{\mathrm{I}} = 0$ (pure in-plane shear mode), $K_{\mathrm{I}} = K_{\mathrm{II}}$ (mixed opening and in-plane shear mode) and $K_{\mathrm{II}} = 0$ (pure opening mode). The minimum stress concentration for mode I loading is at $\theta = 0$; in other words, a fracture under extension will extend in its plane, perpendicular to the least principal stress. For anti-mode I loading, the maximum stress concentration is also perpendicular to the maximum principal stress, suggesting that stylolites under these loading conditions should also propagate in the same plane. For pure mode II loading (the sliding crack model), the minimum value of the stress occurs near the fracture tip across a plane inclined 70.5° to the fracture plane. This predicted orientation for wing cracks conforms to natural examples (Figure 7.38a). The fracture paths predicted by fracture mechanics from the sliding wing crack model (Figure 7.38c) are very close to the experimental results and the field observations. Departures from the 70.5° take-off angle can be explained by envisioning that stresses in the vicinity of the crack tip are sufficiently great to exceed the elastic limit of the fractured rock, giving rise to a "bead" of inelastically deformed material about the fracture tip. Integrating this effect is relatively

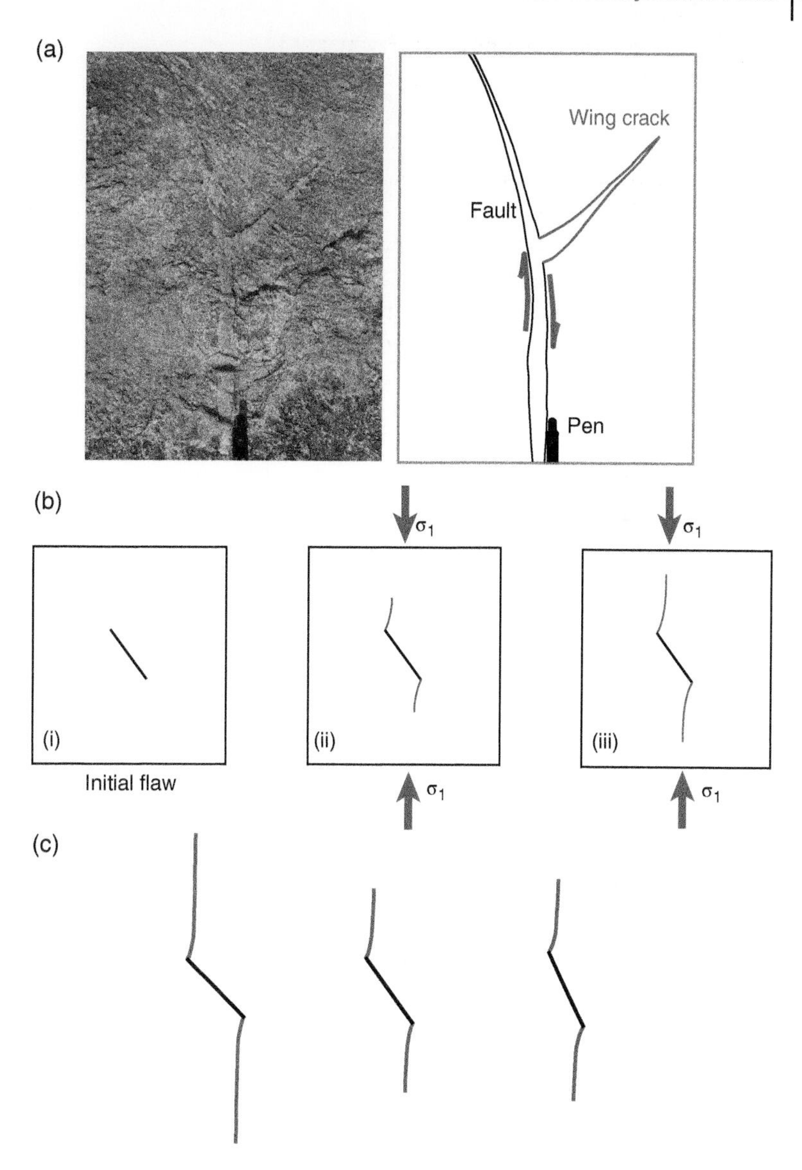

Figure 7.38 (a) Photograph and interpretive sketch of fault in granite with a quartz-filled wing crack. (b) Sketches of three different stages in a model of the formation of a wing crack. (c) Cracks with different orientations relative to principal stresses will develop wing cracks with different geometries.

straightforward provided the size of the inelastic region is small compared to the fracture length. More involved analyses, which are departures from the linearly elastic fracture mechanics approach described here, are required if this is not the case.

The sliding wing crack analysis predicts the most compressive stress occurs on the opposite side of the sliding crack to where the minimum value and extension fracturing occurs. This would be the locus of anti-mode I fracturing, which is the formation of stylolites. Natural examples of

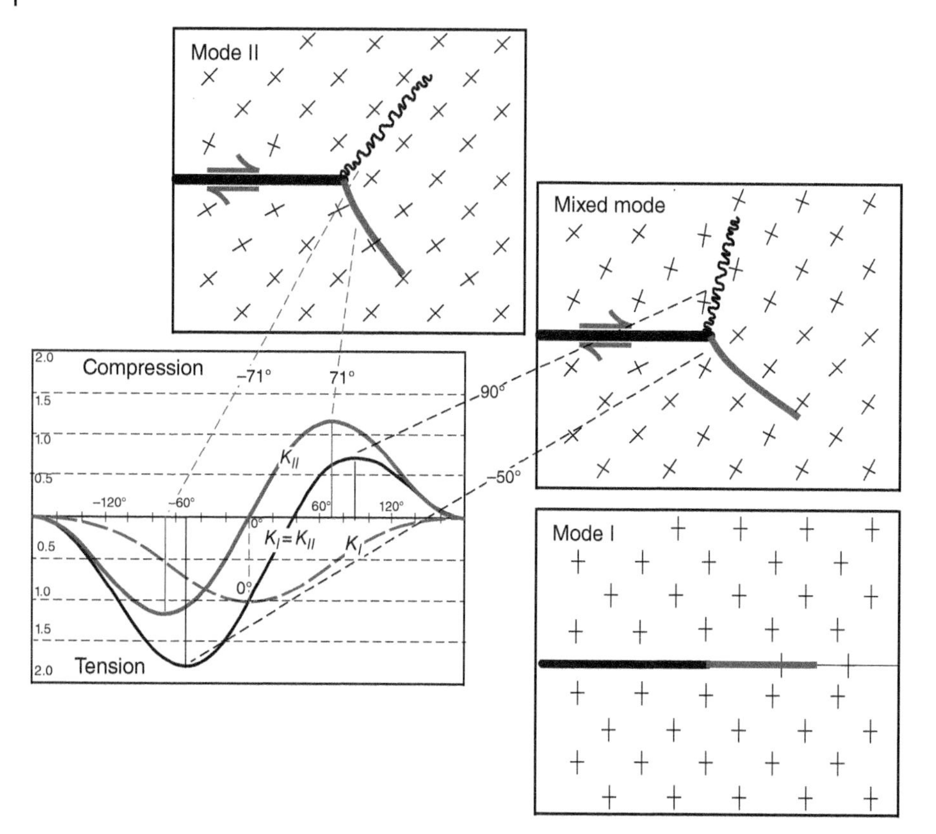

Figure 7.39 Plot showing calculated stress magnitudes at different orientations from crack for an opening mode, an in-plane shearing mode, and a mixed-mode fracture. Small crosses in sketches give orientations of local directions of shortening and elongation. Predicted orientations for fractures (in red) and stylolites (wiggly black lines) are shown in different cases.

mode II cracks with veins on one side and stylolites on the opposite side match the predicted theoretical pattern (Figure 7.40).

The development of wing cracks is especially relevant in understanding the origin of macroscopic shear fractures. In Section 7A.2.2.2, we highlighted the difficulty of extending cracks oblique to the principal directions of the remote, applied stress. The tendency for cracks to curve to an orientation normal to the most tensile stress is a significant factor in suspecting that extensional shear fractures do not exist (see Figure 7.18). Detailed studies of the formation of conjugate faults inclined to the most compressive stress principal direction indicate that it is the linkage, by wing cracks oriented perpendicular to the most tensile principal stress, of pre-existing flaws inclined to the principal stress directions, that leads to the formation of macroscopic faults in the familiar conjugate orientations.

Mode II fractures do not propagate in a plane (neither lab nor theoretical evidence supports this). Faults are not in this sense shear fractures. Indeed, faults in general form by linking of mode I fractures. This is one reason, perhaps the most basic one, why faults are almost never isolated flat shear surfaces, but zones of fracturing.

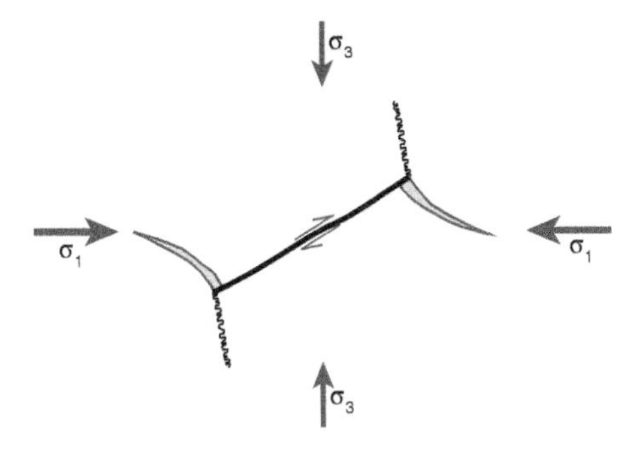

Figure 7.40 Wing cracks and stylolites regularly occur in association with shear fractures, and their orientations indicate the orientations of the principal stress.

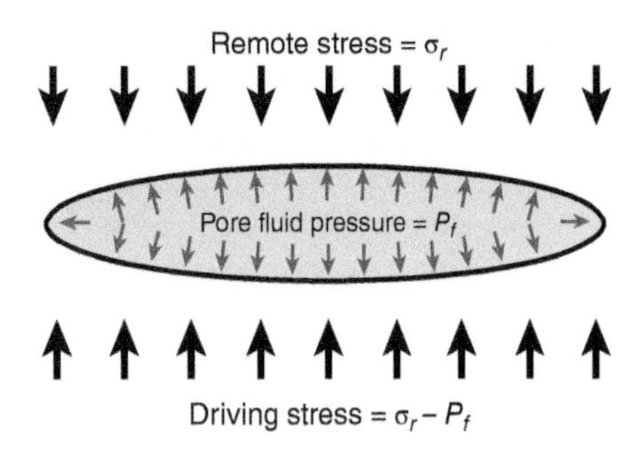

Figure 7.41 Pore fluids can assist fracture propagation because pore fluid pressure can lower the driving stress, the stress required to propagate the crack.

7B.2.2.4 Pore Pressure Effects

Fracture mechanics suggests that fractures under tension can propagate readily because of the stress amplification at their tips. Remote extensional kinematics characterize parts of some divergent tectonic environments such as continental rifts, mid-ocean ridges, the upper parts of elevated plateaus, and the outer parts of folds. In these settings, it is no surprise to find joints, that may be extension fractures. However, extensional fractures are found in all types of tectonic environments. One explanation for the apparent paradox of extension fracture formation within compressional stress fields is the role of fractures that are filled by fluids under pressure (Figure 7.41). The fluid pressure may be great enough to counteract both the compressional stresses and the strength of the rock, causing extensional fracture propagation. The concept of a **fracture driving stress** σ_D (we use an uppercase D for the subscript to distinguish this from the symbol σ_d which is often used to represent deviatoric stress) is useful for understanding the role of pore fluid pressure in extension fracture propagation. Consider a fracture filled by a fluid that exerts a pressure P_f on fracture walls and acted upon by remote stress σ_r perpendicular to the plane of the fracture. The fracture driving stress is defined as the difference between the remote normal stress σ_N and P_f:

$$\sigma_D = \sigma_N - P_f. \tag{7.72}$$

The fracture driving stress is precisely the effective stress on the inside of the fracture. The condition for the propagation of extensional fractures under fluid pressures is that the driving stress is less (more tensional) than the tensile strength of the rock T_0.

$$\sigma_D = \sigma_N - P_f \leq -T_0. \tag{7.73}$$

Clearly, increasing the pore fluid pressure P_f lowers the fracture driving stress, thereby making it more likely that the fracture will propagate. This situation is analogous to that outlined in Section 7A.2.2.4, where we argued that increasing the pore fluid pressure without otherwise changing the differential stresses acting on a rock could transform a stable stress state into an unstable stress state. The process of forming extension fractures in rock due to high fluid pressures is, however, given the special name **hydrofracturing**.

There is considerable evidence for the importance of pore fluid pressure and high fracture driving stresses in the formation of the systematic joint sets seen in a variety of tectonic settings.

For example, unmetamorphosed sedimentary strata, whether deformed or undeformed, regularly exhibit one or more joint sets. High pore fluid pressures are measured in many boreholes in sedimentary basins, where stratigraphic intervals in which the fluid pressures exceed the least principal stress σ_3 are common. High pore fluid pressures occurring at depths and temperatures comparable to those at which joints are inferred to form suggest that pore fluid pressure has a role in joint formation. Additional evidence for the role of pore fluids includes: (1) the presence of incrementally precipitated hydrothermal minerals in extension fractures, implying recurring opening of fractures filled by fluids; (2) fluid inclusions trapped in those minerals preserve evidence that fluids pressures exceeded hydrostatic pressure; and (3) the greater density of jointing in organic-rich shale layers that were the source of hydrocarbons.

7B.2.3 Differential Stress, Pore Fluid Pressure, and Failure Mode

The previous section indicated that increasing pore fluid pressure could lead to extensional failure. The criterion for extensional failure, Equation (7.72), can also be written:

$$P_f \geq \sigma_N + T_0. \tag{7.74}$$

As shown in Figure 7.18b, there is a close relationship between stress and mode of failure. Extensional failure can only occur when the Mohr circle is less than a certain diameter, i.e. differential stresses are low. Once differential stress exceeds a certain value, shear failure (or possibly extensional shear failure) must occur.

The limit of differential stress for extensional failure is readily calculated from Eq. (7.22), $\sigma_2 = -T_0$, which, when substituted into the Griffith criterion, leads to

$$\Delta\sigma = \sigma_1 - \sigma_2 = 4T_0. \tag{7.75}$$

At any value of $\Delta\sigma$ greater than $4T_0$, there will be a component of shear failure.

The relation between stresses and failure is very effectively shown on *failure mode diagrams* which plot pore fluid factor or just pore fluid pressure versus differential stress at failure (Figure 7.42). In these plots, the separate influences of differential stress and pore fluid pressure on failure can be related to extensional and shear modes of failure.

The pore fluid factor λ_v is defined as:

$$\lambda_v = \frac{P_f}{\sigma_v}, \tag{7.76}$$

with P_f as pore fluid pressure and σ_v as the vertical stress. Most failure mode diagrams have used a composite Griffith-Coulomb failure criterion in which the Griffith criterion is assumed for negative normal stresses, and a Coulomb criterion is used for positive normal stress. This failure criterion suggests three types of failure. Extension failure is given by:

$$\lambda_v = \frac{\sigma_3 + T_0}{\sigma_v}. \tag{7.77}$$

Writing the principal stresses as effective stresses allows the relation for extensional shear failure to be written as:

$$\lambda_v = \frac{8T(\sigma_1 + T\sigma_3) - \Delta\sigma^2}{16T\sigma_v}. \tag{7.78}$$

Shear failure is described by (e.g. Cox 2010):

$$\lambda_v = \frac{1}{2\sigma_v}\left[\begin{array}{l} 2C\dfrac{1}{\mu} - \left(\dfrac{1-\cos\left(\tan^{-1}\dfrac{1}{\mu}\right)}{\cos\left(\tan^{-1}\dfrac{1}{\mu}\right)}\right)\sigma_1 \\ + \left(\dfrac{1+\cos\left(\tan^{-1}\dfrac{1}{\mu}\right)}{\cos\left(\tan^{-1}\dfrac{1}{\mu}\right)}\right)\sigma_3 \end{array}\right], \tag{7.79}$$

where μ is the coefficient of internal friction. Figures 7.43a and b show failure mode diagrams

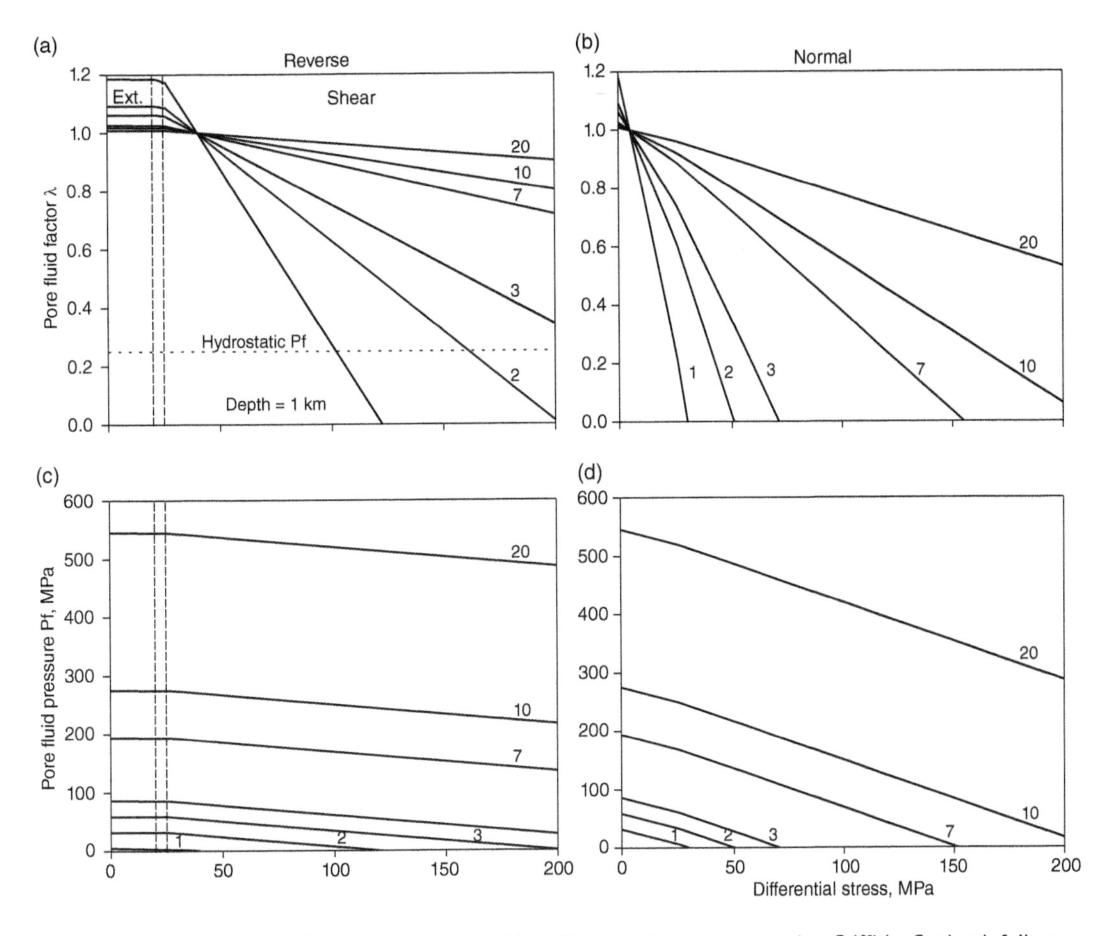

Figure 7.42 Failure mode diagrams for depths of 1 to 20 km in the crust, assuming Griffith–Coulomb failure criteria. Ext denotes the field of extensional failure. Shear denotes the field of shear failure. The two dashed vertical lines enclose the field of extensional-shear failure. (a) and (b): the relationship between pore fluid factor and differential stress for reverse and normal faulting stress conditions respectively; (c) and (d) show pore fluid pressure for the same conditions. T_0 is assumed to be 5 MPa.

based on these relationships for $\mu = 0.75$, $C = 10 = 2T$ for reverse faulting conditions ($\sigma_v = \sigma_3$) and normal faulting ($\sigma_v = \sigma_1$), assuming a vertical stress gradient of 27 MPa/km. The failure envelop for a range of depths from 1 to 20 km is shown: any depth has a unique failure envelop. In general, all the diagrams show an inverse relationship between pore fluid factor and differential stress, indicating a trade-off between these two influences on failure. Diagrams can be constructed to show conditions for reactivation of faults in different orientations.

Failure mode diagrams have two advantages over Mohr diagrams: the distinct roles of pore fluid pressure and differential stress can be clearly visualized,

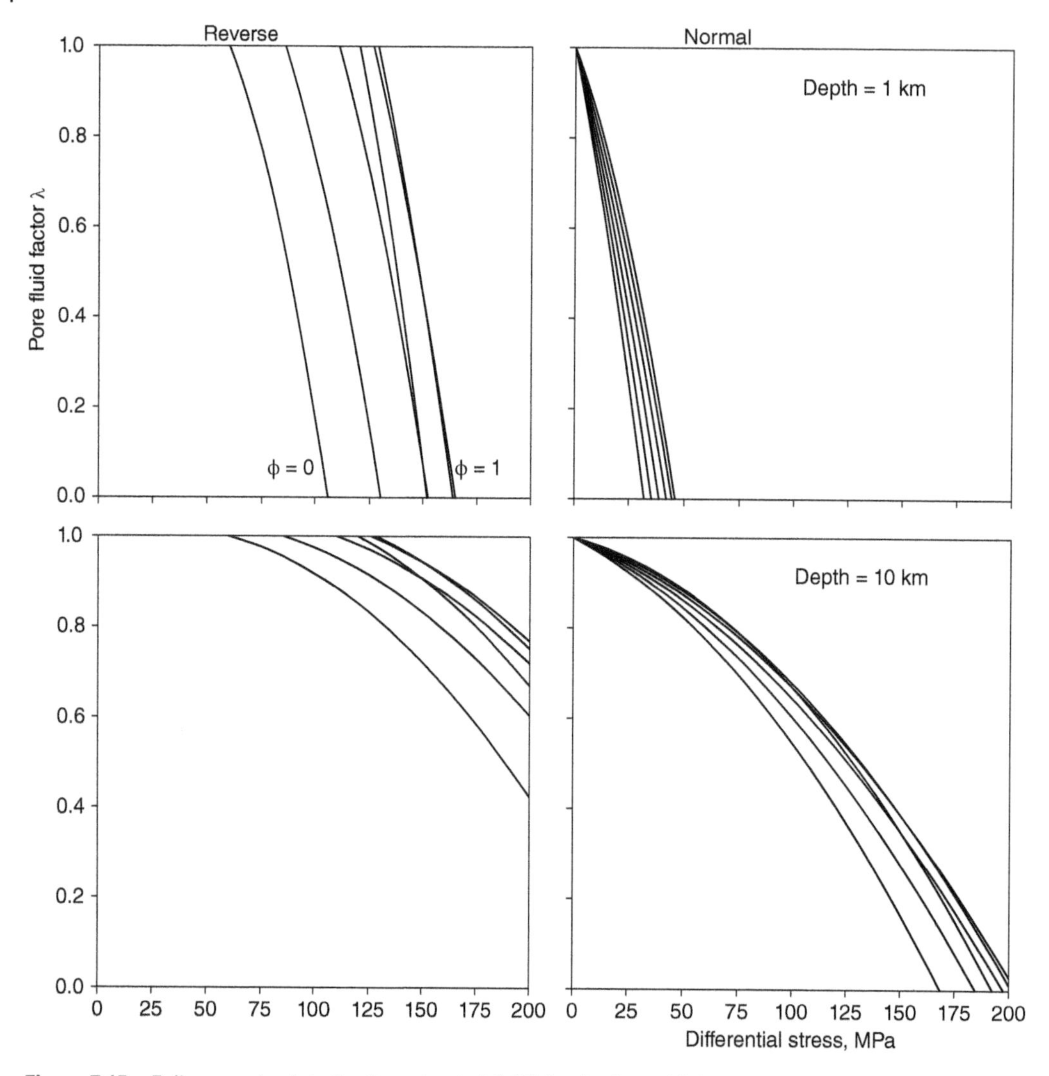

Figure 7.43 Failure mode plots for the extended Griffith criterion, which accounts for the influence of σ_2. Each family of curves is for $\phi = 0, 0.2, 0.4, 0.6, 0.8$ *and* 1.0. $T_0 = 5$ MPa.

and a sequence of stress states is more readily shown as a stress path. However, failure mode diagrams based on the Griffith-Coulomb criterion have some limitations. First, while the pore fluid factor λ_v is a useful concept, it is not a widely understood geological variable, at least outside the structural geology and rock mechanics community. For a more ready appreciation of the mechanics of failure, a simpler plot of pore fluid pressure against differential stress may be better, as illustrated in Figure 7.42c and d for the same parameters as the previous plots. The failure envelops for different depths do not cross over in these plots.

Second, the shear failure envelop for intact failure is nonlinear (e.g. Paterson and Wong 2006, p. 156), and even Byerlee's Law for frictional reactivation requires two linear versions depending on normal stress (Byerlee 1978). Third, the status of extensional shear as a mode of failure is widely accepted as a theoretical possibility, but there is no substantive body of experiments for extensional shear, and the field evidence for this mode of failure is equivocal (Engelder 1999). Moreover, the very limited experimental evidence of extensional shear failure is not consistent with the Griffith extensional-shear failure criterion (Ramsey and Chester 2004).

Most importantly, however, the Griffith-Coulomb failure criterion has no explicit dependence on the intermediate stress σ_2. The influence of σ_2 on failure, as described in Section 7B.2.2.1.6 is currently a subject of active research (Schöpfer et al. 2013; Hackston and Rutter 2016; Ma et al. 2017; Rudnicki 2017). Failure criteria are reviewed in the context of their ability to predict borehole shear failure by Rahimi and Nygaard (2015), with the conclusion that the criteria that account for the effect of σ_2 are superior, although Colmenares and Zoback (2002) have shown that rock types differ considerably in their sensitivity to σ_2.

The extension of the Griffith criterion by Murrell (Murrell 1963) is an obvious place to start investigating the effect of σ_2 since it uses the same single material property (tensile strength) as the Griffith criterion so that it can be compared with the plots in Figure 7.42. The extended Griffith or Griffith-Murrell criterion (Eq. 7.66) can be written as:

$$(\sigma_2-\sigma_3)^2+(\sigma_3-\sigma_1)^2+(\sigma_1-\sigma_2)^2 = 24T(\sigma_1+\sigma_2+\sigma_{13}). \tag{7.80}$$

Using the ratio $\phi = (\sigma_2-\sigma_3)/(\sigma_1-\sigma_3)$ and the differential stress between the principal stress $\Delta\sigma$ leads to:

$$\lambda_v = \frac{1}{\sigma_v}\left[\sigma_1(1-\phi)+3\sigma_3-\frac{\Delta\sigma^2(1+\phi+\phi^2)}{12T}\right] \tag{7.81}$$

This criterion leads to continuous, curved λ_v vs $\Delta\sigma$ relationships, which depend considerably on σ_2 via the ratio of the principal stresses ϕ (Figure 7.43). These graphs are an indication of the possible importance of including σ_2, and do not include the effects of tensile stresses.

7B.2.4 Yield and Creep

One property inherent in most analyses of the flow of isotropic materials is the idea that the principal directions of the incremental strains coincide with the principal directions of the differential stress. This principle was first proposed in the late nineteenth century and re-enumerated in the early twentieth century. It is often expressed mathematically using

$$\frac{\delta e_{xx}-\delta e_{yy}}{(\sigma_{xx}-\sigma_{yy})}=\frac{\delta e_{yy}-\delta e_{zz}}{(\sigma_{yy}-\sigma_{zz})}=\frac{\delta e_{xx}-\delta e_{zz}}{(\sigma_{xx}-\sigma_{zz})} =\frac{\delta e_{xy}}{\sigma_{xy}}=\frac{\delta e_{yz}}{\sigma_{yz}}=\frac{\delta e_{xz}}{\sigma_{xz}}=\delta\lambda, \tag{7.82}$$

which are known as the Levy–Mises relations (see Ford 1963, p. 410). For flows in which volume is conserved, one can write

$$\delta e_{ij}=\delta\lambda\,\delta\sigma'_{ij} \tag{7.83}$$

In Eqs. (7.82) and (7.83), δe_{ij} are the components of the incremental strain tensor, σ_{ij} are the components of the stress tensor, and σ'_{ij} are the components of the deviatoric stress tensor (from Chapter 6, $\sigma'_{ij} = \sigma_{ij}-\sigma_m$, where σ_m is the mean stress). The proportionality constant, $\delta\lambda$, has the "δ" to emphasize the relationships hold for incremental deformations only.

The equivalence of these two formats can be demonstrated by the following reasoning. From Eq. (7.82), one can extract $\delta e_{xx}-\delta e_{yy} = \delta\lambda(\sigma_{xx}-\sigma_{yy})$ and $\delta e_{xx}-\delta e_{zzy} = \delta\lambda(\sigma_{xx}-\sigma_{zzy})$. Adding these two together yields,

$$2\delta\varepsilon_{xx}-\delta e_{yy}-\delta e_{zz}=\delta\lambda(2\sigma_{xx}-\sigma_{yy}-\sigma_{zz}) \tag{7.84}$$

In flows where volume is conserved, $\delta e_{xx}+\delta e_{yy}+\delta e_{zz}=0$ or $\delta e_{xx}=-\delta e_{yy}-\delta e_{zz}$. Thus,

$$\delta e_{xx}=\frac{\delta\lambda\left(2\sigma_{xx}-\sigma_{yy}-\sigma_{zz}\right)}{3} \tag{7.85}$$

It is clear that $(2\sigma_{xx}-\sigma_{yy}-\sigma_{zz})/3=\sigma_{xx}-(\sigma_{xx}+\sigma_{yy}+\sigma_{zz})/3=\sigma_{xx}-\sigma_m=\sigma'_{xx}$. The parallelism of the stress and incremental principal directions and the proportionality of incremental strains and deviatoric stresses are two important links in constitutive relationships for plastic and viscous flows. Another way to illustrate this link is to recognize that the deviatoric stress and incremental strain Mohr circles are geometrically similar.

7B.2.5 Viscous Behavior

In Section 7A.2.6.2, we introduced a generalized version of a constitutive equation for viscous flow that could accommodate linear or nonlinear viscosity:

$$\dot{e}=B\left(\frac{\sigma'}{S_o}\right)^n. \tag{7.31}$$

Without an understanding of the physical and chemical processes entailed in viscous flow, topics that we address in the next chapter, there are limits to how deeply one can explore the properties of this general flow law. This section considers three issues that geologists wishing to understand viscous flow regularly address: (1) the temperature dependence of viscous flow of geological materials; (2) formulating Eq. (7.31) in terms of tensor invariants to take account of the overall character of a viscous flow; and (3) some approaches to address material anisotropy in viscous flow.

7B.2.5.1 Temperature Dependence of Viscosity

Experience, whether it is comparing the relative ease with which a simple syrup pours at different temperatures or the observation that tar distorts readily when warmed by the mid-day sun, demonstrates that viscosities of materials vary with temperature, typically taking lower values at higher temperatures. In geological materials that exhibit linear and nonlinear viscosity, parameter B is a function of several variables. Those variables, and thus the parameter B itself, are thermally activated, meaning that their dependence on temperature can be described by an Arrhenius-type equation. We address the processes involved in temperature dependence in the following chapter. At this point, we take B to be directly proportional to a factor r that measures the rate of those processes. This rate parameter varies with temperature according to

$$r=r_o\exp\left\{-\frac{E_a}{kT}\right\}. \tag{7.86}$$

(In presenting exponential functions, we use the "exp(a)" notation instead of "e^a" in order to avoid confusing the base of the natural logarithm e with the strain measure elongation e introduced in Chapter 4). Here r_o is a limiting value of the rate, E_a is an activation energy for the process, k is the Boltzmann's constant, and T is the temperature in Kelvins. As temperature rises, E_a/kT approaches zero, $\exp\{-E_a/kT\}$ approaches unity, and r approaches r_o. As this rate parameter increases, so does the magnitude of parameter B.

For linearly viscous materials, $S_o/B=\eta$, the viscosity. Because B is inversely proportional to the viscosity, increasing its magnitude lowers the viscosity. Lowering the viscosity means that a given magnitude differential stress is associated with a higher strain rate (see Figure 7.23). Figure 7.44 illustrates how changing the magnitude of B affects materials that exhibit nonlinear viscosity. For both power-law materials with $n=3$ and power-law materials with $n=4$, increasing the magnitude of B has a greater effect at higher differential stresses than at lower differential stresses. Further, increasing the magnitude of B is more significant for larger n values.

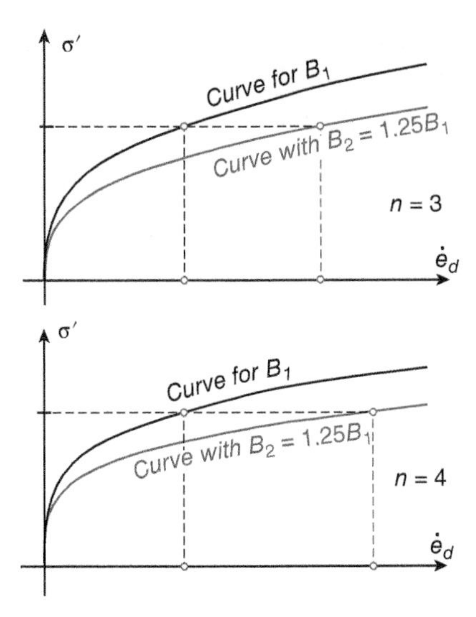

Figure 7.44 Diagram illustrating the impact of temperature increase on power-law materials. Stress-strain rate curves in red (for $n = 3$ and 4) predict higher strain rates at given differential stress for larger values of parameter B.

One can examine this temperature effect using the reciprocal of the viscosity, called the *fluidity*. As reciprocals, low fluidity is equivalent to high viscosity and vice versa. Fluidity is typically denoted by the overworked Greek letter μ; we here add a subscript *FL* to distinguish fluidity from the other two parameters μ often represents in structural geology, the elastic shear modulus and the coefficient of friction. Linear viscosity implies linear fluidity, that is.

$$\dot{e} = \mu_{FL}\,\sigma' \tag{7.87}$$

where the fluidity $\mu_{FL} = B/S_o$. Because fluidity is directly proportional to B, and increasing B leads to greater fluidity. The concept of fluidity is useful in examining the impact of temperature on nonlinearly viscous materials. Comparing Eqs. (7.31) and (7.87), we see that the fluidity of a power-law material

$$\mu_{FL} = B\left[\frac{\sigma'^{(n-1)}}{S_o^{\,n}}\right] \tag{7.88}$$

depends on deviatoric stress in addition to the parameters B and S_o. Still, fluidity is directly proportional to B, so increasing temperature increases fluidity. Thus, in geological settings, both linearly and nonlinearly viscous materials will deform more rapidly where temperatures are higher. Since temperatures increase with depth in the lithosphere, rocks and minerals that exhibit linear or nonlinear viscosity will tend to have lower viscosities at greater depths and will tend to deform more readily at a given magnitude of differential stress.

7B.2.5.2 Writing Flow Laws in Terms of Invariants

Theoretical and experimental analyses indicate that formulations of flow laws written in terms of invariants of the stress, strain, or strain rate tensors are preferred because they take into account the full characteristics of the respective tensor quantity. The most general formulations of flow laws are those written in terms of all three invariants, but in practice, it is acceptable, given the precision of stress, strain, and strain rate measurements, to ignore the variation with the third invariant and focus on formulations written in terms of the first two invariants. In Section 7A.2.4, we noted that there can be no viscous flow in response to the mean stress, so the first invariant is not a direct factor in viscous flow laws. A version of Eq. (7.31) written in terms of the second invariants of the stress and strain rate tensors takes the form

$$\dot{\gamma}_{\text{oct}} = B\left(\frac{\tau_{\text{oct}}}{S_o}\right)^n, \tag{7.89}$$

where $\dot{\gamma}$ is the octahedral shear strain rate, B is a constant with units of strain rate, τ_{oct} is the octahedral shear stress, and as usual, S_o is a constant with units of stress and n is a dimensionless constant. With increasing frequency (for reasons apparent in the following section), analyses of viscous deformation of nonlinear geological materials elect to use fluidity rather than viscosity to relate deviatoric stresses and strain rates. Rewriting Eq. (7.89) in terms of invariants yields

$$\mu_{FL} = B\left[\frac{\tau_{\text{oct}}^{(n-1)}}{S_o^{\,n}}\right]. \tag{7.90}$$

7B.2.5.3 Anisotropic Viscosity

By comparing Eqs. (7.5) and (7.6), which relate deviatoric stresses to deviatoric elastic strains, to Eqs. (7.30a) and (7.30b), which relate deviatoric stresses to strain rates, one can see that there is significant parallelism in the formulations. One can transform the equations referring to elastic distortion into the equations referring to viscous flow by replacing the shear modulus G with the viscosity η and "dotting the *es*," that is replacing each occurrence of a strain tensor component by the analogous strain rate tensor component. Workers have drawn upon the parallelism between these two sets of constitutive equations in addressing materials with anisotropic viscosity. We follow an analysis outlined by Cobbold (1976).

In Section 7B.2.1.2, we showed that in anisotropic materials the three-dimensional stress tensor is connected to the three-dimensional elastic strain tensor by

$$\sigma_{ij} = c_{ijkl} \cdot e_{kl}, \tag{7.91}$$

where the 81 elements c_{ijkl} are elastic stiffnesses, or by

$$e_{ij} = s_{ijkl} \cdot \sigma_{kl}, \tag{7.92}$$

where the 81 elements s_{ijkl} are compliances. By drawing upon the symmetry characteristics of the stress and strain tensors, we noted that the number of independent elements in either the stiffness or compliance tensors can be reduced significantly. For two-dimensional deformations, the number of components in the compliance or stiffness tensors reduces from 3^4 to $2^4 = 32$. Invoking the symmetry of the stress and strain tensors further limits the number of independent components of either tensor to 9, and specifying that the deformation occurs with no change in the volume of the material results in three equations with four independent coefficients (A, B, N, and Q) relating deviatoric stresses and strains (Cobbold 1976):

$$\sigma'_{xx} = 2N\,e_{xx} + 2A\,e_{xy} \tag{7.93a}$$

$$\sigma'_{xy} = 2B\,e_{xx} + 2Q\,e_{xy} = \sigma'_{yx} \tag{7.93b}$$

$$\sigma'_{yy} = 2N\,e_{yy} - 2A\,e_{xy} \tag{7.93c}$$

By assuming that the stress and strain tensors possess orthorhombic symmetry, that is conform with a pure shear deformation, and the plane of anisotropy is parallel to the x or y axis, $A = B = 0$, and Eq. (7.93) devolve to

$$\sigma'_{xx} = 2N\,e_{xx} \tag{7.94a}$$

$$\sigma'_{xy} = 2Q\,e_{xy} = \sigma'_{yx} \tag{7.94b}$$

$$\sigma'_{yy} = 2N\,e_{yy} \tag{7.94c}$$

In this formulation, N is the stiffness for normal stress and Q is the stiffness for shear stress.

Work to date addressing the flow of materials with anisotropic viscosity has focused on viscous flows conforming to the restrictions invoked to derive Eqs. (7.94), specifically two-dimensional, pure-shearing flows with no volume change. The viscous anisotropy is captured, then, in two distinctive viscosities, a viscosity for normal stress N and viscosity for shear stress Q:

$$\sigma'_{xx} = N\,\dot{e}_{xx} \tag{7.95a}$$

$$\sigma'_{xy} = Q\,\dot{e}_{xy} = \sigma'_{yx} \tag{7.95b}$$

$$\sigma'_{yy} = N\,\dot{e}_{yy} \tag{7.95c}$$

Drawing on the results of Section 7B.2.5.2, the viscosities for normal and shear stress are sometimes represented in the following form:

$$N = \eta_N = C_N \left(\dot{\gamma}_{\text{oct}}\right)^{((1/n)-1)} \tag{7.96}$$

$$Q = \eta_S = C_S \left(\dot{\gamma}_{\text{oct}}\right)^{((1/n)-1)} \tag{7.97}$$

Where C_N and C_S are constants describing the viscosity at a particular reference strain rate, $\dot{\gamma}_{\text{oct}}$ is the octahedral shear strain rate, and n is the power-law exponent.

7B.2.6 Plastic Behavior

No Earth materials exhibit perfect plasticity. As noted earlier under, however, rock masses following the Coulomb criterion or Byerlee's law exhibit bulk behavior very much consistent with an elastico-plastic constitutive relation. Similarly, the stress-strain rate curves for nonlinearly viscous materials in which the power exponent n is 3 or greater show that strain rates increase dramatically over a narrow range of deviatoric stress values. As n increases, then, materials approach perfect plasticity.

7B.2.7 Lithospheric Strength Profiles

A fundamental observation familiar to all geologists is that both pressure and temperature increase with depth in the lithosphere. As outlined in Section 7A.1.2 and shown schematically in Figure 7.5, lower temperatures and pressures favor fracturing and faulting, whereas higher temperatures and pressures tend to favor flow. To a first order, then, one can envision that in terms of its resistance to deformation the lithosphere consists of two layers: (1) an upper layer where deformation is dominated by fracturing and faulting; and (2) a lower layer where deformation occurs mainly by flow. Byerlee's law postulates that the differential stress at which Earth materials fail by fracturing and faulting increases linearly with the mean stress or hydrostatic pressure (see Section 7B.2.2.1.2). This observation suggests that lithospheric rocks in the upper layer are stronger at depth than near the surface. Similarly, the temperature dependence of the viscosities of Earth materials (see Section 7B.2.5.1) suggests rocks at higher temperatures are likely to attain a characteristic strain rate at lower flow stress than rocks at lower temperatures. This observation suggests that lithospheric rocks deeper in the lower layer will require a smaller magnitude of differential stress to attain a given strain rate than those nearer the top of that layer. To state the obvious, the strength of the lithosphere must vary with depth.

Figure 7.45 presents an early attempt to use inferences on how temperature and pressure affect rock strength to assess the lithospheric strength at different depths (see Kohlstedt et al. 1995 or Paterson 2001 for classic discussions of this issue; we consider more recent work on this issue at the end of Chapter 9). The figure plots the estimated differential stress magnitude required to cause either (1) faulting in the cooler upper lithosphere; or (2) flow at a characteristic strain rate (here 10^{-15}s^{-1}) in the warmer lower lithosphere. As you might have already surmised, few geologists believed then or now believe that any individual strength versus depth plot will pertain at all or even many points on Earth. Still, geologists continue to use comparable methodologies to define **lithospheric strength profiles** for different locations or for different tectonic provinces.

Spatial variations in rock composition, spatial and temporal variations in the vertical temperature profile, or differences in the tectonic setting guarantee that the precise form of the profile of lithosphere strength versus depth will vary from point to point on Earth. Even if one assumes that the strength versus depth profiles at different points all conformed to the relatively simple "hatchet" shape of the curve shown in Figure 7.45, she or he must anticipate that the precise shape of the curve will vary from location to location. It does not require much imagination to wonder how variations in rock composition between the upper crust and the lower crust or between the

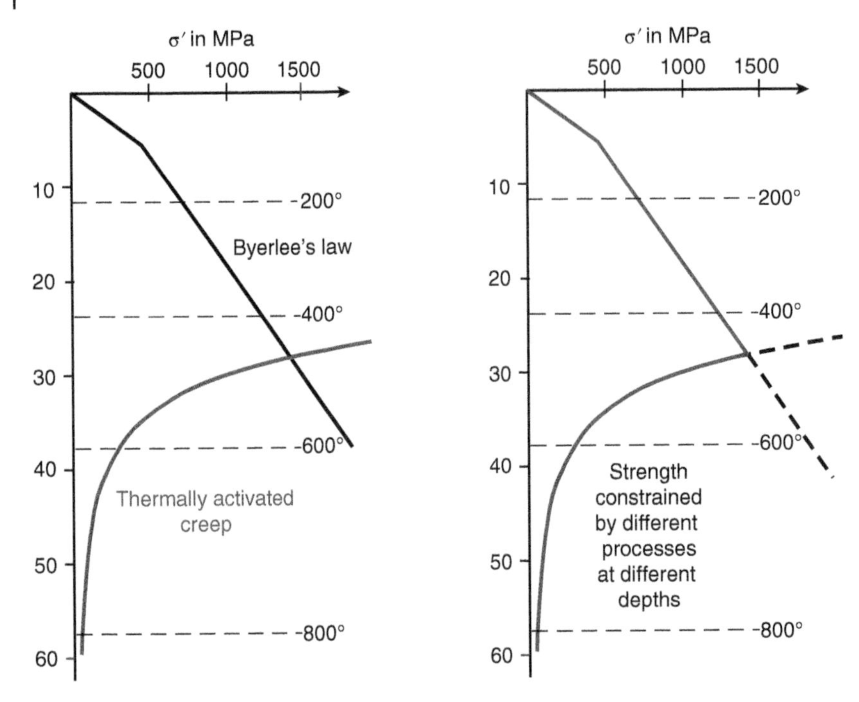

Figure 7.45 At left, the black curve gives of stress needed for faulting at different depths, derived from Byerlee's law. The red curve estimates the stress at which dry quartz-rich rocks creep. Combining the two curves, and taking the lower stress estimate, yields a strength versus depth profile for the upper portion of the continental lithosphere.

lower crust and the upper mantle affect the shape of the strength versus depth curve. Similarly, the lithospheric strength profile for any particular column of rock is likely to change over time, depending upon its proximity to plate boundaries, hot spots, etc. A deeper examination of lithospheric strength profiles requires an understanding of the full range of deformation mechanisms and some understanding of the character of deformation in different characteristic tectonic environments. We return to our consideration of lithospheric strength profiles at the end of Chapter 9.

References

Aharonov, E. and Scholz, C.H. (2017). A physics-based rock friction constitutive law: steady state friction. *Journal of Geophysical Research* **123**: 1591–1614.

Al-Ajmi, A.M. and Zimmerman, R.W. (2005). Relation between the Mogi and Coulomb failure criteria. *International Journal of Rock Mechanics and Mining Sciences* **42**: 431–439.

Brace, W. F. (1964). Brittle fracture of rocks. In: Judd, W. R. (ed.), *State of stress in the Earth's crust*. Elsevier, New York, 111–180.

Byerlee, J. (1978). Friction of rocks. *Pure and Applied Geophysics* **116**: 615–626.

Cobbold, P. (1976). Mechanical Effects of Anisotropy During Large Finite Deformations. *Bulletin Société Geologique De France* **18** (6): 1497–1510.

Colmenares, L.B., Zoback, M.D. 2002. A statistical evaluation of intact rock failure criteria constrained by polyaxial test data for five different rocks. *International Journal of Rock Mechanics and Mining Science* **39**, 695–729. doi:https://doi.org/10.1016/S1365-1609(02)00048-5.

Cox, S.F. 2010. The application of failure mode diagrams for exploring the roles of fluid pressure and stress states in controlling styles of fracture-controlled permeability enhancement in faults and shear zones. *Geofluids* **10**, 217–233. doi:https://doi.org/10.1111/j.1468-8123.2010.00281.x.

Engelder, T. (1999). Transitional-Tensile Fracture Propagation: a Status Report. *Journal of Structural Geology* **21**: 1049–1055.

Ford, H. (1963). *Advanced Mechanics of Materials*. London: Longmans, Green and Co., Ltd.

Griffith, A.A. (1921). The phenomena of rupture and flow in solids. *Philosophical Transactions of the Royal Society of London* **A221**: 163–198.

Griffith, A.A. (1924). The theory of rupture. In: *Proceedings of the 1st International Congress on Applied Mechanics* (ed. C.B. Biezeno and J.M. Burgers), 54–63. Delft: Tech. Boekhandel en Drukkerij J. Walter Jr.

Gundmundsson, A. (2011). *Rock Fractures in Geological Processes*. Cambridge: Cambridge University Press.

Hackston, A. & E. Rutter. 2016. The Mohr–Coulomb criterion for intact rock strength and friction – a re-evaluation and consideration of failure under polyaxial stresses. *Solid Earth* **7** (2), 493–508. doi:https://doi.org/10.5194/se-7-493-2016.

Handin, J. and Hager, R.V. Jr. (1957). Experimental deformation of sedimentary rocks under confining pressure: tests at room temperature on dry samples. *American Association of Petroleum Geologists Volume Studies in Geology* **41** (1): 1–50.

Harper, J., Humphrey, N.F., and Pfeffer, W.T. (1999). Three-dimensional deformation measured in an Alaskan glacier. *Science* **281**: 1340–1342.

Inglis, C.E. (1913). Stresses in a plate due to the presence of cracks and sharp corners. *Royal Institute of Naval Architects Transactions* 219–230.

Jaeger, J.C. (1971). *Elasticity, Fracture, and Flow, with Engineering and Geological Applications*. London: Methuen & Co., Ltd. and Science Paperbacks.

Jaeger, J.C. and Cook, N.G.W. (1971). *Fundamentals of Rock Mechanics*, 2e. London: Chapman & Hall and A Halsted Press Book.

Kohlstedt, D.L., Evans, B., and Mackwell, S.J. (1995). Strength of the lithosphere: Constraints imposed by laboratory experiments. *Journal of Geophysical Research* **100** (B9): 17587–17602.

Lawn, B.R. (1993). *Fracture of Brittle Solids*, 2e. Cambridge: Cambridge University Press.

Lee, Y.-K., S. Pietruszczak, & B.-H. Choi. 2012. Failure criteria for rocks based on smooth approximations to Mohr–Coulomb and Hoek–Brown failure functions. *International Journal of Rock Mechanics and Mining Sciences* **56** (C)., 146–60. doi:https://doi.org/10.1016/j.ijrmms.2012.07.032.

Ma, X., Rudnicki, J.W., Haimson, B.C. 2017. The application of a Matsuoka-Nakai-Lade-Duncan failure criterion to two porous sandstones. *International Journal of Rock Mechanics and Mining Science* **92**, 9–18. doi:https://doi.org/10.1016/j.ijrmms.2016.12.004.

Marone, C. (1998). Laboratory-derived friction laws and their application to seismic faulting. *Annual Reviews of Earth and Planetary Sciences* **26** (May): 643–696.

Mogi, K. (1971a). Effect of the triaxial stress system on fracture and flow of rocks. *Physics of the Earth and Planetary Interiors* **5**: 318–324.

Mogi, K. (1971b). Fracture and flow of rocks under high triaxial compression. *Journal of Geophysical Research* **76**: 1255–1269.

Murrell, S.A.F. (1963). A criterion for brittle fracture of rocks and concrete under triaxial stress and the effect of pore pressure on the criterion. In: *Proceedings of the 5th Symposium on Rock*

Mechanics (ed. C. Fairhurst), 563–577. Minneapolis: University of Minnesota.

Nádai, A. (1950). *Theory of Flow and Fracture of Solids*, vol. 1. New York: McGraw-Hill.

Nye, J.F. (1976). *The Physical Properties of Crystals: Their Representation by Tensors and Matrices – Oxford Science Publications (reprinted)*, 322p. Oxford: Clarendon Press.

Paterson, M. 2001. Relating experimental and geological rheology. *International Journal of Earth Sciences* **90** (1), 157–167. doi:https://doi.org/10.1007/s005310000158.

Paterson, M.S. and Wong, T. (2006). *Experimental Rock Deformation – The Brittle Field.* Springer Verlag.

Rahimi, R., Nygaard, R. 2015. Comparison of rock failure criteria in predicting borehole shear failure. *International Journal of Rock Mechanics and Mining Science* **79**, 29–40. doi:https://doi.org/10.1016/j.ijrmms.2015.08.006.

Ramsey, J. and Chester, F.M. (2004). Hybrid fracture and the transition from extension fracture to shear fracture Jonathan. *Nature* **428**: 63–66.

Rudnicki, J.W., 2017. A three invariant model of failure in true triaxial tests on Castlegate sandstone. *International Journal of Rock Mechanics and Mining Science* **97**, 46–51. doi:https://doi.org/10.1016/j.ijrmms.2017.06.007.

Scholz, C.H. (1998). Earthquakes and friction laws. *Nature* **391**: 37–42.

Scholz, C.H. (2002). *The Mechanics of Earthquakes and Faulting*, 2e. Cambridge: Cambridge University Press.

Schöpfer, M.P.J., Childs, C., Manzocchi, T., 2013. Three-dimensional failure envelopes and the brittle-ductile transition. *Journal of Geophysical Research Solid Earth* **118**, 1378–1392. doi:https://doi.org/10.1002/jgrb.50081.

Twiss, R.J. and Moores, E.M. (2007). *Structural Geology*, 2e. New York: W. H. Freeman & Co.

8

Deformation Mechanisms

8.1 Overview

The term **deformation mechanism** refers to a distinctive combination of physical and chemical processes that work together to accommodate the permanent deformation of minerals and rocks. An additional characteristic of any deformation mechanism is that it conforms to or is consistent with a particular constitutive equation or failure criterion. In this chapter, we examine eight widely recognized deformation mechanisms: (1) *Cataclasis and frictional sliding*; (2) *Coble creep*; (3) *Nabarro–Herring creep*; (4) *Solution transfer;* (5) *Dislocation glide*; (6) *Dislocation creep*; (7) *Diffusion-accommodated grain boundary sliding*; and (8) *Dislocation-accommodated grain boundary sliding* (Figure 8.1). To simplify our discussion of these deformation mechanisms, we group them into four broad deformation mechanism classes: *cataclastic* deformation mechanisms, *diffusional* deformation mechanisms, *dislocational* deformation mechanisms, and *grain boundary sliding* deformation mechanisms (Figure 8.1).

In the previous chapter (Chapter 7 – Rheology), we discussed three physical and chemical processes that contribute to the deformation of minerals and rocks: *elastic deformation*, *fracture formation*, and *frictional sliding*. In this chapter, we introduce six more such physical processes: *diffusion*, *glide of dislocations*, *climb of dislocations*, *intergranular displacement*, *twinning*, and *kinking*. As illustrated schematically in Figure 8.1, multiple **deformation processes** combine in different ways to constitute the different **deformation mechanisms**. The different deformation processes operate at different rates at different temperatures, pressures, differential stress magnitudes, strain rates, and compositions and concentrations of pore fluids. Consequently, a specific deformation process will make greater or lesser contributions to deformation as conditions change. This change in the relative contributions of the different deformation processes is a primary factor responsible for the distinctions between different deformation mechanisms.

The collection of microstructures observed in rocks often constrains which deformation mechanism or combination of deformation mechanisms contributed to the deformation. Syn- or post-deformation effects – which include reactions, recovery (defined in Section 8A.4.3), and recrystallization – create their own distinctive microstructures that sometimes obscure the deformation microstructures. Consequently, geologists rely on the results of theoretical analyses and rock deformation experiments to guide their interpretation of microstructures

An Integrated Framework for Structural Geology: Kinematics, Dynamics, and Rheology of Deformed Rocks,
First Edition. Steven Wojtal, Tom Blenkinsop, and Basil Tikoff.

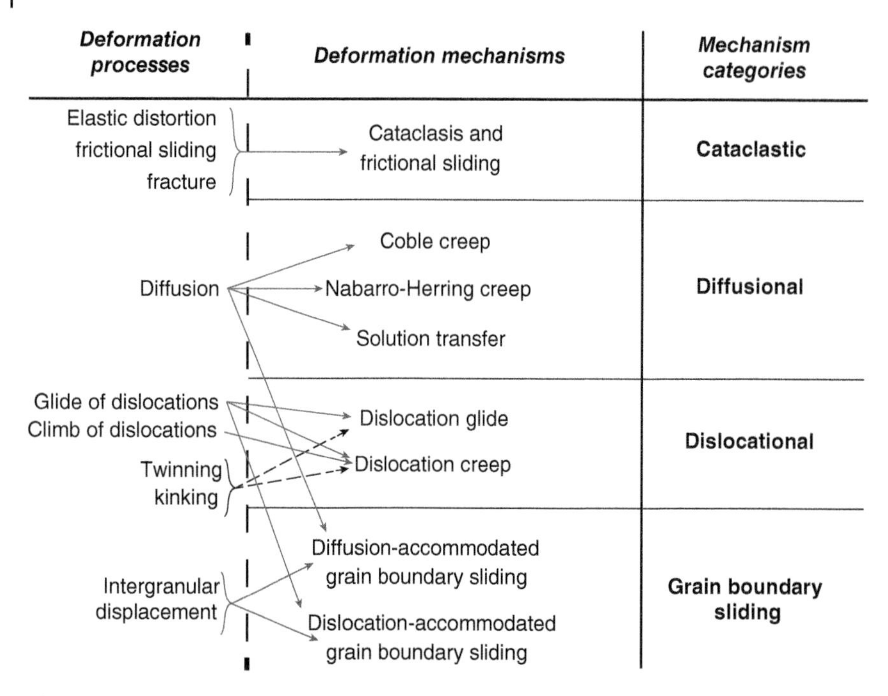

Figure 8.1 A diagrammatic overview of the relationships between deformation processes (left column), deformation mechanisms (center column), and deformation mechanism classes (right column). Solid red arrows from deformation processes to deformation mechanisms indicate that process contributes significantly to the deformation mechanism; dashed black arrows indicate a deformation process contributes to the deformation mechanism in some settings. See text for explanation.

and rock rheology. Theoretical analyses determine the functional form of the constitutive equations associated with the different deformation mechanisms. Rock deformation experiments identify characteristic microstructures and constrain the parameters in theoretically derived constitutive equations.

In most geological settings, two or more different deformation mechanisms operate simultaneously. We introduce and discuss briefly **deformation mechanism maps**, a graphical tool developed by materials scientists to evaluate the relative importance of competing deformation mechanisms. Because deformation mechanism maps are constructed using the constitutive equations associated with different deformation mechanisms, we can use them to analyze graphically how changes in physical conditions impact the relative contributions of different deformation mechanisms. This aspect of deformation mechanism maps contributes to our understanding of tectonic settings. Deformation mechanism maps connect the topics covered in this chapter to the following chapter, which examines case studies where geologists are able to combine data on rheology and deformation mechanisms to understand distributions of strain, strain rates, and stresses in the Earth.

Like the previous four chapters, this chapter is divided into a Conceptual Foundations section and a Comprehensive Treatment section. The Conceptual Foundations section focuses on the physical and chemical processes by which rocks deform and describes how they work together to constitute the main deformation mechanisms. The Comprehensive Treatment section explores in greater detail those deformation processes and examines factors influencing the functional form of the constitutive equations for the different deformation mechanisms. We once again provide citations to other monographs, journal articles, websites, etc., throughout the Comprehensive Treatment section to acknowledge the work we draw upon and to guide interested readers to sources that address the topics covered in this chapter in greater detail. As usual, there is a list of symbols in Table 8.1; this is not completely comprehensive and does not include those in the appendix for reasons of brevity.

Table 8.1 Commonly used symbols in this chapter.

Cataclastic deformation mechanisms			
σ_S	Shear strength	σ_Y	Strength in compression
$N(S)$	Number of particles of size greater than S	D	Fractal dimension in power-law particle-size distribution
$F(s)$	Cumulative fraction > size s in Weibull distribution	s_0	Characteristic particle size in Weibull distribution
β	Shape parameter in Weibull distribution	v	Slip speed
B	Bagnold number	ϕ	Grain fraction
w	Fault zone width	ρ, ρ_m	Density, the density of granular flow
R	Reynolds number	h	Granular flow thickness
U_f	Granular mean flow velocity	μfm	Granular flow viscosity
Diffusional deformation mechanisms			
J	Diffusive flux	c	Concentration
x	Distance		
D_L, D_{GB}, D_{WB}	Nabarro–Herring, Coble, wet grain boundary diffusion coefficients	D_o, D_v, D_{sd}	Limiting value, volume, self-diffusion coefficients
μ_A	Chemical potential of A	A_{NH}, A_C, A_{WB}	Prefactors for Nabarro–Herring, Coble and wet grain boundary constitutive equations
d	Mean grain diameter	δ	Grain boundary width
ΔG_f	Free energy	n_v, n_s, n_o	Number of vacancies, Schottky defects, cations
B	Equilibrium spacing between lattice sites	ϕ	Number of vacancies crossing an area d^2 per second
Dislocational deformation mechanisms			
H	Activation enthalpy	Q	Internal energy
PV	Pressure × Volume	R	Molar gas constant
H^*	Activation enthalpy for dislocation glide	B	Dislocation creep flow law geometry constant
n	Flow law exponent	C	Grain boundary sliding parameters in constitutive equation

(*Continued*)

Table 8.1 (Continued)

ψ	Angle of normal with cylindrical crystal axis	λ	Angle of direction with cylindrical crystal axis
σ_E	External stress	S	Schmid factor
σ_{CR}	Critical resolved shear stress for glide	ρ_d	dislocation density
b	Length of jog	Φ	Number of vacancies diffusing/unit length of dislocation line/unit time
L	Mean distance of dislocations glide	v_j, v_d	Jog, dislocation speed
d	Mean distance between glide planes	t_g, t_c	Time for glide, climb as dislocations move L
γ_{eg}	Shear strain due to, e.g. twinning	M	Number of dislocation sources per unit volume
Γ_{eg}	Tensor shear strain due to, e.g. twinning	ψ	Angular strain due to, e.g. twinning
D	Twin density	I_t	Twinning incidence

8A Deformation Mechanisms: Conceptual Foundation

The first section of the Conceptual Foundations portion of this chapter (Section 8A.1 Elastic Distortion) presents a brief recap of the characteristics of elastic distortion. The next four subdivisions of this chapter address the four main classes of deformation mechanisms: (1) Section 8A.2 Cataclastic deformation mechanisms; (2) Section 8A.3 Diffusional deformation mechanisms; (3) Section 8A.4 Dislocational deformation mechanisms; and (4) Section 8A.5 Grain boundary sliding deformations. For each of these classes, we consider first how the different physical and chemical processes combine to deform rocks. Then we describe briefly the distinctive set of microstructures or structures generated by those deformation mechanisms. Then we draw upon theoretical analyses and experimental data to outline the constitutive relationships or flow laws associated with each mechanism.

In Chapter 7 – Rheology, we focused on the bulk behavior of rocks from a macroscopic perspective, envisioning that rocks are made of continuous, uniform material or, in the case of fracture formation, continuous material with numerous microscopic flaws. This characterization is, of course, a simplification. Sedimentary rocks are polycrystalline precipitates from solutions or aggregates of detrital fragments, which may themselves be single crystals or aggregates of crystals, cemented together. Igneous rocks are collections of interlocking minerals crystallized from melts. Metamorphic rocks are interlocking minerals formed by recrystallization in the solid state. The polycrystalline character of rocks must be addressed in order to understand the mechanisms by which rocks deform. In fact, the crystalline character of individual mineral grains also impacts significantly the mechanisms by which any rock deforms. It is, then, imperative to establish a baseline understanding of crystallinity.

Within any mineral in a rock, atoms, ions, or groups of atoms or ions occur periodically at locations known as *sites* in a regular, three-dimensional array known as a *crystal lattice* (Figure 8.2a). For simplicity in the rest of the chapter, we use the

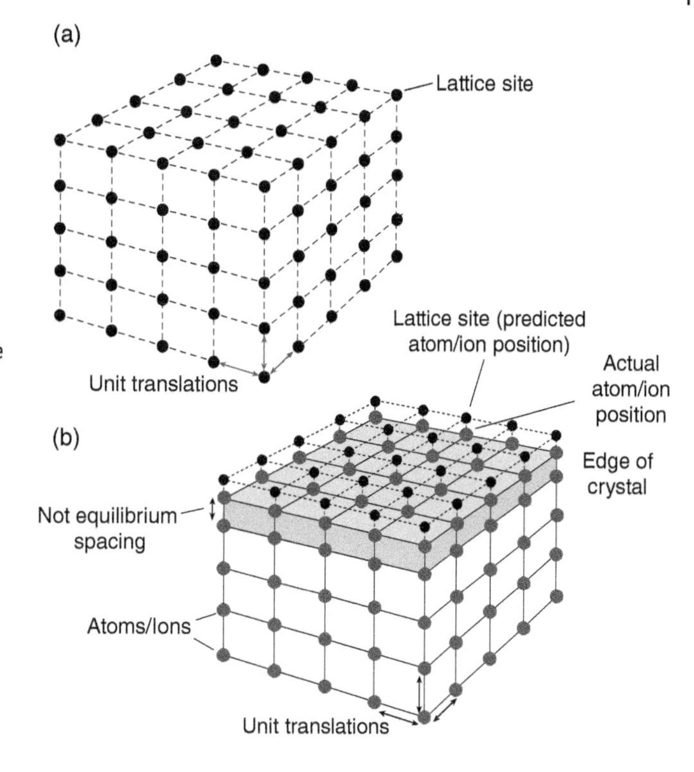

Figure 8.2 (a) Sketch of an idealized isometric or cubic lattice, with lattice sites found at regular intervals along three, mutually perpendicular directions. The atom at each lattice site sits one *unit translation* from its neighbors along any one of those three directions. (b) In the vicinity of an external surface of a crystal, atoms do not occupy positions one unit translation from their neighbors because bonds that draw atoms toward the center of the crystal are not balanced by bonds drawing atoms away from the center of the crystal.

term "atoms" or rarely the phrase "atoms and ions" as shorthand for "atoms, ions, or groups of atoms or ions." Ideally, the periodicity of lattices dictates that identical atoms occur spaced at identical distances, known as *lattice units* or *unit translations*, from each other along characteristic directions, known as *crystallographic axes*, within the crystal. Naturally occurring minerals exhibit a considerable variation in the geometric properties of their lattices and in the configurations of the collections of atoms associated with each lattice site. In this chapter, illustrations depict only lattices with the simple cubic geometry (having three mutually perpendicular crystallographic axes with unit translations of equal length) and with single atoms or ions positioned at each lattice site. Despite this simplification, the illustrations provide significant insight into the processes responsible for rock deformation.

8A.1 Elastic Distortion

Even though we have addressed elastic distortion in the previous chapter, we explicitly consider elastic distortion in this chapter on deformation mechanisms for two reasons. First, elastic distortion plays a role in most deformation mechanisms even though it does not on its own generate permanent deformation. Second, elastic distortion is needed to understand the deformation on the short time scale that occurs in the tectonic plates. For example, geodetic movements associated with plate boundary faults, such as subduction zones, require knowledge of elastic strain buildup and release.

Elastic distortion occurs *as soon as* stresses or surface displacements are imposed on a material – i.e. is *instantaneous* – and persists *only as long as* the stresses or displacements are imposed – i.e.

is *recoverable* (see Section 7A.2.1). These macroscopic characteristics relate directly to effects at the atomic or ionic level (cf. Figures 7.8, 7.9). In the absence of imposed stresses or surface displacements, atoms in crystalline or glassy solids have a spacing dictated by a balance between attractive and repulsive forces. During elastic distortion, imposed stresses or surface displacements cause the bonds between atoms in a crystal lattice to stretch, compress, or shear but not to break. This limits the magnitudes of imposed stresses and displacements that can be accommodated by elastic distortion, especially in the case of the oxides and silicates that compose most rocks.

Recall that elastic distortion adds energy to a deformed material, and the elastic strain energy per unit volume is the product of the stress and strain (Eq. 7.1). Spatial variations in elastic strain energy are common *within* individual crystals. For example, the distorted bonds along the bounding surface of a crystal give rise to measurable surface energy (Figure 8.2b). Similarly, the elastic distortion and elevated stresses in the vicinity of the tip of an incipient fracture store strain energy that can be drawn upon to increase fracture surface area during fracture propagation. That exchange of energy is essential to the development of macroscopic fractures.

Spatial variations in elastic distortion within individual crystals are significant to other important geologic phenomena. In a granular or polycrystalline material, an elastic distortion that appears uniform macroscopically is necessarily variable at the scale of individual grains or crystals. In a granular material with open pores or with pores filled by a phase that is weaker than the grains, uniform stresses or displacements imposed on boundaries of the material do not generate uniform elastic distortion within all grains. This difference is apparent in the two-dimensional model in Figure 8.3a, where certain grain-to-grain contacts experience greater elastic strains and stresses than other grain-to-grain contacts. These variations have two important effects. The first effect is that the portions of grains with relatively high elastic strains and stresses (see Figure 8.3a) have higher energy per unit volume than other grain-to-grain contacts or than the interior of the grains. The second effect, which is particularly significant for cataclastic flow, is the development of *force chains*. Force chains can be visualized in the two-dimensional model by drawing line segments connecting the grain-to-grain contacts that experience high compressive stress and high elastic strain. These line segments define collections of grains that are the primary support of compressive loads or inward displacements imposed on the boundaries of a rock. The networks of force chains together constitute a *load-bearing framework* for the aggregate. Force chains in these networks may form polygons that surround certain grains, isolating them from the effects of the displacements or stresses applied to the boundaries (Figure 8.3b). When a rock is composed of grains with different diameters, force chains almost always pass through larger grains, meaning that grains with larger diameters are more likely to contribute to supporting the boundary displacements or loads.

In rocks composed of interlocking crystals with no open pores, stresses or displacements imposed on the rocks' boundaries may not create the distinctly visible force chains created by the point-to-point contacts seen in Figure 8.3a. Consequently, elastic strains may be more uniformly distributed through a material with interlocking crystals. Nevertheless, variations in elastic strains are likely for two reasons. First, spatial variations in elastic strains within crystals may result from variations in the elastic properties of different mineral species or from variations due to differences in the orientations of neighboring anisotropic grains of the same mineral. Second, grain boundaries at high angles to the direction of maximum shortening experience higher compressive stresses and elastic strains than grain boundaries with different orientations (Figure 8.3c). Both

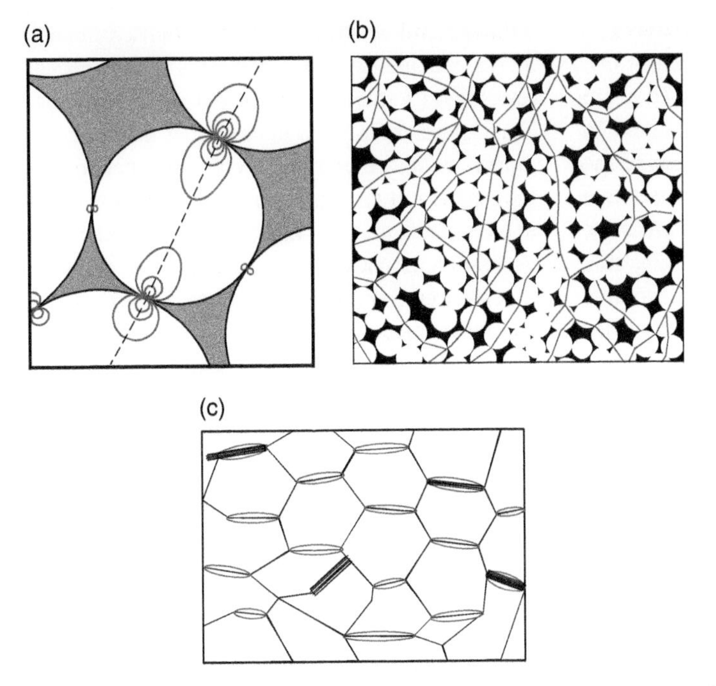

Figure 8.3 (a) Tracing of a photograph of translucent, isotropic circular discs in an array of discs compressed vertically. *Source:* Allersma (2018)/Henderikus Allersma. Where the discs are elastically distorted at disc-to-disc contacts, they become birefringent. The red lines identify successive orders of interference colors, with higher orders of interference colors denoting greater elastic distortion. The dashed lines connect contacts where discs are more highly distorted. (b) Tracing of an array of translucent, isotropic discs compressed vertically where red lines indicate *force chains*. *Source:* Modified from Gendelman et al. (2016). (c) Idealized sketch of mineral grains in a rock with no open pores compressed vertically. Red ellipses identify grain boundaries exhibiting greater elastic distortion and correspondingly higher compressive normal stresses.

effects generate variations in strain energy per unit volume within mineral grains.

As outlined in Section 7A.2.1.1, elastic distortion conforms to a linear relationship between the applied stress and the resulting recoverable strain. In most instances, we treat deforming rock masses as isotropic bodies that follow constitutive equations with the form of Eq. (7.7).

8A.2 Cataclastic Deformation Mechanisms

Three deformation processes – elastic distortion (described earlier), rock fracture, and frictional sliding – combine to define the cataclastic deformation mechanism called cataclasis and frictional sliding.

8A.2.1 Fracture of Geological Materials

Fracture formation contributes to the permanent deformation of rocks in two contexts: (1) settings where fractures occur as isolated features or as widely spaced members of arrays or sets of fractures; and (2) settings where fracturing is pervasive, such that fractures are closely spaced. Section 7A.2.2 examined the conditions under which fractures initiate and propagate. In this section, we consider geological settings where the polycrystalline nature of Earth materials has a significant impact on the formation of both types of fractures.

8A.2.1.1 Characteristics of Individual Fractures

Because rocks are aggregates of differently oriented, often anisotropic mineral grains, they are not the ideally uniform materials envisioned in the discussion of fracture in Section 7A.2.2. In the following section, we consider at the outset two types of natural fractures whose surface areas are many times the dimensions of the mineral grains that compose the rock: transgranular fractures and intergranular fractures. We then consider intragranular fractures, which are confined to individual mineral grains.

Transgranular fractures cut indiscriminately across the grains that compose a rock, typically resulting in a smooth fracture surface (Figure 8.4a). Transgranular fractures indicate that differences in grains' resistance to fracture had little effect on the propagation of the fracture, meaning that the rock approximated the uniform continuum envisioned in Section 7A.2.2. Transgranular fractures form in rocks that are well cemented, consist of interlocking crystals with similar physical properties, or are otherwise highly indurated at the time of fracture. Transgranular fractures are, in essence, the type of fracture expected

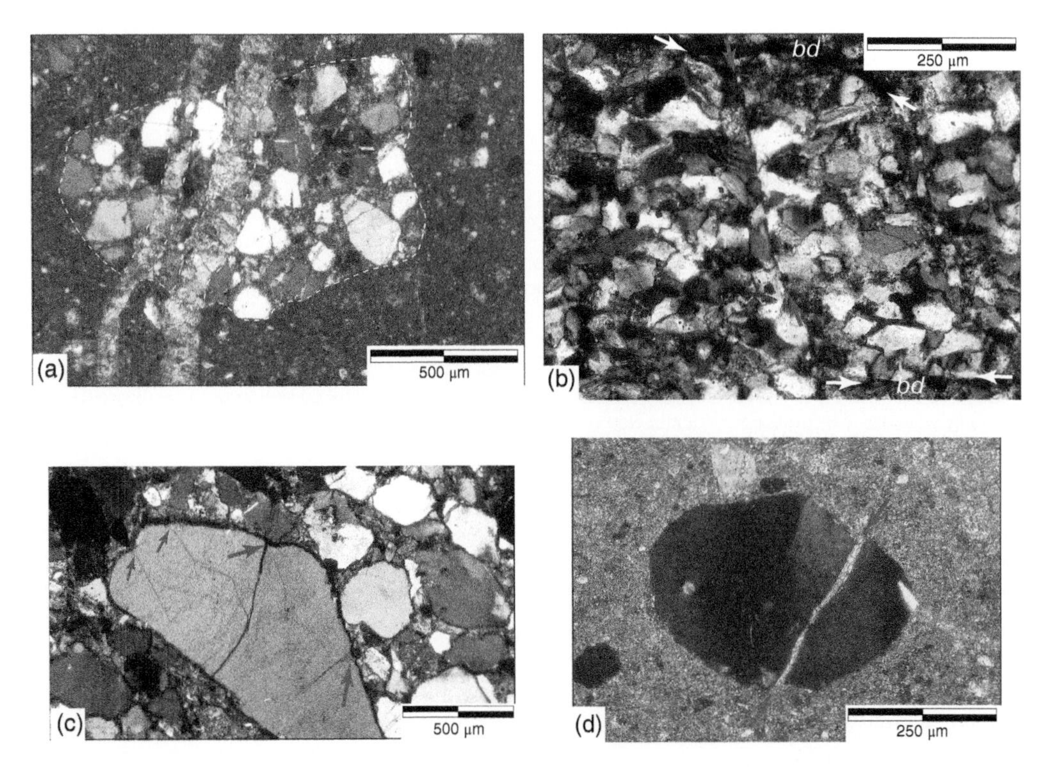

Figure 8.4 Fracture types. (a) Transgranular fracture. Two calcite-filled fractures cut across a quartz sandstone fragment (outlined by a dashed white line) in a banded cataclasite composed of a mixture of fine-grained carbonate and siliciclastic fragments. Note that the calcite-filled fracture at the left splits a quartz clast in the sandstone fragment. (b) Intergranular fracture. A short intergranular fracture (between the two red arrows) along the boundaries of quartz grains within a thin layer in a quartz sandstone; *bd* = mud layers in sandstone. (c) Intragranular fractures in a quartz fragment, indicated by red arrows, caused by impinging neighboring grains. (d) Intragranular fracture. A calcite-filled fracture (between two red arrows) cuts a quartz sandstone fragment in a banded cataclasite. Note that fracture does not extend into the surrounding matrix of ultrafine calcite with small siliciclastic fragments.

from fracture propagation where: (1) stress principal directions have similar orientations at different positions, or vary smoothly with position in the rock; (2) there exists a favorably oriented flaw with a shape that is able to generate sufficiently high stresses to exceed the fracture strength of the rock; and (3) the fracture, whether it is a tensile or shear fracture, maintains a similar orientation relative to the principal stresses as it propagates. Transgranular fractures typically are planar or broadly curved.

Intergranular fractures also have surface areas many times the dimension of the grains that compose a rock, but the fracture follows the boundaries of grains rather than propagating through them (Figure 8.4b). Intergranular fractures, therefore, have a surface that is rough or irregular at the scale of the diameters of the grains composing the rock. Intergranular fractures indicate that the difference in the physical characteristics or orientations of the component crystals altered the fracture propagation. This may occur because rocks are either not well-cemented, cemented together by minerals less resistant to fracture than the mineral grains, or composed of interlocking minerals with significantly different physical properties.

The polycrystalline nature of rock also impacts the formation of the third type of fracture – **intragranular fracture**. Under appropriate conditions, individual fractures initiate within and remain confined to individual mineral grains within the rock (Figure 8.4c). The occurrence of such an intragranular fracture indicates: (1) significant differences in the physical properties of the grain relative to the physical properties of the material surrounding the grains; (2) significant differences in the magnitudes of stresses within the grain relative to those outside the grain; or (3) significant differences between both physical properties and stresses within the grain relative to those outside it. These factors are summarized in the types of dominantly intragranular fractures described in Table 3.1. Some intragranular fractures form when two grains along a force chain *impinge* upon each other (Figure 8.4c). For minerals with well-defined cleavage planes – such as micas, carbonates, and feldspars – intragranular fractures created by grain impingement often follow the mineral's cleavage planes. In other cases, in minerals such as garnet, impingement may form intragranular fractures that follow parting directions in the minerals. Still other minerals, notably quartz, have less preferred directionality in breakage. So, the geometry of the contact, in particular, the curvature of the impinging grains, exerts significant control on the occurrence and subsequent orientations of any intergranular fractures. Intragranular fractures in relatively strong grains found within a matrix composed of weaker materials are thought to develop due to tractions exerted on the periphery of the fractured grain by the deforming matrix (Figure 8.4d).

8A.2.1.2 Pervasive Fracturing and Comminution

Deformation in some settings is accommodated by pervasive arrays of fractures. The component fractures may be subparallel, anastomose, occur in two or more intersecting sets, or have no discernable preferred orientation. Individual fractures may have no discernable offset or exhibit opening or shearing displacements. Pervasive arrays consisting only of opening-mode fractures accommodate volume expansion. In some settings, zones of pervasive, subparallel extension fractures are separated by broad regions with few or no fractures. The rock hosting these *tabular fracture clusters*, as the sets of closely spaced extension fractures are known, often is bleached or otherwise altered, whereas the intervening rock masses are unaltered. These observations suggest that tabular fracture clusters are zones of extensive fluid or volatile transport in addition to being the locus of elongation. Pervasive arrays of shearing-mode fractures accommodate finite shearing offsets. If such an array consists of subparallel fractures, rock between the offset terminations of subparallel fractures often is more extensively fractured or otherwise deformed than rock elsewhere (Figure 8.5a). Many pervasive arrays consist of both opening-mode and shearing-mode fractures.

Intersecting fractures divide an intact rock into fragments. Deformation concurrent with or after the fracture formation may cause neighboring rock fragments: (1) to impinge on each other, leading to the formation of additional fractures and smaller fragments; and (2) to rotate relative to their neighbors or "roll," which often abrades corners off angular fragments, creating rounded fragments. These two effects together constitute the process known as *comminution*. Comminution may generate an incohesive mass of rock fragments known as a *breccia* (if it consists of fragments with a wide range of sizes) or *gouge* (if it consists primarily of fragments of small size). Pervasive fracturing may also generate a variety of cohesive rocks composed of rock and mineral fragments; these rocks are known collectively as *fault rocks* (Section 3.3). Structural geologists distinguish coherent fault rocks on the relative amounts of identifiable rock fragments versus *matrix*, constituents too small to be seen individually without the aid of a hand lens (cf. Figure 2.13). Coherent fault rocks with 0–10% matrix are either breccias or microbreccias. Fault rocks with 10–50% matrix are *protocataclasites* (e.g. Figure 8.5b), those with 50–90% matrix are *cataclasites*, and those with >90% matrix are *ultracataclasites* (e.g. Figure 3.4).

In experiments conducted at low confining pressures, axial shortening of samples regularly forms extension fractures parallel to the most compressive stress (Figure 8.5c *i*; cf. Figures 7.5 and 7.10). In samples shortened at moderate confining pressure, microscopic flaws commonly link together to form fault zones inclined to the direction of greatest compressive stress (Figure 8.5c *ii*; cf. Section 7A.2.2.3 and Figure 7.18). In experiments conducted at higher confining pressures, pre-existing flaws experience sufficiently high stresses near their terminations to propagate as fractures (Figure 8.5c *iii*). The latter two cases can lead to pervasive fracturing of the sample if the propagating fractures intersect. In those instances, the intact sample is divided into angular fragments surrounded by fracture surfaces. As noted earlier, some fragments are likely to form a load-bearing framework. Under continued loading, fragments in the load-bearing framework may experience sufficient stresses to fracture. In cases where the fractured rock experiences shearing displacements, adjacent fragments are likely to rotate relative to each other. The rolling of fragments may abrade corners or point off angular fragments, leading to rounded fragments. The removal of corners or points on rolling fragments contributes to further comminution. Thus, between grain impingement and grain abrasion, the median grain size may be reduced during shearing.

8A.2.2 Frictional Sliding

Section 7A.2.2.1 introduced Amonton's Laws governing frictional sliding on clean planar surfaces:

1) Frictional resistance is proportional to the normal load across a surface, and
2) Frictional resistance is independent of the area of surface contact as long as the normal force remains constant.

Amonton's Laws are thought to arise because macroscopically planar surfaces are necessarily irregular in detail (Figure 8.6a). In those cases: (1) the shearing resistance is a product of the yield strength in shear (σ_S) acting across the true area of contact across the surface; and (2) the true contact area of the surface separating two rock masses depends on the yield strength of the rock in compression (σ_Y). The magnitude of the coefficient of friction of a surface (μ) is the ratio of the

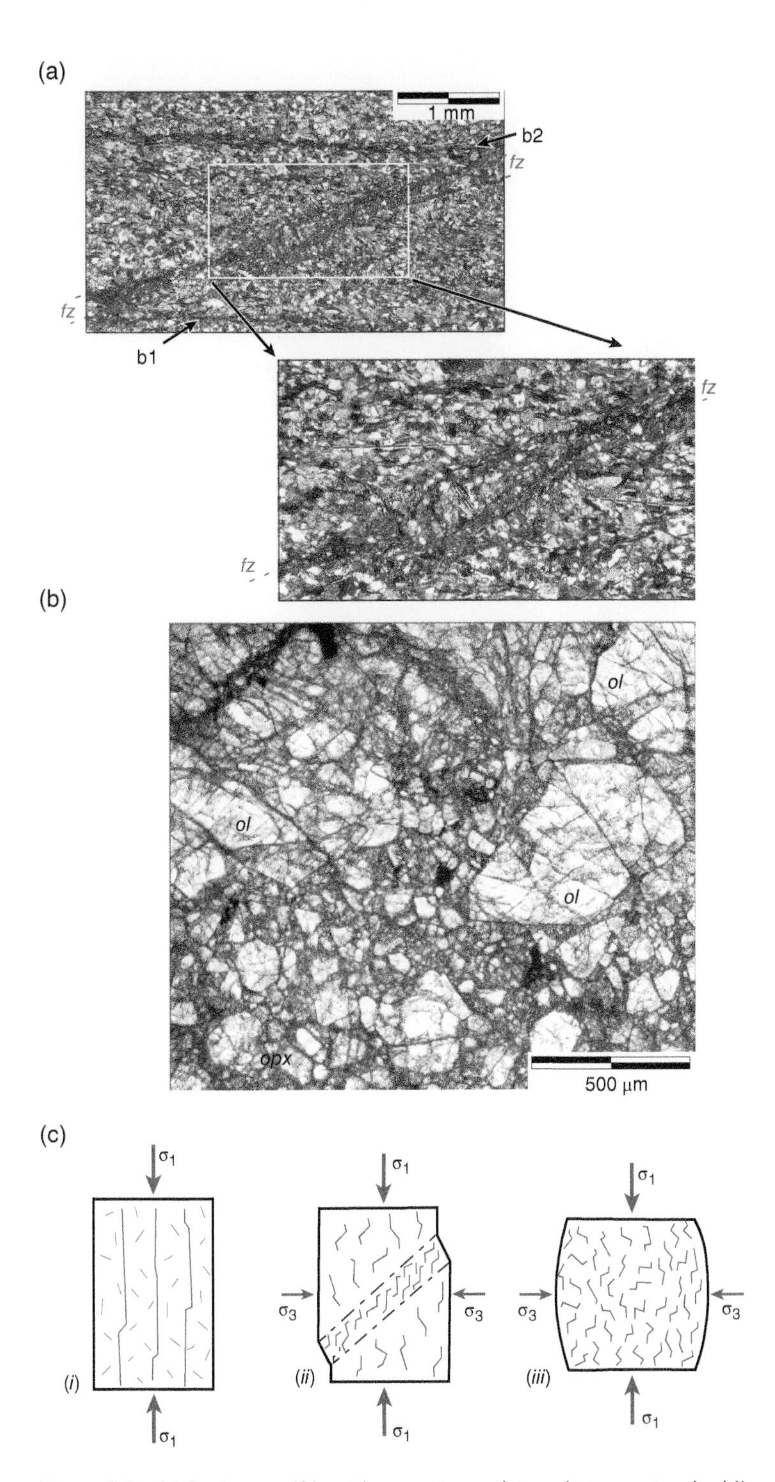

Figure 8.5 (a) Fault zone (*fz*) cutting quartz sandstone between two bedding surfaces (*bd1* and *bd2*). Enlargement shows that two branches of the fault zone surround a small region of rotated bedding (indicated by dashed red lines). (b) Protocataclasite developed in faulted dunite from Twin Sisters Massif, Washington (USA) ol – olivine; opx – orthopyroxene. (c) Idealizations of the observed response to rock samples subjected to compressive loads. *Source:* Sammis et al. (1986)/Springer Nature. At low confining pressure (*i*), sample develops extension fractures. At moderate confining pressures (*ii*) extension fractures link to form a fault zone that cuts obliquely across the sample. At higher confining pressure (*iii*), sample is pervasively fractured.

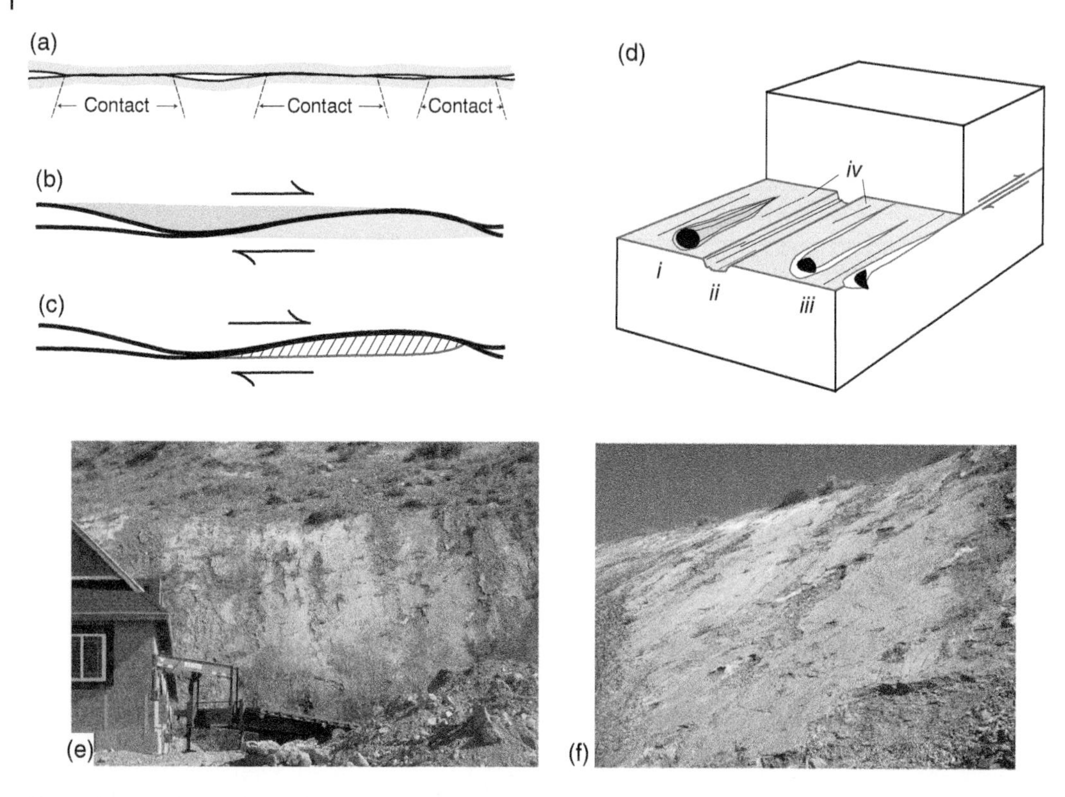

Figure 8.6 (a) Sketch indicating microscopic characteristics of a fault surface, where true contact area is less than the total fault area. (b) Shearing of contact regions may lead to localized continuous deformation of portions of the fault's hanging wall (e.g. the region shaded red) and footwall (e.g. the regions shaded gray). (c) Shearing of contact regions may lead to the formation of a small fault (in red), which can separate rock or mineral fragments, like the ruled area, from the wall rocks. (d) Diagram showing the origin of fault surface markings formed during frictional slip after Power and Tullis (1989) (*i*) Tail of lightly cemented fault gouge (in gray) behind a hard asperity (in black). (*ii*) Interlocking groove and ridge. (*iii*) Spoon-shaped depressions formed around hard particles. (*iv*) Scratches and striae made by small asperities, which probably disintegrated. (e) and (f) Photographs of a portion of the Wasatch Fault exposed at a construction site. Utah, USA. Note the polished appearance of the slip surface and down-dip striations consistent with normal offset on the fault.

rock's strength in shear (σ_S) divided by its strength in compression (σ_Y) (cf. Eq. 7.14):

$$\mu = \tan\frac{\sigma_S}{\sigma_Y}. \tag{8.1}$$

Geological faults rarely resemble the macroscopically planar surfaces assumed in deriving Amonton's Laws. Irregularities with amplitudes on the order of a cm or larger must deform somehow to accommodate fault slip. Fault movement may lead to deformation localized in the vicinity of irregularities in the fault surface (Figure 8.6b) or may nucleate subsidiary slip surfaces that separate the irregularities from the wall rock (Figure 8.6c). In such an instance, intact rock on one side of the fault is separated from intact rock on the opposite side. Smaller irregularities, known as *asperities*, on one side of a fault will also impinge on the opposite wall as slip accrues on the fault. An asperity that is harder than the adjacent wall rock may break off the wall rock and become embedded in it or scratch the opposite side, creating striations on the fault surface (Figure 8.6d). In some instances, abraded surfaces take on a smooth or polished appearance (Figure 8.6e and f). These processes together contribute to frictional *wear* by the distortion and breakage of asperities, leading to the formation of *gouge* or *cataclasite* along a fault surface (Figure 8.7).

Generally, once gouge or cataclasite forms, some component of fault movement is accommodated by deforming the gouge or cataclasite. This deformation need not be uniform. In some instances, adjacent fragments in a gouge or cataclasite will have irregular boundaries that fit together like puzzle pieces, indicating that they were once joined and have moved little since they were formed. In other instances, gouge or cataclasite may consist entirely of highly abraded and rounded fragments. The fragments that compose gouges or cataclasites need not be the same lithologies as the adjacent wall rocks. Other processes, such as the injection of material from different areas of a fault zone, can separate the opposing masses of intact rock. Further, fault zones are often conduits for fluid flow, so precipitation of secondary minerals and/or the alteration or replacement of primary minerals contribute to differences between the wall rocks and the fragments in a fault zone. New minerals created by precipitation, alteration, or replacement may be either stronger or weaker than the minerals created by the comminution of the wall rock.

Despite these difficulties, Amonton's Laws remain useful in describing the stresses at which frictional sliding occurs. This coincidence likely relates to the role of asperities in frictional sliding. As outlined in Section 7A.2.2.1, frictional resistance is rooted in the contact across asperities. Sliding can occur only when shear stresses are sufficiently high to deform permanently the

(a)

(b)

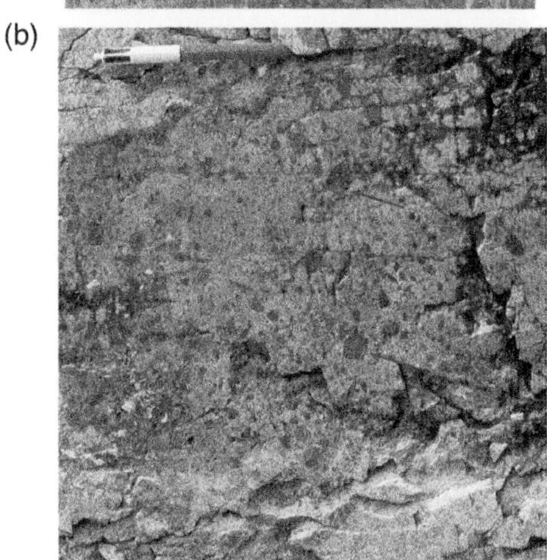

Figure 8.7 (a) Photograph of a portion of the fault surface in Figure 8.6d and e. Light gray material indicated by arrows is cemented cataclasite adhering to fault surface. (b) Fault surface in Cambrian limestones from the Appalachian Valley and Ridge Province, Virginia, USA. Arrows point to a few of the larger abraded limestone fragments visible within the white cataclasite.

material that composes the asperities. Empirically, the deformation of asperities involved in moving fragments past neighboring fragments is sufficiently similar to the deformation of asperities along a single surface that the "rules" for sliding, i.e. Amonton's Laws, pertain in both instances.

8A.2.3 Microstructures Associated with Cataclasis and Frictional Sliding

Microstructural observation typically can recognize the activation of cataclastic deformation processes. Cataclasis and frictional sliding may generate striated surfaces (Figure 8.6d–f), breccias (Figure 8.5), gouge (Figure 3.3), cataclasites (Figures 3.4a, 3.4b, 8.5b, 8.7a, and b), and ultracataclasites (Figure 3.4c).

8A.2.4 Cataclasis and Frictional Sliding as a Deformation Mechanism

Elastic distortion, fracture, and frictional sliding combine to define the deformation mechanism of cataclasis and frictional sliding. Laboratory studies suggest that rock masses deforming by this mechanism conform to a modified Coulomb criterion known as Byerlee's law (see Section 7B.2.2.1.2 for a more thorough discussion). Byerlee's law consists of two empirical relationships:

- For rock bodies deformed at low-to-intermediate magnitudes of normal stress, i.e. 5 MPa $<\sigma_N\leq$ 200 MPa, the tangential stress required for slip is given by

$$\sigma_T = 0.85\sigma_N. \quad (8.2a)$$

- At higher magnitudes of normal stress, i.e. 200 MPa $<\sigma_N\leq$ 2000 MPa, the shear stress required for slip is given by

$$\sigma_T = 50 + 0.6\sigma_N. \quad (8.2b)$$

Thus, deformation (slip within a fault zone) proceeds when the magnitude of differential stress is sufficiently large to reach the shear stresses effectively a yield stress, and the magnitude of the yield stress increases with increasing hydrostatic stress.

8A.3 Diffusional Deformation Mechanisms

Geologists and materials scientists recognize three distinct deformation mechanisms that result from diffusion in rocks. If material diffuses through the crystal lattices of mineral grains (known as *volume diffusion*), the resulting deformation mechanism is called *Nabarro–Herring creep*. If material diffuses primarily along dry grain boundaries separating mineral grains (known as grain boundary diffusion), the resulting deformation mechanism is called *Coble creep*. If there is an intergranular fluid present and diffusion occurs primarily through this intergranular fluid, the resulting deformation mechanism is called *solution transfer*.

8A.3.1 Diffusion

The process of diffusion plays a significant role in the crystallization of igneous rocks, the diagenesis of sedimentary rocks, and the deformation or metamorphism of all varieties of rocks. Understanding the principles that govern diffusion provides an insight into a wide range of geological processes.

8A.3.1.1 Diffusion in Gases and Liquids

The process of diffusion in gases and liquids is familiar to most people. If one adds a drop of a water-soluble dye to a beaker of water, very soon diffusion will eliminate any differences in the concentration of the dye in the beaker. The chemical constituent (the dye) diffuses from regions of high concentration to regions of low concentration, thereby reducing and potentially eliminating differences in the concentration of the constituent. The atoms that constitute the liquid are in constant motion, moving along straight-line paths until they collide with another atom or the wall of their container. Then they rebound and head off in a new direction. More atoms of the constituent leave its source region, where they are highly concentrated than re-enter that region. This effect reduces the concentration of the constituent in the source region and raises its concentration far from the source. With sufficient time, diffusion results in a uniform concentration of the constituent, such that the dye color is uniform in the beaker.

The rate of movement by diffusion of a constituent from one location to another, measured as the mass of the constituent in moles that crosses a unit area per unit time, is the constituent's *diffusive flux*. In the situations outlined here, the diffusive flux of a constituent is proportional to the spatial change in its concentration, i.e. its concentration gradient. That is

$$J \propto \frac{\Delta C}{\Delta x}, \tag{8.3}$$

where J is the diffusive flux, C is the local concentration of the constituent, and x is a linear measure of distance (measured perpendicular to the area across which we measure the flux). Thus, $\Delta C/\Delta x$ is the concentration gradient. Rewriting Eq. (8.3) using a proportionality constant D, called the *diffusion coefficient*, yields

$$J = -D\frac{\Delta C}{\Delta x}. \tag{8.4}$$

Equation (8.4), known as Fick's first law, is a one-dimensional description of diffusion in ideal liquids or gases. The diffusive flux is given as mass in moles per unit area per unit time and has units of mol/l^2t(or mol $l^{-2}t^{-1}$). The concentration gradient has units of mass in moles per unit volume per unit length or mol/l^4(or mol l^{-4}). Thus, the diffusion coefficient must have units of l^2/t, i.e. m^2/s (l^2t^{-1} or m^2s^{-1}). The negative sign in Eq. (8.4) results because material moves from higher concentrations to lower concentrations, i.e. "down the concentration gradient." One can, of course, replace $\Delta C/\Delta x$ by a derivative of the concentration taken relative to the single position coordinate (i.e. dC/dx). Then Fick's first law becomes

$$J = -D\frac{dC}{dx}. \tag{8.5}$$

Diffusion coefficients vary with temperature and pressure, as determined by both theoretical arguments and experimental data. Solely increasing temperature while holding all other factors constant causes the diffusion coefficient to increase, which leads, in turn, to greater diffusive flux. Solely increasing pressure while holding all other factors constant causes the diffusion coefficient to decrease, which leads to lower diffusive flux. The effects of temperature are more pronounced than the effects of pressure.

8A.3.1.2 Diffusion in Crystalline Solids

You might find it counterintuitive, but constituents can diffuse through crystalline solids if there are gradients in their chemical potentials within the crystalline solids. Recall that the fundamental characteristic of crystalline solids is that they consist of regular arrangements of atoms. The positions of the atoms that compose solids are not fixed, although they are generally predictable. That is, atoms oscillate about a point in space with an amplitude and frequency that varies with the temperature. It is only when averaged over time that an atom "occupies" a single position within the solid. Still, the relative anchoring of atoms to specific locations is a significant constraint on diffusion in solids.

In many settings, the concentration (C) of a constituent does not adequately capture the constituent's contribution to the overall physical–chemical state of the system. Rather, we use a parameter called the *chemical potential* to measure a constituent's contribution to the total energy of the system. If the chemical potential of constituent A is denoted by μ_A, then

$$J_A \propto -D_A\frac{d\mu_A}{dx}. \tag{8.6}$$

That is, the diffusive flux of the constituent A is proportional to the product of the spatial gradient of its chemical potential ($d\mu_A/dx$) and a diffusion coefficient (D_A). As is the case for the diffusion in gases and liquids, the negative sign in relation (8.6) means that constituents move from regions of higher chemical potential to regions of lower chemical potential, that is, "down a chemical potential gradient."

Why make the distinction between concentration and chemical potential? As is the case in some solutions, a chemical constituent's concentration may not correspond to its chemical potential (i.e. its contribution to the total energy of the crystal). In these cases, it is not completely accurate to say that a chemical constituent (e.g. dye) moves from high-concentration regions to low-concentration regions. Rather, it is more accurate to state that atoms diffuse from portions of a crystal structure where they have higher chemical potential to portions of the structure where they have lower chemical potential. Thus, diffusion tends to eliminate spatial gradients in chemical potential.

The difference between concentration and chemical potential is highlighted in the case of exsolution. In exsolution, two or more chemical constituents which are evenly distributed throughout a crystal structure at high temperatures segregate as the temperature falls. For example, an alkali feldspar crystallized from a melt with a chemical composition (Na, K)Al_2SiO_4 may, on cooling, separate into two distinct phases to form a perthite where a KAl_2SiO_4 crystal has lamellae of $NaAl_2SiO_4$. The diffusional movement of Na^{+1} and K^{+1} ions creates concentration gradients because those ions' contributions to the total energy of the crystal are lower in separate regions of distinct composition than in a single crystal of uniform composition.

Several types of *defects*, which are imperfections in crystals, facilitate the movement of constituents through solids by diffusion. For this reason, then, we begin by examining the different types of defects that exist in all real crystals.

8A.3.1.3 Defects in Crystalline Solids

In the vicinity of the external surfaces of a crystal or fracture surfaces within the crystal, atoms or ions occupy positions other than those consistent with the regular spacing dictated by the crystal's lattice. This local distortion of the atomic or ionic arrays in the vicinity of fractures or the bounding surfaces of crystals is the reason that both fractures and bounding surfaces have measurable surface energy. These surfaces are not, however, the only places where atoms occupy positions other than those dictated by a crystal's lattice. In all real crystalline solids, there are several types of anomalies where atoms have positions different from those predicted by the lattice. We call those places – where atoms are closer to or farther from their nearest neighbors than expected – *defects* or *imperfections*. A **point defect** is an anomaly at a specific location with a crystal's three-dimensional lattice. Point defects typically alter bond lengths and bond angles within a small volume surrounding the defect, thereby giving rise to highly localized regions of elastic distortion within crystals. This elastic distortion, in turn, raises the energy of the volume of crystal containing the defect relative to a volume of perfect lattice with equal size. Stated another way, the point defect raises the energy per unit volume compared to an equal volume of ideal, undistorted crystal. There are three types of point defect: (1) substitutional; (2) interstitial; and (3) vacancy.

In real crystals, atoms of one chemical species can replace, or substitute for, atoms of a different chemical species in the crystal structure (Figure 8.8). These **substitutionals**, as they are known, typically have an ionic radius or valence similar to the chemical species they replace, but neither similarity is required. For example, Ca^{+2} cations regularly substitute for Mg^{+2} cations, which have considerably smaller ionic radii, or Na^{+1} cations, which have a different valence. Even when the substitution has an ionic radius and valence similar to the expected atom (such as when Fe^{+2} cations substitute for Mg^{+2} cations), the substitutional is

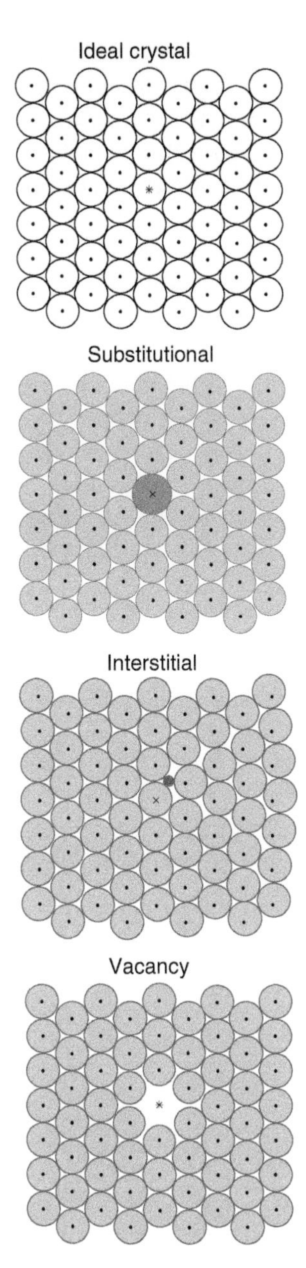

Figure 8.8 An ideal array of atoms or ions in a simple crystal structure (at top) compared with arrays containing substitutionals, interstitials, or vacancies. In each of the three lower images, the array of black dots gives the relative positions of the centers of the atoms or ions in the ideal crystal, and the centers of the atoms or ions in the three arrays do not have the same relative positions.

not identical to the original atom. Differences in the strength and length of its bonds with the surrounding atoms cause local elastic distortions of the lattice. The second type of point defect is an **interstitial** (Figure 8.8). Interstitials occur when an atom is found at a location where, in a perfect lattice, there would be no atom. That is, an interstitial occupies the interstices between atoms at lattice sites. The third type of point defect is a **vacancy** (sometimes called an omission; Figure 8.8), where there is no atom at a place that would be occupied by an atom in a normal lattice. All three types of point defects, particularly vacancies, are critical for facilitating diffusion in crystalline solids.

Real crystals also possess *line* (or *linear*) *defects* known as *dislocations*, which are described in detail in Section 8A.4.1. Dislocations are relevant to understanding diffusion through crystals, however, because they create tubular (i.e. pipe-shaped) volumes of elastically distorted crystal lattice. The distorted lattice results in regions where the spacing of atoms of the crystal lattice is variable. This variation in atom spacing allows faster diffusion along the dislocation than elsewhere in the crystal.

8A.3.1.4 The Role of Crystal Defects in Diffusion in Solids

Diffusion occurs when atoms in a solid crystal lattice move relative to the lattice without distorting the overall lattice. As long as the temperature is above 0 K, atoms in crystalline materials oscillate about equilibrium positions. The amplitudes and frequencies of the oscillations increase with temperature. When temperatures are sufficiently high, these oscillations can create situations in which constituents are able to move from one location in a crystal to another.

The movement of atoms through a crystal typically involves lattice defects. Neighboring atoms in a crystal structure almost never have sufficient vibrational energy to swap positions, even at temperatures close to the crystal's melting temperature. Rather, vacancies play a role in nearly all instances where an atom moves from one location in a lattice to another. That is, an atom with sufficient thermal energy – whether it is a substitutional or a typical atom in the crystal structure – will jump from its position in the lattice to an adjacent vacancy (Figure 8.9a, b). For this reason, materials scientists often refer to the diffusion of vacancies from one portion of a crystal to another rather than envisioning the movement of atoms in the opposite direction. Alternatively, the simultaneous oscillations of an interstitial and an atom in the surrounding crystal lattice may facilitate the interstitial's jump from one interstice to another (Figure 8.9a, b). The frequency of occurrence of these jumps increases with increasing temperature.

In Section 8A.3.1.1, we noted that diffusion will eventually eliminate variations in concentrations of constituents in liquids and gases. An analogous effect occurs in crystals. If a particular constituent occupying the interstices of a crystal, e.g. carbon in iron, has a higher concentration in one portion of the crystal, random movements of these interstitials will lead to a net movement of C atoms away from areas of high concentration, eventually eliminating variations in its concentration across the crystal. Similarly, if a substitutional, e.g. Fe^{+2} substituting for Mg^{+2} in olivine, has a higher concentration in one portion of a crystal than elsewhere in the crystal, diffusion will tend to distribute the substitutional evenly throughout

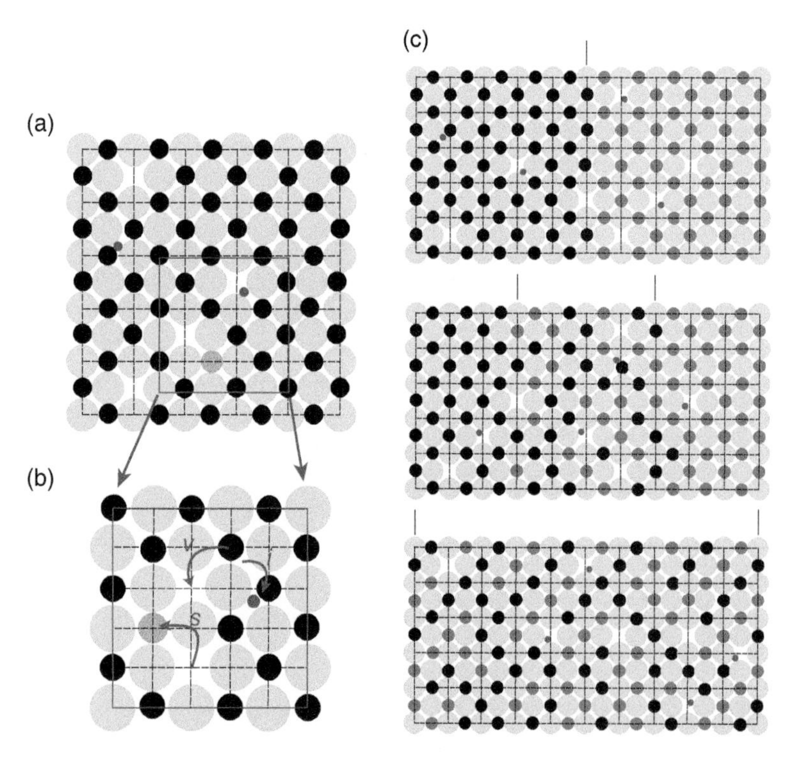

Figure 8.9 (a) A portion of a crystal with a substitutional (in pink), interstitials (in red), and vacancies (blank). (b) Enlarged portion of (a) showing how diffusion within crystals occurs when vacancies and atoms swap positions (*v*), when interstitials move from one interstice to another (*i*), and when substitutionals switch positions with vacancies or other atoms/ions (*s*). (c) Schematic representation of how a chemical gradient in a crystal might, at elevated temperature, drive diffusion that eventually would eliminate any gradient in chemical composition.

the crystal (Figure 8.9c). These effects are comparable to what occurs in liquids and gases. This type of diffusional movement of hydrogen cations (H^+ = protons) and oxygen anions (O^{-2}) through silicate minerals is particularly important for both deformation and metamorphism.

Interstitials, and to a lesser degree substitutionals, raise the internal energy per unit volume of a crystal. However, interstitials and substitutionals in regions of crystals with "perfect" lattice add more to the total energy of the crystal than interstitials and substitutionals in regions where a crystal's lattice is already distorted, such as along dislocations or near grain boundaries. For that reason, in the absence of external concentration gradients, interstitials, and to a lesser degree substitutionals, tend to move toward and cluster: (1) along grain boundaries; or (2) in the pipe-shaped regions surrounding dislocations.

At higher temperatures, crystal structures expand, and the atoms associated with the crystal lattice have greater vibrational kinetic energy. The diffusion of chemical constituents through crystal lattices is, therefore, strongly enhanced by high temperatures.

8A.3.1.5 Diffusion in Geological Solids

Materials scientists distinguish two types of diffusion within polycrystalline solids with no open pores on the basis of the path taken by the diffusing species. When atoms diffuse through the lattices of the crystals that compose the solid, the process is known as **intragranular**, **lattice**, or **volume** diffusion. When atoms diffuse along the boundaries between the crystals that compose the solid, the process is known as **intergranular** or **grain boundary diffusion**. Both types of diffusion occur at the same time, and both are capable of producing the same net effects.

In situations where the polycrystalline solid has open pores, there are two additional potential paths for diffusing species. In addition to lattice and grain boundary diffusion, atoms can diffuse along the surfaces of the crystals surrounding an open pore. This *free surface diffusion* can occur whether the pore is filled by a vapor, fluid, or melt. Chemical species can also diffuse through the vapor, fluid, or melt phases that fill the pore. When the diffusing species move through portions of the vapor, fluid, or melt phase that are beyond the influence of crystals' atomic or ionic charges, we refer to the process as *free-fluid diffusion*.

Materials scientists also distinguish *solution diffusion*, where atoms diffuse through a medium composed of different chemical species from *self-diffusion*, where atoms diffuse through a medium composed of the same chemical species. In geological settings, free-fluid diffusion is usually solution diffusion. The diffusion of hydrogen ions through a silicate mineral mentioned earlier is an example of solution diffusion in solids. Solution diffusion in solids, whether lattice or grain boundary diffusion, is an essential component of metamorphism and metasomatism. Self-diffusion, whether lattice, grain boundary, or free-surface diffusion, is both important and common in the shape changes by diffusion that we consider in the following section.

Experiments indicate that several factors affect the rates of material transfer for the different types of diffusion. Under most conditions, the rate of material transfer is slower for intragranular diffusion and grain boundary diffusion than for free surface or free fluid diffusion. The presence of fluid phases in pores increases the rates of mass transfer significantly. In one experiment, the presence of a pore fluid increased the diffusion rate by a factor of 10^8–10^{10}.

8A.3.2 Grain Shape Change by Diffusion

Diffusion alone can change the shapes of mineral grains in a rock. Recall the force chains that form in an aggregate of grains subjected to stresses or displacements along its boundaries (cf. Figure 8.3a). In the vicinity of the grain-to-grain contacts along a force chain, crystal lattices are elastically distorted (Figure 8.10a) and, in some instances, permanently

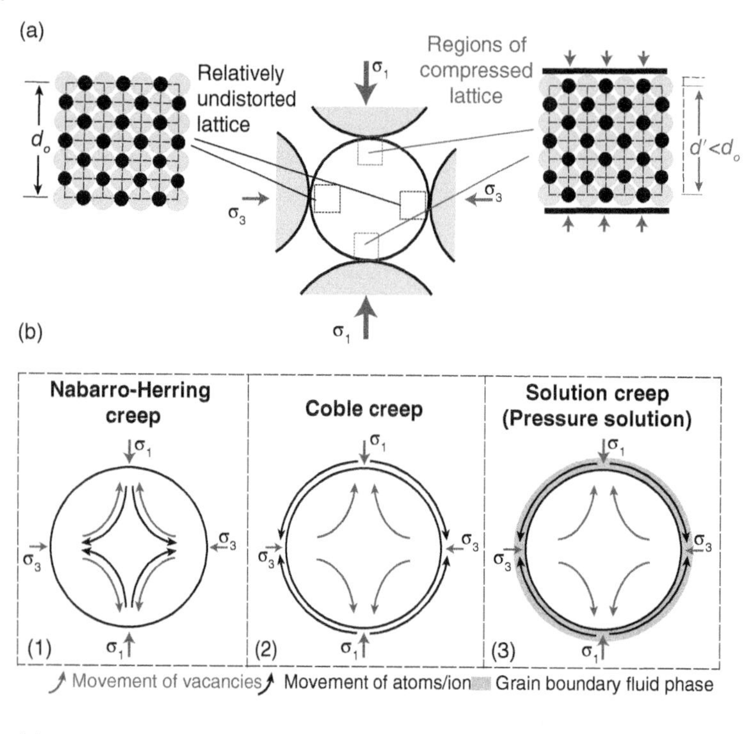

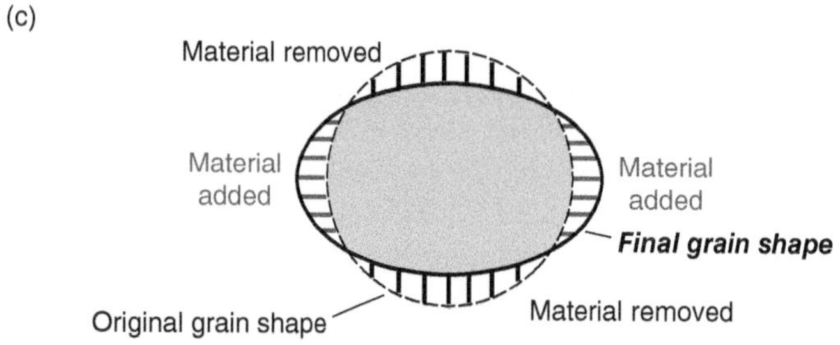

Figure 8.10 (a) Comparison of arrays of atoms or ions at two portions of a grain subjected to shortening. Along the boundaries of the grain parallel to the shortening direction, the lattice is not elastically distorted and stresses are relatively low. Along the boundaries of the grain perpendicular to the shortening direction, the array is shortened, requiring higher stresses to maintain the elastic distortion. (b) The differences in the chemical potential of components will drive diffusion from grain boundaries with greater elastic distortion to regions where elastic distortion is lower, thereby changing the shapes of grains. Diffusion may occur through crystal lattice (1), along a dry grain boundary (2), or through a grain boundary fluid. (c) Mass transfer will change the shape of the grain.

deformed. The elastic distortion near grain-to-grain contacts requires that the stresses in the vicinity of the contact are higher than elsewhere in the grain. Thus, the elastically distorted regions of grains have higher elastic strain energy per unit volume than other regions in the grains.

In the vicinity of the grain-to-grain contacts along a force chain, atoms contribute more to the grain's internal energy than do atoms or ions elsewhere in the grain. Stated in a different way, atoms near grain-to-grain contacts along a force chain have higher chemical potential than atoms elsewhere in the grain.

This spatial gradient in the chemical potential drives diffusion from the vicinity of grain contacts along the force chains to other grain contacts.

Similar processes occur in materials in which there are no open pores (cf. Figure 8.3c). Differential stresses or displacements imposed on the boundaries of an aggregate give rise to different stress magnitudes across grain boundaries with different orientations. In the vicinity of grain boundaries subjected to compressive stresses greater than the hydrostatic stress, elastic strain energy per unit volume is relatively high. In the vicinity of grain boundaries subjected to compressive stresses less than the hydrostatic stress, elastic strain energy per unit volume is relatively low. These variations in elastic strain energy per unit volume give rise to variations in the chemical potential for the atoms that compose the grains. The resulting chemical potential gradients drive the diffusion of atoms away from grain boundaries that experience compressive stresses greater than the hydrostatic stress.

Atoms can diffuse through the crystal lattice and along grain boundaries to areas where they have lower chemical potential and join the crystal lattice there. Importantly, this process changes the shape of individual grains (Figure 8.10c). When diffusive mass transfer changes the shapes of individual grains in a polycrystalline aggregate, it is necessarily accompanied by *intergranular displacement* if there is to be no increase in the rock's volume during deformation by the formation of voids (Figure 8.11).

The movement of atoms through (by volume diffusion) or around (by grain boundary diffusion) grains results in *creep*. Atoms move away from the regions supporting compressive stresses or accommodating imposed convergent displacements and move toward regions not supporting compressive stresses or experiencing divergent displacements. Alternatively, vacancies move toward the regions supporting compressive stresses or accommodating imposed convergent displacements from regions not supporting compressive stresses or experiencing divergent displacements.

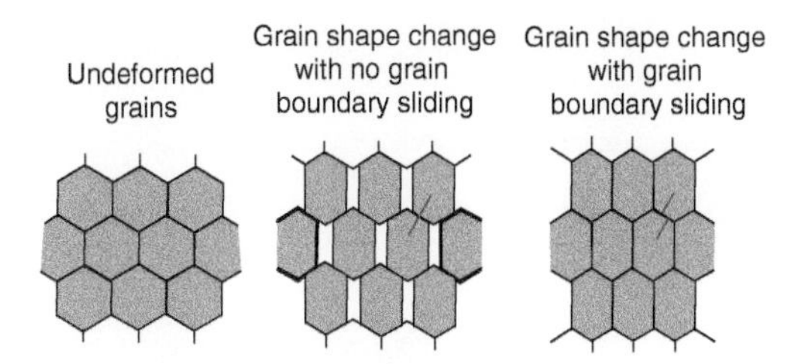

Figure 8.11 If grain boundary sliding is prevented, diffusional removal of material from grains in an aggregate shortened along a left-right axis will generate open pores (center image). Instead, grain boundary sliding accompanies diffusional mass transfer (image at right). *Source:* Modified from McClay (1977).

To summarize, localized distortion of a mineral grain's lattice creates regions of elevated chemical potential, and atoms move from those areas of higher chemical potential toward undistorted portions of the lattice. This results in a change in the shape of the crystal.

8A.3.3 Microstructures Associated with Diffusional Mass Transfer

In a deforming polycrystalline aggregate, the diffusive flux will occur simultaneously along the following paths: (1) through the lattices of the grains that make up the aggregate; (2) along the boundaries between grains in the aggregate; (3) along the interface between a fluid and the mineral grains, if fluid phases exist along the boundaries between grains; and (4) if there are open pores, through the vapor or fluid filling the pores. Since diffusion rates increase dramatically with the addition of a fluid phase, pore fluids will enhance material rearrangement by diffusion. The diffusional changes in the shapes of the mineral grains that compose a rock will necessarily change the shape of the rock. Some of these microstructures have been described in Sections 3.4 and 3.5.

In grain aggregates, atoms diffuse from point contacts, generating grains *truncated* by relatively *planar grain contacts* (Figure 8.12). Atoms diffuse

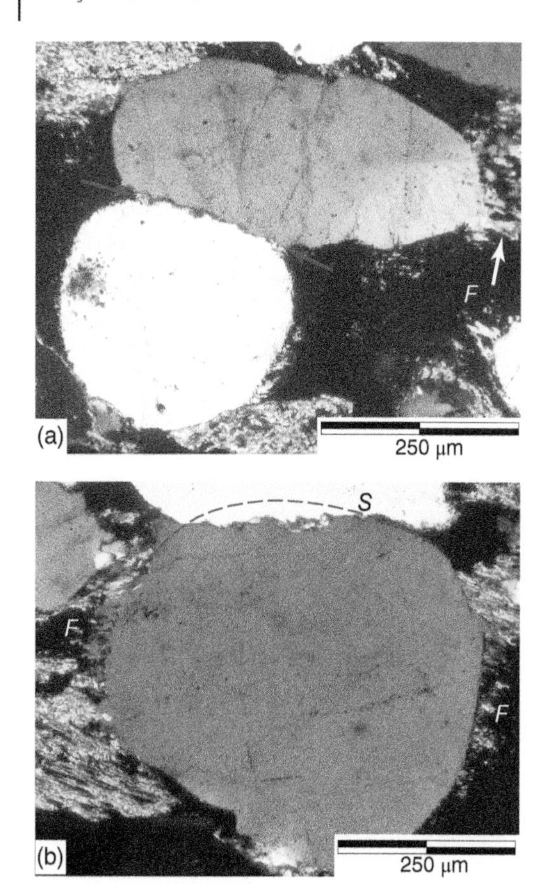

Figure 8.12 Planar grain contacts in rocks deformed by diffusional mass transfer. (a) Two quartz grains in a hematite cemented quartzite have converged across the planar grain boundary indicated by the red arrow, truncating the grain shapes. (b) The rounded detrital quartz grain at the center of the image has relatively planar, truncated contact with white grain near the top of the image; the dashed line shows its inferred initial shape. Note the with a thin seam of selvage *S* decorating the grain contact. Small fibrous overgrowths *F* occur along the left and right sides of the grain.

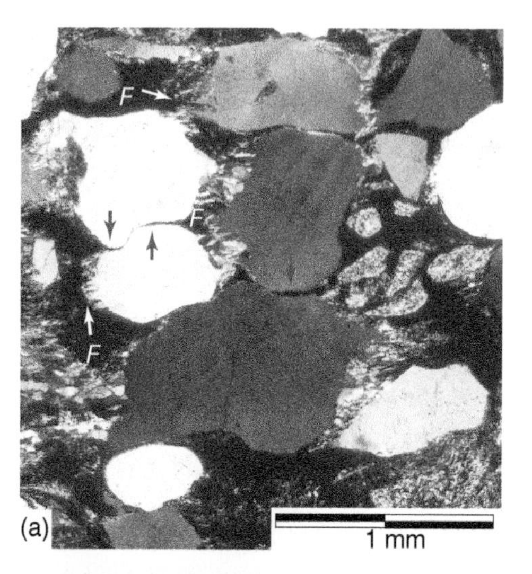

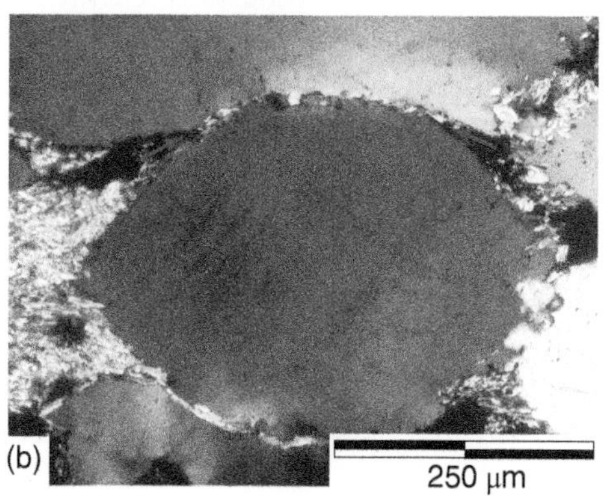

Figure 8.13 Concavo-convex and sutured grain contacts in rocks deformed by diffusional mass transfer. (a) Red arrows indicate locations where two quartz grains have interpenetrated by diffusional removal of quartz, leading to concavo-convex boundaries. Note fibrous growth at locations labeled *F*. (b) Rounded detrital quartz grain at the center of the image has a convex shape, whereas the grain above it has a concave shape. The sutured contact between two grains is indicated by red arrows.

toward open pores in the rock, should they exist, where those constituents add to the grains as *overgrowths* (Figure 8.13). The grain boundaries affected by diffusional removal of atoms are identified by *truncated* or *flattened* grain shapes, *stylolitic grain contacts*, and *concavo-convex grain contacts* without evidence of grain distortion or fracturing (Figures 8.12, 8.13). In many instances, grain boundaries are decorated by *selvage*, which is an accumulation along those grain boundaries of constituents or inclusions within grains that do not diffuse readily. Atoms diffuse toward grain boundaries that experience compressive stresses less than the hydrostatic stress and join the lattice there. There is no need for open cavities into which material grows. New growth will occur along those grain boundaries, and it will make room for itself. These grain boundaries are identified by *overgrowths* of relatively inclusion-free minerals, often with a *fibrous* habit in which the long dimensions of overgrowth parallel to the incremental elongation direction (Figures 8.12, 8.13). Geologists use the terms *pressure fringe*, *pressure shadow*, and *beard* to refer to the regions where mass is added to existing grains.

There are two general types of overgrowth – *syntaxial* and *antitaxial*. A syntaxial overgrowth is one in which the overgrowth is the same mineral as the host grain, and it adds directly to the lattice of the host grain. As a result, segments of the overgrowth closer to the host grain were added to the host grain during earlier deformation increments, and segments of the overgrowth farther from the host grain were added during the later deformation increments (Figures 8.13, 8.14). Thus, the mineral growth is in the direction of elongation. When the overgrowth exhibits a fibrous habit, the characteristics of the overgrowth provide an insight into the deformation path. Coaxial deformations will generate straight fibers, whereas non-coaxial deformations generate smoothly curved fibers or fibers with distinct sectors having different orientations. An antitaxial overgrowth is one in which the overgrowth is a different mineral than the host grain. New mineral growth occurs adjacent to the host grain throughout deformation. So, segments of the overgrowth farther from the host grain were added during earlier deformation increments (Figure 8.15). Segments of the overgrowth closer to the host grain record later deformation increments, and the mineral growth is in the opposite direction

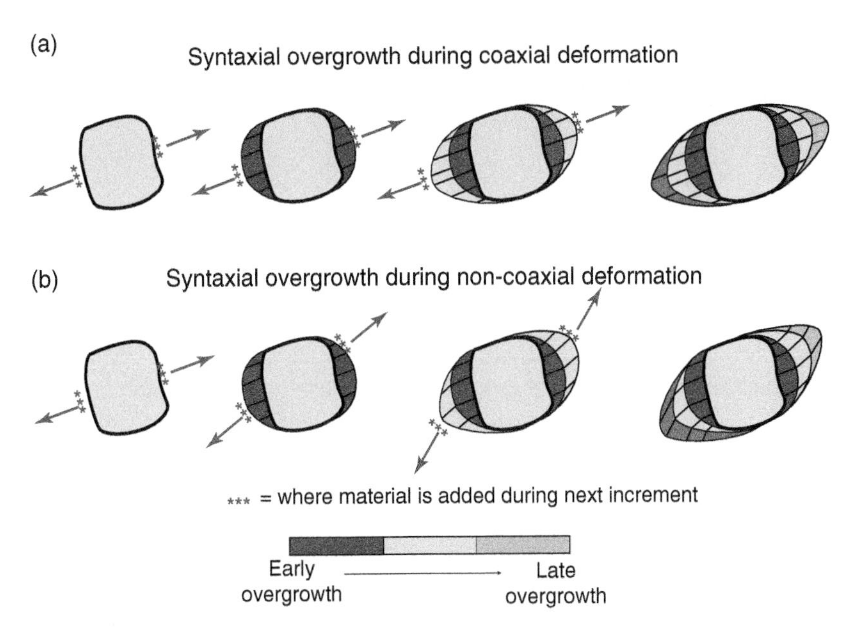

Figure 8.14 Sequential development (from left to right) of syntaxial overgrowths on mineral grains. Red arrows on any image indicate the elongation direction responsible for overgrowth added to create the next image. In syntaxial overgrowths, fibers added during earlier increments are adjacent to host grain, and fibers added during later increments are farther from the host grain. (a) Coaxial deformation with straight, subparallel fibers. (b) Non-coaxial deformation with curved fibers.

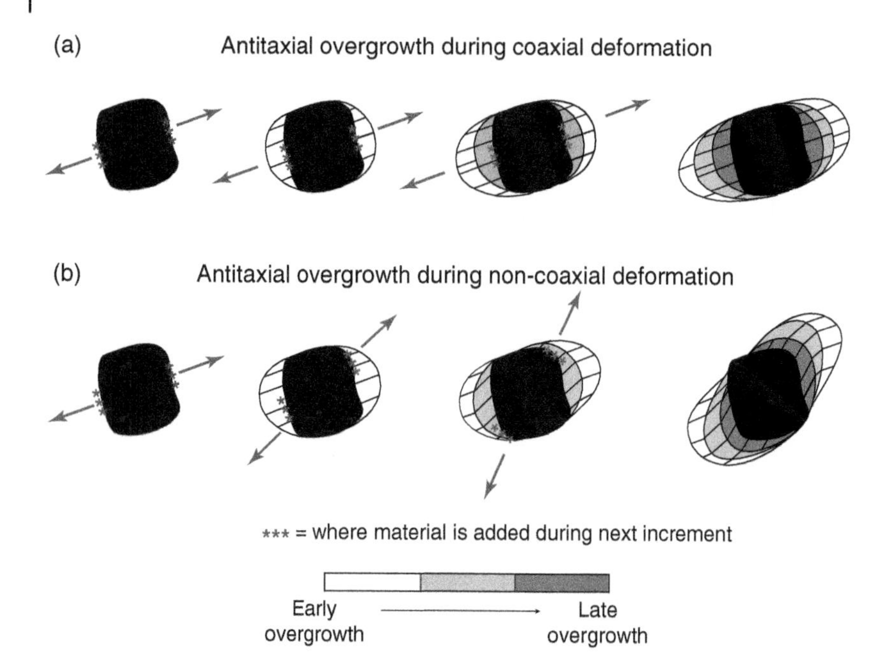

Figure 8.15 Sequential development (from left to right) of antitaxial overgrowths on mineral grains. Arrows on any image indicate elongation direction responsible for overgrowth added to create the next image. In antitaxial overgrowths, fiber segments added during earlier increments are now farther from host grain, and fiber segments added during later increments are adjacent to host grain. (a) Coaxial deformation with straight, subparallel fibers. (b) Non-coaxial deformation with curved fibers.

to the local elongation. Antitaxial overgrowths also provide an insight into the deformation path, with coaxial deformations generating straight fibers and non-coaxial deformations generating smoothly curved fibers or fibers with distinct sectors (Figure 8.16; cf. Figure 8.15). See the Comprehensive Treatment section of this chapter for a discussion of mineral growth in vein fillings.

8A.3.4 Diffusional Mass Transfer as a Deformation Mechanism

The movement of atoms, i.e. *mass transfer*, will occur along all possible diffusional paths *simultaneously*, thereby changing the shapes of individual grains. Changes in the shapes of individual grains in the aggregate will necessarily change the shape of the aggregate. Thus, mass transfer by diffusion can result in finite strain of an aggregate of grains.

If the mass transfer occurs primarily by self-diffusion through the crystal lattices of grains (Figure 8.10a), materials scientists call the process of change of shape by diffusive transfer *Nabarro–Herring creep* after the Frank R.N. Nabarro and Conyers Herring who derived a flow law for deformation by lattice diffusion. If the material transfer occurs primarily by self-diffusion along the boundaries of grains in the aggregate (Figure 8.10b), materials scientists call the process *Coble creep* after Robert L. Coble who derived a flow law for deformation by grain boundary diffusion. If there is an intergranular fluid present, and the material transfer occurs primarily by solution diffusion through this intergranular fluid, geologists sometimes call the process *solution transfer* (or *pressure solution*) (Figure 8.10c).

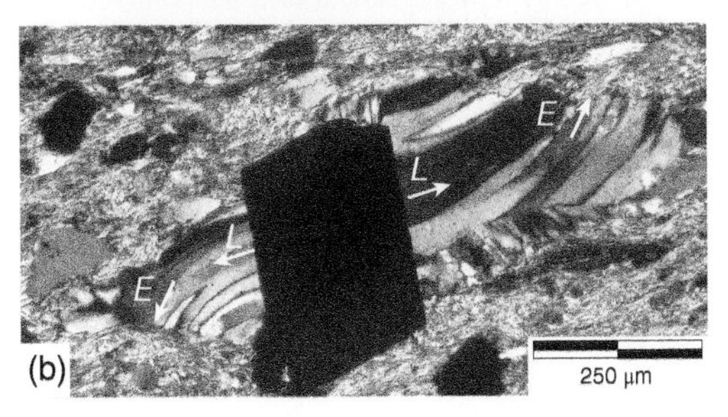

Figure 8.16 Antitaxial overgrowths. (a) Straight quartz fibers in overgrowth on magnetite grain. Note quartz near the ends of the fibers, the older portions of the fibers, have irregular boundaries due to deformation. (b) Curved quartz fibers in overgrowth on magnetite grain. The older ends of the fibers have irregular boundaries and are oblique to the fabric in the surrounding matrix. The younger portions of the fibers, adjacent to the magnetite, are parallel to the fabric.

Because these three processes produce similar microstructures, they cannot be distinguished after the fact. For that reason, most geologists now use the general term **diffusive mass transfer** (and the acronym **DMT**) or the term **diffusional creep** to refer to the process, whereby polycrystalline aggregates change shape by diffusion.

Diffusive transfer *can* and *does* occur without the fluid phase. Nabarro–Herring and Coble creep are known to be important in metallurgy, where they can be experimentally verified. Nabarro–Herring and Coble creep are thought to be important contributors to rock deformation in the lower crust and throughout the mantle, where rocks may be relatively dry. However, the presence of fluids *accelerates* the rate at which diffusive transfer occurs. In upper and middle crustal deformation, diffusive transfer probably occurs as solution diffusion along grain boundaries. The "solvent" is likely an intergranular film of fluid rich in H_2O or CO_2. Evidence supporting those inferences includes: (1) pores and cavities in crustal rocks below the ground water level normally contain aqueous fluids; (2) vacuoles in minerals usually contain a fluid phase of some kind, even minerals found in granulite facies metamorphic rocks where the fluid is commonly rich in CO_2; (3) carbonate minerals and hydrous silicates (clays, micas, and amphiboles) are common constituents of crustal rocks; because they exist in thermodynamic equilibrium with their surroundings, they have carbonic or hydrous phases present along their grain boundaries; and (4) transmission electron microscopy (TEM) reveals minute bubbles interpreted as fluid-filled cavities along dislocation lines and grain boundaries in quartz.

8A.3.5 Flow Laws for Three Diffusional Mass Transfer Deformation Mechanisms

By envisioning that mass transfer occurs along only one path, materials scientists are able to derive constitutive equations for deformation by diffusional mass transfer along a single diffusional pathway (see Section 8B.4). We present simplified versions of the constitutive equations for each of the diffusional deformation mechanisms with the aim of emphasizing the functional relationships between strain rate $\dot{\varepsilon}$, the appropriate diffusion coefficient (D_i), grain diameter d, and deviatoric stress σ'. In each case, we include a prefactor A_i, which is a parameter that depends on the formulation that would vary with temperature and pressure, and ensures both sides of the equation have units of strain rate (t^{-1} or $\sec^{-1}$). A complete description of deformation, in general, by a flow law would require considering the strain rate and deviatoric stress tensors. However, flow laws are commonly expressed in terms of components or scalar values derived from these tensors, as intended here. The stress term in flow laws is often taken as the differential stress, $\Delta\sigma = \sigma_1 - \sigma_3$, and in the axial compression experiments from which many flow laws are derived, the strain rate term refers to the axial strain rate $\dot{\varepsilon}_1$.

For diffusional mass transfer by lattice diffusion, i.e. for Nabarro–Herring creep, the constitutive equation takes the form:

$$\dot{\varepsilon} = A_{NH}\frac{D_L}{d^2}\sigma', \tag{8.7}$$

where D_L is the diffusion coefficient for intragranular or lattice diffusion and d is the mean diameter of grains.

For diffusional mass transfer by grain boundary diffusion, i.e. for Coble creep, the constitutive equation takes the form:

$$\dot{\varepsilon} = A_C\frac{D_{GB}\delta}{d^3}\sigma'. \tag{8.8}$$

In Eq. (8.8), D_{GB} is a diffusion coefficient for grain boundary diffusion, δ is the grain boundary width, and d again is the mean diameter of grains.

For diffusional mass transfer by diffusion along wet grain boundaries, i.e. for solution transfer, the constitutive equation takes the form:

$$\dot{\varepsilon} = A_{ST}\frac{D_{WB}\delta}{d^3}\sigma'. \tag{8.9}$$

In Eq. (8.9), D_{WB} is a diffusion coefficient for wet grain boundary diffusion, δ again is the grain boundary width, and d again is the mean diameter of grains.

There are significant similarities between Eqs. (8.7)–(8.9). In each equation, and, therefore, in any case of diffusion creep, the strain rate $\dot{\varepsilon}$ is linearly related to the deviatoric stress σ'. These relations suggest that all diffusional mass transfer deformation mechanisms – Nabarro–Herring creep, Coble creep, and solution creep – result in linearly viscous rheology. Available experimental data on the deformation of metals and Earth materials confirm this inference. Theory further predicts that the strain rate by diffusional mass transfer is inversely proportional to the mean grain diameter raised to a power. Experimental data also support this inference, indicating that solely reducing the mean grain diameter while holding all other factors constant causes the strain rate by DMT to increase. When the primary diffusional pathway is along a grain boundary, the grain boundary width becomes a factor, and adding this factor to the flow law requires a change in the value of the power to which grain diameter is raised (i.e. d^2–d^3) in order to maintain equivalent units on both sides of the constitutive equation. Note that depending upon the dominant diffusional path, reducing the mean grain diameter by a factor of 2 or 3 should cause strain rates to increase by a factor of 4–8 or 9–27. Diffusional creep is, therefore, very sensitive to the grain size of the rock. Keeping all other conditions the same (e.g. stress, temperature, and rock composition), a rock with a smaller mean grain size can deform more rapidly by DMT than a rock with a larger mean grain size. Diffusion coefficients depend upon temperature. So, the strain rate by any diffusional mass transfer mechanism is dependent on the temperature. If all other factors are held constant and the temperature increases, the strain rate will increase.

For temperatures between 100 and 300 °C, diffusion coefficients for dry grain boundary diffusion are greater than those for lattice diffusion, and diffusion coefficients for wet grain boundary diffusion are 10^8–10^{10} times those for dry grain boundary diffusion. Thus, in the mid- to the upper crust, diffusional mass transfer occurs primarily by diffusion along grain boundaries. If water is present along the grain boundaries, strain rates for diffusional creep are accelerated markedly.

In some geological settings, atoms diffusing from regions of high chemical potential may dissolve into a solute (often an aqueous phase) and be carried away as the solution moves through the rock. A geological example is the removal of material along a stylolite seam and the deposition of that material as a vein in another location. This movement of constituents in a solution that moves from one region to another is called *advection*, i.e. the atoms are *advected* by the moving fluid. Atoms may subsequently precipitate from the solution far from the location where they dissolved. In this way, advection can contribute to wholesale loss of volume in some rock masses or wholesale increase in volume in other rock masses. Further, advection of fluid phases may promote continued rapid removal of atoms at compressed grain boundaries by maintaining low concentrations of dissolved solids in a grain boundary fluid phase. It is important to note, however, that solution transfer can and does occur *without* advection.

8A.4 Dislocational Deformation Mechanisms

At the beginning of the previous section on diffusional mass transfer, we outlined two broad categories of defects found in all real crystals: *point defects* and *line defects*. Both types of defects play significant roles in facilitating diffusion, and therefore diffusional mass transfer, in crystalline solids. The second type of defects, the line defects known as **dislocations**, plays a fundamental role in facilitating the deformation of crystals by two mechanisms, **dislocation glide** and **dislocation creep**. In order to understand these two deformation mechanisms, we need to describe the character of dislocations in greater detail than simply stating that they are pipe-shaped regions of distorted crystal lattices. In this section, we begin by examining the geometric characteristics of dislocations, then consider how dislocations move and show how that motion can lead to the deformation of a crystal, and, finally, examine how dislocation movement contributes to the deformation mechanisms of dislocation glide and dislocation creep.

The deformation mechanism dislocation glide relies entirely on the movement of dislocations through a crystal lattice by a process called the *glide of dislocations*. In an attempt to minimize confusion, in this text, we use "glide of dislocations" to refer to the process by which individual dislocations move through a crystal lattice and "dislocation glide" to refer to the deformation mechanism by which crystals change their shape. Dislocation glide is widely observed in metallurgy, and it can, under appropriate conditions, contribute to deformation in geological settings. Dislocation creep is the dominant deformation mechanism in much of the middle crust, lower crust, and lithospheric mantle. Unlike diffusive mass transfer or dislocation glide, several distinct processes occur concurrently to accommodate dislocation creep. These processes are: (1) the glide of dislocations; (2) the interaction of diffusional processes with dislocations, which enables dislocations to move through a crystal lattice by a process called *dislocation climb*; (3) the interactions between dislocations that increase the net length of, or the number of dislocations, in a crystal; and (4) the tendency for crystals to reduce the net length of, or the number of dislocations, by a collection of processes that fall under the general headings of *recovery* and *recrystallization*.

As part of this section, we also examine the origin and nature of surface imperfections within deformed crystals and use their characteristics to investigate the nature of grain boundaries, i.e. the surfaces that separate different crystals. Finally, we examine briefly the two other processes that contribute to permanent changes in the shapes of some crystalline solids: *mechanical twinning* and *kinking*.

8A.4.1 Dislocations as Elements of Lattice Distortion

8A.4.1.1 Dislocations – Line Defects in Crystals

There are two "end-member" types of dislocations: **edge dislocations** and **screw dislocations**. Each of these end-member types creates tube-shaped volumes of elastically distorted crystal lattice. To visualize line defects in crystals, imagine that one is able to slice halfway through a crystal lattice (Figure 8.17a). An edge dislocation

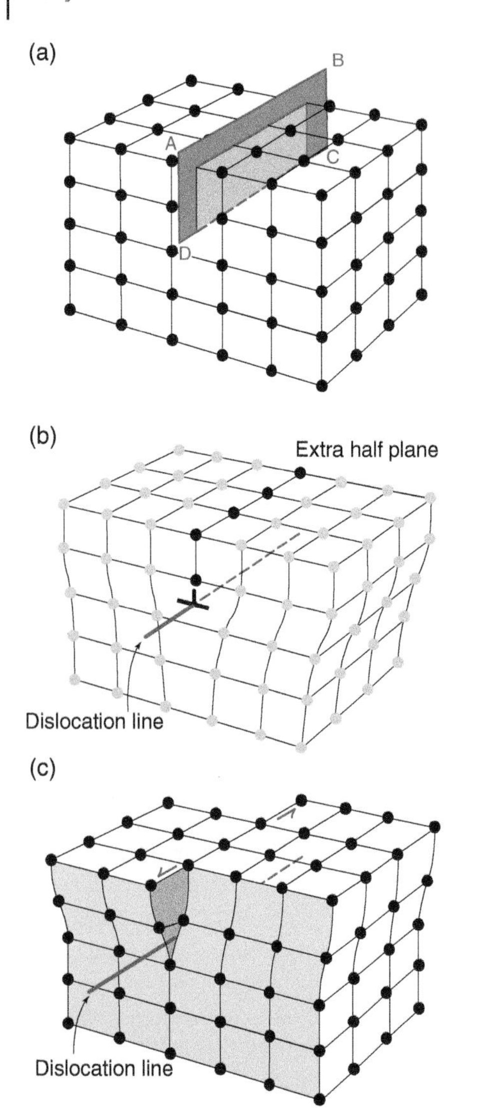

Figure 8.17 (a) A portion of an ideal lattice sliced open along plane *ABCD*. (b) Inserting an extra "half plane" of atoms along the slice creates a region of distorted lattice along the edge of the half plane – an edge dislocation. (c) Displacing a portion of the sliced crystal one unit translation past the other portion of the sliced crystal creates a linear region of distorted lattice – a screw dislocation.

results from pulling apart the two portions separated by the slice and inserting an extra "half plane" of atoms or ions along the slice and allowing bonds to reform (Figure 8.17b). A screw dislocation results from sliding one portion of the crystal one bond length past the other portion and allowing the bonds to reform (Figure 8.17c). In both cases, the result is a tube-shaped region of distorted lattice along the length of the dislocation.

Looking "end on" along an edge dislocation (Figure 8.18a), it is clear that the portion of the crystal lattice in the vicinity of the edge of the extra "half plane" is distorted. Immediately above the plane, whose projection is the line *ab* in Figure 8.18a, atoms in the vicinity of the edge of the half plane are closer together than those elsewhere in the crystal. This portion of the crystal lattice has an elastic strain field characterized by contractional strains and the associated compressional stresses. In the portion of the crystal immediately below the plane whose projection is *ab*, atoms in the vicinity of the edge of the half plane are farther apart than those elsewhere in the crystal. This portion of the crystal lattice has an elastic strain field characterized by extensional strains and associated tensional stresses. Looking down on the plane whose trace is *ab* (Figure 8.18b), elastic distortion is localized along the end of the half plane that defines the dislocation line.

Materials scientists define the character of a line defect by comparing a closed-loop path, called a *Burgers circuit*, around the dislocation line to an identical path within a region of ideal or perfect lattice. Whether one first makes the closed loop in the ideal crystal and compares it to an identical path around the dislocation (as shown in the figure), or makes the closed loop about the dislocation first and compares it with the same loop in ideal crystal, there will be a mismatch between the two paths (Figure 8.18c). A line segment or vector equal in length to the characteristic spacing of atoms in the lattice is needed to close the loop. This vector is known as the *Burgers vector* for that dislocation. For any edge dislocation, the Burgers vector is perpendicular to the

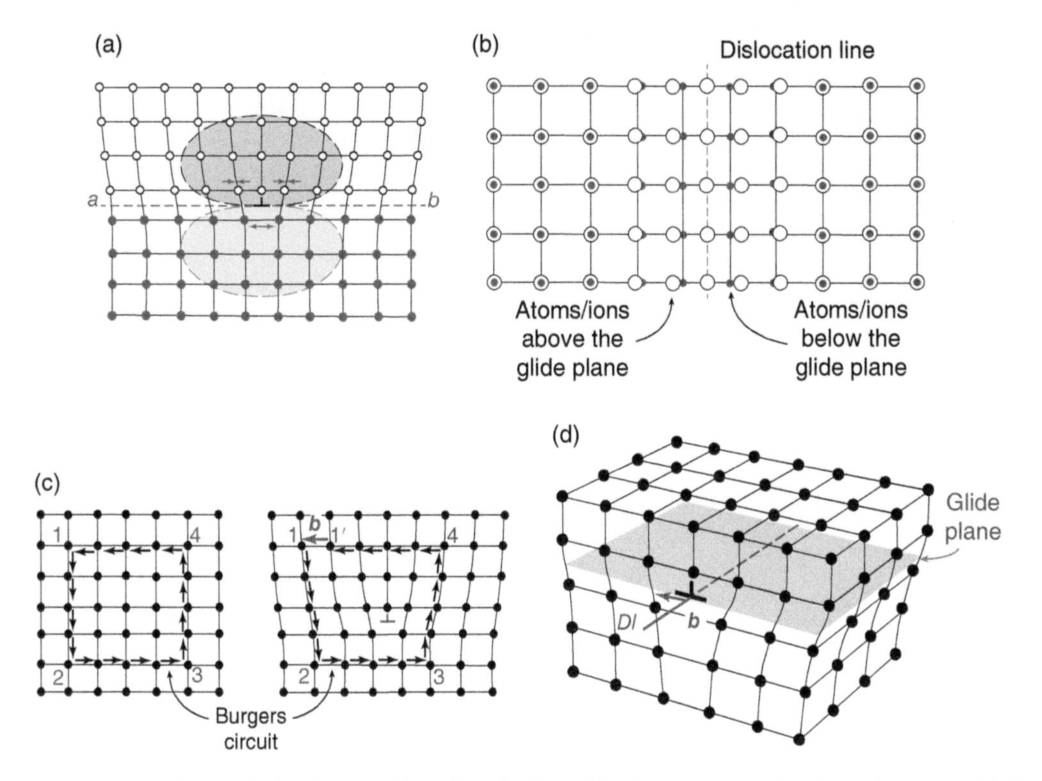

Figure 8.18 Characteristics of edge dislocations. (a) The glide plane *ab*, perpendicular to the extra half plane, separates contracted lattice (in white) from extended lattice (in red). (b) View looking down on plane *ab* showing that the contracted and extended portions of the lattice are localized along the dislocation line that is the edge of the half plane. (c) Comparison of a Burgers circuit in the perfect crystal at left and around the edge dislocation at right. The Burgers vector ***b*** is required to close the Burgers circuit around the edge dislocation. (d) Three-dimensional rendering of an edge dislocation showing that the dislocation line and its mutually perpendicular Burgers vector define the glide plane.

dislocation line (Figure 8.18d). The imaginary plane defined by an edge dislocation line and its perpendicular Burgers vector is known as a **glide plane**.

In the case of a screw dislocation (Figure 8.19), there is no extra half plane. Rather, the lattice planes perpendicular to the screw dislocation line spiral around the line in the manner of a continuous stairway or ramp. The elastic strain field along the length of a screw dislocation is characterized by some stretched bonds, some contracted bonds, and some sheared bonds that accommodate the torsion observed; accordingly the atoms in the vicinity of the dislocation line experience associated compressional, tensional, and shear stresses. Completing the Burgers circuit analysis shows that for any screw dislocation, the Burgers vector is parallel to the dislocation line (Figure 8.19).

Because the edge dislocations in Figures 8.17 and 8.18 have an extra half plane of atoms that extends above the glide plane, materials scientists call them *positive* edge dislocations. A *negative* edge dislocation is one whose extra half plane extends downward from the glide plane (Figure 8.20a). Similarly, if the two portions of the

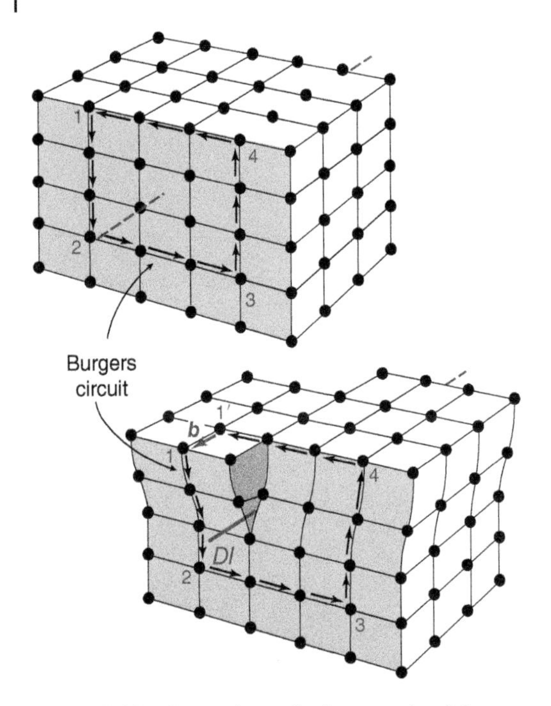

Figure 8.19 Comparison of a Burgers circuit in a region of the perfect crystal lattice and around a screw dislocation. The Burgers vector ***b*** is parallel to the dislocation line *Dl*.

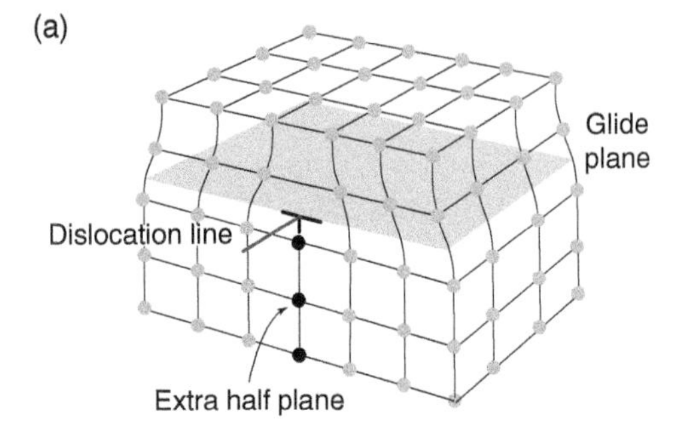

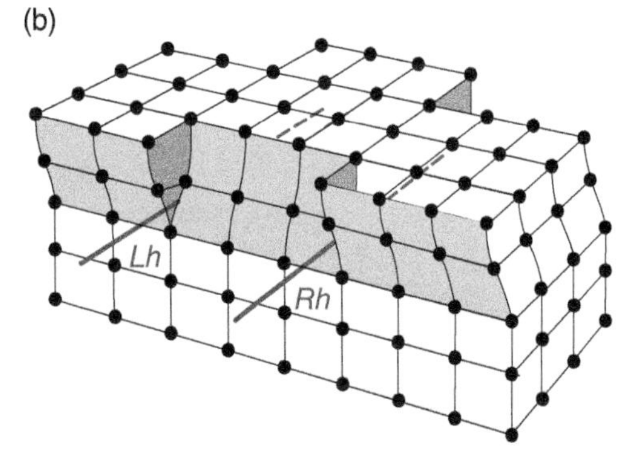

Figure 8.20 (a) A negative edge dislocation. (b) Comparison of a right-handed screw dislocation (*Rh*) and a left-handed screw dislocation (*Lh*).

lattice in Figure 8.17c had been displaced in the opposite directions, lattice planes perpendicular to the screw dislocation line would spiral in the opposite sense. This is the difference between a "right-handed" and "left-handed" screw dislocation (Figure 8.20b).

Neither edge nor screw dislocations can end within a crystal. Rather, all dislocations must extend from one side of a crystal to the other *or* exist as closed loops within a crystal. For example, one can envision an incomplete portion of an extra plane of atoms, or a localized "gap" in a plane of atoms, surrounded by a loop of edge dislocation (Figure 8.21a). More typically, a dislocation line will curve through a crystal, with some segments of the dislocation line having the characteristics of an edge dislocation, other segments of the dislocation line having the characteristics of a screw dislocation, and still other segments sharing the characteristics of both edge and screw dislocations (Figure 8.21b, c). We say the segments of dislocation lines with both edge and screw character are **mixed dislocations**; the Burgers vector of a mixed dislocation is oblique to the dislocation line.

Since the unit translations for crystals are typically on the order of 0.1–1 nm (10^{-10}–10^{-9} m), the lengths of Burgers vectors of dislocations are such that one cannot see the offset of the crystal lattice, even with TEM. TEM can image the distorted lattice surrounding individual dislocations, but dislocations are not visible with optical or scanning electron microscopy. Prior to the widespread access to TEM, materials scientists were able to "see" dislocations in some instances

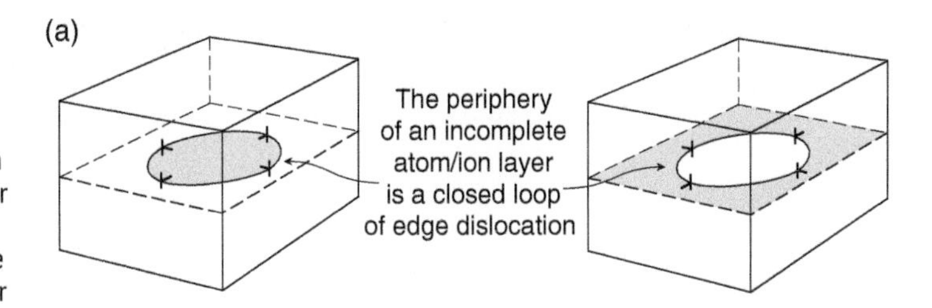

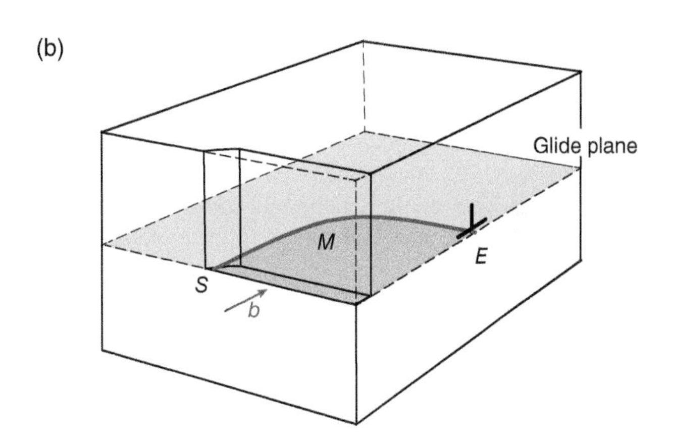

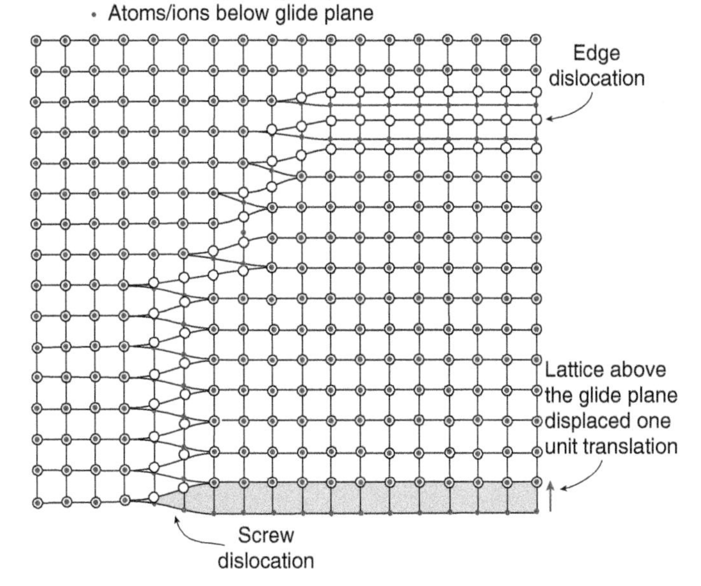

Figure 8.21 (a) A partial layer of atoms or ions or a "hole" in a layer of atoms or ions leads to a closed loop of an edge dislocation. (b) More typically, a dislocation curves through a crystal with segments perpendicular to the Burgers vector (***b***) having edge character (*E*), segments parallel to the Burgers vector having screw character (*S*), and segments oblique to the Burgers vector having mixed character (*M*). (c) View, looking down upon a glide plane, of a curved dislocation line with local edge, mixed and screw character.

by: (1) etching the surfaces of crystals, which will preferentially dissolve the distorted lattice in the vicinity of a dislocation line; or (2) subjecting a crystal to a vapor phase that would diffuse preferentially along the distorted portion of lattice about the dislocation, "decorating" it so it can be seen.

Dislocations occur in all crystalline materials. We use a parameter known as the *dislocation density* to characterize the extent to which dislocations disrupt a crystal. We measure the dislocation density by counting the number of dislocations that intersect a plane of unit area within the crystal or by measuring the total length of dislocation lines per unit volume of the crystal. Dislocation densities have units of a number per unit area, i.e. $10^4\,cm^{-2}$. Crystals grown under the most carefully controlled conditions in laboratories have dislocation densities on the order of $10^3\,cm^{-2}$. Natural crystals have higher dislocation densities, normally on the order of $10^5\,cm^{-2}$.

8A.4.1.2 Movement of Dislocations

Dislocations are critical elements in the permanent deformation of crystalline solids. Stresses or displacements imposed on the boundaries of a crystal that contains dislocations will, under appropriate conditions, cause dislocations to move through the crystal. The movement of a dislocation displaces one portion of a crystal one lattice unit relative to other portions of the crystal without destroying the crystal's lattice. With the movement of a sufficient number of dislocations, crystals are measurably distorted.

Most dislocations consist of segments with edge character, segments with mixed character, and segments with screw character. Because it is easier to visualize how edge dislocations move, we use edge dislocations to illustrate the two types of dislocation movements and to examine interactions between dislocations and other lattice defects.

8A.4.1.2.1 Glide of Dislocations Raising the magnitude of the hydrostatic stresses compresses the portion of a crystal with a dislocation in the same way that increased pressure affects other portions of the crystal. Bonds with different orientations are shortened by comparable amounts. So, the regions of relatively contracted and extended lattices above and below the glide plane of an edge dislocation remain. Elevating hydrostatic stresses does not cause dislocations to move.

Increasing the components of deviatoric stresses acting on a crystal increases the shear stresses acting across all planes in the crystal except the three planes perpendicular to the three principal stresses. Deviatoric stresses cause some bonds to shorten and other bonds to lengthen, depending on their orientation relative to the principal stresses. Similarly, deviatoric stresses will cause changes in bond angle for most bonds. One can visualize the effect of stresses on an edge dislocation by considering the effect of shear stresses resolved on the dislocation's glide plane (Figure 8.22a). In the vicinity of an edge dislocation, shear stresses distort some already stretched bonds enough to break them at the same time as they displace some atoms into positions that they can form bonds with adjacent atoms (Figure 8.22b). As a result, the dislocation line moves one Burgers vector across its glide plane. In its new position, renewed or continued application of shear stress of sufficient magnitude will cause the edge dislocation to continue moving, one lattice unit at a time. Materials scientists call this type of dislocation movement **glide**.

The glide of dislocations is often described as *conservative dislocation motion* because the dislocation moves without changing the number of atoms in the vicinity of the dislocation line. An edge dislocation moving by glide remains in the plane defined by the dislocation line and its Burgers vector – its *glide plane*. An edge dislocation will glide as soon as and as long as the

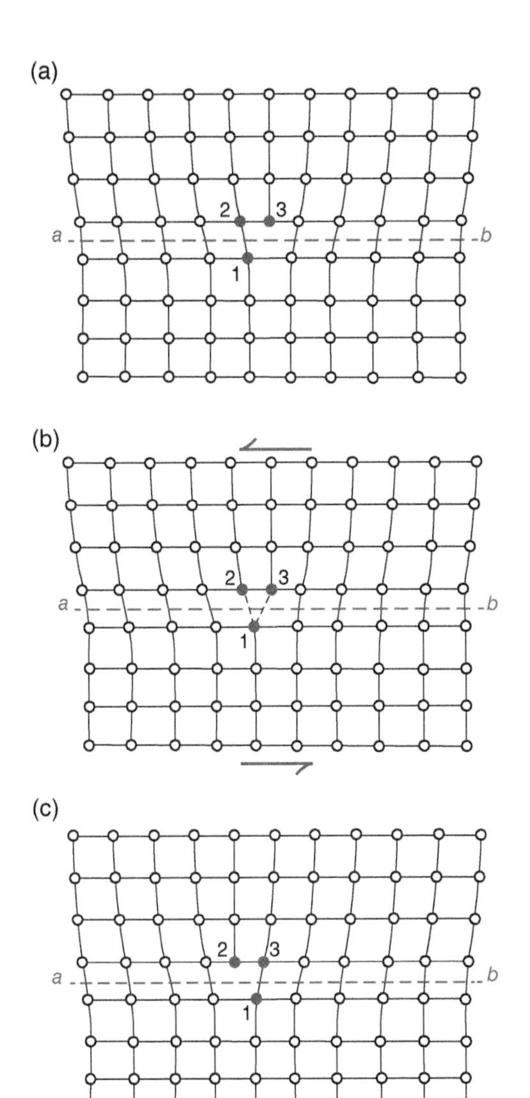

Figure 8.22 Glide or conservative motion of an edge dislocation. (a) A positive edge dislocation with glide plane *ab*. (b) When a critical magnitude of counterclockwise shear stress is applied across the glide plane, the bond connecting atoms #1 and #2 is stretched at the same time as atom #3 (at the end of the half plane) moves closer to atom *#1*. If the shear stress component acting parallel to the Burgers vector is sufficiently large, the #1–#2 bond stretches beyond its breaking point, and atom #1 bonds to atom #3 instead. (c) The net result is the movement of the dislocation line one lattice unit to the left.

magnitude of the shear stress acting on the dislocation's glide plane parallel to the dislocation's Burgers vector is sufficiently large to cause the distortion depicted in Figure 8.22b. This threshold shear stress magnitude is called the *critical resolved shear stress.* A positive edge dislocation will move to the left in response to counterclockwise shear stress, as shown in Figure 8.22. Counterclockwise shear stress would cause a negative edge dislocation to move in the opposite direction. Clockwise shear stress would cause the positive edge dislocation to move to the right and a negative edge dislocation to move to the left.

Shear stresses also distort bonds in the vicinity of screw dislocations and mixed dislocations. Shear stresses equal to the critical resolved shear stress magnitude will break some bonds along the dislocation line as other bonds form, leading to movement, i.e. glide, of the dislocation line. With screw dislocations, glide causes the dislocation line to move in a direction perpendicular to its Burgers vector (Figure 8.23); the same sense of shear stress will cause right-handed and left-handed screw dislocations to move in the opposite direction. Shear stress will cause a mixed dislocation to move oblique to its Burgers vector. Thus, shear stress acting on a glide plane containing a dislocation that exists as a closed loop will cause the loop to expand (Figure 8.23), with the crystal lattice inside the closed loop displaced one unit translation relative to the crystal lattice outside the loop. With sufficient shear stress acting on the glide plane, the loop may expand to the margins of the crystal, resulting in a one-unit translation displacement of the portion of the crystal above the glide plane relative to the portion of the crystal below the glide plane (Figure 8.24). In this way, the glide of dislocations contributes to changing a crystal's shape.

The activation energy for the glide of dislocations is relatively small. One must add only enough strain energy to break bonds in the

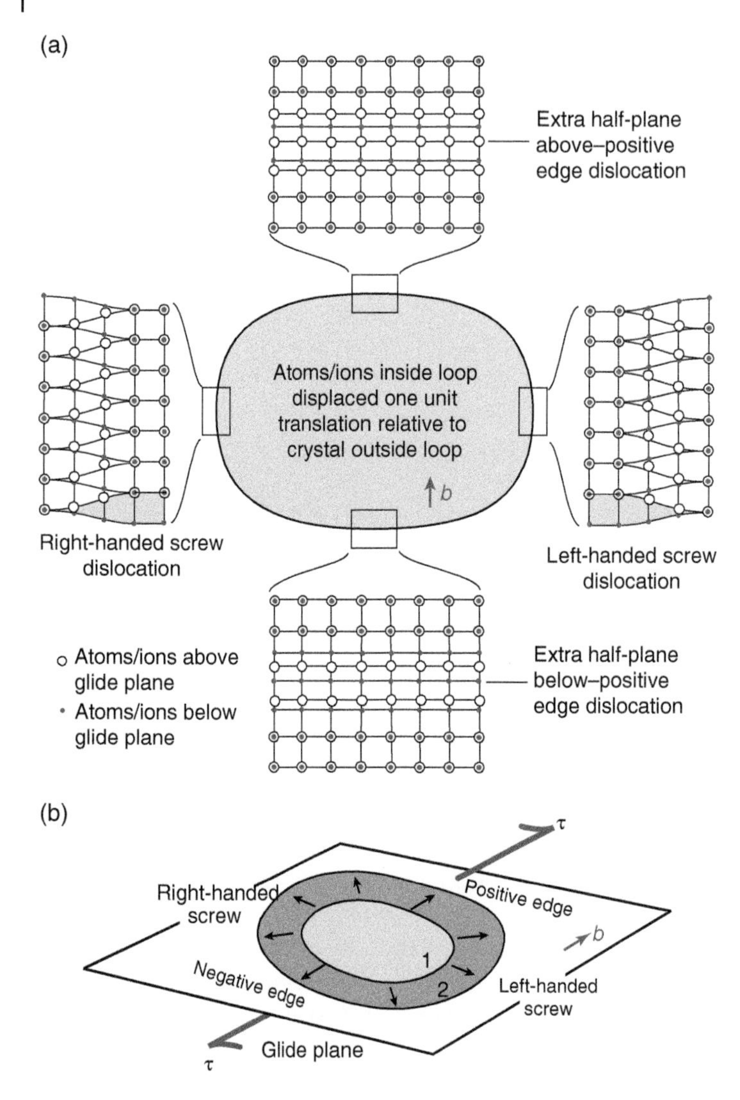

Figure 8.23 (a) A closed dislocation loop has segments of positive and negative edge character, right- and left-handed screw character, and mixed character. (b) A critical magnitude of shear stress resolved on the glide plane will cause the dislocation loop to expand.

vicinity of the dislocation line. Continued addition of these small magnitudes of strain energy can eventually displace all of the crystal above the glide plane one unit translation past all of the crystal below the glide plane. Note also that after a dislocation glides through a portion of the crystal, it leaves "perfect" lattice behind. For this reason, one cannot see the offset within the crystal. Rather, the glide of the dislocation through the entire crystal will produce a unit translation offset of the periphery of the crystal.

Within any crystal, differently oriented lattice planes have different spacings between the atoms that lie in the plane. Thus, on the basis of lattice geometry alone, if dislocations exist in two differently oriented lattice planes that are not

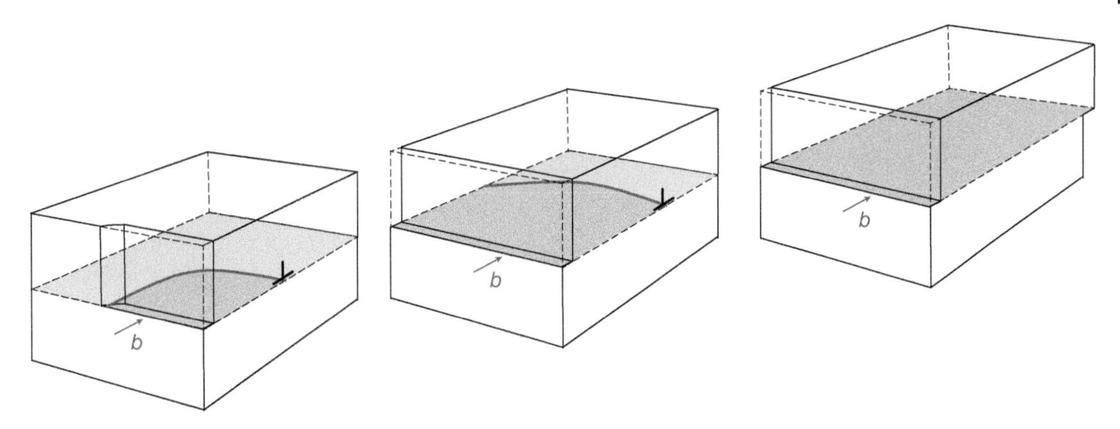

Figure 8.24 Propagation of dislocation through a crystal displaces one portion of the crystal with respect to other portions of the crystal.

symmetrically equivalent, the Burgers vectors of those dislocations will have different lengths. Further, the spacing between atoms will vary with direction across a lattice plane. Thus, the length of potential Burgers vectors will vary with the direction within the lattice plane. We call the combination of a specific Burgers vector direction within a specific lattice plane a *slip system*. In any crystal, slip systems with longer Burgers vectors require greater shear stress magnitudes to initiate glide on them. We observe that only certain slip systems, which are particular combinations of crystallographic planes with characteristic orientations and their corresponding Burgers vectors, will contribute to deformation under particular loading conditions. The dominant slip system for any mineral can change for different T, P, and water content conditions.

8A.4.1.2.2 Climb of Dislocations Edge dislocations also move through a crystal when atoms or ions diffuse into or away from the dislocation line. For any edge dislocation, the diffusion of several vacancies toward the end of the extra half plane will cause a positive edge dislocation to move up one lattice unit and a negative edge dislocation to move down one lattice unit (Figure 8.25). Diffusion of vacancies away from the end of an extra half plane (atoms diffuse toward the end of the half plane) causes a positive edge dislocation to move down and a negative edge dislocation to move up. Because this process enables dislocations to move up or down relative to their glide planes, materials scientists call this type of dislocation movement **climb**. The climb of dislocations is envisioned as a *non-conservative* dislocation movement because it occurs in response to changes in the number of atoms in the vicinity of the dislocation line. The climb of dislocations occurs spontaneously and continuously at a rate controlled by the rate at which vacancies diffuse through a crystal. Changes in the ambient temperatures and pressures, which affect the diffusion rate, may cause climb to speed up or slow down.

8A.4.1.2.3 Comparing Glide and Climb of Dislocations An individual dislocation will glide as long as the magnitude of the shear stress acting on its glide plane equals or exceeds its critical resolved shear stress. Increasing the overall magnitude of the deviatoric stress raises the shear stress magnitudes acting on all lattice planes except those parallel to the stress principal planes

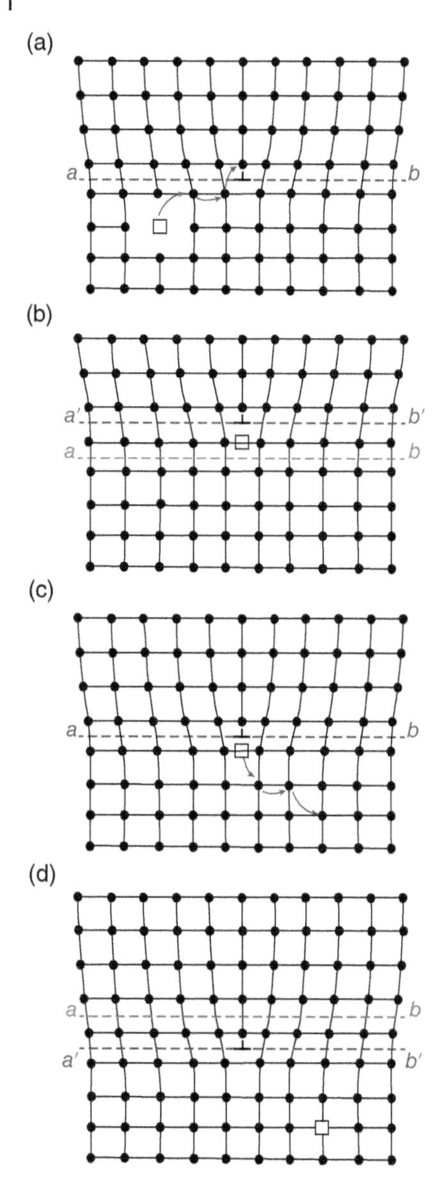

Figure 8.25 Climb or nonconservative motion of a dislocation. (a) and (b) Diffusion of a vacancy toward the end of a half plane causes an edge dislocation to move from its original glide plane *ab* to a glide plane *a′b′* above its original glide plane. (c) and (d) Diffusion of a vacancy away from the end of a half plane causes an edge dislocation to move from its original glide plane *ab* to a glide plane *a′b′* below its original glide plane.

(on which there are no shear stresses). As a consequence, increasing the deviatoric stresses means that dislocations in a wider range of orientations of crystallographic planes might have shear stress values equal to or exceeding their critical resolved shear stress. Stated another way, increasing the deviatoric stress has the potential to activate a larger number of slip systems in a deforming crystal.

To move an edge dislocation a particular distance by climb requires sufficient time for a large enough number of vacancies (or atoms) to diffuse toward or away from the dislocation and accomplish the movement. Climb occurs spontaneously and continuously at a rate that depends upon the ambient temperature and pressure. An increase in temperature causes the lattice to expand and increases the thermal vibration of atoms. These effects raise the rate of material transfer by diffusion, meaning that it takes *less* time for the same number of vacancies (and atoms) to diffuse toward or away from the dislocation and accomplish the movement. An increase in pressure causes the lattice to compress, making it more difficult for atoms to diffuse from one location to another. This lowers the rate of material transfer by diffusion, meaning that it takes *more* time for the same number of atoms or ions to diffuse toward or away from the dislocation and accomplish the movement. The pressure effect is small compared to the temperature effect, however.

For reasons outlined in detail in the following section, the movement of dislocations through a crystal typically requires a combination of glide and climb. The glide of dislocations can occur rapidly if deviatoric stress components are sufficiently large. The climb of dislocations typically requires more time than glide. If the movement of a particular dislocation through a lattice requires *both* glide and climb, the slower process, climb, is the one that controls how quickly that dislocation moves. Because the climb of dislocations typically requires more time than glide, climb is the *rate-limiting step.*

8A.4.2 Dislocation Interactions

In introducing the concept of the glide of dislocations, we envisioned the gliding dislocation moving within a glide plane that extended through a perfect crystal lattice (e.g. Figure 8.24). In reality, a dislocation must glide through a crystal containing numerous imperfections, including point defects, other dislocations in its glide plane, dislocations in glide planes parallel to its glide plane (i.e. dislocations in the same slip system), and dislocations with glide planes oblique to its glide plane (i.e. dislocations in different slip systems). Derek Hull, the author of a classic text on dislocations, offered a visceral image of the potential effects of dislocations oblique to a dislocation's glide plane when he wrote of dislocations needing to move through a "forest" of other dislocations. Each point or line defect is surrounded by elastically distorted lattice. As a dislocation moves, the distorted lattice surrounding it will necessarily begin to overlap with the distorted lattice associated with the other imperfections. Each instance of overlapping lattice distortions has the potential to impede the glide of the dislocation. In the following two sections, we examine briefly the nature and impact of interactions between dislocations and other imperfections.

8A.4.2.1 Interactions that Impede the Glide of Dislocations

Consider first a gliding edge dislocation that approaches a point defect in its glide plane or a few unit translations above or below its glide plane. The region of elastically distorted lattice surrounding and moving with the dislocation will impinge upon the region of elastically distorted lattice surrounding the point defect. If the relative positions of the dislocation line and the point defect are such that the contracted or extended portions of lattice associated with the dislocation attempt to overlap respectively with the contracted or extended portions of the lattice associated with the point defect, that segment of the dislocation will require increased shear stress to move past the point defect (Figure 8.26a). If the externally applied stresses or displacements are sufficiently large, this segment of the dislocation may be driven past the point defect. Otherwise, it will become "pinned" by the point defect (Figure 8.26b). If segments of the dislocation far from the point defect continue to glide, the length of the dislocation line will increase, raising the dislocation density of the deforming crystal and increasing the volume of elastically distorted lattice in it.

A similar effect occurs when two positive edge dislocations (or two negative edge dislocations) in a single glide plane or in closely spaced, parallel glide planes move toward each other (Figure 8.27). As the distance separating the two dislocations decreases, the contracted and extended lattice

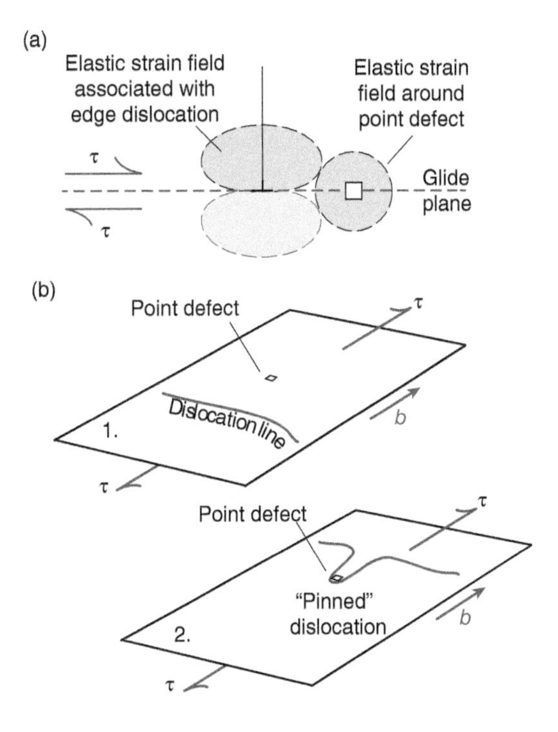

Figure 8.26 (a) A point defect can be an obstacle to a gliding dislocation. (b) 1. A dislocation line approaches a point defect in its glide plane. 2. The point defect may pin a segment of the dislocation. If the rest of the dislocation continues to glide, the dislocation line will lengthen.

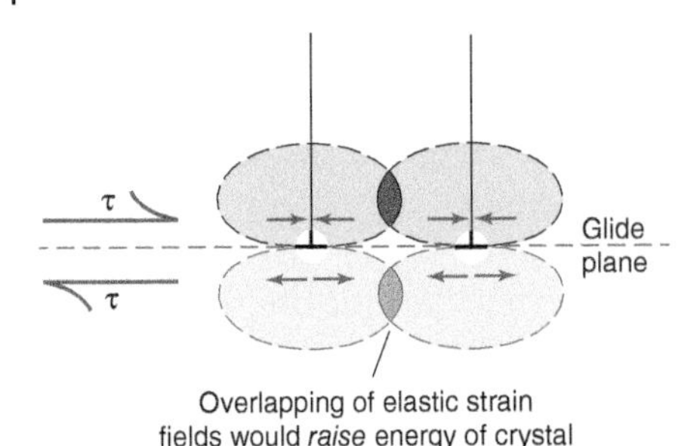

Figure 8.27 Because superposing elastic strain fields raises the internal energy of a crystal, two edge dislocations of the same sign in the same or closely spaced glide planes will tend to repel each other.

regions associated with the first dislocation will begin to overlap respectively the contracted and extended lattice regions associated with the second dislocation. Once again, the overlapping regions of distorted lattice raise the overall elastic strain energy in the crystal, and thus the two dislocations will move into close proximity only if the deviatoric stress components are high enough to drive the overlap. Materials scientists envision that the two defects tend to "repel" each other in order to keep the total elastic strain energy associated with the distorted lattice to a minimum. Comparable repulsion due to overlapping contracted, extended, or sheared regions of lattice can occur with pairs of screw dislocations or with pairs of appropriately oriented edge and screw dislocations.

A different type of interaction occurs when a dislocation attempts to move through the forest of dislocations in glide planes oblique to its glide plane. A dislocation that glides across another dislocation can create local steps or "jogs" in one or both of the dislocations (Figure 8.28a, b). In many cases, the offset of a dislocation line that creates the jog generates a segment of dislocation that must climb in order to move with its host dislocation (Figure 8.28c). Materials scientists infer that such a jog may, if it is sufficiently short, be dragged along by the moving dislocation line, leaving vacancies in its wake. Not surprisingly, the glide of a dislocation that drags along a jog requires higher resolved shear stress than a dislocation without the jog. If a jog has a greater length, it is harder for the dislocation to drag it along. A longer jog will only be able to remain with the rest of the dislocation by the climb, which requires a longer time interval. Such a jog will effectively pin that segment of dislocation, again resulting in an increase in the length of the dislocation (Figure 8.28d).

As the total length of dislocations in a crystal increases, there is a greater likelihood of dislocation interactions. These interactions lead, in turn, to further increases in the total length of dislocations in the crystal, raising the dislocation density in the deforming crystal. There are two important consequences of such an increase in the dislocation density of a crystal. First, continued deformation requires greater magnitudes of deviatoric stress, either to continue the glide of impeded dislocations or to initiate the glide dislocations lying in less favorably oriented glide planes. Both of these factors produce the macroscopic effect that higher values of deviatoric stress components are required to continue changing the crystal's shape. Stated in another way, the crystal's *flow stress* increases, and the material *work hardens*, when dislocations interact. Second, the increased dislocation density means the deformed crystal has higher internal energy than undeformed crystals. As we examine below, this accumulated strain energy fuels significant changes in the microstructure of deformed crystalline aggregates.

8A.4.2.2 Interactions that Facilitate the Glide of Dislocations

Consider again the segment of a gliding edge dislocation that is pinned by a point defect in its glide plane or a few unit translations above or below the glide plane (Figure 8.26a). Because diffusing atoms (or vacancies) continually interact with the dislocation, it is possible that the positive or

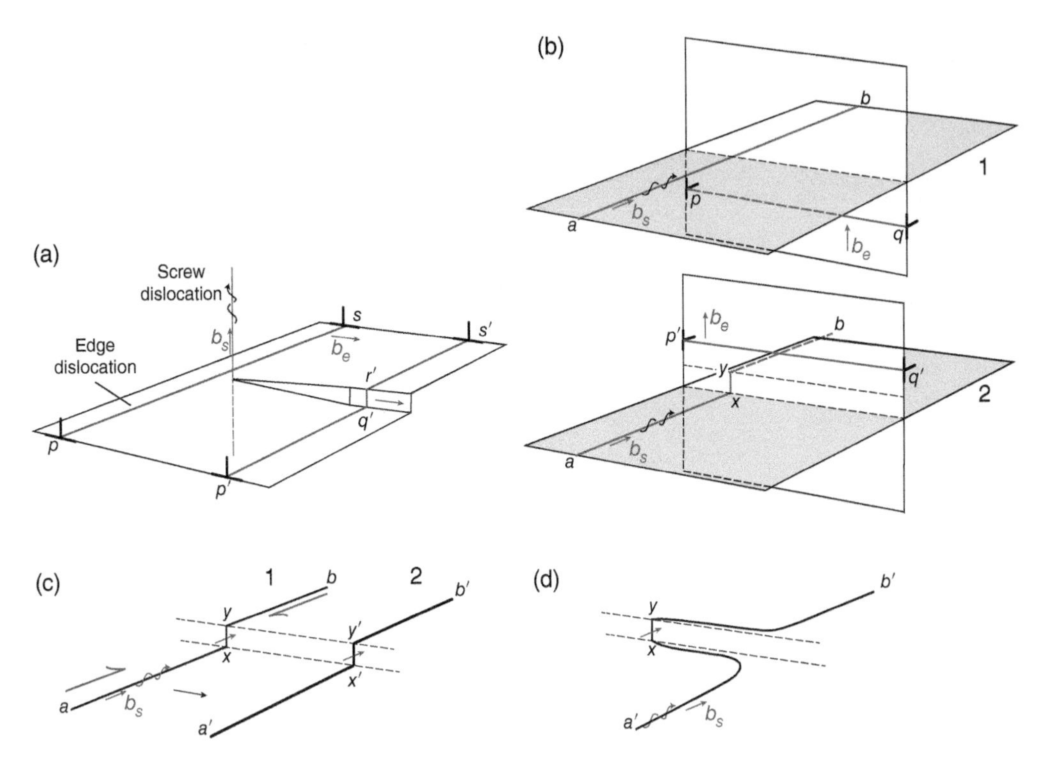

Figure 8.28 (a) After moving edge dislocation *ps* intersects a screw dislocation, it takes the form *p′q′r′s′*, with jog *q′r′*. The dislocation line is incrementally longer. (b) Moving edge dislocation *pq* crosses screw dislocation *ab*, creating the jog *xy*. (c) A critical magnitude of shear stress will induce screw dislocation *axyb* to move to the right. Because the Burgers vector is perpendicular to *xy*, it can move to position *x′y′* only by climb. (d) If climb occurs too slowly, the segment *xy* will be pinned, and movement of the remainder of the dislocation will increase the length of dislocation line in the crystal. *Source:* Modified from Hull and Bacon (1984).

negative climb will enable the pinned segment of the dislocation to "step over" or "slip under" the obstacle in its glide plane and resume gliding. In this instance, the interaction between point defects and a dislocation facilitates rather than impedes the dislocation's motion.

Just as closely spaced dislocations repel each other if elastic strains in the crystal increase as they move toward each other (e.g. Figure 8.27), closely spaced dislocations can be drawn together if elastic strains decrease as they move toward each other. For example, positive and negative edge dislocations in the same glide plane may, if they are sufficiently close to each other, be drawn toward each other until they meet and *annihilate* (i.e. cancel) each other, resulting in defect-free lattice (Figure 8.29). The same is true of right-handed and left-handed screw dislocations – they too can be drawn or driven together and annihilate each other leaving defect-free lattice. The driving energy for dislocation movement in these instances is the lowering of the internal energy of the crystal by eliminating lattice distortions.

8A.4.3 Recovery and Recrystallization

By enabling once-impeded dislocations to move, dislocation climb reduces or eliminates the extent to which increases in the magnitudes of deviatoric stresses are required to drive the glide of

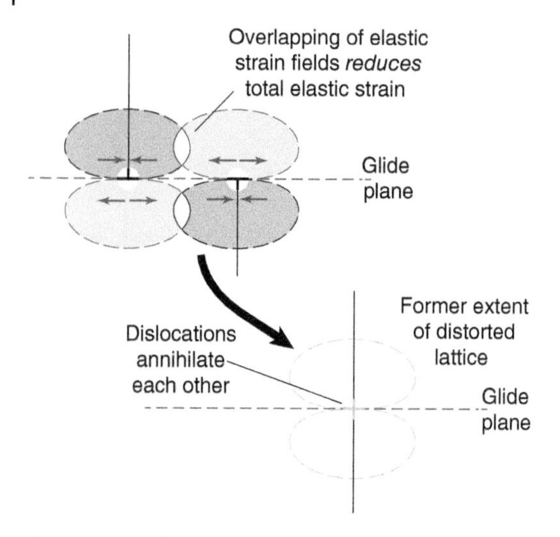

Figure 8.29 Because superposing elastic strain fields lower the internal energy of a crystal, two edge dislocations of the opposite sign in the same or closely spaced glide planes will tend to move two and annihilate each other.

dislocations. Similarly, dislocation annihilation reduces the number of dislocations in a crystal, lowering its internal energy, and reducing incrementally the likelihood of subsequent dislocation interactions. These consequences – reducing or limiting work hardening and lowering the internal energy of deformed crystals – are two effects that fall under the general heading of **recovery**. As Ron Vernon stated succinctly in *Metamorphic Processes: Reactions and Microstructure Development* (1976, p. 166): "Recovery is the set of processes that decreases work (strain) hardening without the development of high angle grain boundaries." A second set of processes is known collectively as **recrystallization**. Recrystallization processes change the microstructure of deformed crystal aggregates by moving the boundaries that separate crystals in an aggregate. Because recrystallization, like recovery, reduces work hardening and lowers the internal energy of deformed crystals, we consider the two processes in the same section. Readers interested in a deeper understanding of these processes and their effects should consult the Comprehensive Treatment section of this chapter.

8A.4.3.1 Recovery

Processes that bring individual deformed crystals closer to their initial, lower energy state are *recovery processes*. Thus, dislocation climb and dislocation annihilation are recovery processes, as is the movement of dislocations to the periphery of the crystal. In order to envision another important recovery process, consider two edge dislocations of the same sign lying in parallel glide planes (Figure 8.30a). If the spacing of the glide planes is appropriate, the two dislocations will be drawn into an arrangement where the contracted-lattice region associated with one overlaps the extended-lattice region associated with the other (Figure 8.30b). Such an arrangement is favored because it reduces the volume of elastically distorted lattice and lowers the internal energy of the crystal.

Should large numbers of edge dislocations move into such a lower energy configuration, as illustrated in Figure 8.31a, they form a planar array called a *low-angle boundary* or *tilt boundary*. The movement of dislocations into a tilt boundary like that illustrated in Figure 8.31a typically creates adjacent regions of the crystal lattice that are relatively dislocation-free. Further, the extra half planes associated with edge dislocations in the boundary combine to produce a misorientation of the portions of a crystal lattice separated by the tilt boundary (Figure 8.31b). Other types of tilt boundaries may be three-dimensional arrangements of edge dislocations in different glide planes (Figure 8.31c), screw dislocations, or combinations of edge and screw dislocations. In all cases, the tilt boundaries divide a distorted single crystal into separate, small, relatively dislocation-free (i.e. undistorted) regions called *subgrains* because their lattices are slightly misoriented relative to each other (Figure 8.32). As subgrains form, i.e. as dislocations move into

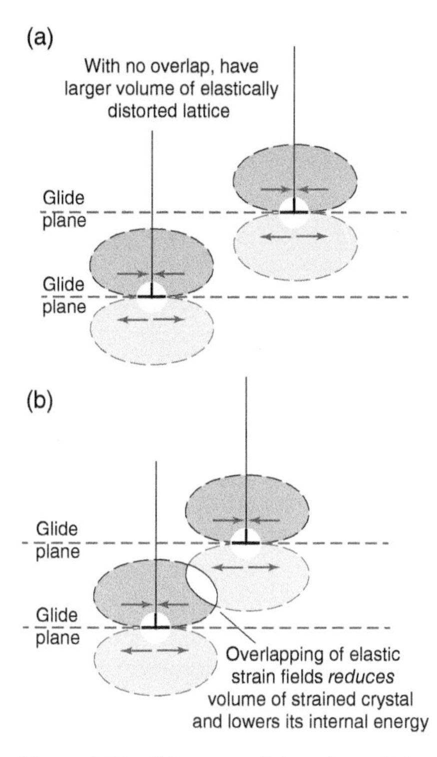

Figure 8.30 When two dislocations of the same sign are in glide planes with an appropriate spacing (a), the contracted lattice associated with one dislocation is drawn to overlap the extended lattice associated with the other dislocation (b). Because the crystal has lower total energy when the dislocations take this configuration, the configuration is stable.

tilt boundaries, they lower the overall internal energy of the crystal.

8A.4.3.2 Effects of Recovery

Individual dislocations distort the crystal lattice around them. When dislocation densities are high, the lattice distortion may be sufficiently great to broaden peaks in an x-ray diffraction pattern or locally alter a crystal's optical properties. High dislocation densities can, for example, cause isotropic minerals to become birefringent. Experimentally deformed quartz and olivine regularly possess

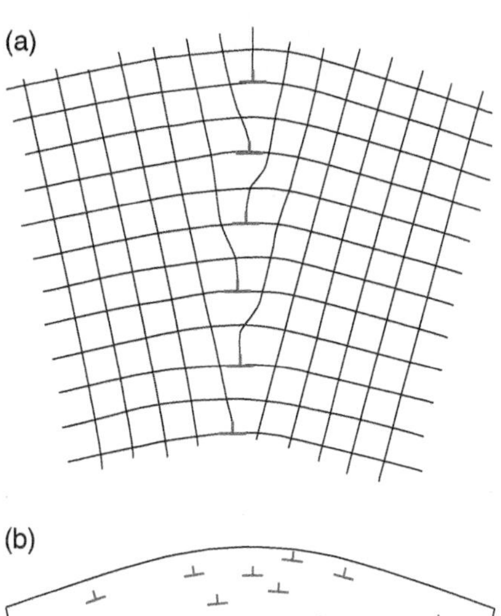

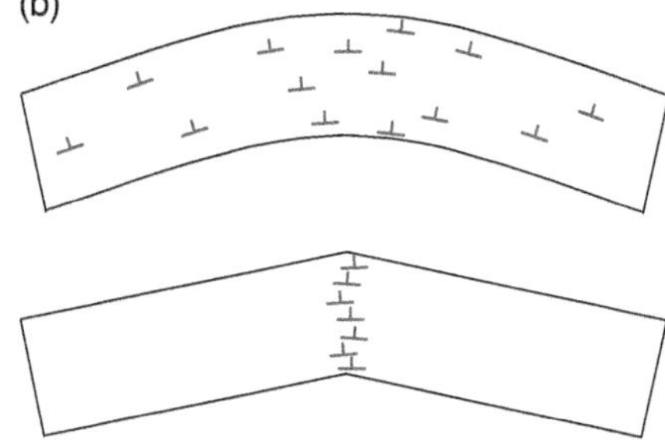

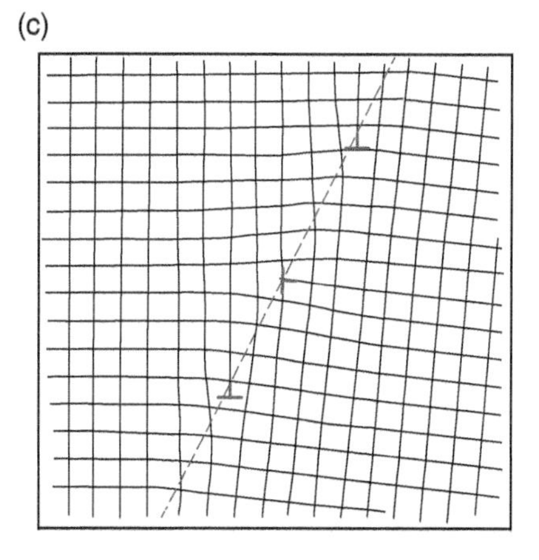

Figure 8.31 Low angle tilt boundaries. (a) Positive edge dislocations positioned in a stable 3D array. (b) Edge dislocations in a deformed crystal move into a low angle tilt boundary separating two relatively dislocation free portions of the crystal. (c) A low angle tilt boundary consisting of edge dislocations in different glide planes. Movement of this boundary will require both the glide and climb of dislocations.

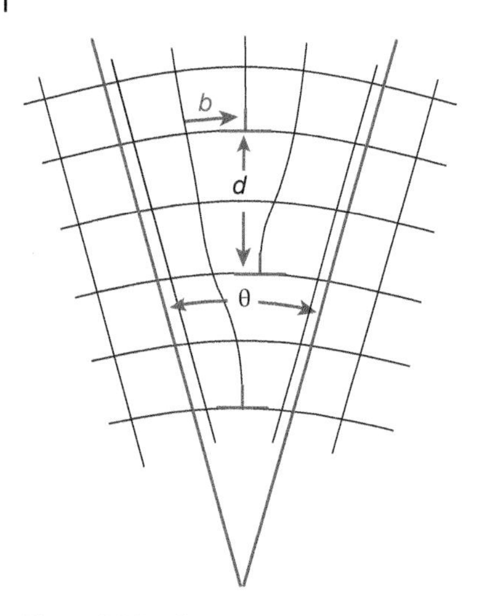

Figure 8.32 Enlargement of a portion of the low angle tilt boundary in Figure 8.31. The tilt angle θ is directly proportional to b, the length of the Burgers vector, and inversely proportional to d, the spacing between dislocations.

deformation lamellae, which are thin, tabular regions with indices of refraction different from that of the host grain. Examination with TEM shows that these regions have extremely high dislocation densities.

As a result of dislocation climb, dislocation annihilation, and moving dislocations to the periphery of a crystal or into tilt boundaries, a deformed crystal may *recover* its sharp x-ray diffraction peaks. Similarly, individual subgrains *recover* their uniform optical properties. In addition, the crystal may regain its lower (*i.e.* initial) flow stresses, i.e. the effects of work hardening are reduced or eliminated. Climb is instrumental to recovery because it enables dislocations to move past obstacles, allowing them to move into appropriate glide planes where they can move freely. Complete recovery requires elevated temperatures *or* very long time intervals.

8A.4.3.3 Recrystallization

In polycrystalline aggregates, recovery often generates situations where a relatively strain-free grain is adjacent to a highly strained grain. This sets the stage for a more dramatic reordering of the microstructure known as *recrystallization*. The *migration* of grain boundaries is integral to and characteristic of recrystallization. Grain boundary migration is driven by differences in the internal energy between less-strained grains or strain-free subgrains and their highly strained neighbors. As a grain boundary migrates across the more highly strained grain and the relatively strain-free grain grows, the internal energy of the aggregate is lowered.

In order to understand grain boundary migration, one must have some understanding of the nature of boundaries separating different grains in polycrystalline aggregates. Grain boundaries are thought to be tabular volumes where atoms and ions do not conform to a lattice. The movement of such a tabular volume of atoms that lack a coherent crystal structure requires: (1) atoms to leave the more highly strained crystal lattice; (2) join or diffuse across the grain boundary; and (3) move to and occupy sites in the lattice of the less-strained grain. *Grain boundary mobility* is, therefore, dependent upon diffusion, which is, in turn, strongly dependent upon temperature.

Ambient conditions dictate whether and the degree to which grain boundaries are *mobile*. At lower temperatures, grain boundaries are not mobile, so only recovery processes operate. At somewhat higher temperatures, grain boundaries become mobile, and a deforming aggregate will begin to recrystallize. If temperatures are just high enough for grain boundaries to exhibit limited mobility, microstructures are dominated by small embayments in grain boundaries, where a relatively strain-free crystal has expanded into and replaced a small portion of a more highly strained neighboring grain. At even higher temperatures, grain boundaries are able to sweep across larger portions of highly distorted grains.

8A.4.3.4 Post-tectonic (Annealing) versus Syntectonic (Dynamic) Recovery and Recrystallization

In metallurgical applications and in some geological settings, recovery or recrystallization occurs after the deformation. Metallurgists call post-deformation recovery and recrystallization *annealing*. In most geological settings, the processes of recovery and recrystallization – annihilation of dislocations, movement of dislocations into stable three-dimensional arrays, i.e. low-angle tilt boundaries, and the mobility of low- and high-angle grain boundaries – occur concurrently with deformation. Geologists sometimes use the terms *dynamic* or *syntectonic* to refer to recovery or recrystallization that occurs at the same time as deformation accumulates.

8A.4.4 Microstructures Indicative of Dislocation-Accommodated Deformation

The glide of dislocations through a crystal, whether assisted by dislocation climb or not, displaces one portion of a crystal lattice relative to the rest of the lattice (cf. Figures 8.21, 8.23, 8.24). Each glide plane is comparable to a microscopic fault or a thin shear zone, with one portion of a crystal displaced relative to the remainder of the crystal by a multiple of Burgers vector length. Thus, the glide of dislocations creates *displacement gradients* in the crystal. In the same way that the slip of individual cards past each other can change the shape of a card deck, the glide of numerous dislocations on different glide planes enables minerals to change shape. This occurs without destroying the mineral's crystallinity. Thus, if one sees that mineral grains in a rock have developed shape-preferred orientations without evidence for grain fracture or diffusional mass transfer, it is reasonable to propose that dislocation movement contributed to the shape change (*strain*).

Individual dislocations are too small to be seen individually using optical or scanning electron microscopy, and, at present, most structural geologists do not have regular access to the TEM that can confirm the role of dislocations. How then can one confirm that dislocations have contributed to the deformation of grains? Recall that as dislocation movement accomplishes a change in the shape of mineral grains, dislocations interact and dislocation densities multiply. As dislocation densities increase, dislocations tend to move into distinctive geometric arrays like low-angle tilt boundaries, where their combined effects create visible microstructures. It is the collective effects of large numbers of dislocations that generate the suite of microstructures we use to confirm that dislocation movement contributed to grain shape change. In the following paragraphs, we describe several optical microstructures that indicate that dislocation movement contributed to deformation. The specific character of the microstructures depends on strain magnitude, deformation rate, and the temperature at which deformation proceeded. In this section, we outline the most prominent dislocation microstructures; Section 8B.3.7 discusses in greater detail the differences in microstructures developed under different deformation conditions. The microstructures previously described in Section 3.7 are dislocation microstructures; they are interpreted here in detail.

One common optical microstructure that develops at low-strain magnitudes over a broad range of deformation temperatures is **undulose extinction**, where different portions of an individual grain go extinct at different times during stage rotation (Figures 8.33a, 3.15a). One might imagine that, as is shown schematically in the upper image of Figure 8.31b, this microstructure arises because the many dislocations in a crystal create smoothly curved lattice planes across a typical mineral grain. This might occur in experimentally deformed crystals, but TEM studies of naturally deformed minerals suggest that undulose extinction results from numerous low-angle tilt boundaries that together create sufficient misorientation

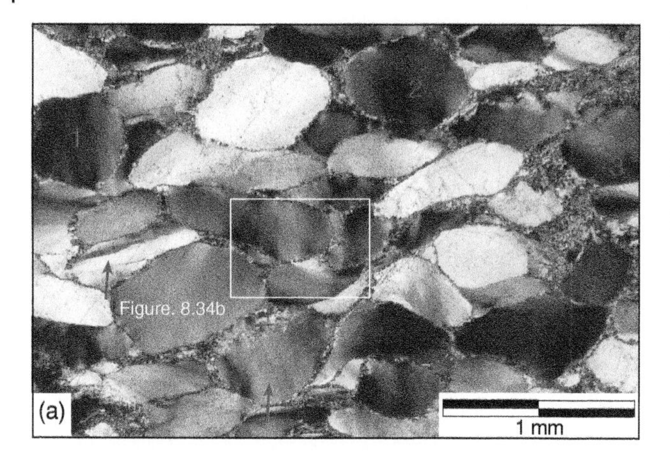

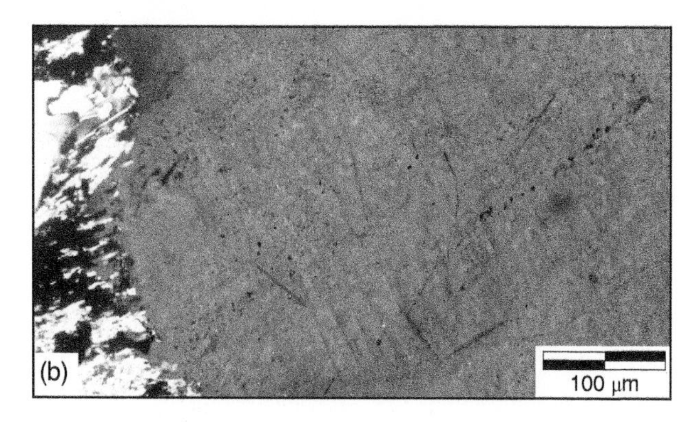

Figure 8.33 (a) Elongate quartz grains in a deformed quartzite. Grains 1, 2, 3, and others exhibit undulose extinction. Grains indicated by red arrows have developed patchy to banded extinction. The white box indicates the field of view of Figure 8.34b. (b) Deformed quartz grain with deformation lamellae indicated by a red arrow.

of crystal lattices to generate the optically visible microstructure.

Higher strain magnitudes require the movement of larger numbers of dislocations. Thus, continued deformation, particularly at low-to-moderate temperatures, is accompanied by further increases in dislocation density. As a greater number of dislocations move into subgrain arrays, we see both the formation of a greater number of subgrain boundaries and an increase in the angular misorientation across individual subgrain boundaries. Individual original grains may exhibit patchy extinction or *deformation bands*, the latter of which are lenticular regions in grain that lack distinct boundaries but that go extinct at a different point during stage rotation than neighboring regions (Figure 8.34a).

In rocks where deformation accrued at a high rate relative to the rate of recovery (this may result from high strain rates at moderate temperatures or at moderate strain rates at low temperatures), deformed grains will develop **deformation lamellae**. Deformation lamellae are thin, tabular regions apparent under plane-polarized light because they have an index of refraction different from that of the host grain. Deformation lamellae may also be apparent under cross nicols (Figures 8.33b, 3.16b). As noted earlier, TEM studies indicate that deformation lamellae in experimentally deformed samples are regions of extremely high dislocation density, whereas, in naturally deformed samples, they are localized regions that are sufficiently misoriented from the host grain to exhibit a different index of refraction.

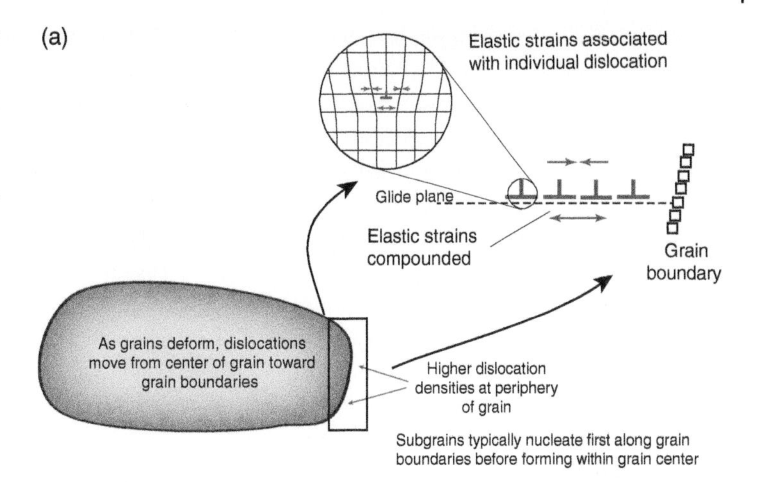

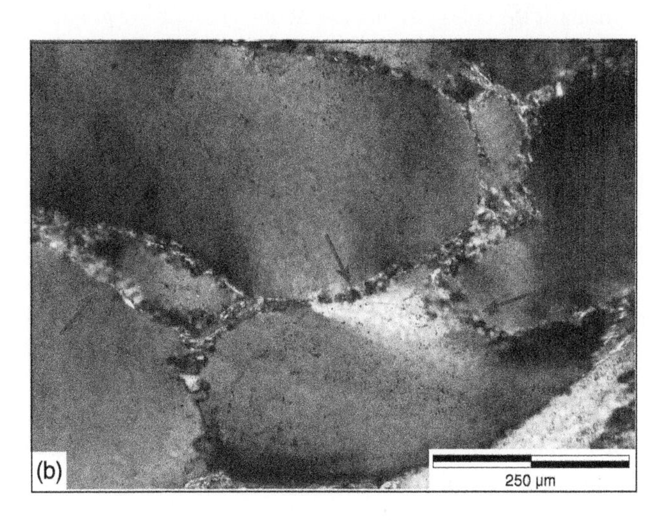

Figure 8.34 (a) Dislocations pile up in the vicinity of grain boundaries. Thus, dislocation densities typically are higher near peripheries of grains than in centers of grains. (b) Large quartz grains exhibiting undulose extinction. Arrows point to subgrains and new grains formed along the margins of deformed grains.

As dislocations glide toward grain boundaries, the "halos" of elastic distortion surrounding dislocations interact with the distorted lattice associated with the grain boundaries. Only if stresses are sufficiently high are dislocations forced to glide into positions where their elastic strain field overlaps with the elastic distortion near the grain boundary. Otherwise, the grain boundaries repel dislocations attempting to glide toward the periphery of grain, and dislocations *pile up* in the vicinity of grain boundaries. Observations show that the dislocation densities in the vicinities of the boundaries of deformed grains are higher than those in the centers of grains. For this reason, recovery processes, driven by the increased internal energy due to high dislocation densities, typically initiate or occur earlier near grain boundaries than in grain centers (Figure 8.34a). Generally, low-angle tilt boundaries and eventually subgrains form first along the margins of grains (Figure 8.34b).

As grains continue to deform, subgrains develop greater angular misorientations between their lattices and that of the host. Recovery and minor grain boundary mobility may lead to the formation of small distinct regions along the original grain boundaries whose angular misorientation relative to the host grains exceeds 10°. We refer to those regions as **new grains**. The result is an aggregate that exhibits a **core-and-mantle texture**, where "core" refers to the deformed remnants of the original grains with undulose extinction and deformation lamellae and "mantle" refers to the collection of relatively strain-free subgrains and new grains developed near the periphery of the original grains (Figures 8.35a, 3.17b).

Spatial variations in dislocation densities, due to variations in the progress of recovery, generate differences in the internal energy of portions of adjacent grains. If the energy differences across grain or subgrain boundaries are sufficiently high, they can drive the migration of grain boundaries. This process is facilitated by slower strain rates, at higher temperatures, and the presence of sufficient fluid phases to enhance diffusion. Under appropriate conditions, such as at slow strain rates or at elevated temperatures, subgrains and new grains may grow by grain boundary migration. If this process proceeds sufficiently far, the distorted cores of grains may be replaced completely by subgrains and/or new grains, generating a fully recrystallized rock.

8A.4.4.1 Development of Crystallographic Preferred Orientations

As dislocations sweep across grains within a polycrystalline aggregate and the individual grains change their shapes, grains must move past their neighbors in order to prevent the creation of cavities in the aggregate (cf. Figure 8.11). In addition, for many grains, the change in shape will be accompanied by a reorientation, relative to a fixed external reference direction, of their lattices. One can begin to understand this effect by focusing on two grains within an aggregate undergoing coaxial shortening (i.e. pure shear deformation) by dislocation movement (Figure 8.36). Grain boundaries that are approximately parallel to the strain principal planes will be stable or will only slowly reorient during deformation. Most rock-forming minerals have limited numbers of slip systems that operate within a particular range of temperatures and differential stresses, and often an individual mineral grain will be oriented so that only one of those slip systems is active. Figure 8.36a illustrates such a situation where the grain boundary between the shaded and unshaded grains is approximately parallel to a principal plane and only one set of glide planes in the unshaded grain is favorably oriented for slip.

The unshaded grain will change its shape by slip on those favorably oriented glide planes. Slip on those glide planes will be accompanied by reorientation of both the glide planes and the crystallographic axes of the unshaded grain. In order to understand why that reorientation occurs, think of the grain boundary as akin to a bookshelf and the glide planes as akin to the surfaces along which book covers meet. If the books were to slump by slip between their covers, the spines of the books would reorient from orientations perpendicular to the bookshelf to orientations oblique or even parallel to the bookshelf. The shape of the aggregate of several books would change from a rectangular shape for the upright books to a rhomb shape for the slumped books. In an analogous manner, deformation of grains leads to reorientation, relative to an external reference, of crystallographic directions.

In pure shear, crystallographic directions perpendicular to glide planes in deforming grains tend to "rotate" toward the shortening direction. In simple shear, crystallographic directions perpendicular to glide planes tend to rotate toward the direction perpendicular to the overall shear plane. Similarly, if a mineral has more than one active set of glide planes, the observed patterns will be more varied. In any case, as deformation proceeds, these processes contribute to the formation of *crystallographic preferred orientations* (*CPOs*) or *lattice-preferred orientations* (*LPOs*) in the aggregate (Section 3.11).

The slip systems capable of accommodating deformation vary from mineral to mineral, depending on

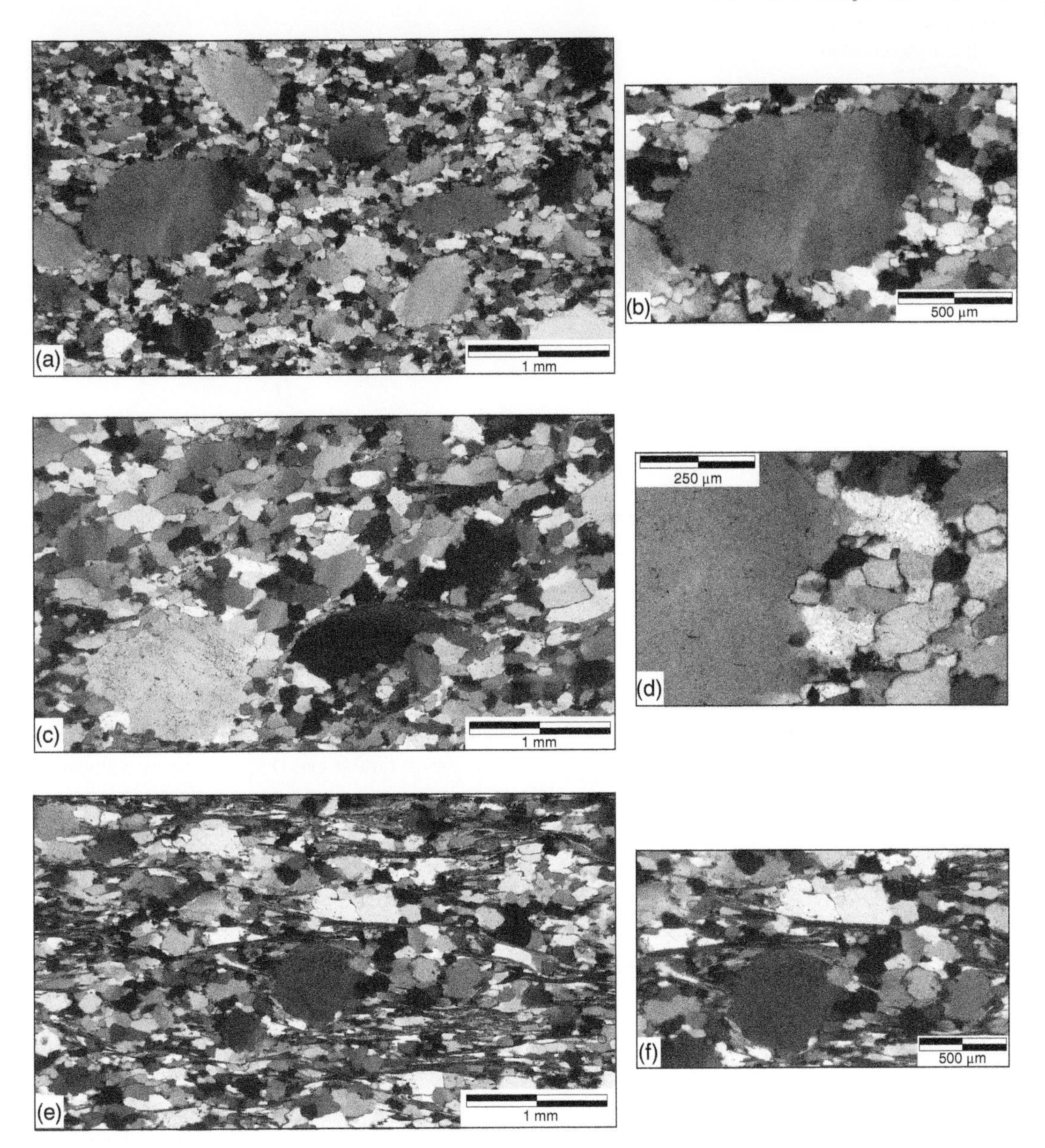

Figure 8.35 (a) and (b) Deformed quartzite where recrystallization has progressed farther than in Figure 8.34b. Note large remnants of deformed detrital grains exhibiting undulose or banded extinction surrounded and separated by subgrains and new grains. (c) and (d) In the sample where recrystallization has proceeded even farther, relict detrital grains are smaller and more widely separated because more of host grains have been replaced by subgrains or new grains. Note the significant angular misorientation between the new grains. (e) and (f) With even greater recrystallization, very few relict grains are apparent. New grains have grown, and some are nearly as large as the remnants of the detrital grains.

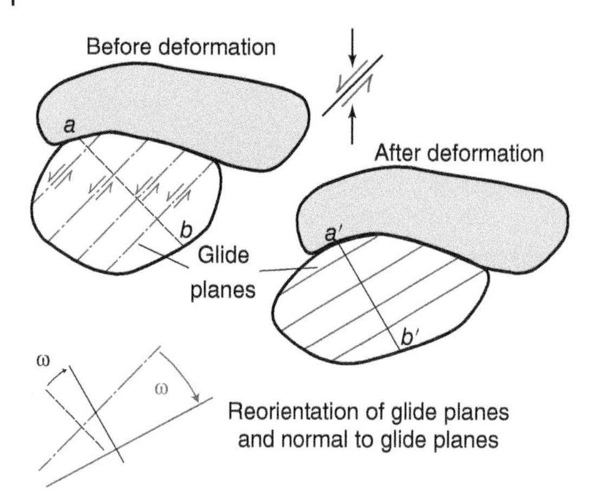

Figure 8.36 The boundary separating these two mineral grains, which is perpendicular to the direction of maximum shortening, has a stable orientation. Distortion of the unshaded mineral grain by the glide of dislocations on favorably oriented planes causes the crystallographic direction *ab* to reorient to orientation *a′b′*, i.e. to rotate toward the direction of maximum shortening.

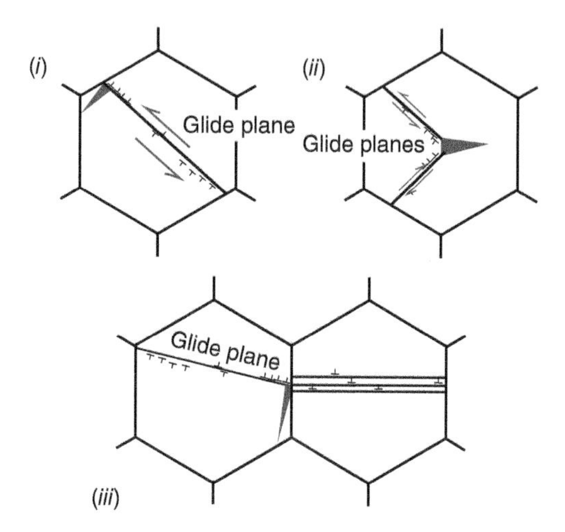

Figure 8.37 As deformation by dislocation glide proceeds, buildup of numbers of dislocations on glide planes within individual crystals may lead to the initiation of fractures within grains favorably oriented glide planes meet grain boundaries (*i*), where two active glide planes intersect (*ii*), and where glide from one crystal imposes stresses on neighboring crystal (*iii*).

mineral symmetry and the relative strengths of differently oriented bonds. For any individual mineral, different slip systems operate at different ambient temperatures, differential stress levels, water contents, strain geometries, and strain rates. Still, the CPO that develops typically reflects the strain path of the deformation, with coaxial strain paths generating CPOs with orthorhombic symmetry and non-coaxial strain paths generating CPOs with monoclinic symmetry. CPOs created during shape changes persist through recovery and recrystallization.

8A.4.5 Dislocation Glide: A Deformation Mechanism

In deformation occurring at very high strain rates and low temperatures, distortion of polycrystalline metals by dislocation movement may occur with neither recovery nor recrystallization. In these instances, distortion is accomplished by the deformation mechanism called **dislocation glide**. The lack of recovery means that deformation occurs with undiminished numbers of dislocations, extensive dislocation interactions, dramatic increases in dislocation densities, and significant increases in flow stress. Deformation by dislocation glide is accompanied by work hardening. Metallurgists know that deformation by dislocation glide often leads to the embrittlement of metal, that is, the metal crystals fracture during deformation because increases in dislocation density generate pile-ups of a sufficient number of dislocations to initiate fractures within crystals (Figure 8.37).

In most geological settings, however, either temperatures are sufficiently high or the rates of deformation are sufficiently slow that recovery and recrystallization play a role in nearly all deformation accommodated by dislocation movement. One setting where the dislocation glide deformation mechanism may occur is in the high strain rate deformation associated with geological fault zones. The embrittlement observed due to dislocation pile-ups in metals is similar to the plastic deformation in fault zone rocks that leads to comminution by grain fracture.

8A.4.6 Flow Law for Dislocation Glide

Activation enthalpy H is a measure of the total internal energy of the system. H has two components: Q = the internal energy of the system, and PV = the pressure–volume work done on or by the system. That is, $H = Q + PV$. Flow laws for deformation by dislocation glide are determined by assuming that the activation enthalpy is a function of the deviatoric stress σ', that is, activation enthalpy is $H(\sigma')$. Empirically, activation enthalpy is seen to decrease with increasing differential stress, and there is a differential stress magnitude at which the activation enthalpy goes to zero. At that point, deformation by dislocation glide proceeds without thermal activation. Taking H^* as the activation enthalpy for dislocation glide, these conditions yield constitutive relationships of the form

$$\dot{\varepsilon} \propto \exp\left\{\frac{-H^*(\sigma')}{RT}\right\} \tag{8.10a}$$

or

$$\dot{\varepsilon} \propto \sigma^2 \exp\left\{\frac{-H^*(\sigma')}{RT}\right\}. \tag{8.10b}$$

Both of these relationships indicate that strain rate is exponentially related to differential stress. The magnitude of the differential stress required for deformation varies weakly with the strain rate or temperature. Thus, the flow law approximates that of a perfectly plastic material.

8A.4.7 Dislocation Creep: A Deformation Mechanism

Single crystals and polycrystalline aggregates change shape by the combination of: (1) the glide and climb of dislocations; (2) the dislocation interactions associated with dislocation movement; and (3) the recovery and recrystallization processes outlined earlier. Materials scientists and geologists apply the name **dislocation creep** to the deformation mechanism that results from this combination of processes. Steady-state deformation by dislocation creep is governed by a single constitutive equation.

In dislocation creep, the glide of dislocations is responsible for changing the shapes of individual mineral grains, but the rate of dislocation climb exerts a strong influence on the flow stress required for deformation and, to some degree, the rate at which deformation may proceed. That is, deformation by dislocation creep entails competing work-hardening and work-softening processes. Work hardening results because the glide of dislocations often leads to dislocation interactions that impede the glide of some dislocations, increasing dislocation densities and flow strengths. Work softening operates through the collection of processes known as recovery. Specifically, the combined glide and climb of dislocations into low-angle grain boundaries lowers the internal energy of deformed crystals or grain aggregates and enables dislocations to glide "unencumbered" through relatively dislocation-free subgrains separated by low-angle grain boundaries.

8A.4.8 Flow Law for Dislocation Creep

The strain and the specific microstructural features observed in a polycrystalline aggregate deformed by dislocation creep depend upon the rate of deformation ($\dot{\varepsilon}$) or the magnitude of the deviatoric stress (σ'), which are functionally related by the constitutive equation for dislocation creep:

$$\dot{\varepsilon} = BD_L\left(\frac{\sigma'}{G}\right)^n. \tag{8.11}$$

In Eq. (8.11), B is a parameter that depends on the geometry of deformation, the crystal's elastic parameters, and has units that ensure both sides of the equation have units of strain rate (t^{-1} or sec^{-1}). As noted earlier D_L is the diffusion coefficient for lattice diffusion. G is the crystal's shear modulus, and n is a dimensionless parameter. The functional dependence on the parameters B, D_L, and G is consistent with theory, and for many rock-forming minerals, their values are constrained experimentally.

Equation (8.11) indicates that the strain rate is proportional to a normalized value of components of the deviatoric stress tensor raised to a power n. Using the normalized deviatoric stress, i.e. σ'/G, instead of deviatoric stress alone keeps the flow law dimensionally sound, that is, it eliminates problems associated with raising a variable with units of stress to a power greater than unity. For most crystalline materials, the exponent n takes a value of 3–5. Typically, this flow law holds for differential stresses in the range $0 \leq \sigma'/G \leq 0.3$. As was the case for diffusional mass transfer, the strain rate for dislocation creep is directly proportional to the diffusion coefficient. Increasing temperature while holding all other factors constant causes the diffusive flux to increase and thereby increases the strain rate by dislocation creep. Increasing pressure while holding all other factors constant decreases the diffusive flux and the strain rate due to dislocation creep will decrease. As noted before, the effects of increasing temperature are more pronounced than the effects of increasing pressure.

8A.4.9 Other Lattice Deformation Processes – Twinning and Kinking

There are two other deformation processes, deformation twinning and kinking, that contribute to the deformation of some crystalline materials. We describe these processes in this section on dislocation creep because both can accompany deformation by dislocation creep and may contribute to the internal energy of a deformed crystal that drives recovery and recrystallization. Neither process, however, is itself a deformation mechanism because: (1) the amount of deformation accommodated by the process is limited; and (2) because the process is not associated with a distinctive flow law or yield criterion.

8A.4.9.1 Deformation or Mechanical Twinning

Twins are two or more crystals separated by a special kind of grain boundary called a *composition surface*. The composition surface is special because the atoms along the surface occupy positions consistent with the lattices of the crystals on both sides of the surface (Figure 8.38a). For this reason, the crystallographic axes of adjacent twins have well-defined geometric orientations relative to each other defined by a symmetry operation called a *twin law*. For macroscopic twins, corresponding crystal faces have predictable orientations relative to each other. In thin sections of many anisotropic minerals, twinning is apparent because the extinction positions in adjacent twins have specific orientations relative to each other and relative to the composition surface that separates them (the optical appearance of twins was described in Section 3.8).

Twinned crystals separated by an irregular composition surface usually are *penetration twins*. *Contact twins* have a planar composition surface. Three or more crystals twinned by the same twin law are *repeated* or *multiple twins*. If the composition surfaces in a multiple twin are parallel, the twins are *polysynthetic*. When successive composition surfaces in a multiple twin are not parallel, the twins are *cyclic*. Thin, plate-like twins within a larger, untwinned crystal are *twin lamellae*.

Growth twins originate in some minerals as "accidents" during crystal growth. Twins in other minerals form during cooling when minerals revert from a higher symmetry structure characteristic of higher temperatures to a lower symmetry structure stable at lower temperatures. In several common minerals, twins occur as a result of deformation. This type of twinning, known as *mechanical twinning*, *deformation twinning*, *glide twinning*, or *stress-induced twinning*, is a mechanism by which the mineral changes shape. Mechanical twinning

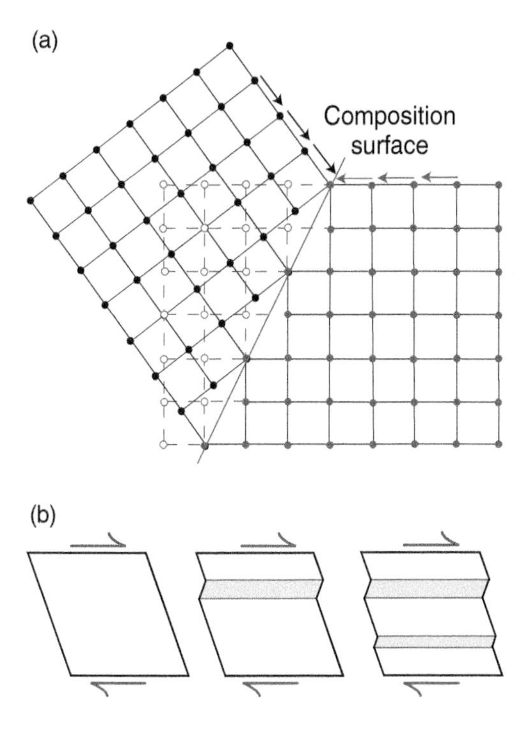

Figure 8.38 (a) A twin boundary is a rational boundary, i.e. atoms and ions fall at positions consistent with the lattices on either side of the boundary. (b) Deformation or mechanical twinning accommodates a change in the shape of a crystal. (c) A deformed marble with calcite grains exhibiting polysynthetic and cyclic twinning. Twins are apparent because they extinguish at different positions during stage rotation.

accommodates small finite displacement of one portion of a crystal relative to the rest of the crystal through a characteristic reflection or rotation of the lattice across a composition surface (Figure 8.38b). Mechanical twinning occurs regularly in calcite, dolomite, plagioclase feldspar, quartz (where it generates Brazil and Dauphine twins not visible under the petrographic microscope), and orthopyroxene. In calcite, where mechanical twinning is common (Figure 8.38c), the maximum shortening accommodated by twinning is 15%.

8A.4.9.2 Kinking of Crystals

A **kink** is a highly angular fold possessing: (1) a *kink plane*, an imaginary plane that intersects the folded layers at an angle; and (2) a *kink axis*, an imaginary line that lies in the kink plane about which the folded layers rotate. The rotation of the layering is accommodated by interlayer slip (Figures 3.16a, 8.39a). **Kink bands** form where two subparallel kink planes have kink axes that also are subparallel, but the layers rotate in the opposite senses about the two kink axes.

Under appropriate conditions, kink folds or kink bands may form by rotating the lattice within an individual mineral grain (Figures 8.39b, 3.16a). The minerals capable of deforming by kinking possesses either lattice planes with especially weak bonds or a single slip system with very low critical resolved shear stress relative to the mineral's other slip systems. In sheet silicates, for example, the weak bonds across the basal planes in sheet silicates may be exploited as cleavage planes or as a weak slip system, depending upon the orientation of the grain in a deforming polycrystalline aggregate. If the mineral is shortened along its basal plane, the grain may accommodate the shortening by kinking (Figure 8.39c). Similarly, slip on the basal planes in ice crystals occurs readily, and ice crystals shortened along their basal planes regularly kink. Kinks also commonly form in amphiboles (such as hornblendes, etc.) and in pyroxenes (in both orthopyroxene and clinopyroxene). Like twinning, this deformation process typically is restricted to low strain values.

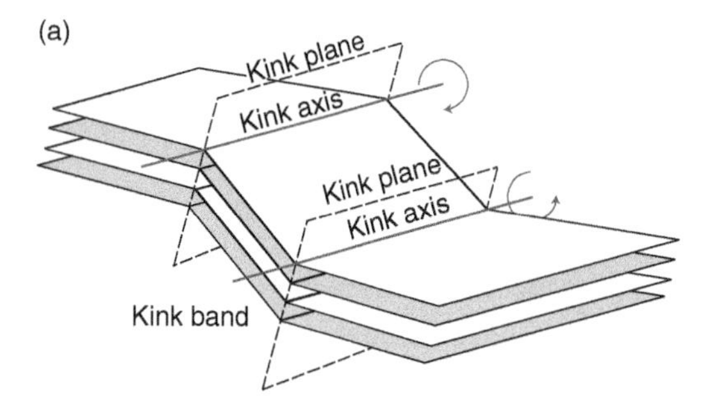

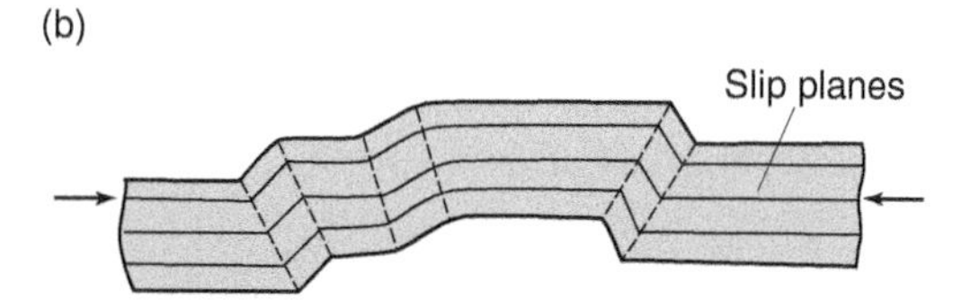

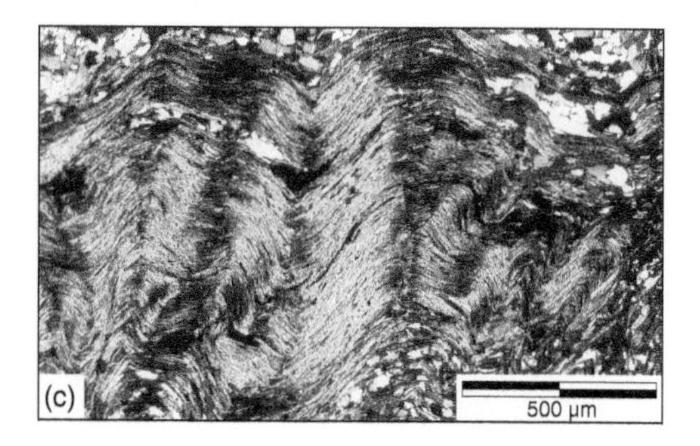

Figure 8.39 (a) A kink band. (b) Sketch of kinked mineral, where interlay slip is accommodated by slip on numerous slip planes or glide planes. (c) Photomicrograph of kinked micas in deformed schist.

8A.5 Diffusion- and/or Dislocation-Accommodated Grain Boundary Sliding

When polycrystalline aggregates deform by diffusional or dislocational creep, mineral grains must move past each other if the changes in the shapes of individual mineral grains are to contribute to the deformation of the aggregate without creating voids. In those cases, the displacements of grains relative to their neighbors are subsidiary to the changes in the shapes of the grains. That is, the amount that a mineral grain moves relative to its neighboring grains is dictated by the degree to which the grain changes its shape, and it is unusual for grains to switch neighbors.

Under appropriate conditions, typically when grain sizes are small, polycrystalline aggregates deform by a different mode in which in most of the deformation of the aggregate occurs because individual grains switch their neighbors. Individual mineral grains change their shapes to enable grains to move past their neighbors, but the strains experienced by individual grains are small relative to the overall strain experienced by the aggregate. The net changes in the shapes of grains may be sufficiently small that the aggregate exhibits no, or at best a weak, measurable dimensional or shape-preferred orientation. In some instances, grains in the aggregate may exhibit a weak CPO, but this is not always observed.

Materials scientists call this mechanism for deforming polycrystalline aggregates **grain boundary sliding**. Some distortion of grain shapes is necessary in order for grains to switch neighbors (Figure 8.40), and the relatively minor changes to the shapes of individual mineral grains may be accomplished either by dislocation

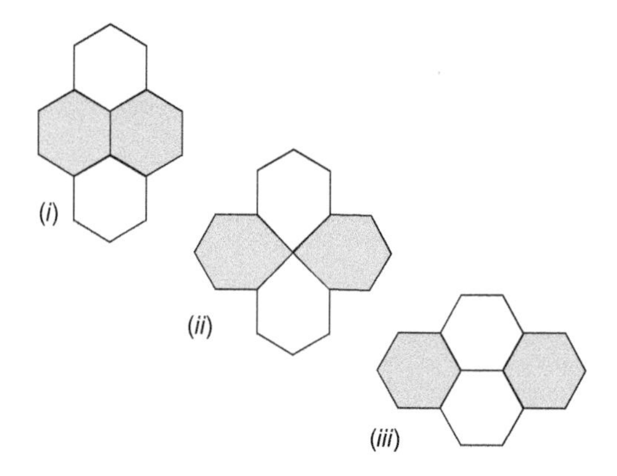

Figure 8.40 Sketch of a neighbor switching event of the sort expected in grain boundary sliding deformation. Neighbor-switching requires mineral grains in *i.* to distort slightly to a configuration similar to that shown in *ii.*, and then deform again to the configuration shown in *iii.*

movement within grains or by diffusional mass transfer. Thus, materials scientists distinguish between *diffusion-accommodated grain boundary sliding* and *dislocation-accommodated grain boundary sliding.*

Laboratory studies, supported by theoretical analyses, indicate the grain boundary sliding conforms to a constitutive equation of the form:

$$\dot{\varepsilon} = C\left[\frac{(\sigma')^n}{d^p}\right]. \quad (8.12)$$

In Eq. (8.12), the strain rate $\dot{\varepsilon}$ is directly proportional to the deviatoric stress components σ' raised to a power n and inversely proportional to the grain diameter d raised to a power p. The stress exponent n has a value between 1 and 2, and the grain size sensitivity exponent p typically falls between 1 and 3. C is a factor that encompasses material parameters and the activation enthalpy for creep.

In metallurgy, deformation experiments in which samples exhibited extreme ductility with minor strain hardening, said to exhibit *superplasticity*, typically generated a microstructure consisting of fine, generally equant grains that have little or no lattice-preferred orientation. Drawing an analogy between the metallurgical and geological microstructures, some geologists equate grain boundary sliding with superplastic flow. In our view, it is better to avoid using the term "superplastic flow," which carries specific connotations about deformation conditions and behavior, in favor of using the term "grain boundary sliding."

8A.6 Deformation Mechanism Maps

In most geological settings, two or more of the distinctive deformation mechanisms defined earlier contribute to the total deformation. The different deformation mechanisms can and, in many cases, do operate concurrently and competitively, although one deformation mechanism might predominate under different temperature, pressure, and strain rate/deviatoric stress conditions, depending upon the mineralogical composition of the rock.

Two concurrent and competitive processes A and B are *independent* of each other if they occur with no special sequence and each occurs at its own distinct rate. Both diffusional mass transfer and dislocation creep are concurrent and competitive deformation mechanisms, and geologists have identified different settings in which one or the other was the predominant mechanism. This situation contrasts with the operation of *dependent* processes, which must occur in a specific order. If process A must occur before process B can occur, A and B are dependent. For example, the climb and glide of dislocations in steady-state dislocation creep are dependent processes. A sufficient amount of climb must occur in order to allow dislocation glide to proceed at a particular differential stress level, and because climb occurs at a slower rate, it is the rate-limiting step in the process. We underscore this dependent relationship by including climb and glide as linked steps in the single deformation mechanism dislocation creep.

Under conditions where two or more competing deformation mechanisms contribute to the deformation of a rock, the mechanism that makes the largest contribution to the overall strain rate will predominate. Using experimental data to calibrate the constitutive equations or flow laws for the different deformation mechanism, one can predict when a specific deformation mechanism predominates under a particular set of conditions.

Four critical variables are functionally related by constitutive equations: strain rate $\dot{\varepsilon}$, deviatoric stress σ', temperature T (often indirectly through the material's diffusion coefficient or activation enthalpy), and mean grain diameter d:

$$\dot{\varepsilon}_{\text{Total}} = F_{\text{Total}}\left(\sigma', T, d\right). \quad (8.13)$$

If two independent mechanisms (A and B) contribute to the deformation, the total deformation rate is the sum of the contributions of the two mechanisms:

$$\dot{\varepsilon}_{\text{Total}} = \dot{\varepsilon}_A + \dot{\varepsilon}_B \tag{8.14}$$

or

$$\dot{\varepsilon}_{\text{Total}} = F_A\left(\sigma', T, d\right) + F_B\left(\sigma', T, d\right). \tag{8.15}$$

For any combination of deviatoric stress, temperature, and mean grain diameter, one can illustrate Eq. (8.15) graphically: the strain rate due to each deformation mechanism is a line segment and the total strain rate is the length of the line segments laid end to end (Figure 8.41). As the parameters deviatoric stress σ', temperature T, and mean grain diameter d change, the relative contribution of the two mechanisms to the overall strain rate will change. Figure 8.41 shows an example of deformation accommodated by a combination of dislocation creep and diffusive mass transfer in which temperature and grain diameter are held constant but the components of the deviatoric stress are varied. As the magnitudes of the deviatoric stress components increase, the total strain rate increases *and* the relative contribution of dislocation creep increases. This example provides insight into why dislocation creep is predominant in laboratory deformation experiments, which must be conducted at sufficiently high strain rates to deform samples in days or weeks.

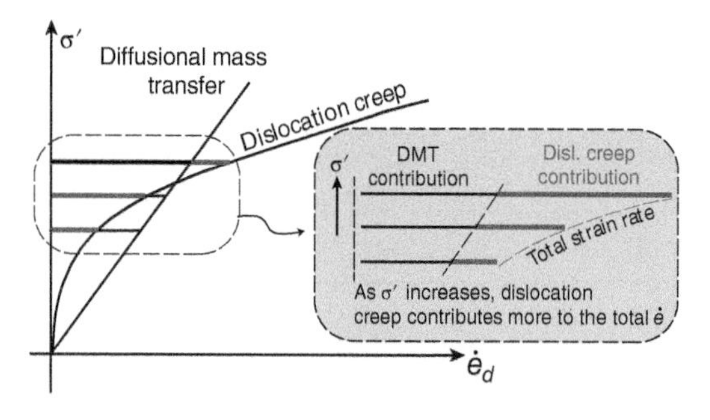

Figure 8.41 Relative magnitudes of strain rates accommodated by diffusional mass transfer and dislocation creep at three differential stress magnitudes. At lower differential stresses, the total strain rate is lower but diffusional mass transfer contributes more to the total strain rate. At the highest differential stresses, the total strain rate is higher and dislocation creep contributes more to the total strain rate.

Materials scientists and geologists use plots, called **deformation mechanism maps**, that display graphically the magnitudes of strain rates, computed using this sort of analysis, for a range of values of the critical parameters. In deformation mechanism maps, the strain rate $\dot{\varepsilon}$ is taken to be the dependent parameter that is a function of three independent variables, deviatoric stress σ', temperature T, and mean grain diameter d. One would need a four-dimensional plot to show the relationships. By holding one of the independent variables constant, it is possible to plot strain rate $\dot{\varepsilon}$ as a function of the other two (Figure 8.42a). Using contours of constant strain rate, one can portray the same information on a two-dimensional diagram. When two or more competing deformation mechanisms contribute to the total strain rate, regions where one strain rate surface extends above the others indicate values of independent variables where that deformation mechanism predominates and intersections of strain rate surfaces indicate where two mechanisms make equal contributions (Figure 8.42b). The resulting two-dimensional plot is essentially a contour map of the total strain rate surface overlain by lines that indicate where two deformation mechanisms contribute equally to the overall strain rate and separating fields where an individual deformation mechanism predominates.

Two types of deformation mechanism maps are commonly used in geology. The first type shows strain rate $\dot{\varepsilon}$ as a function of deviatoric stress σ' and temperature T, with mean grain diameter d held constant (Figure 8.42c). A stress-temperature

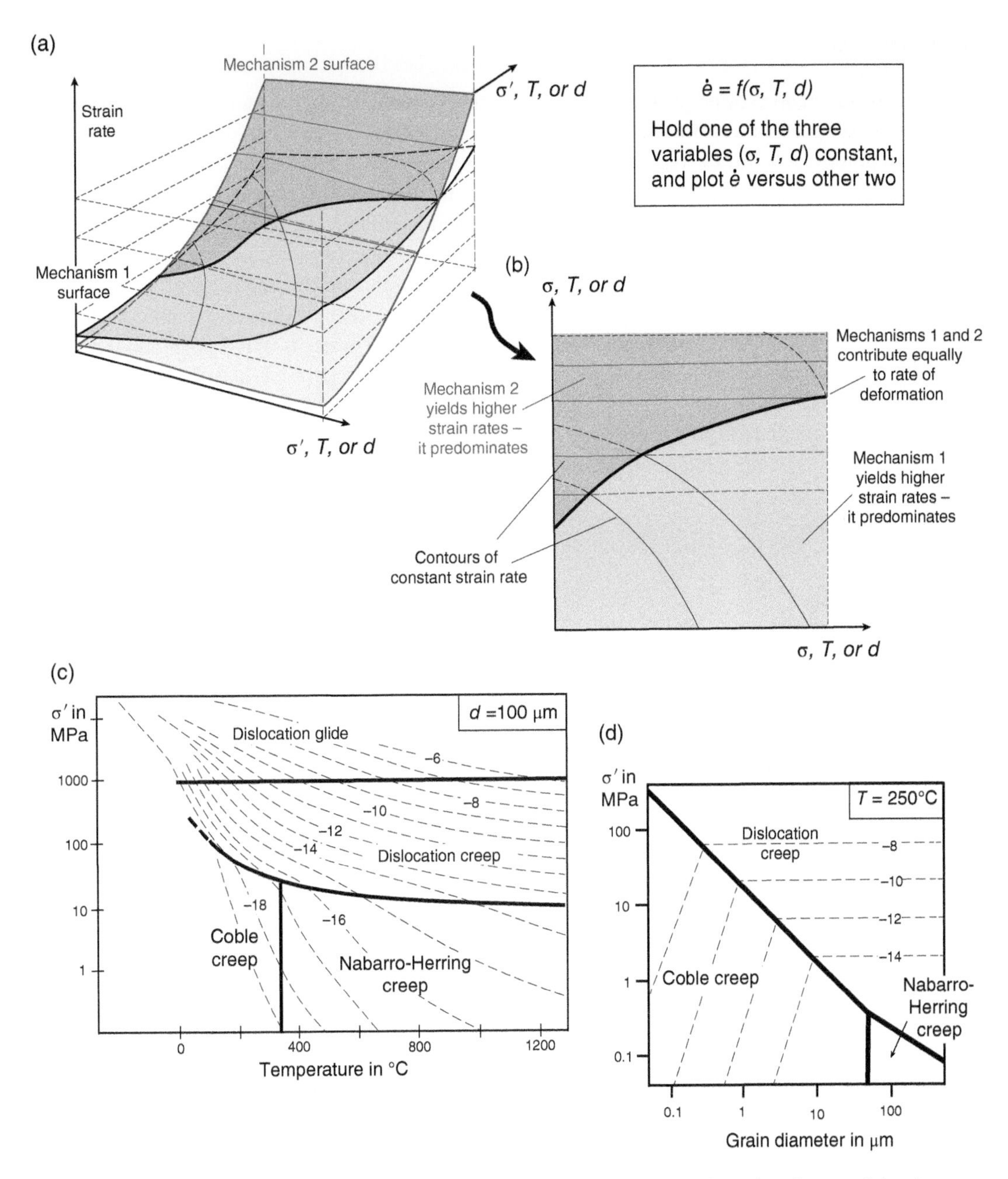

Figure 8.42 Deformation mechanism maps. (a) Diagram showing strain rate plotted again two of the three variables differential stress (σ'), temperature (T), or grain diameter (d). Note that the strain rate surface for deformation mechanism 1 is higher closer to the origin, i.e. mechanism 1 generates greater strain rates there, whereas the strain rate surface for deformation mechanism 2 is higher farther from the origin. The two mechanisms contribute equally to the strain rate where the two surfaces intersect. (b) A two-dimensional rendering of (a), showing a field where the mechanism 1 surface is higher, a field where the mechanism 2 surface is higher, and the line where the two surfaces intersect. (c) A differential stress-temperature deformation mechanism map. *Source:* Modified from Ashby (1972). The example here, for quartz with a grain size of 100 μm, from Rutter (1976). (d) A differential stress-grain diameter deformation mechanism map. *Source:* Modified from Mohamed and Langdon (1974). The example here, for quartz at 250 °C, from Mitra (1984).

deformation mechanism plot might be used, for example, to predict which deformation mechanism will predominate within a particular rock type (with a particular grain size) at different depths within the lithosphere, where temperatures and deviatoric stress levels change predictably. The second type shows rate $\dot{\varepsilon}$ as a function of deviatoric stress σ' and mean grain diameter d, with temperature T held constant (Figure 8.42d). A stress-grain size deformation mechanism plot might be used, for example, to predict which deformation mechanism will predominate in a particular rock type in a deformation setting as its microstructure changes.

8A.7 Summary

Several physical-chemical processes are responsible for the deformation of geological materials: elastic distortion of crystal lattices; fracture; frictional sliding; diffusion; permanent distortion of lattices by dislocation movement, twinning, or kinking, and grain displacement. Two or more of these deformation processes may, depending upon conditions, operate simultaneously to accommodate the deformation of rock masses. Specific combinations of deformation processes link together to constitute a deformation mechanism, which can be characterized by a distinctive yield criterion or constitutive equation. Generally, the different deformation mechanisms generate different characteristic collections of structures or microstructures.

- Cataclasis and frictional sliding is the deformation mechanism in which elastic distortion of crystal lattices, fracture, and frictional sliding combine to enable rock masses to deform. Distinctive microstructures are fractured grains, comminution, and the formation of breccias or cataclasites. Cataclasis and frictional sliding is associated with macroscopic plastic rheology.
- Diffusional mass transfer is the deformation mechanism in which rock deformation is accommodated by solid-state diffusion and grain displacement. Distinctive microstructures include truncated grains, pressure shadows or overgrowths, stylolites, and mineral-filled veins. Diffusional mass transfer is associated with a macroscopic linearly viscous rheology.
- Dislocation creep is the deformation mechanism in which permanent distortion of crystals occurs by dislocation movement, lattice diffusion, elastic deformation of lattices, and grain displacement. Microstructures indicative of dislocation creep include undulose extinction, deformation lamellae, deformation bands, development of subgrains, deformation-induced recrystallization, and crystallographic or lattice-preferred orientations. Dislocation creep is associated with a power-law viscous rheology.
- Grain displacement combined with diffusion or permanent deformation of lattices by dislocation movement, twinning, or kinking constitute the deformation mechanism of grain boundary sliding. Grain boundary sliding is difficult to recognize, but it is invoked when a rock shows evidence of macroscopic ductility and locally aligned grain boundaries but lacks grain-shape fabrics (dimensional preferred orientations) and CPO. Grain boundary sliding is associated with a slightly nonlinear viscous rheology.

Two or more deformation mechanisms may contribute competitively but independently to the overall deformation. When that occurs, deformation mechanism maps provide a way to use the collection of microstructures in a deformed rock to determine what deformation mechanisms were responsible for the deformation, and by assessing the relative contributions of the deformation mechanisms to assess the conditions under which the deformation occurred.

8B Deformation Mechanisms – Comprehensive Treatment

Like the Comprehensive Treatment sections of previous chapters, this section addresses the topics covered in the Conceptual Foundations section in greater detail and with greater reliance on physical and mathematical reasoning. We consider the four deformation mechanism classes – cataclastic deformation mechanisms, diffusional deformation mechanisms, dislocational deformation mechanisms, and grain boundary sliding deformation mechanisms – in the same order as in the Conceptual Foundations section. Our more detailed discussion of each of the four deformation mechanism classes necessarily refers to and draws upon an understanding of other deformation mechanism classes because: (1) deformation mechanisms from two or more classes regularly operate simultaneously and competitively; or (2) deformation mechanisms from two or more classes often operate sequentially with structural or fabric changes wrought by the first facilitating the operation of the second. Deformation mechanism maps, introduced at the end of the *Conceptual Foundations* section, appear throughout the *Comprehensive Treatment* section because they are an effective way to evaluate the relative contributions of competing deformation mechanisms *and* to show transitions in sequential mechanisms and their underlying causes in natural deformations.

8B.1 Cataclastic Deformation Mechanisms

Chapter 2 outlined the general characteristics of fractures and faults, but a more detailed examination is necessary here in order to support a more comprehensive analysis of the processes, settings, and constitutive relationships of cataclastic deformation mechanisms. Fractures and faults share many characteristics. Most fundamentally, they are surfaces or tabular zones of discontinuity in rock masses; primary features do not extend or continue across them without disruption. Yet there are important differences between these two types of structures, arising from differences in the types of relative displacements accommodated by the structures and in their typical habit as "surfaces or tabular zones" of discontinuity. In some instances, fractures and faults have demonstrable genetic relationships.

8B.1.1 Joints, Fractures, and Mesoscopic Faults

8B.1.1.1 Characteristics and Origin of Joints

Joints are discrete surfaces that cut across primary and/or secondary fabrics in rocks, but across which there is no visible offset of those earlier features despite the fact that there is no longer cohesion across the surface (Figure 2.9). Individual joints typically are planar or broadly curved, and the rock immediately adjacent to the joints rarely is more deformed than rock far from the fracture. Joints regularly occur in flat-lying, unmetamorphosed sedimentary rocks and in unaltered and unmetamorphosed igneous rocks, where they may be the only features indicating that the rocks are deformed. Deformed but unmetamorphosed sedimentary and igneous rocks regularly exhibit *systematic* arrays of joints (Figure 8.43a–c). Joints also commonly occur in metamorphic rocks, where they are inferred to develop either near the end of the deformation event that coincided with metamorphism, after the rocks have cooled significantly, or during separate deformation events. Exposed joint surfaces may display a variety of surface markings consistent with formation by episodic propagation of an individual fracture (Figure 8.43d–f). These observations are consistent with the inference that such joints initiate and propagate as nearly ideal, mode I brittle fractures to accommodate small extensional strains. In this model, rock masses were distorted elastically, the

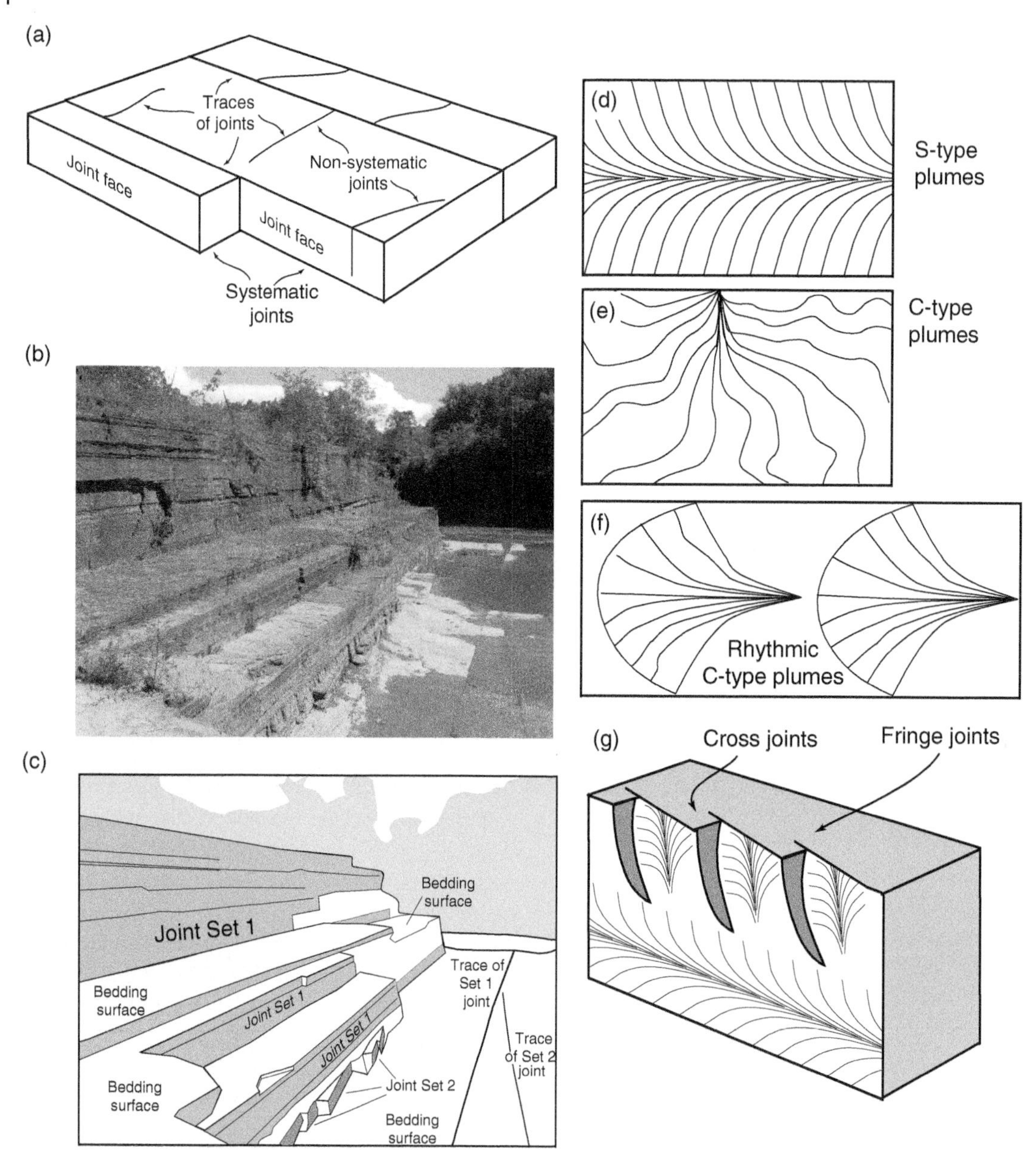

Figure 8.43 Joints. (a) Block diagram showing schematic relationships between systematic and non-systematic joints. (b) and (c) Photograph and interpretive sketch of joints in Devonian strata in the Appalachian Plateau province of central New York state, USA. (d), (e), (f), and (g) Surface markings on joints. (d) S-type plumes indicating joint propagation along a layer from right to left. (e) C-type plume indicating joint propagating from top to bottom of a layer. (f) Rhythmic C-type plumes indicating episodic joint propagation from right to left across the layer. (g) Fringing joints at the margin of a jointed layer.

release of stored elastic strain energy drove the propagation of the fracture, and the opening or aperture of the fracture is minimal (Pollard and Aydin 1988). Linear elastic fracture mechanics provides a solid explanation for the nucleation and growth of joints formed in this manner, which likely characterizes the majority of joints.

8B.1.1.2 Characteristics and Origin of Extension Fractures and Veins

Extension fractures, where the fracture walls have moved apart and across which primary features are displaced perpendicular to fracture surfaces, record elongation of rock masses greater than that responsible for joint formation. Extension fractures are commonly in-filled by secondary minerals, creating *extension veins* (Figures 8.44a, 3.8). The *apertures*, or maximum openings, of extension fractures and extension veins typically occur near the midpoint of the length of the fracture or vein (Vermilye and Scholz 1995) (Figure 8.44b), and the magnitude of opening typically is small relative to the fracture or vein length, on the order 0.01–0.001×√(fracture length) (Schultz et al. 2008). With their finite displacement of the fracture/vein walls, extension fractures and extension veins are sometimes described as "tabular" features. Despite the tabular geometry of extension veins, the similarity of shapes for veins of different sizes, the similarity in the patterns of displacements in the vicinity of extension fractures and veins, the lack of distortion of rock in the vicinity of both extension fractures and veins suggest that extension fractures and many veins also develop as ideal mode I fractures (Figure 8.44c).

In naturally deformed rocks, fractures or veins with smaller lengths or apertures are more numerous than those with large lengths or apertures. In fact, quantitative studies of fracture size distributions (e.g. Hooker et al. 2014 and references therein) demonstrate that populations of fractures and/or veins often adhere to power-law distributions, where the number of fractures and/or veins with a particular dimension (i.e. fracture length or fracture aperture) is inversely proportional to that dimension raised to a power. Closer examination of measured fracture/vein size distributions often

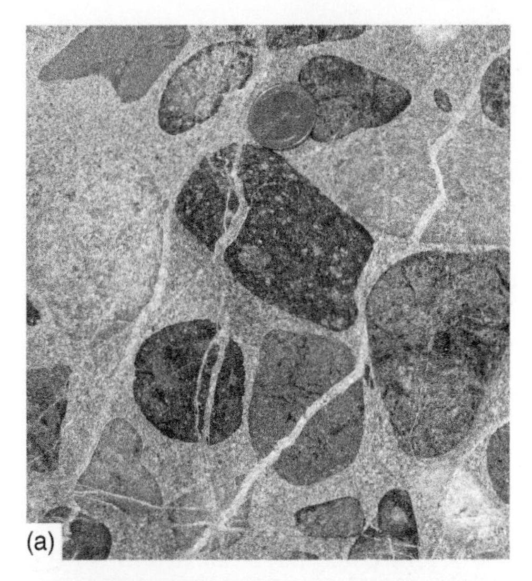

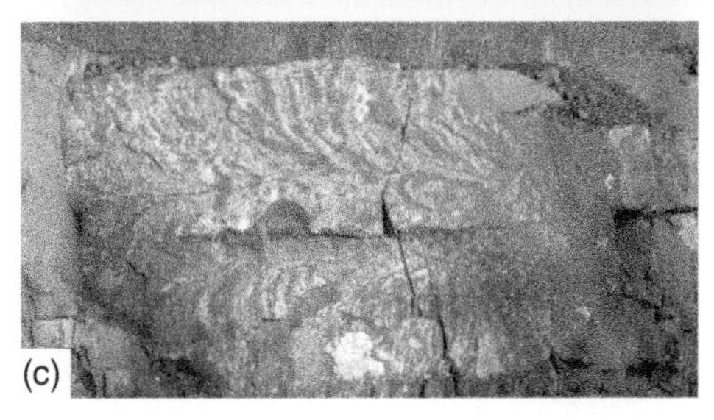

Figure 8.44 (a) Conglomerate cut by several mineral-filled extension veins. Movement of wall rock normal to vein is apparent where veins cut across individual cobbles. Note coin for scale. (b) A single mineral-filled vein with maximum aperture near the midpoint of fracture length. (c) Mineral coating highlights S-type plumes on this fracture surface. The fractured layer is ~50 cm thick.

reveals, however, departures from the power-law distributions, with fewer than expected fractures or veins with the smallest *and* the largest dimensions. Under-sampling of the smallest elements is commonly attributed to a sampling bias known as *truncation*, in which "undercounting" the smallest fractures or veins at any scale of observation occurs because the smallest features are difficult to detect. Undercounting of the large features may result from a different sampling bias known as *censoring*; the finite sample size collected along a finite scan line or within a finite map area or sample volume cannot provide an adequate picture of the largest individual features.

Departures from the power-law scaling for features of large size may not be an artifact of sampling biases, but instead indicate a change in the fracture/vein growth process. For example, the lengths of bedding-normal veins recording extension of an individual sedimentary stratum may be limited by the thickness of the stratum. So, the apertures of the largest veins increase without comparable lengthening of the vein. Alternatively, the overburden may limit the growth of the apertures but not the propagation of subhorizontal fractures, producing populations of veins where the longest veins have narrower-than-expected apertures. Real departures from power-law distribution could also indicate that factors not incorporated into linear elastic fracture mechanics, such as inelastic behavior of the rock mass or processes other than elastic strain and fracture surface formation (such as the need to transport constituents to be precipitated to the veins) impact the development of the fractures and veins. For example, in systems of linked extension veins and stylolites (Figure 8.45), the geometry – orientation and spacing – of veins and solution seams within the overall displacement field influence the propagation and eventual growth of the veins (Seyum and Pollard 2016).

The morphology of vein fillings can be an important source of information on the relative rates of

Figure 8.45 Arrays of stylolites and mineral-filled veins. The mutually crosscutting characters of veins and stylolites indicate they formed contemporaneously.

fracture opening versus the rate at which the constituents to be precipitated reach the fracture. Veins filled by continuous mineral fibers suggest that mineral growth was contemporaneous with and kept pace with the fracture opening. As mentioned in Section 3.4, important observations for interpreting how vein filling occurred include: (1) whether visible discontinuities occur along either of the two vein walls where the vein minerals contact the host rock; (2) whether visible discontinuities near the center of the vein filling, known as *median sutures*, exist; (3) whether fibers in veins widen in a particular direction across the vein or have constant width; and (4) whether fibers have symmetrical or asymmetric geometries. These observations allow five vein-filling processes to be distinguished: **Syntaxial**, **Antitaxial**, **Composite**, **Ataxial**, **Monotaxial**, and **Composite**. The features that identify these types of fibrous vein fillings, and their interpretations, are given in Table 8.2.

Syntaxial vein fillings consist of mineral(s) that match those found in the fracture wall, and vein fibers extend directly from grains along the fracture walls (Figure 8.46a). In an ideal, symmetrical syntaxial vein, the oldest segments of fibers are

Table 8.2 Fibrous vein-filling textures and their interpretation.

Texture	Diagnostic features	Common	Interpretation
Syntaxial	Two bands of fibers No continuity of fibers across suture if present Fiber orientations often symmetric across median line Fiber widths may increase toward center of vein	Vein filling similar to wall rock Vein filling in optical continuity with wall rock	Fibers grew toward center of vein from both walls
Antitaxial	One band of fibers Fibers continuous across median line Fiber shapes and orientations are symmetric across median line Fiber widths often increase toward walls	Vein filling different to wall rock	Fibers grew toward both walls of vein from center
Composite	Both syn- and antitaxial portions in the same vein		Alternating syn- and antitaxial growth
Monotaxial	One band of fibers No median line or suture Fibers widen from one wall to the opposite wall		Fibers grew from one wall to the opposite wall
Ataxial	Fibers have even widths across vein	Interlocking teeth on fiber edges	Random refracturing

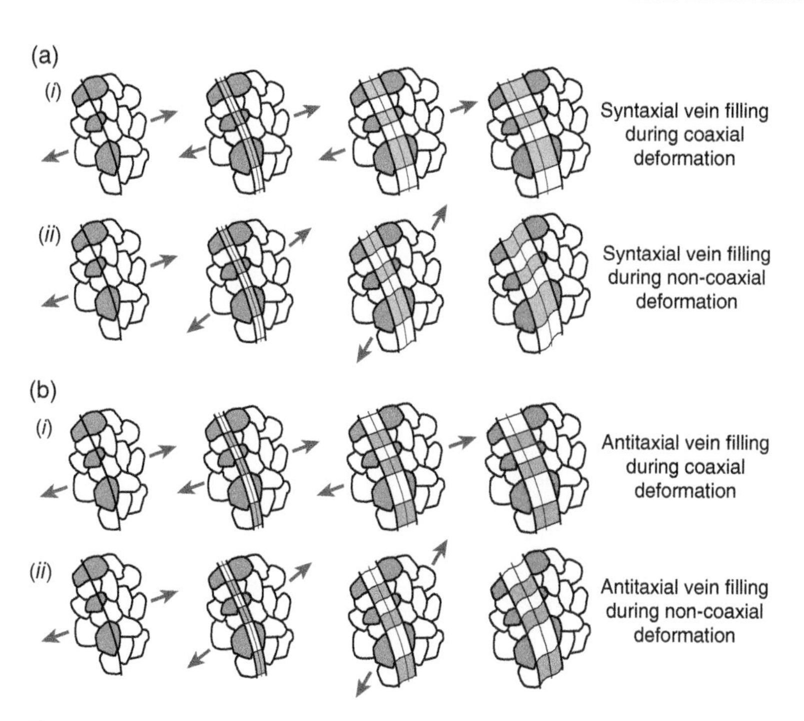

Figure 8.46 Sequential development (from left to right) of syntaxial and antitaxial vein fillings in coaxial (*i.*) and non-coaxial (*ii.*) deformations. Arrows on any image indicate elongation direction responsible for overgrowth added to create the next image. (a) In syntaxial veins, fibers grow by adding new material along the medial line of fracture. Older segments of fibers are along margins of vein, and younger segments are in center of vein. (b) In antitaxial veins, fibers grow by adding new material along vein walls. Older segments of fibers are in the center of the vein, and younger segments are along the vein walls.

attached directly to the lattices of grains along the fracture wall. Younger segments of fibers are added along the *median surface* of the vein, far from the original fracture wall. *Tracking fibers* (Section 3.4) grow parallel to the incremental elongation direction. Thus, syntaxial fibers precipitated during coaxial deformation are straight, whereas those precipitated during non-coaxial deformation are curved. Antitaxial fibers consist of minerals not found in the fracture walls (Figure 8.46b). In ideal, symmetrical antitaxial veins, fibers initially precipitate along what will become the median surface of the vein and grow by adding to both ends of the initial precipitate. Antitaxial fibers will be straight or curved depending on whether the deformation path is coaxial or non-coaxial. In either case, the younger segments of antitaxial fibers are found along the margins of the vein filling, and the earliest fiber segments are in the vicinity of the median surface of the vein (Figure 8.47a, b). Some veins have both antitaxial and syntaxial portions in the same vein: these are called **composite** veins, and may record a history of at least two different vein-filling events, likely with different fluid chemistries.

In syntaxial and antitaxial vein fillings, the locus of fiber growth does not change over time. In some settings, however, the surface along which new material is added jumps from one position to another over time. The resulting veins

Figure 8.47 Antitaxial vein filling. (a) Straight, parallel, antitaxial calcite fibers in adjacent veins formed during coaxial deformation. (b) Antitaxial calcite fibers indicate the vein walls separated normal to fracture walls, then moved obliquely to fracture walls (with right-lateral offset), and finally moved normal to fracture walls. Calcite fibers near center of vein, which precipitated during earlier deformation increments, are twinned. (c) Vein with segmented quartz fibers precipitated by the crack-seal process. Arrows show bounding surfaces of fracture segments.

typically have fibers with approximately constant width across the vein, but the fibers may show a complex interdigitation so that individual fibers appear to be built out of "tablets" or "blocks." Such fibers are called ataxial or non-systematic.

Syntaxial, antitaxial and ataxial vein fillings all have a degree of symmetry about the center of the vein. However, some vein fillings consist of continuous fibers that show a systematic widening from one side of the vein to the other, implying that growth occurred in one direction. This is monotaxial growth, with the fibers growing in the direction of widening.

In some instances, vein fillings have a fibrous habit, but close examination of fibers indicates that they have numerous planar arrays of inclusions, fluid inclusions, or voids that divide fibers into segments (Figure 8.47c). The arrays of inclusions were described as inclusion bands (parallel to the vein margin) or inclusion trails (at some angle to the vein margin) in Chapter 3. These veins are inferred to form during alternating episodes of crack growth (increasing both crack length and crack aperture) and mineral in-filling, creating *crack-seal veins* (Ramsay 1980). Alzayer et al. (2015) analyzed in detail the dimensions and types of filling in a large number of crack-seal veins. In the populations they studied, the fracture aperture was a smaller fraction of fracture length in newly nucleated fractures, suggesting fractures lengthened with little increase in aperture during the early stages of their growth. The apertures of fractures with evidence for many crack-seal events are a greater fraction of the fracture length, suggesting that later in the growth history, fracture aperture increases with little or no increase in fracture length. These data support an inference that deformation-related changes in rock properties or that the later increments of fracture growth do not conform fully to linear elastic fracture mechanics models of fracture growth in a homogeneous medium.

In some veins, minerals exhibit coarsening-inward structure or microstructure and/or possess well-defined crystal terminations (euhedral vein fillings: Table 3.2). This morphology indicates the minerals grew into open voids, i.e. that filling occurred after the fracture opened. The time interval between opening and filling may be long. Based on the morphology and mineral composition of vein fillings and on major element, minor element, and isotopic analyses, Evans et al. (2014) showed that vein fillings precipitated from fluids with markedly different compositions and at different temperatures and pressures over time periods of millions of years. In this and in many other instances, it is clear that chemical constituents dissolved in fluid phases were carried through the rock as the solution moved from one region to another. Further, compositional and isotopic variations in vein fillings indicate that the temperatures at which vein fillings precipitated changed over time in ways that indicate that fluid flow along fractures *advected* both chemical constituents and heat. This sort of advection of fluids with dissolved chemical constituents can lead to wholesale loss of volume in some rock masses or wholesale increase in volume in other rock masses.

Finally, many extension veins are filled by blocky mineral growth that provides little insight into its origin. In some instances, careful examination reveals remnants of fibrous or coarsening inward fabrics, indicating that static recrystallization or the action of a later fluid phase has promoted recrystallization that obliterated a record of earlier deformation increments.

8B.1.1.3 Relationship between Extension Fractures and Stylolites

A *stylolite* is a generally planar discontinuity whose walls have moved toward each other; the interpenetration of the opposing walls of the discontinuity is accommodated by the removal of mass by diffusion (Figures 2.9, 8.48a). Macroscopic stylolites share some characteristics with joints, extension fractures, or extension veins. Like joints, fractures, and veins, stylolites are surfaces

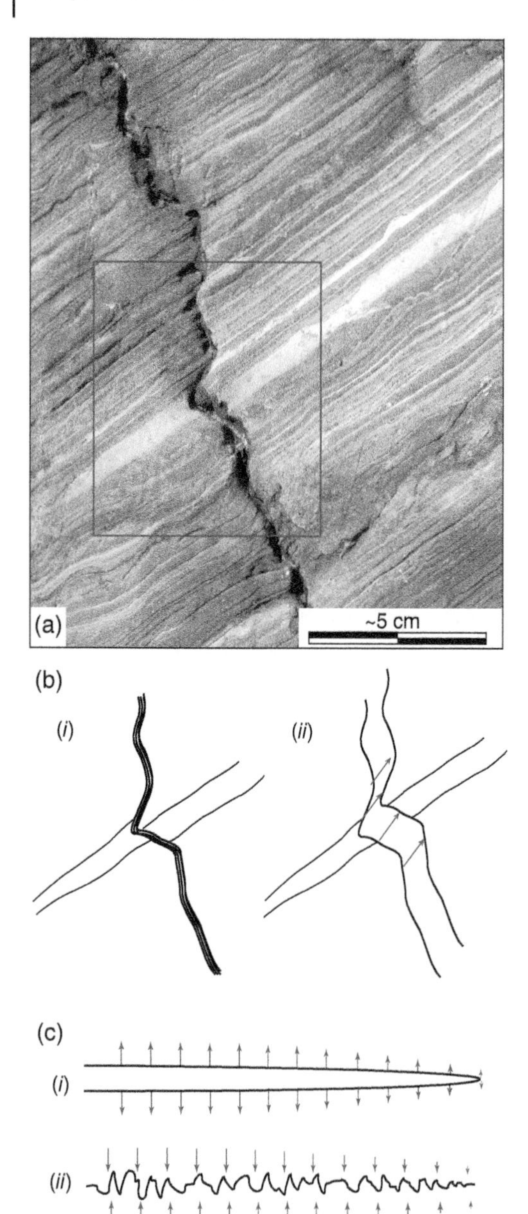

Figure 8.48 (a) Photograph of a bedding-normal stylolite in fine-grained limestone. (b) At the left, the sketch of the stylolite in (a) showing with the apparent offset of a layer. At the right, the image showing the inferred pre-deformation state. (c) Sketches showing displacement patterns for opening-mode crack (*i*) and stylolite (*ii*).

or tabular zones of discontinuity in rock masses – primary features do not extend or continue across stylolites without disruption. In addition, as is true of joints and extension fractures, rock immediately adjacent to stylolites typically is not more deformed than rock far from the feature. Finally, stylolites commonly occur associated with extension fractures, extension veins, and faults (cf. Figure 8.45). For these reasons, some structural geologists classify stylolites as a type of fracture – a *stylolitic fracture*.

The view of stylolites as a type of fracture derives from a comparison of the displacement fields in the vicinity of fracture and stylolite terminations: if a fracture is a Mode 1 opening crack, then a stylolite is a Mode 1 closing crack – an *anticrack* (Figure 8.48b, c). The similarities between the displacement fields of opening and closing cracks do not necessarily carry over to the distributions in the stresses in the vicinity of the two types of features. Opening mode fractures typically are a sort of stress discontinuity. There need not be a comparable stress discontinuity in the vicinity of a stylolite (cf. Koehn et al. 2007; Aharonov and Karcz 2019). For this reason, we conclude that stylolites are sufficiently different from fractures that it is more sensible to address their development in Section 8B.2 on diffusional mass transfer than in this section.

8B.1.1.4 Oblique Extension Fractures, Shear Fractures, and Mesoscopic Faults

Discontinuity surfaces in deformed rock masses sometimes are the locus of finite displacement both perpendicular *and* parallel to the surface. *Oblique extension fractures* result when rock masses simultaneously move away from and past each other (Figure 2.9) (*slickolites*, which we also defer to discuss until Section 8B.2 on diffusional mass transfer, result when rock masses simultaneously move toward and past each other). For many mixed-mode fractures, the component of movement parallel to the surface is on same order as the component of movement perpendicular to the surface.

When wall rocks are displaced *only* parallel to an individual discontinuity surface, as mode II or mode III fractures, the structure is a *shear fracture*. As shown in Figure 7.39, linear elastic fracture mechanics prescribes the orientations and magnitudes of elastic strains and stresses in the vicinity of planar fractures subjected to mode II/III (shearing) or mixed-mode (oblique extension) conditions as well as those subjected to mode I (opening) conditions, and these calculated strain magnitudes conform closely to strains observed near the fractures on which offsets are small. As slip accrues on a shear fracture and the magnitude of rock distortion increases, it is reasonable to ask if there are limits to the applicability of the linear elastic fracture mechanics approach.

Many geologists use the terms "shear fracture" and "slip surface" (cf. Section 2.1.3) interchangeably. In this book, we use "shear fracture" to refer to relatively small, discrete surfaces on which the magnitude of shearing offset is comparable to the offsets on extension or stylolitic fractures, i.e. on the order of mms to cms depending on the length of the feature (slip magnitudes $\leq 0.01 \times \sqrt{[\text{fracture length}]}$). Shear fractures defined this way often are sufficiently small to be contained within an individual thin section or hand sample, and they are nearly always confined to a single outcrop. Offsets of this magnitude generate permanent strains in the rock surrounding the fracture that is sufficiently small to be approximated readily by the elastic strains in fracture mechanics models. We use "slip surface" to refer to larger discrete surfaces, such as those visible in outcrop, on which slip magnitudes are $\geq 0.01 \times \sqrt{[\text{fracture length}]}$ (Figures 8.49a, b). The larger the displacement on a slip surface, the more likely it is that permanent deformation of the rock adjacent to slip surfaces exceeds that modeled using linear elastic fracture mechanics. As shearing displacement increases, so does the likelihood that a fault is no longer a single surface (cf. Figure 8.49c, d), but a fault zone (Figures 2.10, 8.50, and 8.51).

8B.1.2 Fault Zones

With the variety of rheological properties exhibited by different rock types and the range of physical conditions under which faults form, it is unlikely that any two fault zones will display the same collection and organization of structures and fabrics. Still, many fault zones have common characteristics. Shear displacement typically is concentrated in a narrow *fault core*, which is nested within a thicker *damage zone* (Figure 2.10; Caine et al. 1996). Commonly, one can identify a single discrete surface within the fault core, called a *principal slip surface*, across which displaced wall rocks are juxtaposed. Even in those instances, where the opposing wall rocks are in contact across the principal slip surface, the wall rocks typically are visibly deformed (Figures 2.10, 8.49c, and d). In some fault zones, layers, lenses, or pods of *fault rocks* – physically and/or chemically altered rocks derived from the wall rocks – decorate the principal slip surface or separate the rock masses juxtaposed across the fault zone (Figures 8.50, 8.51). The damage zone that surrounds the fault core is apparent as: (1) deflected bedding, foliation, or banding in the wall rock; (2) arrays of subsidiary faults, which typically are not parallel to the principal slip surface and which may or may not splay from it; (3) arrays of deformation bands parallel or oblique to the principal slip surface; (4) arrays of macroscopic fractures, mineral-filled veins, or stylolites; (5) gouge, breccia, or cataclasite cut by many microfractures; (6) locally developed cleavage or foliation; (7) layers or zones of fine-grained fault rocks not associated with the principal slip surface; and (8) chemical and mineral alteration of the wall rock (Engelder 1974; Brock and Engelder 1977; Harris and Milici 1977; House and Gray 1982;

Figure 8.49 (a) Small slip surface in Carboniferous siltstones, Cumberland Plateau, Tennessee, USA. Compass is ~7 cm across. Arrows indicate tip points of the trace of the slip surface. (b) Slip surface cutting and offsetting chalk layers, Flamborough Head, UK. (c) Linkage of two slip surfaces inclined from upper left to lower right is accommodated by a series of subsidiary slip surfaces and accommodated by other slip surfaces inclined from upper right to lower left. Flamborough Head, UK. (d) Prominent slip surface offsetting Carboniferous siltstones and sandstones Cumberland Plateau, Tennessee, USA. Arrows indicate hanging wall and footwall cutoffs of the base of a sandstone layer. Note subsidiary faults developed in the hanging wall.

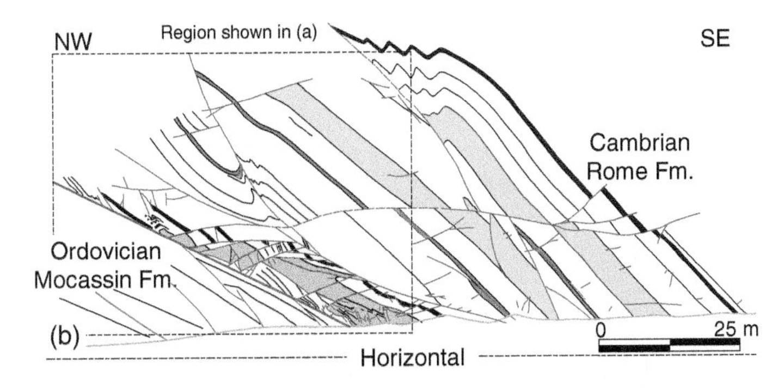

Figure 8.50 Structures apparent at an exposure of the Copper Creek thrust fault, Valley and Ridge Province, Tennessee, USA. (a) Photograph of roadside exposure (now overgrown) with Cambrian Rome Formation in the hangingwall above the principal slip surface and Ordovician Moccasin Formation in the footwall. (b) Section, drawn on plane perpendicular to thrust and parallel to thrust slip, of the portion of the damage zone seen in (a). The full section shown in Figure 9.11. (c) Close-up of the principal slip surface. Arrows indicate the layer of banded cataclasite that decorates the principal slip surface.

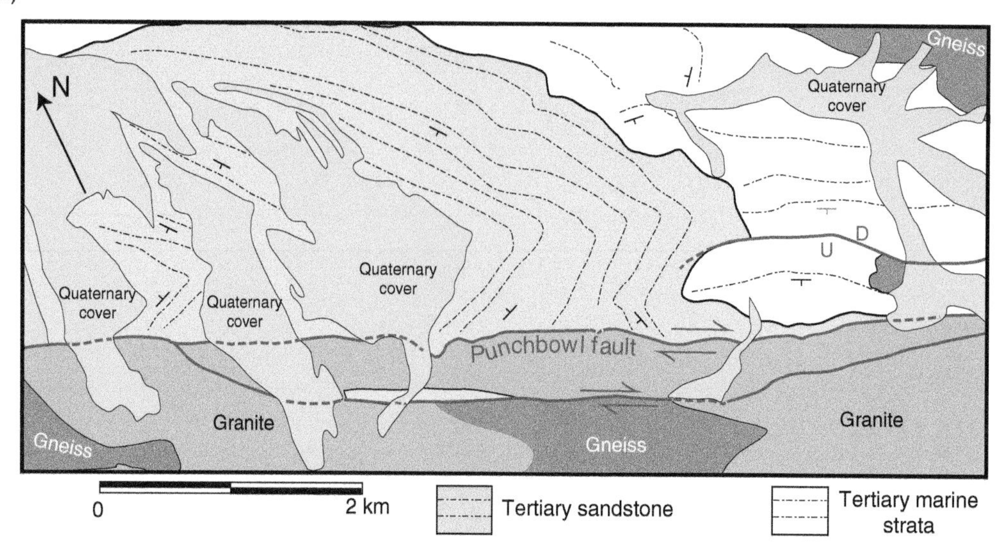

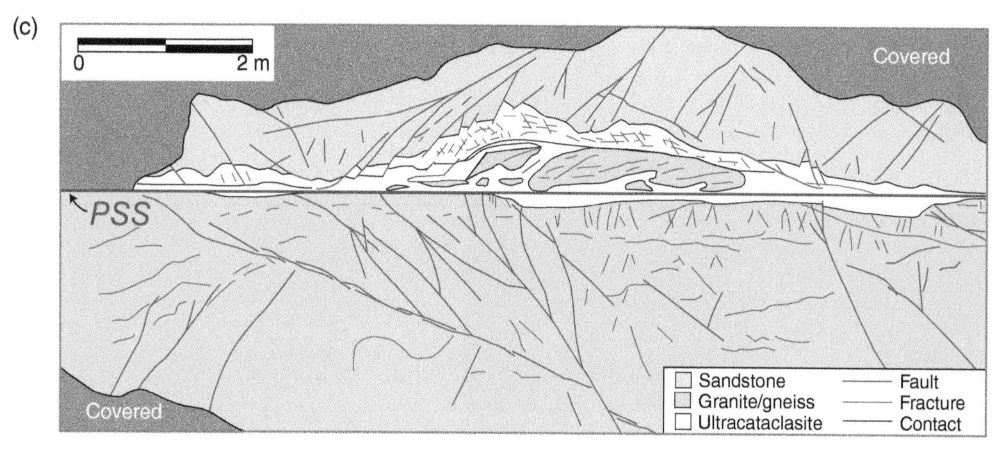

Figure 8.51 The Punchbowl fault, San Gabriel Mountain, California, USA. (a) Generalized geologic map of the Devil's Punchbowl area. *Source:* Dibblee (1987)/GeoScienceWorld. (b) Photograph looking NW, with brown-to-tan folded sandstones of the Miocene Punchbowl Formation juxtaposed against white granitic gneisses across the Punchbowl fault. (c) Outcrop map of the Punchbowl fault. *Source:* Modified from Chester et al. (2004). PSS = principal slip surface.

Mitra 1984, 1992, 1993; Chester and Logan 1986, 1987; Wojtal and Mitra 1986, Chester et al. 1993; Erickson 1994; Biegel and Sammis 2004; Chester and Chester 2004; and many others).

Any measure of deformation in the vicinity of the fault is made in comparison to the "background" deformation magnitudes observed far from the fault. In the examples illustrated in Figures 8.50 and 8.51, and in many other instances, damage zones have diffuse boundaries, where the densities of minor faults, fractures, or veins, the intensity of cleavage, or the extent of microfracturing gradually approach background levels with increasing distance from the fault core (Gilotti and Kumpulainen 1986; Wojtal and Mitra 1986; Chester et al. 1993, 2004; Mitra 1993; Erickson 1994; Biegel and Sammis 2004; Anders et al. 2014). The damage zones associated with some faults have abrupt boundaries. For example, the prominent faults defining the boundaries of the linked systems of faults known as duplex zones may delimit the extent of a complex fault zone (Boyer and Elliott 1982; Fermor and Price 1987) (Figure 8.52).

Large displacement faults juxtapose rock masses from widely separated locations that may have been subjected to distinctly different physical conditions and histories. Consequently, the damage zones rarely are symmetrical and typically have distinctly different thicknesses, measured perpendicular to the fault surface, on opposite sides of a fault.

8B.1.2.1 The Character of Principal Slip Surfaces

Our knowledge of the character of principal slip surfaces is far from complete. Studies of three-dimensional exposures of principal slip surfaces are limited to near-surface exposures, typically of active or recently active extension faults or recently exhumed strike-slip or normal faults (cf. Figure 8.6a, b). Over observation scales ranging from fractions of a mm to 10s of m, near-surface principal slip surfaces are "rough," i.e. depart from being planar. Early work characterizing fault surface roughness argued that the root mean square of the heights of asperities, which measures the average deviation from a planar surface, and the wavelengths of those asperities vary with observation scale length in a manner consistent with the inference that fault surfaces are *self-similar* (Power et al. 1988). The profile of a self-similar fault surface taken at one scale can be enlarged or shrunk without distortion to describe the shape of the fault surface at a different length scale. This early work also argued that asperities have greater relief and the wavelengths of asperities are shorter along lines perpendicular to the slip direction when compared to lines parallel to the slip direction (Power and Tullis 1989, 1991). More recent analyses of the shapes of fault surfaces confirm some conclusions of the early work but refute the inference that fault surfaces have self-similar shapes. Sagy et al. (2007), for example, used Lidar to characterize precisely the shapes of several well-exposed faults

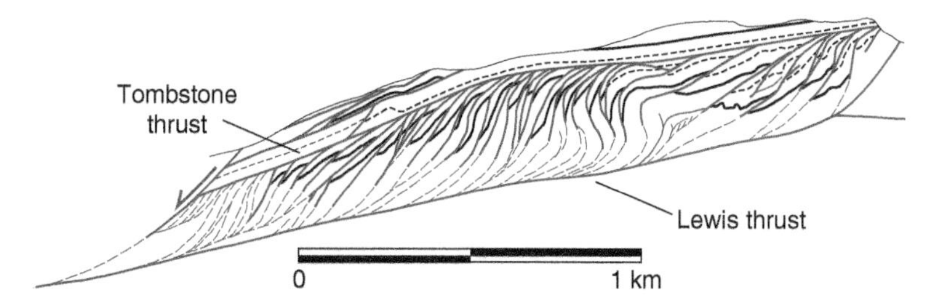

Figure 8.52 Cross section of the Haigh Brook window of the Lewis thrust, Canadian Rocky Mountains, Alberta, Canada. *Source:* Modified from Fermor and Price (1987).

with a range of offsets. Their measurements confirm that fault surfaces exhibit different geometric characteristics when measured parallel and perpendicular to the inferred slip direction. Variations in fault surface topography along lines parallel to the slip have longer wavelengths and slightly lower amplitudes than those observed along lines perpendicular to the slip. In addition, their data indicate that small-offset faults tend to be rough at all observation scales, whereas large-offset faults regularly have relatively long wavelength bumps or depressions but are smooth over short length scales. Sagy et al. (2007) concluded that fault shapes smooth as slip on the fault accrues.

Studies of two particularly large and well-exposed faults provide an additional insight into the geometric characteristics of fault surfaces. Quarrying operations in central Greece have revealed a spectacular 300 m-long and 30–50 m-tall exposure of the Arkitsa normal fault (Jackson and McKenzie 1999). This is one of the largest exposed fault surfaces to have been characterized in detail. The fault surface exhibits prominent down-dip corrugations with meter-scale relief about a plane, and analysis of the shape of the fault surface indicates that lines of no-surface curvature parallel the long dimension of the corrugations (Resor and Meer 2009). The fault surface also displays several types of surface striae or *slickensides*, including scratches, striations, grooves and ridges, and/or gouge trails inferred to form when a resistant asperity on one side of the fault (sometimes called a "tool") ploughed through less-resistant rock on the opposing side of the fault (Candela and Brodsky et al. 2016; Candela and Brodsky 2016; Petit 1987; Power et al. 1988; Power and Tullis 1989, 1991) (Figure 8.6d). These smaller surface markings are thought to originate during individual slip events, some of which are associated with earthquakes on this seismically active fault. The striae vary slightly but significantly in orientation relative to the corrugations, suggesting that there may have been a recent change in the direction of slip on this fault. The eastern sides of the corrugations on this surface are polished, whereas the western sides are not. This suggests that the fault slip now has a component perpendicular to the corrugations and that corrugations now play a role in resisting fault slip (Resor and Meer 2009).

Analyses of a second, iconic near-surface exposure of a fault, the 60 m by 15 m exposure of the Corona Heights fault in California, provide additional insight into the geometric character of fault surfaces and the physical processes by which fault slip occurs (Kirkpatrick and Brodsky 2014). This fault surface also exhibits corrugations with amplitudes on the order of 1 m. On this fault surface, local variations in the orientations of striae suggest that differential stress orientations varied or wall rocks experienced local inelastic distortion about irregularities in the fault surface. In addition, regions of cataclastically deformed wall rocks have traces on the fault surface parallel to the corrugations and thicknesses on the order of the amplitudes of fault roughness, suggesting a systematic relationship between relief on the fault and distortion of the wall rock. These observations, together with an analysis showing that large displacement faults have lower surface roughness than small displacement faults, supports the inference that faults and wall rocks undergo *wear*, i.e. become smoother as slip accrues (Brodsky et al. 2011, 2016).

Cataclasites observed on near-surface exposures of active faults (cf. Figure 8.7b) are similar to those observed on ancient, inactive, and exhumed faults (Figure 8.7c), leading many to infer that similar processes operate on near-surface faults and those at depth. In particular, many workers infer that striae and grooves (plowed by tools) and layers of cataclastically deformed wall rock are consistent with frictional slip on faults that is episodic and possibly seismic. Frictional slip occurs at a critical magnitude of shear stress, is largely independent of the rate of slip, and conforms to plastic behavior. There are, however, reasons to consider that for all the insight geologists have garnered from studying near-surface fault surfaces, these studies do not automatically constrain the geometry or behavior of principal slip surfaces at crustal depths where most major earthquakes initiate.

Three-dimensional exposures of principal slip surfaces in exhumed fault zones are rare, and the traces of principal fault surfaces rarely extend more than several 10s of meters before encountering a branch with a comparable principal slip surface. Individual principal slip surfaces often have relatively straight or, at most, broadly curved traces, even when the wall rocks are extensively fractured, faulted, or folded (Figures 8.50a, b, and 8.51c). Further, many small-displacement faults and some large-displacement faults exhumed from middle to lower crustal levels lack cataclastic deformation of the wall rocks and lack breccia or cataclasites. Instead, faults are coated by arrays of mineral fibers inferred to form by precipitation during *pressure-solution slip* (Arthaud and Mattauer 1969; Elliott 1976), where diffusional mass transfer facilitates removal of material on the sides of asperities that face toward the movement of the opposing wall and mineral growth occurs on sides that face away from the movement of the opposing side (Figure 8.53). This mechanism of fault slip may occur at stress levels dependent upon the rate of fault slip, i.e. conform more to viscous behavior.

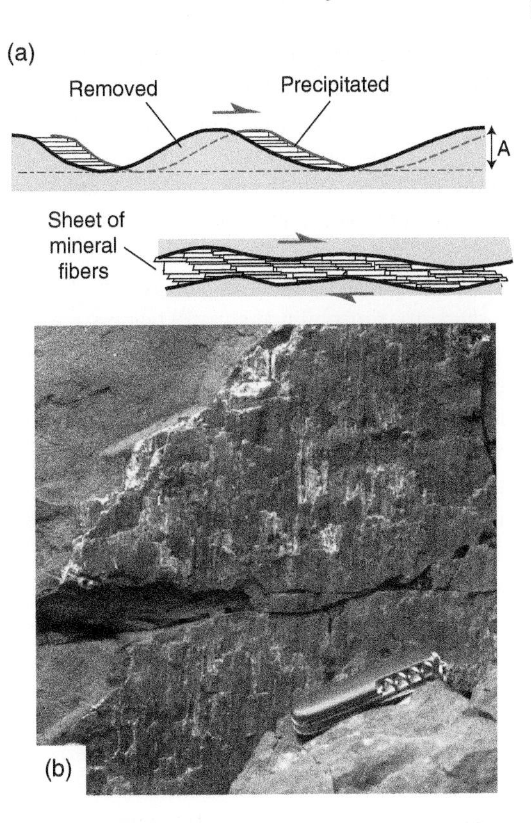

Figure 8.53 (a) Formation of sheets of mineral fibers by "pressure-solution slip." *Source:* Modified from Elliott (1976). *A* is the amplitude of irregularities on the fault surface. The black curve shows the fault surface, and the red line shows the position of the upper plate after fault movement. Note that the area of material that must be removed by diffusional mass transfer corresponds to the area of the mineral fibers precipitated. The net result of movement is a sheet of mineral fibers on a fault surface. (b) Quartz and chlorite mineral fibers coating a minor fault surface in red Devonian sandstone, Valley and Ridge Province, Maryland, USA.

8B.1.2.2 Fault Wear and the Generation of Cataclasites

Section 8A.2.1 and Figure 8.5a drew upon deformation experiments conducted at different confining pressures to outline the range processes by which the rock along and adjacent to a fault might deform during fault slip (Sammis et al. 1986). To recap briefly, at low confining pressures, experimental samples typically develop mode I fractures aligned parallel to the direction of shortening. Extrapolating to fault slip at low confining pressure, one might anticipate "slabbing," where wall rock is split by a family of fractures (Figure 8.5c *i*). Under moderate confining pressure, cracks inclined to the direction of the most compressive stress experience minute shear displacements. Elastic strain fields surrounding the tips of cracks that experienced such minute shear displacements provide the energy for those cracks to lengthen, typically along trajectories that curve so crack tips are opening mode fractures (Figure 8.5c *ii*). These fractures can link together to create macroscopic slip surfaces; such a process may lead to shearing off asperities in the wall rock during fault slip. Under high confining pressure, opening mode crack growth is suppressed, and clusters of small cracks link together to form slip surfaces (Figure 8.5c *iii*). With increasing confining pressure, the probability that rocks possess numerous small slip surfaces increases. When rocks are penetratively fractured, they deform as three-dimensional aggregates of elastic particles of different sizes that impinge upon each other. Bulk, permanent deformation of the aggregate accrues as these elastic particles rotate and/or move past their neighbors. The rotation and relative movement of particles typically require continued fracturing of particles, especially if the volume of the aggregate is constrained to be constant. Thus, finite bulk deformation occurs by *comminution*. This section considers in greater detail the deformation by cataclasis within or adjacent to fault zones.

Early experimental studies of comminution suggested that the particle-size distributions generated by cataclastic deformation were log normal and that the median diameter of the distributions was smaller for distributions generated at higher confining pressure (Sammis et al. 1986; see also Chester et al. 2005). Studies of cataclastic rocks adjacent to and within natural fault zones reveal, however, that particle-size distributions differ from log-normal particle-size distributions and often conform to power-law distributions (Figure 8.54). In power-law particle-size distributions, the number of particles of mass greater than M is given by

$$N(M) = KM^{-f}, \qquad (8.16)$$

where K and f are constants, or alternatively the number of particles greater than dimension S is given by

$$N(S) = KS^{-h}, \qquad (8.17)$$

where $h = 3(f-1)$ (c.f. Sammis et al. 1987; Blenkinsop 1991). Particle-size distributions that adhere to Eqs. (8.18) or (8.19) are sometimes called *fractal*, denoted by

$$N(S) \sim S^{-D}, \qquad (8.18)$$

where D is the fractal dimension. D is the slope of a graph of log $N(S)$ versus log S (Figure 8.54b), and $D = h$. Analyses of two-dimensional sections of natural fault gouges indicate that $D \sim 1.6$. If gouges are isotropic, the fractal dimension of the corresponding three-dimensional particle array would be greater by one, i.e. $D \sim 2.6$.

Sammis et al. (1987) outlined a mechanism capable of producing power-law particle size distributions. They noted that in real particle aggregates, particles with large diameters are likely to be surrounded by neighboring grains with smaller diameters. An individual particle with a large diameter, as shown in Figure 8.55, is likely to contact several neighboring particles with smaller diameters, with each of those grain-to-grain contacts supporting a fraction of the force transmitted through the large particle. With smaller forces transmitted across at each of these contact points, the large diameter fragment is less likely to fracture. In this sense, the smaller, neighboring particles "cushion" the larger fragment. Should the smaller particles contact neighbors of similar size, they are more likely to fracture, generating an array of even smaller particles as they or their neighbors fracture. Once one of the smaller diameter fragments itself is surrounded by even smaller neighboring particles, it too will be cushioned by those even smaller particles. The net result of the fracturing of similar-sized neighbors and the cushioning effects of smaller neighbors is that the aggregate evolves toward an arrangement in which: (1) particles tend to be surrounded by particles of smaller diameters; and (2) particles of all diameters have the same statistical probability of fracturing. Sammis et al. (1987) note that an

(a)

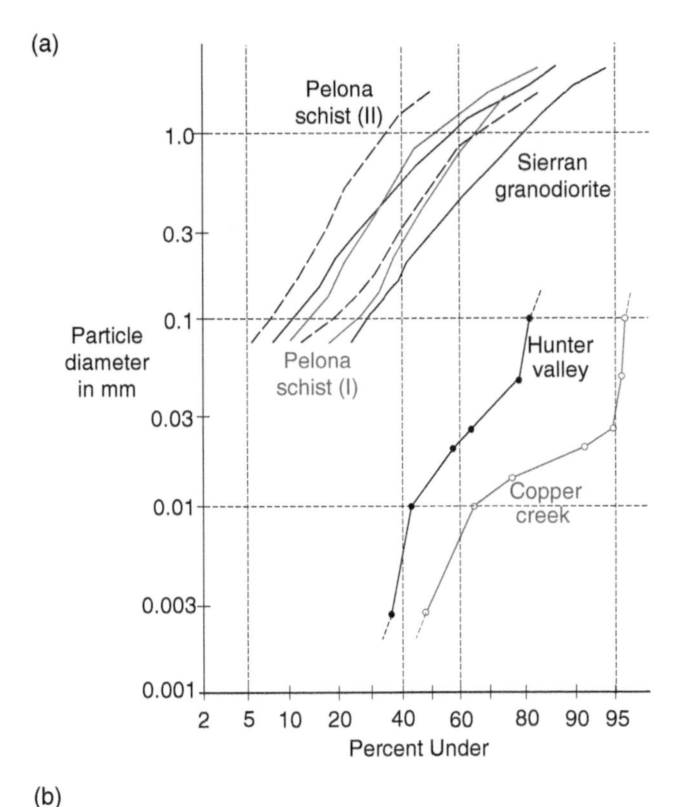

(b)

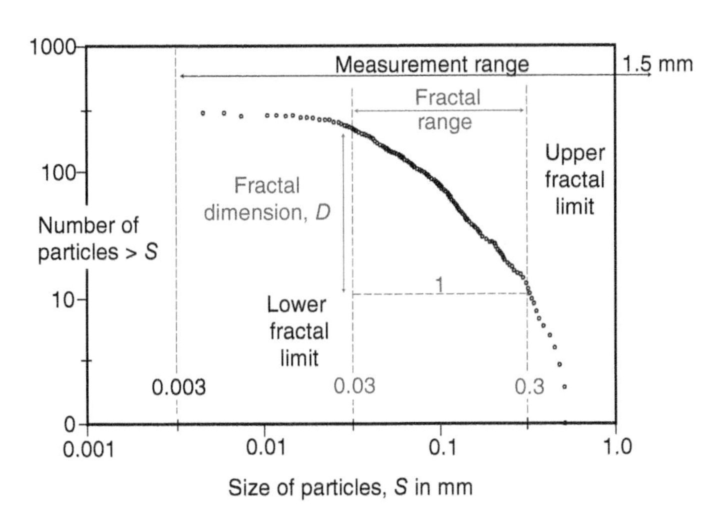

Figure 8.54 (a) Plots after Sammis et al. (1987) giving bounding lines for particle size distributions for experimentally generated cataclasites of Sierran granodiorite (solid black lines) and Pelona Schist shortened perpendicular (solid red lines) and parallel (dashed black lines) to foliation. Also shown are particle size distributions from banded cataclasites from the southern Appalachian Hunter Valley and Copper Creek thrust faults after Wojtal and Mitra (1986). (b) Particle size distribution of sample from Cajon Pass (southern California, USA) drill hole. *Source:* Modified from Blenkinsop (1991).

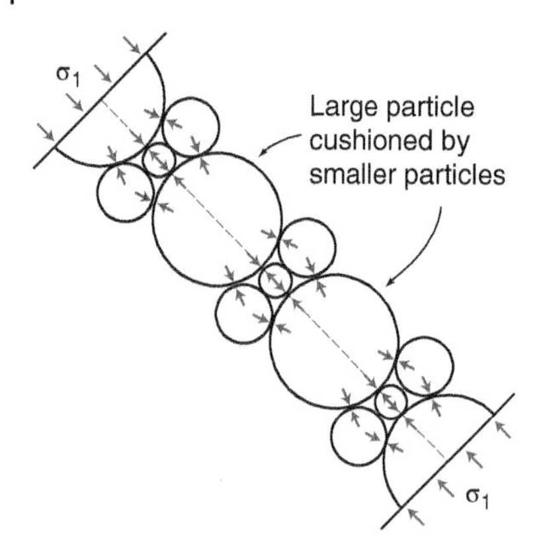

Figure 8.55 Schematic diagram showing "beams" composed of particles along force chains where larger particles are "cushioned" by smaller particles. Redrawn from Sammis et al. (1987).

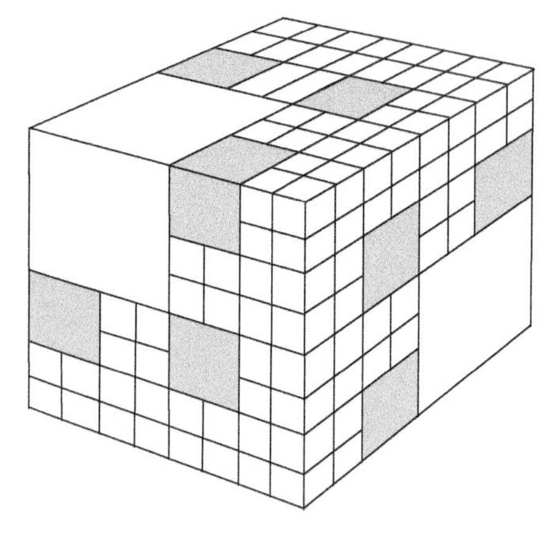

Figure 8.56 Representation of a fractal particle size distribution. *Source:* Sammis et al. (1987)/ Springer Nature.

aggregate of particles where no two particles of the same size are in contact is necessarily *self-similar* and will have a fractal dimension comparable to those measured for many natural fault gouges and cataclasites. Further, once this self-similar fabric develops, it is likely to persist.

The self-similar geometry of cataclastic rocks implies that their particle-size distributions can be directly related to the total strain due to cataclasis. If elastic processes make no finite contribution to shearing and fragments do not deform internally, strain results only from the relative displacement of particles across fractures. To evaluate the strain accommodated by a self-similar cataclasite, consider the idealized volume of deformed rock shown in Figure 8.56, i.e. a cube with edge length = $L(0)$ and volume $V(0) = [L(0)]^3$ (Sammis et al. 1987). Cataclasis has subdivided the cube by fractures so that the largest particle has an edge length $L(0)/2$, the edge length of successively smaller particles $L(n) = L(0)/2^n$, and each fracture event divides a cube of order n into 8 subcubes of order $n + 1$. From Eq. (8.18), the fractal dimension is

$$D = \frac{\log N(n)}{\log\left[1/L(n)\right]}, \tag{8.19}$$

and inspection of Figure 8.56 shows that this self-similar array of cubes and subcubes has 2 subcubes with edge length $L(0)/2$, and two cubes with edge length $L(0)/4$ in each of the six cubic regions with volume $[L(0)/2]^3$, and so on. Thus, $D = \log 6/\log 2 = 2.58$, similar to the value inferred in several studies (e.g. Sammis et al. 1987; Marone and Scholz 1989; Blenkinsop 1991; Hadizadeh et al. 2010).

However, Phillips and Williams (2021) argue that the data used to show power-law size distributions is better fitted by log-normal relationships. One aspect of obtaining the best particle size distribution is to understand how much, if any, of the measured distribution is affected by truncation and censoring effects similar to those described for fracture length and aperture distributions in Section 8B.1.1.2. Clearly, any distribution is likely to become more linear and thus well fitted to a power law as the particle size range is restricted. Phillips and Williams argue that even allowing for a restricted range of particle sizes to accommodate truncation and censoring, log-normal distributions are a better fit to a large collection of data (61 naturally and experimentally deformed rocks).

Log-normal grain size distributions are predicted by some models of cataclasis (Kolmogorov 1941; Epstein 1947), in which the probability of particle fracturing is random, or independent of particle size, in contrast to the Sammis model. Particle size evolution can be modelled by specifying a relationship between one increment of fragmentation and the next, such as the random relationship in the Kolmogorov and Epstein models. However, if a fragmentation episode is related to the previous one by a power-law function of grain size breakage, the probability density function of particle sizes will follow a "Generalized Gamma" (GG) distribution (Ord et al. 2022), one form of which gives the Weibull distribution:

$$F(s) = 1 - \exp\left[-\left(\frac{s}{s_0}\right)^{\beta}\right], \tag{8.20}$$

where $F(s)$ is the cumulative fraction greater than size s, s_0 is a characteristic size, and β is a shape parameter. The Weibull distribution has been widely used to describe fragment distributions.

For a given data set, several alternative mathematical functions may be close fits to the data, and distinctions between them may be too fine to be significant. In ascribing a power-law relation to particle size distributions, fine particle sizes are commonly assumed to be an artefact of truncation, but there are increasing hints that a power law does not fit the smaller particle size distributions. Careful and comprehensive measurements of fine particle sizes are, therefore, necessary. None of these idealized mathematical models takes into account the polymineralic nature of most cataclastic rocks, which is likely to mean that different minerals will have different size distributions. Other processes that could affect particle size distributions in cataclastic rocks are not accounted for in the models, such as mineral alteration, dissolution, and melting. An example of the effect of such processes on grain sizes is described in Section 8B.1.2.4.

8B.1.2.3 Granular Flow

Fault rocks often show evidence that they contained variable mixtures of grains and fluids and that these mixtures flowed under a range of conditions and at different fault slip speeds (Figure 8.57a, b). It is, therefore, appropriate to outline what is known about granular flows as a starting point for drawing inferences about fault zone behavior. Experimental and theoretical results show that granular flows are complex phenomena. Granular flows vary in: (1) grain/fluid ratios; (2) grain sizes, size distributions, and shapes; and (3) grain and fluid velocities, viscosities, and temperatures. Variations in these properties give rise to the complexities of granular flow rheology. Moreover, flow properties typically evolve during their movement (Iverson and Vallance 2001). The involvement of granular flow in the deformation of fault gouges is intriguing because granular flows can have a variety of rheologies that could relate to different types of fault behavior.

8B.1.2.3.1 Physics of Granular Flows Granular flows have two primary components: relatively rigid grains and a surrounding viscous medium. The physical behavior of a collection of grains within a fluid depends on the proportion of momentum transferred by grain inertia compared to that transferred by the viscous flow of the fluid. Grain concentration N, equal to the ratio of solid to fluid mass per unit volume, is, therefore, a key parameter affecting granular flow. Granular flows belong to one of three regimes (Bagnold 1954): (1) *grain inertial*; (2) *transitional*; and (3) *macroviscous* (Figure 8.57c). In the transitional and macroviscous regimes, grain concentrations are lower ($N < 1$), and the rheology of the mixture is determined by fluid viscosity and independent of grain density. In grain inertial flows, where $N > 1$, momentum transfer is dominated by grains (Iverson and Vallance 2001), and collisions between grains

Figure 8.57 Fault rocks. (a) Carbonate cataclasite near Cospedal village, Castilla y Leon, showing the granular nature of cataclasite. The pen diameter is approximately 3 mm. (b) The Portizuelo fault (Spain) showing a marked grain size variation across the 20 cm wide yellow fault gouge. Note quartzite fragments, which could be due to segregation effects during granular flow. (c) Granular flow regimes, and their relationship to the Bagnold number. Arrows indicate the effects of decreasing grain fraction ϕ and grain size d, and increasing fluid viscosity η, on transitions between the regimes. (d) Schematic evolution of gouges and cataclasites, shown as changes in average grain size and fractal dimension of grain size distribution. Microstructural mechanisms are shown on the left of the curve, and granular flow regime to the right. ϕ/ϕ^* is the ratio between grain fraction and the maximum possible packing of grains.

increase the viscosity of the grain-fluid mixture over the viscosity of the fluid alone. In the context of fault rocks, a different regime can be identified in which rheology is largely independent of fluids and dominated by grain friction and comminution: this is the *frictional regime* (Monzawa and Otsuki 2003) (Figure 8.57c).

Numerical and laboratory experiments reveal that even in the frictional regime, granular flows can exhibit diverse behaviors. Simulated gouges with angular fragments have coefficients of sliding friction (μ) similar to those measured for many natural samples ($\mu \approx 0.6$, i.e. equivalent to "Byerlee's Law"), whereas gouges composed of spherical particles have coefficients of friction as low as 0.1 (Abe and Mair 2009). The experiments suggest that comminution occurs in fault gouges by two mechanisms: particles may *split*, or merely *abrade* (creating small fragments and rounding grains) (Mair and Abe 2011). Abrasion dominates at low normal stresses and in the later stages of gouge evolution, and could be one reason fractal dimensions of particle sizes have been measured (e.g. Blenkinsop 1991) with high values since large numbers of small grains are created. Grain splitting in the experiments leads to fractal particle size distributions, but abrasion eventually creates a bimodal size distribution, for which a fractal or power-law fit to the particle size distribution is evidently unsatisfactory. The roughness of fault walls is another important factor in gouge evolution, which promotes grain splitting.

Transitions between the fluidized regimes are specified by the Bagnold number B. Slip speed v in a fault zone of width w can be expressed as a function of the four variables: grain fraction ϕ, fluid viscosity η and density ρ, and grain size d:

$$v = \frac{wB\eta f(\phi)}{\rho d^2}, \tag{8.21}$$

where $f(\phi) = \sqrt{(\phi^{-1/3} - 1)}$. Transitions between the regimes are sensitive to these variables as shown by the arrows in Figure 8.57c. The transition from frictional to fluidized flow is predicted to occur as grain fractions and size decrease, and as fluid viscosity increases. These factors should also favor transitions from the grain inertial to transitional and macroviscous regimes. Despite some limitations to applying the Bagnold analysis to fault rocks (for example, a single grain size is assumed, and roughness of fault zone walls or clasts is overlooked), the approach suggests that the rheology of a fault that is governed by granular flow could vary temporally and spatially.

There are two fundamentally different types of granular flow. A dimensionless parameter, the Reynolds number R distinguishes between *turbulent* (high R) and *laminar* (low R) flows:

$$R = \frac{\rho_m U_f h}{\mu_{fm}}, \tag{8.22}$$

where ρ_m is the density of the grain-fluid mixture, U_f is the mean flow velocity, h is the flow thickness, and μ_{fm} is the flow viscosity.

Granular flows can lead to segregation of particles, mainly because of differences in particle sizes and densities (Tunuguntla et al. 2014), though the shape and frictional properties may be relevant (Gillemot et al. 2017). Segregation of larger particles to the bases or tops of flows is known as *normal* or *inverse* grading, respectively; the latter is also familiar as the Brazil nut effect (in which Brazil nuts, the largest nuts, end up near the surface of a package of mixed nuts). Normal grading is a characteristic product of turbulent flows; most laminar flows lead to inverse grading. Even for granular flows in air studied in industrial settings, segregation mechanisms are not yet understood in all conditions (McCarthy 2009). One segregation mechanism commonly considered to apply to geological flows is grain dispersive pressure – the Bagnold effect (Bagnold 1954) – that is thought to fractionate larger grains to the top of gravity flows. Alternative segregation mechanisms proposed for grain inertia-dominated flows include fluidization, convection, trajectory segregation, and gradients in granular temperature (thermal diffusion) (Ehrichs et al. 1995; Schröter et al. 2006). Sorting can also be due to shear-induced segregation or *kinetic sieving*, where smaller particles concentrate at the bottom of flows by falling through gaps between larger particles (Savage and Lun 1988; Gray and Thornton 2005; Gray 2013). This mechanism is particularly relevant to particles with ratios of large to small particles of less than two (Savage and Lun 1988). A complimentary mechanism that promotes upward movement of large particles called *squeeze expulsion* also occurs in these flows (Jing et al. 2017). Segregation due to these mechanisms increases with flow distance and density contrast between grains and fluid, and decreases with flow thickness (Thornton et al. 2006).

8B.1.2.3.2 Granular Flows in Fault Rocks: Evidence and Implications Explicit lines of evidence for recognizing fluidized granular flow in fault zones have been proposed, such as observing fault breccia grains that appear to be completely surrounded by matrix and thus are apparently unsupported (Fondriest et al. 2012). The proportion of grains that can be recognized as being derived from a parent grain, at any given grain size, could be diagnostic for fluidization (Monzawa and Otsuki 2003; Otsuki 2003). Very low values of this proportion imply fluidization. Monzawa and Otsuki were able to demonstrate that some fault gouges have this property. The viscosity of the granular flow is a function of the volume fraction of the grains ϕ, normalized by the maximum volume fraction possible. As the particle size distribution evolves, and this ratio decreases from 1, the viscosity of the granular flow is also dramatically reduced. Fluidized fault rocks are "nearly frictionless" (Monzawa and Otsuki 2003).

Several of the segregation mechanisms discussed earlier do not depend on gravity and might be expected to occur in granular flows along any fault zone. Figure 8.57b shows an example of size segregation in a fault gouge in a vertical strike-slip fault. Such gradational layering is quite persuasive evidence for granular flow in fault zones.

Some features of granular fault rocks indicate a direct link to slow slip phenomena. The obvious role of fluids in fault rocks, evident from the abundance of veins and alteration, corresponds with the inferred role of fluids in slow slip events. Vein textures that record sub-millimeter slip may be consistent with the scale of movements in episodic tremor associated with slow slip events (Fagereng et al. 2011).

8B.1.2.3.3 Synthesis Considering the evidence from natural fault gouges, experiments, and theory together suggest a possible sequence of events for gouges and cataclasites (Figure 8.57d). Intact rock is deformed at first by extension microfractures (see Sections 7A.2.2.3, 7B.2.2.3). At this stage, there may be a very little matrix or fluid, and the rock is well within the frictional regime. This stage evolves into grain splitting and further into grain abrasion, which creates smaller and rounder grains and leads to macroviscous granular flow, with potentially very low viscosities/coefficients of friction.

8B.1.2.4 Growth of Fault Zones

Faults do not appear instantaneously with finite lengths and displacements. Like all structural elements, they nucleate at small sizes and grow over time (Douglas 1958; Elliott 1976). In laboratory deformation experiments, some cracks experience shear displacements, and these shear fractures lengthen by opening tensile fractures at their terminations. With continued shortening or loading of experimental samples, numerous shear fractures link together to form macroscopic slip surfaces. Size limitations make it difficult to study the growth of large slip surfaces in the laboratory. So, geologists turn to the field and theoretical studies for insight on macroscopic fault growth. As the magnitude of slip on a surface grows, the rocks surrounding the slip surface typically develop finite strains, thus departing from conditions governed by linear elastic fracture mechanics models. Modifications of linear elastic fracture mechanics yield models for fault growth that incorporate regions of permanent, inelastic deformation near the fault terminations (Cowie and Scholz 1992; Cowie and Shipton 1998).

Examinations of the characteristics of populations of faults also provide an insight into fault growth. Measurements of the displacements on separate faults in a population and their dimensions have been used to develop phenomenological models for the growth of individual faults (Watterson 1986; Walsh and Watterson 1988) and for assessing the impact of the linkage of two or more slip surfaces to create large faults (Segall and Pollard 1983; Dawers and Anders 1995; Peacock and Sanderson 1995). Studies of the character of deformation within fault zones and their thickness have played a prominent role in evaluating fault growth. So, a brief survey of potential sources of deformation with fault damage zones is warranted.

1) Structures in the damage zone can originate as the fault surface propagates through a rock mass (Elliott 1976; Cowie and Scholz 1992).
2) The roughness of a principal slip surface affects the resistance to slip on the fault, leading to

wear, i.e. the generation of breccia or gouge in the fault core, as wall rocks deform during slip over the rough surface (Power et al. 1988; Power and Tullis 1989, 1991).

3) At a larger scale, most fault zones are not straight and many exhibit distinct bends; movement of wall rock past curves or bends in fault zones requires a zone of wall rock to deform in the same way that the movement of a glacier over an irregular bed requires a basal layer of ice with finite thickness to distort. Structure in the wider damage zone may therefore be a product of the geometry of the fault (e.g. Childs et al. 2008).
4) Drawing another analogy to glacier flow, over time ice above a glacier's basal layer accrues deformation – straight boreholes in glaciers become curved – by integrating strain-rate gradients associated with the macroscopic viscous flow of the glacier (Figure 7.22b). Comparable integration of strain-rate gradients associated with the viscous flow of rock masses adjacent to a fault may contribute to the development of structures in the damage zone associated with that fault (Wojtal 1992).

It is likely that combinations of fault propagation, fault roughness, macroscopic geometry, and/or macroscopic viscous flow contribute to the formation of the damage zone of an individual fault. Cross-cutting relationships provide indications of the relative ages of structures at individual locations within the damage zone, but determining precisely the time of origin of structures in the damage zone is rarely possible. If fault displacement carries rock originally from one lithospheric level to another, structures developed under one set of physical conditions overprint those developed under different conditions. In those cases, one can infer that the overprinting structures are related to fault slip rather than fault surface propagation (Wojtal and Mitra 1988).

Hull (1988) used a plot of the logarithm of fault displacement (d) versus the logarithm of fault zone thickness (T) (Figure 8.58a) as the starting point of a discussion of the factors affecting the relationship between fault displacement and fault zone thickness. Hull compiled fault zone thickness and

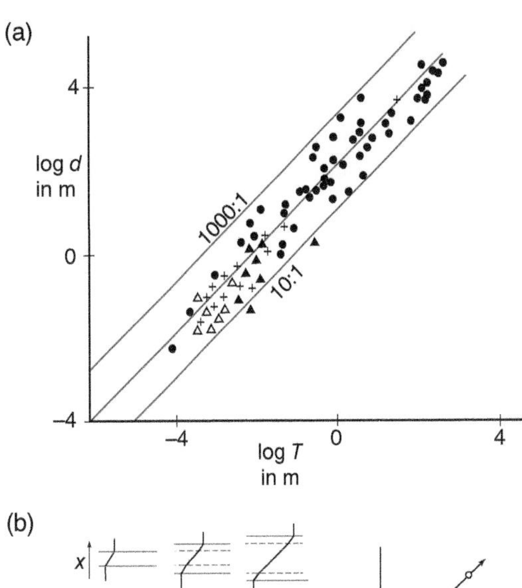

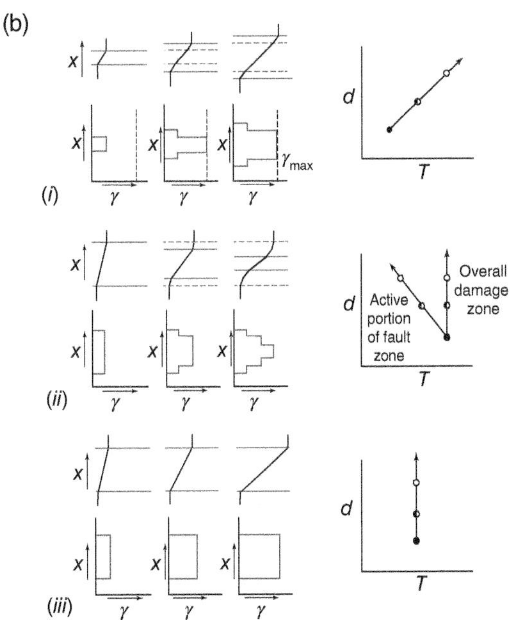

Figure 8.58 (a) A plot of fault displacement (d) versus fault zone thickness (T). (b) Diagrams modeling fault growth for fault zones that (*i*) widen with increasing displacement, (*ii*) have an active zone that narrows with increasing displacement, and (*iii*) maintain constant width with increasing displacement. *Source:* Both (a) and (b) Hull (1988)/Elsevier.

displacement data from a few comprehensive studies of populations of faults in which researchers working in a particular tectonic setting measured the displacements and fault zone thicknesses of faults with range sizes. He argued that if fault size is a proxy for fault age, the observed relationship between fault zone thickness and fault displacement suggests that damage zones develop by deforming as-yet-undeformed "shoulders" of the fault zone as displacement increases (Figure 8.58b *i*). Hull concluded that the observed fault zone thickness-fault displacement scaling is evidence that a single deformation process – cataclasis and frictional sliding – prevailed as faults grew. A corollary to this conclusion is the inference that the bulk rheology of rock masses would have been similar at different times during the growth of the fault.

Hull's analysis provoked published discussions by Blenkinsop (1989) and Evans (1990), and examining the criticisms raised in those discussions facilitates a closer look at the implications of the observed fault zone thickness-fault displacement data. First, Blenkinsop presented data showing two orders of magnitude variation in the ratio of gouge thickness T to displacement d (from ~ 0.002 to 0.04) within a short distance along a small, strike-slip fault; he noted that this variation questions the notion that there is a characteristic scaling between d and T during fault growth. Second, as noted earlier, only by compiling data from fault populations formed in different rock types and under different physical conditions are there sufficient data to support statistical analyses. Both Blenkinsop and Evans questioned the validity of combining data on faults formed under different physical conditions and in different rock types. Because Hull's analysis depended upon data from faults with a range of magnitudes of fault zone thicknesses and fault zone displacements, Hull needed to present the data on log-log plots. Blenkinsop and Evans both noted that log-log plots tend to minimize departures from trends. Evans replotted subsets of Hull's data selected for similarities in rock type or similarities in physical conditions using linear (not logarithmic) axes; the individual plots do not support so clearly conclusions drawn from the composite plot.

Figures 8.58b *ii* and *iii* illustrate two alternative models for fault zone development that Hull presented. In the alternative model for fault zone development shown in Figure 8.58b *ii*, shearing is concentrated in a zone that narrows over time. Schmid (1982), Mitra (1984), and Wojtal and Mitra (1986, 1988) and others noted that large-displacement faults regularly are decorated by thin layers of distinctive fine-grained, often banded, fault rocks. The particle size distributions in these fault rocks derived from unmetamorphosed to weakly metamorphosed sedimentary rocks do not conform to Eq. (8.18); they lack large fragments expected of a fractal particle size distribution and often have a preponderance of mineral grains with diameters on the order of 1–10μ (Figure 8.54a). Structures and microstructures in these rocks indicate that: (1) dislocation creep and, where appropriate, deformation twinning operated concurrently with and contributed to particle fracture; and (2) that diffusional mass transfer contributed to deformation once the particle size was sufficiently reduced. Due to the role of fracture in the formation of these rocks, they are typically called "cataclasites." Even at the temperatures found in the middle and upper crust, however, deformation mechanism maps suggest aggregates of extremely fine-grained particles are capable of deforming by diffusional mass transfer at high strain rates under moderate differential stresses or moderate strain rates at low differential stresses (Mitra 1984; Wojtal and Mitra 1986, 1988). Such a proposed change in the dominant deformation mechanism would be associated with a pronounced change in the rheology of these rocks. If differential stress levels in the fault zone remained roughly constant, layers of fine-grained fault rocks would accrue more deformation per unit time than the surrounding coarser-grained rock. In order for strain rates in the fine-grained layers to remain roughly constant, differential stresses in the fault rocks would need to fall, and as differential stress levels fell, coarser-grained rock units would accrue less deformation. Either result would lead to significant localization of shearing deformation within the fault rock layer. More recent work has demonstrated that deformation in these fine-grained fault rock layers is more complex than the two-step sequence outlined here; we discuss the role of transitions in deformation mechanisms in accommodating fault-related shearing and in determining the scale of the deformation of wall rock adjacent to major faults in Chapter 9.

A different sequence of processes can lead to a comparable transition in the strength of fault rocks derived from granitic basement protoliths at middle to lower crust levels. Grain size reduction, whether due to cataclasis, cleavage of minerals, plasticity-induced fracture, or development of subgrains and new grains, increases the ratio of surface area to volume for minerals and facilitates retrograde metamorphism in fluid-rich fault zones, often producing fine-grained, quartz-phyllosilicate fault rocks (Wojtal and Mitra 1988; Newman and Mitra 1993; Wibberley 2005; Bhattacharyya and Mitra 2011). A transition to diffusional mass transfer as the dominant deformation mechanism, along with the geometric effect of phyllosilicate minerals forming with their weak basal planes in orientations favorable for fault-parallel shear, helps localize deformation within the fault rock layers.

Wibberley (2005) compiled fault-zone thickness to fault displacement ratios for faults with fault rocks exhibiting evidence for a transition in dominant deformation mechanism. On a plot of log T (fault zone thickness) versus log d (fault displacement), these faults plot to the left of Hull's log T versus log d data, i.e. with lower T/d ratios. A pattern of thicker fault zones for faults with larger displacements is nonetheless apparent. The transition in dominant deformation mechanism to diffusional mass transfer within a layer of fault rocks does not, then, preclude the possibility that fault zone thickness grows at fault displacement increases. Some analyses of fault zone scaling have considered damage zone thickness and fault core thickness separately (Shipton et al. 2006; Wibberley et al. 2008), reviving the questions of whether there are generally applicable scaling relationships and what they imply about deformation mechanisms in fault zones.

In the third alternative model for fault zone development shown in Figure 8.58b *iii*, shearing magnitude increases within a fault zone of uniform thickness. Such a model would pertain to a fault where displacement accrues by the mechanism of pressure-solution slip described in Section 8B.1.2.1. This model is also applicable to faults that develop in rocks with pronounced weak layers. The shearing associated with detachment or décollement zones, whether developed in extensional or contractional settings, is often confined to evaporite (halite or gypsum-anhydrite) layers (Hudec and Jackson 2007; Bartel et al. 2014) or shales (Kehle 1970; Morley et al. 2017, 2018). Shearing within shale layers may be facilitated by elevated pore fluid pressure (Hubbert and Rubey 1959; see Aydin and Engelder 2014 for a recent analysis). Shearing in shales, particularly those with pore fluids present, may occur by a combination of slip-on basal planes of sheet silicates and diffusional mass transfer of non-phyllosilicate detrital grains (Gratier et al. 2011). This combination of slip on the basal planes of phyllosilicates and diffusional mass transfer in rocks with moderate-to-high pore fluid pressures may be a significant mode of deformation in many subduction zone settings. Modeling by Fagereng and den Hartog (2016) suggests that mélange zones deforming by this combination of mechanisms may exhibit steady-state creep over a broad range of strain rates at low effective friction coefficients, provided pore fluid pressures are maintained and the margins of the mélange zones are smooth. Such a mélange can, in principle, undergo continued shearing within the zone of constant thickness. In their models, local narrowing of the mélange zone due to irregularities in either the overriding or underthrust plate can cause sufficient increases in the strain rate to cause velocity strengthening, which they argue may, in turn, lead to abrupt, possibly seismic, slip events. In some detachment zone settings, cleavage seams and arrays of crack-seal veins formed during the shearing of shale layers record temporal variations in the relative rates of dissolution, diffusion, and precipitation, which could reflect intrinsic variations in the rates of dissolution, diffusion, and precipitation or extrinsic variations in stress and/or rate of shearing (Fisher and Brantley 2014). Modeling underscores the role of ambient temperature in controlling the relative rates at which these processes contribute to the shearing of shale horizons, particularly in subduction zone settings (Fisher et al. 2019).

To summarize, analyses of fault zone thickness versus fault displacement data have continued relevance to understanding faulting, especially when they are combined with examinations of deformation mechanisms in fault zones or fault rocks or with models of deformation mechanics. Taking a broader look over Section 8B.1, the structures and microstructures in fault zones and fault rocks, and related modeling or analytical studies, underscore that frictional sliding and cataclasis occur concurrently and in competition with other deformation mechanisms. Consequently, it is important to examine the competing processes in greater detail.

8B.2 Diffusional Deformation Mechanisms

The organization of this section differs from that of Section 8A.3. In particular, Section 8B.2.1 on the characteristics of stylolites is intended to draw upon and complement material addressed in the discussion of fractures and veins in Section 8B.1 – Cataclastic Deformation Mechanisms. This discussion of diffusional mass transfer structures appears at the beginning of this section so that it might link, with little intervening material, more effectively with the discussion in Section 8B.1. Subsequent sections then: (1) analyze the deformation processes that contribute to diffusional deformation mechanisms in greater depth than the discussion in Section 8A.3; and (2) outline derivations of the constitutive relationships for the different diffusional deformation mechanisms.

8B.2.1 Diffusional Mass Transfer Structures

Section 8A.3.2 defined and described the two main microstructures that indicate that diffusional mass transfer contributed to a rock's change in shape – truncated grains, often with sutured boundaries and fibrous mineral overgrowths (Sections 3.4, 3.5, Figures 8.10, 8.12, 8.13, and 8.16). Diffusional mass transfer operating at the scale of individual grains can produce strong shape-preferred orientations (Figure 8.59a). The same processes are also capable of generating shape-preferred orientations of macroscopic objects visible in outcrop, such as cobbles in deformed conglomerates (Figure 8.59b). *Stylolites*, discontinuity surfaces whose walls have moved toward each other, are another product of diffusional mass transfer. Diffusional removal of mass accommodates the interpenetration of the opposing walls of the discontinuity surface (Figure 8.60). Because stylolites are important outcrop-scale deformation elements in many deformation settings, and because they are important in controlling rock permeability, it is appropriate to examine their characteristics in some detail.

8B.2.1.1 Characteristics of Stylolites

Stylolites are distinctive nonplanar or wave-like surfaces across which primary features such as bedding planes or fossils are truncated and/or offset (Section 3.4, Figures 8.46, 8.61). In their classic form, stylolites are surfaces across which short,

Figure 8.59 (a) Detrital quartz pebbles exhibit a pronounced grain shape fabric. Diffusional mass transfer contributed significantly to fabric development. Individual grains are elongate roughly perpendicular to bedding, which is indicated by the layer of finer sand grains between the arrows. Stylolite surfaces (denoted *s*), decorated by lichen, highlight the margins of grains, and emphasize the elongated grain fabric. Grain *x* has distinct stylolitic contacts and fibrous overgrowths. (b) Deformed quartzite cobbles exhibit a clear shape preferred orientation. The contribution of diffusional mass transfer is apparent at contact indicated by arrows, where rounded cobble has impinged upon elongate neighboring cobble. Local development of mineral fibers (*f*) is also evidence for diffusional mass transfer.

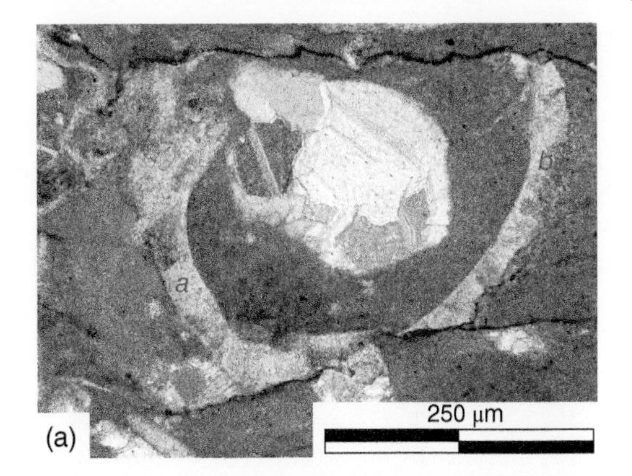

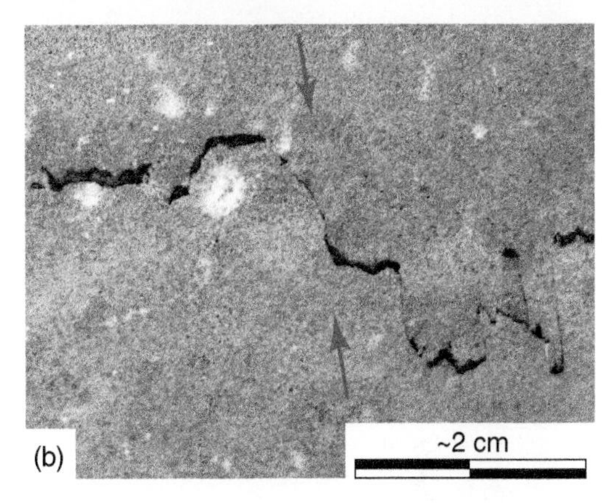

Figure 8.60 (a) Photomicrograph of stylolites in deformed limestone. Two segments of a fossil shell (labeled *a* and *b*) are truncated against distinct stylolites (denoted *s*) decorated by selvage. (b) Photograph of a single stylolite developed in oolitic limestone. Segments of the stylolite parallel to the arrows are thin, whereas segments roughly normal to arrows have thick selvage accumulations. Arrows give inferred directions of movement across the stylolite surface.

stubby columns of rock interpenetrate like meshing teeth (the root word for "stylolite" is *stylos*, Greek for column or pillar). When the columns or teeth are perpendicular to the overall trend of the stylolitic surface, the trace of the stylolite resembles the graph of a noisy but quasiperiodic signal. Although individual stylolites are flat at a large scale, they are rough at a small scale (Ebner et al. 2009a, b, and 2010; Koehn et al. 2012, 2016). In other instances, a macroscopic stylolite may consist of a collection of wispy, irregular surfaces that separate and rejoin (Figure 8.61b). Citing the roughness of individual surfaces or noting that collections of irregular, wispy stylolites occur in zones that have a finite width, many geologists categorize stylolites as "tabular" features. The amplitudes of the teeth or columns on any individual stylolite

Figure 8.61 (a) Three subparallel stylolites, part of an array of stylolites defining a spaced cleavage in this limestone. Arrow points to an apparent offset in a lamina due to the removal of material across the stylolite. (b) Two generations of stylolites in limestone. The pen points to one of two subparallel stylolites that extend across the field of beds at high angles to layering. The arrows point to anastomosing small stylolites that link to form arrays inclined at a moderate angle to layering. (c) Prominent stylolitic seams, with planar boundaries and mm-thick layers of selvage.

and the widths of the zone containing wisps that together define a single stylolite generally are small relative to the lengths of the feature, typically on the order of $0.01 \times \sqrt{(\text{stylolite length})}$ (Benedicto and Schultz 2010). It is worth noting, however, that the teeth with the greatest amplitude on any stylolite are not necessarily found midway along its length; rather they can occur anywhere along the trace of the feature. Regardless of their shape, stylolites usually are marked by an accumulation of microscopic iron oxides, sulfides, or clay mineral grains called *selvage*. The thickness of selvage along the stylolite surface can be highly variable or even be discontinuous, but generally greater selvage thickness indicates a greater magnitude of interpenetration across the surface. If the thickness of selvage reaches 2–3 mm, some geologists call the feature a *stylolite seam* (Figure 8.61c).

Arrays of subparallel stylolites that are oblique and/or normal to bedding define macroscopic foliations that may accommodate pre-folding layer-parallel shortening, contribute to deformation in the hinges of folded layers, and/or accommodate layer-parallel shearing during folding or adjacent to faults (Figure 8.61). In some instances, the arrays of stylolites are penetrative at the scale of a hand sample or thin section and accommodate homogeneous strains (Geiser 1974; Henderson et al. 1986). In other settings, collections of parallel stylolite surfaces or stylolite seams define *spaced* or *domainal* cleavages (Section 3.9), in which cleavage *domains* characterized by individual *disjunctive*, or crack-like, stylolites, individual seams with *diffuse* or gradational boundaries, or anastomosing seams are separated by *microlithons*, regions lacking stylolites or seams. Arrays of spaced stylolites accommodate deformation that is discontinuous, i.e. the particle displacements change abruptly across stylolites, but, in many cases, the overall deformation is mesoscopically homogeneous (Markley and Wojtal 1996). In these situations, the role of diffusion in the development of stylolites imposes a rate-dependence on the bulk behavior of rock that can be captured by viscous rheology (Gratier et al. 1999, 2014). Further support for inferring that deformation by arrays of stylolites correlates with bulk viscous rheology comes from examinations of folding; both the distribution of strains accommodated by stylolites in the folded layers and the shapes of the folds are consistent with viscous rheology for the folded layers (Hudleston and Treagus 2010).

8B.2.1.2 Arrays of Veins and Stylolites

Stylolites often co-occur with extension veins, with constituents that diffuse away from stylolites in order to accommodate interpenetration of their opposing walls subsequently precipitated in extension veins (Figures 8.43, 8.61c). In these linked systems, the geometry – orientation and spacing – of veins and stylolites within the overall displacement field influence the propagation and eventual growth of the veins (Seyum and Pollard 2016). A different result of stylolite and vein geometry is apparent in other settings. As deformation accrues in sedimentary rocks, arrays of stylolites oblique or normal to bedding provide pathways for advecting fluids, enhancing the transport of chemical constituents between layers. In cases where well-developed stylolites cut across multiple beds, stylolites often accommodate sizeable offsets and contribute to significant local volume losses or gains due to secondary mineral deposition in veins (Engelder and Marshak 1985; Marshak and Engelder 1985; Bell and Cuff 1989; Markley and Wojtal 1996) (Figure 8.61b and c). Volume loss or gain during the formation of stylolitic cleavages is not universal, however. In some deformed and weakly metamorphosed siliciclastic rocks, the volume of material precipitated in veins or as overgrowths on minerals in microlithons is approximately equal to that removed along stylolites (Erslev 1998). In cases where stylolites accommodate significant finite strains in rocks, whether advection contributes to local changes in the volume of the deforming rock or not, the net rheology of the rock may transition from one dominated by the propagation of fractures to one constrained by the solution removal and precipitation of phases.

8B.2.1.3 Origin and Development of Stylolites

Stylolites may initiate within uniform rock via a type of mode I fracture propagation where diffusional mass transfer facilitates the interpenetration of fracture walls across an "anti-crack," a surface analogous to an extensional crack but with displacements normal to the surface having negative rather than positive magnitudes (Fletcher and Pollard 1981). Other factors are likely to figure in the initiation of most stylolites, however. For example, bedding plane stylolites typically initiate along surfaces that juxtapose sediments with different mineral constituents (i.e. clays in shale versus calcite limestone or quartz in sandstone) or beds with different mean grain sizes (i.e. fine-grained limestone versus coarse-grained limestone) (Koehn et al. 2007). Similarly, stylolites normal or oblique to bedding often initiate along pre-existing joints (Geiser and Sansone 1981) or along the margins of mineral-filled veins (Mullenax and Gray 1984). The presence of pore fluids along a pre-existing joint will enhance localized diffusional mass transfer, and could lead to the utilization of the joint as a locus of diffusional removal of mass. Similarly, the juxtaposition of different minerals across the wall of a vein may contribute to the initiation of stylolite along the margin. Even if the vein filling consists of the same mineral as the host rock (i.e. quartz veins in a quartzite or calcite veins in a limestone), the presence of fluid phases along the vein margin *and* the juxtaposition of mineral grains with different sizes can facilitate the initiation of a stylolite along the vein margin.

The role of juxtaposing different minerals in the formation and development of stylolites warrants further discussion. Trurnit (1968) determined empirically a "solubility" series, a list ranking minerals on the likelihood that they would or would not diffuse from a grain-to-grain contact (Table 8.3). Minerals at the top of the solubility series, like halite and calcite, tend to diffuse from contact with minerals near the bottom of the series, such as quartz, micas, clays, sulfides, or iron oxides. The juxtaposition of detrital layers composed mainly of calcite or quartz grains against thin shale layers composed of clay minerals and iron oxides is, then, a factor in the formation of bedding plane stylolites in limestones and sandstones as a result of compaction. Bedding-normal stylolites in limestones often nucleate at the ends of chert nodules, where calcite, near the top of Trurnit's solubility list, is in contact with silica, nearer the bottom of the solubility list. In fact, empirical studies of stylolite morphology (Ebner et al. 2010) and numerical modeling of the origin of stylolitic fractures (Koehn et al. 2007, 2012, 2016) indicate that the irregular distribution of minerals up or down Trurnit's series is fundamental to the origin of the roughness of stylolites.

Table 8.3 Trurnit's "Solubility" Series from most likely to least likely to diffuse from a grain contact.

1	Halite and potassium salts
2	Calcite
3	Dolomite
4	Gypsum
5	Anhydrite
6	Amphibole and pyroxene
7	Chert
8	Quartzite
9	Quartz, glauconite, rutile and hematite
10	Feldspars and cassiterite
11	Micas and clay minerals
12	Arsenopyrite
13	Tourmaline and sphene
14	Pyrite
15	Zircon
16	Chromite

Since interpenetration across any stylolite is accommodated by diffusional mass transfer, the development of the stylolite depends on the rates of diffusion of the chemical constituents in the minerals in rocks as well as the stress and displacement patterns in the vicinity of the stylolite. The particular minerals present in a rock thus exert strong control on the likelihood that stylolites will contribute to deformation and on the character of the stylolites that develop. For example, classic, toothed stylolites oblique to bedding regularly develop in limestones with relatively low clay content, whereas in unmetamorphosed shaly limestones, thick solution seams or collections of anastomosing thin stylolites are more common (Alvarez et al. 1978; Marshak and Engelder 1985). The development of stylolites in siliciclastic shales typically requires higher temperature settings, such as those associated with anchimetamorphism or lower greenschist grade metamorphism. Also, analyses of populations of stylolites indicate that the amplitude of roughness and volume of selvage increase with increasing interpenetration across stylolite surfaces, in good agreement with models of the formation of stylolites (Koehn et al. 2012, 2016; Peacock et al. 2017).

8B.2.2 Understanding Diffusion Through Crystalline Materials

In Section 8A.3.1, we introduced the deformation process diffusion by considering situations where concentration gradients exist in an ideal gas or liquid – a pungent cheese placed in a room or a drop of dye introduced into a beaker of water. The odor of the cheese spreads throughout the room or the dye spreads throughout a beaker of water because random movements in the gas or liquid tend to eliminate concentration gradients. In most deformation settings, however, the concentration (C) of a constituent does not adequately capture its contribution to the overall physical-chemical state of the system, and it is the *chemical potential* that measures constituents' contributions to the total energy of the system. For this reason, Section 8B.2.2.1 examines the relationship between the concentration of a chemical constituent, its chemical potential, and its diffusive flux. Section 8B.2.2.2 considers briefly the factors that affect the magnitudes of diffusion coefficients, and Section 8B.2.2.3 examines the role of vacancies in diffusion in crystalline materials.

8B.2.2.1 Diffusion in NonIdeal Gases and Liquids

In nonideal solutions, the chemical potential of a constituent A is given by

$$\mu_A = \mu_A^o + RT \ln C_A, \tag{8.23}$$

where μ_A^o is a reference value of the constituent A's chemical potential, R is the gas constant, T is the absolute temperature, and C_A is the concentration of constituent A. Taking the derivative with respect to the spatial coordinate x of both sides of Eq. (8.23) yields

$$\frac{d\mu_A}{dx} = \frac{d\left(\mu_A^o + RT \ln C_A\right)}{dx} = RT \frac{1}{C_A} \frac{dC_A}{dx}. \tag{8.24a}$$

Rearranging this equation yields

$$\frac{dC_A}{dx} = \frac{C_A}{RT} \frac{d\mu_A}{dx}. \tag{8.24b}$$

Substituting into Eq. (8.5) yields

$$J_A = -\frac{D_A C_A}{RT} \frac{d\mu_A}{dx}, \tag{8.25}$$

where J_A is the flux of constituent A and D_A is the diffusion coefficient for that constituent. The relationship (8.6), presented in Section 8A.3.1.2, is a simplified representation of Eq. (8.25). Chemical constituents in nonideal liquids move from regions of higher chemical potential to regions of lower chemical potential, that is they move "down a chemical potential gradient."

8B.2.2.2 Diffusion Coefficients

Both theoretical arguments and experimental data indicate that diffusion coefficients vary with temperature. Theoretical arguments indicate that the functional variation of the diffusion coefficient is

$$D = D_o \left\{ \exp \frac{(-Q - PV)}{kT} \right\}, \tag{8.26}$$

where D_o is a limiting value for the diffusion coefficient, Q is the activation energy for diffusion, P is the pressure, V is the molar volume, k is the Boltzmann constant, and T is the absolute temperature. As the temperature increases, the absolute value of $[(-Q - PV)/kT]$ decreases, $\exp[(-Q - PV)/kT]$ approaches unity, and the diffusion coefficient approaches more closely its limiting value D_o. Thus, as stated in the Conceptual Foundations section, increasing temperature while holding all other factors constant causes the diffusive flux to increase. As pressure increases, the absolute value of $[(-Q-PV)/kT]$ increases, $\exp[(-Q-PV)/kT]$ becomes smaller, and D becomes a smaller fraction of the limiting value D_0. Thus, increasing pressure while holding all other factors constant decreases the diffusive flux. The effects of temperature are more pronounced than the effects of pressure.

8B.2.2.3 Mechanisms of Diffusion in Crystals

The diffusion of chemical species through a crystal lattice, whether it is self-diffusion or solution-diffusion, almost always involves vacancies. The vibrational energy needed to move atoms or ions in a crystal from one lattice site or interstitial position to another by exchanging positions with a vacancy (cf. Figure 8.8) is so much lower than exchanges with no vacancy involvement that the latter rarely occurs. Given the essential role of vacancies in diffusion in crystalline materials, it is appropriate to examine their origin and persistence. The occurrence of vacancies in crystals may be surprising, but as Cottrell (1964, p. 69) noted, "Just as matter 'dissolves' into space at high temperatures, i.e. forms a vapor, so also does space dissolve in matter. . ." Materials scientists are able to show, using fundamental principles of physical chemistry, why and how "space dissolves into matter" and to derive expressions giving the concentration of vacancies in crystals as a function of measurable parameters.

The discussion of vacancies in Section 8A.3.1.3 focused on the distortions they create in perfect crystals, noting that vacancies increase the internal energy of crystals. In order to understand why vacancies exist even though they do locally distort a crystal lattice, we begin with a perfect crystal in equilibrium with its vapor. Imagine that an atom from the interior of the crystal moves from its lattice site, leaving a vacancy, and attaches to a step on a free surface of the crystal. Because the atom attaches to a step in the crystal's surface, the external surface area of the crystal remains the same. The free energy of the crystal increases, however, because the number of bonds broken in removing the atom from its initial site is larger than the number of bonds reformed in attaching the atom to the surface. At the same time, the entropy of the crystal changes because the frequency of vibration of the atoms in the vicinity of the vacancy changes. In addition, the volume of the crystal changes by an amount that is the sum of two terms: (1) adding the atom to the free surface of the crystal increases the volume of the crystal; and (2) the crystal's volume changes as atoms surrounding the vacancy move in response to its formation. This movement of atoms surrounding the vacancy is called "relaxation" with the notion that atoms surrounding the vacancy "relax" inward. It is important to note, however, that atoms may "relax" outward. As illustrated in Figures 8.8 and 8.9, for example, the removal of a positive (or negative) ion in an ionic crystal typically means that the surrounding negative (or positive) ions no longer are drawn toward the now-vacated lattice site and move away from that site. The Gibbs energy of formation of an individual vacancy equals the increase in internal energy plus the pressure-volume work associated with the change in volume due to vacancy formation minus the change entropy associated with vacancy formation (See Appendix 8-I).

Although every individual vacancy increases the free enthalpy of a crystal, at any temperature above 0 K, there is a configurational entropy contribution that more than compensates for the increase in free enthalpy associated with forming the vacancy. Configurational entropy depends upon the positions of the constituents in a crystal, and it is a function of the number of configurations possible at a particular energy level. For small numbers of vacancies relative to the number of lattice sites (i.e. there are large numbers of possible configurations), the crystal's configurational entropy term exceeds the total free energy of vacancy formation.

Thus, crystals have lower Gibbs energy with vacancies than in a vacancy-free state.

The equilibrium number of vacancies varies with temperature. Taking n_v as the number of vacancies in an atom composed of n atoms,

$$n_v = n \exp\left(-\frac{\Delta G_f}{RT}\right), \tag{8.27}$$

where ΔG_f is the free energy (in joules per mole) associated with the formation of a vacancy, R is the gas constant, and T is the absolute temperature (see Appendix 8-I). Nicolas and Poirier (1976) calculate that equilibrium concentrations of vacancies $N_v = n_v/n$ in metals range from $\sim 10^{-18}$ at low homologous temperatures (homologous temperature = ambient temperature divided by melting temperature) to $\sim 10^{-4}$ as homologous temperatures approach unity.

Most rock-forming minerals and many other crystalline solids consist of three-dimensional arrangements of positively and negatively charged ions held together by bonds with strong ionic character. In ionic solids, individual missing cations or anions lead to unsustainable charge imbalances. The simplest model for vacancy formation in ionic crystals envisions that: (1) vacant cation sites are balanced by an equivalent number of vacant anion sites; and (2) the two types of vacancies need not be directly associated with each other. Even with the degree of independence required to meet condition (2), materials scientists consider that a pair of vacant cation and anion sites together constitute a single *Schottky defect*. If the equilibrium number of Schottky defects is n_s, there are n_s cation vacancies and n_s anion vacancies to be distributed among N lattice sites. Materials scientists have determined (Appendix 8-I) that the equilibrium number of Schottky defects in an ionic crystal is

$$n_s \sim N \exp\left(-\frac{\Delta G_f}{2RT}\right). \tag{8.28}$$

Equation (8.28) is derived using the assumption that the cation and anion vacancies are independent of each other, but that condition may not hold. An alternative model for vacancy formation in ionic crystals, for example, involves an ion, typically a cation because it has a smaller ionic radius, leaving its lattice site to become an interstitial. Such an interstitial–vacancy pair, called a *Frenkel defect*, maintains charge neutrality in a crystal, but the two are not independent of each other. Nevertheless, the number of vacancies is given by a relation comparable to Eq. (8.28). In addition, many rock-forming minerals consist of arrays of ions and covalently bound ionic groups, such as SiO_4^{-4}, CO_3^{-2}, or SO_4^{-2}. This analysis does not address the complexity of vacancy formation in crystals with different bond types and strengths. Nevertheless, one solid "take away" from the analyses outlined earlier (and presented in Appendix 8-I) is that crystals have equilibrium concentrations of vacancies at all temperatures.

8B.2.3 The Effect of Differential Stress

The analyses in the previous section provide a foundation for an alternative approach to understanding what drives diffusional mass transfer. We begin by envisioning a cube-shaped crystal subjected to biaxial stresses (Figure 8.62). The most compressive principal stress component, which has magnitude $\sigma_1 = P+\sigma$, acts on the crystal surface perpendicular to the x_1 axis. The least compressive principal stress component, which has magnitude $\sigma_3 = P-\sigma$, acts on the crystal surface perpendicular to the x_3 axis. The intermediate principal stress, which has magnitude $\sigma_2 = P$, acts on the crystal surface perpendicular to the x_2 axis. The hydrostatic stress $\sigma_m = (\sigma_1+\sigma_2+\sigma_3)/3 = (P-\Delta\sigma+P+\Delta\sigma+P)/3 = P$. Recall that $\Delta\sigma$ denotes the differential stress. The preceding section examined the formation of a vacancy by envisioning the removal of an atom from the center of the crystal to a crystal face subjected to a pressure P. Differential stress either increases or decreases the force per unit

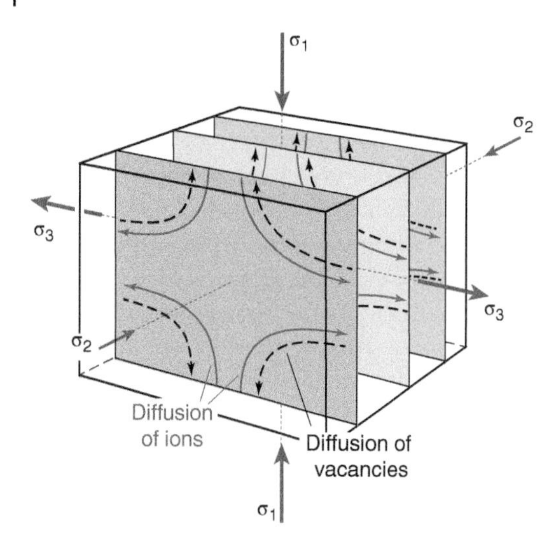

Figure 8.62 Schematic image of crystal undergoing deformation by diffusional mass transfer in response to applied differential stresses. Vacancies diffuse from the surface subjected to the least compressive to the surface subjected to the most compressive stress, whereas atoms/ions diffuse from the surface subjected to the most compressive to the surface subjected to the least compressive stress.

area acting on crystal surfaces, which increases or decreases the magnitude of the pressure–volume work associated with creating a vacancy. For this reason, the equilibrium number of vacancies in the vicinity of the crystal surface perpendicular to the x_1 axis will differ from the equilibrium number of vacancies in the vicinity of the crystal surface perpendicular to the x_3 axis.

If b is the equilibrium spacing between lattice sites and b^3 is the volume of a unit cell of the crystal, then one can determine the equilibrium concentration of vacancies in a crystal subjected to hydrostatic stresses by dividing Eqs. (8.27) or (8.28) by b^3:

$$C_o = \frac{1}{b^3}\exp\left(-\frac{\Delta G_f}{kT}\right). \tag{8.29}$$

Under non-hydrostatic stresses, the pressure–volume work contribution to ΔG_f is increased (for stresses more compressive than the mean stress) or reduced (for stresses less compressive than the mean stress) by an amount $\Delta\sigma\Delta V = \sigma b^3$. If C_i (where $i = 1$ or 3) is the concentration of vacancies near the crystal surface perpendicular to the x_i axis,

$$C_i = \frac{1}{b^3}\exp\left(-\frac{\Delta G_f \pm \Delta\sigma b^3}{kT}\right) = \frac{1}{b^3}\exp\left(-\frac{\Delta G_f}{kT}\right) \times \exp\left(-\frac{\pm\Delta\sigma b^3}{kT}\right) = C_o \exp\left(-\frac{\pm\Delta\sigma b^3}{kT}\right). \tag{8.30}$$

For $i = 1$, the most compressive normal stress acting on the crystal surface $\sigma_1 = P+\Delta\sigma > P$, $\exp(-\Delta\sigma b^3/kT) < 1$, and $C_1 < C_o$. For $i = 3$, the least compressive normal stress acting on the crystal surface $\sigma_3 = P-\Delta\sigma < P$, $\exp(\Delta\sigma b^3/kT) > 1$, and $C_3 > C_o$. The concentration gradient will drive the diffusion of vacancies from the crystal surface perpendicular x_3 axis to the surface perpendicular to the x_1 axis. An accumulation of vacancies adjacent to the surface perpendicular to σ_1 leads to a net removal of mass from this portion of the crystal, and depletion of vacancies adjacent to the surface perpendicular to σ_1 leads to a net addition of mass to this portion of the crystal.

8B.2.4 Flow Laws for Diffusional Deformation Mechanisms

Starting with the inference that vacancies diffuse at a rate proportional to gradients in their concentration, Nicolas and Poirier (1976) used the following series of steps to derive a flow law for diffusional mass transfer. In a cube-shaped crystal with edge length d, the concentration gradient is proportional to $(C_3-C_1)/d$, and the flux of vacancies is

$$J = D_v \operatorname{grad} C, \tag{8.31}$$

where D_v is the diffusion coefficient for vacancies and grad C is the three-dimensional spatial gradient of the concentration of vacancies. The number of vacancies crossing an area d^2 per second is

$$\phi = J d^2. \tag{8.32}$$

Removing ϕ vacancies from the surface of a crystal corresponds to adding mass with a volume of ϕb^3 to that surface. If that volume is spread over the surface, this corresponds to adding a layer of thickness $\phi b^3/d^2$. With this added mass, the length of the crystal grows from an initial length l_o to a final length l' or experiences a stretch of $T = l'/l_o$ each second. Using ε instead of T to denote the strain accomplished by this mass movement, the amount that the crystal is elongated each second is

$$\varepsilon = \frac{l'}{l_o} = \frac{\left(\phi b^3 / d^2\right)}{d}. \tag{8.33}$$

Thus, the strain rate is

$$\dot{\varepsilon} = \phi \frac{b^3}{d^3}. \tag{8.34}$$

Substituting from Eqs. (8.30) and (8.31), Equation (8.32) becomes

$$\begin{aligned} \phi &= \left(-D_v \text{ grad } C\right)d^2 \\ &\propto dD_v C_o \left[\exp\left(\frac{\Delta\sigma b^3}{kT}\right) - \exp\left(\frac{-\Delta\sigma b^3}{kT}\right)\right] \end{aligned} \tag{8.35}$$

Combining Eqs. (8.34) and (8.35) yields an equation that relates strain rate and stress. Before doing so, however, we use three steps to simplify the right-hand side of Equation (8.35).

First, because $(e^x - e^{-x})/2 = \sinh x$, the right-side term can be rewritten using $\sinh(\Delta\sigma b^3/kT)$. Second, because $\Delta\sigma b^3 << kT$ and $\sinh x \approx x$ for small x, that term can be reduced to a function of $(\Delta\sigma b^3/kT)$ only. Finally, by inserting a proportionality constant $1/\alpha$ that varies with crystal shape and the stress state, Equation (8.35) becomes

$$\phi = \frac{dD_v C_o}{\alpha} \cdot \frac{\Delta\sigma b^3}{kT} \tag{8.36}$$

and from Eq. (8.34)

$$\dot{\varepsilon} = \left[\frac{dC_o D_v}{\alpha} \cdot \frac{\Delta\sigma b^3}{kT}\right]\frac{b^3}{d^3} = \frac{b^3 C_o D_v}{\alpha d^2} \cdot \frac{\Delta\sigma b^3}{kT}. \tag{8.37}$$

Normally, one would want the constitutive equation written in terms of a diffusion coefficient for self-diffusion. The diffusion of vacancies through a crystal is clearly related to the diffusion of atoms through the crystal, but the diffusion coefficient for self-diffusion D_{sd} is not equal to the diffusion coefficient for vacancies. The movement of an atom from one position to another is not, like the movement of a vacancy, independent of its surroundings. Instead, the likelihood that an atom moves from its lattice site depends upon the likelihood that it is adjacent to a vacancy. Thus, the diffusion coefficient for self-diffusion is a product of a fraction reflecting the likelihood that a lattice site is vacant ($N_v = n_v/n$) and the diffusion coefficient for vacancies, i.e. $D_{sd} = N_v D_v = C_o D_v b^3$ and $D_v = D_{sd}/C_o b^3$. Substituting for D_v in Eq. (8.36) and then substituting the resulting expression for ϕ into Eq. (8.34) yields

$$\dot{\varepsilon} = \frac{D_{sd} b^3}{\alpha d^2 kT} \Delta\sigma. \tag{8.38}$$

By setting $b^3/\alpha kT = A_{NH}$, recalling that $\Delta\sigma$ refers to differential stress, and recognizing that the diffusion coefficient for self-diffusion is precisely equivalent to D_L, the diffusion coefficient for lattice diffusion in Section 8A.3.5, it is apparent that Eq. (8.38) is equivalent to Eq. (8.7), the constitutive equation for Nabarro–Herring creep.

In Section 8A.3.5, we indicated that the strain rate and stress terms in flow laws are commonly the axial strain rate $\dot{\varepsilon}_1$ and differential stress ($\Delta\sigma = \sigma_1 - \sigma_3$) because these are the quantities that are measured in laboratory experiments to obtain flow laws. The experiments and the theoretical approach discussed earlier show that the geometry of flow will affect the precise meaning of the flow law strain rate and stress terms. It is quite hard to extrapolate expressions such as Eq. (8.38) to natural flows, a problem considered by Paterson (2001).

Stepping back from the details of the mathematics presented in the two previous sections, it is worthwhile to consider the general form of

Equation (8.38), the constitutive equation for mass transfer via lattice diffusion. First of all, the strain rate $\dot{\varepsilon}$ is linearly related to the differential stress. The theoretical analysis indicates that material deforming by Nabarro–Herring creep should exhibit linearly viscous behavior, a result that has been experimentally verified. Further, the strain rate is inversely proportional to the square of the mean grain diameter. This too is seen empirically to hold true. Finally, it is the diffusion coefficient for self-diffusion, which is strongly dependent upon the absolute temperature, not the inverse proportionality to kT, that is responsible for higher strain rates at higher temperatures. There is, then, solid theoretical, experimental, and empirical evidence indicating that diffusional mass transfer by lattice diffusion leads to linear viscosity.

8B.2.5 Paths of Rapid Diffusion – Dislocations and Grain Boundaries

Along the tubular volumes surrounding dislocation lines and within the tabular volumes of grain boundaries, atoms do not occupy positions predicted by or expected of a crystal lattice. The irregularity in the locations of atoms near dislocations and within grain boundaries is one factor responsible for the high diffusivity of some chemical species along dislocations and grain boundaries. Dislocations and grain boundaries alter the overall rate of diffusion through a polycrystalline aggregate in proportion to the volume that the dislocations' tubes and/or grain boundaries' tabular regions occupy.

Envisioning a dislocation as a pipe through a crystal, the effective diffusion coefficient D_{eff} for the crystal must incorporate a term measuring the contributions of diffusion along the pipes associated with a dislocation, i.e.

$$D_{eff} = D_L + \pi\rho_d b^2 D_{disl} \tag{8.39}$$

where D_L is the diffusion coefficient for diffusion through the crystal lattice, ρ_d is the dislocation density, b is the Burgers vector of the dislocation, and D_{disl} is the diffusion coefficient for diffusion along the dislocation (Karato 2008). Recall that dislocation densities are given as a number per unit area. The cross-sectional area of the dislocation is approximated as πb^2. So, $\pi\rho_d b^2$ is a measure of the fraction of a surface across which vacancies or atoms diffuse that has the diffusion coefficient D_{disl} rather than D_L.

A similar expression pertains to diffusion along grain boundaries. That is

$$D_{eff} = D_L + \frac{\pi\delta}{d} D_{GB} \tag{8.40}$$

where D_L again is the diffusion coefficient for diffusion through the crystal lattice, δ is the grain boundary width, d is the grain diameter, and D_{GB} is the diffusion coefficient for diffusion along the grain boundary (Karato 2008). The functional form of the coefficient of the D_{GB}, i.e. direct proportionality to the grain boundary width and inverse proportionality to the grain diameter, results from determining the relative sizes of the grain boundary volume (which is proportional to $\delta \times d^2$) and the volume of the grain (which is proportional to d^3). The specific scalar factor included in the expression, here equal to π, depends on the shape of the grains and varies in different analyses.

The activation energies for diffusion along dislocations and along grain boundaries are small when compared to the activation energies for lattice diffusion, especially at relatively low temperatures and for small grain diameters. In those situations,

$$\frac{\pi\delta}{d} D_{GB} \gg D_v \text{ and } D_{eff} \approx \frac{\pi\delta}{d} D_{GB}.$$

Replacing the diffusion coefficient for self-diffusion D_{sd} in Equation (8.38) by D_{eff} yields a constitutive equation for diffusional mass transfer by grain boundary diffusion:

$$\dot{\varepsilon} = \frac{\pi b^3 \delta D_{GB}}{\alpha d^3 KT} \Delta\sigma. \tag{8.41}$$

By setting $\pi b^3/\alpha\ kT = A_C$ and recalling that $\Delta\sigma$ refers to differential stress, it is apparent that Eq. (8.41) is equivalent to Eq. (8.8), the constitutive equation for Coble creep.

Reviewing the general form of Eq. (8.41), we see that the strain rate for mass transfer via grain boundary diffusion is linearly related to the differential stress, i.e. Coble creep gives rise to linearly viscous rheology. The strain rate is inversely proportional to the cube of the mean grain diameter, a result of significant importance in nature. The diffusion coefficient for grain boundary diffusion exhibits the typical strong dependence on the absolute temperature, and the activation energy for grain boundary diffusion is relatively low. Thus, deformation by grain boundary diffusion often is a significant contributor to the linear viscous rheology of crustal rocks in a variety of natural deformation settings.

8B.2.6 The Effect of Fluid Phases Along Grain Boundaries

The common occurrence of diffusional mass transfer structures and microstructures in rocks deformed at low temperatures is widely cited as support for the inference that the presence of fluids in rocks enhances diffusional mass transfer. In Section 8A.3.5, we argued that mass transfer via diffusion along wet grain boundaries is a distinct deformation mechanism: *solution transfer* or *pressure solution*.

As noted in Eqs. (8.6) and (8.25), the diffusive flux is driven by the gradient in the chemical potential of the diffusing constituents. In the case of solution transfer, the value of the chemical potential of any mobile constituent depends upon the local chemical equilibrium across the interface between the solid phases and the aqueous solution (Rutter 1976, 1983). Further, mass flux occurs via a three-step process: (1) constituents dissolve one location; (2) constituents diffuse through an aqueous fluid phase along the grain boundaries; and (3) constituents reprecipitate at a different location. Thus, the diffusive flux varies with the molar volume (V) of any diffusing constituent and depends on the constituent's concentration (C_o) in the solution coating the grain boundaries. Accounting for these additional conditions results in a constitutive equation of the form:

$$\dot{\varepsilon} = A\frac{C_o D_{WB} V \delta}{d^3}\Delta\sigma, \tag{8.42}$$

where C_o is the concentration of the solution outside of the grain boundary, D_{WB} is a diffusion coefficient for diffusion through the solution, V is the molar volume of the crystalline solid, δ is the grain boundary width, and d is the mean diameter of grains. The constant A varies with the geometry of the fluid-crystal interface. By setting $AC_oV = A_{ST}$ and recalling that $\Delta\sigma$ refers to differential stress, Equation (8.42) is equivalent to Eq. (8.9), the constitutive equation for solution transfer.

Despite the widespread recognition of solution transfer as a distinct deformation mechanism, there is no consensus regarding the details of the processes responsible. This is due, in large measure, to uncertainty about the character and distribution of fluid phases, especially aqueous fluids, along grain boundaries. One question concerns how "wet" are grain boundaries, i.e. whether or not the fluid phase completely covers the contact between adjacent mineral grains. There are two factors at play here. First, experimental data suggest that aqueous fluids do not "wet" in a chemical sense the surfaces of silicate minerals effectively. One can assess the degree of chemical interaction of a fluid and a solid by measuring the wetting angle, which, in turn, depends upon the relative importance of the cohesion of the liquid and its adhesion to the surface. A low wetting angle indicates relatively high adhesion and relatively greater interaction between the fluid and the surface. A high wetting angle indicates relatively high cohesion and relatively lesser interaction between the fluid and the surface. Think, for example, of the tendency of water to "bead" on surfaces that resist wetting. Aqueous fluids on silicate surfaces tend to have wetting angles greater than 45°, which suggests a low degree of chemical interaction.

The second question concerns what percentage of the grain boundary surface area is coated by fluid, as more grain boundary area coated by fluid yields a greater volume of high diffusivity material. Compressing an aggregate can spread the high diffusivity fluid across a greater fraction of the mineral surface. Fully wetted grain boundaries, which would constitute a continuous high diffusivity network, are sometimes invoked as a requirement for solution transfer. Since fluids cannot support differential stresses, however, it is not clear how fully wetted, continuous grain boundaries coexist with grains supporting differential stresses. Fully wetted and continuous grain boundaries would seem to require sizable gradients in fluid pressures across relatively short distances. Would that cause fluid flow? It has been proposed that fluids under pressure have sufficiently high viscosity that they can support differential stresses, but that is thought likely to impede diffusional transport. Alternatively, it is possible that fluid phases do not fully coat the grain boundary, leaving "islands" or patches of mineral-to-mineral contact that transmit differential stresses from one crystal to another. These patches would necessarily be regions of slower transport by diffusion, but the distorted crystal lattice in the vicinity of the patches would dissolve more readily due to its relatively high strain energy.

With fluids along grain boundaries, minerals dissolve into and precipitate from the fluid. The rate of mineral dissolution along a grain boundary will depend upon interfacial reactions. Those reactions are likely to depend upon many variables, including mineral composition and structure, temperature and pressure, and concentration of dissolved components in the fluid (the latter two affect the mineral viscosity). These variables may depend, in turn, on whether the system is closed, i.e. where fluids are confined to a limited region where their temperature and concentration of dissolved components tend to equilibrate with their surroundings, or open, i.e. fluid temperatures and dissolved components do not equilibrate with the rock as they flow through it. Open systems, where advection can increase or decrease rock mass and add or subtract chemical constituents, require interconnected pathways for fluid flow. As we noted in Section 8B.1.1.3, there are feedbacks between structures and fabrics generated during earlier strain increments and which deformation mechanisms predominate during later increments. In particular, stylolite surfaces marked by thick accumulations of selvage, on the order of 2–3 mm, often serve channels for fluid flow. This is especially important in sedimentary rocks where thin shale beds may act as impermeable layers that restrict fluid movement between adjacent strata. The channelized flow often gives rise to spatially varying amounts of shortening or elongation.

8B.3 Dislocational Deformation Mechanisms

The organization of this section differs from that in Section 8A.4 – Dislocational Deformation Mechanisms. Instead of the single section on the movement of dislocations (Section 8A.4.1.2), this Comprehensive Treatment portion of the chapter contains separate sections addressing the glide of dislocations (Section 8B.3.2) and the climb of dislocations (Section 8B.3.3). Section 8B.3.4 then examines how dislocation interactions impact dislocation movement and dislocation densities. Section 8B.3.5 derives expressions concerning the stresses associated with dislocations and the strains accommodated by dislocation movement. Those relationships provide a foundation for the derivation of the constitutive equation for dislocation creep *in* Section 8B.3.6. The final two sections (Sections 8B.3.7, 8B.3.8) address recovery and recrystallization mechanisms and strains due to twinning and kinking.

8B.3.1 Origin of Dislocations

Dislocations, like point defects, raise the energy of a crystal. It is, then, appropriate to ask why dislocations form and persist in crystals. Materials scientists have proposed a number of scenarios proposed to account for the existence of dislocations in crystals, and we outline a few of those scenarios here.

In the first, consider a crystal that grows at a rate of 1–2 mm/yr. The crystal must add a layer of atoms to its growth surface every 1–2 seconds. For a crystal just 1 mm across, that entails millions or tens of millions of atoms attaching to the growth surface each second. Slowing the rate of growth to 1–2 mm/thousand yr still requires thousands to tens of thousands of atoms to attach to the growth surface each second. Errors in organization, that is accidents during the crystal growth, can be expected. It is easy to imagine how those errors generate substitutions, omissions, or interstitials. The incomplete addition of a plane of atoms would create an edge dislocation extending across the crystal (Figure 8.63a) or a closed loop of edge dislocation surrounding an incomplete plane (Figure 8.21a). Hypotheses for the origin of screw dislocations are more involved. One such hypothesis envisions a partial mismatch of lattices when two growing crystallites coalesce (Figure 8.63b). Another hypothesis proposes that variations in the concentrations of impurities across the surfaces of crystal platelets cause them to deform and nucleate a screw dislocation. Screw dislocations themselves, once they exist, are thought to facilitate crystal growth. Instead of needing to precipitate a small portion of a completely new plane of atoms on an existing growth surface, a process that is energetically unfavored, atoms can adhere to step of the "spiral staircase" of a screw dislocation, thereby extending the crystal along the screw dislocation line (Figure 8.63c).

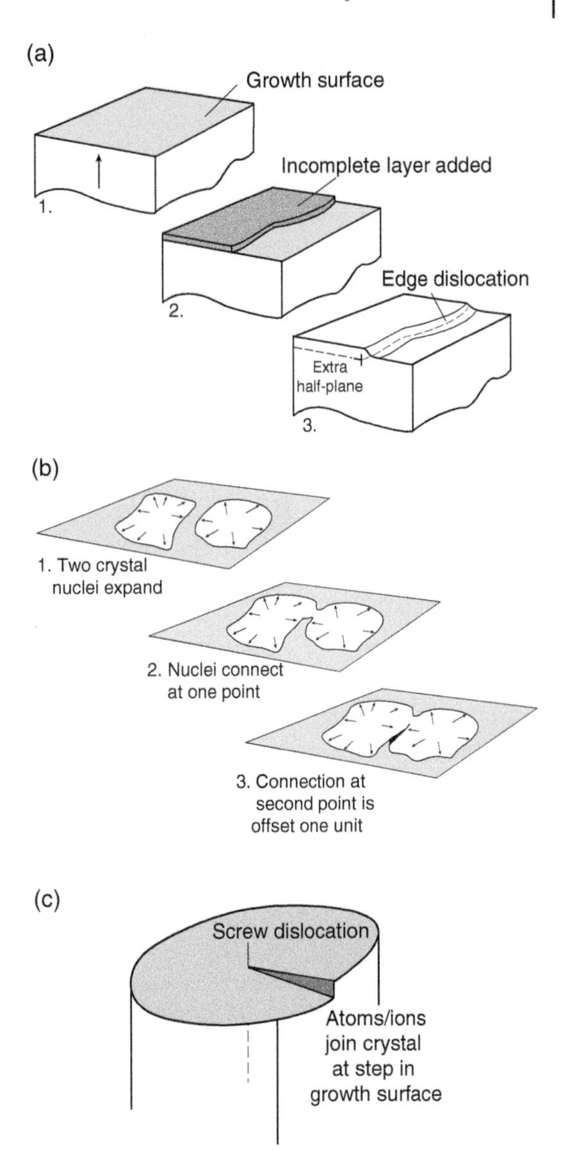

Figure 8.63 (a) Origin of an edge dislocation by incomplete addition of a later of atoms or ions. (b) Origin of a screw dislocation by coalescence of two crystal nuclei. (c) Crystals often grow by adding atoms or ions to the step in a lattice associated with a screw dislocation.

8B.3.2 Dislocation Movement

8B.3.2.1 Glide of Dislocations

8B.3.2.1.1 Slip Systems In describing the glide of dislocations in Section 8A.4.1.2, we introduced the concept of *slip systems*, which are defined by the orientations of the crystallographic plane or planes in which dislocations lie and the crystallographic direction or directions of the dislocations' Burgers vectors. Materials scientists use Miller index notation to identify the particular crystallographic planes in which dislocations glide and to indicate the orientations of the dislocations' Burgers vectors. Indices in parentheses, such as (*hkl*) or (*hkil*) for trigonal and hexagonal minerals, specify the crystallographic orientation of individual sets of planes containing dislocations. Indices in square brackets, such as [*uvw*] or [*uvtw*] for trigonal and hexagonal minerals, identify the crystallographic direction within those planes of the dislocations' Burgers vectors. Indices in braces or curly brackets, such as {*hkl*} or {*hkil*} for hexagonal or trigonal minerals, denote crystallographic *forms*, which consist of two or more sets of differently oriented but symmetrically equivalent planes. Indices in pointy or angle brackets, such as <*uvw*> (or <*uvtw*>), denote two or more distinct but symmetrically equivalent linear directions within crystals.

Some minerals possess slip systems defined by slip in one direction on a family of parallel planes. For example, the (100)[001] slip system in the orthorhombic mineral olivine denotes slip parallel to olivine's *c* axis on planes perpendicular to the mineral's *a* axis and parallel to its *b* and *c* axes. Other minerals have slip systems defined by slip in symmetrically equivalent directions on two or more distinct but symmetrically equivalent families of planes. Thus, the $\{110\}<1\bar{1}0>$ slip system in the isometric mineral halite denotes slip on any of the six planes that are parallel to one *a* axis and inclined 45° to the two other *a* axes, in directions inclined 45° to the one *a* axis that is parallel to the slip plane. It is possible to have slip systems that require a mix of the notation for individual planes or directions and symmetrically equivalent planes or directions. Thus, the $(0001)<11\bar{2}0>$ slip system in the trigonal mineral quartz refers to slip in one of three *a*-axis directions on the mineral's basal plane.

8B.3.2.1.2 Critical Resolved Shear Stress Most minerals possess two or more potential slip systems (Table 8.4). The main criterion determining whether or not a particular slip system contributes to deformation is, as noted in Section 8A.4.1.2, whether the shear stress acting on the crystallographic planes containing the dislocations and resolved in the direction parallel to their Burgers vectors reaches the threshold magnitude of the *critical resolved shear stress*, σ_{CR}. This is the stress required to initiate the glide of dislocations in that particular direction on any of the planes in the slip system. In any crystal with more than one slip system, the different slip systems will have different critical resolved shear (tangential) stress magnitudes. A primary factor determining the critical resolved shear stress for a particular slip system is

Table 8.4 Slip systems of common minerals.

Mineral	*Slip plane*	*Slip direction*
Quartz	(0001)	$\langle 11\bar{2}0\rangle$
	$\{10\bar{1}0\}$	(0001)
	$\{10\bar{1}0\}$	$\langle 1\bar{2}10\rangle$
	$\{10\bar{1}0\}$	$\langle 1\bar{2}13\rangle$
	{0111}	$\langle 11\bar{2}0\rangle$
Calcite	$\{\bar{1}2\bar{1}0\}$	$\langle \bar{2}021\rangle$
	(0001)	$\langle \bar{1}\bar{1}20\rangle$
	(0001)	$\langle \bar{1}2\bar{1}0\rangle$
	$\{\bar{1}012\}$	$\langle 2\bar{2}01\rangle$
	$\{\bar{1}012\}$	$\langle 10\bar{1}1\rangle$
	$\{10\bar{1}0\}$	$\langle \bar{1}2\bar{1}0\rangle$
	$\{10\bar{1}4\}$	$\{10\bar{1}0\}$
	$\{10\bar{1}4\}$	$\langle \bar{2}021\rangle$
Olivine	(010)	[100]
	(0kl)	[100]
	(001)	(100)
Orthopyroxene	(100)	[001]
	(100)	[010]
	(010)	[001]
Micas	(001)	<110>
	(001)	[100]

the packing of atoms within that system's crystallographic planes. Glide is easier on crystallographic planes with more dense packing of atoms. First, denser packing keeps the length of dislocation's Burgers vectors to a minimum, and the work required for glide scales directly with the distance atoms must move during each glide step. Second, in any volume in a crystal containing an arbitrary number of atoms, if certain planes have greater numbers of atoms per unit area, those planes will necessarily be more widely spaced, meaning that those planes are less strongly bound to each other (Nicolas and Poirier 1976, p. 40). Both effects also pertain to the more open structures found in rock-forming minerals.

Another important factor determining the magnitude of σ_{CR} is the intrinsic strength of the bonds within a crystal. To illustrate this point, consider the ionic solid LiF and the minerals halite (NaCl) and periclase (MgO). All are isostructural, i.e. crystallize with geometrically similar, isometric or cubic arrays of cations and anions. Their crystal structures consist of arrays of cations (Li^{+}, Na^{+}, or Mg^{+2}, respectively) coordinated with six near-neighbor anions (F^{-}, Cl^{-}, or O^{-2}, respectively). Although these three solids have the same crystal structure, they exhibit different physical properties due to differences in the strengths of the ionic bonds in the different compounds. Ionic bond strength is directly proportional to the product of the cation and anion valences. If all other factors were equal, the bond between divalent Mg and O ions (product of the valences = −4) would be stronger than the bonds between univalent Na and Cl or Li and F ions (product of the valences = −1). Ionic bond strength is also inversely proportional to the square of the interionic distance. The ionic radius of Mg^{+2} < ionic radius of Na^{+} < ionic radius of Li^{+}, and the ionic radius of O^{-2} < ionic radius of Cl^{-} < ionic radius of F^{-}. On the basis of interionic distance alone, the relatively short-length Mg-O bonds would be stronger than the intermediate-length Na-Cl bonds, which would be stronger than the longer Li-F bonds. The combination of these factors makes the Mg-O bond considerably stronger than the Na-Cl bond, which is, in turn, stronger than the Li-F bond. In all three solids, $\{110\}\langle 1\bar{1}0\rangle$ is the slip system on which glide occurs most readily, but the differences in the strengths of the bonds in the three minerals lead to significant differences in the σ_{CR} required to activate the slip system in the different minerals. Not surprisingly, at room temperatures, deformation of MgO by dislocation movement requires differential stresses of several hundred MPa, whereas halite deforms at differential stresses on the order of 10 MPa. LiF crystals have sufficiently low critical resolved shear stresses that one can cause dislocations to glide by scratching the surface of a crystal (Gilman and Johnston 1962).

8B.3.2.1.3 The Schmid Factor In a deforming crystal, the magnitudes of the stresses resolved in arbitrary directions on planes with an arbitrary orientation within a crystal are determined by the magnitudes of the principal stresses and their orientation relative to the planes in question. To illustrate this relationship, consider a cylindrical crystal with across sectional area A_o (Figure 8.64). A force of magnitude F acts across the circular end of the crystal. The plane P within the crystal is inclined to the axis of the cylindrical crystal; its normal makes an angle ψ with the axis of the cylinder. A direction $\boldsymbol{d}$ lying in the plane P makes at an angle λ to the axis of the cylindrical crystal. Geometry determines the magnitude of the stress component acting parallel to $\boldsymbol{d}$ on the plane P, i.e. it is a function of the force F, the angle ψ, and the angle λ. The external stress $\boldsymbol{\sigma}_E$ acting parallel to the axis of the crystal has a magnitude $\sigma_E = F/A_o$. The area of the plane P is $A_o/\cos\psi$. So, the magnitude of the stress vector acting on plane P is

$$\sigma_P = \left(\frac{F}{A_o}\right)\cos\psi = \sigma_E \cos\psi. \tag{8.43}$$

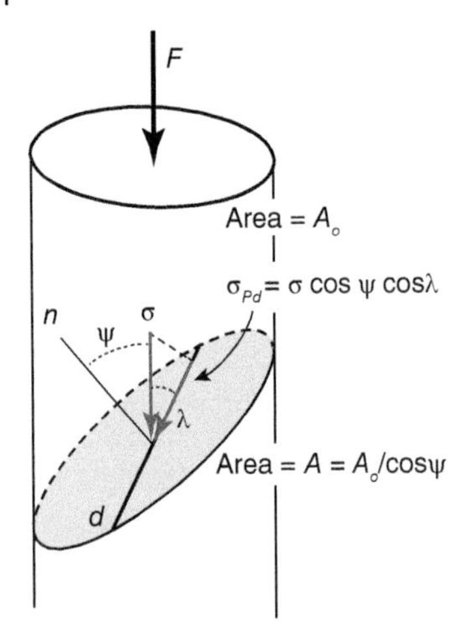

Figure 8.64 A cylindrical crystal with area A_o has an end load ***F***. Shear stress resolved parallel to direction *d* on plane *A*, which inclined to the end load, is σ_{pd} cosψ cosλ. See text for explanation.

The magnitude of the component of $\boldsymbol{\sigma}_E$ acting parallel to ***d*** is

$$\sigma_{Pd} = \sigma_P \cos\ \lambda. \tag{8.44}$$

Combining Eqs. (8.43) and (8.44), the magnitude of the component of the stress vector acting parallel to ***d*** on plane *P* is

$$\sigma_{Pd} = \left(\frac{F}{A_o}\right)\cos\psi\cos\lambda = \sigma_E\cos\psi\cos\lambda = S\sigma_E. \tag{8.45}$$

The variable *S* is called the *Schmid factor* after the scientist who first applied this calculation to study slip systems. For any angle ϕ, $\cos\phi \leq 1$ for all ϕ. So, $S \leq 1$ for all values of angles ψ and λ (in fact, $S \leq 1/2$ for all values of ψ and λ). Thus, $\sigma_{Pd} \leq \sigma_E$ for any plane *P* and linear direction ***d***.

If we consider the plane *P* and direction ***d*** to denote the sole slip system in the crystal, dislocations with Burgers vectors parallel to ***d*** will glide on planes parallel to *P* only when the magnitude of $\boldsymbol{\sigma}_E$ is sufficiently large that $\sigma_{Pd} = \sigma_{CR}$, the critical resolved shear stress for glide, i.e.

$$\sigma_E = \frac{\sigma_{Pd}}{S} = \frac{\sigma_{Pd}}{\cos\psi\cos\lambda} = \frac{\sigma_{CR}}{\cos\psi\cos\lambda}. \tag{8.46}$$

As the end load is increased from zero, the crystal will distort elastically until σ_E reaches the value specified in Eq. (8.46). Note that the orientations of the plane *P* and direction ***d*** determine the magnitude of the external stress $\boldsymbol{\sigma}_E$, at which dislocations will begin to glide on plane *P*. The onset of plastic deformation will occur at different magnitudes of $\boldsymbol{\sigma}_E$ depending upon the values of ψ and λ. This means that the elastic limit of this crystal varies with the orientations of *P* and ***d*** relative to the axis of the cylindrical crystal.

If a crystal has several symmetrically equivalent combinations of slip planes and Burgers vectors' directions, the factor *S* determines which, if any, of those slip planes would become active under a particular imposed external stress $\boldsymbol{\sigma}_E$. When a crystal has two or more distinct slip systems, the factor *S* determines whether either or both of those slip systems are activated under a given orientation and magnitude of $\boldsymbol{\sigma}_E$.

Beginning with an undeformed crystal, as the magnitude of $\boldsymbol{\sigma}_E$ increases, dislocations in the glide plane(s) for which product $S\sigma_E \geq \sigma_{CR}$ will begin to glide. In nature, those dislocations interact with other defects in the crystal, developing the jogs, pile-ups, etc., that cause the magnitude of the critical resolved shear stress σ_{CR} on glide planes to rise. The continued glide of dislocations on those glide planes requires either that: (1) the dislocations climb past obstacles; or (2) the magnitude of the external stress $\boldsymbol{\sigma}_E$ increases. If more dislocations are pinned than can climb past obstacles, the critical resolved shear stress rises, and continued deformation requires $\boldsymbol{\sigma}_E$ to increase. As $\boldsymbol{\sigma}_E$ increases, there is a greater likelihood that: (1) the product $S\boldsymbol{\sigma}_E$ equals the critical resolved shear stress for other, symmetrically equivalent glide planes, i.e. members of the slip system that were not favorably oriented for glide at the outset; or (2) the product $S\boldsymbol{\sigma}_E$ equals the critical resolved shear stress for other slip systems with intrinsically higher values of critical resolved shear stress. More dislocations gliding may lead to more dislocation interactions, which typically further raises the external stress required for continued deformation.

8B.3.2.1.4 Complexities Related to Ionic-Covalent Bonding in Minerals

Most rock-forming minerals differ from the idealized cubic arrays of atoms used thus far to illustrate the character of dislocations. In common mineral groups like the silicates, carbonates, or sulfates, for example, cations are bound to and coordinated with anionic groups that themselves are held together by strong partially ionic and partially covalent bonds. The more complex atomic-scale structure of most rock-forming minerals affects the glide of dislocations in two ways. First, even in lattice planes with densely packed atoms, there may be variations in the strengths of bonds between atoms that affect the possible directions for the glide of dislocations. The glide of dislocations with Burgers vectors parallel to weaker bonds is more likely than the glide of dislocations with Burgers vectors parallel to stronger bonds. Second, in more complex crystal structures, it is more likely that equivalent atoms are more widely spaced and are separated by nonequivalent atoms, so that equivalent atoms effectively occur only in every other plane of atoms.

One result commonly observed in more complex crystal structures is the *dissocation* of a *dislocation* into two or more *partial dislocations*. In such cases, instead of a single Burgers vector connecting symmetrically equivalent atoms across a single dislocation line, the crystal lattice on one side of a glide plane is displaced relative to the lattice on the opposite side by the combined effect of movement across two or more dislocation lines – the partial dislocations (Figure 8.65). Each partial dislocation has its own associated Burgers vector, and the sequential displacement associated with the passage of the two or more partial dislocations is required to yield a total Burgers, whose direction and magnitude give the displacement of symmetrically equivalent atoms in the complex crystal structure.

(a)

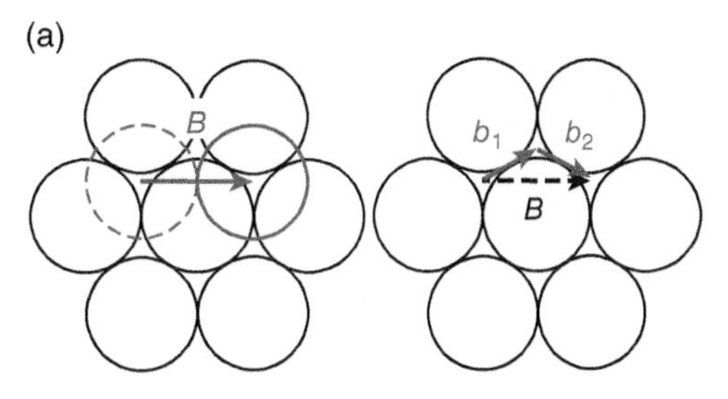

(b)

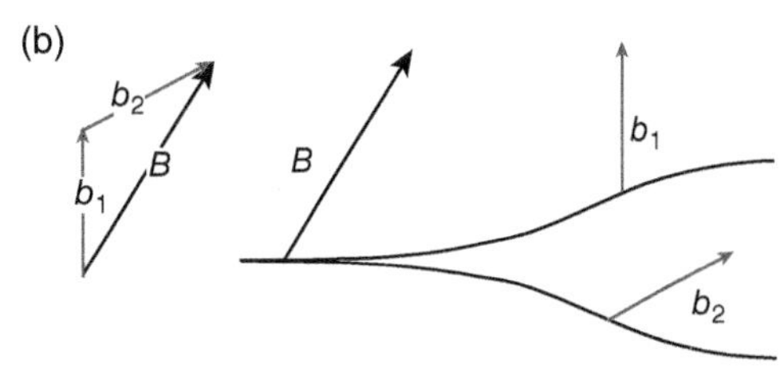

Figure 8.65 (a) A portion of a layer of closest-pack atoms (in black) with an atom in the overlying layer in red. The dashed circle shows the original position of an atom, and solid circle shows the position after passage of dislocation with Burgers vector ***B***. Moving atom along path $b_1 + b_2$ is easier than along path ***B***. (b) Dislocation with Burgers vector ***B*** dissociates into two partial dislocations with Burgers vectors b_1 and b_2, whose sequential passage accomplishes the lattice offset associated with the original dislocation.

8B.3.2.2 Climb of Dislocations

8B.3.2.2.1 The Role of Vacancies in Climb

Figure 8.66a is a depiction of the extra half plane of an edge dislocation with a "ragged" or "stepped" configuration due to the presence of vacancies. Vacancies occur along the dislocation line for the same reason that they exist in crystals. Their presence increases the number of possible configurations for atoms or ions and vacancies, which, in turn, increases the configurational entropy contribution to the Gibbs energy or free enthalpy of the dislocation sufficiently to exceed the added energy associated with the increased length of the dislocation line. Thus, edge dislocations have

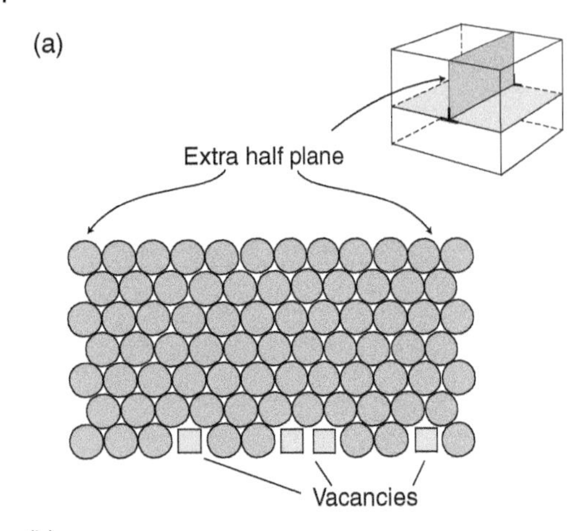

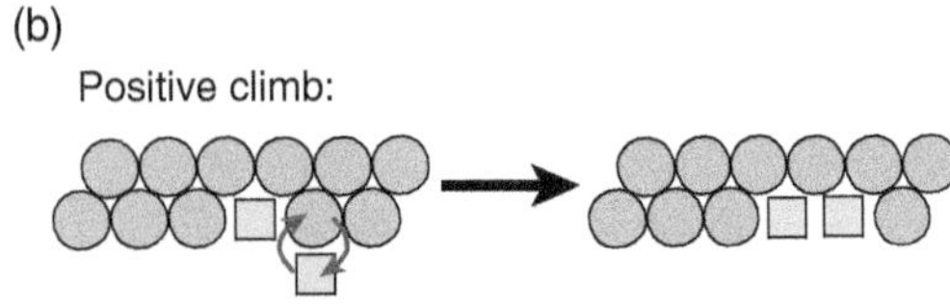

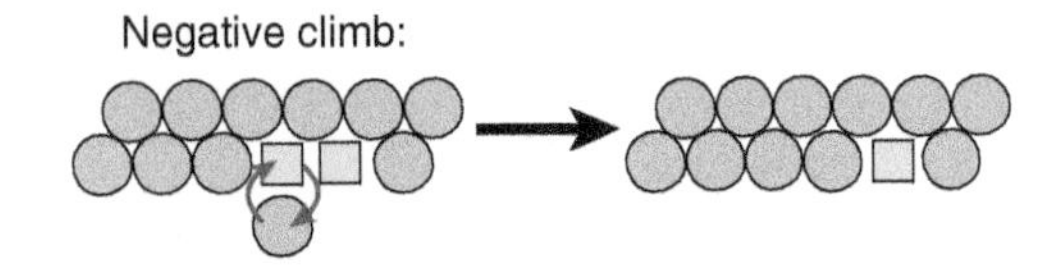

Figure 8.66 (a) Schematic view showing the "ragged" edge of the extra half plane of an edge dislocation formed by the occurrence of vacancies along the dislocation line. (b) Two ways that diffusion of vacancies or atoms/ions are likely to contribute to the ragged character of the end of the half plane, and enable the dislocation to climb.

lower Gibbs energy or free enthalpy with steps than without them.

Positive or negative climb of the edge dislocation occurs by respectively absorbing (i.e. adding) vacancies to the extra half plane or emitting (i.e. subtracting) vacancies from the extra half plane at one of the steps in the dislocation (Figure 8.66b). From a mechanistic point of view, adding a vacancy to a step or subtracting one from a step in a dislocation line causes relatively little to change the overall state of the crystal, especially compared to adding or removing an atom in the middle of a dislocation line. Under most circumstances, the diffusion of vacancies toward or away from the edge dislocation controls the rate at which an edge dislocation climbs, not the process of emitting or absorbing the vacancy.

8B.3.2.2.2 Factors Controlling the Rate of Climb

Since the extra half plane of a typical edge dislocation possesses jogs or has a stepped shape, the velocity at which the dislocation climbs is approximately equal to the velocity at which individual jogs move along the dislocation line. If b is the spacing of atoms along the edge dislocation and Φ is the number of vacancies arriving via diffusion per unit length of dislocation line per unit time, then the number of vacancies that arrive at the jog per unit time is the product of Φ and b, the length of the jog. The product Φb has units of the number of vacancies absorbed per unit time. Since each vacancy absorbed moves the jog a distance b, the velocity of the jog v_j is

$$v_j = \Phi b^2. \tag{8.47a}$$

We take velocity at which the dislocation climbs v_c to be equal to the velocity of the jog along the dislocation, i.e.

$$v_j = \Phi b^2 = v_c. \tag{8.47b}$$

Absorbing or emitting a vacancy along the edge of dislocation is comparable to creating or absorbing a vacancy at a free surface parallel to the extra half plane. Section 8B.2.3 showed that variations in the magnitudes of stress acting on free surfaces of crystals alter the local concentrations of vacancies. The lattice distortion in the vicinity of the edge of an extra half plane gives rise to a local increase in differential stress $\Delta\sigma$, which, in turn, lowers the concentration of vacancies in the vicinity of the edge dislocation (see Appendix 8-II). Vacancies then diffuse toward the edge of the dislocation at a rate dependent upon the gradient in the concentration of vacancies. If L is the average distance between the dislocation line, where the concentration of vacancies is C, and portions of the crystal with the equilibrium concentration of vacancies (C_o), then the diffusive flux of vacancies (in number of vacancies per unit area per unit time) is

$$J = -D_v \frac{dC}{dx} = C_o D_v \frac{\Delta\sigma b^3}{kTL}, \tag{8.48}$$

where D_v is the diffusion coefficient for vacancies and $\Delta\sigma$ is the stress in the vicinity of the dislocation (see Appendix 8-II). To reach the edge of the extra half plane, vacancies must diffuse across the surface of an imaginary cylinder with radius b surrounding the dislocation line. Since the cylinder's surface area per unit length of the dislocation line is $2\pi b$, the number of vacancies that reach the dislocation line per unit time $\Phi = 2\pi bJ$ or

$$\Phi \propto bC_o D_v \frac{\Delta\sigma b^3}{kTL}. \tag{8.49a}$$

From Section 8B.2.3, the diffusion coefficient for self-diffusion $D_{SD} = b^3 C_o D_v$, so

$$\Phi \propto \frac{D_{SD}}{b^2} \frac{\Delta\sigma b^3}{kTL} \tag{8.49b}$$

and the diffusion-controlled climb velocity v_c is

$$v_c = \Phi b^2 \propto \frac{D_{SD} \Delta\sigma b^3}{kTL}. \tag{8.50}$$

The important result here is that the rate of climb of dislocation is dependent upon the magnitude of the diffusion coefficient for self-diffusion.

8B.3.3 Dislocation Interactions

8B.3.3.1 Dislocation Interactions that Increase Dislocation Density

A gliding edge dislocation is constrained to a particular crystallographic plane – its *glide planes* – because the Burgers vectors of the edge dislocation are perpendicular to the dislocation line. The Burgers vector of a pure screw dislocation is parallel to the dislocation line. So, pure screw dislocations do not have well-defined glide planes. If crystal symmetry is such that there are multiple lattice planes with equivalent atom densities that could contain symmetrically equivalent dislocations, a gliding pure screw dislocation line can "step" from one such lattice plane to another symmetrically equivalent plane. This process is known as *cross slip* (Figure 8.67a). Cross slip creates closed dislocation loops that are not confined to a single, continuous lattice plane, and it regularly leads to an increase in the total length of dislocations. In minerals with sufficiently complex crystal structures that dislocations dissociate into partial dislocations, cross slip is a multiple-step process. Until all of the steps are completed, the dislocation line may be pinned at the intersection of the glide planes (Figure 8.67b).

Consider the situation in which a segment of an edge dislocation is pinned, either due to the creation of jogs or due to cross slip, at two points along its length. If the length of the jogs or the cross-slipped segments are sufficiently long and the shear stress resolved on the glide plane containing the edge dislocation equals or exceeds the critical resolved shear stress, the finite length of edge dislocation may develop into a *Frank-Read source.* Figure 8.68 illustrates the sequence of events that can occur repeatedly at Frank-Read sources. At the outset (Figure 8.68a), the segment of edge dislocation pinned at points a and b lies in a glide plane. Shear stress on the glide plane equal to the critical resolved shear stress causes the dislocation to glide. By time t_2, movement of the dislocation has swung the dislocation line segments immediately adjacent to the pinning points a and b into orientations parallel to the Burgers vector. These segments now have right- or left-handed screw characters and move to the left or right respectively in response to the applied shear stress. By time t_3, segments of the dislocation near points a and b have once again swung into

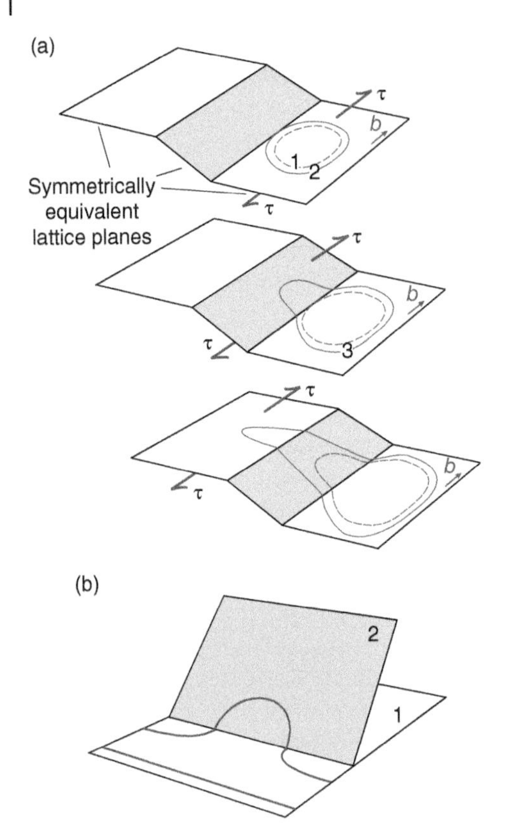

Figure 8.67 (a) As a closed dislocation loop expands (shown at times 1, 2, and 3), a pure screw segment of the dislocation (where the dislocation line is parallel to the Burgers vector) is free to move from one glide plane to a symmetrically equivalent plane. (b) If dislocation dissociates into two partial dislocations, complications associated with the cross slip of partial dislocations from glide plane 1 to glide plane 2 can slow or pin dislocations at intersections of two glide planes.

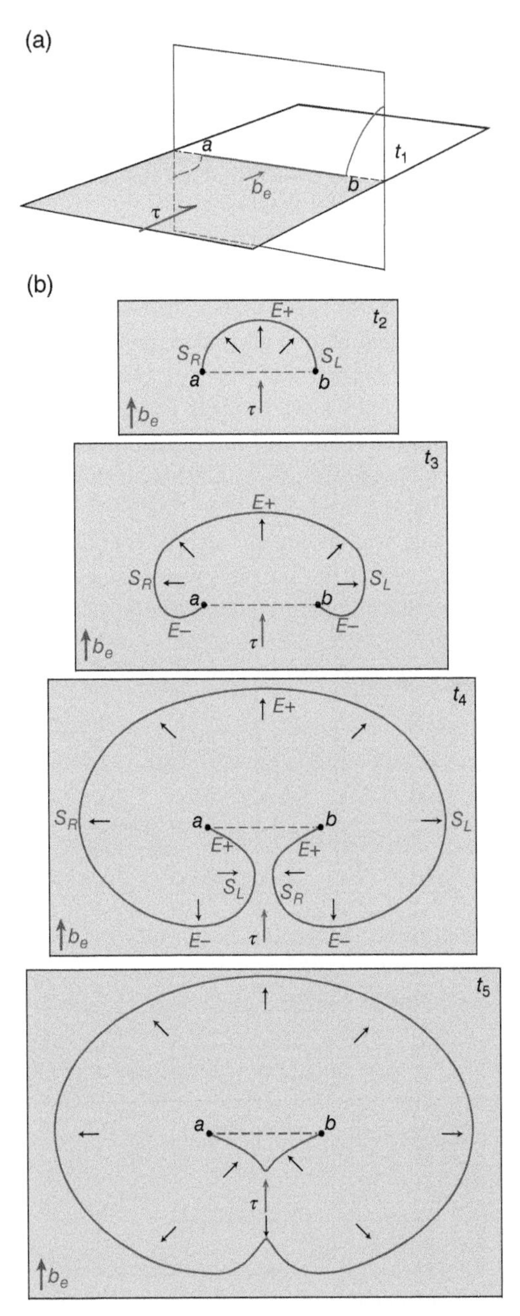

Figure 8.68 A Frank-Read source. (a) Glide plane with a segment of edge dislocation with pinned points a and b at time t_1. (b) View showing successive times t_2 to t_5, during which the dislocation loop expands and intersects itself. See text for additional information.

orientations approximately perpendicular to the Burgers vector. At that time, however, they have negative edge character and so move in a direction 180° from the positive edge segments. By time t_4, continued slip of the different segments has created two new segments of dislocation line with screw character "behind" pinning points a and b. These two newest segments of screw dislocation move toward each other as the rest of the dislocation loop expands. Eventually, the two screw-dislocation segments meet and annihilate each other, leaving: (1) a closed loop of dislocation that can continue to expand as long as shear stresses equal or exceed the critical resolved shear stress value: and (2) a new segment of positive edge dislocation between points *a* and *b*. The segment of edge dislocation between *a* and *b* may itself begin gliding and eventually exhibit a comparable evolutionary sequence as the original, pinned dislocation. This sequence of events can occur over and over, leading some materials scientists to describe Frank-Read sources as "mills" for generating dislocations and increasing the density of dislocations in a deforming crystal.

As dislocation densities increase due to dislocation interactions, there is a greater likelihood of additional dislocation interactions. This, in turn, leads to further increases in the dislocation density, etc. The net result is a significant increase in the number of dislocations in a deforming mineral. Recall that dislocation densities in a typical undeformed crystal are on the order of $10^4 cm^{-2}$. Dislocation densities of $10^{12} cm^{-2}$ have been measured in experimentally deformed single crystals – an increase in dislocation density by a factor of up to a billion over the dislocation density at the outset of deformation. This degree of increase of dislocation density gives an indication of the sort of changes that are wrought by deformation.

8B.3.3.2 Effects of Dislocation Interactions

Some dislocations with jogs are still able to glide, albeit only at higher values of critical resolved shear stress to glide (see Section 8A.4.2.1). Other pinned dislocations can no longer glide. Both factors may lead to the activation of dislocations that lie in less favorably oriented glide planes, i.e. slip systems that require higher values of the critical resolved shear stress to move. As a result, higher values of deviatoric stresses are required to continue changing a crystal's shape, i.e. the crystals *work harden*. Increased dislocation densities are also directly responsible for broadening peaks in an x-ray diffraction pattern, locally altering minerals' optical properties, and the appearance of *deformation lamellae*. Finally, the increased volume of distorted lattice associated with high dislocation densities raises the internal energy of that crystal and provides the impetus to drive the recovery and recrystallization (see Section 8A.4.3).

We observe that impurities concentrate along grain boundaries because they add less to a crystal aggregate's total internal energy by residing in an already distorted lattice than by sitting in and locally distorting perfect crystal lattice. Concentrating impurities along grain boundaries tends to reduce even further the "bondedness" of atoms and ions in grain boundaries and further enhances the mobility of atoms and ions along grain boundaries. Grain boundaries are, then, important paths for components to diffuse through an aggregate (cf. Section 8B.2.5), whether that diffusion is self-diffusion (where a component diffuses through a volume composed of the same chemical component) or solution diffusion (where an "impurity" diffuses through a "host solution").

Diffusion also occurs relatively readily along dislocations because: (1) the distorted crystal lattice in the vicinity of the dislocation has space for additional atoms or ions; and (2) because the addition of an "impurity" generates a proportionally smaller increase in internal energy. Dislocations that intersect the grain boundaries provide a relatively easy path for impurities to diffuse from the grain boundary into the center of a grain. Increases in dislocation density multiply the number of paths of rapid diffusion into the center of a deforming crystal. This effect may explain why deformation enhances metamorphism.

One situation where such diffusional effects are very important is in the *hydrolytic weakening* of silicates. Unless H_2O can be incorporated into the lattice of one of the minerals in a silicate rock, water concentrates along the boundaries between grains in the rock. From its position along grain boundaries, water can diffuse along dislocations. Water along dislocations is not "free" water, it reacts with the silicate. That is, Si-O-Si bridges in a silicate structure, which are strong covalent-ionic bonds, are replaced by Si-OH···HO-Si bridges, held together by a weaker hydrogen bond. The critical resolved shear stress needed to break the hydrogen bond is lower than that needed to break the Si-O-Si bond, meaning that dislocations move at lower differential stresses. H_2O probably moves with the moving dislocation, because it adds less energy to the lattice by staying in the distorted lattice in the vicinity of the dislocation. The overall effect is *lower flow stress* for wet silicates deforming by dislocation movement than for dry silicates.

8B.3.4 Stresses Associated with Dislocations

The atoms in the vicinity of a dislocation are displaced from their equilibrium positions, defining a local elastic strain field that surrounds, is tied to, and moves with the dislocation. Along a pure screw dislocation, atoms are offset distance b, the length of the Burgers vector, parallel to the dislocation line. Offsets, and, therefore, elastic shear strains, decrease with distance from the dislocation line at a rate proportional to $1/r$, where r is the radial distance from the dislocation line (Nicolas and Poirier 1976, pp. 80–85; Hull and Bacon 1984, pp. 75–77). The magnitudes of the stress tensor components in the elastically distorted crystal are proportional to the respective strain components, whose magnitudes vary with distance from the dislocation line. In the vicinity of a screw dislocation then, the magnitudes of individual components of stress tensor vary as

$$\sigma \sim \frac{Gb}{r}, \tag{8.51}$$

where G is the shear modulus of the crystal.

Algebraic expressions specifying the displacement and strain fields in the vicinity pure edge dislocation are more involved (Hull and Bacon 1984, pp. 77–78), with elastic strains due to shortened interatomic or interionic distances on the side of the glide plane with the extra half plane of atoms/ions and elastic strains due to elongated interatomic or interionic distances on the opposite side of the glide plane. Thus, the elastic stresses have opposite signs on either side of the edge dislocation's glide plane – compressive on the side with the extra half plane and tensile on the opposite side of the glide plane. Nevertheless, expressions for stresses near a pure edge dislocation are similar to Eq. (8.51) in that stress magnitudes are directly proportional to G and b and decrease at a rate proportional to $1/r$.

In steady-state flow, the applied stress and the stress in the vicinity of a dislocation line are generally balanced, i.e.

$$\frac{Gb}{r} \sim S\sigma_E. \tag{8.52a}$$

In such a situation, we envision that the dislocation density during steady-state flow is one dislocation per square with area r^2, where r is given by

$$r \sim \frac{Gb}{S\sigma_E}. \tag{8.52b}$$

Substituting from Equation (8.52b), the dislocation density ρ_d in steady-state flow is

$$\rho_d = \frac{1}{r^2} \sim \left(\frac{S}{b}\right)^2 \left(\frac{\sigma_E}{G}\right)^2. \tag{8.53}$$

Empirical studies confirm the functional form of this relationship, i.e. that the dislocation density is proportional to the applied stress squared.

8B.3.5 Strains Accommodated by the Glide of Dislocations

8B.3.5.1 Estimating Magnitudes of Strain Due to the Glide of Dislocations

In order to visualize the relationship between dislocation glide and strain, consider a crystal with the

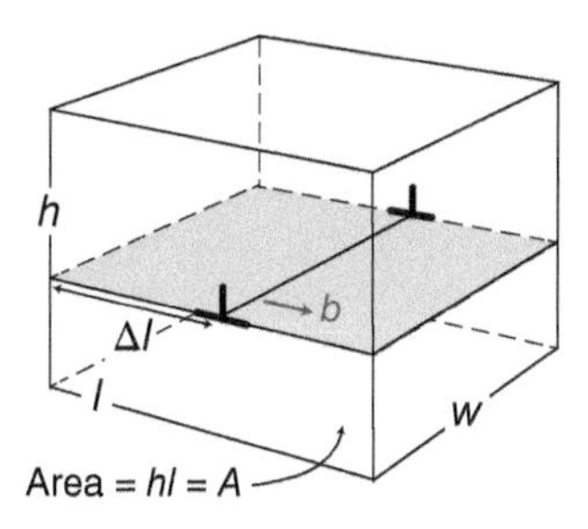

Figure 8.69 An edge dislocation with Burgers vector b in a crystal with length l, width w, and height h.

shape of a rectangular parallelepiped with edges h (height), l (length), and w (width) (Figure 8.69). A single, straight edge dislocation lies in a glide plane perpendicular to h and parallel to l and w. The length of the dislocation is w; its Burgers vector is aligned parallel to the crystal's length l and has a magnitude of b. The glide of that dislocation across the entire glide plane produces a shear strain $\gamma = b/h$. If the dislocation does not glide completely across the glide plane, but traverses a distance $\Delta l < l$, it generates an average shear strain of

$$\gamma = \frac{b}{h}\frac{\Delta l}{l}. \tag{8.54}$$

If there are N such dislocations in the crystal (in different but parallel glide planes) and if the average distance the dislocations have traversed across their respective glide planes is Δl, the total shear strain is

$$\gamma = N\frac{b}{h}\frac{\Delta l}{l}. \tag{8.55}$$

Recall that dislocation densities are specified as the net length of dislocation per unit volume or the number of dislocations per unit area. With N dislocations like the one shown in Figure 8.69, the total length of dislocation line is Nw. The area of the front face A in the figure is hl, and the volume of the crystal is $V = hlw = Aw$. The dislocation density ρ_d in this crystal is then

$$\rho_d = \frac{Nw}{V} = \frac{N}{A} = \frac{N}{hl}, \tag{8.56}$$

and Eq. (8.54) becomes

$$\gamma = \rho_d b \Delta l. \tag{8.57a}$$

Assuming that similar results would obtain from a population of curved dislocations with density $\bar{\rho}_d$ and average slipped distance $\overline{\Delta l}$, and replacing γ by the more general label for strain ε, the strain generated by dislocation glide is

$$\varepsilon = \bar{\rho}_d b \overline{\Delta l}. \tag{8.57b}$$

For a crystal 10^{-2} m = 1 cm across with a dislocation density typical of undeformed crystals $\approx 10^5\,\text{cm}^{-2}$, $b \sim 1\,\text{Å} = 10^{-8}$ cm, and $\overline{\Delta l}$ taken to be on the order of 0.5 cm ($\approx$ the diameter of a typical mineral grain in the rock), Equation (8.57b) gives a shear strain of 5×10^{-3}, which is far too small to be measured. In Section 8B.3.3.1, we noted that dislocation densities of $10^{12}\,\text{cm}^{-2}$ have been measured in experimentally deformed crystals. Substituting that value of dislocation density into Eq. (8.57b) gives a shear strain of 5×10^4. In experimentally deformed crystals, dislocation densities of $10^{12}\,\text{cm}^{-2}$ are not found throughout the sample; those dislocation densities occur within deformation lamellae. Thus, this extreme value of shear strain is likely to be generated only locally within a crystal. Still, this "back-of-the-envelope" calculation demonstrates that measurable shear strains are readily attainable by dislocation glide.

8B.3.5.2 Dislocation Glide and Three-Dimensional Strains

The state of strain accommodated by the glide of dislocations described in the preceding section and quantified in Eq. (8.57) is a three-dimensional simple shear, i.e. a plane strain with the maximum elongation and maximum shortening axes lying within a plane perpendicular to the glide plane lw in Figure 8.69. The position gradient matrix for three-dimensional simple shear is, in its simplest format (where the local x_2 axis is perpendicular to the shear plane and the local x_1 axis is parallel to the shear direction), specified by defining a single parameter – the shear strain parallel to the shear axis = γ defined in Eq. (8.57a) (see Section 5B.4.14):

$$\mathbf{F} = \left[\mathrm{F}\right] = \begin{vmatrix} 1 & \gamma & 0 \\ 0 & 1 & 0 \\ 0 & 0 & 1 \end{vmatrix}$$

One can use the tensor transformation law to determine the components of the equivalent position gradient matrix referred to a different coordinate frame with axes appropriate to the deformation of the aggregate, but that position gradient matrix still has only a single independent parameter. Similarly, the Finger tensor = $\mathbf{F{\cdot}F}^{\mathrm{T}}$, the symmetric 3×3 matrix that characterizes fully the change in shape, has a single independent parameter.

The boundary conditions of deformation regularly dictate non-plane strain states or plane strain states that cannot be attained by simple shear on the planes oriented parallel to a crystal's slip system. The most general form for the Finger tensor, one that can describe fully any change in the shape of the crystal, has six independent components. If the deformation occurred with no change in volume, a reasonable inference in the case of dislocation glide, the number of independent components required to specify the matrix is five. The requirement of five independent slip systems to accomplish a general 3D deformation is called the *von Mises condition*, after Richard von Mises who first enumerated it.

There is a profound disjunction between what is required to define an arbitrary, general change of shape and what can be attained by dislocation glide on a single set of parallel crystallographic planes system. In terms of the mathematical description of the strain state, only by combining five distinct simple shear position gradient matrices, each corresponding to simple shear in a unique direction on uniquely oriented planes, can one generate an arbitrary, general three-dimensional strain state. From this, materials scientists conclude that five independent slip systems are required to accomplish a general three-dimensional shape change of a crystal. In this context, an "independent" slip system is one whose effect cannot be reproduced by a linear combination of other slip systems. Thus, two slip systems, i.e. $(110)[1\bar{1}0]$ and $(101)[01\bar{1}]$ in isometric minerals, can be independent even though they are symmetrically equivalent. On the other hand, in the mineral quartz only two of the three slip directions in $(0001)\langle 11\bar{2}0\rangle$ slip system are independent; the third can be expressed as a linear combination of the other two.

Many rock-forming minerals do not have the requisite five independent slip systems. Instead, these minerals regularly develop features that compensate for the fact that dislocation glide alone cannot accommodate the strain. The accommodation features include bending of glide planes, kinking of glide planes, and opening of fractures between glide planes or along the boundaries between grains (cf. Figure 8.37).

One might envision that a collection of randomly oriented grains in a deforming polycrystalline aggregate could together accommodate any arbitrary, three-dimensional change of shape. In a deforming polycrystalline aggregate, however, individual crystals are not free to change shape independently. Simple shear on glide planes with different orientations in different crystals can lead to incompatible changes in the shapes for neighboring crystals. Distorting crystals may slide past their neighbors on relatively weak grain boundaries and reorient slightly to accommodate these differences during the initial deformation increments, but the amounts of grain boundary sliding and grain rotation are themselves limited. Any sizable distortion of individual crystals would tend to open gaps along grain boundaries or cause crystals to attempt to overlap. These effects prevent the aggregate from undergoing finite changes of shape. As a result, in addition to the intracrystalline compensation structures mentioned earlier, mobile grain boundaries contribute to accommodating grain impingement and minimizing gaps or holes in deforming aggregates.

8B.3.6 Constitutive Equations for Dislocation Creep

Dislocation creep requires both the glide and climb of dislocations. The glide of dislocations alone accomplishes the change of shape of crystals. In all but the most restrictive settings, a general change of shape requires that either: (1) the glide of dislocations in multiple slip systems; or (2) the combination of glide on limited numbers of slip systems and subgrain or grain boundary mobility. Interactions between gliding dislocations, particularly when several slip systems are active but also when only one glide system is active, impedes continued glide. Dislocation climb, which occurs spontaneously, frees once-impeded dislocations to glide again and accommodate additional strain. Similarly, the mobility of subgrain and grain boundaries requires the climb of dislocations. In essentially all conceptualizations of dislocation creep, the glide of unencumbered dislocations is envisioned to be sufficiently rapid that dislocations glide instantaneously between impediments, where they are held until the climb releases them again. There are, however, different models for conceptualizing how the glide and climb of dislocations combine, and these different models lead to different formulations of the constitutive equation for dislocation creep. In this section, we outline one of the models used to determine a constitutive equation for dislocation creep as an example. A fundamental assumption in formulating this model is the assumption that climb is the "rate-limiting step" for dislocation movement.

Equation (8.57b) relates the shear strain experienced by a crystal to the mean dislocation density $\bar{\rho}$, Burgers vector length b, and mean distance $\overline{\Delta l}$ that dislocations glide. Taking L to be equal to $\overline{\Delta l}$, the mean distance that dislocations glide, and differentiating Eq. (8.57b) with respect to time yields an expression relating the velocity of dislocations to the macroscopic strain rate.

$$\frac{d\varepsilon}{dt} = \frac{d\bar{\rho}_d}{dt} bL + \bar{\rho}_d b \frac{dL}{dt}. \tag{8.58a}$$

Assuming that deformation is steady-state and that the dislocation density does not vary with time during steady-state flow, i.e. $\frac{d\bar{\rho}_d}{dt} = 0$,

$$\dot{\varepsilon} = \bar{\rho}_d b v_d, \tag{8.58b}$$

where $\dot{\varepsilon}$ is the strain rate and v_d is the mean velocity of dislocations. Equation (8.60b) is known as Orowan's equation, after the physicist/metallurgist Egon Orowan who first derived it. The mean velocity of dislocation is

$$v_d = \frac{L}{t_g + t_c}, \tag{8.59}$$

where t_g is the time required for glide and t_c is the time required for climb as the dislocation moves a distance L. The assumption that glide occurs rapidly and climb occurs slowly – climb is the "rate-limiting step" – means that $t_g << t_c$ and $v_d \approx L/t_c$. If d is the mean distance between glide planes, i.e. the distance that dislocation must climb, and v_c is again the climb velocity, then

$$v_d = \frac{L}{t_c} = \frac{L}{d} v_c. \tag{8.60}$$

Substituting this result into Orowan's equation yields

$$\dot{\varepsilon} = \bar{\rho}_d b \frac{L}{d} v_c. \tag{8.61}$$

Although the dislocation density does not vary with time in steady-state flow, we argued in

Section 8B.3.4 that the mean dislocation density varies with the differential stress magnitude at which steady-state flow occurs. The expression relating applied stress to dislocation density, i.e. Equation (8.53), is modified slightly in the steady-state creep model of Weertman (1968).

Weertman envisioned crystals with a series of glide planes distance d apart. Dislocations in those glide planes are closed loops of diameter L (Figure 8.70a). The continued glide of dislocations needed to accommodate the finite strain of a crystal is possible because dislocations are generated by sources located within the closed loops. However, dislocations on adjacent glide planes interact to form dislocation pairs called "dislocation dipoles" (Figure 8.70b). Once a dislocation in one glide plane is linked to another dislocation in a dipole, it no longer glides. Other dislocations "piled up" behind it in the plane are able to glide again only after climb annihilates the two dislocations in the dipole.

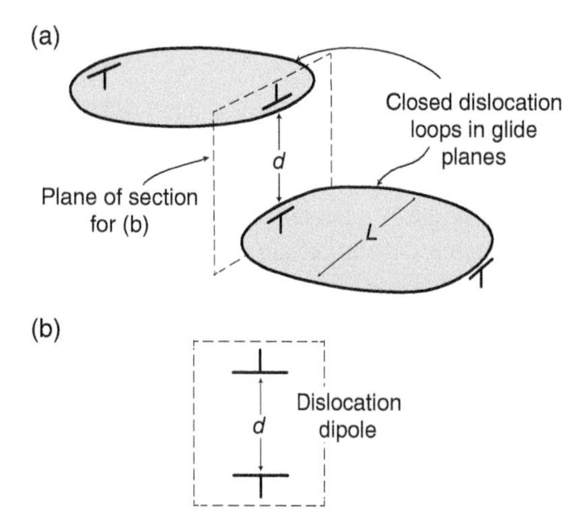

Figure 8.70 (a) Closed dislocation loops on parallel glide planes distance d apart. The positive edge dislocation on one glide plane and the negative edge dislocation on the other glide plane together form a dislocation dipole. (b) Schematic view of the dislocation dipole.

The density of those *mobile* dislocations is the product of the number of dislocation sources per unit volume × the length of dislocations per dipole × the number of dipoles for one source. Weertman proposed that there is one dislocation source in every cylindrical volume of size $\pi L^2 d$, i.e. the number of dislocation sources per unit volume is

$$M = \frac{1}{\pi L^2 d}. \tag{8.62}$$

The stress due to neighboring dislocations within a glide plane is proportional to $1/L$, and the stress between two dislocations on adjacent glide planes is proportional to $1/d$. Thus, the number of dipoles per source is ~ L/d. Combining these results, the dislocation density $\bar{\rho}_d$ is

$$\bar{\rho}_d = M \cdot \pi L \cdot \frac{L}{d} = \frac{\pi M L^2}{d}. \tag{8.63}$$

Weertman used Eqs. (8.62) and (8.63) to derive two expressions that are substituted for terms in Orowan's equation. In addition, by drawing upon the expressions relating the Burgers vector to the elastic stress in the vicinity of a dislocation (Eq. 8.53), he was able to derive an expression outlining the functional relationship between stress and strain rate, i.e. a constitutive relationship.

First, for each source, the length of dislocation is the circumference of the dislocation loop, i.e. πL. To estimate the number of dipoles for each source, Weertman argued that the stresses acting on each dislocation dipole balance. Combining Eqs. (8.62) and (8.63) indicates that $\bar{\rho} = 1/d^2$. Substituting this result into Orowan's Eq. (8.58b) yields

$$\dot{\varepsilon} = b\frac{L}{d^3}v_c. \tag{8.64}$$

Rearranging Eq. (8.64) yields

$$L = \left(\frac{1}{\pi M d}\right)^{1/2}. \tag{8.65a}$$

Replacing L in Eq. (8.64) gives

$$\dot{\varepsilon} = b\frac{1}{(\pi M d)^{1/2}}\frac{1}{d^3}v_c = \frac{b}{d^{7/2}M^{1/2}}v_c. \quad (8.65b)$$

Substituting the spacing between glide planes d for the distance r in Eq. (8.52) from Section 8B.3.4, $d \sim b(G/\Delta\sigma)$ and

$$\dot{\varepsilon} \sim \left(\frac{\Delta\sigma}{G}\right)^{7/2}\frac{b}{b^{7/2}M^{1/2}}v_c. \quad (8.66)$$

Finally, using the climb velocity derived in Section 8B.3.2.2.2 and given in Eq. (8.50),

$$\dot{\varepsilon} \sim \left(\frac{\Delta\sigma}{G}\right)^{9/2}\frac{D_{SD}}{M^{1/2}}. \quad (8.67)$$

Expression (8.67) has the same general form as the general flow law for dislocation creep, Eq. (8.11), given in Section 8A.4.8. Both indicate that the strain rate is directly proportional to the differential stress raised to a power $n > 1$, directly proportional to a diffusion coefficient for the diffusion of species through a crystal lattice, and independent of the grain diameter. Expression (8.67), derived in part by assuming that the density of dislocation sources, M, is independent of stress, predicts $n = 4.5$. If the density of dislocation sources depends upon stress, expression (8.67) predicts different power-law exponents (Karato 2008, pp. 158–159). For example, taking the dislocation source density to be related to the stress by $M \propto \bar{\rho}$, whereas above $\bar{\rho} \propto (\Delta\sigma)^2$, then expression (8.67) predicts a power-law exponent $n = 3.5$. Alternatively, if $M \propto \bar{\rho}/d \propto (\Delta\sigma)^3$, expression (8.67) predicts power-law exponent $n = 3$.

Materials scientists have envisioned alternatives to the dislocation dipole mechanism for coupling the glide and climb of dislocations. Those alternatives lead to different assumptions for the density of dislocation sources, the rates of dislocation production, and the distances over which dislocations must climb to facilitate continued dislocation glide. Those alternatives also predict flow laws for dislocation creep in which strain rate is directly proportional to the differential stress raised to a power $n > 1$, directly proportional to a diffusion coefficient for the diffusion of species through a crystal lattice, and independent of the grain diameter.

8B.3.7 Recovery, Recrystallization, and Dislocation Creep Regimes

The previous sections underscore the importance of dislocation climb in facilitating deformation by dislocation creep. Dislocation climb, of course, requires diffusion. Similarly, grain boundaries migrate primarily by the diffusion of atoms, ions, and vacancies across the tabular volumes that constitute grain boundaries (see Section 8A.4.3.3), which are a non-crystalline region. In short, diffusion is essential to dislocation movement, recovery, and the degree to which recrystallization contributes to and facilitates dislocation creep. The rates of diffusional mass transfer depend strongly upon temperature, and as a result, the temperature is an important factor determining the degree to which recovery and recrystallization occur during deformation by dislocation creep. Absolute temperature is not the sole determining factor, however. There is an interplay between the rate of deformation and the ambient temperature. Higher deformation rates tend to increase dislocation densities and are associated with higher flow stresses, whereas higher temperatures tend to lower dislocation densities and are associated with lower flow stresses. While acknowledging that it is not temperature alone that is the sole determinant, structural geologists identify suites of optical and electron microstructures that characterize deformation by dislocation creep at low, moderate, and high-temperature ranges. In what follows, we focus on optical microstructures indicative of the character of deformation. With widespread access to TEM in the late 1960s and early 1970s, workers were able to understand more fully the submicroscopic differences between the microstructures associated with dislocation creep at low-to-moderate temperatures and the microstructures associated with dislocation creep at higher temperatures in a number of common minerals. We refer interested readers to White (1977), Nicolas and Poirier (1976), Barber (1985), Hirth and Tullis (1992), or Passchier and Trouw (2005) for details of the electron microscope substructures associated with the optical microstructures we describe.

At low temperatures, the glide of dislocations generates small strains in the original grains in a polycrystalline aggregate. The climb of dislocations is limited. So, only dislocations along the margins of the original grains are able to move into grain boundaries. Small subgrains form along the margin of the original grains, either by intergranular nucleation or when a grain boundary migrates across a very small volume of highly distorted crystal leaving a small, relatively strain-free grain. The grain boundary migration is, due to the lower temperatures and limited diffusion, severely limited. Further, the small strain-free grains, which have low dislocation densities, distort at lower differential stresses than the surrounding grains with higher dislocation densities and greater numbers of impeded dislocations. Typically, the initially strain-free small grains accrue an increment of "easy" deformation, and they soon acquire dislocation densities and impeded dislocation arrays comparable to their host grains. With total internal energies comparable to the host grains, there is no longer any impetus for grain boundaries to migrate. This also militates against the recrystallized grains growing at the expense of their hosts. Thus, subgrains remain very small, usually about 1–5 μ in diameter (cf. Figures 8.33a, 8.34b). Any small recrystallized grain may, however, be replaced by relatively strain-free (i.e. dislocation-free) subgrain, should one nucleate in the vicinity. This generates a microstructure composed of little-deformed original grains surrounded by and progressively replaced by very small subgrains and new grains. A similar suite of microstructures may result from moderate temperature deformation at high strain rates, where the critical factor is that the rate of climb is unable to keep pace with the rates at which dislocations interact, become pinned, etc.

Because dislocations climb readily at moderate-to-high temperatures, climb counteracts to a significant degree the interactions between dislocations, reduces the numbers of pinned dislocations, and lessens the associated increases in the dislocation density that occur at lower temperatures. Since dislocations have greater mobility, original grains accrue large strain magnitudes more readily than grains deformed at lower temperatures. In addition, the combination of climb and glide of dislocations enables them to move into stable, low energy, three-dimensional arrays during deformation, and dislocation densities in the original grains remain lower than those observed in grains deformed at lower temperatures. Lower dislocation densities mean that individual dislocations are likely to move across an entire grain without a significant increase in their critical resolved shear stress. Experimental samples deformed at moderate to high temperatures typically deform at roughly constant flow stress. The degree to which these processes occur varies depending upon whether temperatures are moderate or high.

At moderate temperatures and moderate strain rates, original grains usually strain homogeneously (i.e. flatten or elongate uniformly). The centers or *cores* of original grains exhibit undulose extinction and may have large subgrains. Along the margins of original grains, and at other grain locations where dislocations have piled up, such as in deformation bands or near fractures in grains, numerous small *polygonal subgrains* form. These subgrains form by progressive rotation of crystal lattice by adding dislocations to low-angle tilt boundaries. This type of recrystallization is, therefore, called *progressive subgrain rotation recrystallization*. When the misorientation of grains exceeds 10°, the resulting grains are classified as *new grains*. Subgrains and new grains typically have lower internal energy than the parent grains, but they may not grow at the expense of old grains because diffusion rates are insufficient to enable grain boundaries to sweep across the distorted, original grains leaving less-distorted, new crystal lattice behind. At moderate temperatures, where grain boundary mobility is lower, the ultimate size of the recrystallized grains is only slightly larger than those formed at lower temperatures (Figure 8.35a, b).

At high temperatures, climb occurs easily, so dislocations are highly mobile and the average dislocation density remains low during deformation, especially when strain rates are moderate. Further, grain boundaries are highly mobile. Even though the differences in dislocation density (and, therefore, internal energy) between highly distorted grains and less distorted grains are relatively low, grain boundaries routinely sweep across the more distorted grains. We call this process of replacing original grains by recrystallized grains *grain boundary migration recrystallization*. This process occurs repeatedly, with recrystallized grains themselves replaced after they are slightly deformed (Figure 8.35c–f). Recrystallization by the combined action of subgrain rotation and grain boundary migration keeps pace with deformation, so an aggregate may look undistorted despite the fact that it accommodated sizable strains. We noted earlier that grain boundaries themselves are regions of relatively high internal energy. Thus, when temperatures are sufficiently high, grain aggregates reduce their overall internal energy by developing straight grain boundaries that meet at 120° triple junctions. At such elevated temperatures, recrystallization may proceed to an *exaggerated grain growth* stage, resulting in larger grains than the original undeformed rock. This texture, which may occur during deformation (i.e. syntectonically), resembles an *annealing* microstructure.

In a study that combined analyses of experimentally and naturally deformed quartzites, Hirth and Tullis (1992) distinguished distinct dislocation creep *regimes* that correspond to the low, moderate, and high-temperature settings outlined earlier. They termed the low temperature/high strain rate creep regime *recrystallization-accommodated dislocation creep* or *regime 1 dislocation creep*. Hirth and Tullis distinguished two dislocation creep regimes in which climb is important. In their *progressive subgrain rotation dislocation creep* or *regime 2 dislocation creep*, which occurs at moderate temperatures/moderate strain rates, original grains usually strain homogeneously (i.e. are flattened or elongated uniformly), develop subgrains of moderate size, and recrystallization proceeds by progressive subgrain rotation. In *grain boundary migration dislocation creep* or *regime 3 dislocation creep* of Hirth and Tullis, which occurs at high temperatures/slow strain rates, aggregates undergo extensive *recrystallization by grain boundary migration*, which enables the aggregate to deform with little increase in flow stress. Other common minerals, particularly carbonates, do not exhibit the same progression of microstructures, but there are comparable temperature and strain rate-related variations in their deformation microstructures.

8B.3.8 Twinning and Kinking

8B.3.8.1 Mechanical Twinning in Calcite

Mechanical or deformation twinning occurs in a wide variety of minerals. Mechanical twins are an especially common product of deformation in the carbonate minerals calcite and dolomite, particularly at low to moderate temperatures. In this section, we use studies of calcite twinning to illustrate how analyzing mechanical twinning contributes to understanding deformation. Our focus on twinning in calcite in this section also mirrors to some degree the structural geology literature, where published studies of mechanical twinning in calcite outnumber published studies of twinning in other minerals by a significant degree. The reasons for this emphasis on deformation twinning in calcite include: (1) calcite is nearly ubiquitous in sequences of sedimentary and metamorphic rocks and is a common secondary mineral in many igneous rocks; (2) twins in calcite are readily recognized using straightforward optical techniques; (3) because calcite belongs to the trigonal crystal system and has three, symmetrically equivalent crystallographic planes along which twinning regularly occurs, grains may have a variety of orientations and still be favorably oriented to form mechanical twins; and (4) the differential stresses required to form deformation twins are relatively low. So, deformation twins form regularly. These factors are among the reasons that structural geologists have developed techniques for using mechanical twins in calcite to determine the orientations and magnitudes of strain principal directions *and* to estimate the orientations and magnitudes of the most and least compressive stresses.

In the first two of the following sections, we briefly describe the geometric basis of techniques to measure strains due to mechanical twinning in calcite or assess stress principal directions and outline an empirical approach to estimate differential stress magnitudes. In a brief subsequent section, we describe how the morphology of twins in calcite provides an insight into the temperature at which the minerals were deformed. In order to use twinned calcite to assess strains or stress in a rock, one must first demonstrate that the twins are a product of deformation and not growth twins. Also, one must assume that the twinned crystals record accurately the boundary conditions to which the entire rock was subjected, i.e. that twinned crystals experienced the same displacement gradients as the entire rock and exhibited the same intrinsic material properties as the aggregate. Finally, since the techniques for determining strain or stress from twinned calcite require measurements from collections of grains, the techniques assume that the strains or stresses are homogeneous within the volume or area encompassing the collection of grains.

8B.3.8.2 Measuring the Strain Accommodated by Twinning in Calcite

Figure 8.38b, which depicts schematically the formation of a mechanical twins in calcite, illustrates two generalizations of mechanical twins in calcite: (1) twin formation generates in an increment of simple shear parallel to the twin plane; and (2) because the magnitude of angular shear is fixed by the calcite twin law, larger shear strain magnitudes require thicker twins. The figure does not address directly, however, significant characteristics of twinning in calcite. In calcite, the most common mechanical twins form along one of three symmetrically equivalent $\{01\bar{1}2\}$ planes known as *e* planes (Turner 1953; Groshong 1972) (Figure 8.71a). For *e* twinning, the direction of shearing, known as the *glide direction* g, is parallel to the intersection of the *e* plane and one of three rhomb $\{10\bar{1}1\}$ planes. The pole to an individual twin plane e_i, the glide direction g defined by the twin plane's intersection with a rhomb plane r_j, and the host crystal's *c* axis all fall on a single great circle on a stereographic projection. Thus, they all lie in a single plane. Using a universal stage to view the twins in a section plane containing *c*, e_i, and g, one finds that twinning reorients the edge of a calcite crystal through an angle $\alpha = 38°17'$ (Figure 8.71b). An individual twin lamella of thickness t_i accomplishes a shear displacement of u_i (the shear displacement in Figure 8.71b is positive-valued). By inspection, $u_i = 2\,t_i \cdot \tan(\alpha/2) = 0.69t_i$.

A typical twinned calcite grain will possess numerous twin lamellae (Figure 8.71b), i.e. twinning is polysynthetic. The shear strain accomplished by *n* parallel twin lamellae of thicknesses t_i is

$$\gamma_{eg} = \frac{1}{t}\sum_{i=1}^{n} u_i = \frac{2}{t}\sum_{i=1}^{n} t_i \tan\left(\frac{\alpha}{2}\right), \tag{8.68}$$

where *t* is the thickness of the grain measured perpendicular to the twin plane (Groshong 1972). Substituting for α and simplifying, the shear strain γ_{eg} is

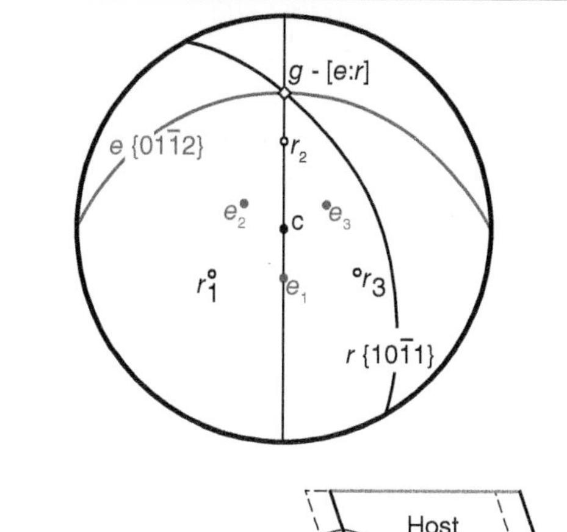

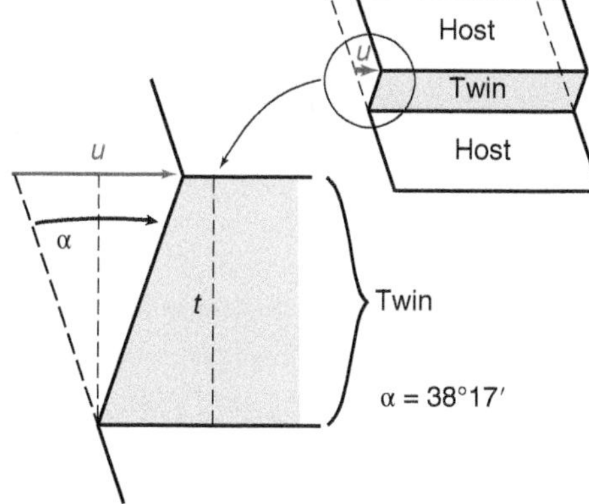

Figure 8.71 Lower hemisphere stereographic projection showing the geometry of the *e* twin planes and *r* rhomb planes relative to the *c* axis of calcite. The intersection of an *e* plane and *r* plane define a glide direction *g*. Sketched section parallel to the *e-g* plane showing schematically the shear displacement *u* accomplished by twinning. Enlargement illustrates the geometry relating shear displacement *u*, twin thickness *t*, and the twin angle α.

$$\gamma_{eg} = \frac{0.69}{t}\sum_{i=1}^{n} t_i. \tag{8.69}$$

and the angular shear accommodated by twinning in this grain is

$$\psi = \tan^{-1}\gamma_{eg} = \tan^{-1}\left\{\frac{0.69}{t}\sum_{i=1}^{n} t_i\right\} \tag{8.70}$$

Conel (1962, cited in Groshong 1972) used the angular shear accomplished by any twin set to calculate the principal values of infinitesimal strain in a Cartesian coordinate frame with axes parallel to g and e_i and normal to the plane containing those lines

$$\varepsilon' = \begin{bmatrix} \frac{1}{2}\tan\psi & 0 & 0 \\ 0 & \frac{-1}{2}\tan\psi & 0 \\ 0 & 0 & 0 \end{bmatrix} \tag{8.71}$$

He then used the tensor transformation law to calculate infinitesimal strain components referred to thin-section wide Cartesian coordinate axes *x-y-z*. Conel multiplied the transformed strain components from each grain by the ratio of the grain area divided by the total area of all grains measured, and then added them to determine a bulk strain value.

Groshong (1972) noted three shortcomings to Conel's method. First, one can calculate a "bulk strain" value using a single twin set from a single grain. Second, and more fundamentally, the strain components in Eq. (8.71) are a function of the orientation of the twins relative to the section-wide coordinate frame. Third, the method provides no statistical measure of variation in the results. Groshong (1972) proposed an alternative method for calculating strain that treats individual twinned grains as a strain gauges with different orientations and uses them to calculate strain magnitudes.

Groshong's method begins with the same raw data – measured widths of twins referred to a coordinate frame with axes parallel to the twins' shear direction g, the pole to the twin planes e_i, and the normal to the plane they define. Using Eq. (8.69) to calculate γ_{eg} (shear strain in the *e-g* coordinate frame) and noting that the tensor shear strain $\Gamma_{eg} = \gamma_{eg}/2$, the data from individual grains are combined to solve simultaneously an equation that relates the strain in individual grains to the total strain (Jaeger and Cook 1976, p. 44).

$$\Gamma_{eg} = (l_e l_g)\varepsilon_x + (m_e m_g)\varepsilon_y + (n_e n_g)\varepsilon_z + (l_e m_g + m_e l_g)\Gamma_{xy} + (m_e n_g + n_e m_g)\Gamma_{yz} + (n_e l_g + l_e n_g)\Gamma_{zx} \quad (8.72a)$$

Because the shear displacement constrains only deviatoric strains, the three direct strains sum to zero, i.e. $\varepsilon_x + \varepsilon_y + \varepsilon_z = 0$ and $\varepsilon_z = -(\varepsilon_x + \varepsilon_y)$, reducing this number of variables in this equation to five:

$$\Gamma_{eg} = (l_e l_g - n_e n_g)\varepsilon_x + (m_e m_g - n_e n_g)\varepsilon_y + (l_e m_g + m_e l_g)\Gamma_{xy} + (m_e n_g + n_e m_g)\Gamma_{yz} + (n_e l_g + l_e n_g)\Gamma_{zx} \quad (8.72b)$$

In this equation, the measured tensor shear strain Γ_{eg}, the direction cosines l_e, m_e, and n_e of the individual grain's *e* axis with respect to the section-wide *x-y-z* axes, and the direction cosines l_g, m_g, and n_g of the individual grain's g axis with respect to the section-wide *x-y-z* axes are known. The components of the overall strain, ε_x, ε_y, Γ_{xy}, Γ_{yz}, and Γ_{zx} are unknowns to be determined. Clearly, five twin set measurements are sufficient to determine a unique solution, but additional precision and rigor accrue with more measurements. Groshong (1972) outlined a least-squares method for solving simultaneously the overdetermined situation to yield strain values with standard errors.

This technique relies upon accurate measurement of twin thickness, which, in turn, requires that twin lamellae are clearly distinguished from the host grain. For that reason, the technique is reliable for grains that are less than half twinned, which corresponds to a tensor shear strain $\Gamma_{eg} = \gamma_{eg}/2 = 0.17$. Also, like all strain measurement techniques, this technique has potential shortcomings. Only a fraction of favorably oriented grains develop mechanical twins under any set of deformation conditions. Examinations of the size of twinned carbonate grains indicate that grains with smaller diameters are less likely to twin than comparably oriented grains with larger diameters; the effect is particularly pronounced for large-diameter grains surrounded by small-diameter grains (highly likely to twin) and small-diameter grains surrounded by large-diameter grains (highly unlikely to twin) (Spiers 1979; Newman and Mitra 1993). Further, twins that do form often record higher shear strains than might be expected for planes parallel to the *e* plane; thus measured strains probably over overestimate strain magnitudes somewhat. Still, this technique has proven successful at providing quantitative strain data in a variety of tectonic settings, particularly studies of weakly deformed rocks in the foreland regions of mountain belts (e.g. Groshong et al. 1984; Groshong 1988; Craddock et al. 1993; Ong et al. 2007; Hnat and van der Pluijm 2011).

8B.3.8.3 Using Twinned Calcite to Assess Stresses

The simple shear accommodated by any calcite twin produces an incremental plane strain with directions of maximum incremental shortening and maximum incremental elongation lying in the *e*-g plane and oriented at 45° angles to the *e* plane (Figure 8.72). One can envision the ideal plane stress system responsible for this incremental plane strain: the most compressive (σ_1) and most tensile (σ_3) principal stresses lie in the *e*-g plane, with the σ_1 parallel to the direction of maximum incremental shortening and σ_3 parallel to the direction of maximum incremental elongation. In a rock with variably oriented calcite grains possessing deformation twins, it is likely that some grains have their *e* planes oriented so that they conform to this ideal loading geometry, i.e. the *e* planes in those grains are parallel to the maximum shear stress associated with the imposed stress. For most grains, however, the incremental shortening and elongation directions associated with their twins, denoted, respectively the *C* and *T* axes, are not parallel to σ_1 and σ_3. Still, collectively the mechanical twins exhibit significant geometric coherence (Turner 1953): (1) the glide directions g (or [*e*:*r*] in Turner's notation) exhibit a strong preferred orientation; (2) the segments of great circles containing

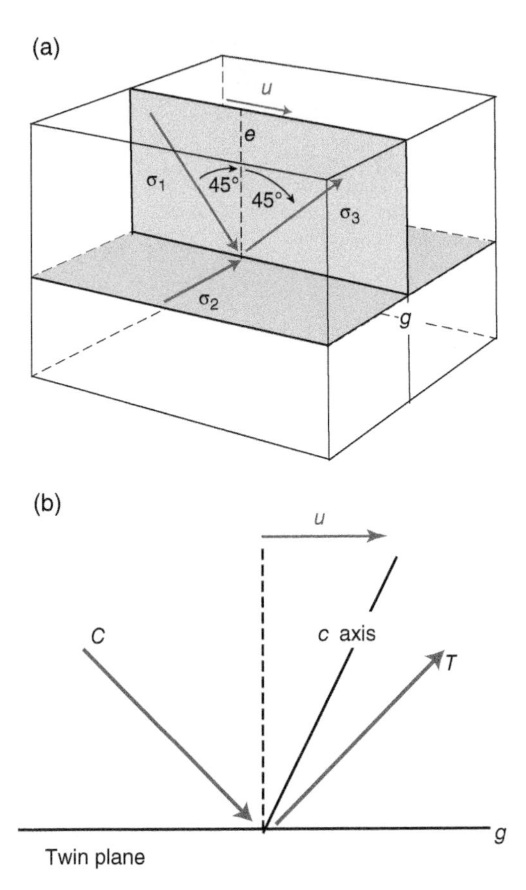

Figure 8.72 (a) Idealized biaxial stress state responsible for generating the shear displacement *u* parallel to *g*. (b) Sketch of the *e*-*g* plane showing the orientations of the *C* and *T* axes associated with shear displacement *u*.

the *c* axes of a twin and of its host, or great circles containing the pole to *e* and the *c* axis of the host, either of which defines the *e*-*g* planes of twins, are highly ordered; and (3) plots of the *C* and *T* axes for all twins tend to define clusters, the centers of which are interpreted to indicate σ_1 and σ_3, respectively. Plots of *C* and *T* axes are the foundation of Turner's *dynamic analysis of twinning* to determine the orientations of principal stresses.

Rowe and Rutter (1989) drew upon deformation experiments to define two ways to use deformation twins in calcite to constrain differential stress magnitudes. The first utilizes measurements of *twinning incidence* I_t, which is the percentage of grains within a particular grain size class that contain optically visible grains. Their analysis of deformation experiments indicates that twinning incidence depends upon stress. For any grain size, twinning incidence is greater at higher differential stress levels. They determined empirically the relationship between twinning incidence and grain size to be:

$$\sigma' = 523 + 2.13 I_t - 204 \log d, \tag{8.73}$$

where σ is stress in MPa, I_t is the twinning incidence in percent, and *d* is grain size in microns (μm). The second method uses *twin density D*. Twin density is derived from measurements of the number of twins per mm in grains of different sizes. The twin density *D* is the slope of the best fit line of a plot of the number of twins per mm versus grain diameter in microns. Rowe and Rutter determined that twin density varies with differential stress according to

$$\sigma' = -52 + 171 \log D, \tag{8.74}$$

where σ again is stress in MPa and *D* is twin density measured in a number of twins per mm per micron.

8B.3.8.4 Temperature-related Effects on Deformation Twins and Kinks

Twin boundaries, like grain boundaries, are able to migrate provided that the temperature is sufficiently high to activate lattice diffusion in the host crystal and the lattices separated by the boundary have sufficiently different internal energies. Burkhard (1993) identified four calcite twin morphology types (Figure 8.73) and demonstrated that their occurrences vary systematically with temperature. Ferrill et al. (2004) compiled twinning data from several field areas to calibrate the temperatures at which the different twin morphologies develop. To distinguish Type I and Type II twins, both of which have planar boundaries, one must

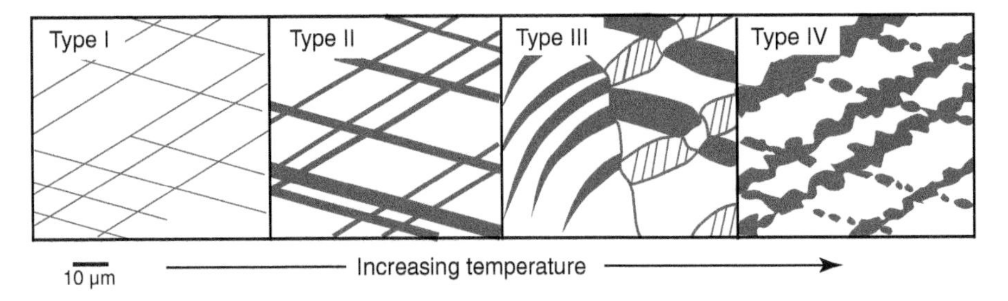

Figure 8.73 Diagram illustrating changes in calcite twin morphology correlated with temperature. See text for explanation. *Source:* Burkhard (1993)/Elsevier.

determine the *twinning intensity*, which is the number of lamellae from a twin set per mm, and the *mean thickness* of lamellae in a set. At temperatures <170 °C, mean twin widths typically are less than 1 μm, but twin intensities will vary with strain magnitude. In these low-temperature settings, calcite tends to accrue strain by increasing the number of twins per mm rather than increasing the thickness of twins. At temperatures >200 °C, mean twin widths typically are greater than 1 μm, with thicker twins associated with higher strain magnitudes. At these moderate temperatures, calcite tends to accrue strain by increasing the thickness of twins rather than forming new twins. At higher temperatures, twins develop curved boundaries, with twins in cross-cutting sets taking on lens-shaped forms. The highest temperatures, above 250 °C, twin boundaries become mobile resulting in patchy development of twinning.

Kink bands in crystals are regions where lattices are highly strained. As a result, kinks often are the locus of recrystallization. One can watch this process occur in a collection of spectacular videos of deforming ice crystals (Wilson and Marmo 2000). Similar processes occur in many deformed minerals.

8B.4 Grain Boundary Sliding and Superplasticity

The term "superplasticity" was introduced briefly in Section 8A.4: it is useful to add some detail since the significance of superplasticity has been recently re-evaluated. The original use of the term in the metallurgical literature was to describe the results of experiments, which led to very high strains in extension without failure. The experiments were conducted at typically high homologous temperatures, with fine grain sizes, and the rheology consisted of a power law with a large value of the stress exponent m. Microstructures of equant grains with a weak CPO resulted, in which the majority of the strain was contributed by grain boundary sliding. Geological applications of the term "superplasticity" have generalized these specific circumstances, for example, by considering that superplasticity applies in other loading configurations than extension, or that superplasticity is synonymous with grain boundary sliding, or that "phenomenological superplasticity" can be defined as very high continuous strains (Gilotti and Hull 1990), such as those that are common in some mylonites. The latter term does not describe a deformation mechanism, and indeed several mechanisms can contribute to such a state of strain.

A model for superplasticity as a deformation mechanism suggests that grains change shape relatively little but slide past each other (Ashby and Verrall 1973) (Figure 8.74). This proposed mechanism is similar to Nabarro–Herring creep, but has less grain shape change and more grain boundary sliding. Further, the predicted strain rates are faster than Nabarro–Herring creep by a factor of 5. Experimental evidence for the operation of this classic mechanism has been found in forsterite and diopside aggregates. The results of these experiments show that a CPO developed by sliding on grain boundaries that are also crystallographic planes and call into question the diagnostic criteria for distinguishing diffusion creep from dislocation creep (Maruyama and Hiraga 2017). Superplasticity and Nabarro–Herring Creep result in similar microstructures (Figure 8.74) of equant grains, and this distinguishes them from dislocation creep microstructures which generally have some clearly inequant grains, depending on the amount of recovery (Figure 8.74).

Recently, it has been suggested that diffusional mechanisms may be important even at seismic slip velocities (Verberne et al. 2014). In experiments and on natural fault surfaces in carbonates, shiny "mirrorlike" fault surfaces have been found to be composed of nano-sized carbonate particles, arranged in fibers in the slip direction. These fibers seem to evolve from discrete, equant particles, which slide past each other on grain boundaries. Contacts between adjacent strings of a particle are joined by sintering (a diffusional process described in Section 8A.2.2) to form the fibers. Diffusion can potentially occur at seismic

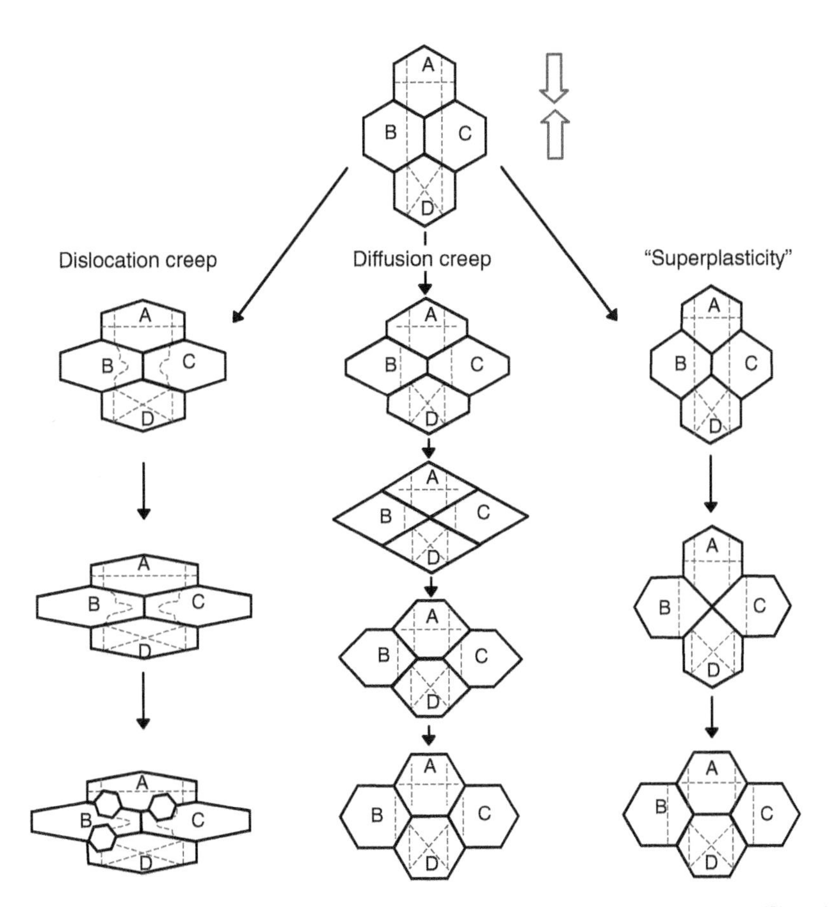

Figure 8.74 The differences between dislocation creep (1), diffusional creep (2), and grain boundary sliding (3) (accommodated by diffusional creep). *Source:* Maruyama and Hiraga (2017)/John Wiley & Sons. Grain shapes change profoundly in the first two cases; the latter is dominated by relative movement of grains (neighbor switching).

velocities because of the exceptionally fine grain size, promoted by higher temperatures, and possibly some fluid-assisted diffusion.

The combination of grain boundary sliding and diffusion is characteristic of superplasticity (the deformation mechanism) as described in the materials science literature. However, in terms of the classification of deformation mechanisms used in this book, grain boundary sliding is a better term. We can envisage a division of dominant non-cataclastic deformation mechanisms into dislocation creep, diffusion creep and grain boundary sliding. Figure 8.74 clarifies that the main difference between diffusion creep and grain boundary sliding (accommodated by diffusion) is that grain shapes are changed less profoundly in the latter, and more sliding occurs along grain boundaries. The balance between the two is therefore related to microstructure.

Appendix 8-I

To understand why and how vacancies exist in crystals, we draw upon analyses in Nicolas and Poirier (1976, pp. 53–64), Poirier (1985, pp. 40–50), Putnis (1992, pp. 186–188 and pp. 235–236), and Karato (2008, pp. 76–80).

Envision that an atom moves from its lattice site in the interior of the crystal, creating a vacancy, and attaches to a step on a free surface of the crystal. Because the atom attaches to a step in the crystal's surface, the external surface area of the crystal remains the same. The free energy of the crystal is increased, however, because the number of bonds broken in removing the atom from its initial site is larger than the number of bonds reformed in attaching the atom to the surface. At the same time, the entropy of the crystal is changed because the frequency of vibration of the atoms in the vicinity of the vacancy has changed. If ΔE_f is the net work done in breaking and reforming bonds, and ΔS_f is the change in entropy of the crystal, the free energy of forming an individual vacancy ΔF_{fi} (the subscripts "*fi*" refer to "formation" and "individual," respectively) is

$$\Delta F_{fi} = \Delta E_f - \mathrm{T}\ \Delta S_f. \tag{8I.1}$$

The volume of the crystal changes by an amount that is the sum of two terms. First, adding the atom to the free surface of the crystal increases the volume of the crystal by an amount ΔV_a. At the same time, the crystal's volume changes by an amount ΔV_r (smaller than ΔV_a) as atoms surrounding the vacancy move, i.e. "relax," in response to its formation. If atoms surrounding the vacancy "relax" inward, ΔV_r is negative. If atoms "relax" outward, ΔV_r is positive. Regardless of the sign of ΔV_r, the change in volume of formation of the individual vacancy is

$$\Delta V_{fi} = \Delta V_a + \Delta V_r. \tag{8I.2}$$

From Eq. (8I.2), the enthalpy of vacancy formation is the sum of the term related to breaking and reforming bonds and a term related to the work against pressure due to the change in the volume:

$$\Delta H_{fi} = \Delta E_f + P\Delta V_{fi}, \tag{8I.3}$$

where P is the hydrostatic stress or pressure. The Gibbs energy or free enthalpy of formation of an individual vacancy is

$$\Delta G_{fi} = \Delta H_{fi} - \mathrm{T}\Delta S_{fi} = \Delta E_f + P\Delta V_{fi} - \mathrm{T}\Delta S_{fi}. \tag{8I.4}$$

Individual vacancies increase the free enthalpy of a crystal, but there is a *configurational entropy* contribution that more than compensates for that increase. Configurational entropy is a function of the number of configurations possible at a particular energy level. For small numbers of vacancies relative to the number of lattice sites, there are large numbers of possible configurations and, thus, a large configurational entropy contribution. This is the reason that crystals have lower Gibbs energy or free enthalpy with vacancies than in a vacancy free state. One can use this reasoning to infer that there is an equilibrium concentration of vacancies the magnitude of which varies with temperature.

Taking n_v as the number of vacancies, the free enthalpy of formation for an individual vacancy given by Eq. (8I.4), and S_c as the configurational entropy, the change in free enthalpy resulting from vacancy formation is

$$\Delta G = n_v \Delta G_{fi} - T\, S_c. \qquad (8I.5)$$

If there are W possible configurations for the n_v vacancies, Boltzmann's entropy formula gives

$$S_c = k \ln W, \qquad (8I.6)$$

where k is the Boltzmann constant. Equation (8I.5) then becomes

$$\Delta G = n_v\ \Delta G_{fi} - k\ T \ln W. \qquad (8I.7)$$

The number of possible configurations W is directly proportional to the product of factors giving the numbers of possible sites for each vacancy. In a crystal composed of n atoms, there are n possible sites in which to put the first vacancy, $(n-1)$ sites in which to put the second vacancy, $(n-2)$ sites in which to put the third vacancy, and so on to the last vacancy $(n-n_v+1)$. However, since all vacancies are identical, it makes no difference in what order they occupy the sites. Therefore, the product of these factors must be divided by the number of permutations of n_v, i.e. $n_v!$, to determine the configurational entropy. Thus,

$$W = \frac{n(n-1)(n-2)\ldots(n-n_v-1)}{n_v!} = \frac{n!}{(n-n_v)!n_v!} \qquad (8I.8)$$

and

$$\ln W = \ln(n!) - \ln[(n-n_v)!] - \ln(n_v!). \qquad (8I.9)$$

Using the approximation known as Stirling's formula, i.e. $\ln(X!) \approx X \ln X - X$, Equation (8I.9) reduces to

$$\ln W = n_v \ln\left(\frac{n}{n_v} - 1\right) - n \ln\left(1 - \frac{n_v}{n}\right). \qquad (8I.10)$$

Because n/n_v is much greater than 1,

$$\ln\left(\frac{n}{n_v} - 1\right) \approx \ln\frac{n}{n_v} \quad \text{and} \quad \ln\left(1 - \frac{n_v}{n}\right) \approx -\frac{n_v}{n},$$

so Eq. (8I.7) becomes

$$\Delta G = n_v\ \Delta G_{fi} - k\mathrm{T}\left(n_v \ln\frac{n}{n_v} + n_v\right). \qquad (8I.11)$$

To define the minimum of the free enthalpy as a function of the number of vacancies, one must find where the derivative of ΔG with respect to n_v equals zero, i.e.

$$\frac{\partial \Delta G}{\partial n_v} = \Delta G_{fi} - kT \ln\left(\frac{n}{n_v}\right) = 0. \qquad (8I.12)$$

Rearranging and taking the fraction $n_v/n = N_v$ to be the crystal-wide fraction of vacant lattice sites at temperature T,

$$N_v = \frac{n_v}{n} = \exp\left(-\frac{\Delta G_{fi}}{kT}\right). \qquad (8I.13)$$

The Boltzmann constant k has units joules $\deg^{-1}$. So, the formulation in Eq. (8I.13) yields ΔG_{fi} in terms of joules per defect. Expressing Gibbs free energy or free enthalpy of vacancy formation in terms of joules per mole is more desirable (Karato 2008, p. 77). In order to do so, one must replace the Boltzmann constant with the gas constant $R = k \cdot N_A$, where Avogadro's number $N_A = 6.02 \times 10^{23}$. In this case, Equation (8I.13) becomes

$$\frac{n_v}{n} = \exp\left(-\frac{\Delta G_f}{RT}\right), \qquad (8I.14)$$

where ΔG_f in joules per mole $= N_A \cdot \Delta G_{fi}$, and

$$n_v = n \exp\left(-\frac{\Delta G_f}{RT}\right). \qquad (8.27)$$

One must modify this derivation in order to calculate the equilibrium number of Schottky defects in an ionic crystal because the requirement of equal numbers of cation and anion vacancies alters the calculation of configurational entropy. If the equilibrium number of Schottky defects is n_s, there are n_s cation vacancies and n_s anion vacancies to be distributed among N lattice sites.

There are, then, both cation and anion contributions to the configurational entropy:

$$W_c = \frac{N!}{(N - n_s)!n_s!} \tag{8I.15a}$$

for cations, and

$$W_a = \frac{N!}{(N - n_s)!n_s!} \tag{8I.15b}$$

for anions. With the assumption that the cation and anion vacancies are independent of each other, the total number of configurations is $W_c \cdot W_a = W^2$. Thus, the expression for ΔG of Schottky defects that accounts for the number of equivalent configurations (i.e. comparable to Eq. 8I.11) is

$$\Delta G = n_s \Delta G_{fi} - 2k\mathrm{T}\begin{bmatrix} N \ln N - (N - n_s) \\ \ln(N - n_s) - n_s \ln n_s \end{bmatrix}, \tag{8I.16}$$

and the expression for the change in ΔG with the change in the number of Schottky defects (i.e. comparable to Eq. 8I.12) is

$$\frac{\partial \Delta G}{\partial n_s} = \Delta G_{fi} - 2kT \ln\left[\ln(N - n_s) - \ln n_s\right] = 0, \tag{8I.17a}$$

or

$$\Delta G_{fi} = 2kT \ln\left[\frac{N - n_s}{n_s}\right]. \tag{8I.17b}$$

Rearranging this equation yields

$$\frac{n_s}{N - n_s} = \exp\left(-\frac{\Delta G_{fi}}{2kT}\right). \tag{8I.18}$$

Since $N >> n_s$,

$$n_s = (N - n_s)\exp\left(-\frac{\Delta G_{fi}}{2kT}\right) \approx N \exp\left(-\frac{\Delta G_{fi}}{2kT}\right) \tag{8I.19}$$

Here too, in order to express ΔG_f in terms of joules per mole, replacing the Boltzmann constant with the gas constant R yields

$$n_s \sim N \exp\left(-\frac{\Delta G_f}{2RT}\right) \tag{8.28}$$

Appendix 8-II

Each vacancy absorbed at a jog in a dislocation moves the jog a distance b along the dislocation and raises the dislocation a distance b. In Section 8B.3.2.2.2, this observation was the basis of the inference that the climb velocity of a dislocation (v_c) is equal to the velocity at which the jog moves along the dislocation line (v_j), i.e.

$$v_j = \Phi b^2 = v_c. \tag{8.49b}$$

Absorbing or emitting a vacancy along the edge of the dislocation is comparable to creating or absorbing a vacancy at a free surface parallel to the extra half plane. The distortion of the crystal lattice in the vicinity of a dislocation gives rise to a local increase in stress σ. As is the case at the surface of a crystal, variations in stress in the vicinity of dislocations affect the local concentrations of vacancies. Utilizing Eqs. (8.29) and (8.30) from Section 8B2.3, the concentration of vacancies in the vicinity of the edge dislocation C is related to the equilibrium concentration of vacancies within the crystal C_o by

$$C = C_o \exp\left(-\frac{\Delta\sigma b^3}{kT}\right). \tag{8II.1}$$

$C < C_o$. So, vacancies will tend to diffuse toward the edge of the dislocation at a rate dependent upon the magnitude of the concentration gradient. If L is the average distance between the dislocation line and a portion of the crystal with the equilibrium concentration of vacancies, then

$$\frac{\mathrm{d}C}{\mathrm{d}x} = \frac{C - C_o}{L} = \frac{1}{L}\left\{\left[\left(C_o \exp\left(-\frac{-\Delta\sigma b^3}{kT}\right)\right)\right] - C_o\right\}$$
$$= \frac{C_o}{L}\left[\exp\left(\frac{-\Delta\sigma b^3}{kT}\right) - 1\right]. \tag{8II.2}$$

Using a Taylor expansion for $e^x = 1+x+x^2/2! + x^3/3! + \cdots$ where $x = (-\Delta\sigma b^3/kT)$, and noting that $\Delta\sigma b^3 << kT$ so that $(-\Delta\sigma b^3/kT)^n/n! \sim 0$ for $n \geq 2$, Equation (8II.2) reduces to

$$\frac{dC}{dX} \simeq -C_o \frac{\Delta\sigma b^3}{kTL}. \quad (8II.3)$$

The diffusive flux of vacancies (in number of vacancies per unit area per unit time) then is

$$J = -D_v \frac{dC}{dx} = C_o D_v \frac{\Delta\sigma b^3}{kTL}, \quad (8.48)$$

where D_v is the diffusion coefficient for vacancies.

References

Abe, S., & K. Mair. 2009. Effects of gouge fragment shape on fault friction: New 3D modelling results. *Geophysical Research Letters* **36**, 2–5. doi:https://doi.org/10.1029/2009GL040684

Aharonov, E. and Karcz, Z. (2019). How stylolite tips crack rock. *Journal of Structural Geology* **118**: 299–307.

Allersma, H.G.B. (2018). http://hgballersma.net/tudweb/hgball.htm#optical http://homepages.ipact.nl/~allersma/album.htm.

Alvarez, W., Engelder, T., and Geiser, P. (1978). Classification of solution cleavage in pelagic limestones. *Geology* **6**: 263–266.

Alzayer, Y., P. Eichhubl, & S. E. Laubach. 2015. Non-linear growth kinematics of opening-mode fractures. *Journal of Structural Geology* **74**, 31–44. doi:https://doi.org/10.1016/j.jsg.2015.02.003.

Anders, M.H., Laubach, S.E., and Scholz, S.H. (2014). Microfractures: a review. *Journal of Structural Geology* **69**: 377–394.

Arthaud, F. and Mattauer, M. (1969). Exemples de stylolites d'origine tectonique dans le Languedoc, leurs rélations avec la tectonique cassante. *Société Géologique de France Bulletin* **7** (11): 738–744.

Ashby, M.F. (1972). A first report on deformation-mechanism maps. *Acta Metallurgica* **20**: 887–897.

Ashby, M.F. and Verrall, R.A. (1973). Diffusion-accommodated flow and superplasticity. *Acta Metallurgica* **21**: 149–163.

Aydin, M. G., & T. Engelder. 2014. Revisiting the Hubbert-Rubey pore pressure model for overthrust faulting: inferences from bedding-parallel detachment surfaces within middle devonian gas shale, the Appalachian Basin, USA. *Journal of Structural Geology* **69**, 519–537. doi:https://doi.org/10.1016/j.jsg.2014.07.010.

Bagnold, R. 1954. Experiments on a gravity-free dispersion of large solid spheres in a Newtonian fluid under shear. *Proceedings of the Royal Society* **225**, 49–63. doi:https://doi.org/10.1098/rspa.1954.0186.

Barber, D.J. (1985). Dislocations and microstructures. In: *Preferred Orientation in Deformed Minerals and Rocks: An Introduction to Modern Texture Analysis* (ed. H.R. Wenk), 149–182. Orlando: Academic Press.

Bartel, E.M., Neubauer, F., Heberer, B., and Genser, J. (2014). A low-temperature ductile shear zone: the gypsum-dominated western extension of the brittle Fella-Sava Fault, Southern Alps. *Journal of Structural Geology* **69**: 18–31.

Bell, T.H. and Cuff, C. (1989). Dissolution, solution transfer, diffusion vs. fluid flow and volume loss during deformation/metamorphism. *Journal of Metamorphic Geology* **7**: 425–447.

Benedicto, A., & R.A. Schultz. 2010. Stylolites in limestone: magnitude of contractional strain accommodated and scaling relationships. *Journal of Structural Geology* **32**/9, 1250–1256. doi:https://doi.org/10.1016/j.jsg.2009.04.020.

Bhattacharyya, K. and Mitra, G. (2011). Strain softening along the MCT zone from the Sikkim Himalaya: relative roles of quartz and micas. *Journal of Structural Geology* **33**: 1105–1121.

Biegel, R.L. and Sammis, C.G. (2004). Relating fault mechanics to fault zone structure. *Advances in Geophysics* **47**: 65–111.

Blenkinsop, T.G. (1989). Thickness-displacement relationships for deformation zones: discussion. *Journal of Structural Geology* **11** (8): 1051–1054.

Blenkinsop, T.G. (1991). Cataclasis and processes of particle size reduction. *Pure and Applied Geophysics* **136** (1): 59–86.

Boyer, S.E. and Elliott, D. (1982). Thrust systems. *American Association of Petroleum Geologists Bulletin* **66**: 1196–1230.

Brock, W.G. and Engelder, T. (1977). Deformation associated with the movement of the Muddy Mountain overthrust in the Buffington window, southeastern Nevada. *Geological Society of America Bulletin* **88**: 1667–1677.

Brodsky, E.E., Gilchrist, J.J., Sagy, A., and Gollettini, C. (2011). Faults smooth gradually as a function of slip. *Earth and Planetary Science Letters* **302**: 185–193.

Brodsky, E. E., J. D. Kirkpatrick, & T. Candela. 2016. Constraints from fault roughness on the scale-dependent strength of rocks. *Geology* **44/1**, 19–22. doi:https://doi.org/10.1130/G37206.1.

Burkhard, M. (1993). Calcite twins, their geometry, appearance, and significance as stress-strain markers and indicators of tectonic regime: a review. *Journal of Structural Geology* **15** (3–5): 351–368.

Caine, J.S., Evans, J.P., and Forster, C.B. (1996). Fault zone architecture and permeability structure. *Geology* **24** (11): 1025–1028.

Candela, T., & E. E. Brodsky. 2016. The minimum scale of grooving on faults. *Geology* 44/8, 603–606. doi:https://doi.org/10.1130/G37934.1.

Chester, F.M. and Chester, J.S. (2004). Stress and deformation along wavy frictional faults. *Journal of Geophysical Research* **105-B10** (23): 421–423. 430.

Chester, F.M. and Logan, J.M. (1986). Implications for mechanical properties of brittle faults from observations of the Punchbowl fault zone, California. *Pure and Applied Geophysics* **124** (1–2): 79–106.

Chester, F.M. and Logan, J.M. (1987). Composite planar fabric of gouge from the Punchbowl fault, California. *Journal of Structural Geology* **9** (5–6): 621–634.

Chester, F.M., Evans, J.P., and Biegel, R.L. (1993). Internal structure and weakening mechanisms of the San Andreas fault. *Journal of Geophysical Research* **98** (B1): 771–786.

Chester, F.M., Chester, J.S., Kirschner, D.L. et al. (2004). Structure of Large-Displacement, Strike-Slip Fault Zones in the Brittle Continental Crust. In: *Rheology and Deformation of the Lithosphere at Continental Margins* (ed. G.D. Karner, B. Taylor, N.W. Driscoll and D.L. Kohlstedt), 223–260. New York: Columbia University Press.

Chester, J.S., Chester, F.M., and Kronenberg, A.K. (2005). Fracture surface energy of the Punchbowl fault, San Andreas system. *Nature* **437** (1): 133–136.

Childs, C., T. Manzocchi, J. J. Walsh, C. G. Bonson, A. Nicol, & M. P. J. Schopfer. 2008. A geometric model of fault zone and fault rock thickness variations. *Journal of Structural Geology* **31**/9, 1–48. doi:https://doi.org/10.1016/j.jsg.2008.08.009.

Conel, J. E. (1962). Studies of the development of fabrics in some naturally deformed limestones. PhD dissertation. California Institution of Technology, Pasadena 257 p.

Cottrell, A.H. (1964). *The Mechanical Properties of Matter*. New York: Wiley.

Cowie, P.A. and Scholz, C.H. (1992). Physical explanation for the displacement-length relationship of faults using a post-yield fracture mechanics model. *Journal of Structural Geology* **14/10**: 1133–1148.

Cowie, P.A. and Shipton, Z.K. (1998). Fault tip displacement gradients and process zone dimensions. *Journal of Structural Geology* **20** (8): 983–997.

Craddock, J.P., Jackson, M., van der Pluijm, B.A., and Versical, R.T. (1993). Regional shortening fabrics in the eastern North America: far-field stress transmission from the Appalachian-Ouachita orogenic belt. *Tectonics* **12** (1): 257–264.

Dawers, N. H., & M. H. Anders. 1995. Displacement-length scaling and fault linkage. *Journal of Structural Geology* **17**/5, 607–614. doi:https://doi.org/10.1016/0191-8141(94)00091-d.

Dibblee, T.W. Jr. (1987). Geology of the Devil's Punchbowl, Los Angeles County, California. In: *Geological Society of America Centennial Field Guide 6 – Cordilleran Section* (ed. M.L. Hill), 207–210. Geological Society of America.

Douglas, R.J.W. (1958). Mount head map-area, Alberta. *Geological Survey of Canada Memoire* **291**: 241.

Ebner, M., D. Koehn, R. Toussaint, & F. Renard. 2009a. The influence of rock heterogeneity on the scaling properties of simulated and natural stylolites. *Journal of Structural Geology* **31**/1, 72–82. doi:https://doi.org/10.1016/j.jsg.2008.10.004.

Ebner, M., D. Koehn, R. Toussaint, F. Renard, & J. Schmittbuhl. 2009b. Stress sensitivity of stylolite morphology. *Earth and Planetary Science Letters* **277**/3–4, 394–398. doi:https://doi.org/10.1016/j.epsl.2008.11.001.

Ebner, M., R. Toussaint, J. Schmittbuhl, D. Koehn, & P. Bons. 2010. Anisotropic scaling of tectonic stylolites: a fossilized signature of the stress field? *Journal of Geophysical Research* 115, B06403. doi:https://doi.org/10.1029/2009JB006649.

Ehrichs, E. E., H. M. Jaeger, G. S. Karczmar, J. B. Knight, V. Y. Kuperman, & S. R. Nagel. 1995. Granular convection observed by magnetic resonance imaging. *Science* **267**, 1632–1634. doi:https://doi.org/10.1126/science.267.5204.1632.

Elliott, D. (1976). The energy balance and deformation mechanisms of thrust sheets. *Philosophical Transactions of the Royal Society of London* **A283**: 289–312.

Engelder, J.T. (1974). Cataclasis and the generation of fault gouge. *Geological Society of America Bulletin* **85**: 1515–1522.

Engelder, T. and Marshak, S. (1985). Disjunctive cleavage formed at shallow depths in sedimentary rocks. *Journal of Structural Geology* **7**: 327–343.

Epstein, B. (1947). The mathematical description of certain breakage mechanisms leading to the logarithmico-normal distribution. *Journal of the Franklin Institute* **244**: 471–477.

Erickson, S.G. (1994). Deformation of shale and dolomite in the Lewis thrust fault zone: Northwest Montana, USA. *Canadian Journal of Earth Sciences* **31** (5): 1440–1448.

Erslev, E.A. (1998). Limited, localized nonvolatile element flux and volume change in Appalachian slates. *Geological Society of America Bulletin* **110** (7): 900–915.

Evans, J.P. (1990). Thickness-displacement relationships for fault zones. *Journal of Structural Geology* **12** (8): 1061–1065.

Evans, M.A., DeLisle, A., Leo, J., and Lafonte, C.J. (2014). Deformation conditions for fracturing in the Middle Devonian sequence of the central Appalachians during the Late Paleozoic Alleghenian orogeny. *American Association of Petroleum Geologists Bulletin* **98** (11): 2263–2299.

Fagereng, Å. and Den Hartog, S.A.M. (2016). Subduction megathrust creep governed by pressure solution and frictional-viscous flow. *Nature Geoscience* **10**: 51–57.

Fagereng, Å., F. Remitti, & R. H. Sibson. 2011. Incrementally developed slickenfibers – Geological record of repeating low stress-drop seismic events? *Tectonophysics* **510**, 381–386. doi:https://doi.org/10.1016/j.tecto.2011.08.015.

Fermor, P.R. and Price, R.A. (1987). Multiduplex structure along the base of the Lewis thrust sheet in the southern Canadian Rockies. *Bulletin of Canadian Petroleum Geology* **35** (2): 159–185.

Ferrill, D. A., A. P. Morris, M. A. Evans, M. Burkhard, R. H. Groshong, & C. M. Onasch. 2004. Calcite twin morphology: a low-temperature deformation geothermometer. *Journal of Structural Geology* **26**/8, 1521–1529. doi:https://doi.org/10.1016/j.jsg.2003.11.028.

Fisher, D.M. and Brantley, S.L. (2014). The role of silica redistribution in slip instabilities along convergent margins, Kodiak, Alaska. *Journal of Structural Geology* **69**: 395–414.

Fisher, D.M., Smye, A.J., Marone, C. et al. (2019). Kinetic models for healing of the subduction interface based on observations of ancient accretionary complexes. *Geochemistry, Geophysics, Geosystems* **20**: 3431–3449.

Fletcher, R.C. and Pollard, D.D. (1981). Anticrack model for pressure solution surfaces. *Geology* **9** (9): 419–424.

Fondriest, M., S. A. F. Smith, G. Di Toro, D. Zampieri, & S. Mittempergher. 2012. Fault zone structure and seismic slip localization in dolostones, an example from the Southern Alps, Italy. *Journal of Structural Geology* **45**, 52–67. doi:https://doi.org/10.1016/j.jsg.2012.06.014.

Geiser, P.A. (1974). Cleavage in some sedimentary rocks of the central Valley and Ridge Province, Maryland. *Geological Society of America Bulletin* **85**: 1399–1412.

Geiser, P.A. and Sansone, S. (1981). Joints, microfractures, and the formation of solution cleavage in limestone. *Geology* **9** (6): 280–285.

Gendelman, O., Pollack, Y.G., Procaccia, I. et al. (2016). What determines the static force chains in stressed granular media? *Physical Review Letters* **116**: 1–5. doi: 0031-9007=16=116(7)=078001(5).

Gillemot, K.A., E. Somfai, & T Börzsönyi. 2017. Shear-driven segregation of dry granular materials with different friction coefficients. *Soft Matter* 415–420. doi:https://doi.org/10.1039/C6SM01946C.

Gilman, J.J. and Johnston, W.G. (1962). Dislocations in lithium fluoride crystals. *Solid State Physics* **13**: 147–222.

Gilotti, J.A. and Hull, J.M. (1990). Phenomenological superplasticity in rocks. In: *Deformation Mechanisms, Rheology, and Tectonics*, vol. **54** (ed. R.J. Knipe and E.H. Rutter), 229–240. Geological Society Special Publication.

Gilotti, J.A. and Kumpulainen, R. (1986). Strain softening induced ductile flow in the Sarv thrust sheet, Scandinavian Caledonides. *Journal of Structural Geology* **8** (3–4): 441–455.

Gratier, J.-P., Renard, F., and Labaume, P. (1999). How pressure solution creep and fracturing processes interact in the upper crust to make it behave in both a brittle and viscous manner. *Journal of Structural Geology* **21** (7): 1189–1197.

Gratier, J.-P., J. Richard, F. Renard, S. Mittempergher, M. L. Doan, G. DiToro, & A. M. Boullier. 2011. Aseismic sliding of active faults by pressure solution creep: evidence from the San Andreas fault observatory at depth. *Geology* **39/**12, 1131–1134. doi:https://doi.org/10.1130/G32073.1;3figures;DataRepositoryitem2011336.

Gratier, J.-P., Renard, F., and Vial, B. (2014). Postseismic pressure solution creep: evidence and time-dependent change from dynamic indenting experiments. *Journal of Geophysical Research* **119**: 2764–2779. doi: 10:1002/2013JB010768.

Gray, J.M.N.T., 2013. A hierarchy of particle-size segregation models: from polydisperse mixtures to depth-averaged theories. *AIP Conference Proceedings* 1542, 66–73. doi:https://doi.org/10.1063/1.4811869.

Gray, J.M.N. & A. R. Thornton. 2005. A theory for particle size segregation in shallow granular free-surface flows. *Proceeding of the Royal Society* **A461**, 1447–1473. doi:https://doi.org/10.1098/rspa.2004.1420.

Groshong, R. H. 1988. Low-temperature deformation mechanisms and their interpretation. *Geological Society of America Bulletin* **100/**9, 1329–1360. doi:https://doi.org/10.1130/0016-7606(1988)100<1329:LTDMAT>2.3.CO;2.

Groshong, R.H. Jr. (1972). Strain calculated from twinning in calcite. *Geological Society of America Bulletin* **83** (7): 2025–2038.

Groshong, R.H. Jr., Pfiffner, O.A., and Pringle, L.R. (1984). Strain partitioning in the Helvetic thrust belt of eastern Switzerland from the leading edge to the internal zone. *Journal of Structural Geology* **6** (1–2): 5–18.

Hadizadeh, J., Sehhati, R., and Tullis, T. (2010). Porosity and particle shape changes lead- ing to shear localization in small-displacement faults. *Journal of Structural Geology* **32**: 1712–1720. https://doi.org/10.1016/j.jsg.2010.09.010.

Harris, L.D. and Milici, R.C. (1977). Characteristics of thin-skinned style of deformation in the southern Appalachians and potential hydrocarbon traps. *U. S. Geological Survey Professional Paper* **1018**: 1–40.

Henderson, J. R., T. O. Wright, & M. N. Henderson. 1986. A history of cleavage and folding: an example from the goldenville formation, Nova Scotia. *Geological Society of America Bulletin* **97**/11, 1354–1414. doi:https://doi.org/10.1130/0016-7606(1986)97<1354:AHOCAF>2.0.CO;2.

Hirth, G. and Tullis, J. (1992). Dislocation creep regimes in quartz aggregates. *Journal of Structural Geology* **14**: 145–159.

Hnat, J.S. and van der Pluijm, B.A. (2011). Foreland signature of indenter tectonics: insights from calcite twinning analysis in the Tennessee salient of the Southern Appalachians, USA. *Lithosphere* **3** (5): 317–327.

Hooker, J. N., S. E. Laubach, & R. Marrett. 2014. A universal power-law scaling exponent for fracture apertures in sandstones. *Geological Society of America Bulletin* 126 (9–10): 1340–1362. doi:https://doi.org/10.1130/B30945.1.

House, W.M. and Gray, D.R. (1982). Cataclasites along the Saltville thrusts, USA and their implications for thrust sheet emplacement. *Journal of Structural Geology* **4**: 257–269.

Hubbert, M.K. and Rubey, W.W. (1959). Role of fluid pressure in mechanics of overthrust faulting. *Geological Society of America Bulletin* **70**: 115–166.

Hudec, M. R, & M. P. A. Jackson. 2007. Terra infirma: understanding salt tectonics. *Earth-Science Reviews* **82**/1–2, 1–28. doi:https://doi.org/10.1016/j.earscirev.2007.01.001.

Hudleston, P. J., & S. H. Treagus. 2010. Information from folds: a review. *Journal of Structural Geology* **32**, 2042–2071. doi:https://doi.org/10.1016/j.jsg.2010.08.011.

Hull, J. (1988). Thickness-displacement relationships for deformation zones. *Journal of Structural Geology* **10** (4): 31–35.

Hull, D. and Bacon, D.J. (1984). *Introduction to Dislocations*, 3e. New York: Pergamon Press.

Iverson, R. M., Vallance, J. W., 2001. New views of granular mass flows. *Geology* **29**, 115–118. doi:https://doi.org/10.1130/0091-7613(2001)029<0115:NVOGMF>2.0.CO.

Jackson, J. and McKenzie, D. (1999). A hectare of fresh striations on the Arkitsa Fault, central Greece. *Journal of Structural Geology* **21**: 1–6.

Jaeger, J.C. and Cook, N.G.W. (1976). *Fundamentals of Rock Mechanics*, 2e, 585. London: Chapman and Hall.

Jing, L., Kwok, C.Y., Leung, Y. F., 2017. Micromechanical origin of particle size segregation. *Physical Review Letters* **118**, 118001. doi:https://doi.org/10.1103/PhysRevLett.118.118001.

Karato, S.-I. (2008). *Deformation of Earth Materials: An Introduction to the Rheology of Solid Earth*, 463. Cambridge: Cambridge University Press.

Kehle, R. O. 1970. Analysis of gravity sliding and orogenic translation. *Geological Society of America Bulletin* **81**/6, 1641–1624. doi:https://doi.org/10.1130/0016-7606(1970)81[1641:AOGSAO]2.0.CO;2.

Kirkpatrick, J.D. and Brodsky, E.E. (2014). Slickenline orientations as a record of fault rock rheology. *Earth and Planetary Science Letters* **408**: 24–34.

Koehn, D., Renard, F., Toussaint, R., and Passchier, C.W. (2007). Growth of stylolite teeth patterns depending on normal stress and finite compaction. *Earth and Planetary Science Letters* **257**: 582–595.

Koehn, D., M. Ebner, F. Renard, R. Toussaint, & C. W. Passchier. 2012. Modelling of stylolite geometries and stress scaling. *Earth and Planetary Science Letters* **341–344**. 104–113. doi:https://doi.org/10.1016/j.epsl.2012.04.046.

Koehn, D., Rood, M.P., Beaudoin, N. et al. (2016). A new stylolite classification scheme to estimate compaction and local permeability variations. *Sedimentary Geology* **346**: 60–71.

Kolmogorov, A.N. (1941). Über das logarithmisch normale Verteilungsgesetz der Dimensionen der Teilchen bei Zerstückelung. *Doklady Akademii Nauk SSSR* 31: 99–101. Translated as: On the logarithmic normal distribution of particle sizes under grinding. Paper 29, In: Selected Works of A. N. Kolmogorov, Volume II. 1992. Editor: A. N. Shiryayev. Springer.

Mair, K.,& S. Abe. 2011. Breaking up: comminution mechanisms in sheared simulated fault gouge. *Pure and Applied Geophysics* **168**, 2277–2288. doi:https://doi.org/10.1007/s00024-011-0266-6.

Markley, M. and Wojtal, S. (1996). Mesoscopic structure, strain, and volume loss in folded cover strata, Valley and Ridge Province, Maryland. *American Journal of Science* **296**: 23–57.

Marone, C. and Scholz, C.H. (1989). Particle-size distribution and microstructures within simulated fault gouge. *Journal of Structural Geology* **11**: 799–814. https://doi.org/10.1016/0191-8141(89)90099-0.

Marshak, S. and Engelder, T. (1985). Development of cleavage in limestones of a fold-thrust belt in eastern New York. *Journal of Structural Geology* **7**: 345–359.

Maruyama, G. and Hiraga, T. (2017). Grain- to multiple-grain-scale deformation processes during diffusion creep of forsterite + diopside aggregate: 1. Direct observations. *Journal of Geophysical Research* **B122**: 5890–5915.

McCarthy, J. J., 2009. Turning the corner in segregation. *Powder Technology* **192**, 137–142. doi:https://doi.org/10.1016/j.powtec.2008.12.008.

McClay, K.R. (1977). Pressure solution and Coble creep in rocks and minerals: a review. *Journal of the Geological Society of London* **134**: 57–70.

Mitra, G. (1984). Brittle to ductile transition due to large strains along the White Rock thrust, Wind River Mountains, Wyoming. *Journal of Structural Geology* **6**: 51–61.

Mitra, G. (1992). Deformation of granitic basement rocks along fault zones at shallow to intermediate crustal levels. In: *Structural Geology of Fold and thrust Belts* (ed. S. Mitra and G.W. Fisher), 123–144. The Johns Hopkins Studies in Earth and Space Sciences.

Mitra, G. (1993). Deformation processes in Brittle deformation zones in granitic basement rocks. a case study from the Torrey Creek area, Wind River Mountains. In: *Basement behavior in Rocky Mountain foreland structure*, vol. **280** (ed. C. Schmidt, R. Chase and E. Erslev), 177–195. Geological Society of America Special Paper.

Mohamed, F.A. and Langdon, T.G. (1974). Deformation mechanism maps based on grain-size. *Transactions of the Metallurgical Society* **5**: 2330–2345.

Monzawa, N., & K. Otsuki,. 2003. Comminution and fluidization of granular fault materials: implications for fault slip behavior. *Tectonophysics* **367**, 127–143. doi:https://doi.org/10.1016/S0040-1951(03)00133-1.

Morley, C.K., von Hagke, C., Hansbery, R.L. et al. (2017). Review of major shale-dominated detachment and thrust characteristics in the diagenetic zone: part I, meso- and macro-scopic scale. *Earth Science Reviews* **173**: 168–228.

Morley, C.K., von Hagke, C., Hansbery, R.L. et al. (2018). Review of major shale-dominated detachment and thrust characteristics in the diagenetic zone: part II, rock mechanics and microscopic scale. *Earth Science Reviews* **176**: 19–50.

Mullenax, A.C. and Gray, D.R. (1984). Interaction of bed-parallel stylolites and extension veins in boudinage. *Journal of Structural Geology* **6** (1–2): 63–71.

Newman, J. and Mitra, G. (1993). Lateral variations in mylonite zone thickness as influenced by fluid-rock interactions, Linville Falls fault, North Carolina. *Journal of Structural Geology* **15** (7): 849–863.

Nicolas, A. and Poirier, J.P. (1976). *Crystalline Plasticity and Solid State Flow in Metamorphic Rocks*. London: Wiley.

Ong, P.F., van der Pluijm, B.A., and Van der Voo, R. (2007). Early rotation and late folding in the

Pennsylvania salient (U.S. Appalachians): evidence from calcite-twinning analysis of Paleozoic carbonates. *Geological Society of America Bulletin* **119** (7–8): 796–804.

Ord, A., Blenkinsop, T., Hobbs, B. (2022). Fragment size distributions in brittle deformed rocks. *Journal of Structural Geology* **154**, 104496. doi:https://doi.org/10.1016/j.jsg.2021.104496.

Otsuki, K., 2003. Fluidization and melting of fault gouge during seismic slip: identification in the Nojima fault zone and implications for focal earthquake mechanisms. *Journal of Geophysical Research* **108**, 2192. doi:https://doi.org/10.1029/2001JB001711.

Passchier, C.W. and Trouw, R.A.J. (2005). *Micro-Tectonics*, 2e. Berlin: Springer-Verlag.

Paterson, M.S. (2001). Relating experimental and geological rheology. *International Journal of the Earth Sciences (Geol Rundsch)* **90**: 157–167.

Peacock, D.C.P. and Sanderson, D.J. (1995). Strike-slip relay ramps. *Journal of Structural Geology* **17** (10): 1351–1360.

Peacock, D.C.P., Korneva, I., Nixon, C.W., and Rotevatn, A. (2017). Changes of scaling relationships in an evolving population: the example of "sedimentary" stylolites. *Journal of Structural Geology* **96**: 118–133.

Petit, J.P. (1987). Criteria for the sense of movement of fault surfaces in brittle rocks. *Journal of Structural Geology* **9** (5–6): 597–608.

Phillips, N.J. and Williams, R.T. (2021). To D or not to D? Re-evaluating particle-size distributions in natural and experimental fault rocks. *Earth and Planetary Science Letters 553*: 116635. https://doi.org/10.1016/j.epsl.2020.116635.

Poirier, J.-P. (1985). *Creep of Crystals: High-Temperature Deformation Processes in Metals, Ceramics, and Minerals*. Cambridge: Cambridge University Press, 260 p.

Pollard, D.D. and Aydin, A. (1988). Progress in understanding jointing over the past century. *Geological Society of America Bulletin* **100**: 1181–1204.

Power, W.L. and Tullis, T.E. (1989). The relationship between slickenside surfaces in fine-grained quartz and the seismic cycle. *Journal of Structural Geology* **11** (7): 879–893.

Power, W.L. and Tullis, T.E. (1991). Euclidean and fractal models for the description of rock surface roughness. *Journal of Geophysical Research* **96** (B1): 415–424.

Power, W.L., Tullis, T.E., and Weeks, J.D. (1988). Roughness and wear during brittle faulting. *Journal of Geophysical Research* **93** (B12): 12278–15268.

Putnis, A. (1992). *Introduction to Mineral Sciences*, 457. Cambridge: Cambridge University Press.

Ramsay, J.G. (1980). The crack-seal mechanism of rock deformation. *Nature* **284**: 135–139.

Resor, P. G. & V. E. Meer. 2009. Slip heterogeneity on a corrugated fault. *Earth and Planetary Science Letters* **288**,483–491. doi:https://doi.org/10.1016/j.epsl.2009.10.010.

Rutter, E.H. (1983). Pressure solution in nature, theory and experiment. *Journal of the Geological Society of London* **140**: 725–740.

Rowe, K.J. and Rutter, E.H. (1989). Paleostress estimation using calcite twinning: experimental calibration and application to nature. *Journal of Structural Geology* **12** (1): 1–17.

Rutter, E.H. (1976). The kinetics of rock deformation by pressure solution [and discussion]. *Philosophical Transactions of the Royal Society of London* **A283**: 203–219.

Sagy, A., E. E Brodsky, & G. J. Axen. 2007. Evolution of fault-surface roughness with slip. *Geology* **35**, 283–284. doi:https://doi.org/10.1130/G23235A.1.

Sammis, C.G., Osborne, R.H., Anderson, J.L. et al. (1986). Self-similar cataclasis in the formation of fault gouge. *Pure and Applied Geophysics* **124**: 53–78.

Sammis, C., King, G., and Biegel, R. (1987). The kinematics of gouge deformation. *Pure and Applied Geophysics* **125**: 777–812.

Savage, S. B. & C. K. K. Lun. 1988. Particle size segregation in inclined chute flow of dry cohesionless granular solids. *Journal of Fluid Mechanics* **189**, 311–335. doi:https://doi.org/10.1017/S002211208800103X.

Schmid, S.M. (1982). Microfabric studies as indicators of deformation mechanisms and flow laws operative in mountain building. In: *Mountain Building Processes* (ed. K.J. Hsu), 95–110. London: Academic Press.

Schröter, M., S. Ulrich, J. Kreft, J. B. Swift, & H. L. Swinney, 2006. Mechanisms in the size segregation of a binary granular mixture. *Physical Review* **E74**, 1–14. doi:https://doi.org/10.1103/PhysRevE.74.011307.

Schultz, R. A, R. Soliva, H. Fossen, C. H. Okubo, & D. M. Reeves. 2008. Dependence of displacement – length scaling relations for fractures and deformation bands on the volumetric changes across them. *Journal of Structural Geology* **30**, 1405–1411. doi:https://doi.org/10.1016/j.jsg.2008.08.001.

Segall, P. and Pollard, D.D. (1983). Nucleation and growth of strike slip faults in granite. *Journal of Geophysical Research* **B88** (1): 555–568.

Seyum, S. & D. D. Pollard. 2016. The mechanics of intersecting echelon veins and pressure solution seams in limestone. *Journal of Structural Geology* **89**, 250–263. doi:https://doi.org/10.1016/j.jsg.2016.06.009.

Shipton, Z.K., Soden, A.M., Kirkpatrick, J.D. et al. (2006). How thick is a fault? Fault-displacement-thickness scaling revisited. In: *Earthquakes: Radiated Energy and the Physics of Faulting*, Geophysical Monograph Series, vol. 170 (ed. R. Abercrombie, A. McGarr, G. Di Toro and H. Kanamori), 193–198. American Geophysical Union.

Spiers, C.J. (1979). Fabric development in calcite polycrystals deformed at 400°C. *Bulletin Mineralogie* **102**: 282–289.

Thornton, A.R., J. M. N. Gray, & A. J. Hogg. 2006. A three-phase mixture theory for particle size segregation in shallow granular free-surface flow. *Journal of Fluid Mechanics* **550**, 1–25. doi:https://doi.org/10.1098/rspa.2004.1420.

Trurnit, P. (1968). Pressure solution phenomena in detrital rocks. *Sedimentary Geology* **2**: 89–11114.

Tunuguntla, D.R., O. Bokhove, & A. R. Thornton. 2014. A mixture theory for size and density segregation in shallow granular free-surface flows. *Journal of Fluid Mechanics* **749**, 99–112. doi:https://doi.org/10.1017/jfm.2014.223.

Turner, F.J. (1953). Nature and dynamic interpretation of deformation lamellae in calcite of three marbles. *American Journal of Science* **251**: 276–298.

Verberne, B.A., Spiers, C.J., Niemeijer, A.R. et al. (2014). Frictional properties and microstructure of calcite-rich fault gouges sheared at sub-seismic sliding velocities. *Pure and Applied Geophysics* **171**: 2617–2640.

Vermilye, J.A. and Scholz, C.H. (1995). Relation between vein length and aperture. *Journal of Structural Geology* **17** (3): 423–434.

Vernon, R. (1976). *Metamorphic Processes: Reactions and Microstructure Development*. Netherlands: Springer.

Walsh, J.J. and Watterson, J. (1988). Analysis of the relationship between displacements and dimensions of faults. *Journal of Structural Geology* **10** (3): 239–247.

Watterson, J. (1986). Fault dimensions, displacements and growth. *Pure and Applied Geophysics* **124** (1/2): 365–373.

Weertman, J. (1968). Dislocation climb theory of steady-state creep. *Transactions of the American Society of Metals* **61**: 681–694.

White, S.H. (1977). The geological significance of recovery and recrystallization in quartz. *Tectonophysics* **39**: 143–170.

Wibberley, C.A.J. (2005). Initiation of basement thrust detachments by fault-zone reaction weakening. In: *High Strain Zones: Structure and Physical Properties*, vol. **245** (ed. D. Bruhn and L. Burlini), 347–372. Geological Society, London, Special Publications.

Wibberley, C. A. J., G. Yielding, & G. Di Toro. 2008. Recent advances in the understanding of fault zone internal structure: a review. *Geological Society, London, Special Publications* **299**, 5–33. doi:https://doi.org/10.1144/SP299.2.

Wilson, C. and Marmo, B. (2000). Flow in polycrystalline ice: Part 1 – Examples of microscopic flow. http://www.tectonique.net/MeansCD/contribs/wilson/introduction.html.

Wojtal, S. (1992). One-dimensional models for plane and non-plane power-law flow in shortening and elongating thrust zones. In: *Thrust Tectonics* (ed. K. McClay), 41–52. London: Chapman & Hall.

Wojtal, S. and Mitra, G. (1986). Strain hardening and strain softening in fault zones from foreland thrusts. *Geological Society of America Bulletin* **97**: 674–687.

Wojtal, S. and Mitra, G. (1988). Nature of deformation in some fault rocks from Appalachian thrusts. In: *Geometries and Mechanisms of Thrusting, with Special Reference to the Appalachians*, vol. **222** (ed. G. Mitra and S. Wojtal), 17–33. *Geological Society of America Special Paper*.

9

Case Studies of Deformation and Rheology

9.1 Overview

Chapters 4 through 8: (1) introduced the fundamental concepts structural geologists use to assess the kinematics and intensity of deformation in rocks; (2) presented basic principles of rock mechanics and rock rheology; and (3) reviewed laboratory and field data constraining the behavior of rock deformed under different temperature, pressure, and strain rate conditions. In this final chapter, we use three case studies to illustrate how those concepts and principles can be applied in different contexts. Two sections of this chapter focus on using rock structures and fabrics to constrain inferences on the character and rheology of rock deformation under different geologic conditions. The first case study in Section 9.2 focuses on structures and deformation fabrics characteristic of the mid- to lower crust, and in Section 9.3 the second considers deformation at shallower crustal levels. The third case study presented in Section 9.4 outlines how an understanding of deformation and rheology provides insight to a practical question of concern to society.

The foundation of the first case study is an analysis of deformation microstructures in quartzites from thrust sheets in the Ruby Gap duplex of the Redbank thrust zone in central Australia. By using data on the composition and geochronology of sheet silicates in those thrust sheets to augment the quartzite microstructural analysis, workers have demonstrated that deformation occurred at temperatures of 350–400° C and moderate pressures. The data and analyses examined here, drawn from several publications, provide insight into deformation rates, cooling rates, and temporal changes in rock rheology in this mid- to lower crustal setting. The second case study draws upon research into the kinematics and intensity of mesoscopic deformation in a thrust sheet cut by an array of minor faults, and the microstructural character and inferred behavior of fine-grained fault rocks along the fault surface. The interplay between grain-scale deformation localized within the thin layer of fault rocks along the Copper Creek thrust in the southern Appalachian (the United States) fold-thrust belt and the mesoscopic deformation of the overlying unmetamorphosed to weakly metamorphosed sedimentary rocks provides constraints on the rheological behavior of rocks at temperatures of ~200° C and under moderate differential stresses, i.e. in mid- to upper crustal settings. The third case study examines the important issue of induced seismicity in upper crustal settings, where human intervention leads to earthquakes in tectonically active settings.

In Section 9.5, we consider how those three case studies, when considered in the context of other

An Integrated Framework for Structural Geology: Kinematics, Dynamics, and Rheology of Deformed Rocks, First Edition. Steven Wojtal, Tom Blenkinsop, and Basil Tikoff.

published work, help us to refine inferences, outlined in Chapters 7 and 8, on the general character of the rheology of the lithosphere.

9.2 Integrating Structural Geology and Geochronology: Ruby Gap Duplex, Redbank Thrust Zone, Australia

9.2.1 Geological Setting and Deformation Character

The Arunta Inlier is an east–west trending, 200–300 km wide belt of deformed and metamorphosed rocks that extends more than 800 km across central Australia. The belt exposes a sequence of igneous and sedimentary rocks that were deformed and metamorphosed around 1800 Ma, intruded by a dike swarm at circa 900 Ma, and then covered by sedimentary rocks of Late Proterozoic to Paleozoic ages. The basal unit of this sedimentary sequence, the Heavitree Quartzite, is recognized in the Wiso and Georgina Basins to the north, the Amadeus Basin to the south, and in the narrow Ngalia Basin that divides the Arunta into North and South Blocks (Figure 9.1a). The widespread occurrence of this fine-grained and well-sorted quartz arenite indicates that a broad area in what is now central Australia was a tectonically stable region with little vertical relief at the end of the Proterozoic. Continued, largely conformable sedimentation into the early Paleozoic across this broad area demonstrates that stable conditions and low relief persisted across it into the Paleozoic (Shaw et al. 1991). The uplift responsible for the current exposure of the Proterozoic basement rocks in the Arunta Block is the product of crustal shortening during the Devonian-Carboniferous (400–300 Ma) Alice Springs orogeny (Figure 9.1a).

Reconstructions indicate that crustal shortening during the Alice Springs orogen developed in an intraplate, intracratonic setting, not the more typical setting for crustal shortening, i.e. convergence across a plate boundary (Lambeck 1986; Shaw et al. 1991, 1992). That is to say, the Alice Springs orogen was neither associated with the subduction of oceanic lithosphere like most convergent orogens nor due to transpression across a San Andreas-type plate boundary. Nevertheless, the crustal shortening during the Alice Springs orogen led to the development of large-scale deformation features entirely comparable to those associated with convergent plate boundaries. In particular, the shortening of the Proterozoic basement and Proterozoic to Paleozoic cover strata along the southern margin of the Arunta Block led to the formation of a complex array of linked subhorizontal detachment faults and more steeply inclined reverse faults like those seen in other zones of crustal convergence (Teyssier 1985; Collins and Teyssier 1989; Dunlap et al. 1995; Pfiffner 2017) (Figure 9.1b, c).

In a variety of tectonic settings, subhorizontal shortening is accommodated by reverse slip on inclined faults and/or by the reverse component of oblique slip on inclined faults (cf. Figure 2.12). Reverse faulting, as shown in Figure 2.12, juxtaposes rock from structurally lower positions in the hanging wall against the rock in structurally higher positions in the footwall. The illustrations of fault slip in Figure 2.12 are, of course, highly simplified and therefore do not capture fully the character of crustal shortening by faulting. First, individual faults are rarely planar; more typically fault surfaces are broadly curved or consist of relatively planar segments connecting more highly curved regions. Second, displaced rock masses are not rigid bodies; faulting must be accompanied by distortion of the wall rock where faults are curved. Where curved or step-shaped faults cut through stratified rocks, fault slip duplicates layers, leading to folded layers on one or both sides of the fault (Figure 9.2a). Third, it is unusual to find a single fault in isolation. Deformation by faulting

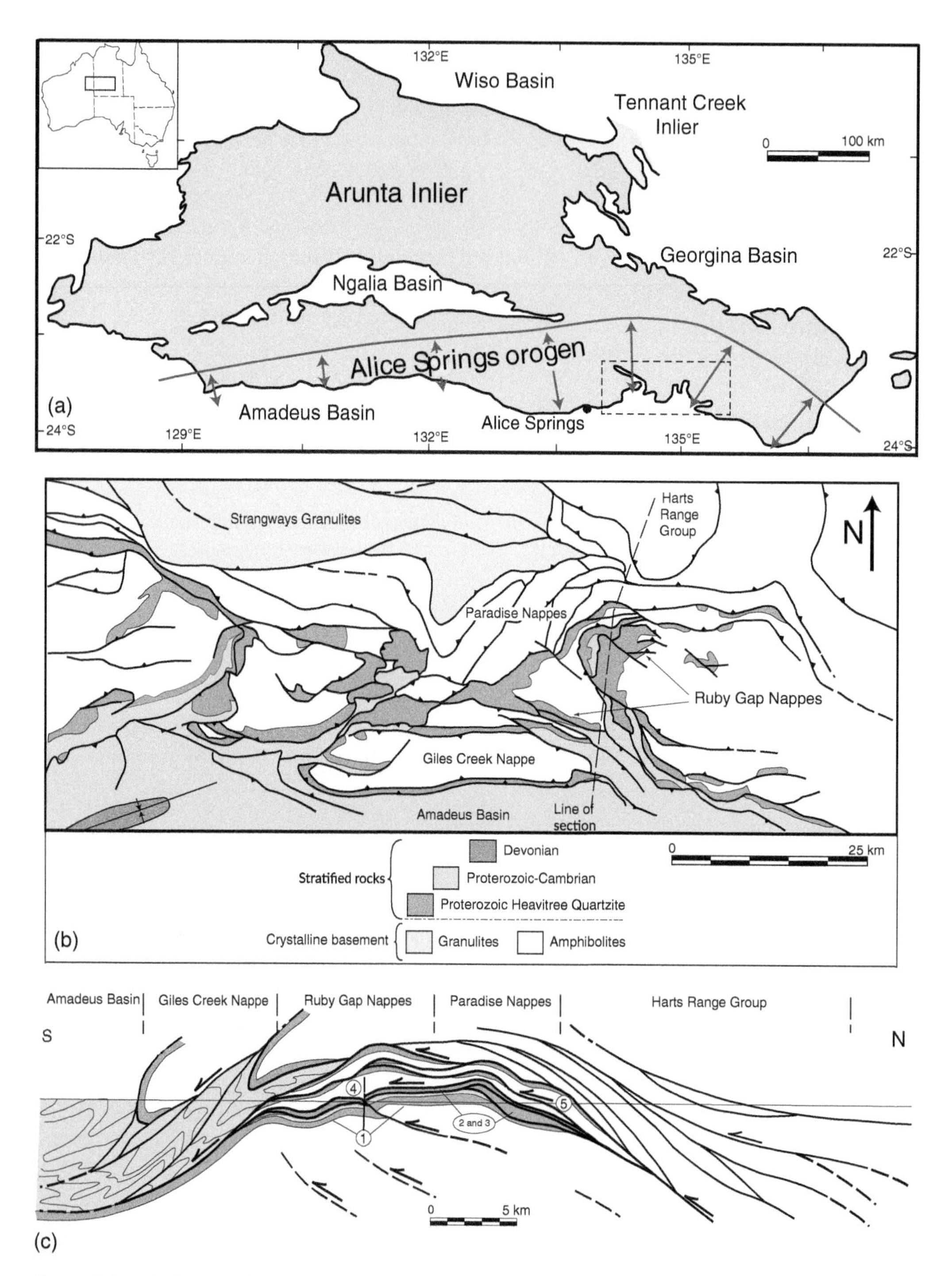

Figure 9.1 (a) Map showing the setting and extent of the Alice Springs orogen along the southern margin of the Arunta Inlier in central Australia. *Source:* After Shaw et al. 1991. The dashed box in the lower right shows the location of (b). (b) Geologic map showing main structural elements of the Ruby Gap duplex. The dashed line shows the line of section in (c). (c) Structure section across the Ruby Gap duplex. Circled numbers denote the five horses in the duplex described in the text. *Source:* (b) and (c) After Dunlap et al. (1995).

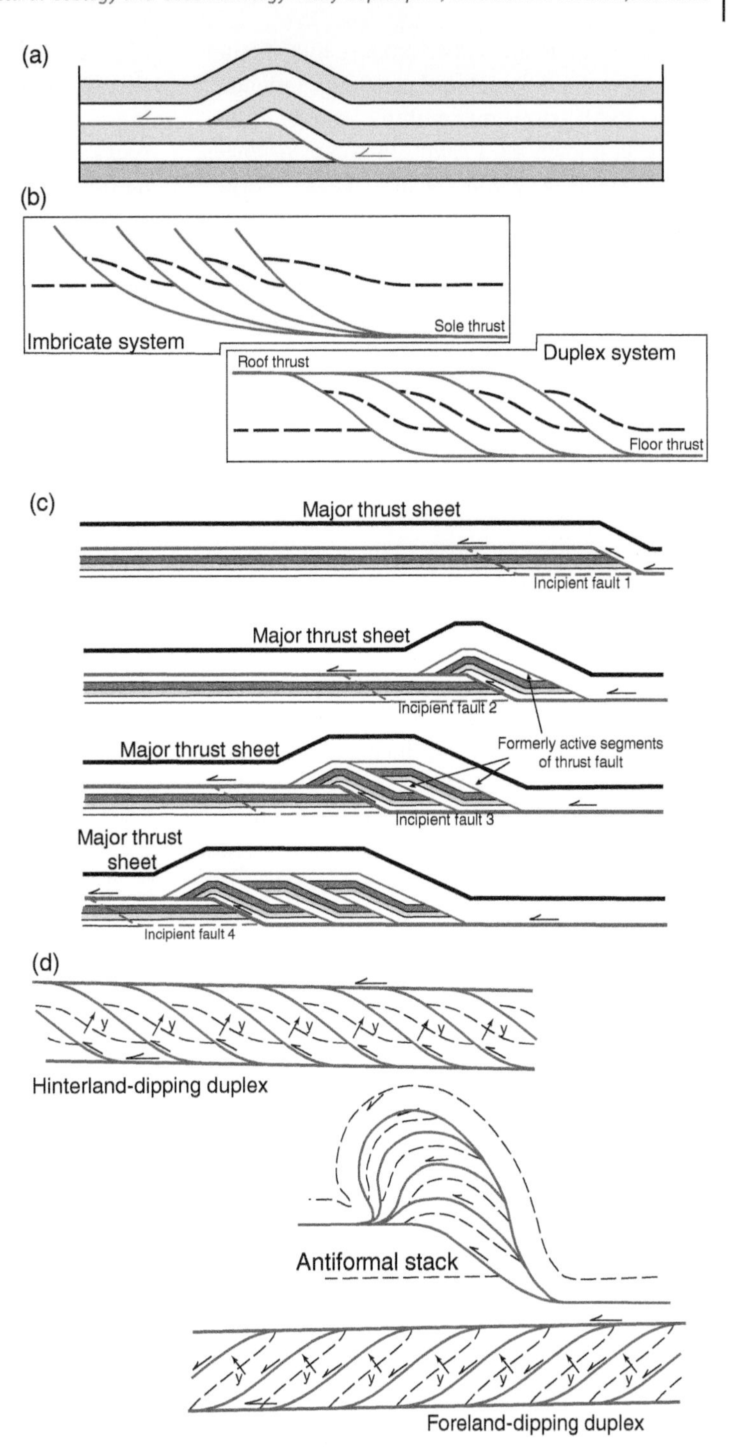

Figure 9.2 (a) Movement of a thrust sheet over a step-shaped thrust fault forms an anticline above the thrust ramp. (b) Illustrations of imbricate and duplex thrust systems. (c) Sequential development of a duplex by footwall imbrication. (d) Illustrations of hinterland-dipping duplexes, antiformal stacks, and foreland-dipping duplexes. *y* denotes younging direction for strata in horses. *Source:* (b–d) Redrawn from Boyer and Elliott (1982).

normally entails slip on two or more linked, intersecting, or otherwise interrelated faults. Individual faults often separate into multiple strands, each of which is known as a *splay*. In some instances, splays extend away from the primary fault surface and terminate in the rock masses on either side of it. In other instances, splays separate and then rejoin the primary fault or another major fault, creating fault-bounded blocks known as *lenses*, *lozenges*, or *horses*. In settings that have experienced subhorizontal shortening, both of these geometries are common, generating, respectively, *imbricate systems* and *duplex systems* (Boyer and Elliott 1982) (Figure 9.2b). In imbricate systems, the fault surface with the greatest slip magnitude may be the lowest imbricate in the stack of imbricates, creating a *hanging wall imbricate fan*, or the highest imbricate in the stack, creating a *footwall imbricate fan*. In duplex systems, two or more splays connect a subhorizontal or shallowly inclined lower fault segment, the *floor thrust*, to a subhorizontal or shallowly inclined upper fault segment, the *roof thrust*. Boyer and Elliott (1982) noted that duplex systems are effectively "herd(s) of horses." The two subhorizontal faults accommodate slip magnitudes greater than the slip magnitudes on any of the individual splays, but the magnitude and relative timing of slip on individual splays can vary depending on the kinematics of duplex formation.

Duplex systems regularly form in stratified rocks when a subhorizontal fault segment at one stratigraphic level "ramps up" to a subhorizontal fault segment at a higher stratigraphic level (Figure 9.2c). In those instances, successive splays typically form when slip initiates in the previously unfaulted footwall on an extension of the lower subhorizontal fault segment (i.e. the floor thrust) and on an inclined fault segment connecting the extension of the floor thrust to the upper subhorizontal fault segment (i.e. the roof thrust). This process is called footwall imbrication. Steps 1 through 4 in Figure 9.2c illustrate the formation of a duplex by successive footwall imbrication. In this situation: (1) splays carry older or structurally lower rocks over younger or structurally higher rocks, (2) rocks within the horses are weakly deformed, so while layers in individual horses typically are folded one can infer their original shape from their final shape, and (3) slip often concentrates on the youngest splay with earlier formed splays becoming mostly inactive. Duplex systems formed this way exhibit different overall geometries depending on the magnitude of slip on the latest active splay (Figure 9.2d). *Hinterland-dipping duplexes* form when the slip on successive splays is less than the length of the splays. *Antiformal stacks* form when the magnitude of slip on successive splays is comparable to the length of splays, and *foreland-dipping duplexes* form when the slip on successive splays is greater than the length of the splay.

The footwall imbrication mechanism for duplex formation is sufficiently common in stratified rocks in the mid- to upper crust that most workers use it as their working hypothesis for duplex formation unless there is clear evidence to the contrary. Duplex systems also regularly develop in both stratified and unstratified (i.e. crystalline basement) rocks in the mid- to lower-crust, but it is not clear whether footwall imbrication is as common there as it is in upper crustal settings. Extensive penetrative deformation developed prior to or accompanying fault slip in mid- to lower-crustal settings can impact the evolution of duplex systems by altering fault geometries and the distribution of slip on them. For example, shortening a rock mass prior to or during the development of a floor thrust, roof thrust, or splay may mean that those faults may cut across earlier faults, folds, or deformation fabrics. In those instances, fault slip may place younger or structurally higher rocks on older or structurally lower rocks (Figure 9.3a, b). Since the rocks above or below any fault may contain truncated folds, there will be no direct correlation between bends in

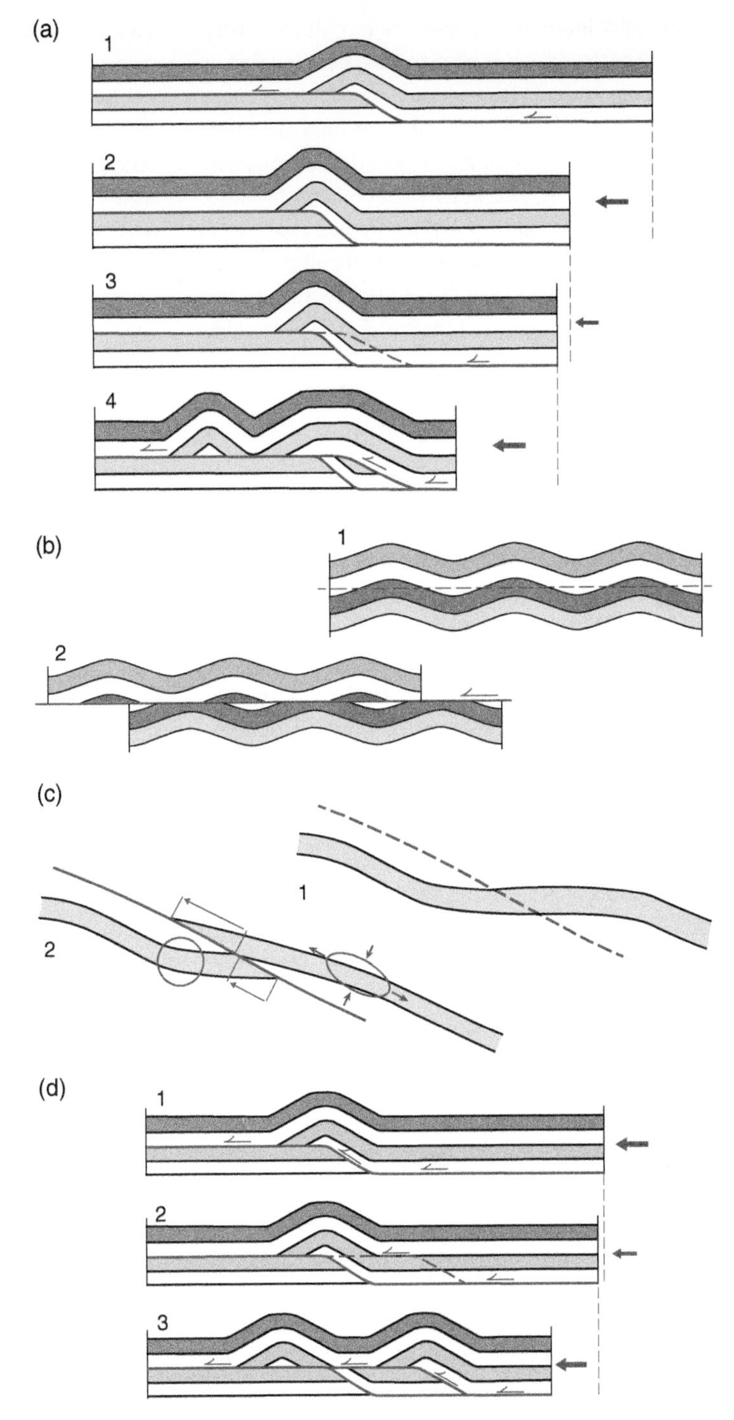

Figure 9.3 Structures formed by departures from the footwall imbrication paradigm. (a) Ramp anticline formed in 1, shortened and steepened in 2, and then cut by an out-of-sequence thrust in 3 to form structures in 4. (b) Folded strata cut by planar thrust. (c) Pronounced variation of displacement along fault due to distortion of hanging wall in stretching fault. Contrast the small displacement arrow shown on the footwall with the longer arrow on the hanging wall: the extra displacement is due to stretching in the hanging wall. (d) Truncated folds formed by hanging wall imbrication.

faults and folds in the wall rocks. Alternatively, floor thrusts, splays or roof thrusts can develop as *stretching faults* (Means 1989), where differences in the finite strain magnitudes on opposite sides of a fault lead to significant variations in the displacement along the fault surface (Figure 9.3c). Finally, deformation of the wall rock may oversteepen, fold, or otherwise impede slip on the structurally lowest splay in a duplex, setting the stage for hanging wall imbrication that crosscuts or bypasses the earlier splays in the duplex (Figure 9.3a, d).

Although the crustal shortening associated with Alice Springs orogen occurred in an intraplate rather than plate boundary tectonic setting, Collins and Teyssier (1989) demonstrated that the faults along the southern margin of the Arunta block define imbricate systems and duplex systems like those found in other, more typical convergent orogens. In this section, we outline the work of Dunlap and Teyssier (1995) and Dunlap et al. (1995, 1997), who analyzed the development of a duplex formed in the amphibolite-grade basement and Heavitree Quartzite in the vicinity of Ruby Gap, northeast of Alice Springs. Their analysis of the Ruby Gap duplex used the character of dislocational deformation microstructures in quartzites and the geochemistry and geochronology of micas formed during deformation to analyze the development of the Ruby Gap duplex and drew upon thermal modeling to understand the temporal and structural development of the Alice Springs orogen.

Figure 9.1c is a cross section depicting the Ruby Gap duplex (after Dunlap et al. 1997). A weakly to moderately deformed stratigraphic sequence of basement gneisses overlain by Heavitree Quartzite and Bitter Springs Formation (labeled 1 in Figure 9.1c) lies beneath the basal thrust of the duplex. Two lower horses in the duplex (labeled 2 and 3 in Figure 9.1c) are thin sheets composed solely of deformed cover strata. Two upper horses (labeled 4 and 5 in Figure 9.1c) consist of basement gneisses overlain by strongly deformed quartzite. The roof of the duplex, the Ruby thrust, carried a complex of highly deformed basement gneisses currently exposed north of Ruby Gap (the Paradise Nappes and the Harts Range Group) and nappes composed of a deformed basement and cover strata exposed south of Ruby Gap (the Giles Creek Nappe) (Figure 9.1c). The entire complex has a broadly antiformal shape inferred to be due to thrust faulting beneath the floor thrust. Even if the Ruby thrust is restored to a subhorizontal orientation, however, the Ruby Gap Nappes and Giles Creek Nappe have downward-facing geometries (i.e. they are synformal anticlines). Thus, the overall structure of the Ruby Gap duplex is an antiformal stack. As detailed later and in the following sections, deformation intensity and metamorphic grade increase from south to north over this structure.

9.2.2 Microstructures and Deformation Mechanisms

9.2.2.1 Microstructures

Deformation microstructures in the Heavitree Quartzite vary systematically with a position in the duplex (Dunlap et al. 1991, 1997). Along the southern margin of the duplex, quartzites beneath the floor thrust (from sheet 1 in Figure 9.1c) are undeformed or weakly deformed. Weakly deformed quartzites from sheet 1 exhibit fabrics defined by dimensional preferred orientations of large (~750-μm diameter), somewhat elongate (axial ratio ≤ 2:1), detrital quartz grains. In thin section, rounded detrital grains regularly exhibit undulose extinction (Figure 9.4a). Pore spaces between large quartz grains are filled by mixtures of irregular to rounded quartz grains 50–100 μm across and detrital muscovite fragments 100–200 μm in length (mu in Figure 9.4a). Both the smaller detrital quartz and detrital muscovite grains also exhibit undulose extinction. Some large detrital quartz grains indent their neighbors across

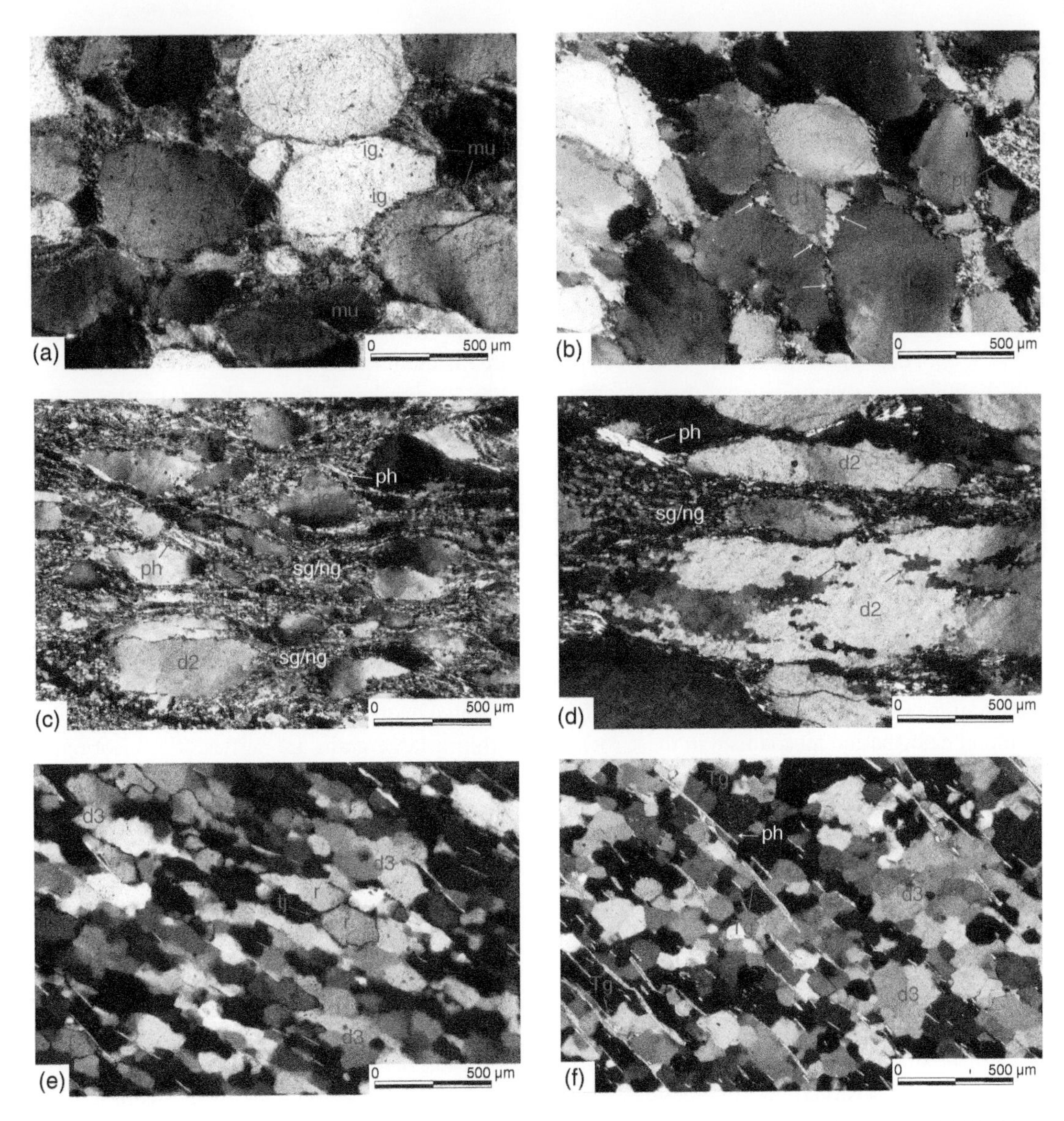

Figure 9.4 Deformation microstructures in quartzites at different positions in the Ruby Gap duplex. (a) Weakly deformed detrital quartz grains in a sample from the footwall along the southern margin of the duplex. ig = indented grains; og = quartz and muscovite overgrowth; mu = detrital muscovite; sz = selvage. (b) Quartzite from sheet 1 near center of the duplex. d1 = quartz grains exhibiting deformation bands and deformation lamellae; sg = quartz grains with well-developed subgrains; ph = aggregates of neocrystallized phengite. White arrows indicate bands of subgrains and new grains along the margins of original grains. (c) and (d) Moderately recrystallized quartzites from the northern end of sheet 1 and the main body of sheet 2. d2 = deformed cores of detrital grains exhibiting undulose extinction, deformation bands, deformation lamellae, and subgrain arrays; sg/ng = subgrains and new grains in recrystallized quartz aggregate; ph = phengite. (e) and (f) Fully recrystallized quartzites from sheets 4 and 5. d3 = quartz grains subdivided by numerous subgrains; r = quartz grains with distinctive grain boundaries. Tg = T grain boundaries; tj = three- or four-grain junctions where grain boundaries meet at approximately 120° angles.

contacts subparallel to the grains' long dimensions (i.e. ig in Figure 9.4a). In other instances, contacts subparallel to the long dimensions of quartz grains are straight and marked by selvage composed of extremely fine quartz, opaque minerals, and/or detrital sheet silicates (sz in Figure 9.4a). Contacts at high angles to the long dimensions of grains exhibit overgrowths composed of mixtures of uniformly extinguishing, fine-grained quartz and white mica (i.e. og in Figure 9.4a). These microstructures are consistent with the inference that diffusional mass transfer was the dominant deformation mechanism in the southern section of sheet 1, with the removal of quartz and muscovite along the longer planar or indented contacts and addition of quartz and white mica across the shorter contacts.

In sheet 1 quartzites from the central portion of the duplex, quartz grains up to ~500 μm in length again have elongate shapes (with axial ratios ≤ 2:1) that define foliation and lineation. A distinctive suite of microstructures indicates that the dominant deformation mechanism in these rocks differs from that observed to the south. Most large quartz grains exhibit undulose extinction, often together with deformation bands and deformation lamellae (e.g. grains labeled d1 in Figure 9.4b). Some grains are subdivided into subgrains 20–50 μm across (sg in Figure 9.4b). The margins of these original grains possess narrow bands of small (10–50 μm diameter), relatively strain-free quartz grains inferred to be subgrains and new grains (indicated by white arrows in Figure 9.4b). This suite of microstructures is consistent with the deformation of quartz by dislocation creep with relatively minor recovery and recrystallization. These samples, like those from locations to the south, contain detrital muscovite grains, which here typically exhibit undulose extinction and often have a frayed appearance. These samples also contain fine, dispersed flakes of white mica with phengitic compositions in aggregates mixed with fine-grained quartz or along the margins of large quartz grains (ph in Figure 9.4b). In many instances, the long axes of the fine mica grains are parallel to the long dimensions of the deformed quartz grains. These neocrystallized phengites, like the recovery microstructures in quartz, indicate deformation at higher temperatures than in samples farther south.

In sheet 1 quartzites exposed beneath northern sections of the duplex and quartzites from sheet 2 or portions of sheet 3 within the duplex, a dimensional preferred orientation of large (200–500-μm-long), flattened (axial ratios ≤ 3:1) quartz grains defines prominent foliation and lineation (Figure 9.1c, d). These large quartz grains are the remnants of original detrital grains, now separated from each other by arrays of fine-grained (10–50-μm diameter) quartz intermixed with white mica. Large quartz grains exhibit undulose extinction, deformation bands, deformation lamellae, and arrays of subgrains (i.e. grains labeled d2 in Figure 9.4c, d). In some instances (i.e. where indicated by red arrows in Figure 9.4d), subgrains within a large grain are sufficiently misoriented from their hosts to be classified as new grains. These remnants of detrital grains have irregular or serrated boundaries separating them from the aggregates of fine-grained, uniformly extinguishing quartz and irregularly shaped white mica grains that surround them. The serrated boundaries of the large quartz grains and the occurrence of new grains within and around the remnants of the original grains support the inference that the aggregates of fine-grained quartz are composed of subgrains and new grains (sg/ng in Figure 9.4c, d). This collection of microstructures is consistent with deformation by dislocation creep accompanied by moderate recovery and subgrain rotation recrystallization. Phengites in these samples (labeled ph in Figure 9.4c, d) typically are larger than those in the weakly deformed samples farther south, exhibit more uniform extinction, and exhibit a strong dimensional preferred orientation parallel to the long axes of the quartz grains. Both of these observations suggest that white micas grew or were thoroughly recrystallized during deformation.

Strongly foliated and lineated samples from sheet 1 beneath the most northerly portions of the floor thrust, from the structurally higher portions of sheet 3, and from quartzites in sheets 4 or 5 within the duplex exhibit microstructures significantly different from those described earlier. The high strain inferred for these quartzites is not apparent in the shape of quartz grains. Figure 9.4e shows that there are few if any remnants of original detrital grains and no clear distinction between original grains and matrix. Instead, the microstructure is defined by an interlocking array of quartz grains and less common, thin, well-aligned white mica grains. Some individual, uniformly extinguishing quartz grains are surrounded by distinct grain boundaries (e.g. grains labeled r in Figure 9.4e). More typical are 200–500 μm long and 100–200 μm wide individual grains (labeled d3 in Figure 9.4e) containing distinctive subgrains. Both individual grains and subgrains in larger grains extinguish uniformly. Contacts parallel to the long dimensions of quartz grains are composed of short straight segments slightly oblique to each other, giving the boundaries a serrated look. Contacts defining the ends of these elongate grains typically consist of one or two straight segments inclined approximately 60° angles to the long dimension of the grains. This sometimes creates three- or four-grain junctions where grain boundaries meet at approximately 120° angles (tj in Figure 9.4e). White mica grains 100–300 μm long and ~5 μm wide define the foliation in these rocks. Individual white mica grains extinguish uniformly, indicating that they are either neocrystallized or recrystallized. The microstructure in more sheet silicate rich samples from these portions of the duplex is generally similar. The framework of recrystallized quartz again consists of both uniformly extinguishing individual grains and grains containing distinctive subgrains (r and d3, respectively, in Figure 9.4f). Phengites again exhibit a strong dimensional preferred orientation, but individual phengite grains are both more numerous and larger, up to ~500 μm long and ~20 μm across. With more white mica, T grain boundaries, where the quartz–quartz grain boundaries are pinned by the mica, are as or more common than the serrated boundaries seen in Figure 9.4e. These collections of microstructures are consistent with deformation by dislocation creep accompanied by extensive, syntectonic recovery and recrystallization by grain boundary migration.

Dunlap et al. (1997) also recognized localized zones where microstructures like undulose extinction, deformation bands, and deformation lamellae developed in quartzites that had previously undergone extensive recrystallization by grain boundary migration. Similarly, in the southern part of the duplex, they identified zones of intense cataclastic deformation along exposures of thrust surfaces separating sheets 1, 2, and 3. Exposures of those thrusts to the north lack evidence for cataclastic deformation, as do exposures of the thrusts beneath sheets 4 and 5.

9.2.2.2 Deformation Mechanisms

As noted in the previous section, quartzite samples from the southern portion of the footwall of the Ruby Gap duplex are either undeformed or exhibit microstructures indicating deformation by diffusional mass transfer. Footwall samples beneath the northern portion of the duplex and from the different sheets within the duplex exhibit a range of microstructures consistent with deformation by dislocation creep. In Section 8A.4 Dislocational Deformation Mechanisms, we emphasized the role of dislocation climb in accommodating finite changes in shapes of crystals by dislocation creep. In particular, climb (1) enables gliding dislocations to move past obstacles, thereby facilitating changes in grain shape; (2) enables dislocations to move readily into stable 3D arrays (i.e. low-angle tilt boundaries) surrounding relatively dislocation-free subgrains, thereby lowering the total internal energy of deformed grains; and (3) enables low- and high-angle grain boundaries to migrate, replacing highly strained lattice by relatively strain-free lattice, thereby lowering the internal energy of deformed polycrystalline aggregates. The rate of climb depends on the rate that atoms and ions diffuse through crystals, which is, in turn, controlled by the ambient temperature. Syntectonic recovery (the movement of dislocations into stable 3D arrays) and recrystallization (via the migration of grain boundaries) facilitate continued deformation of polycrystalline aggregates by limiting strain hardening, and both recovery and recrystallization proceed more effectively at higher temperatures. Section 8B.3.7 examined in greater detail different recovery and recrystallization mechanisms and introduced the notion, articulated fully by Hirth and Tullis (1992), that characteristic suites of microstructures develop at different combinations of strain rate and ambient temperature. In particular, Hirth and Tullis argued for three distinctive suites of dislocation creep microstructures, each characteristic of a *dislocation creep regime*. The systematic differences in dislocation microstructures found in samples at different positions within the duplex indicate that different dislocation creep regimes prevailed at different locations within the duplex.

At lower temperatures and/or higher strain rates, the rate of climb of dislocations is slow compared to other processes contributing to deformation by dislocation movement. Under these conditions, some dislocations initially glide readily, but interactions between dislocations either: (1) impede glide on those dislocations so they glide only at higher differential stresses; or (2) halt it entirely. Higher differential stress levels are required to continue deformation, either by enabling impeded dislocations to begin gliding again or to initiate glide on other, less favorably oriented dislocations. Thus, the material work hardens. Since more dislocations are now active, dislocation interactions increase and dislocation densities continue to increase. As deformation proceeds, dislocation densities increase most dramatically in the vicinity of grain boundaries. The high dislocation densities there raise the strain energy of those portions of the grains' lattices sufficiently to drive the development of optically visible microstructure changes. Because there is little dislocation climb, there is limited movement of dislocations into subgrain boundaries and severely limited grain boundary migration. Small subgrains form along grain boundaries by either intergranular nucleation or when grain boundaries migrate across small volumes of a highly distorted lattice. Subgrain growth is severely limited, however, resulting in a microstructure that consists of (1) distorted original grains, which may exhibit undulose or patchy extinction and/or deformation bands; and (2) numerous small (1–5-μm diameter), relatively uniformly extinguishing subgrains and new grains that decorate the boundaries of the distorted original grains. This is the *Regime 1 microstructure* of Hirth and Tullis (1992). In laboratory deformation experiments, the Regime 1 microstructure is associated with significant strain hardening and higher flow stresses. Deformed Heavitree Quartzite in much of sheet 1 beneath the floor thrust of the Ruby Gap duplex and the southernmost portions of sheets 3 and 4 within the duplex exhibit the Regime 1 microstructure (Figure 9.5).

At moderate temperatures and/or at slower strain rates, dislocations climb at rates capable of significantly reducing dislocation interactions and increasing dislocation mobility. Due to reduced numbers of pinned dislocations, dislocations regularly sweep through grains toward grain boundaries. The increased mobility of dislocations in grain centers enables them to move readily into stable 3D arrays that give rise to undulose extinction. In the vicinity of grain boundaries, the

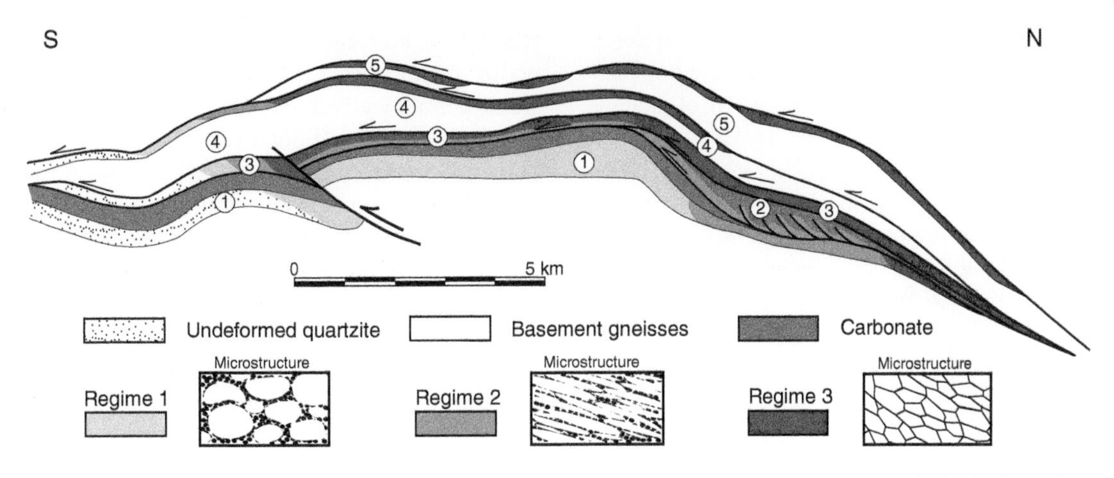

Figure 9.5 Cross section indicating location and extent of quartz dislocation creep regimes within the Ruby Gap duplex. Circled numbers denote the five horses in the duplex (see Fig. 9.1). *Source:* Dunlap et al. (1997) / John Wiley & Sons.

higher dislocation densities lead to the formation of tilt boundaries that surround relatively dislocation-free polygonal subgrains. As grains continue to distort, ever greater numbers of dislocations move to the vicinity of grain boundaries, where they add to existing low-angle tilt boundaries resulting in subgrain rotation recrystallization. The resulting microstructure, with distorted centers of grains exhibiting undulose extinction and deformation bands and margins of grains marked by arrays of subgrains and new grains, is the *Regime 2 microstructure* of Hirth and Tullis (1992). In laboratory deformation experiments, the Regime 2 microstructure forms in experiments in which the rates of recovery were sufficiently high to keep flow stresses from rising dramatically, and deformation proceeded at moderate flow stresses. Quartzites in the northernmost portion of sheet 1 beneath the floor thrust of the Ruby Gap duplex and in the southern and central portions of sheet 2 and in the southern portions of sheets 3 and 4 of the duplex exhibit the Regime 2 microstructure (Figure 9.5).

At high temperatures and/or at slow strain rates, dislocation climb occurs at rates that keep the numbers of pinned dislocations low. As grain distort, dislocations readily move to the vicinity of grain boundaries, where they either form stable 3D arrays or add to existing low-angle tilt boundaries. Relatively unstrained subgrains form and grow readily, and so they regularly decorate the former boundaries of the original grains. Both individual dislocations and grain boundaries are sufficiently mobile that even small differences in the internal energy of neighboring grains are sufficient to drive grain boundary migration. Relatively strain-free subgrains and new grains expand at the expense of neighboring more highly strained lattice, resulting in wholesale recrystallization by grain boundary migration. In this instance, distorted original grains are replaced by large, relatively strain-free subgrains and new grains separated by relatively straight grain boundaries that meet at 120° angles. As the originally strain-free subgrains and new grains subsequently distort, they are, in turn, replaced by successive generations of subgrains and new grains formed by continued grain boundary migration recrystallization. The resulting *Regime 3 microstructure* (Hirth and Tullis 1992) develops without strain hardening in deformation experiments and is inferred to indicate steady-state flow in natural settings. In the northernmost portions of sheet 2, most of sheet 3 and all of sheets 4 and 5 of the duplex, quartzites are completely recrystallized and exhibit dislocation of Regime 3 microstructure (Figure 9.5).

9.2.3 Rheological Analysis Using Microstructures by Comparison to Experimental Deformation

With an understanding of the characteristics of deformation in the different dislocation creep regimes, one can draw general inferences about deformation conditions at different positions within the duplex depicted in Figure 9.5. Regime 1 microstructures in southern portions of the footwall of the duplex and in the southernmost portions of horses 1, 3, and 4 suggest that deformation proceeded at relatively low temperatures and probably at relatively high flow stresses. Regime 2 microstructures in most of sheet 2 and in the southern portions of sheets 3 and 4 suggest that deformation proceeded at moderate temperatures and moderate flow stresses. Regime 3 microstructures in sheets 4 and 5 are consistent with deformation at moderate to high temperatures and relatively low flow stresses.

As noted in Sections 6A.11.2.5.3 and 8B.3.7, the size of the subgrains or recrystallized grains formed during steady-state flow are functionally related to the magnitudes of the differential stresses responsible for the deformation. Dunlap et al. (1997) measured size distributions of subgrains and new grains produced during dynamic recrystallization in samples across the duplex. Quartzites from the Ruby Gap duplex have recrystallized grain sizes that range from ~20 μm in Regime 1 quartzites from the structurally lowest and southern portions of the Ruby Gap duplex to >160 μm in Regime 3 quartzites in the structurally higher and northern portions of the duplex. As outlined in Chapter 6, the relationship

$$\Delta\sigma = cd^{-v}, \tag{9.1}$$

where d is the measured mean diameter of subgrains, c and v are experimentally derived constants defined by Twiss (1977), provides a means to calculate $\Delta\sigma$, the differential stress magnitude at which deformation proceeded. Figure 9.6a, redrawn from Dunlap et al. (1997), indicates that differential stress magnitudes varied systematically with position during the development of the Ruby Gap duplex. Using the values of the parameters c and v defined by Twiss (1977), Dunlap et al. inferred that differential stresses in the southern segments of the duplex were on the order of hundreds of MPa, whereas those prevailing in the northern portions of the structure were less than or equal to 50 MPa (Figure 9.6b).

9.2.4 Geochronology

Sheet silicates in quartzites from the Ruby Gap duplex belong to one of two distinct populations (Dunlap et al. 1991, 1997): (1) detrital muscovite; or (2) neocrystallized phengite. Deformed quartzites from the footwall of the duplex and some Regime 1 quartzites within the duplex possess large muscovite grains that exhibit undulose extinction and often have a frayed appearance. In some Regime 2 quartzites, aggregates of fine-grained sheet silicates contain small muscovite grains that exhibit undulose extinction. Muscovites exhibiting undulose extension or possessing a frayed appearance are interpreted as detrital remnants. More commonly, the aggregates of fine-grained sheet silicates found in Regime 2 quartzites are composed of uniformly extinguishing phengite. Similarly, Regime 3 quartzites from the northernmost portion of the footwall of the duplex and from the several horses that compose the duplex regularly contain individual, uniformly extinguishing phengites crystals ~500 μm long and ~20 μm across. Dunlap et al. (1991) noted that fine-grained phengite does not occur as overgrowths on detrital muscovite and inferred that phengite is neocrystallized.

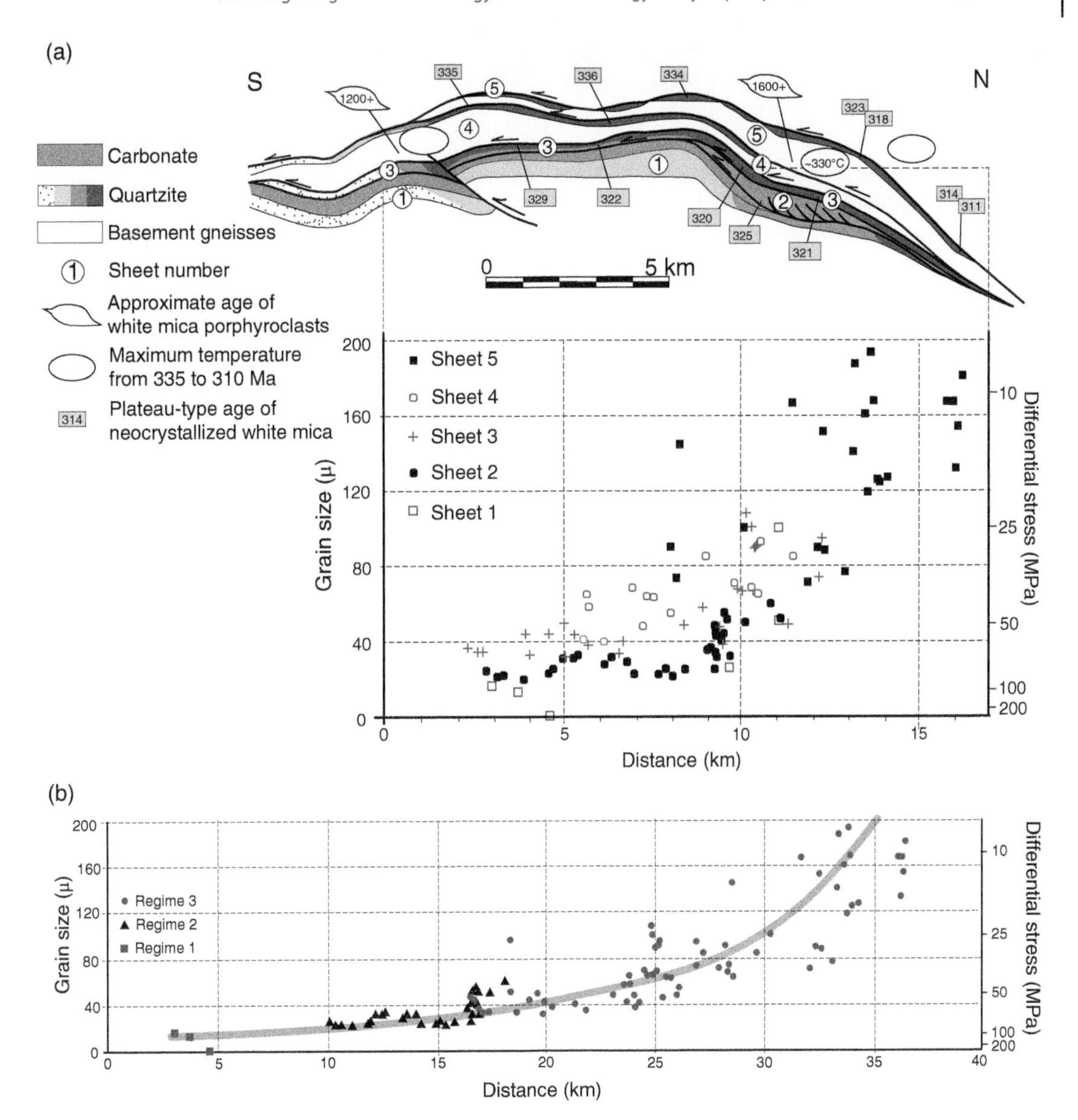

Figure 9.6 (a) Plot of mean recrystallized grain size and inferred differential stress at different positions along a south–north transect across the Ruby Gap duplex. Note different symbols denote data from different horses in the duplex. (b) Grain size and differential stress data from (a), here differentiated by dislocation creep regime, versus restored distance from the southern end of the duplex. Curve depicts estimated smooth variation in differential stress across orogen. *Source:* Both (a) and (b) Dunlap et al. (1997) / John Wiley & Sons.

Geochronology confirmed that inference; they reported K/Ar ages of approximately: (1) 1569 Ma for relatively large, detrital muscovite grains from sheet 1 quartzites beneath the floor thrust, consistent with having been eroded from a Middle Proterozoic source; and (2) 376 Ma for fine-grained phengites from quartzites in the northern portion of sheet 1, consistent with crystallization and growth during mid-Paleozoic deformation.

Ar^{40}/Ar^{39} dating studies revised slightly and refined inferences concerning the mid-Paleozoic ages of phengites in Regime 2 and 3 quartzite samples from different sheets in the duplex (Dunlap 1997; Dunlap et al. 1997). Phengites from Regime 2 quartzites exhibit discordant age spectra inferred to be due to the incorporation of argon from pre-existing muscovite. So, they are less useful in interpreting the deformation history. Phengites in the fully recrystallized Regime 3 quartzites yield flat age spectra consistent with neocrystallization during syntectonic recrystallization. This interpretation is based on their clean, uniform extinction and their association with quartzite microstructures indicative of syntectonic recrystallization (not post-tectonic annealing recrystallization). Moreover, phengites in individual samples yield narrow ranges of Ar^{40}/Ar^{39} ages, and there are systematic differences in the ages of phengite from samples at different positions within the duplex (Figure 9.5a). At each structural level along a south-to-north transect across the duplex, phengites from quartzite samples in the south yield older ages than those from samples in the north. Further, in the northern portion of the duplex, phengites in quartzites at lower structural levels in the duplex generally yield older ages than those at structurally higher positions in the duplex. Considering collectively the microstructural characteristics of phengites in regime 3 quartzites, Dunlap and collaborators inferred that phengite ages in Figure 9.6a are neocrystallization ages and not cooling ages.

9.2.5 Evaluating Displacement Through Time

Dunlap et al. (1997) combined:

1) observations on the deformation mechanisms and rheological behavior of quartzites outlined in Section 9.2.3;
2) geochronologic data derived from neocrystallized phengite in those rocks discussed in Section 9.2.4; and
3) temperature-time paths for potassium feldspars in sheets 4 and 5 derived from thermal modeling (Dunlap et al. 1995; Dunlap 1997)

to infer an unusually well-constrained view of the development of the Ruby Gap duplex. Figure 9.7a presents their restoration of the Ruby Gap duplex at ~311 Ma ago, just before the formation of the late footwall structures that warped the duplex into its current antiformal geometry. This figure presents Ar^{40}/Ar^{39} data from neocrystallized phengite at different positions in the duplex and the character of the quartzite deformation microstructures at different positions across most of the structure. The lines labeled *1*, *2*, and *3 DCRT* denote the traces of surfaces across which quartzites transitioned from one dislocation creep regime to another, determined using temperature conditions inferred from thermal modeling and estimates of the transition temperatures extrapolated from experimental deformation data. At the time depicted in this section, rocks below line *3 DCRT* (shaded gray) would have been at temperatures above the closure temperature for phengite, and quartzites would have been actively deforming under Regime 3 conditions. As the reverse movement on faults in the duplex carried rocks across the *DCRT* surfaces, ambient temperatures would have fallen below the inferred blocking temperature for the Ar^{40}/Ar^{39} dating of phengite as Regime 3 microstructures “froze” into deformed quartzites.

Figure 9.7a–d depict sequential restorations of the Ruby Gap duplex consistent with thermal

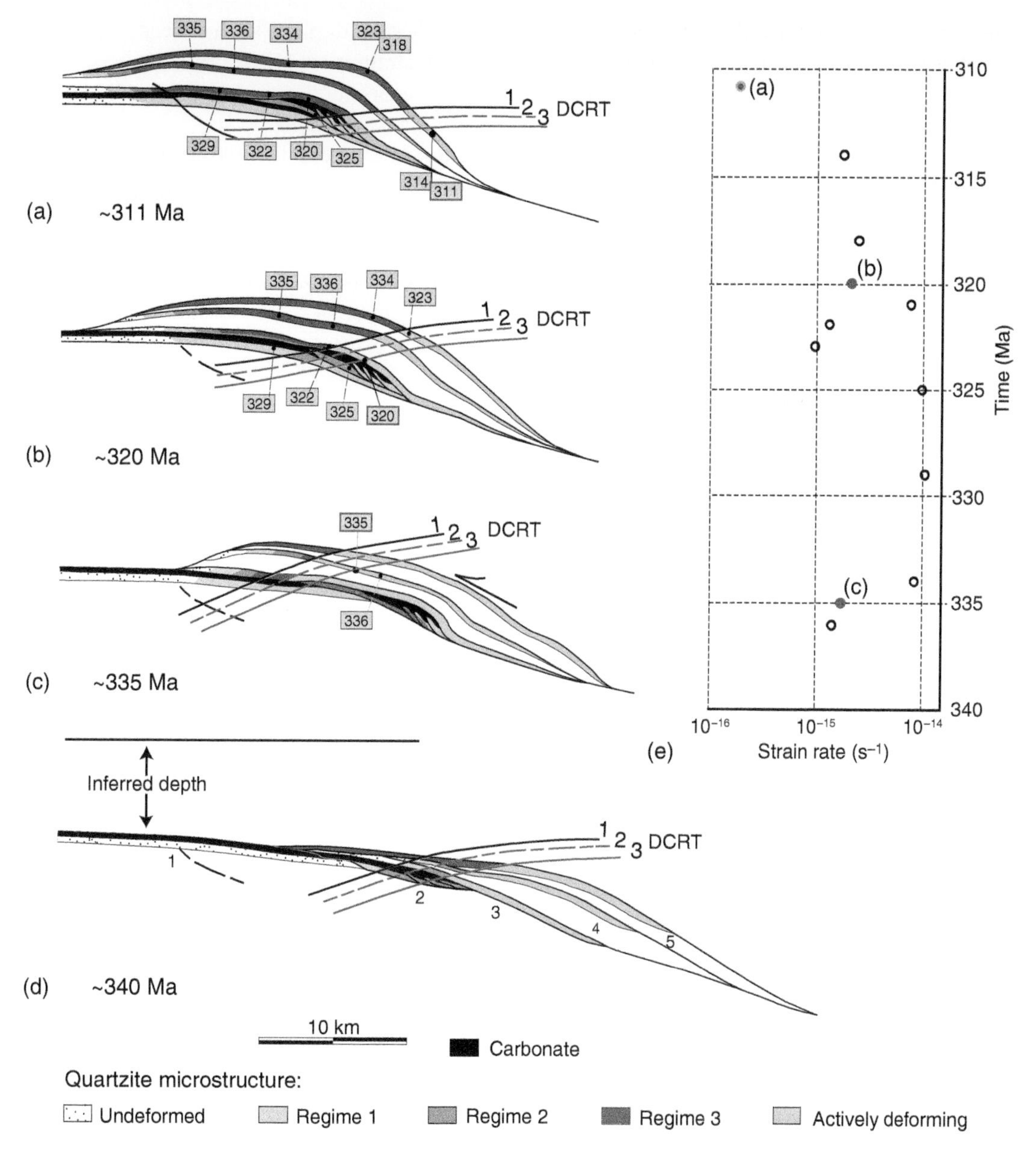

Figure 9.7 (a–d) Sequential restoration of Ruby Gap duplex from ~340 to ~311 Ma after Dunlap et al. (1997) / John Wiley & Sons. Restorations constrained by combining interpretations of quartz deformation microstructures, $^{40}Ar/^{39}Ar$ dating of phengites, and thermal modeling. (e) Plot of inferred strain rates in Heavitree Quartzite in the Ruby Gap duplex versus time. The lines DCRT 1–3 indicate the inferred depths at which rocks undergo a dislocation creep regime transition, i.e. transition from one dislocation creep regime to another. See text for explanation.

modeling and Ar^{40}/Ar^{39} age data for neocrystallized phengite (Dunlap et al. 1997). For example, in Figure 9.7b the duplex is restored to the inferred configuration of the structure at ~320 Ma ago, just prior to the formation of the latest structures that warped the duplex into its current antiformal geometry. Figure 9.7c depicts the inferred geometry of the duplex at ~335 Ma, and Figure 9.7d shows the inferred geometry and relative position of the duplex early in the development of the Alice Springs orogen. Reversing the sequence to view Figure 9.7a–d suggests that uplift associated with thrusting was accompanied by cooling, which, in turn, fixed Ar^{40}/Ar^{39} phengite ages and "froze" quartzite deformation microstructures in portions of the duplex that crossed the *DCRT* contours.

9.2.6 Orogenic Development Through Time

Figure 9.7b–d, and a indicates that Ar^{40}/Ar^{39} phengite ages and quartzite deformation microstructures were fixed in segments of the duplex that lie progressively closer to the leading edge of the duplex over the time interval from ~340 to ~310 Ma. Dunlap et al. (1997) estimated the strain rates at which deformation proceeded at each step by substituting differential stress values derived from measured recrystallized grain sizes and temperature estimates from feldspar compositional variations and thermal models into the dislocation creep flow laws for wet quartzite from Paterson and Luan (1990). Figure 9.7e plots variations in the calculated strain rate magnitudes in Ruby Gap quartzites over the time interval during which the duplex developed. The calculated values indicate increasing strain rates for deformation accommodated by dislocation creep as horses 3, 4, and 5 begin to travel over horse 2, with continued deformation by dislocation creep at high strain rates as continued thrusting carries those sheets over the footwall (labeled 1 in Figures 9.5–9.7). During the final stages of the growth of the duplex, i.e. between ~320 and ~310 Ma, strain rates for deformation accommodated by dislocation creep waned as the horses were uplifted and cooled.

Figure 9.8a places the reconstruction of the duplex at ~340 Ma into the regional context of the Alice Springs orogeny. Figures 9.8b and c sequentially restore the horses in the Ruby Gap duplex to their pre-orogenic configuration, indicating that the emplacement of the Paradise nappes over distorted strata of the Amadeus cover accommodated in excess of 50 km of crustal shortening. Dunlap et al. argued that the nucleation of thrust faults in the duplex and initiation of slip on those faults, responsible for the transition from Figure 9.8c to Figure 9.8b, occurred at the differential stresses dictated by the Mohr–Coulomb criterion or Byerlee's Law. This early deformation, probably at moderate to high differential stresses, would have brought about crustal thickening, resulting in greater overburden for strata in the duplex (e.g. Figure 9.8b). With the change in ambient conditions (elevated temperatures and higher pressures), continued deformation of quartzites in the duplex would occur by dislocation creep (e.g. Figures 9.7 and 9.8a). With subsequent erosional unroofing, lowering of ambient temperatures and pressures would lead to a return of deformation by pressure solution, cataclasis, and sliding on discrete faults, likely at elevated levels of differential stress.

9.2.7 Summarizing Deformation in the Ruby Gap Duplex

Dunlap et al. (1997) combined:

1) an analysis of spatial variations in the character of the microstructure of deformed Heavitree Quartzite;
2) geochronological data on the timing of recrystallization of feldspar in basement and phengite in quartzites; and
3) thermal modeling of orogenic heating

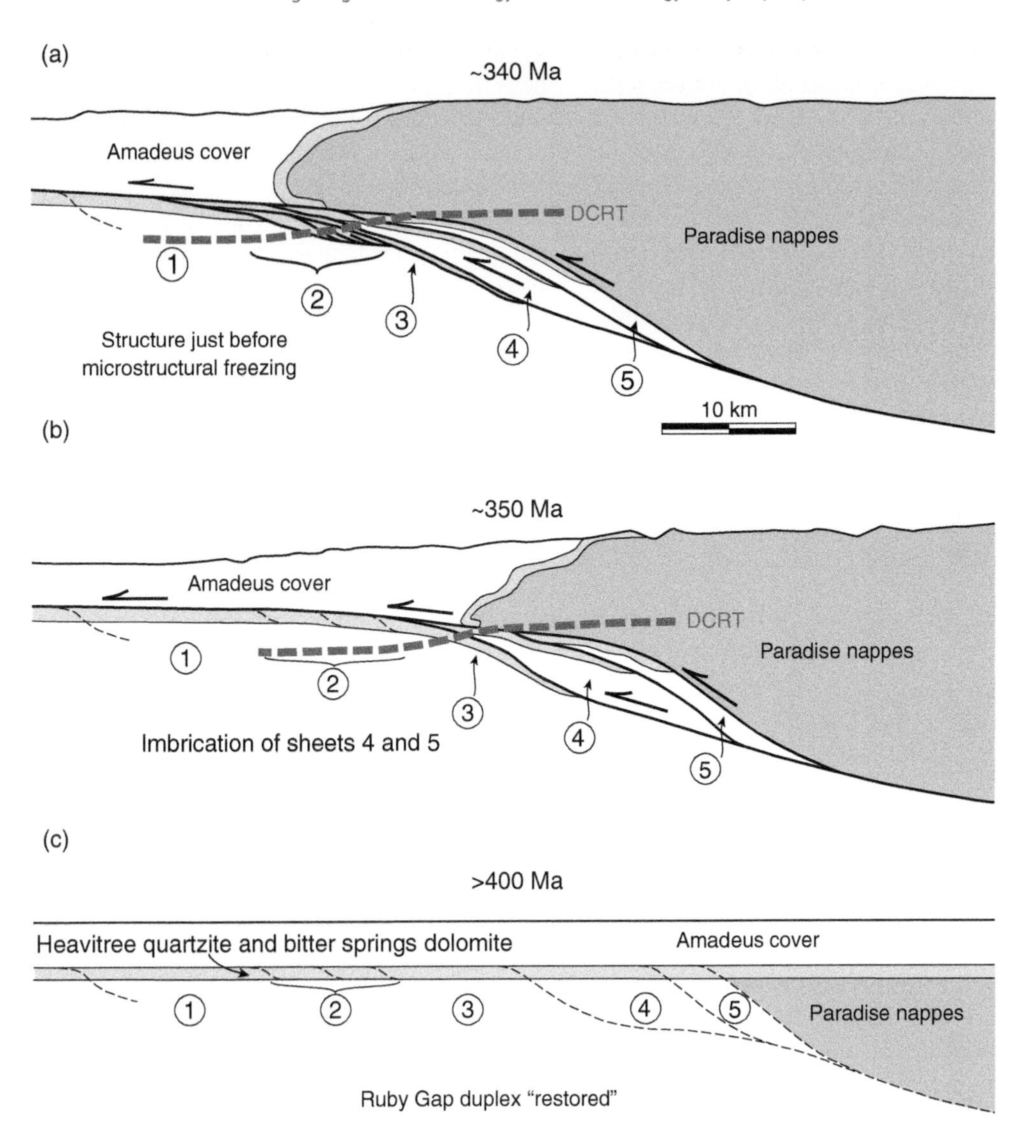

Figure 9.8 Final steps in the restoration of the Ruby Gap duplex. *Source:* Dunlap et al. (1997) / John Wiley & Sons. (a) Reconstruction of the Alice Springs orogen at the stage depicted in Figure 9.7d, e, i at ~340 Ma. (b) Reconstruction of the Alice Springs orogen at ~350 Ma. (c) Restored positions of horses in the Ruby Gap duplex prior to the onset of deformation, i.e. at ~400 Ma.

to constrain a detailed history of the development of the mid-crustal Ruby Gap duplex during ~30 Ma of the waning stages of the Alice Springs orogeny. Their reconstruction, focused on a period beginning about 340 Ma ago, ties variations in quartzite microstructure to spatial and temporal variations in temperature and differential stress. Selecting reasonable values for parameters in flow laws for dislocation creep for wet quartzite, they determine that this mid-crustal, intracratonic duplex likely developed under conditions comparable to those typically found in convergent plate boundary settings.

By linking their microstructural analysis to the Ar^{40}/Ar^{39} ages of phengite grown in Regime 3 and to a lesser degree Regime 2 quartzites, Dunlap et al. demonstrate that a footwall imbrication mechanism is the preferred kinematic model for the development of the Ruby Gap duplex. Unlike the standard models for duplex formation, where rock masses in individual horses bend and flex during their emplacement but are otherwise weakly deformed, the horses in the Ruby Gap duplex were elongated and thinned significantly during their formation and emplacement.

9.3 The Interplay of Deformation Mechanisms and Rheologies in the Mid-Crust: Copper Creek Thrust Sheet, Appalachian Valley and Ridge, Tennessee, United States

9.3.1 Introduction

Fold-thrust belts are characteristic features of post-Paleoproterozoic convergent orogens, and comparable structural elements may have developed associated with crustal convergence during the Archean. In the well-preserved Paleozoic Appalachian-Caledonian, Mesozoic-Cenozoic Cordilleran, and Cenozoic Alpine-Himalayan orogens, belts of thrusted and folded unmetamorphosed to variably metamorphosed continental margin and synorogenic sedimentary strata are hundreds of kilometers across. Toward the interior of these orogenic belts are collections of thrust sheets that consist of deformed and metamorphosed crystalline basement and cover or deformed and metamorphosed accreted terranes. Clearly, the details of the geometry, timing, and evolution of the fold-thrust structures in these different belts are unique, yet there are sufficient similarities between the features in the different belts that it is reasonable to consider that similar underlying mechanics govern the development of different fold-thrust belts. This section aims to use the characteristics of thrust structures in the southern Appalachians of North America as a guide to understanding the development and mechanics of the mid- to upper-crustal levels of fold-thrust belts. Thus, this section complements the discussion of mid- to lower-crustal deformation products of convergence in Section 9.2.

9.3.2 General Characteristics of the Southern Appalachian Fold-Thrust Belt

The Appalachian orogenic belt stretches more than 3000 km along the southeastern margin of the North American craton (Figure 9.9a). Structures in this belt define four broad, arcuate *salients* convex to the northwest – the Newfoundland, Northern Appalachian, Central Appalachian, and Southern Appalachian salients. The St. Lawrence, New York, and Virgina *recesses*, where structural trends define tight arcs concave to the northwest, separate the salients. This orogen-scale geometry is likely a product of the general character of the late Proterozoic rifting along the southeastern margin of North America (Rankin 1976; Thomas 1977). There is considerable evidence for the onset of deformation and metamorphism relatively soon after the late Proterozoic rifting, but most North American workers recognize three main phases of orogenic activity in the Appalachians: a Middle to Late Ordovician Taconic phase, an Early to Middle Devonian Acadian phase, and a Carboniferous to Permian Alleghanian phase. Taconic and Acadian deformation and metamorphism affect most prominently the Newfoundland and Northern Appalachian salients, though there is a record of Taconic and Acadian activity in both the unmetamorphosed sedimentary rocks in the foreland and crystalline rocks of the hinterland in both the

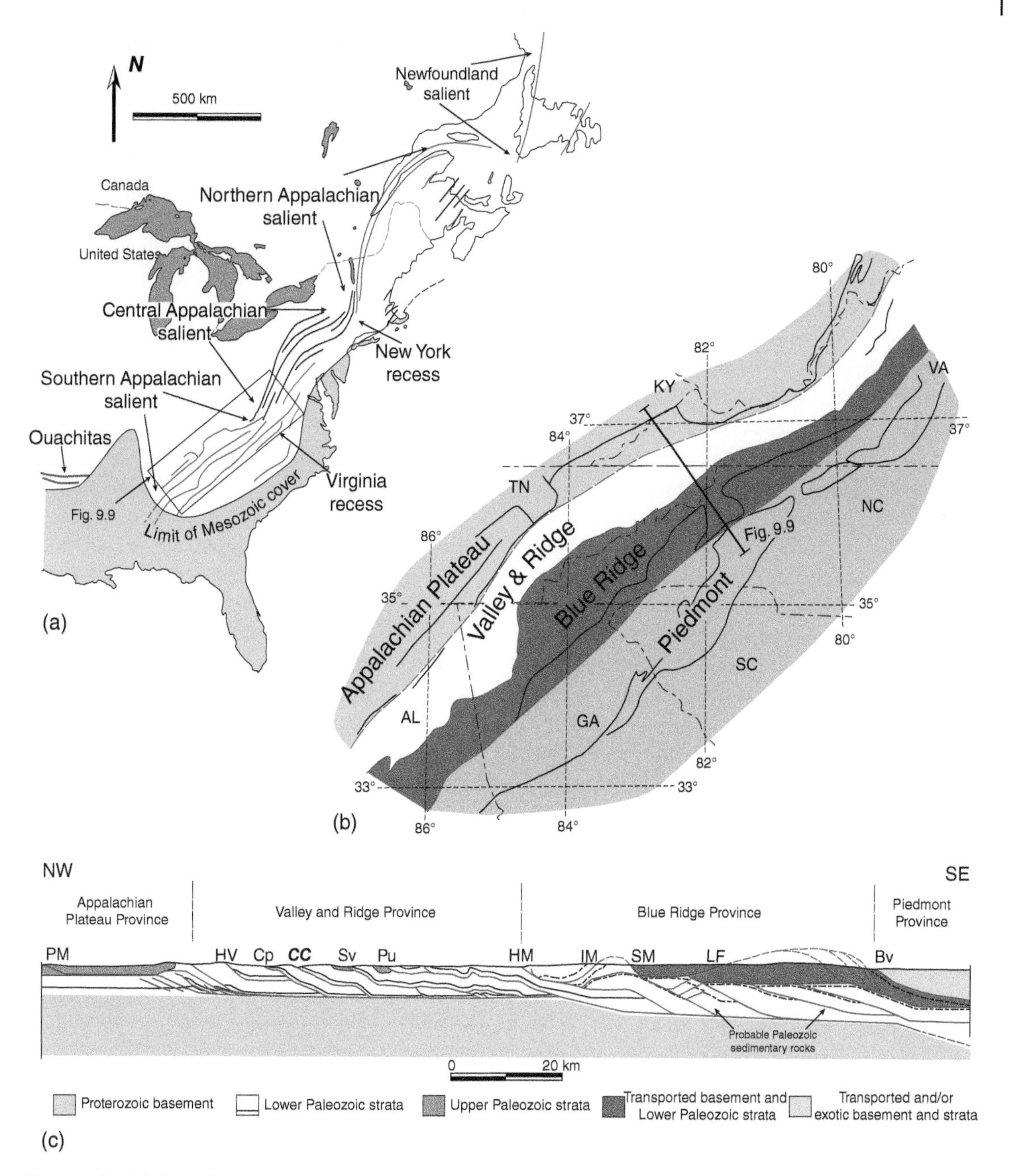

Figure 9.9 (a) The salients and recesses of the Appalachian orogenic belt along the SE margin of North America. Traces of major faults in red, and axes of major folds in black. The light red rectangle shows the area of the map in (b). (b) Map of the Southern Appalachian salient, showing the extent of the Plateau, Valley and Ridge, Blue Ridge, and Piedmont Provinces and the location of the section in (c). AL = Alabama; GA = Georgia; SC = South Carolina; NC = North Carolina; VA = Virginia; KY = Kentucky; TN = Tennessee. (c) Generalized structure section across the Southern Appalachian salient; line of section given in (b). *Source:* (b) and (c) Modified from Hatcher et al. (2007).

Central and Southern Appalachian salients. To a first order, these two orogenic phases are products of subduction along the margin of North American craton followed by a collision with the European craton. Alleghanian deformation and metamorphism are focused in the Central and Southern Appalachian salients, but there is also clear evidence for post-Acadian deformation and metamorphism in the two northern salients. The Alleghanian orogeny in the Central and Southern Appalachian salients is inferred to result from a collision between North American and African cratons. Broadly contemporaneous orogenic belts, the Ouachitas in the states of Arkansas and Oklahoma and the Marathon Mountains in the state of Texas, are likely southwest continuations of the Appalachian orogenic belt. Similarly, the Caledonides of East Greenland, Scotland, and Scandinavia extend this orogenic belt to the northeast. Reconstructions suggest that these belts together were comparable in length, geologic complexity, and temporal duration to the modern Cordilleran orogenic along the western margins of the North and South American continents or the Alpine-Himalayan belt and its continuation through southeast Asia to Indonesia.

Our focus in this chapter is the Alleghanian deformation in the Southern Appalachian salient. North American geologists recognize four physiographic provinces from NW to SE across the strike of the Southern Appalachians (Figure 9.9b): (1) the Appalachian Plateau Province underlain by flat-lying to gently inclined upper Paleozoic clastic wedge strata; (2) the Valley and Ridge Province in which folded and thrust-faulted lower to mid-Paleozoic continental margin strata define linear ridges and subparallel valleys; (3) the Blue Ridge Province, where deformed and metamorphosed siliciclastic lower Paleozoic to Proterozoic continental margin strata and Proterozoic crystalline basement support rugged topography at generally high elevation; and (4) the Piedmont Province, a region of generally low elevation underlain by deformed metasedimentary rocks and crystalline basement with both North American and exotic affinities. A collection of ESE-dipping thrust faults defines the overall structure of the Southern Appalachians (Figure 9.9c). In the Appalachian Plateau and Valley and Ridge Provinces to the north and west, thrusts carry sheets of unmetamorphosed to variably metamorphosed continental margin sedimentary strata to the WNW. In the Blue Ridge and Piedmont Provinces to the south and east, thrust sheets consist of deformed and metamorphosed North American crystalline basement and cover and, farther to the SE, deformed and metamorphosed accreted terranes.

The different physiographic provinces exhibit some differences in their underlying structure. In both the Plateau and Valley and Ridge Provinces, thrust faults extend down-dip to a regional décollement horizon near the base of the Cambrian Rome Formation. Beneath this regional décollement, undeformed Late Proterozoic to Cambrian strata lying on Proterozoic crystalline basement dip gently (1–3°) to the SE (Figure 9.9c). To the west, in the Plateau Province, this regional décollement steps up to laterally extensive detachment horizons in Silurian and/or Carboniferous strata and eventually emerges along the NW boundary of the Pine Mountain block in Kentucky and Virginia and along the west side of the Sequatchie Valley in the Cumberland Plateau of Tennessee (Figures 9.9c and 9.10). To the southeast, the first thrust fault to bring resistant siliciclastic strata from the Rome Formation to the surface marks the transition to the Valley and Ridge Province. A transect across the Valley and Ridge Province typically encounters five to eight individual thrust faults or systems of linked thrust faults that bring strata from the regional décollement to the surface (Figure 9.9c). In the Valley and Ridge Province, shorter distances between exposures of thrusts and steeper inclinations of the thrusts combine to create the distinctive pattern of linear

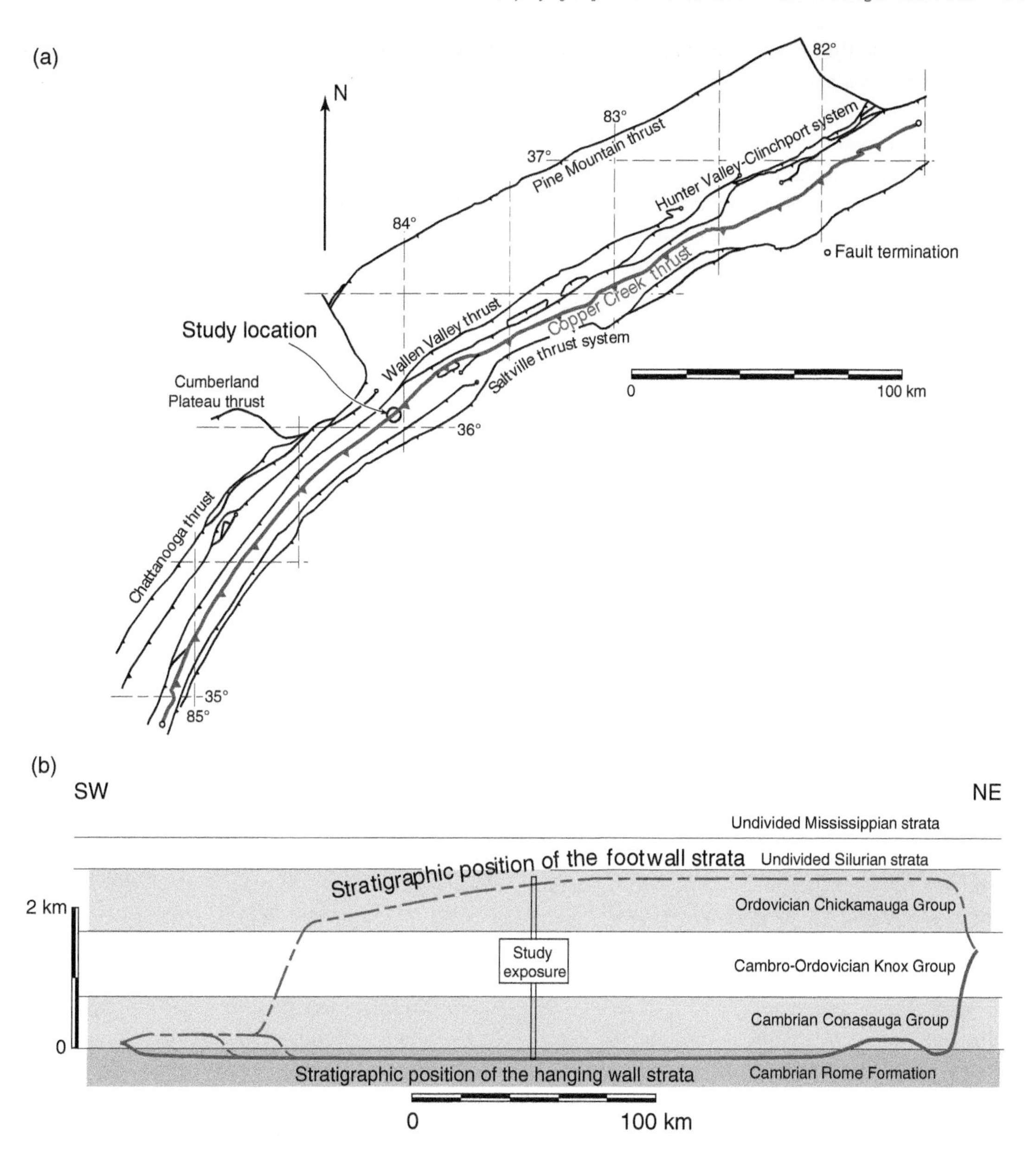

Figure 9.10 (a) Map showing the main thrust faults of the Appalachian Plateau and Valley and Ridge Provinces of southwest Virginia and northeast Tennessee. The Copper Creek thrust shown in red. (b) A stratigraphic separation diagram, indicating the stratigraphic position of hanging wall and footwall strata at different locations along the trace of the thrust exposure.

ridges and valleys that gives the province its name. In the central and southeastern portions of this Province, some thrust faults are folded, and transported sedimentary rocks regularly exhibit mappable mesoscopic fabrics.

In the Blue Ridge Province farther to the SE, the geometries and topologies of thrust faults resemble those of thrusts in the Valley and Ridge Province. The higher elevation and rugged topography of the Blue Ridge Province arise because thrusts carry erosionally resistant late Proterozoic to early Cambrian strata and Proterozoic crystalline rocks. Sedimentary rocks in Blue Ridge thrust sheets generally exhibit well-developed deformation fabrics and are variably metamorphosed. Crystalline rocks in these thrust sheets typically have experienced retrograde metamorphism and exhibit strong fabrics associated with thrusting. Along the SE edge of the Blue Ridge, an escarpment marks the NW margin of the Piedmont Province. This escarpment is an erosional boundary, but its structural significance is not clear. The macroscopic structure and general deformational character of rocks in the Piedmont Province are similar to those in the Blue Ridge, with metamorphosed sedimentary and crystalline rocks in the Piedmont exhibiting fabrics developed during and associated with transport to north and west above gently southeast-dipping thrust faults.

This section examines thrust-related deformation in the western portion of the Valley and Ridge Province. Restored sections of this portion of the Valley and Ridge suggest overburdens of 5 to 7 km on the regional detachment during emplacement. Assuming a geothermal gradient of ~30° C/km, deformation temperatures were unlikely to have exceeded 200° C significantly, an estimate that is consistent with available indicators of paleotemperature such as conodont color alteration indices across the belt. Thus, the overall deformation conditions in the western portion of the Valley and Ridge conform to the mid- to upper crust. The sedimentary rocks that compose the bulk of the thrust sheets in this portion of the province are neither metamorphosed nor have they developed penetrative deformation fabrics during emplacement. Instead, minor faults, folds, and/or arrays of variably spaced stylolites and fractures accommodated gradients in the displacement field. These deformation elements are concentrated in damage zones, which typically are hundreds of meters thick, adjacent to thrust surfaces. The cores of these thrust fault zones, in which fault rocks exhibit deformation fabrics apparent at the scale of a hand sample, rarely are thicker than tens of centimeters. The aim of this section is to examine the mechanics of this type of deformation in the mid-crust.

9.3.3 Deformation of the Copper Creek Thrust Sheet

The specific focus of this section is the deformation associated with the emplacement of the Copper Creek thrust sheet, a regionally significant sheet composed of unmetamorphosed sedimentary rocks exposed in the western portion of Valley and Ridge Province. The thrust underlying this sheet has a trace length in excess of 300 km, stratigraphic separations of 2–3 km along most of its length, and estimates of net slip that range from 15 to 50 km (Hatcher et al. 2007) (Figure 9.10), typical of many Southern Appalachian thrusts. A spectacular roadcut exposure near the midpoint of the mapped trace of this fault provides a unique picture of the mesoscopic structures associated with the emplacement of this thrust (Harris and Milici 1977) (Figure 9.11). At this location, a nearly planar principal slip surface separates dolostones, shales, and sandstones of the Cambrian Rome Formation from shales and limestones of the Ordovician Moccasin Formation (near the top of the Chickamauga Group). Strata above and below the principal slip surface are faulted, but none of these minor faults cut across the principal slip surface at this exposure. We consider here two

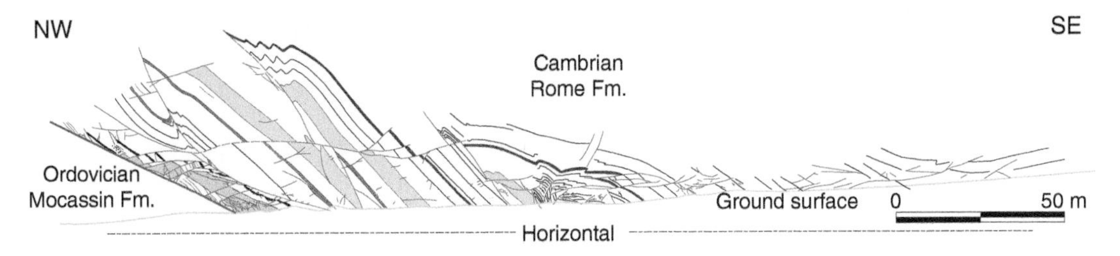

Figure 9.11 Profile of an exposure of the Copper Creek thrust fault in NE Tennessee (see Figure 9.10 for location), showing the minor faults and folds in Cambrian Rome Formation strata in the hanging wall of the thrust.

elements of the emplacement-related fabric associated with this thrust: (1) the array of regularly oriented minor faults and related folds in deformed hanging wall strata within approximately 150 m of the principal slip surface; and (2) the distinctive layer of fault rocks found along the principal slip surface. These elements share characteristics with elements associated with other exposures of the Copper Creek thrust, other thrust faults in the southern Appalachians, and thrust faults in other fold-thrust belts.

9.3.3.1 Mesoscopic Structural Elements

We consider first the characteristics of the minor fault array. Structural geologists have devised several methods to relate the geometry of mesoscopic faults, i.e. the orientations of the fault surface and any lineations on it (Figure 9.12a), to the general character of deformation. Figure 9.12b depicts an idealized planar fault whose slip direction is specified by a slickenline. The pole to the fault and slickenline together define an imaginary plane, the *m plane* of Arthaud (1969). A section parallel to the *m* plane viewed along the *slip normal*, an imaginary linear element that lies in the fault plane and is perpendicular to the *m* plane, best displays the fault offset. Note that the *m* plane contains the shortening (or P) and extension (or T) axes defined by some workers (cf. Marrett and Allmendinger 1990) (Figure 9.12b). Figure 9.12c shows the geometric relationships between the fault pole, the slickenline, and the slip normal. Since the slip normal is perpendicular to the plane containing the local shortening and elongation directions, it defines a tectonic *b* axis for the local deformation accommodated by the fault and is akin to the hinge lines or axis of a fold associated with fault movement (Figure 9.12d).

Nearly all faults in the arrays found within 150 m of the principal slip surface are readily assigned to one of two subpopulations. The first subpopulation consists of faults that cut bedding at low angles ($<45°$). Throughout the hanging wall side of the principal slip surface, low-angle faults have strikes roughly parallel to the strike of the underlying thrust and offsets that shorten sedimentary layering (Figures 9.11 and 9.13a). Calcite or quartz slickenfibers on low-angle faults typically are nearly parallel to the fault dip (most are within 20° of the fault's dip direction, i.e. have rakes $\geq 70°$; Figure 9.12a). The second subpopulation consists of faults that cut bedding at high angles ($\geq 45°$). In strata within 10 m of the principal slip surface, high-angle faults have strikes approximately parallel to the thrust strike and offsets that typically extend sedimentary layering. Calcite or quartz slickenfibers on high-angle faults near the principal slip surface typically have rakes $\leq 45°$. Farther from the principal slip surface (between 10 and ~150 m above the principal slip surface), the strikes of high-angle faults are more variable (Figure 9.13b), and slickenfibers on high-angle faults farther from the principal slip surface exhibit a range of orientations from nearly

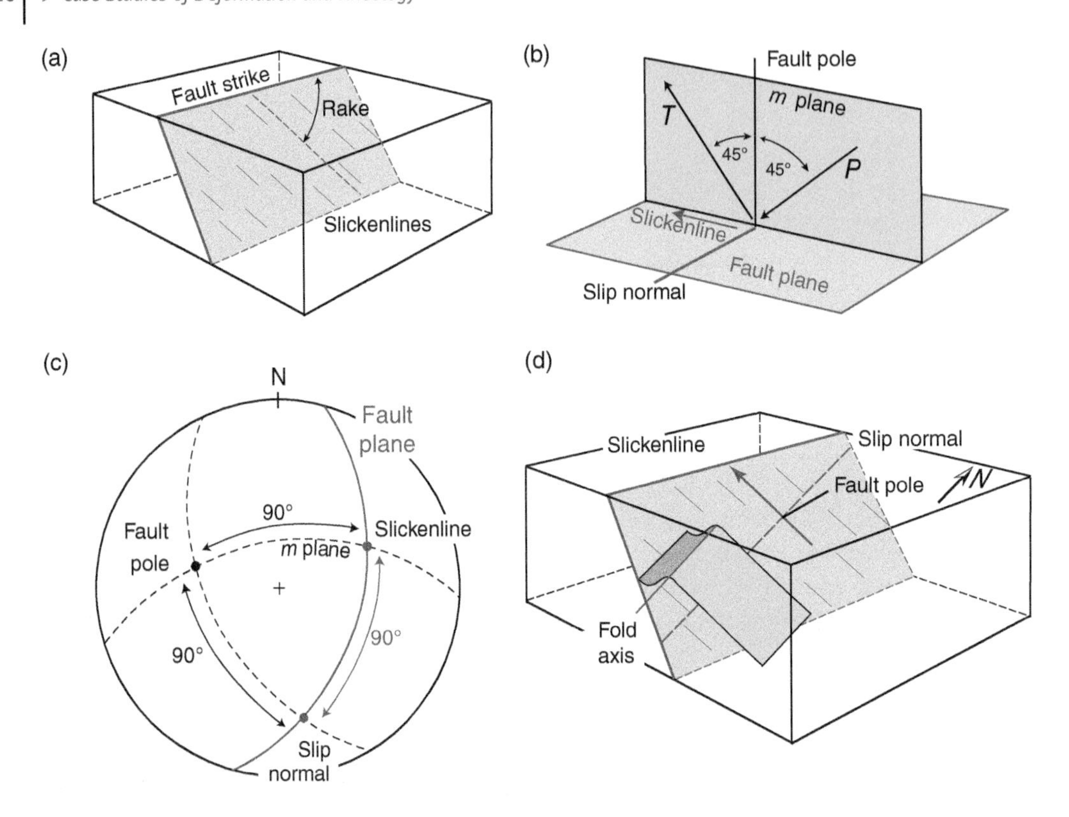

Figure 9.12 (a) Block diagram of fault with slickenlines. The *rake* of a slickenline is the angle measured between the strike of the fault and slickenline. (b) The slickenline on and pole to a planar fault define the *m plane*, which contains the shortening or P axis and elongation or T axis associated with fault slip. The slip normal is perpendicular to the *m* plane. (c) Stereographic projection showing the geometric relationships between the fault plane, slickenline, fault pole, and slip normal. (d) Schematic block diagram illustrating the fault geometry illustrated in (c). Note that the slip normal is parallel to and comparable to the fold axis of an associated fold.

strike-parallel to down-dip. The offsets on most high-angle faults elongate bedding (Figure 9.11).

Figures 9.13c are equal-area projections of slip normals to minor faults with slickenfibers. Within 10 m of the principal slip surface: (1) most slip normals to low-angle faults cluster at a broad point maximum parallel to the strike of the principal slip surface, consistent with subhorizontal shortening and subvertical elongation; (2) the small numbers of low-angle fault slip normals not associated with the point maximum fall along a diffuse great circle girdle (dashed line on equal-area projection) that strikes parallel to the principal slip surface but is inclined to it; and (3) slip normals to high-angle faults define a point maximum approximately perpendicular to the weakly defined great circle girdle. Farther from the principal slip surface (between 10 and ~150 m above it), fewer low-angle faults preserve slickenfibers, and the equal area projection is dominated by steeply inclined slip normals associated with high-angle faults. The high-angle faults in these strata mainly are oblique to the principal slip surface, and their slip normals define a point

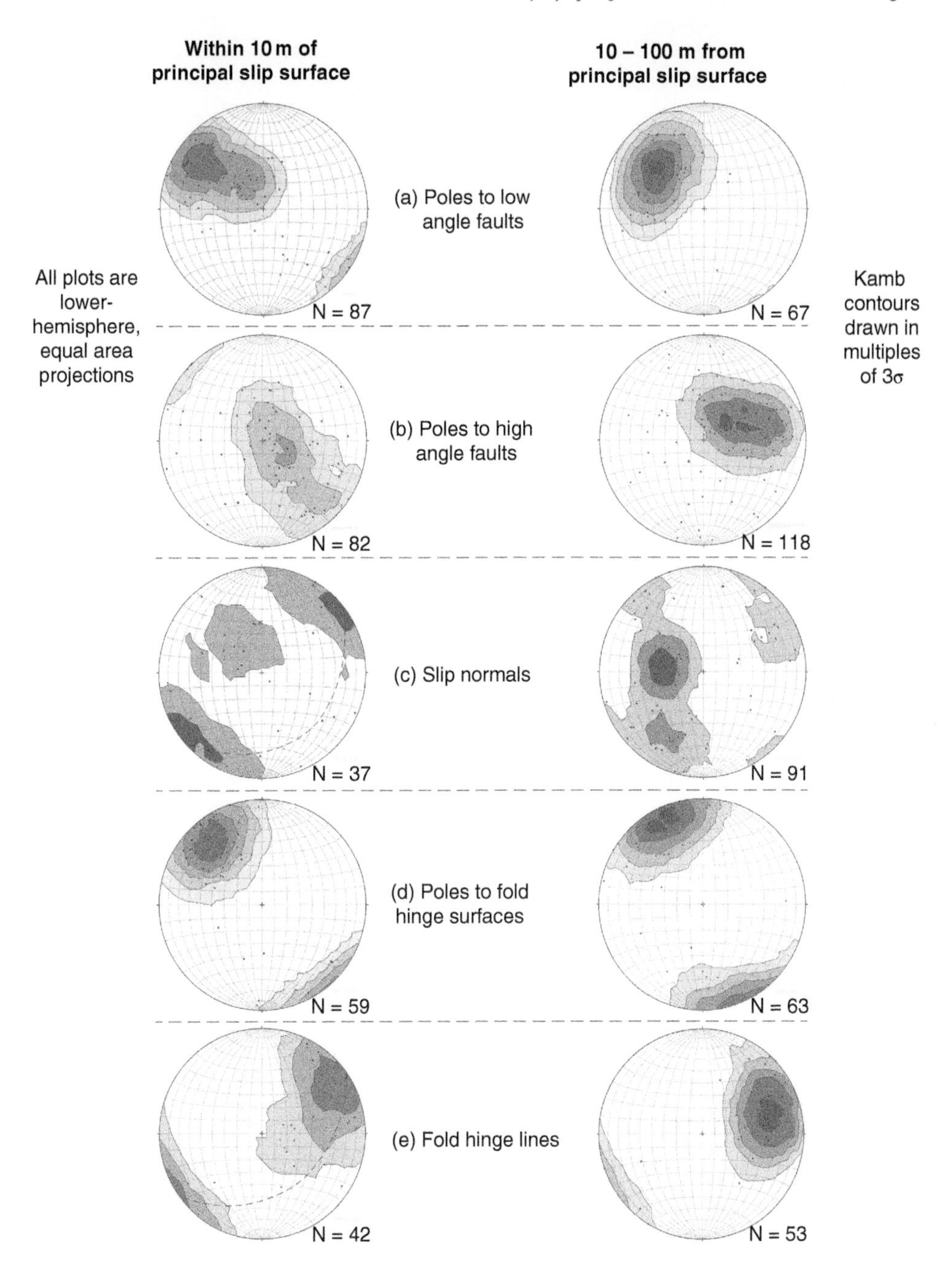

Figure 9.13 Lower hemisphere equal-area projections of structural elements in the deformed Rome Formation strata at the study site. Plots in the column at left depict structural elements observed within 10 m of the principal slip surface. Plots in the column at right depict structural elements observed between 10 and ~150 m above the principal slip surface. (a) Poles to low angle faults. (b) Poles to high angle faults. (c) Slip normals for slickensided faults. (d) Poles to fold hinge surfaces. (e) Fold hinge lines.

maximum slightly oblique to that defined by high-angle faults within 10 m of the principal slip surface.

In the exposure of the Copper Creek thrust damage zone described here, small folds occur adjacent to and are genetically related to minor faults (e.g. Figure 9.12d). More common are asymmetric folds associated with reverse faults. Typically these fault-fold associations resemble idealized fault-bend folds or fault-propagation folds. Less common are drape folds or roll-overs associated with extensional faults. In aggregate, the hinge surfaces of folds tend strike subparallel to the strike of the thrust (Figure 9.13d). A sizable fraction of folds have hinge lines subparallel to the strike of the fold hinge surfaces, and by extension parallel to the strike of the principal slip surface. A number of folds have hinge lines that plunge moderately or steeply, and within 10 m of the principal slip surface fold hinge lines define a great circle inclined approximately 30° to the principal slip surface (dashed line in Figure 9.13e).

9.3.3.2 Kinematics of Mesoscopic Deformation

Within 10 m of the principal slip surface, both low- and high-angle faults are common. High-angle faults regularly cut and offset low-angle faults and their associated folds; only rarely do low-angle faults cut and offset high-angle faults. The distances separating mesoscopic faults are relatively small. Measured values of fault surface area per unit volume for faults with visible traces >1 m are on the order $S/V \approx 1.0\,m^2/m^3$ within 10 m of the principal slip surface. Farther from the principal slip surface, low-angle faults are more common than high-angle faults; there are sizable regions with no high-angle faults. Where both occur, the crosscutting relationships observed near the principal slip surface tend to hold, but values of mesoscopic fault S/V are significantly lower (between 0.1 and $0.5\,m^2/m^3$) farther from the principal slip surface. At this exposure, there is a region ~150 m above the principal slip surface in which high-angle faults outnumber low-angle faults, although both have very small offsets.

Crosscutting relationships within the damage zone above the principal slip surface of the Copper Creek thrust indicate that: (1) low-angle faults typically form during earlier deformation increments, shortening strata in the direction of sheet transport; and (2) high-angle faults typically form during later deformation increments, elongating strata parallel and perpendicular to the direction of sheet transport near the principal slip surface and elongating strata parallel to sheet transport farther from the principal slip surface. Comparable arrays of mesoscopic faults and folds occur in the damage zones adjacent to other Appalachian thrust faults with displacements of tens of kilometers, and these arrays exhibit similar deformation kinematics. The damage zone adjacent to the Cumberland Plateau thrust, which has an inferred displacement of at most a few kilometers, exhibits a similar peak value of S/V, but the thickness of the regions exhibiting the peak S/V values is greater in sheets that have greater displacements.

9.3.3.3 Estimating Strain Magnitudes Within the Damage Zone

The slip on any individual fault in a mesoscopic array contributes to a displacement field in which net displacement varies with position. In some instances, spatial gradients of the next displacement due to fault slip are approximately constant when measured over length scales greater than the mean spacing between faults. Such a deformation is *mesoscopically homogeneous*, and one can determine displacement gradient values, calculate the components of two- or three-dimensional strain matrices, and thereby determine an ellipse or ellipsoid depicting the fault-related strain (Wojtal 1989; Horsman and Tikoff 2005). These methods calculate strain magnitudes using the reciprocal quadratic elongation, a strain measure that is rarely employed despite its utility.

A significant advantage of the reciprocal quadratic elongation is that determines strain values referred to the current or deformed state. For any deformed material line, the reciprocal quadratic elongation $\lambda' = (l_o/l')^2$, where l_o is the original length of the line and l' is the current or deformed length of the line. Note that $1<\lambda'<\infty$ for lines that are shortened, and $0<\lambda'<1$ for lines that are lengthened. In the matrix or tensor form of the reciprocal quadratic elongation λ', the diagonal components λ'_{ii} refer to longitudinal strains in the x'_i coordinate direction. Off-diagonal components λ'_{ij} refer to shear strains in the x'_j coordinate direction on a plane whose normal is the x'_i coordinate direction. Off-diagonal components are related to measured angular shear values ψ_{ij} by the formula $\lambda'_{ij} = \lambda'_{ii}(\tan\psi_{ij})$.

In the portion of the hanging wall damage zone within 10 m of the principal slip surface, offsets on the relatively late high-angle faults are sufficiently large that it is difficult to determine reliably the offsets on the relatively early low-angle faults. As a result, one cannot reconstruct fully the displacement field in the rocks immediately above the principal slip surface. Far above the principal slip surface, faults are widely spaced, offset on faults are small, and strain magnitudes are low. In the portion of the damage zone between 10 and ~75 m above the principal slip surface, the geometry and offsets of faults are clear, fault offsets are significant, and the overall deformation is more amenable to strain measurement. For this reason, we consider first the deformed strata between 10 and ~75 m above the principal slip surface.

Figure 9.14a is a mirror image of Figure 9.11, the cross section of the damage zone in the hanging wall of the Copper Creek thrust. The mirror image is used so that the deformation can be analyzed readily using right-handed coordinate axes in familiar orientations. Thus, superposed on a portion of this section between 10 and ~75 m above the principal slip surface are right-handed Cartesian coordinate axes x'_1 and x'_2. Points A'–D' along the x'_1 axis and J'–M' along the x'_2 axis identify locations near the centers of fault-bounded blocks in this portion of the damage zone. The distances between A' and B', C', and D' give the x'_1 coordinates of the latter three points, and the distances between J' and K', L', and M' give the x'_2 coordinates of those three points. Holding A' fixed, one can measure the net displacements required to restore B', C', and D' to their pre-faulting positions relative to A' (i.e. so that the beds are aligned). This defines the reciprocal displacements $\boldsymbol{u}'$ of those points (cf. Figure 9.14b). Figure 9.15a, b give the magnitudes of the u'_1 and u'_2 components of the reciprocal displacements at different x'_1 values. These plots constrain the range of values for the reciprocal displacement gradients $\Delta u'_1/\Delta x_1$ and $\Delta u'_2/\Delta x'_1$. Similarly, holding J' fixed, one can estimate the reciprocal displacements $\boldsymbol{u}'$ required to restore K', L', and M' to their pre-faulting positions with respect to J'. Figure 9.15c, d give the magnitudes of the u'_1 and u'_2 components of the reciprocal displacements at different x'_2 values and constrain the range of values of the reciprocal displacement gradients $\Delta u'_1/\Delta x'_2$ and $\Delta u'_2/\Delta x'_2$. In each of those graphs, the dashed black lines indicate the maximum and minimum values of the gradients determined from the measured reciprocal displacement values. The solid red line is the bisector of the angle between the two dashed lines, taken here to estimate a sort of average gradient of the reciprocal displacements.

Following Wojtal (1989), we use the reciprocal displacement gradients to determine a two-dimensional reciprocal deformation matrix **E** for the deformation, where $E_{ij} = \Delta u'_i/\Delta x'_j + \delta_{ij}$. The reciprocal quadratic elongation matrix $\boldsymbol{\lambda}'$ is the product of the transpose of **E** times **E**, i.e. $\boldsymbol{\lambda}' = \mathbf{E}^T\mathbf{E}$. Using a Mohr diagram for the reciprocal quadratic elongation, one can determine graphically the magnitudes and orientations of the principal strains for this fault-accommodated strain.

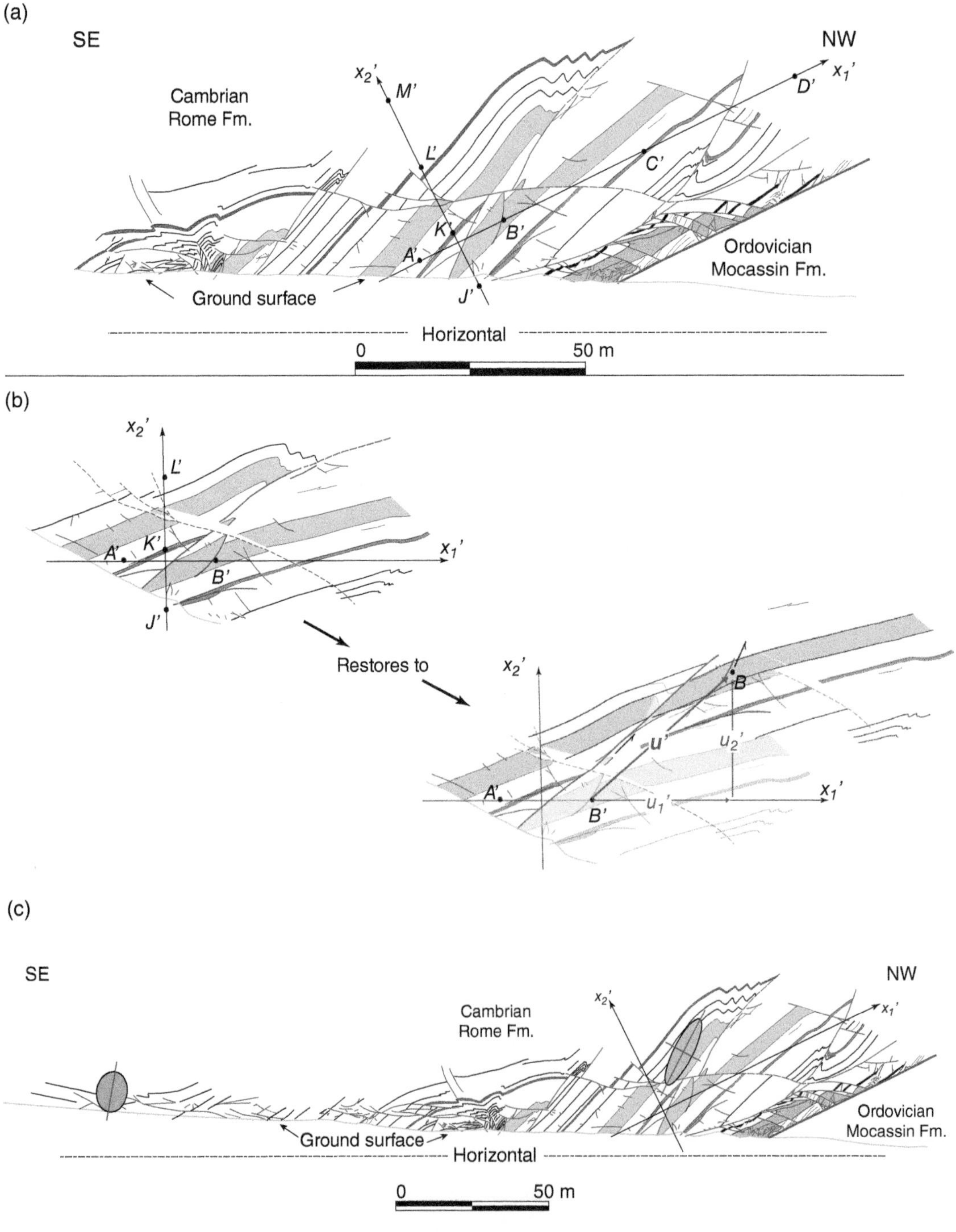

Figure 9.14 (a) Mirror image of a portion of Figure 9.11 with right-handed coordinate axes superposed on the section. (b) Illustration of the the method used to determine the reciprocal displacements of rock masses.
(c) Ellipses superposed on the mirror image of Figure 9.11 depict axial ratios and orientations of finite strains measured in two regions above the thrust.

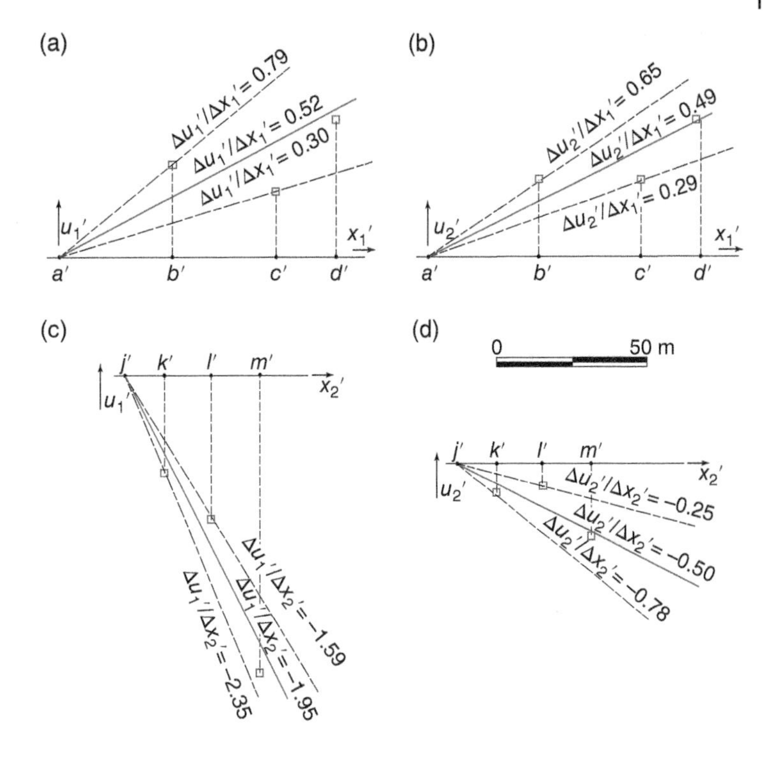

Figure 9.15 Plots of components of the reciprocal displacements of rock masses (u_i') versus the rock masses' position coordinates (x_j') used to determine reciprocal displacement gradients. (a) u_1' versus x_1' plot. (b) u_2' versus x_1' plot. (c) u_1' versus x_2' plot. (d) u_2' versus x_2' plot.

Taking the mean values reciprocal displacement gradients, the reciprocal deformation matrix is

$$\mathbf{E} \approx \begin{vmatrix} 1.5 & -2 \\ 0.5 & -0.5 \end{vmatrix} \tag{9.2}$$

and the reciprocal quadratic elongation matrix λ' is

$$\lambda' \approx \begin{vmatrix} 2.5 & -2.75 \\ -2.75 & 4.25 \end{vmatrix} \tag{9.3}$$

Figure 9.16 is a Mohr diagram drawn using the reciprocal quadratic elongation values given in Equation (9.3). The Mohr diagram constrains the magnitudes and orientations of the principal strains for the fault-related strain. The diagram indicates that the directions of maximum elongation and shortening are inclined ~36° and ~54°, respectively, to the principal slip surface. Using principal values read directly from the Mohr diagram, the axial ratio of the fault-accommodated mesoscopic strain is $\sqrt{\lambda'_{\mathrm{II}}/\lambda'_{\mathrm{I}}} = \sqrt{6.27/0.47} \approx 3.7:1$. The ratio of the current area of the plane of the section versus its original area is $1/\sqrt{(\lambda'_{\mathrm{II}} \times \lambda'_{\mathrm{I}})} = 1/\sqrt{(6.27 \times 0.47)} \approx 0.6$, suggesting significant elongation normal to the plane of the section in this portion of the thrust sheet.

The magnitudes of the reciprocal displacement gradients are approximate, and thus the calculated reciprocal quadratic elongation values may not represent the strain magnitudes precisely. The λ'_{II} value of ~6.3 corresponds to the shortening of material lines to approximately 40% of their original lengths. Shortening of that magnitude is generally consistent with structures seen in the segment of the damage zone between 10 and ~75 m above the principal slip surface. The λ'_{I} value of ~0.5 corresponds to lengthening material lines to approximately 140% of their original lengths. If deformation occurred with little

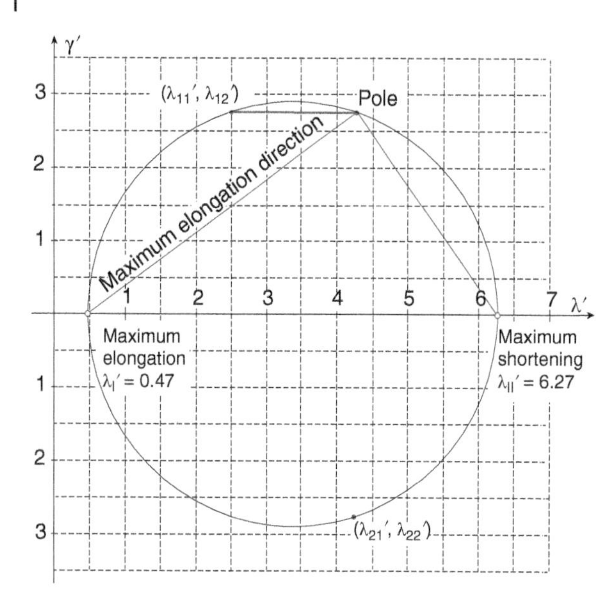

Figure 9.16 Reciprocal quadratic elongation Mohr diagram for mesoscopic deformation in strata 10–75 m above the principal slip surface.

volume change, which can be anticipated in deformation accommodated by minor faults, the measured magnitude of subvertical elongation and measured area ratio imply considerable elongation normal to the plane of section analyzed. Many faults have lineations with rakes <45°, consistent with the movement of material out of the plane of the section, so some elongation normal to the plane of section is consistent with the observed structures. Further, such elongation can be anticipated. Consider, for example, the leading edge of a thrust sheet. After the sheet has moved, the leading edge has an arcuate shape between the thrust's tip points, and the arc length is longer than the original linear distance between those tip points by a factor of 10–15%. In such a case, an area ratio of 0.85–0.9 might be anticipated. The calculated elongations normal to the plane of section exceed those values significantly, however. Holding λ'_{II} constant at ~6.3, a λ'_{I} value closer to ~0.2 (corresponding to lengthening material lines to ≥200% of their original lengths during vertical thickening) is required to bring the area ratio closer to an anticipated magnitude. Alternatively, holding λ'_{I} constant at ~0.5, a λ'_{II} value closer to ~2.75 (corresponding to shortening material lines to 60% of their original length) is required to bring the area ratio closer to an anticipated magnitude. We report here the measured values while acknowledging possible errors in the precise magnitudes of the measured strains. We infer, however, that the strain ellipse drawn in Figure 9.14a is a first-order estimate of the magnitudes and principal directions of bulk strains in strata between 10 and ~75 m above the principal slip surface.

Farther than ~75 m from the thrust surface, faults and folds are very widely spaced, and offsets on them are smaller than the offsets on faults closer to the thrust. In addition, a change in the orientation of the enveloping bedding makes it difficult to assess the original orientations of structures relative to the principal slip surface. Crude estimates of the reciprocal displacement gradients yield a reciprocal deformation matrix of

$$\mathbf{E} \approx \begin{vmatrix} 1.05 & -0.05 \\ -0.1 & 0.8 \end{vmatrix} \tag{9.4}$$

and a reciprocal quadratic elongation matrix λ' of

$$\lambda' \approx \begin{vmatrix} 1.1 & -0.13 \\ -0.13 & 0.64 \end{vmatrix} \qquad (9.5)$$

These values correspond to principal values of $\lambda'_{II} \approx 1.24$ and $\lambda'_{I} \approx 0.76$, an axial ratio of the fault-accommodated mesoscopic strain of $\sqrt{\lambda'_{II}/\lambda'_{I}} = \sqrt{1.24/0.76} \approx 1.3:1$, a ratio of the current area of the plane of the section to the original area of $1/\sqrt{(\lambda'_{II} \times \lambda'_{I})} = 1/\sqrt{(1.24 \times 0.76)} \approx 1.03$. Such a strain state is depicted schematically by the strain ellipse at the SE end of the section in Figure 9.14a.

Returning to the deformation within 10 m of the principal slip surface, the faulted strata there share many characteristics with the deformation in strata 10 to ~75 m above the principal slip surface. One characteristic that distinguishes this 10 m thick layer from the overlying strata is the number of closely spaced high-angle faults with moderate to large displacements (Figures 9.12 and 9.14a). The high-angle faults cut and offset low-angle faults, accommodate shearing relative to the principal slip surface, and elongate previously faulted rock parallel to the transport direction. Restoring those faults reveals a structure that resembles that seen in strata 10 to ~75 m above the principal slip surface (Figure 9.17a, b). Figure 9.17c and d illustrate schematically the effect of slip on the late, high-angle faults, and Figure 9.17e and f show semiquantitatively that slip on these faults result in total strain characterized by an axial ratio exceeding that in strata 10 to ~75 m above the principal slip surface and a principal elongation direction at a lower angle to the principal slip surface.

Combining inferences on the fault-accommodated deformation at different distances above the principal slip surface at this location yields insight to the deformation gradient within the Copper Creek thrust sheet. Across a layer a few hundred meters thick, the principal elongation direction curves from orientations at high angles to principal slip surface farther from that surface to orientations at low angles to the principal slip surface nearer to it. In addition, the strain magnitudes increase significantly across this layer (Figure 9.18). The deformation gradient across this layer is akin to those apparent in curved cleavage trajectories, deformed dikes, and variations in fold geometries adjacent to sheets emplaced at higher temperatures and pressures (Mitra and Elliott 1980; Ramsay et al. 1983; Gilotti and Kumpulanien 1986; Murphy 1987; Mitra 1993; Handschy 1998).

9.3.3.4 Character of Fault Rocks Developed Along the Principal Slip Surface

At the Copper Creek thrust exposure considered here, coherent shales, sandstones, and dolostones cut by minor faults and calcite-filled veins in the hanging wall are separated from coherent shales cut by minor faults, calcite-filled veins, and bedding-normal stylolites in the footwall by a thin layer of distinctive, carbonate-rich fault rocks. In hand sample and thin section, these fault rocks exhibit bands defined by differences in the minerals that compose them, differences in the sizes of mineral grains, or combinations of different minerals and grain sizes. Broadly similar fault rocks decorate other exposures of the Copper Creek thrust, other southern Appalachian thrust faults, and thrust faults in other thrust belts (Wojtal and Mitra 1986; Woodward et al. 1988; Kennedy and Logan 1997, 1998; Kennedy and White 2001; Wells 2017). The character of the bands in aggregate and the internal characteristics of individual bands indicate that the deformation processes operating within these fault rocks are distinct from those responsible for the development of the mesoscopic or macroscopic deformation of the thrust sheets above or below the principal slip surfaces of thrust faults.

9.3.3.4.1 Optical Observations Drawing upon an optical microscopy analysis of fault rocks from

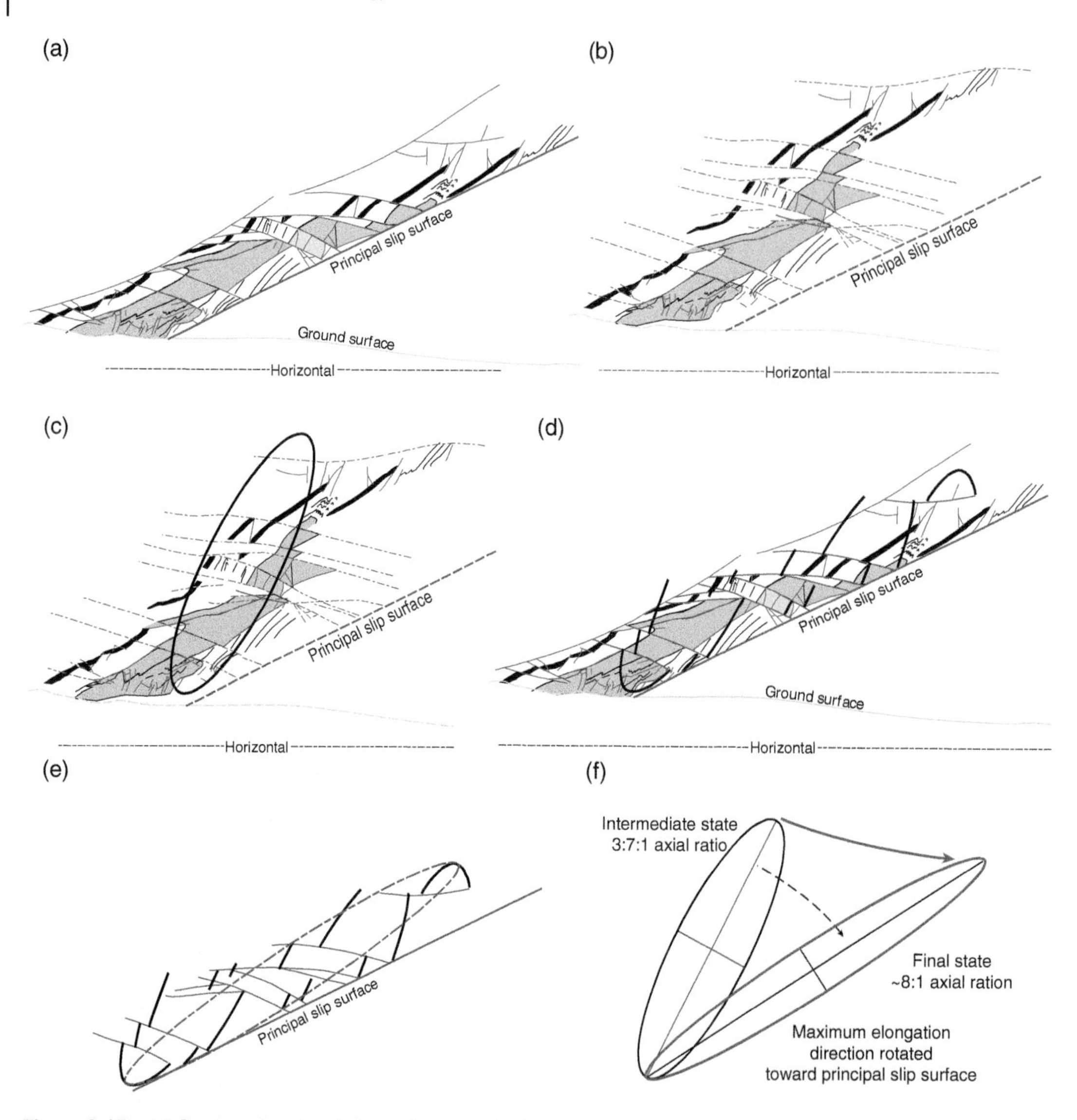

Figure 9.17 (a) Section showing deformed strata within 10 m of the principal slip surface. (b) Inferred geometry for strata within 10 m of the principal slip surface after restoring slip on late high-angle faults. (c) Intermediate strain ellipse, comparable to that in the overlying strata, superposed on the restored section from (b). (d) Displacement of segments of intermediate strain ellipse by late high-angle faults. (e) Estimate of total strain ellipse for strata within 10 m of the principal slip surface. (f) Comparison of intermediate and final strain ellipses for strata within 10 m of the principal slip surface.

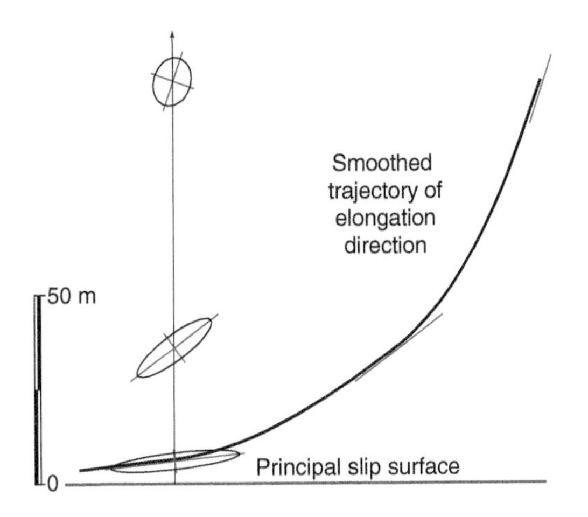

Figure 9.18 Schematic deformation profile for deformed strata above the Copper Creek thrust.

two southern Appalachian thrust faults, Wojtal and Mitra (1986) called these banded, carbonate fault rocks "foliated" cataclasites and ultracataclasites. They noted that rounded and angular fragments of wall rock, mineral-filled veins, and preexisting cataclasite sit in a matrix composed mainly of equant calcite grains with diameters less than 10 μm (Figure 9.19). Fragments are often cut by fractures or calcite-filled veins that do not extend into the surrounding fine-grained matrix. The fractures cutting fragments have no preferential orientation, consistent with an inference that fragments rotated during shearing of the fine-grained matrix. Matrix grains exhibit no discernable shape preferred orientation. The macroscopic "foliation" in these carbonate fault rocks is defined by alternating light and dark bands, with greater and lesser matrix fractions, aligned parallel to thrust-zone boundaries. Bands with high matrix fractions are often nearly planar and extend tens of centimeters across a hand sample or outcrop. Wojtal and Mitra argued that rock and then mineral fracture (comminution) was responsible for the initial grain size reduction, with creep by diffusional mass transfer accommodating significant

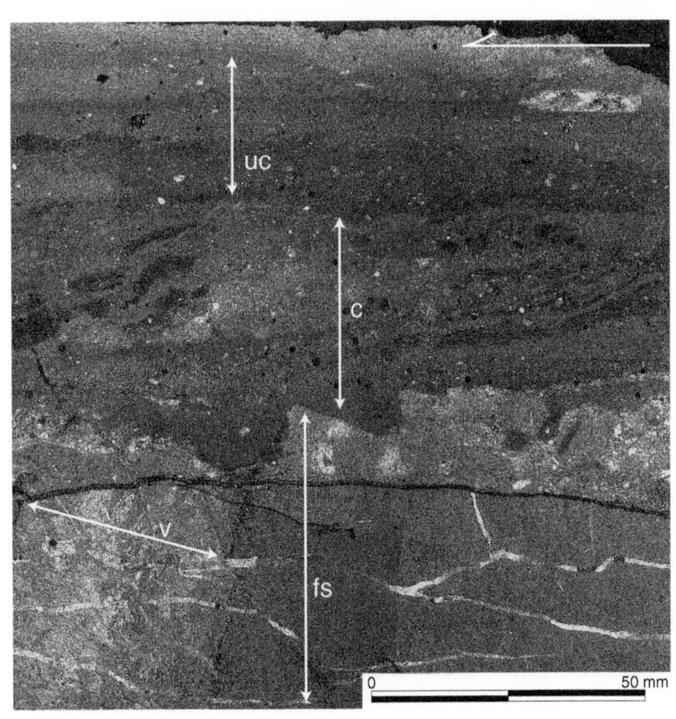

Figure 9.19 Photomicrograph of Copper Creek fault rocks. FS = fractured footwall shale with an irregular upper surface. v = calcite-filled vein cutting footwall shale, truncated by cataclasite layer (c) and ultracataclasite layer (uc).

shear strain once sufficient fine-grained, carbonate matrix formed. The presence of fragments of earlier formed cataclasite indicate cycling between cataclastic processes and creep over the time required to emplace the sheet. Wojtal (1992a) cited similarities in the color and intensity of cathodoluminescence exhibited by calcite in the ground mass of the cataclasites and veins cutting the wall rocks to infer that fine-grained calcite in the matrix is, like the calcite in veins, secondary, or that at least both types of calcite equilibrated with fluid phases with similar compositions. These observations support the inference of the importance of diffusional mass transfer in the deformation of these fault rocks.

9.3.3.4.2 SEM and TEM Observations Scanning electron microscope (SEM) and transmission electron microscope (TEM) analyses of the microstructures in banded Copper Creek fault rocks indicate that the deformation sequence inferred from optical observations requires additions and emendations (Wells et al. 2014; Wells 2017). Three distinctive layers are apparent in fault rocks above the lower contact with the underlying footwall shale: (1) a 500 μm thick band of *densely veined siliciclastic shale*; (2) a 300–350 μm thick *calcite vein layer* containing fragments of the footwall shale; and (3) a band at least 1 cm thick of ultrafine calcite grains that she called, for reasons outlined later, *the shear zone* (Figure 9.20a, b)

The footwall shale is locally cut by calcite-filled veins that are mainly oriented approximately parallel to bedding; these veins cut across less common veins oriented approximately normal to bedding. The mean diameter of calcite grains is as large as ~75 μm in undeformed segments of veins, but the grain size is reduced to <1 μm where veins intersect. The overlying densely veined shale and calcite vein layer are both characterized by more numerous calcite-filled veins. In the densely veined shale, calcite-filled veins cut the siliciclastic shale at low angles to bedding and at low angles to the principal slip surface. The mean diameter of calcite grains in the veins in shale is ~9 μm, but vein intersections are common and calcite grain diameters are significantly smaller (~1 μm or smaller) in intersections. In the calcite vein layer, isolated small fragments of shale sit in a matrix of vein calcite. Layers of calcite with mean grain diameters of 5–30 μm alternate with discontinuous layers composed of ultrafine calcite grains whose mean diameters of <1 μm.

In the densely veined shale and calcite vein layer, both coarse- and fine-grained vein calcite are deformed. Coarser calcite grains typically are twinned, often with trains of open pores along the twin boundaries, at twin-twin intersections, and at twin-grain boundary intersections (Figure 9.20c, d). Coarser calcite grains exhibit a strong lattice-preferred orientation (LPO), with *c*-axes at a high angle to the foliation or banding in the fault rocks. Ultrafine calcite (with grain diameters of ~0.9–0.34 μm) occurs in the discontinuous layers within veins. Twinning is not common in ultrafine calcite grains, and neighboring grains regularly exhibit interpenetrating grain boundaries. Ultrafine aggregates locally have up to 6% grain boundary porosity. In addition, linear features defined by aligned grain boundaries extend across several grains and four-grain junctions are relatively common, and grains sometimes exhibit inter- and intragranular fractures (Figure 9.20e, f). The ultrafine calcite aggregates exhibit no LPO.

The shear zone is a layer at least 1 cm thick in which clasts up to 400 μm across consisting of polycrystalline aggregates of ultrafine calcite or of footwall shale occur within and are surrounded by the ultrafine-grained matrix (Figure 9.20g). The matrix of the shear zone exhibits a microstructure similar to the discontinuous, fine-grained layers in the complex calcite vein. The mean grain size of calcite grains in the shear zone is only slightly smaller than that in the complex calcite vein band (mean diameter is 0.31 μm), but very few grains

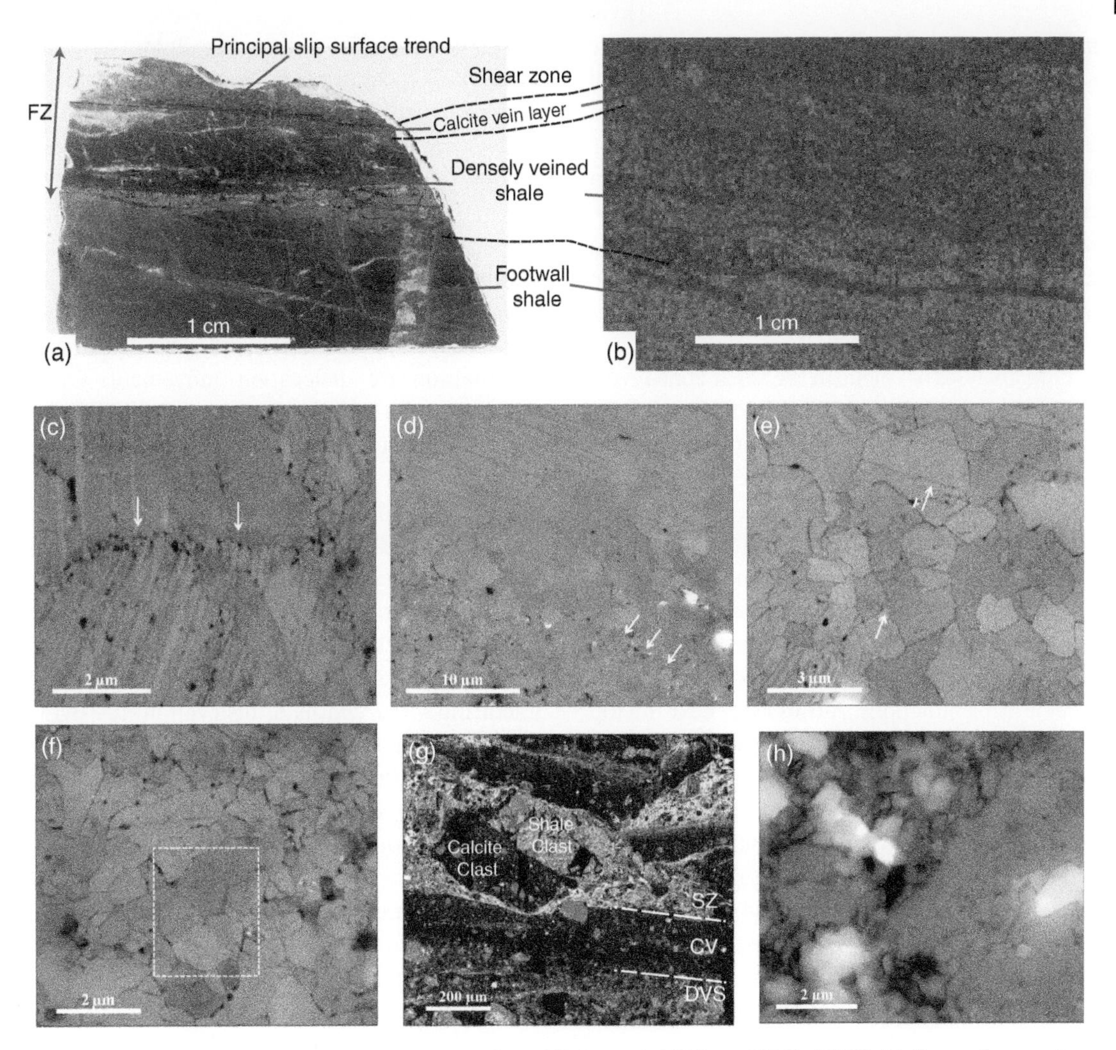

Figure 9.20 Images of Copper Creek fault rocks from (Wells et al. 2014) and Wells (2017). (a) Photomicrograph showing footwall shale and the adjacent fault zone (FZ) consisting of densely veined shale, calcite vein layer, and shear zone. (b) X-ray fluorescence map of shale to shear zone transition. Calcite veins (in blue) are oriented parallel or perpendicular to layering in shale. Al = red; Si = green, Ca = blue. (c–f) Back-scatter electron (BSE) images of fault rocks. (c) Pores (in black) and ultrafine grains along twin-twin and twin-grain boundary intersections. (d) Boundary between coarse calcite grain (above) and ultrafine calcite grains (below). Note aligned grain boundaries (indicated by arrows) in the ultrafine calcite grains. (e) Intragranular fracture (indicated by F), interpenetrating grain boundaries, and regular grid pattern (indicated by arrow) in ultrafine calcite grains. (f) Four grain junction in ultrafine calcite. (g) Energy-dispersive spectrometry map of fault rocks in (a). SZ = shear zone, CV = complex calcite vein, and DVS = densely veined shale. In this image, Si = red, Na = green, K = blue, Al = white, and Ca = black. Interconnected zones of ultrafine calcite surround clasts composed of shale and an aggregate of coarse and fine calcite. (h) BSE image of the edge of clast composed of an aggregate of calcite grains and a shale fragment (C). The left side of the image is ultrafine calcite and small shale fragments.

have diameters as large as 1 μm. These ultrafine calcite grains are elongated, with a mean axial ratio of 2 : 1, but the long axes of grains exhibit a wide range of orientations (Figure 9.20h). Ultrafine calcite grains in the shear zone again exhibit four-grain junctions, inter-and intragranular fractures, and linear features defined by aligned grain boundaries. Ultrafine calcite aggregates also have interpenetrating grain boundaries and pores along grain boundaries.

TEM analyses of calcite in the shear zone reveal that coarser grains (with diameters >10 μm) are commonly vesicular, with rounded vesicles observed along twin and grain boundaries. In addition, the lattice within twins in coarser grains is relatively dislocation free, but dislocations are abundant along twin boundaries and at twin-twin intersections. Ultrafine grains typically are largely devoid of dislocations and are separated by relatively straight, vesicle-free grain boundaries that meet at triple junctions. A thin layer near the top of the shear zone exhibits elongate grains that define a lineation parallel to the thrust slip direction. Within this thin layer, localized clusters of angular nanograins (with diameters as small as 7 nm but typically ~30 nm) occur along the grain boundaries separating some ultrafine grains. Some of the large clasts within the shear zone are partially coated by distinct layers composed of ultrafine calcite; these subtle coatings are most clearly visible in cathodoluminescence.

9.3.3.4.3 Inferences from SEM and TEM Observations

The fault rocks separating the hanging wall and footwall are composed largely of calcite, a mineral not common in the strata immediately above or below the fault. As noted earlier, the calcite in the fault rocks and in veins cutting the footwall exhibit similar cathodoluminescence, suggesting that they precipitated from or equilibrated with a common pore fluid source. Given the relatively low ambient temperatures (<200° C), precipitation from rather than equilibration with a common fluid source is more likely. In addition, microstructures observed in the footwall shale and the densely veined shale underscore that the calcite in the fault rocks originated as vein fills. The least deformed calcite in the fault rocks is the coarse-grained filling in veins. Coarse calcite exhibits deformation twinning, with locally high dislocation densities along twin boundaries, and a strong LPO. These microstructural features are consistent with deformation by dislocation movement. Open pores at twin–twin and twin–grain boundary intersections suggest that twin boundaries were obstacles to twin boundary glide and grain boundary mobility. It is likely that the pores at twin–twin and twin–grain boundary intersections were sites of stress concentration, and the stress concentrations likely led to grain fracture (cf. Figure 8.37). Twin geometry may have exerted control on the shape and size of the products by fracturing – the widths of dislocation-free twins correspond to the diameters of dislocation-free ultrafine calcite grains. There is little evidence that bulging grain boundary or subgrain rotation recrystallization accompanied deformation by twin gliding or contributed to grain size reduction in the deformed vein calcite.

Microstructures in the aggregates of ultrafine calcite suggest different deformation mechanisms predominated in those rocks. Most ultrafine calcite grains are untwinned and dislocation-free. Neighboring grains are sometimes separated by interpenetrating grain boundaries, suggesting some grain boundary mobility due to diffusion or dislocation movement. In other instances, grains are separated by relatively straight grain boundaries, and the boundaries of adjacent grains align to form planar boundaries that extend across several grains. In addition, four-grain junctions are common, and the aggregates exhibit no LPO. Together, these characteristics are consistent with deformation by grain boundary sliding, likely accommodated by diffusional mass transfer.

The microstructures in the Copper Creek fault rocks suggest that fault zone movement alternated between steady, aseismic creep and rapid, potentially seismic, slip events. There are two types of fragments or clasts within the ultrafine-grained matrix: (1) optically visible fragments or clasts; and (2) nanograins, coated clasts, and vesicular calcite grains apparent in TEM. The internal microstructure of fragments visible optically resemble portions of the calcite vein layer or the ultrafine-grained matrix itself, confirming that the fault rocks cycled between (1) periods when microstructures consistent with dislocation creep in coarse-grained calcite and grain boundary sliding in ultrafine-grained calcite develop; and (2) periods when fracturing or cataclasis generated the fragments or clasts. Regarding the features apparent only in TEM, numerous authors have associated the presence of nanograins and vesicular calcite with dissociation of calcite due to frictional heating, and the presence of coated clasts with rapid, presumably seismic slip rates (Han et al. 2007; Boutareaud et al. 2008; Fondriest et al. 2012; Han and Hirose 2012; Niemeijer et al. 2012; Violay et al. 2013; Collettini et al. 2014; Rempe et al. 2014; Spagnuolo et al. 2015; Smeraglia et al. 2017).

Figure 9.21a is stress versus grain size deformation mechanism map for calcite at 180° C after Wells et al. (2014). At this ambient temperature, deformation of calcite at strain rates faster than 10^{-9} s^{-1} is predominately by fracture consistent with Byerlee's law at moderate to high differential stresses. Deformation of coarse-grained calcite at strain rates between 10^{-9} and 10^{-13} s^{-1} can occur by twin gliding at moderate differential stresses. Deformation of ultrafine calcite at those strain rates can occur by diffusion-accommodated grain-size sensitive creep (combinations of diffusion creep and grain boundary sliding) at moderate to low differential stress magnitudes. Figure 9.21b outlines possible sequences of deformation events inferred from overprinting microstructures in the Copper Creek fault rocks (Wells 2017). The oldest microstructures observed in these fault rocks are the arrays of calcite-filled veins cutting the footwall shale. Fracture of the footwall shale is likely to have occurred during events ranging from accelerated creep episodes to seismic slip events. Slip at seismic or near seismic rates is required to generate the nanograins, vesicular calcite, and coated clasts observed in these fault rocks. It is not clear, however, what fraction of the total slip of this fault occurred by this mechanism. Slip at sub-seismic rates, which would still likely lead to the formation of fractures, accounts for an unknown magnitude of fault movement. Several microstructures indicate, however, that other mechanisms that contributed to sub-seismic slip by deformation distributed within the fault rocks. Calcite precipitated as fracture filling is likely to be relatively strain-free and able to deform readily, creating the collection of microstructures observed in the coarse calcite, i.e. deformation twinning, elevated dislocation densities along twin boundaries, open pores at twin–twin and twin–grain boundary intersections, and the development of LPO. In regions where grain size was reduced sufficiently, grain size-sensitive slip (either diffusion-accommodated or dislocation-accommodated grain boundary sliding) could accommodate localized shearing, potentially at low differential stress magnitudes. Grain growth due to recrystallization could close off pore space, creating impermeable layers or regions. Pore fluids in the rock between impermeable layers or surrounded by impermeable regions would experience increased pore fluid pressure enhancing the likelihood of an unstable, probably seismic slip or fracture formation during stable slip. Figure 9.21c indicates that the formation of the array of microstructures preserved in the Copper Creek fault rocks is likely to have been accompanied by significant variations in the rate at which fault movement occurred and in the magnitude of the differential stress required to drive fault movement (Wells 2017).

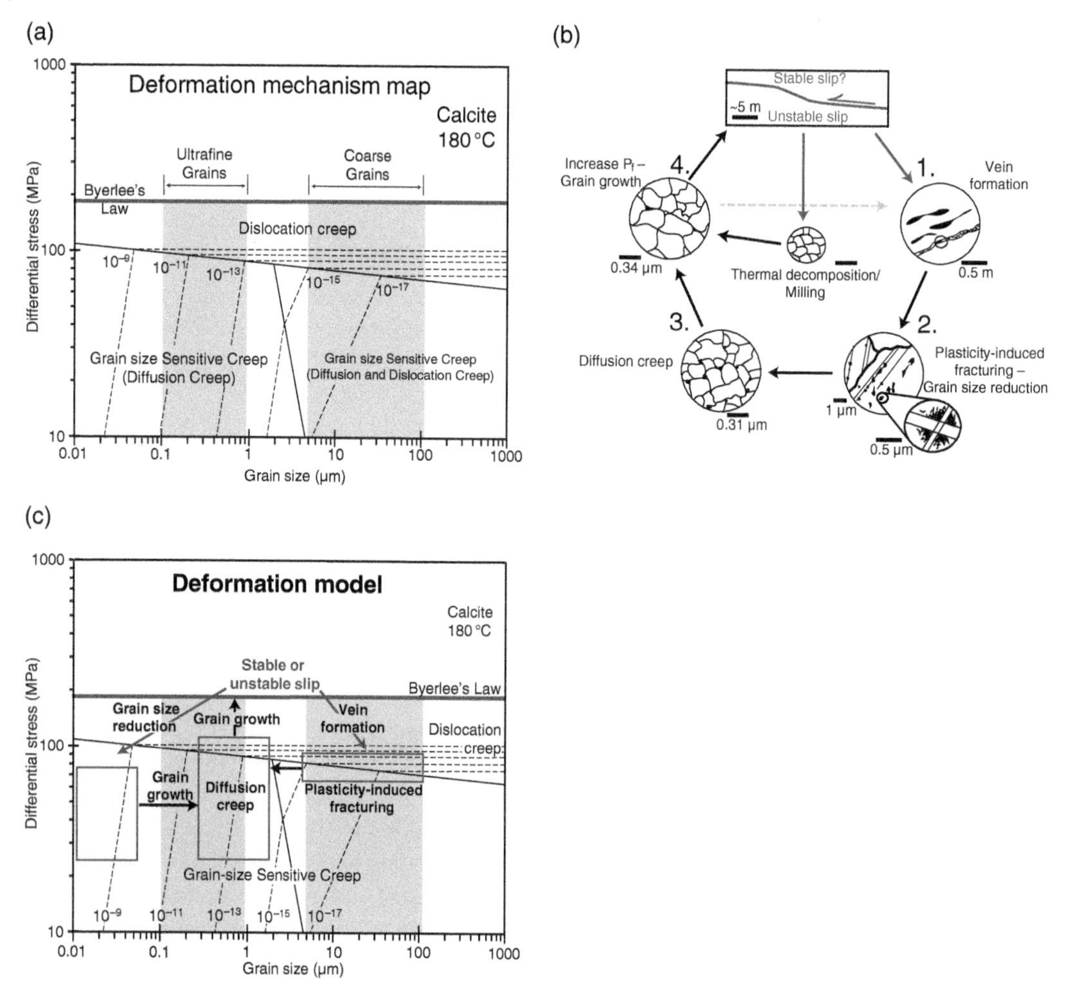

Figure 9.21 Interpretations of fault rock deformation after Wells (2017). (a) Differential stress versus grain size deformation mechanism map for calcite deforming at 180° C. Bands in red indicate ranges of calcite grain sizes in Copper Creek fault rocks. (b) and (c) Diagrams indicating deformation model for Copper Creek fault rocks outlined using deformation mechanism map from (a). Stable or unstable slip creates veins and facilitates extreme grain size reduction and possible thermal decomposition. Plastic deformation of vein calcite and grain growth of milled calcite yields ultrafine calcite that deforms plastically. Recrystallization promotes increased pore fluid pressure, setting the stage for stable or unstable slip episode.

9.3.4 Summarizing Deformation of the Copper Creek Thrust Sheet

Two components constitute the deformation profile developed during emplacement of the Copper Creek thrust sheet: (1) an array of minor faults and folds that accommodate mesoscopic strains within a layer 100s of meter thick along the base of the thrust sheet; and (2) distinctive fault rocks in a layer centimeters to perhaps a meter thick that decorates the principal slip surface. These two components are, to a first order, analogous to the structures seen along the base of glacial ice masses (Woodward et al. 1988; Wojtal 1992a and b), where basal slip and internal deformation contribute to glacier movement. Clearly different mechanical processes are responsible for the basal slip and internal distortion components of the deformation profiles in thrust sheets and glacial ice masses. Further, in glacial ice masses, observations demonstrate that the basal slip and internal deformation components develop simultaneously, although it has long been known that the relative contributions of the two components vary with position within a single glacier and probably vary over time at any position within a glacier (Savage and Paterson 1963).

The character of the fault rocks outlined earlier argues convincingly that both rapid, probably seismic slip episodes and periods of steady, aseismic slip both contributed to the movement of the sheet. It is easy to speculate and difficult to evaluate the relative contributions of creep and slip to the emplacement of any thrust sheet. Geodetic measurements of active faults confirm that aseismic creep, accelerated creep events, and seismic slip work together to accomplish fault slip, but geologists and geophysicists are only beginning to understand the spatial, temporal, and spatiotemporal variations found on active faults (e.g. Khoshmanesh and Shirzaei 2018; Bletery and Nocquet 2020, and references therein). The relatively thin coating of fault rocks found at this exposure of the Copper Creek thrust sheet might be read as an indication that seismic slip predominated during sheet emplacement. Other thrust exposures of thrust faults developed in unmetamorphosed sedimentary rocks display, however, considerably thicker layers of fine to ultrafine-grained fault rocks (cf. Woodward et al. 1988), and sheets emplaced at greater depths and higher temperatures display considerably thicker zones of emplacement-related deformation adjacent to their principal slip surfaces. Further work is required to determine the relative importance of slip on discrete surfaces, which probably was relatively rapid, and deformation distributed across thick layers, perhaps accruing slowly.

9.4 Induced Seismicity

9.4.1 Overview of Induced Seismicity

Examples of seismicity that coincides temporally and spatially with engineering projects such as mining, injection or withdrawal of fluids into the crust, and dam filling have been known for at least a century and from many places in the world. Some of these seismic events have caused loss of life, and they have collectively caused millions of dollars' worth of damage. This seismicity continues to be a source of alarm, exacerbated when earthquakes occur in areas of high population density. Because the origin of some of this seismicity is debated, it is particularly important to follow an inductive approach in analyzing such seismicity before it can be ascribed to human activity. The concept of *induced* seismicity is controversial, and we make some recommendations for its definition toward the end of the chapter, after taking a descriptive approach to examples of seismicity that are *associated* with anthropogenic activities.

This section of the chapter differs from the previous sections in that integration of displacements, strain, stress and rheology is not emphasized. We focus instead on stress and failure in the crust, led by a description of examples that show the importance of some fundamental concepts in the book to applied geology.

The seismicity discussed here commonly occurs in stable, intraplate regions, and has been associated with four broad categories of human activity (e.g. Doglioni 2018; Keranen and Weingarten 2018):

1) Mining, including quarrying, open pit and underground operations (e.g. McGarr 2002);
2) Fluid injection, including low-pressure waste injection, high-pressure waste injection, and fracking (e.g. Ellsworth 2013);
3) Fluid withdrawal, for example during hydrocarbon production (e.g. van Thienen-Visser and Breunese 2015); and
4) Reservoir construction (e.g. Talwani 1997a; Gupta 2002)

In each case, the data can be regarded as the results of an experiment in which the natural stresses, fluid pressures and seismicity of the Earth may have been altered by humankind. The close study of these phenomena offers great potential for learning about natural processes and conditions in the crust. This section does not cover seismicity associated with armaments testing. Earthquake magnitudes measured on different scales are notoriously difficult to unify: in the following, we have tried as much as possible to use the earthquake moment magnitude (M_w) scale (which relates directly to the slip magnitude, slipped area, and physical properties of the faulted rocks), but this has not been possible in all cases.

9.4.2 Earthquakes in the Witwatersrand Basin, South Africa

9.4.2.1 Background

Gold in the Witwatersrand basin (Figure 9.22) is dominantly hosted in siliciclastic rocks of the Central Rand Group deposited in the Witwatersrand basin at approximately 2.8 Ga. This extraordinary enrichment of gold in the Earth's crust has provided between a third and a half of gold mined on Earth (Frimmel 2008). Mining has taken

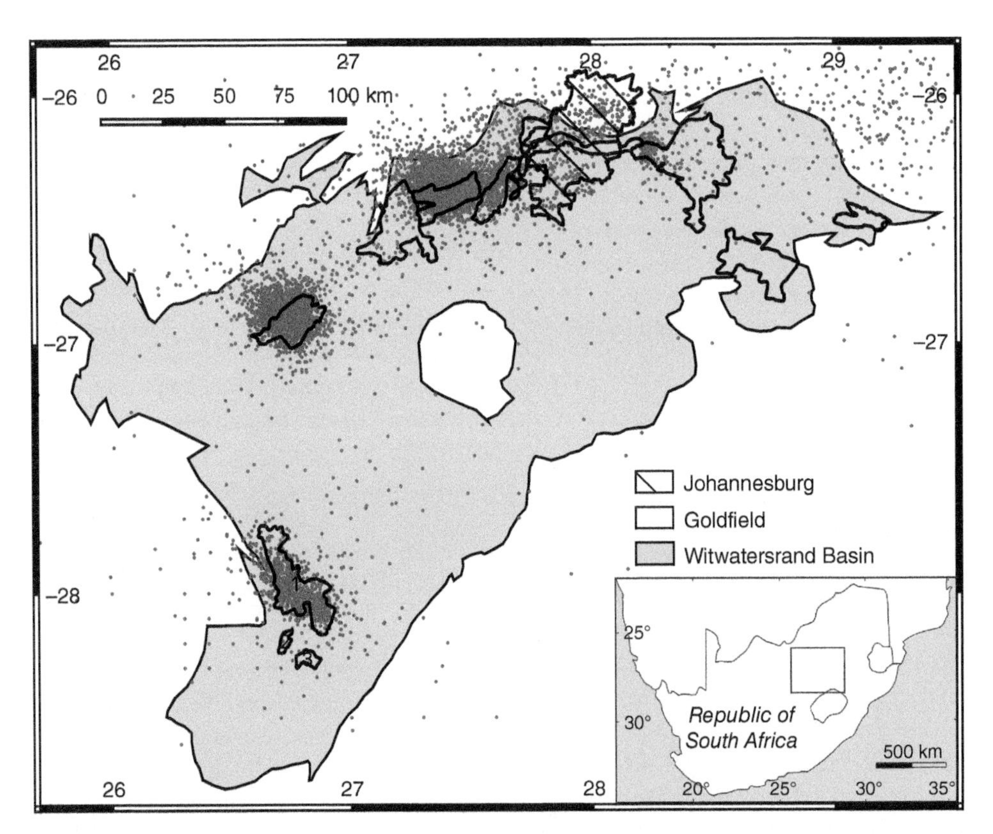

Figure 9.22 Seismicity of the Witwatersrand Basin. Epicentres from the reviewed ISC bulletin from 1964 to 2016 (International Seismological Centre, *On-line Bulletin*, http://www.isc.ac.uk, Internatl. Seismol. Cent., Thatcham, United Kingdom, 2016. http://doi.org/10.31905/D808B830). Outlines of the nine Witwatersrand goldfields and Johannesburg shown on top of the outcrop of the Witwatersrand basin (shaded). Latitude and Longitude in WGS 84.

place in nine major goldfields around the basin (Figure 9.22), which occupies an area of 300 km × 150 km. The gold occurs in "reefs" which are tabular bodies of mineralization largely controlled by bedding or surfaces subparallel to bedding in quartzites and conglomerates. Mining occurs by extraction of subhorizontal "stopes" with horizontal dimensions between 100 and 1000 m and heights of 1–2 m. The majority of current mining occurs at depths of 2–4 km; some mines at depths of >4 km are among the deepest in the world.

The ambient stress in the Witwatersrand basin has a vertical maximum principal stress σ_1 of about 26.5 MPa/km, which is about twice the measured horizontal stress (McGarr and Gay 1978). Pore fluid pressures are close to hydrostatic. This stress state is close to failure. However, mine water is drained during mining, reducing pore fluid pressures to zero.

Gold mining has occurred since the discovery of the deposits in 1886; for most of that history, seismicity in the immediate vicinity of the mines has been detected at rates of as much as 1000 events/day by seismic networks installed around the mines. Earthquakes as large as M_w 5.2 have caused extensive surface and underground damage, and seismic activity has been associated with the dangerous occurrence of rockbursts underground. Apart from fatalities and injuries, the deformation associated with mining has caused the loss of many millions of dollars in just one of the nine goldfields.

9.4.2.2 Seismicity

Detailed studies of seismicity using geophones installed in mines in the Witwatersrand Basin have provided a rich source of data on seismicity. From these studies, some of which involved over half a million events, it is evident that there are two types of seismicity (Richardson and Jordan 2002). By far the majority of events in both types have hypocentral depths between 2 and 4 km, which is the range of depths of mining during the monitoring period.

Type A. Low magnitude, ($-0.4 < M_w \leq 0$). Most of these events follow blasting within seconds to minutes and are located near mining operations. They show an evolving spatial distribution with time that reflects the progress of mining. Their source characteristics as suggested by low S to P wave velocity ratios include components of mode I fracturing at low normal stresses.

Type B. Higher magnitude ($M_w > 1$). These events do not correlate directly spatially or temporally with day-to-day mining operations, although they still occur within the mining area. The seismological data allow a calculation of the critical slip distance *Dc* for Type B events (see Section 7B.2.2.1.5): the value of ca. 1 mm is between values reported in laboratory experiments and for earthquakes.

Detailed analysis of source mechanisms confirms that there are at least two common failure modes for Witwatersrand seismicity. Some source mechanisms have a significant component of implosion. Other events have shear failure mechanisms. There are also rare recordings of explosional sources which are not due to blasting.

9.4.2.3 Interpretation

The effect of drainage in the Witwatersrand mines is to remove pore fluid pressure and thus make failure less likely. However, excavation increases stress around the stopes where the ore is extracted because one principal stress is effectively reduced to zero on the face of the open stope (Figure 9.23). There is an extreme stress concentration at the end of the stope, as well as enhanced stress in the areas ahead of the stope, both above and below. These stresses are sufficient to exceed the failure criterion for the siliciclastic host rocks, causing shear failure and collapse. Type A events are interpreted to represent primary, grain-scale failure in response to such blasting-induced stresses. Shear failure and collapse can be associated if a shear failure intersects mine workings (Figure 9.24).

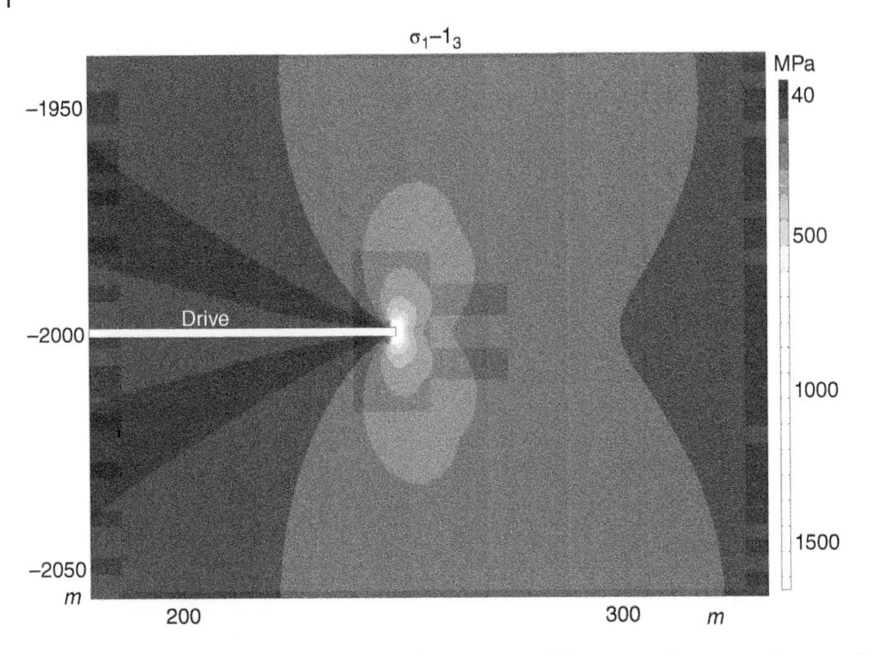

Figure 9.23 Stress concentration around a stope. Von Mises stress due to gravity around a 2 m high stope with a length of 300 m at a depth of 3 km. The rock is modeled with the properties of a quartzite, assuming linear elasticity. The Von Mises stress is: $\sqrt{\dfrac{(\sigma_1-\sigma_1)^2+(\sigma_2-\sigma_3)^2+(\sigma_3-\sigma_1)^2}{2}}$

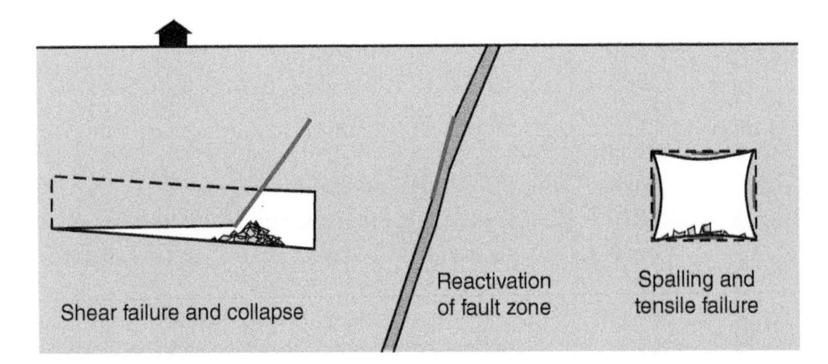

Figure 9.24 Failure mechanisms in the Witwatersrand goldfields. Shear failure that intersects workings can also be associated with collapse. Shear failure may occur by reactivation of older fault zones or other discontinuities. Tensile failure may occur around openings (Spalling).

Shear failure without implosion is attributed to reactivation of pre-existing planes of weakness such as joints, fractures, faults, bedding planes and dikes (Figure 9.24). Type B events are attributed to this failure mechanism because their foci are located on known faults, and failure resulting from reactivation of old fault zones has been observed underground. A slip weakening model, similar to that devised for natural earthquakes, seems applicable to these larger events. Spalling, which is tensile failure induced by the compressive stress in the absence of confining pressure, accounts for the explosional seismic sources (Figure 9.24).

9.4.3 Basel, Switzerland

9.4.3.1 Background

Extraction of geothermal energy using Enhanced Geothermal Systems (EGS) involves pumping fluids at high pressures through an injection well into relatively impermeable, hot rocks to create a reservoir of hot water by hydraulic fracturing. The hot water is pumped to the surface via a production well. Small earthquakes are generated by the hydraulic fracturing process, and the increases in fluid pressures also have the potential to induce larger earthquakes on pre-existing structural weaknesses.

Basel in Switzerland was the site of an EGS experiment started in 1996 (Deichmann and Giardini 2009). Basel is situated at the junction of the Rhine graben and the Jura Mountains of the Alps. Like most of Europe, the Rhine graben has maximum principal stress in the NW direction, related to the convergence of the African and European plates and the opening of the Atlantic.

In a now-infamous case, seismicity following the start of fluid injection at Basel in 2006 caused the closure of the experiment, as well as damage estimated at US$7 m. No previous seismic event had been detected in the 1 km^3 of the volume of seismicity around the injection well. Breakouts in the injection well show maximum horizontal stress oriented at 144°. Focal mechanisms in the region suggest a dominantly strike-slip stress field with some normal faulting.

9.4.3.2 Seismicity

Injection in the Basel experiment started on 2 December 2006 into a 5 km deep well on the outskirts of Basel. Seismic monitoring from surface and borehole seismometers provided a detailed record of more than 11,200 events (Bachmann et al. 2011). Flow rates and well pressures were increased up to 0.05 m^3/s and 29.6 MPa, respectively, with an accompanying increase in frequency and magnitude of events (Figure 9.25) clustered around the injection well, and up to one km away. Seismicity rates of several hundred events per day were recorded, with the catalogue complete down to magnitudes $M_w > 0.9$. A magnitude $M_w \sim 2.2$ event occurred in the morning of 8 December, as a result of which the injection pressure was reduced and eventually pumping was stopped. However, an $M_w \sim 3.1$ event occurred 5 hours later, which was widely felt. The pressure in the well was reduced to hydrostatic over 4 days, and seismicity decreased. The experiment was halted due to public concern, but seismicity continues at low levels.

An analysis of 195 of the larger events shows that they occurred on a near-vertical zone striking NNW, at depths between 4 and 5 km (Deichmann and Giardini 2009). Focal mechanisms of these events show NS and EW subvertical planes, consistent with sinistral strike-slip motions on NS planes, and comparable to focal mechanisms of tectonic faults. However, the largest shock (M_w 3.1) and 8 preceding events probably occurred on a steeply dipping WNW fault with dextral motion. The seismicity migrated away from the injection well with time.

9.4.3.3 Interpretation

The larger earthquakes at Basel are considered to be due to slip on pre-existing faults caused by a reduction in normal stress due to the pore fluid pressure following injection (Deichmann and Giardini 2009). This conclusion is also supported by the observation that most of the seismicity occurred before and after maximum injection pressure, which did not reach the value of the minimum horizontal stress. Migration of the seismicity is consistent with the idea that it was related to the diffusion of fluid away from the borehole. The main shock and foreshocks represent the reactivation of a dextral WNW fault. NS faults and possibly EW faults were also reactivated: these orientations have high slip tendencies given a horizontal NW maximum principal stress.

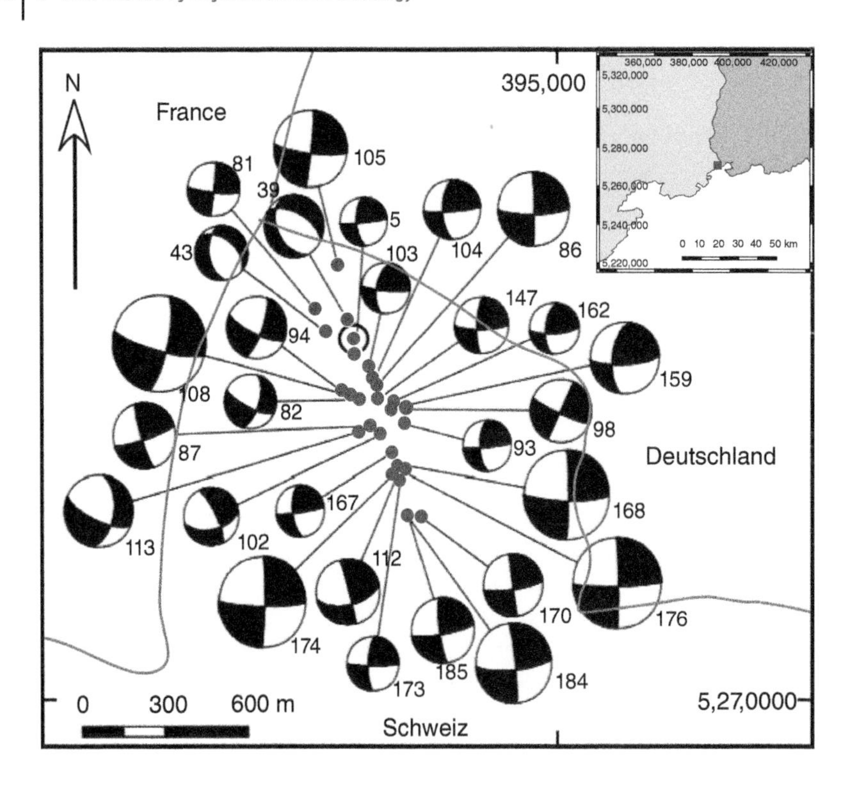

Figure 9.25 Epicentres and focal mechanisms for Basel events (Modified from Deichman and Giardini 2009). Diameters of beachballs are proportional to magnitudes. Most events are compatible with sinistral strike-slip movement on NS panes and dextral on EW planes. Some events have normal components of movement. The mainshock is event 108. Inset shows the location of the figure, under central Basel at the junction of Switzerland, France and Germany, with a WGS84/Pseudomercator coordinate reference system.

9.4.4 Blackpool, United Kingdom

9.4.4.1 Background

The application of hydrofracturing (fracking) to extract shale gas has led to dramatic changes in the world's energy supply. Both the extraction of shale gas and the technique of fracking itself have been highly controversial, on the grounds that hydrocarbon extraction and utilization will contribute to global climate change and that fracking has adverse environmental effects. These concerns have been heightened in the United Kingdom by seismicity associated with the development of the United Kingdom's first shale gas well near Blackpool; seismicity associated with fracking has also been reported from the United States and Canada.

The shale gas target is a sequence of Carboniferous shales/mudstones deposited widely in basins across northern England and potentially representing an extensive shale gas resource. These basins were extended during sedimentation and locally inverted during the Variscan orogeny, creating networks of fractures and faults, which may have been further reactivated in younger deformation events. In March 2011, fracking operations began at Preese Hall near Blackpool in Lancashire (Figure 9.26), but were suspended following seismicity that was widely felt in April and May of that year. Fracking began in the nearby location of Preston New Road, 3 km south of Preese Hall, in October 2018 (Figure 9.26), but was again halted following seismicity in October. Operations resumed in 2019 from another well drilled at the same site offset by approximately 200 m from the original well, but were again suspended after widely felt seismicity in August 2019.

The azimuth of the maximum horizontal stress was constrained by drilling-induced tensile

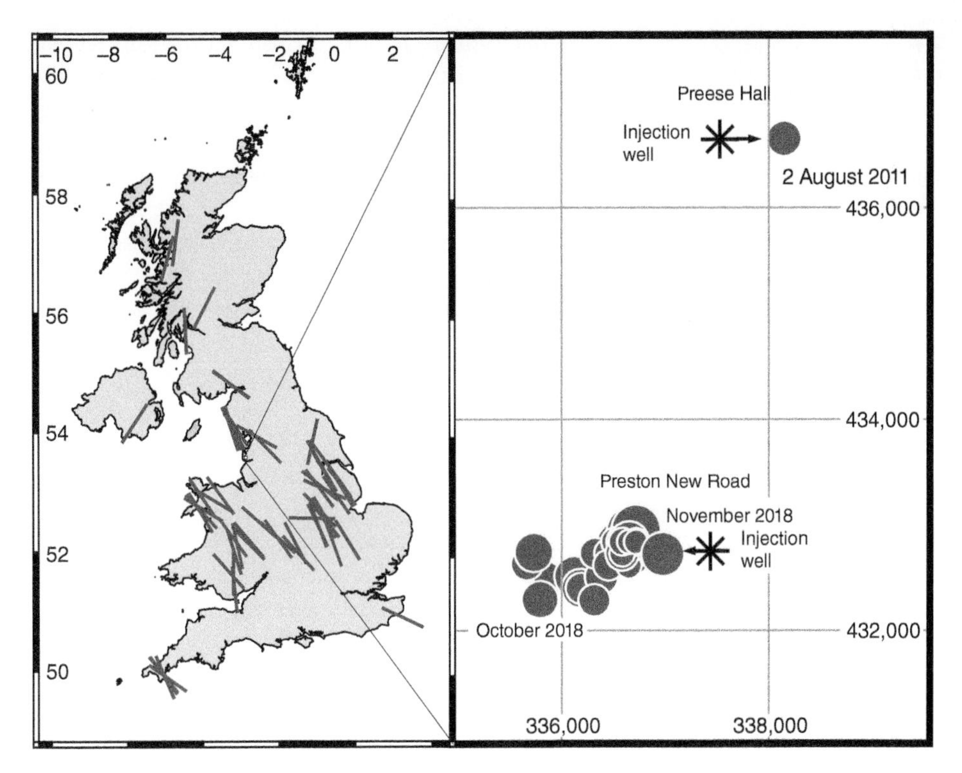

Figure 9.26 Orientation of maximum horizontal stress in UK (red bars). Inset shows epicentres for seismicity associated with fracking in Lancashire. Preese Hall injection well (arrow shows well deviation direction), and well-located event in August 2011 (Clarke et al. 2014), which is representative of the felt events in April and May. Epicentres of October–November seismicity relative to the Preston New Road injection well. Diameters of events proportional to magnitude. Locations from https://earthquakes.bgs.ac.uk/earthquakes/dataSearch.html. All coordinates British National Grid.

fractures at 007°, a similar orientation to that determined by other nearby in situ stress measurements (Figure 9.26). Estimates of the principal stresses, made from boreholes, are: σ_1 = 73 – 64 MPa, σ_2 = 62 – 55 MPa, σ_3 = 44 – 40 MPa, with the intermediate principal stress vertical, i.e. a strike-slip stress field, and the value of the stress ratio $\Phi = (\sigma_{2.} - \sigma_3)_/ (\sigma_2 - \sigma_3)$ is 0.63. Perhaps uniquely for seismicity associated with hydrofracking, a very detailed 3D seismic survey was carried out around the injection well at Preese Hall. This revealed NE–SW striking, steep faults, which are also seen elsewhere at outcrop. These faults are considered to have been inherited from basement structures.

Only one other seismic event has been detected within 5 km of either fracking site since 1970. A simple interpolation of the United Kingdom's background rate of seismicity suggests that one might expect less than one earthquake with magnitude $M_L > 1.3$ every 30 years in an area of 25 km^2. Because earthquake magnitudes are especially sensitive in this context, the M_L scale in which the original values were reported is retained in this section. A calibration of M_w to the M_L scale is however available, showing that $M_L = -0.2$ corresponds to $M_w = 0.8$ and $M_L = 2.3$ corresponds to $M_w = 2.4$. M_L is consistently less than M_w in this calibration. The following account of the 2011 seismicity bear Blackpool is based on Clarke et al. (2014).

9.4.4.2 Seismicity

Two seismic events were felt on 1 April ($M_L = 2.3$) and 27 May ($M_L = 1.5$) 2011, both of them within tens of minutes of injection of large volumes of fracking fluid at the Preese Hall site (Figure 9.27). A number of smaller events with similar waveforms were detected on a regional network and on a local network that was installed following the first felt event. At least 6 of these preceded the $M_L = 2.3$ event.

A subsequent event on 2 August 2011 with $M_L = -0.2$ was located 300–400 m east and 330–360 m below where fluid injection occurred (Figures 9.26 and 9.27). A focal mechanism for this event shows steeply dipping NE striking (sinistral) and NW striking (dextral) planes with strike-slip mechanisms. Given the similarity in waveforms, this event is assumed to be similar to the previous ones. The focus of the August event may lie on a NE–SW fault imaged by the seismic survey.

Another sequence of seismicity near Preston New Road began in early October 2018. Epicenters of events with M_L in the range −0.1 to 1.1, determined by the British Geological Survey, show a progression from West to East during the month of October (Figure 9.26). The seismicity includes the M_L 0.8, 0.8 and 1.1 events on 26, 27 and 29 October, respectively, which exceeded the regulatory threshold of M_L 0.5 for the suspension of fracking operations; these were subsequently halted. By the end of December 2018, 57 events had been detected by a local network. Earthquakes with magnitudes up to $M_L = 2.9$ occurred in August 2019, leading to another suspension of operations at the Preston New Road site. An additional 135 events ($-1.7 \leq M_L \leq 2.9$) were associated with this phase of seismicity.

9.4.4.3 Interpretation

The sequence of events in March–April 2011 near Blackpool can be interpreted in some detail by analyzing pressures in the injection well. It seems that on initial injection, a hydrofracture propagated until it reached a pre-existing fault, after which fluid escaped into the fault. Subsequently

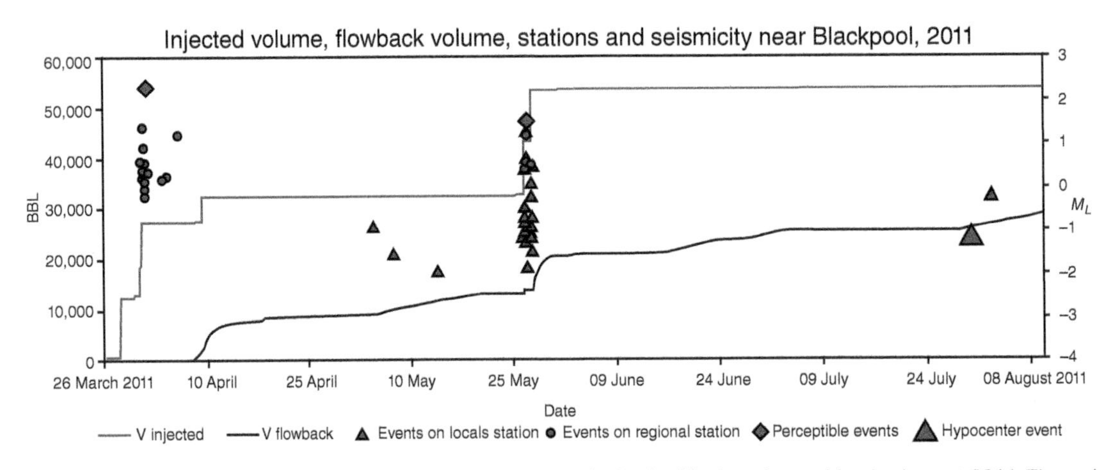

Figure 9.27 Injected volume, flowback volume and seismicity in the Blackpool area, March–August 2011. The red curve represents injected volume (v injected), and the black curve represents flow back volume (v flowback) from the wellhead (BBL represents US barrels). Circles represent seismic events detected on regional seismic stations (more than 80 km away), the triangles represent events detected on two local stations. M_L is a local magnitude relative to the two largest events detected on the regional network, shown as diamonds. The event that provided the source mechanism and reliable hypocenter location is shown as a large red triangle. *Source:* Redrawn from Clarke et al. (2014).

pore fluid pressure rose on the NE-striking fault which is favorably oriented in the strike-slip stress field with σ_1 in a NS horizontal direction, generating precursory activity and finally the $M_L \sim 2.3$ event. For both the October 2018 and the August 2019 seismicity, slip on critically stressed faults was also deemed to be responsible for the largest seismic events. An interesting codicil to this history is that the United Kingdom government announced a moratorium on high-volume hydraulic fracturing in November 2019.

9.4.5 Oklahoma, United States

9.4.5.1 Background

The association between injection of wastewater and seismicity has been widely appreciated since studies carried out in an oil field at Rangely, Colorado, where controlled injection and withdrawal of fluids created corresponding variations in seismicity (Rayleigh et al. 1972, 2013). These early experiments have been thrown into sharp focus by recent developments, particularly in the United States. Unconventional hydrocarbon production generates vast quantities of wastewater, which are disposed of by injection into underground aquifers. Since 2008, unconventional hydrocarbon production has led to large-scale wastewater disposal in the mid-continent of the United States: simultaneously, seismicity rates have risen greatly, although the seismicity is sometimes many km away from the injection wells, leading to controversy about its cause.

This study focuses on perhaps the most widespread case in Oklahoma, but similar scenarios are known in the United States from Arkansas, Colorado, New Mexico, Ohio and Texas, and in Italy. Wastewater disposal in Oklahoma nearly doubled between 2004 and 2008 (Keranen et al. 2014), with the first pressurized injection of water in 2005. Although total seismicity and wastewater injection have decreased since 2015, the largest earthquake was an M_w 5.8 event near Pawnee in 2016, which was one of three earthquakes with $M_w > 5$ in that year.

The salty wastewater in Oklahoma is generated by separation from oil in "dewatering plays," which produce up to 200 times more water than oil. High volume injection wells can handle $164 \times 10^6\,m^3$/month. The wastewater is injected into a dolomitized carbonate, the Arbuckle Group, which has exceptionally high porosity due to karst features and is isolated from near-surface reservoirs. The Arbuckle Group overlies a Precambrian igneous and metamorphic basement, which is connected hydraulically to the Arbuckle Group by fractures and faults. Within the study area, the maximum horizontal stress is oriented quite uniformly 083–084°, and the stress state is strike-slip, with a stress ratio $\Phi \sim 0.62$ (Walsh and Zoback 2016). The background seismicity is estimated to be between 0.5 and 1 $M_w \geq 3$ events per year.

9.4.5.2 Seismicity

Seismicity in Oklahoma increased by a factor of 40 during 2008–2013 compared to 1976–2007. The seismicity tends to occur in swarms of earthquakes that are closely associated spatially over a period of several years. One particularly instructive swarm occurred near the city of Jones, Oklahoma, starting in 2008. In 2011, an M_w 5.7 earthquake associated with this swarm damaged the town of Prague, Oklahoma, 50 km to the east of Jones.

Curiously, the majority of over 700 events in the swarm between 2010 and 2013 occurred between the high injection rate wells to the SE of Jones, and some lower rate wells to the NE (Figure 9.28). These events occurred at depths of between 2 km, corresponding to the Arbuckle Group, and about 4 km, well into the basement. The swarm activity generally spread away from the high injection rate wells over this time at a rate of 100–150 m/day. Aftershock sequences of individual earthquakes form lineaments that correspond to faults with steeply dipping nodal planes in favorable

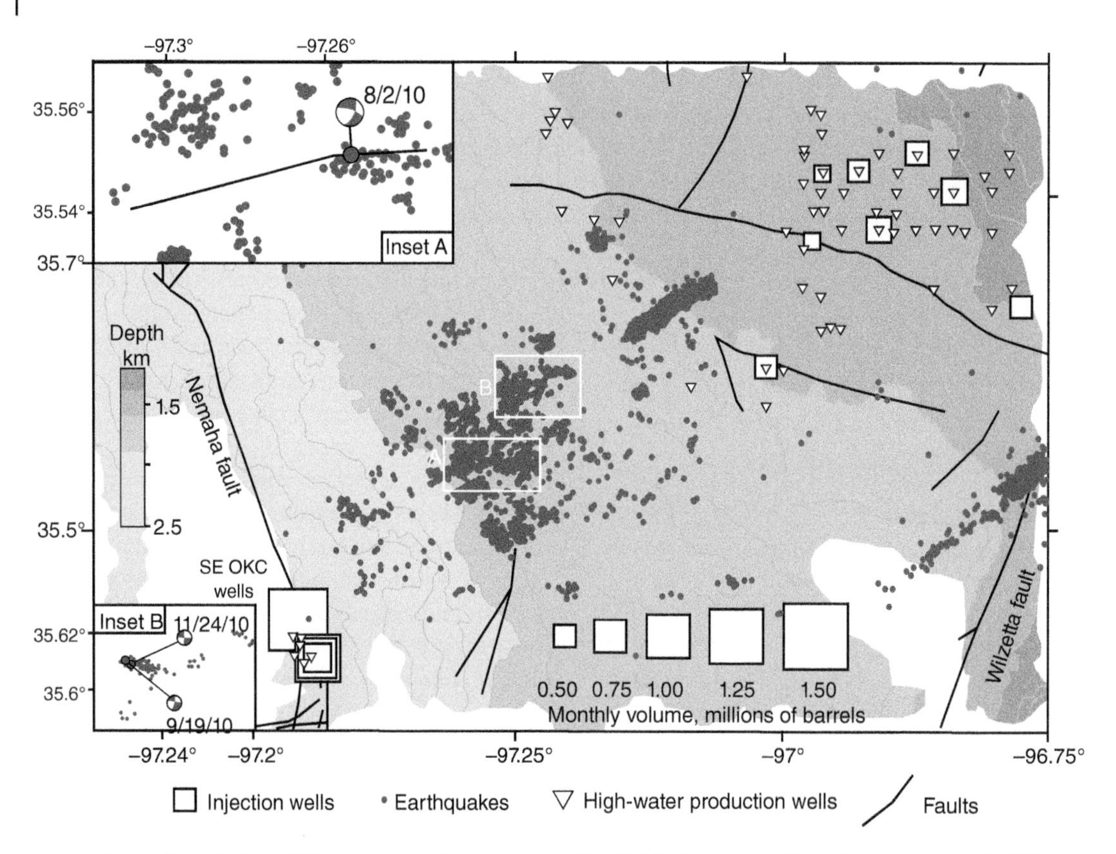

Figure 9.28 Seismicity in Oklahoma from Keranen et al. (2014)/AAAS. Jones earthquake catalog March 2010 to March 2013 using local stations. Squares are injection wells operating at an average rate ≥400,000 barrels/month; triangles are high-water production wells. Background shading and contours represent the depth to the top of the Hunton Group. The Hunton Group is higher in section than the Arbuckle Group but has more data on formation depth.

orientations for slip in the ENE stress field (Figure 9.28). The seismological characteristics of these earthquakes are indistinguishable from natural seismic events.

9.4.5.3 Interpretation

A hydrogeological model of the pore fluid pressures created by injection shows that pore fluid pressure in the Arbuckle and basement rocks would have migrated to the NE from the high injection rate wells over a period of four years (Keranen et al. 2014). A pore fluid pressure increase of 0.07 MPa correlates with where earthquakes occurred. In this case study, it is evident that the seismicity is closely associated with very small pore fluid pressure increases on pre-existing faults in favorable orientations. One of the most surprising results is that the seismicity is not in the area of the highest pore fluid pressure, perhaps because there is a lack of suitably oriented faults, but can occur up to 35 km away from the injection wells (Figure 9.28).

The importance of local geological controls on seismicity is also emphasized by other studies of seismicity in Oklahoma, which show that moment release depends mainly on the injection depth, with injection rates becoming more important as injection depths approach the basement interface (Hincks et al. 2018).

9.4.6 Koyna and Warna, India

9.4.6.1 Background

Seismicity associated with reservoir impoundment has a long observational history, including, for example, earthquakes in the United States from 1936 following the damming of the Colorado River and the filling of Lake Mead, Egypt in 1936 associated with the Aswan High Dam on the Nile River, and Zimbabwe in 1962–1963 following the damming of the Zambezi River and the filling of Lake Kariba (Gupta 2002). At least 168 instances of reservoir-induced seismicity (RIS) have been documented worldwide (Foulger et al. 2018). These examples show that such events may have magnitudes up to $M_W \sim 6$, which are clearly a major concern for dam stability. Nevertheless, there has been considerable debate about both the attribution of the earthquakes to reservoir filling, and the detailed mechanisms that potentially relate the two. The M_W 7.9 Wenchuan earthquake in China – which caused the loss of 200,000 lives – has been controversially related to the filling of the Zipingpu reservoir (Klose 2012).

Relationships between reservoir levels and seismicity are particularly interesting because they have fundamental implications for the permeability and mechanics of the crust, as well as having important risk and hazard connotations.

The 103 m high Koyna dam is located in the Deccan Volcanic Province, close to the Western Ghats Escarpment on the western side of India (Figure 9.29) (Talwani 1997b). Basalts cover the surface in a layer between 400 and 1600 m thick over a granitic basement. The dam was initially impounded in 1961, following which seismicity began immediately. The largest event of M 6.3 occurred in 1967, killing 200, injuring 1500 and damaging the Koyna dam. The 80-m-high Warna dam was completed in 1993, 35 km SSE of Koyna, and filled to a depth of 60 m. The seismicity discussed here occurred initially near the Koyna dam and subsequently closer to the Warna dam. Before dam construction, there is a record of only two felt events in the region.

Both dams, used for hydroelectricity generation, are in the tropics, and the reservoirs behind the dams refill rapidly following monsoonal rain in June and July to reach maximum levels in August. Most of the volume of the reservoirs (1.9 km^3 for Koyna and 0.54 km^3 for Warna) is drained by June. There is up to 42% annual variation in the maximum volume in each reservoir.

Several faults/shear zones in the area have been delineated from Landsat and Lidar imagery (Figure 9.29). NNE-trending deformation zones include the Koyna River Fault Zone (KRFZ) and the Patan Fault (Catchings et al. 2015). Lidar imagery reveals a NS lineament 35 km long south of the Koyna Dam (Arora et al. 2017). The lidar imagery also reveals the trace of the Donachiwada fault, a NS vertical structure, which showed fissures after the 1967 earthquake. Magneto-telluric data suggest a moderately conductive vertical feature extending up to 8 km depth where seismicity is found at that depth. The low resistivity of the feature is attributed to the presence of fluids in a fault zone.

9.4.6.2 Seismicity

The seismicity at Koyna and Warna is notable for its high magnitudes, considerable depths and very protracted history (over 50 years). Seismicity started in 1961. Between 1967 and 2016, there were 22 events with a magnitude greater than 5 (including the M 6.3 event) and 200 events with $M \geq 4$. Seismicity occurred at depths between 4 and 16 km. Three changes in geographic concentrations of seismicity can be identified. Between 1961 and 1967, seismicity was mainly located north of the Koyna dam. Events between 1967 and 1992 were generally in a 10 km wide NS zone

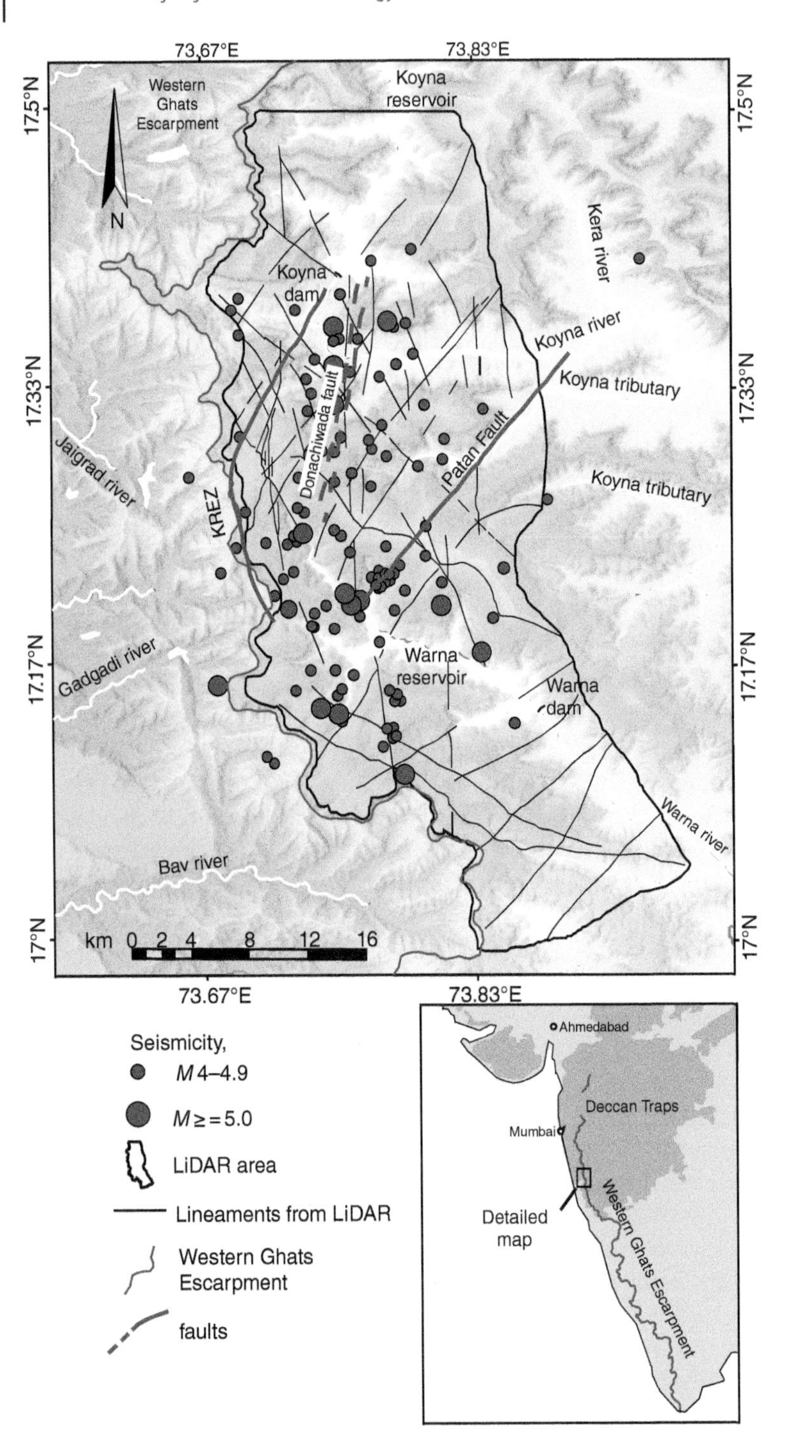

Figure 9.29 Koyna and Warna epicentral map. Epicentral distribution of $M \geq 4.0$ earthquakes for the period 1968–2016 redrawn from Arora et al. (2017). A broad 10 km wide N–S zone is demarcated between Koyna and Warna reservoirs along which most of the earthquakes occur. The zone shows LiDAR-derived digital elevation model (DEM) and lineaments in three directions but N–S fractures control seismic activity.

extending 20 km south of the dam with deeper events near the reservoir (Figure 9.29). Eight $M \geq 5$ earthquakes are close to a NS lineament within this zone. Seismic activity migrated southward from the NS zone to the vicinity of the Warna dam following its construction in 1993. These later events were upstream of the Warna dam, within the NS zone (Figure 9.29).

Reservoir water levels have variable relationships to seismicity between Koyna and Warna. Before the construction of the Warna dam in 1993, a September peak in $M \geq 4$ events followed the August maximum in Koyna dam reservoir levels (Figure 9.30). After 1993, the peak in $M \geq 4$ events occurred in March, during the draw-down phase. However, the patterns are quite variable from year to year.

Relations between seismicity and the rate of water level change, the maximum water level reached, and the duration of high water levels have been proposed. For example, it has been suggested that a change in water level of 12 m/week generates the loading rate necessary for $M \geq 5$ events to occur. An interesting feature of events triggered by variations in the rates of water level changes is that the triggered events occur within days, by comparison to the intervals of several months that relate peak seismic activity to water level maxima. Thus, there are variable time scales for the response of seismicity to water levels.

Seismic activity increases when the highest reservoir level exceeds the level of the previous year. For example, the largest earthquake occurred in December 1963 when the highest water level was reached after impoundment in 1962. From 1967 to 1973, the maximum water levels were kept well below this maximum, and magnitudes decreased. The next $M > 5$ event occurred in October 1973 following an increase in the reservoir level to 1 m higher than 1967 levels.

Focal mechanisms for most events indicate normal faulting, although the M 6.3 event has a strike slip solution with one nodal plane (sinistral) striking at 026°, approximately parallel to the KRFZ. The maximum horizontal stress orientation suggested by the normal faults is north–south, similar to the nearest independent stress determination 200 km to the northwest of Koyna from borehole breakouts.

Coseismic changes in water levels were detected in some observation boreholes, in some cases up to 24 km from the earthquakes. The observed water level changes could be satisfactorily accounted for by calculated co-seismic volume changes. Pre-seismic water level variations, and changes due to distant earthquakes, were also detected.

9.4.6.3 Interpretation

The concentration of seismic activity at Koyna and Warna into a N–S corridor to the south of the Koyna Dam, which is also the locus of several structures visible in Lidar and surface geology, illustrates a strong structural control on the seismicity (Figure 9.29). Focal mechanisms and lineaments suggest both NS and NW–SE structures. Earthquakes with focal depths far below the ~1 km thickness of basalts must represent movement on basement structures, which have presumably reactivated to give surface manifestations in the Lidar data. The duration and high magnitudes of the seismic record at Koyna and Warna may relate to the prominence of these structures. It is interesting that the N–S structures are parallel to the coast of India here, and to the prominent Western Ghats Escarpment that was created following the breakup of Gondwana at the time of eruption of the Deccan traps (65 Ma).

The variable time scales for the relationship between seismicity and lake levels at Koyna can be separated into short times (days) with an *immediate* effect, and a longer term effect, sometimes called *protracted* seismicity. The immediate effect has been attributed due to an increase in pore

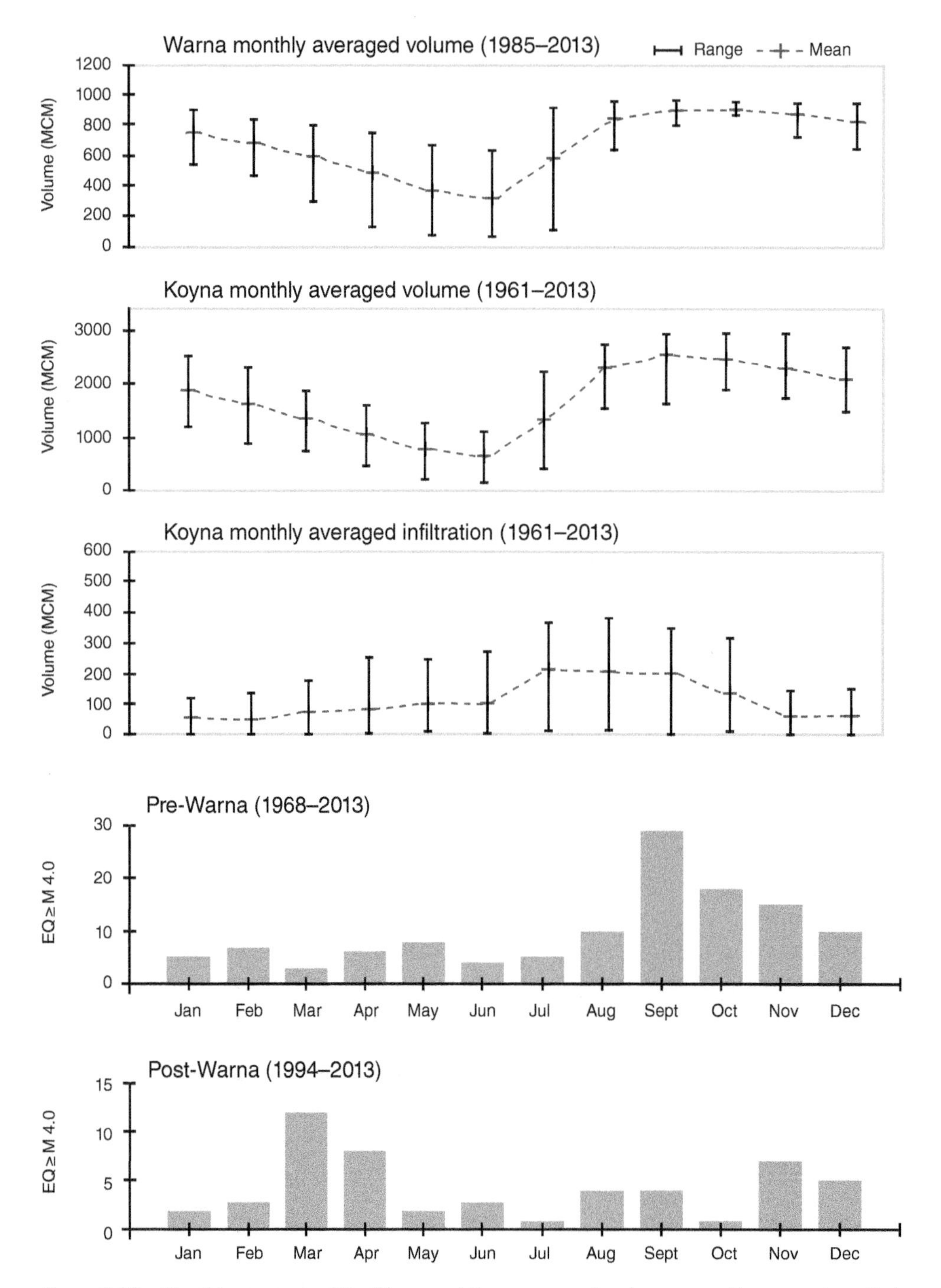

Figure 9.30 Monthly averages of the Warna and Koyna reservoir volumes over the years since their respective impoundment and monthly averages of the infiltration volume of the Koyna reservoir. The monthly distribution of earthquakes (for $M \geq 4.0$) during pre-Warna time shows a prominent peak during September, and during post Warna time, a March–April peak becomes prominent with diminished peaks in September and December. *Source:* Arora et al. (2017) / Springer Nature.

fluid pressure as loading compresses rock around a fixed volume of fluid. As when a wet sponge is squeezed, the fluid pressure builds up until failure occurs, a behavior known as *poroelasticity*. This behaviour is also observed in laboratory experiments in which fluid is confined within a rock sample when it is compressed: such experiments are referred to as "undrained." Earthquakes that occurred immediately on impoundment of the dams, and events that may be triggered by changes in rates of water level change, can be attributed to the immediate effect. The short time scales argue that an elastic volume change has occurred. Poroelasticity is also implicated in the coseismic water level changes, which were responding to volume changes induced by the passage of seismic waves.

The observation that seismicity is subdued until loading to higher lake levels than previously also has a laboratory analogue. It is known as the *Kaiser effect*, which is observed in acoustic emission experiments, when samples are cyclically stressed to progressively higher stresses. Acoustic emissions begin when stresses exceed the level of the previous cycle because fractures do not propagate or form until this condition is reached. This effect seems to apply to the crust at Koyna and Warna (Yadav et al. 2016).

The longer time scales of protracted seismicity may be caused by increased pore fluid pressure from the reservoir water, which takes time to diffuse through the crust to critically stressed faults. The NS orientation of the maximum horizontal stress at Koyna makes the NS structures favorably oriented for normal movement, as observed in many focal mechanisms, and NW–SE structures are well-oriented for strike slip movement, also as observed. Most of the large events at Koyna and Warna, occurring in the months after maximum lake levels, can be attributed to the diffusion of pore fluid pressure. Some of the pre-1993 events occurred as much as 20 km south of the Koyna Dam, indicating the scale of the diffusion.

9.4.7 A Framework for Understanding Induced Seismicity

9.4.7.1 Relating Anthropogenic Activity and Seismicity

Building on the methodology used in climate change, three steps are necessary when relating seismicity to anthropogenic activities:

Detection. Seismicity needs to be demonstrated to be abnormal compared to natural "background" seismicity if it is to be ascribed to anthropogenic activity.

Association. Spatial and temporal correlations between anomalous seismicity and anthropogenic activities need to be investigated on defined scales.

Attribution. Seismicity needs to be tested for a causal link to anthropogenic activity because correlation alone is not causation. Usually, attribution involves developing a model that can explain the observed correlations, for example, that Coulomb stress changes or pore fluid pressures have increased slip or dilation tendencies.

Examining the earlier case studies using these steps, Table 9.1 shows clearly that the seismicity described in each case can be ascribed to anthropogenic activities.

There has been some discussion about the correct terminology to use for seismicity in relation to anthropogenic activity, with some advocating distinctions between induced and triggered seismicity. However, such distinctions rely on advanced analysis of seismic signals, so we suggest a simpler vocabulary:

Associated Seismicity: Seismicity that occurs within a defined distance or time of an engineered event

Induced Seismicity (synonyms: related, triggered): Seismicity that is:

1) Abnormal compared to background (i.e. pre-anthropogenic activity) seismicity
2) Associated spatially and temporally with anthropogenic activity on defined scales
3) Attributable by a mechanistic model

Table 9.1 Detection, association, and attribution of seismic events in the case studies discussed here.

	Detection	Association		Attribution
Case study		**Spatial correlation**	**Temporal correlation**	**Causal relation**
Witwatersrand Basin	Orders of magnitude more frequent seismicity than adjacent areas of the Kaapvaal craton	All Events within mining area: events have hypocentral depths of 2–4 km, the same as mining	Type A events within seconds to minutes of blasting; Type B events within mining period	Stress changes induced by stoping are appropriate
Basel	Seismicity rates 3 orders of magnitude greater than backgrounds	Within 1 km of injection well;	Many events coincide with injection, but a long "aftershock" sequence	Migration of activity away from well corresponds to the diffusion of pore fluid pressure
Blackpool	Seismicity rate several times background	All events within 3 km of injection well	Events correspond within hours to days of injection	Migration of events corresponds to fracking sequence
Oklahoma	Seismicity rates ~1.5 orders of magnitude greater than background	Seismicity up to 35 km from injection well	Strong correlation on yearly time scale	Modeling predicts observed migration of seismic activity by diffusion of pore pressure
Koyna and Warna	Almost no seismicity known before reservoir filling	Seismicity in a zone extending km away from the Dam	Strong annual periodicity in lake level and seismicity	Coseismic water level changes directly detected

Acceptance of these criteria could lead to the resolution of much controversy. The study of deformation and rheology is a key part of the third step.

Since the cases studied earlier are all clearly both anomalous in terms of their seismicity and attributable to the anthropogenic activity, they correspond to mining-induced seismicity (Witwatersrand basin), injection-induced seismicity (IIS) (Basel and Blackpool), and reservoir-induced sesimicity RIS (Koyna and Warna).

9.4.7.2 Mechanics of Induced Seismicity

Different modes of failure and different mechanics operate in the various types of induced seismicity. For example, one of the most significant differences between the mining-induced seismicity in the Witwatersrand basin and the other cases of induced seismicity is that pore fluids are not involved in the former, because fluids are drained from the mines. The Witwatersrand earthquakes involve shear failure caused by differential stress, and implosion and tensile failure due to spalling. The roles of differential stress vs. pore fluid in failure differences can be explored very well on a Failure Mode diagram.

The schematic Failure Mode diagram in Figure 9.31 uses a Griffiths failure criterion for tensile normal stresses and a Mohr–Coulomb criterion for compressive normal stresses to identify the limits of fields of stable stress states. Along the abscissa, where pore fluid factor λ_v is zero, the differential stresses for failure in compression or tension are greatest. As pore fluid increases, the differential stresses required for failure decrease, meaning that there is a smaller range of stable differential stresses. This diagram can show several of the types of failure referred to in the earlier examples:

1) Mining-induced seismicity (MIS). Mining must involve draining pore fluids so that the first part of the path will be toward zero pore fluid factor (Figure 9.31). Earthquakes with shear components are likely to be due to reactivation. Failure with volumetric components is not distinguished on this diagram.
2) Injection-induced seismicity (IIS) involves increasing the pore fluid factor, but differential stress changes very little, only due to localized increases in vertical load due to the injected fluid. When the tensile failure criterion: $P_f \geq \sigma_N + T_0$ is reached, failure will occur by either tensile or shear failure, depending on the differential stress (Chapter 7). P_f will necessarily be suprahydrostatic for reverse stress conditions and intact failure. Most injection-induced earthquakes for which source mechanisms have been determined do not have distinguishable sources from seismic events, implying that they are largely due to shear reactivation of discontinuities.
3) Withdrawal-induced seismicity (WIS) may decrease fluid pressure and reduce vertical loading due to fluid withdrawal. There is an interesting difference in the effect of reducing the vertical stress σ_v between normal and reverse stress regimes. In reverse regimes where $\sigma_v = \sigma_3$, reducing σ_v will increase $\Delta\sigma = \sigma_1 - \sigma_3$, if σ_1 remains constant (Figures 9.31 and 9.32). However, the dominant effect of fluid withdrawal is to cause subsidence and horizontal contraction over the center of the evacuated reservoir. This leads to increases in horizontal and differential stresses, driving the stress regime toward a reverse state (Figure 9.32), potentially even from a normal state of stress. Above the peripheries of the reservoir, the subsidence creates a normal state of stress.
4) Reservoir-induced seismicity (RIS). The poroelastic effect increases fluid pressures on loading, and is implicated by the “immediate” seismicity on reservoir impoundment. However, in this case there is also a small change in differential stress due to the additional load of water in the reservoir, so the path on the failure mode diagram is not vertical. The difference between the

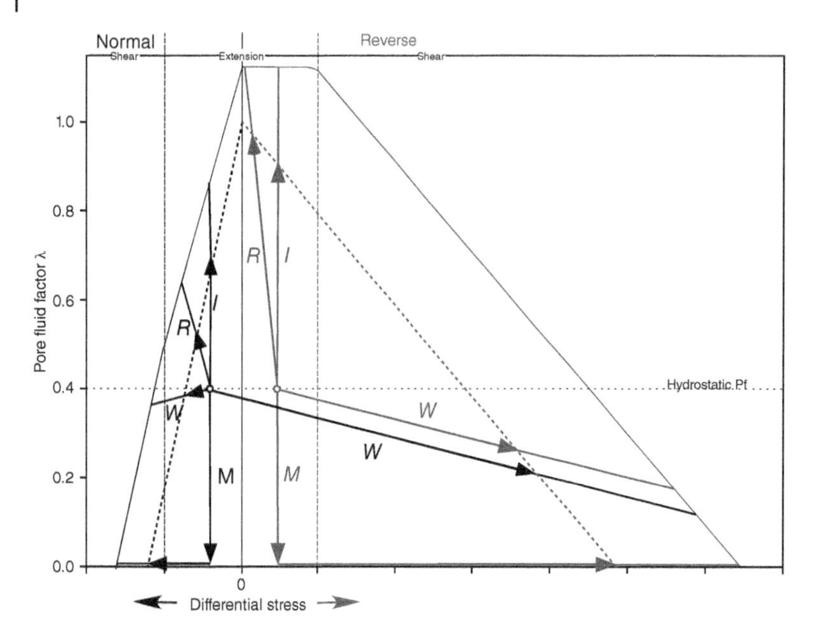

Figure 9.31 The mechanics of induced seismicity, illustrated on a schematic failure mode diagram. The diagram shows pore fluid factor λ plotted against differential stress $\Delta\sigma$. Failure envelopes are constructed for a Griffith failure criterion where $\sigma_n < 0$, and a Coulomb criterion for $\sigma_n > 0$. Solid lines give conditions for failure of intact rock; dashed lines for reactivation (very commonly inferred for induced seismicity). The left side of the diagram shows normal faulting stress regimes; the right-side reverse faulting. In both cases, changes in λ and $\Delta\sigma$ are shown for three types of induced seismicity to an initial condition in the crust of λ and $\Delta\sigma$ which is below failure.

1) M is mining-induced seismicity. During mining, pore fluids are drained, but large changes in $\Delta\sigma$ are caused by mining.
2) I is injection-induced seismicity. λ increases without change in $\Delta\sigma$.
3) R is reservoir-induced seismicity. Pore fluid factor λ increases due to the poroelastic effect, and $\Delta\sigma$ also changes because the vertical load increases. In the normal faulting regime, this causes an increase in $\Delta\sigma$, but in the reverse faulting regime, $\Delta\sigma$ decreases. The change in $\Delta\sigma$ is very small: it is likely to be exaggerated in this diagram.

The three types of induced seismicity are clearly distinguished by their different effects on λ and $\Delta\sigma$.

reverse and normal stress regimes is just the opposite of the case for WIS earlier. In a reverse faulting stress field, the additional load increases the least principal stress σ_3 so that it draws closer to the maximum principal stress σ_1 and $\Delta\sigma$ decreases (Figure 9.31 and 9.32). Whereas in a normal faulting stress field, the additional load increases σ_1 and $\Delta\sigma$ increases. A fuller analysis of the changes in stress can account for the fact that loading an elastic solid changes both principal stresses, but the qualitative result is the same. Less pore fluid pressure increase is therefore required for failure in a normal fault stress field than in a reverse fault stress field (Figure 9.31). Protracted RIS is explained by diffusion of pore fluid pressure: the observed velocities are within the range of those possible for the crust.

In all these types of induced seismicity, a common theme is the reactivation of discontinuities rather than the failure of intact rock. This distinction has

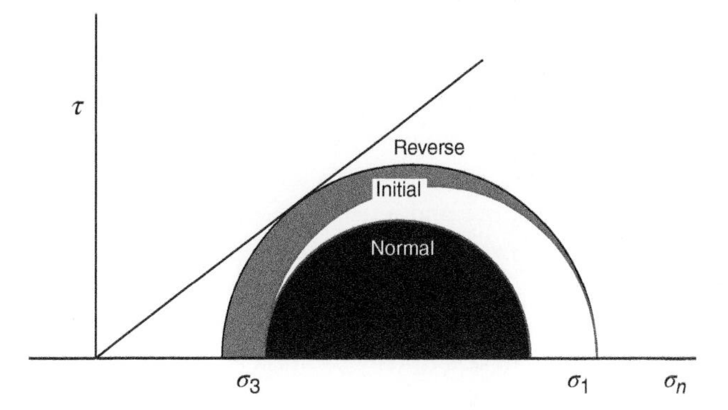

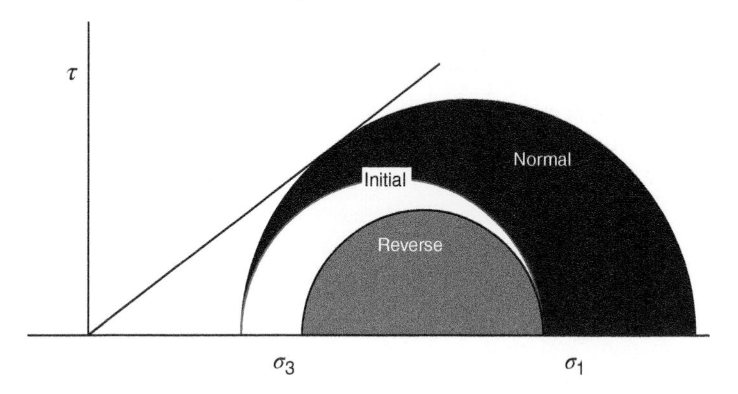

Figure 9.32 Mohr diagrams showing the effect of reducing (a) and increasing (b) vertical load in normal and reverse faulting regimes. In (a), which corresponds to fluid withdrawal, reducing the vertical load brings reverse faulting regimes closer to failure, and stabilizes normal faulting. Fluid injection is modeled by (b), which has just the opposite effects. Note that pore fluid pressure changes are not included in this diagram.

important implications for the range of potential magnitudes of events, and for the largest possible magnitude of an induced event. The possibility of reactivation will be determined by the orientation of the discontinuities, the magnitudes of the stresses (slip and dilation tendency are useful measures of these variables) and the frictional properties of the rocks. However, the range of magnitudes is likely to depend largely on the size range of reactivated discontinuities, which could, for example, follow typical fault power-law scaling (cf. Section 8B.1.2.4). The presence of abundant large discontinuities at Koyna and Warna may be a factor in the unusually large magnitudes of RIS there.

Considering collectively all examples of induced seismicity, there are variable time relations between anthropogenic activity and seismicity. Variation in the time lag between anthropogenic activity and seismicity depends on the mechanism inducing the seismicity. Poroelastic mechanisms, expected to act immediately or over very short time scales, are inferred to have operated in Koyna and Warna. The Blackpool example indicates that other mechanisms may give rise to responses over short time scales. In that setting, fracture propagation speed is inferred to have an effect: the felt seismicity is interpreted to have begun after a fracture propagated at rates of cms^{-1}. In the Oklahoma case of IIS, there is no direct correlation between seismicity and the rate of fluid injected. Increases in pore fluid pressure leading to seismicity can have highly variable time scales depending on the permeability of rocks in the vicinity of human activity. Variations in the rate of diffusion may allow for seismicity to occur after tens of years. Permeability is affected both by faults and fractures and intrinsic formation permeability. For example, the presence of carbonate reefs affects the localization of earthquakes in limestones.

Induced seismicity can be surprisingly deep. At Koyna and Warna, some events occurred at depths of 30 km. Lower crustal seismicity was also associated with the filling of the Aswan dam in Egypt. These observations demonstrate that hydraulic connectivity may exist throughout the crust, and that basement rocks can be reactivated in induced seismicity.

Table 9.2 compares important parameters for the various case studies. The stress state for all these case studies is normal or strike slip or a combination. Figure 9.31 shows that normal faulting requires less differential stress for failure at any pore fluid pressure factor.

Figure 9.33 attempts to show schematically some of the mechanisms identified with each cause of induced seismicity. The case studies show that a particular anthropogenic cause of induced seismicity can be due to different mechanical effects. For example, RIS can be due to both the poroelastic effect and the diffusion of pore fluids. Some mechanisms are common to different types of seismicity: pore fluid diffusion is applicable to both injection and RIS. There is also abundant evidence that induced seismicity is inherently related to the ambient stress state, as well as the presence of favorably orientated pre-existing discontinuities. These relationships are the key to predicting and ameliorating induced seismicity.

9.4.7.3 The Critically Stressed Crust

One of the most puzzling features about much induced seismicity is that pore pressure and load changes induced by anthropogenic activities are very small in relation to geological stresses as calculated from lithospheric strength profiles. For example, typical annual lake level changes at Koyna are 35 m, which corresponds to an increase in vertical load and pore fluid pressure change of just 0.35 MPa. Yet this is sufficient to cause both immediate and protracted seismicity. This is one piece of evidence that the crust in areas of induced seismicity is in a *critically stressed state*, or very close to failure, so that small changes in load or pore fluid pressure can cause failure (Townend and Zoback 2000).

Several other lines of argument suggest that the crust is critically stressed. Compilations of data

Table 9.2 Summary of stress states and seismic characteristics of the case studies discussed here.

Case study	Stress regime	ΔPf MPa	σ_1 MPa	σ_2 MPa	σ_3 MPa	$\Delta\sigma$ MPa	Φ	Typical depth km	Depth range km
Witwatersrand	Normal	0	80	40	40	40	0	3	2–4
Basel	Strike-slip/ normal	29.6						4.5	3.5–5
Blackpool	Strike slip	20	64	55	40	24	0.63	2.9	
Oklahoma	Strike slip	0.07	155	125	77	65	0.61	5–6	5–6
Koyna and Warna	Strike slip/ normal								4–16

(a) Mining-induced seismicity

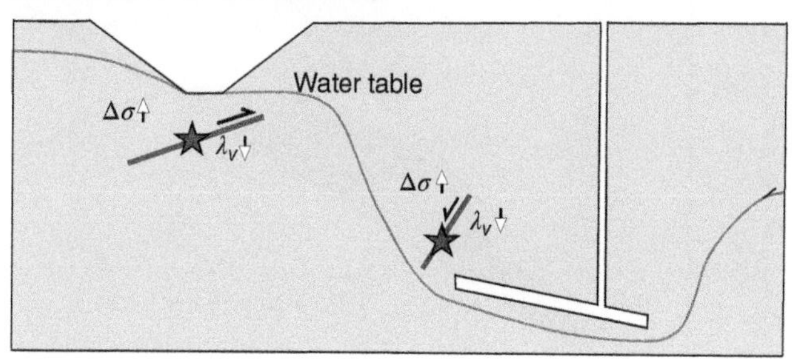

(b) Injection-induced seismicity

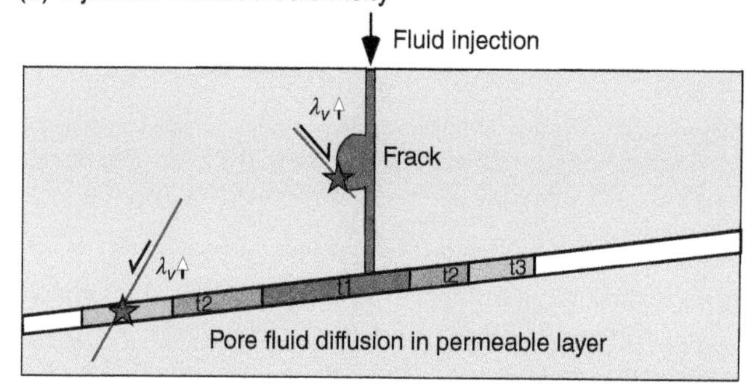

(c) Withdrawal-induced seismicity

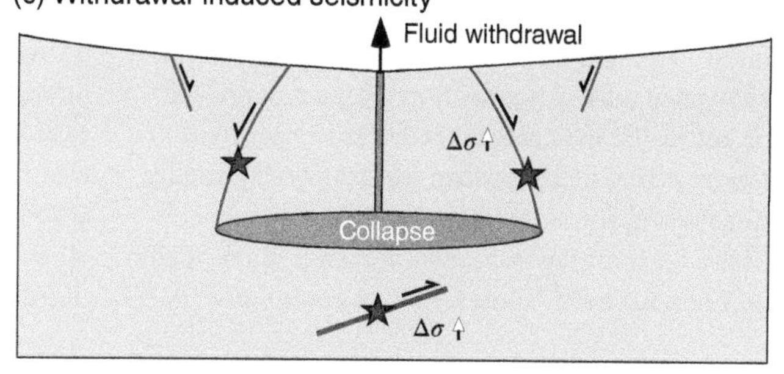

(d) Reservoir-induced seismicity

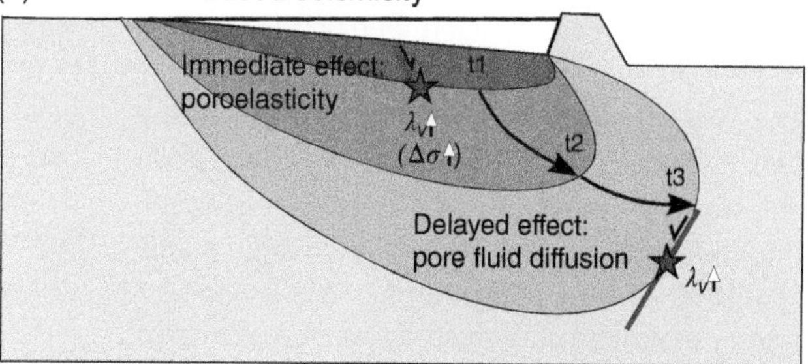

Figure 9.33 Summary of the mechanism of induced seismicity for the four categories of a) mining, b) injection, c) withdrawal and d) reservoir-induced seismicity. A particular category can have different mechanisms.

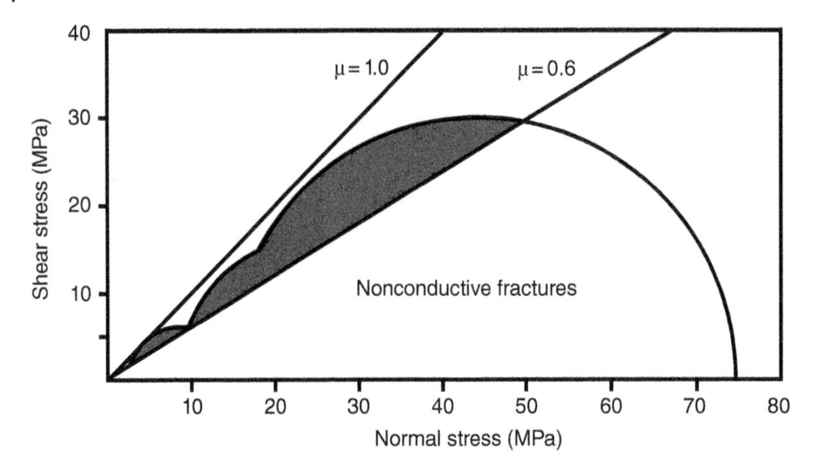

Figure 9.34 Shear and normal stresses on fractures identified with borehole imaging techniques in Cajon Pass, Long Valley, and Nevada Test Site boreholes. The filled area represents hydraulically conductive fractures and faults, and the open area represents nonconductive fractures. *Source:* Based on Townend and Zoback (2000).

from drill holes show that fractures which produce water are almost all in orientations that have shear and normal stresses above the failure criterion (Figure 9.34). Many examples are now known of earthquakes triggered by static stress changes induced by slip on faults. These *Coulomb stress changes* in many cases are only on the order of MPa or less, again indicating that the crust was close to failure before the seismicity was triggered. Finally, direct stress measurements by the hydrofrac technique for many intracontinental areas show high differential stress – close to failure conditions.

9.5 Using Case Studies to Assess Lithospheric Strength Profiles

The case studies presented in Sections 9.2 and 9.3 illustrate the character of deformation in the mid- to lower and mid- to upper levels, respectively, of the continental crust. Since the geometry and evolution of the deformation structures and fabrics described in each study are characteristic of those found in comparable tectonic settings elsewhere, Sections 9.2 and 9.3 provide clues to the rheological behavior of rock at those depths within the continental crust. Section 9.4 takes a different tack for analyzing the rheology of continental crust, using seismicity at a range of depths in a variety of tectonic settings to constrain inferences on rock rheology at different levels of the continental crust. In this section, we endeavor to provide a more general context for the results of the case studies and use them as points of departure in addressing the broad issue, raised in Section 9.4.7.3, of the state of stress at different levels in the lithosphere.

9.5.1 Lithospheric Strength Profiles

9.5.1.1 Defining Lithospheric Strength Profiles

One approach to analyzing tectonic stresses is to determine limiting values for differential stresses, that is to characterize rock strength, at different depths. In Section 7B.2.7, we presented a plot of inferred rock strength versus depth for hypothetical continental lithosphere whose mechanical behavior is everywhere equivalent to that of diopsidite, a rock composed only of the mineral diopside. Figure 9.35a reproduces that plot of the

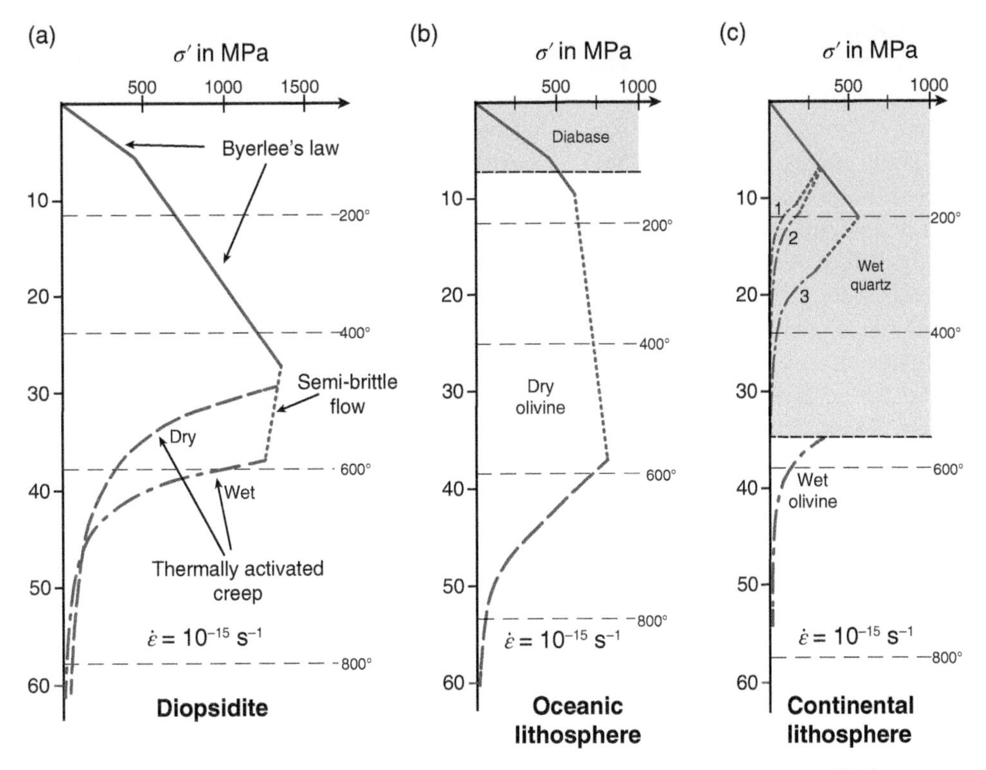

Figure 9.35 Idealized lithospheric strength profiles, calculated for strain rates of 10^{-15} s^{-1}. (a) Strength profile for lithosphere composed of diopsidite; strength at depth depends upon whether the rock is wet or dry. (b) Strength profile for oceanic lithosphere with diabase crust and dry olivine mantle. (c) Strength profile for the continental crust. The strength of 35 km thick crust is controlled by Byerlee's law and one of three flow laws for quartz: 1 = Kronenberg and Tullis (1984); 2 = Luan and Paterson (1992); 3 = Gleason and Tullis (1994). Mantle strength from Chopra and Paterson (1984).

differential stress magnitude required to drive deformation at a rate of 10^{-15} s^{-1} at different depths. This plot, taken from Kohlstedt et al. (1995), was constructed using experimentally constrained flow laws for the deformation mechanisms inferred to pertain at different depths within the lithosphere. Its distinctive "hatchet" or "meat cleaver" shape reflects the inferences that: (1) deformation in the upper portion of the profile (depicted by the solid line in Figure 9.35a) conforms to Bylerlee's law (cf. Section 8A.2.4), that occurs mainly by slip on faults at differential stress values proportional to the magnitude of hydrostatic stress which, in turn, increases linearly depth; (2) deformation in the lowest portion of the profile (depicted by the dashed and dashed-dotted lines in Figure 9.35a) occurs by the thermally activated mechanisms dislocation creep (cf. Section 8A.4.7) and to a lesser degree diffusional creep (cf. Section 8A.3.5), which proceed at higher rates as temperatures increase with depth; and (3) deformation at intervening depths (depicted by the dotted line in Figure 9.35a) occurs by a semi-brittle flow that adheres to the relation known as Goetze's criterion:

$$\sigma_1 - \sigma_3 = \sigma_3' \tag{9.6}$$

where σ_1 and σ_3 are, respectively, the most compressive and least compressive principal stress magnitudes and the effective confining pressure $\sigma_3' = \sigma_3 - P_f$ (the least compressive principal stress minus the pore fluid pressure).

The precise shape of this plot, or of similar plots estimating the strength of rock along vertical profiles through the lithosphere, depends significantly on which minerals are inferred to compose the rock at different depths, the geothermal gradient assumed for the region, and the presence or absence of fluid phases in the rocks. For example, Kohlstedt et al. (1995) also estimated rock strength at different depths for an idealized dry oceanic lithosphere (Figure 9.35b). In constructing that profile, Kohlstedt et al. inferred that deformation: (1) adheres to Byerlee's law in the upper ~10 km of the profile; (2) follows a temperature-dependent, power-law flow law for olivine at depths greater than ~35 km; and (3) adheres to Goetze's criterion for dry olivine in the intervening portion of the upper mantle. Figure 9.35c plots Kohlstedt et al.'s estimated rock strength at different depths for wet continental lithosphere. This profile reflects the inferences that rock: (1) adheres to Byerlee's law for wet, quartz-rich rock in the upper ~12 km of the crust; (2) follows one of three temperature-dependent, power-law relationships for wet, quartz-rich rock in the lower crust; (3) has sufficient strength in the uppermost mantle to accrue stresses required to generate earthquakes (cf. Chen and Molnar 1983); and (4) follows a temperature-dependent, power-law for wet olivine-rich rock in the subjacent lithospheric mantle.

In characterizing rock strength along profiles through the lithosphere, by general consensus "strong" segments of the lithosphere require differential stresses on the order of 100 MPa to deform, whereas "weak" segments of the lithosphere deform at differential stresses ≤10 MPa (Thatcher and Pollitz 2008). Using those characterizations of relative strength, the lithospheric strength profile in Figure 9.35c has a strong mid- to upper crust, a weak lower crust, and a strong upper mantle. Strength profiles of this sort are known colloquially as "jelly sandwich" profiles – where the weak "jelly" of the lower crust is sandwiched between the strong upper crust and upper mantle (Figure 9.36a; note that the estimated flow stress magnitudes in Figure 9.36 are those needed to drive deformation at strain rates of 10^{-14} not the 10^{-15} s^{-1} strain rates of Figure 9.35).

9.5.1.2 Refining Lithospheric Strength Profiles

Jackson (2002) was among the first to take issue with jelly sandwich-type continental lithospheric strength profiles. Analyzing seismic and gravity data from the southern margin of the Himalayas, Jackson argued that: (1) earthquakes beneath Tibet previously interpreted to have occurred in the upper mantle are more reasonably reinterpreted to have occurred within the lower crust of the subducting Indian subcontinent; and (2) the thickness of the portion of the lithosphere whose elastic behavior determines the configuration of the subducted Indian plate (the elastic layer thickness = T_e), calculated from free-air gravity profiles across the southern margin of the Himalayas, is ~30 km. This calculated value of the elastic layer thickness is approximately equal to but systematically slightly less than the observed thickness of the portion of the lithosphere where earthquakes occur (the seismogenic layer = T_s). Jackson concluded that relocating earthquake foci in the Indian subcontinental crust refuted an important line of evidence arguing for a strong upper mantle in the continental lithosphere. Noting that calculated T_e values ≈ 30 km ≈ observed T_s values, Jackson concluded that the continental lithosphere is characterized by a single strong layer, similar to albeit thinner than the oceanic lithosphere depicted in Figure 9.35b.

Burov and Watts (2006) utilized a different approach to assess which sort of lithospheric strength profile is most appropriately applied to continents. They considered two broad categories of models for continental lithospheric strength:

(1) jelly sandwich models, where moderately strong upper crust and strong upper mantle are separated by weak lower crust, akin to the continental lithospheric strength profile envisioned by Kohlstedt et al. (1995) and illustrated by the lithospheric strength profile in Figure 9.36a; and (2) "crème brûlée" models, in which the principal strong layer in the lithosphere is the mid to upper crust, akin to the model espoused by Jackson (2002) and illustrated by the lithospheric strength profile in Figure 9.36b. Burov and Watts also interpreted gravity data from the southern margin of the Himalaya, but their analysis of Bouguer gravity anomalies suggested an elastic layer thickness T_e of ~70 km for the subducted Indian plate. Burov and Watts then used numerical modeling to evaluate the physical characteristics of the lithosphere in two dynamic situations: (1) a stability test that examined the longevity of a mountain range 3 km high and 200 km across, bounded on both sides by 36 km thick crust and subjected to continuous, slow (5 mm/a) horizontal shortening; and (2) a collision test where a continent-continent convergence at 60 mm/a is driven by the subduction of an oceanic plate with 7 km thick serpentinized oceanic crust. After 10 Ma of model time, the jelly sandwich models had geometries generally comparable to profiles of actual continental mountain ranges or modern subduction zones, whereas the crème brûlée models developed unusual lithospheric structures that bear little or no resemblance to actual mountain ranges or subduction zones. Burov and Watts concluded, therefore, that the jelly sandwich models, where moderately strong upper crust and upper mantle are separated by weak lower crust, best define the long term (~10 Ma) character of the continental lithosphere.

Data from rock deformation experiments underscore that the parameters discussed earlier – which minerals compose the rock, what are the ambient temperatures, and whether pore fluids

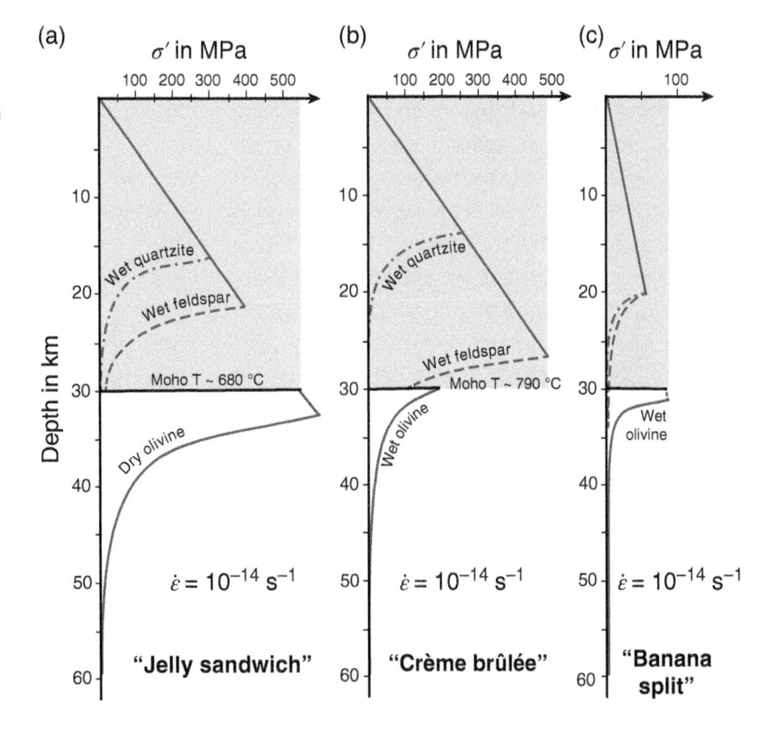

Figure 9.36 The "menu" of lithospheric strength profiles, based on Bürgmann and Dresen (2008). (a) "Jelly sandwich" strength profile, with weak lower crust sandwiched between the strong upper crust and upper mantle. (b) "Crème brûlée" strength profile, with the moderately strong upper crust and weak lower crust and upper mantle. (c) "Banana split" strength profile, with weak crust and mantle; note difference in scale for differential stress values.

are present or absent – affect significantly rock strength. Deformation experiments and theoretically derived flow laws also identify other factors that affect significantly rock strength: (1) the grain size of the minerals that compose rocks; (2) the composition of any fluid phases present; and (3) the presence or absence of melt (Bürgmann and Dresen 2008). The effects of varying the mineralogic composition of rock are readily apparent in Figure 9.35c. Under very similar physical conditions the quartz-rich lower crust has relatively low strength compared to the immediately subjacent olivine-rich upper mantle. The decreases in rock strength with increasing depth across the lower crust and within the upper mantle in this lithospheric profile reflect the effects of increased ambient temperature. Pore fluids directly affect rock strength in three ways: (1) the effective stress required to drive fault slip decreases as the magnitude of the pore fluid pressure increases (see Section 7A.2.2.4); (2) diffusion occurs much more rapidly along "wet" grain boundaries than along "dry" grain boundaries (see Section 8A.3.1), so deformation by diffusional mass transfer occurs at significantly higher rates when pore fluids are present; and (3) dislocation creep of most silicate minerals proceeds at lower flow stresses when pore fluids are present due to hydrolytic weakening (see Section 8B.3.3). The presence of fluids also affects rock strength indirectly by facilitating retrograde metamorphic or metasomatic reactions that replace stronger mineral constituents with weaker mineral constituents (e.g. when feldspar plus water react to form quartz and mica). Deformation by diffusional mass transfer is strongly dependent up the mean grain diameter of grains (see Section 8A.3.5). If those retrograde metamorphic or metasomatic reactions, or any other processes, lead to a reduction of grain size, there is likely to be an increase in the rate of deformation (see Section 9.3.3.4.3).

Where more than one factor is at play, positive feedbacks can enhance changes in rock strength. As an example, fluid infiltration might lead to retrograde metamorphic/metasomatic reactions that replace strong minerals with weaker minerals while also reducing the mean grain size. At the same time, the fluid phases would facilitate the diffusion of chemical species along grain boundaries. The combined effects of these changes would be a pronounced acceleration of the rate of deformation by diffusional mass transfer, potentially weakening the rock dramatically. Bürgmann and Dresen (2008) noted that these sorts of combinations of effects are especially likely in plate boundary zones. They argued, therefore, that a third selection should be added to the menu of possible lithospheric strength profile types – a "banana split" model where both the crust and the upper mantle are weak (Figure 9.36c). The banana split lithospheric strength profile is characterized by: (1) a wet upper and middle crust that strengthens with increasing depth but is nevertheless not capable of supporting differential stresses greater than ~75 MPa; (2) a wet and warm, and therefore very weak, lower crust; and (3) a wet upper mantle only capable of supporting differential stresses of ~100 MPa. Bürgmann and Dresen argue that this sort of lithospheric strength profile pertains to the Pacific-North American plate boundary zone and is likely to develop in other plate boundary settings.

9.5.1.3 The Role of Time in Lithospheric Strength

Thatcher and Pollitz (2008) underscored the significance of the time scale of loading in characterizing the lithospheric strength profile at any location. For example, when considering rock response to the short-duration, transient distortion associated with seismic waves traveling through them, the upper crust, lower crust, and upper mantle appear equally strong. As the length of the observation time increases, differences emerge in the responses of the upper crust, the lower crust, and the upper mantle. Considering the earliest deformation that occurs over time spans characteristic of post-seismic (1–10 a) to glacio-isostatic responses (10^3–10^4 a), both the upper and lower crust are relatively strong, and it is only in the mantle that flow accrues sufficient deformation to accommodate the loading. Over the time spans of longer duration associated with lithosphere isostatic responses (10^6–10^7 a), the response to applied loads is viscous in the lower crust and mantle, and those layers are more appropriately viewed as weak.

9.5.1.4 The Effect of Rock Composition on Lithospheric Strength

Figures 9.35 and 9.36 clearly illustrate the impact of varying the chemical/mineralogic composition of rock on lithospheric strength profiles. Figure 9.37, redrawn from figure 4 in Thatcher and Pollitz, presents an alternative way to depict the effects of differences in the composition of the lower crust and upper mantle. The figure plots effective viscosities of different rock compositions in the lower crust, calculated at the temperatures indicated along the right side of the diagram. In addition, the figure gives calculated viscosities for wet and dry olivine for the upper mantle.

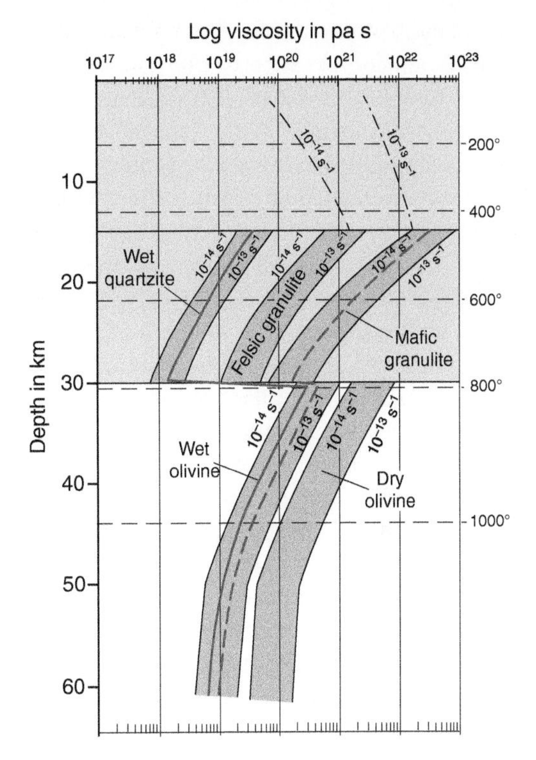

Figure 9.37 Plot of estimated viscosity versus depth for three compositions of continental crust and wet and dry mantle. *Source:* Modified from Thatcher and Pollitz 2008.

In any viscous flow, the differential stress σ' is related to the strain rate $\dot{e}$ by

$$\sigma' = \eta \dot{e}, \tag{9.7}$$

where η is the viscosity of the deforming material. In a lithospheric column where the lower crust and upper mantle deform at a uniform strain rate, the differential stress σ' at any depth is directly proportional to the viscosity η of the rock at that depth. If the rheology of wet quartzite determines the strength of the lower crust and the rheology of dry olivine determines the strength of the upper mantle, the solid red line in Figure 9.37 depicts schematically the magnitude of the viscosity at different depths. Since differential stresses are directly proportional to the viscosity, differential stress magnitudes in such a lithospheric column are likely to increase with depth to the top of the lower crust, then follow a trajectory estimated by the solid red line, increase abruptly at the crust-mantle boundary, and subsequently decrease with depth. Such a variation in differential stresses with depth corresponds to the jelly sandwich strength profile. Alternatively, in a lithospheric column where a lower crust composed of mafic granulites overlies a wet mantle, the dashed red line in Figure 9.37 gives the viscosity at different depths. Again taking the differential stresses to be proportional to viscosity, differential stresses would vary with depth in a manner depicted by the dashed red line. Such a variation of differential stresses with depth corresponds to a crème brûlée strength profile.

An analysis of post-earthquake "afterslip" data and measurements of viscoelastic relaxation yields viscosity values for the lower crust and mantle that are largely consistent with the plot of viscosity versus depth given in Figure 9.37 (Pollitz 2019). That study also calculated effective viscosities for the deformation of the upper crust at strain rates of 10^{-14}–10^{-13} s^{-1}; they are shown as dashed and dash-dotted black lines, respectively, in Figure 9.37. This analysis supports the inference that lithospheric strength profiles for quartz-dominated lower crust generally conform to jelly sandwich strength profiles and granulite lower crust generally conforms to crème brûlée strength profiles.

The several lithospheric strength profiles considered to this point have relied upon relatively simplistic models for the composition of the continental crust. An evaluation of the impact of more realistic variations in the chemical and mineralogic composition of the lower crust is beyond the scope of this discussion. It is worthy to note, however, that one such analysis by Shinevar et al. (2018) yielded results that very generally confirm the conclusions of analyses of relatively simplistic crustal compositions. In particular, the results indicate: (1) higher viscosities in the shallower portions and lower viscosities in the deeper portions of the lower crust; (2) generally lower viscosities at most depths in the vicinity of a major plate boundary (the San Andreas fault); and (3) a transition toward crustal behavior consistent with the crème brûlée strength profile farther from the plate boundary.

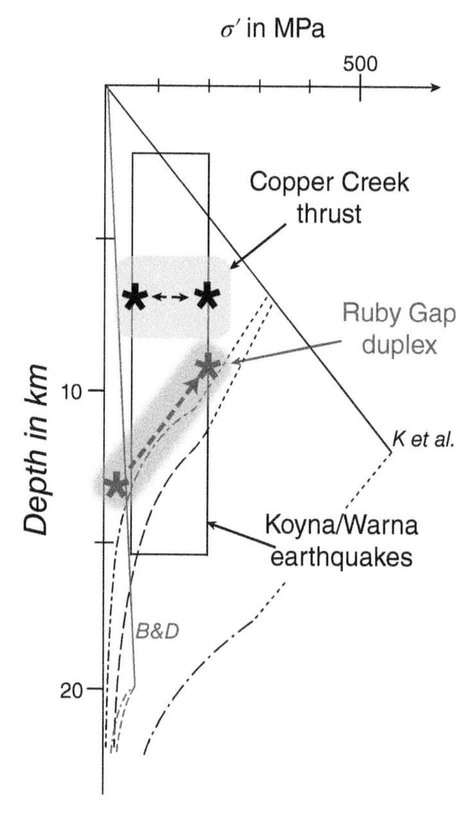

Figure 9.38 Differential stress magnitudes determined from the three case studies plotted against portions of the lithospheric strength profiles of Kohlstedt et al. (1995) and Bürgmann and Dresen (2008).

9.5.2 Comparing Stress Magnitudes Inferred from the Case Studies to Lithospheric Strength Profiles

Figure 9.38 plots the differential stress magnitudes inferred from the structural and microstructural studies outlined in Sections 9.2–9.4 against a backdrop of different continental lithospheric strength profiles. We consider first differential stress data from the Ruby Gap duplex presented in Section 9.2. Dunlap et al. (1997) determined the differential stress magnitudes at different positions within the duplex by: (1) measuring the size distributions of subgrains and new grains developed by dynamic recrystallization during deformation by dislocation creep; and (2) using the experimentally constrained recrystallized subgrain size piezometer (Twiss 1977). The recrystallized grain size piezometric technique is widely used and is inferred to be reliable, but one must take care in interpreting the results presented here. Figure 9.6 shows that the lowest differential stress magnitudes are inferred from quartzite microstructures developed at higher temperatures and pressures, reflecting deformation at greater depths north of Ruby Gap, whereas the lower temperature microstructures developed at higher differential stresses closer to the southern margin of the Capricorn orogen. The two red asterisks in Figure 9.38 depict the two end-member combinations of measured differential stresses and inferred depth of formation values for these deformed Heavitree Quartzites. Within the Ruby Gap duplex, quartzites exhibiting higher temperature microstructures were uplifted as they were transported toward the south and often overprinted by lower temperatures microstructures. The red arrow connecting asterisks in Figure 9.38 depicts schematically that anticipated differential stress-depth path. The two "end-member" differential stress-depth pairs and the line connecting them essentially fall on one of the "jelly sandwich" differential stress versus depth curves from Figure 9.35c. We, therefore, infer that the quartzite microstructure data from the southern margin of the Capricorn orogen conform best with the jelly sandwich strength profile.

The differential stress magnitudes associated with the emplacement of the Copper Creek thrust sheet are constrained only by the deformation mechanism map for the calcite-rich fault rocks in Figure 9.21. As noted in Section 9.3.4 and indicated by Figure 9.21, deformation in these fault rocks is likely to have cycled between two modes: (1) a higher differential stress mode, approaching or equaling magnitudes consistent with deformation by Byerlee's law; and (2) a low differential stress mode accommodated by diffusional mass transfer and grain boundary sliding in the fine-grained fault rocks. These two modes are again depicted schematically in Figure 9.38 by asterisks within a shaded region demarcating a range of inferred differential stress magnitudes. The height of the shaded region is intended to depict the uncertainty in fixing the depth at which deformation occurred; the width of the region encompasses the range of differential stresses associated with the different deformation modes. The estimated maximum differential stress is generally consistent with laboratory deformation experiments at confining pressures comparable to those inferred for deformation at depths of 7–10 km. The minimum differential stress reflects the inference, supported by deformation microstructures in the fine-grained fault rocks, that fault movement was accommodated intermittently by diffusional mass transfer and grain boundary sliding in ultrafine-grained, calcite-rich fault rocks. Microstructures in the fault rocks are consistent with repeated cycling between the higher and lower differential stress deformation modes. Deformation of strata above and below the thrust surface, accommodated by mesoscopic faults and folds, is assumed to have accrued at differential stress levels within this range of differential stresses. In this instance, temporal variations in stresses are again likely, although they are likely to have fluctuated many times within the range indicated rather than having undergone, as inferred for the Ruby Gap quartzites, a single episode of reducing differential stress, followed by a period of steady differential stress levels, and ending with a gradual increase in differential stresses as rocks cooled. If the emplacement of the Copper Creek sheet occurred mainly at the higher stress levels, this example too would be consistent with either the jelly sandwich or crème brulée profiles. On the other hand, if the emplacement of the Copper Creek sheet occurred mainly at the lower stress, this example would be more compatible with the banana split lithospheric strength profile.

It is more straightforward to draw inferences on the state of stress from induced seismicity even though the actual path from stability to failure is different in the four classes of induced seismicity discussed in Section 9.4. Of the four classes of induced seismicity, MIS is likely to be a product of localized perturbations of the stress state, where, for example, removal of rock reduces stresses acting on a mine wall sufficiently to cause rock failure.

Nevertheless, the seismicity does provide insight into differential stress levels. Despite differences in the precise causes of seismicity in extraction induced, injection-induced, and reservoir-induced cases, human activity alters the loads in relatively large volumes of rock, creating changes to the stress state that may resemble more nearly the changes in stresses responsible for natural deformation. The resulting seismicity, therefore, provides somewhat greater insight to stress states for a greater range of lithospheric depths. Earthquakes in Oklahoma (the United States), with focal depths up to 25 km have stress drops ranging from ~1 to ~100 MPa (Huang et al. 2017; Wu et al. 2018). In Basel, induced earthquakes are shallower (typically at 3–5 km depths) but stress drops cluster around 2 MPa, with maximum stress drops of ~100 MPa (Goertz-Allmann et al. 2011). The seismicity in Koyna and Warna discussed in Section 9.4.6 occurs mainly at moderate depths, between 4 and 16 km although earthquakes with focal depths nearly twice as deep are known from that area. The differential stresses inferred from stress drops associated with these earthquakes are small to moderate (Talwani 1997b). The magnitudes of stress drops associated with these induced earthquakes are comparable to those associated with tectonic earthquakes (Hardebeck and Aron 2009; Hardebeck and Okada 2017; Huang et al. 2017). There is no consensus whether earthquake-related slip dissipates only a fraction or nearly all of the stored elastic energy associated with an active fault. In cases where coseismic slip is significant, however, subsequent seismicity occurs mainly in the volume of rock surrounding the slipped area, suggesting that fault slip can, in fact, lead to a significant reduction in shear stress on a fault (e.g. Wetzler et al. 2018). If that is the case, then earthquake-related stress drops measure, albeit imprecisely, differential stresses in the volume of faulted rock. Taking this to be the case, the stress drops associated with shallower foci induced earthquakes are consistent with either jelly sandwich or crème brûlée lithospheric strength profiles. The stress drops associated with induced earthquakes with deeper foci are more consistent with the banana split models of the lithospheric strength, however.

9.5.3 Recap

The three examples considered in this chapter generally conform with aspects of existing work on variations of rock strength. Although the data presented in this chapter are far from exhaustive, they suggest that differential stress levels early in deformation history conform with the jelly sandwich strength profiles initially described by Kohlstedt et al. (1995). We note, however, that because our case studies are limited to examples in crustal settings, we cannot determine whether the crème brûlée model of Jackson (2002) or the jelly sandwich model of Burov and Watts (2006) better describes the variation of differential stresses in the crust. The examples considered here do show that rock strength decreases as deformation accrues, as suggested by Bürgmann and Dresen (2008). As such, the examples do not resolve the question of how differential stresses vary in the lithosphere.

Rather, the case studies show the limitations of envisaging that simplistic models can be generalized to pertain to any one region for long time periods or to pertain to broad geographic areas at any single time. Natural stress states change on time scales from those of seismic cycles to millions of years. Regional variations in crustal rheology, perhaps largely dictated by compositional variations and strain accumulation, are to be expected. The case studies show, however, that careful analysis of individual settings, employing geometric, kinematic and dynamic analyses and using both inductive and deductive reasoning as described in Chapter 1, can provide data critical to answering general questions important to geologists, geophysicists, and other Earth scientists. These data have important practical applications, as well as illuminating how the Earth works.

9.6 Broader Horizons

We have endeavored to show in this book how the principles of continuum mechanics (Chapters 4–7) and a material science approach to microstructures and deformation mechanisms (Chapters 3 and 8) can be applied to structures observed in rocks to understand the deformation of the Earth. This process ideally involves geometric, kinematic (displacement and strain) and dynamic (stress) analyses that lead to a quantitative evaluation of the rheology of the Earth at different levels in the crust and mantle for both modern and past orogenies. Stated in this way, the approach appears to be deductive, but, in fact, structural geologists commonly start to solve problems by making observations that are described by a well-established vocabulary from over a century of detailed structural studies, which is an inductive approach. This inductive approach is also how structural geology is typically, and appropriately, introduced at an undergraduate level. We have only included an elementary framework for structural observations (Chapter 2), but hope to present such introductory material in the future, notwithstanding the several textbooks on the market that do this very well.

Structural geologists are in great demand to solve practical problems in society. For many years one of the largest markets for their employment has been the oil and gas industry, but fortunately, that era is now drawing to a close. Nevertheless, energy will continue to be a major concern of structural geology, including the problems of CO_2 capture and storage and geothermal energy. The use of underground resources for other aspects of waste disposal, in particular radioactive materials, is another area demanding structural geology expertise.

As the world transitions into a green economy, forecasts suggest that vast increases in the supply of both traditional and more unusual elements will be needed to build the transport and energy systems of the future. Exploration for such resources is proceeding apace. The locations and geometry of ore bodies containing these resources may be directly controlled by structures in rocks, and also may have been affected by deformation events after ore formation. In both cases, it is necessary to apply structural geology to locate and mine the ore. Two factors are changing the nature of the structural analysis required in the exploration and mining industry: the increasing difficulty of finding new ore bodies, which involves exploring under cover rocks and at greater depths, and the imperative to explore more efficiently due to the environmental impacts of exploration.

Structural geology is relevant to several other aspects of the environment. The induced seismicity case study shows how activities such as dam construction, mining, fracking, fluid extraction and wastewater disposal may create seismic activity with important economic consequences. Understanding natural seismic hazards also requires a knowledge of active and potentially reactivated fault systems. Many semi-arid parts of the world such as much of southern Africa depend on water in crystalline basement aquifers, which is stored and accessed through fracture systems. Locations of highly productive water wells are controlled by the geometry of these fracture systems and the in-situ stress. Landslide and volcanic hazards can involve problems amenable to a geometry → kinematics → dynamics approach.

Finally, we take a speculative look at the direction in which structural geology may travel in the future. Active tectonic regions are likely to continue to fascinate structural geologists because of the insights that they afford into the mechanics of the Earth today. A particularly interesting branch of these studies is how surface and deep Earth processes may be linked through erosion, exhumation, uplift and crustal and mantle deformation. Geodetic studies in active tectonic regions give tremendous insight into short term, present-day deformation of the Earth, and a challenge is to link these results to longer time scales.

Links between structural and geochronological studies are one way to tackle this challenge, and recent advances in dating fault rocks have made impressive forays in this direction. There is a fascinating interaction between deformation and mineral chemistry that may affect radiometric ages. Such studies are at the forefront of material science, which continues to reveal aspects of microscale deformation through the increasing application of techniques such as scanning electron microscopy and electron backscatter diffraction.

We finish with two especially broad outlooks on the future of structural geology. Deformation, stress, rheology, heat flow, fluid flow and chemical reactions can perhaps all be comprehended in one framework using generalized thermodynamic approaches, including nonequilibrium thermodynamics. This topic is in its infancy but has the potential to offer a holistic view of the entire range of processes that occur in a deforming rock. Most of these processes are overlooked by conventional approaches, despite the obvious feedbacks that exist between them.

Structural geology has been applied to understand planetary geology for the past 30 years, but will likely continue with increasing data from other planetary bodies and with advances in understanding meteorite impacts. These studies have an interesting potential to inform understanding of the Earth, particularly with respect to early Earth processes.

References

Arora, K., Chadha, R.K., Srinu, Y. et al. (2017). Lineament fabric from airborne LiDAR and its influence on triggered earthquakes in the Koyna-Warna region, western India. *Journal of the Geological Society of India* 90: 670–677. https://doi.org/10.1007/s12594-017-0774-9.

Arthaud, F. (1969). Méthode de détermination graphique des directions de raccourrassement, d'allongement et intermediare d'une population des failles. *Bulletin Societé géologique de France, 7e Serie* 11: 729–737.

Bachmann, C.E., Wiemer, S., Woessner, J., and Hainzl, S. (2011). Statistical analysis of the induced Basel 2006 earthquake sequence: introducing a probability-based monitoring approach for enhanced geothermal systems. *Geophysical Journal International* 186: 793–807. https://doi.org/10.1111/j.1365-246X.2011.05068.x.

Bletery, Q. and Nocquet, J.-M. (2020). Slip bursts during coalescence of slow slip events in Cascadia. *Nature Communications* 11: 2159. https://doi.org/10.1038/s41467-020-15494-4.

Boutareaud, S., Calugaru, D.G., Han, R. et al. (2008). Clay-clast aggregates: a new textural evidence for seismic fault sliding? *Geophysical Research Letters* 35: 1–5.

Boyer, S. and Elliott, D. (1982). Thrust systems. *American Association of Petroleum Geologists Bulletin* 66 (9): 1196–1230.

Bürgmann, R. and Dresen, G. (2008). Rheology of the lower crust and upper mantle: evidence from rock mechanics, geodesy, and field observations. *Annual Reviews of Earth and Planetary Sciences* 36: 531–567.

Burov, E.B. and Watts, A.B. (2006). The long-term strength of continental lithosphere: "jelly sandwich" or "crème brûlée"? *GSA Today* 16: 4–10. https://doi.org/10.1130/1052-5173(2006)016<4.TLTSOC>2.0CO;2.

Catchings, R.D., Dixit, M.M., Goldman, M.R., and Kuma, S. (2015). Structure of the Koyna-Warna Seismic Zone, Maharashtra, India: a possible model for large induced earthquakes elsewhere. *Journal of Geophysical Research* 120: 3479–3506. https://doi.org/10.1002/2014JB011695.

Chen, W.-P. and Molnar, P. (1983). Focal depths of intracontinental and intraplate earthquakes and their implications for thermal and mechanical properties of the lithosphere. *Journal of Geophysical Research* 88: 4183–4214.

Chopra, P.N. and Paterson, M.S. (1984). The role of water in the deformation of dunite. *Journal of Geophysical Research* 89: 7861–7876.

Clarke, H., Eisner, L., Styles, P., and Turner, P. (2014). Felt seismicity associated with shale gas hydraulic fracturing: the first documented example in Europe. *Geophysical Research Letters* 41: 8308–8314. https://doi.org/10.1002/2014GL062047.

Collettini, C., Carpenter, B.M., Viti, C. et al. (2014). Fault structure and slip localization in carbonate-bearing normal faults: an example from the Northern Apennines of Italy. *Journal of Structural Geology* 67: 154–166.

Collins, W.J. and Teyssier, C. (1989). Crustal scale ductile fault systems in the Arunta Inlier, central Australia. *Tectonophysics* 158: 49–66.

Deichmann, N. and Giardini, D. (2009). Enhanced geothermal system below Basel. *Seismological Research Letters* 80: 784–798. https://doi.org/10.1785/gssrl.80.5.784.

Doglioni, C. (2018). A classification of induced seismicity. *Geoscience Frontiers* 9: 1903–1909. https://doi.org/10.1016/j.gsf.2017.11.015.

Dunlap, W.J. (1997). Neocrystallization or cooling? $^{40}Ar/^{39}Ar$ ages of which micas from low grade mylonites. *Chemical Geology* 143: 181–203.

Dunlap, W.J. and Teyssier, C. (1995). Paleozoic deformation and isotopic disturbance in the southeastern Arunta Block, central Australia. *Precambrian Research* 71: 229–250.

Dunlap, W.J., Teyssier, C., McDougall, I., and Baldwin, S. (1991). Ages of deformation from $^{40}Ar/^{39}Ar$ dating of white micas. *Geology* 19: 1213–1216.

Dunlap, W.J., Teyssier, C., McDougall, I., and Baldwin, S. (1995). Thermal and structural evolution of the intracratonic Arltunga Nappe Complex, central Australia. *Tectonics* 14: 1182–1204.

Dunlap, W.J., Hirth, G., and Teyssier, C. (1997). Thermomechanical evolution of a ductile duplex. *Tectonics* 16 (6): 983–1000.

Ellsworth, W.L. (2013). Injection-induced earthquakes. *Science* 341: 1225942–1225942. https://doi.org/10.1126/science.1225942.

Fondriest, M., Smith, S.A.F., Di Toro, G. et al. (2012). Fault zone structure and seismic slip localization in dolostones, an example from the Southern Alps, Italy. *Journal of Structural Geology* 45: 52–67.

Foulger, G.R., Wilson, M.P., Gluyas, J.G. et al. (2018). Global review of human-induced earthquakes. *Earth-Science Reviews* 178: 438–514. https://doi.org/10.1016/j.earscirev.2017.07.008.

Frimmel, H.E. (2008). Earth's continental crustal gold endowment. *Earth and Planetary Science Letters* 267: 45–55. https://doi.org/10.1016/j.epsl.2007.11.022.

Gilotti, J. and Kumpulanien, R. (1986). Strain-softening induced ductile flow in the Särv thrust sheet, Scandinavian Caledonides: a description. *Journal of Structural Geology* 8: 441–455.

Gleason, G.C. and Tullis, J. (1994). A flow law for dislocation creep of quartz aggregates determined with the molten salt cell. *Tectonophysics* 247: 1–23.

Goertz-Allmann, B.P., Goertz, A., and Weimer, S. (2011). Stress drop variations of induced earthquakes at the Basel geothermal site. *Geophysical Research Letters* 38: L09308. https://doi.org/10.1029/2011GL47498.

Gupta, H.K. (2002). A review of recent studies of triggered earthquakes by artificial water reservoirs with special emphasis on earthquakes in Koyna, India. *Earth-Science Reviews* 58: 279–310. https://doi.org/10.1016/S0012-8252(02)00063-6.

Han, R. and Hirose, T. (2012). Clay-clast aggregates in fault gouge: an unequivocal indicator of seismic faulting at shallow depths? *Journal of Structural Geology* 43: 92–99.

Han, R., Shimamoto, T., Hirose, T. et al. (2007). Ultralow friction of carbonate faults caused by thermal decomposition. *Science* 316: 878–881.

Handschy, J.W. (1998). Spatial variation in structural style, Endicott Mountains allochthon, central Brooks Range, Alaska. In: *Architecture of the Central Brooks Range Fold and Thrust Belt, Artic Alaska*, Geological Society of America Special Paper, vol. 324 (ed. J.S. Oldow and H.G.A. Lallemant), 33–50. Geological Society of America.

Hardebeck, J.L. and Aron, A. (2009). Earthquake stress drops and inferred fault strength on the Hayward fault, East San Francisco Bay, California. *Bulletin of the Seismological Society of America* 99 (3): 1801–1814. https://doi.org/10.1785/0120080242.

Hardebeck, J.L. and Okada, T. (2017). Temporal stress changes caused by earthquakes: a review. *Journal of Geophysical Research* 123: 1350–1365. https://doi.org/10.0102/2017JB014617.

Harris, L.D. and Milici, R.C. (1977). Characteristics of thin-skinned style of deformation in the southern Appalachians and potential hydrocarbon traps. *U. S. Geological Survey Professional Paper* 1018: 1–40.

Hatcher, R.D., Lemiszki, P.J., and Whisner, J.B. (2007). Character of rigid boundaries and internal deformation of southern Appalachian foreland fold-thrust belt. In: *Whence the Mountains? Inquiries into the Evolution of Orogenic Systems: A Volume in Honor of Raymond A. Price*, Geological Society of America Special Paper, vol. 433 (ed. J.W. Sears, T.A. Harms and C.A. Evenchick), 243–276.

Hincks, T., Aspinall, W., Cooke, R., and Gernon, T. (2018). Injection depth. *Science* 1255: 1251–1255. https://doi.org/10.1126/science.aap7911.

Hirth, G. and Tullis, J. (1992). Dislocation creep regimes in quartz aggregates. *Journal of Structural Geology* 14: 145–160.

Horsman, E. and Tikoff, B. (2005). Quantifying simultaneous discrete and distributed deformation. *Journal of Structural Geology* 27: 1168–1189.

Huang, Y., Ellsworth, W.L., and Beroza, G.C. (2017). Stress drops of induced and tectonic earthquakes in the central United States are indistinguishable. *Science Advances* 3: e1700772.

Jackson, J. (2002). Strength of the continental lithosphere: time to abandon the jelly sandwich? *GSA Today* 12: 4–10.

Kennedy, L.A. and Logan, J.M. (1997). The role of veining and dissolution in the evolution of fine-grained mylonites: the McConnell thrust, Alberta. *Journal of Structural Geology* 19: 785–797.

Kennedy, L.A. and Logan, J.M. (1998). Microstructures of cataclasites in a limestone-on-shale thrust fault. *Tectonophysics* 295: 167–186.

Kennedy, L.A. and White, J.C. (2001). Low-temperature recrystallization in calcite: mechanisms and consequences. *Geology* 29: 1027–1030.

Keranen, K.M. and Weingarten, M. (2018). Induced seismicity. *Annual Review of Earth & Planetary Sciences* 46: 149–174. https://doi.org/10.1146/annurev-earth-082517-010054.

Keranen, K.M., Weingarten, M., Abers, G.A. et al. (2014). Sharp increase in central Oklahoma seismicity since 2008 induced by massive wastewater injection. *Science* 345: 448–451. https://doi.org/10.1126/science.1255802.

Khoshmanesh, M. and Shirzaei, M. (2018). Multiscale dynamics of aseismic slip on central San Andreas fault. *Geophysical Research Letters* 85: 2274–2282.

Klose, C.D. (2012). Evidence for anthropogenic surface loading as trigger mechanism of the 2008 Wenchuan earthquake. *Environmental Earth Sciences* 66: 1439–1447. https://doi.org/10.1007/s12665-011-1355-7.

Kohlstedt, D.L., Evans, B., and Mackwell, S.J. (1995). Strength of the lithosphere: constraints imposed by laboratory experiments. *Journal of Geophysical Research* 100: 17587–17602.

Kronenberg, A.K. and Tullis, J. (1984). Flow strengths of quartz aggregates: grain size and pressure effects due to hydrolytic weakening. *Journal of Geophysical Research* 89: 4281–4297.

Lambeck, K. (1986). Crustal structure and evolution of the central Australian basins. In: *The Nature of the Lower Continental Crust*, Geological Society Special Publication, vol. 24 (ed. J.B. Dawson, D.A. Carswell, J. Hall and K.H. Wedepohl), 133–145.

Luan, F. and Paterson, M.S. (1992). Preparation and deformation of synthetic aggregates of quartz. *Journal of Geophysical Research* 97: 301–320.

Marrett, R. and Allmendinger, R.W. (1990). Kinematic analysis of fault-slip data. *Journal of Structural Geology* 12 (8): 973–986.

McGarr, A. (2002). Control of strong ground motion of mining-induced earthquakes by the strength of the seismogenic rock mass. *Journal of the South African Institute of Mining and Metallurgy* 102: 225–229.

McGarr, A. and Gay, N.C. (1978). State of stress in the Earth. *Annual Reveiws of Earth and Planetary Science* 6: 405–436.

Means, W.D. (1989). Stretching faults. *Geology* 17: 983–896.

Mitra, G. (1993). Deformation processes in Brittle deformation zones in granitic basement rocks. A case study from the Torrey Creek area, Wind River Mountains. In: *Basement behavior in Rocky Mountain Foreland Structure*, Geological Society of America Special Paper, vol. 280 (ed. C. Schmidt, R. Chase and E. Erslev), 177–195. Geological Society of America.

Mitra, G. and Elliott, D. (1980). Deformation of basement in the Blue Ridge and the development of the South Mountain cleavage. In: *The Caledonides in the USA*, vol. 2 (ed. D.R. Wones), 307–312. Va. Polytech. Inst. and State Univ. Memoir.

Murphy, D.C. (1987). Suprastructure/infrastructure transition, east-central Cariboo Mountains, British Columbia: geometry, kinematics, and tectonic implications. *Journal of Structural Geology* 9: 13–29.

Niemeijer, A., Di Toro, G., Griffith, W.A. et al. (2012). Inferring earthquake physics and chemistry using an integrated field and laboratory approach. *Journal of Structural Geology* 39: 2–36.

Paterson, M.S. and Luan, F.C. (1990). Quartzite rheology under geological conditions. In: *Deformation Mechanisms, Rheology, and Tectonics*, Special Publication Geological Society of London, vol. 54 (ed. R. Knipe and E.H. Rutter), 299–307.

Pfiffner, O.A. (2017). Thick-skinned and thin-skinned tectonics: a global perspective. *Geosciences* 7: 71. https://doi.org/10.3390/geosciences7030071.

Pollitz, F.F. (2019). Lithosphere and shallow asthenosphere rheology from observations of post-earthquake relaxation. *Physics of the Earth and Planetary Interiors* 293: 106271. https://doi.org/10.1016/j.pepi.2019.106271.

Raleigh, C.B., Healy, J.H., and Bredehoeft, J.D. (1972). Faulting and crustal stress at Rangely, Colorado. In: *Flow and Fracture of Rocks, Geophysical Monograph*, vol. 16 (ed. H.C. Heard, I.Y. Borg, N.L. Carter and C.B. Raleigh), 275–284. https://doi.org/10.1029/GM016p0275.

Raleigh, C., Healy, J., and Bredehoeft, J. (2013). Faulting and Crustal Stress at Rangely, Colorado. In: *Flow and Fracture of Rocks*, 275–284. American Geophysical Union (AGU) https://doi.org/10.1029/GM016p0275.

Ramsay, J.G., Casey, M., and Kligfield, R. (1983). Role of shear in development of the Helvetic fold-thrust belt of Switzerland. *Geology* 11: 439–442.

Rankin, D.W. (1976). Appalachian salients and recesses: late Precambrian continental breakup and the opening of the Iapetus Ocean. *Journal of Geophysical Research* 81 (32): 5605–5619.

Rempe, M., Smith, S.A.F., Ferri, F. et al. (2014). Clast-cortex aggregates in experimental and natural calcite-bearing fault zones. *Journal of Structural Geology* 68: 142–157.

Richardson, E. and Jordan, T.H. (2002). Seismicity in deep gold mines of South Africa: implications for tectonic earthquakes. *Bulletin of the Seismological Society of America* 92: 1766–1782. https://doi.org/10.1785/0120000226.

Savage, J.C. and Paterson, W.S.B. (1963). Borehole measurements in the Athabasca Glacier. *Journal of Geophysical Research* 68: 4521–4536.

Shaw, R.D., Etheridge, M.A., and Lambeck, K. (1991). Development of the Late Proterozoic to Mid-Paleozoic intracratonic Amadeus Basin in central Australia: a key to understanding tectonic forces in plate interiors. *Tectonics* 10 (4): 688–721.

Shaw, R.D., Zeitler, P.K., McDougall, I., and Tingate, P.R. (1992). The Palaeozoic history of an unusual intracratonic thrust belt in central Australia based on ^{40}Ar-^{39}Ar, K-Ar and fission track dating. *Journal of the Geological Society, London* 149: 937–954.

Shinevar, W.J., Behn, M.D., Hirth, G., and Jagoutz, O. (2018). Inferring crustal viscosity from seismic velocity: application to the lower crust of Southern California. *Earth & Planetary Science Letters* 494: 83–19. https://doi.org/10.1016/j.epsl.2018.04.055.

Smeraglia, L., Bettucci, A., Billi, A. et al. (2017). Microstructural evidence for seismic and aseismic slips along clay-bearing, carbonate faults. *Journal of Geophysical Research* 122: 3895–3915. https://doi.org/10.1002/2017JB014042.

Spagnuolo, E., Plümper, O., Violay, M. et al. (2015). Fast-moving dislocations trigger flash weakening in carbonate-bearing faults during earthquakes. *Scientific Reports* 5: 16112.

Talwani, P. (1997a). On the nature of reservoir-induced seismicity. *Pure & Applied Geophysics* 150: 473–492. https://doi.org/10.1007/s000240050089.

Talwani, P. (1997b). Seismotectonics of the Koyna-Warna area, India. *Pure & Applied Geophysics* 150: 511–550.

Teyssier, C. (1985). A crustal thrust system in an intracratonic tectonic environment. *Journal of Structural Geology* 7: 689–700.

Thatcher, W. and Pollitz, F.F. (2008). Temporal evolution of continental lithospheric strength in actively deforming regions. *GSA Today* 18: 4–11. https://doi.org/10.1130/GSAT01804-5A.1.

van Thienen-Visser, K. and Breunese, J.N. (2015). Induced seismicity of the Groningen gas field: history and recent developments. *The Leading Edge* 34: 664–671. https://doi.org/10.1190/tle34060664.1.

Thomas, W.A. (1977). Evolution of the Appalachian-Ouchita salients and recesses from reentrants and promontories in the continental margin. *American Journal of Science* 277: 1233–1278.

Townend, J. and Zoback, M.D. (2000). How faulting keeps the crust strong. *Geology* 28: 399–402. https://doi.org/10.1130/0091-7613(2000)28<399:HFKTCS>2.0.CO;2.

Twiss, R.J. (1977). Theory and applicability of a recrystallized grain-size paleopiezometer. *Pure & Applied Geophysics* 115: 228–244.

Violay, M., Nielsen, S., Spagnuolo, E. et al. (2013). Pore fluid in experimental calcite-bearing faults: abrupt weakening and geochemical signature of co-seismic processes. *Earth and Planetary Science Letters* 361: 74–84.

Walsh, F. R., & Zoback, M. D. (2016). Probabilistic assessment of potential fault slip related to injection- induced earthquakes : Application to north-central Oklahoma , *USA. Geology*, 44, 991–994. https://doi.org/10.1130/G38275.1

Wells, R.K. (2017). Rheologic evolution of carbonates and shale along thrust faults. Ph.D. dissertation. Texas A. & M. University.

Wells, R.K., Newman, J., and Wojtal, S. (2014). Microstructures and rheology of a calcite-shale thrust fault. *Journal of Structural Geology* 65: 69–81.

Wetzler, N., Lay, T., Brodsky, E., and Kanamori, H. (2018). Systematic deficiency of aftershocks in areas of high coseismic slip for large subduction zone earthquakes. *Science. Advances* 4: eaao3225.

Wojtal, S. (1989). Measuring displacement gradients and strain in faulted rocks. *Journal of Structural Geology* 11: 669–678.

Wojtal, S. (1992a). One-dimensional models for plane and non-plane power-law flow in shortening and elongating thrust zones. In: *Thrust Tectonics* (ed. K. McClay), 41–52. London: Chapman & Hall.

Wojtal, S. (1992b). Shortening and elongation of thrust zones within the Appalachian fold-thrust belt. In: *Structural Geology of Fold and Thrust Belts (Elliott Volume)* (ed. S. Mitra and G. Fisher), 93–103. Baltimore: Johns Hopkins University Press.

Wojtal, S. and Mitra, G. (1986). Strain hardening and strain softening in fault zones from foreland thrusts. *Geological Society of America Bulletin* 97: 674–687.

Woodward, N.B., Wojtal, S., Paul, J.B., and Zadins, Z. (1988). Partitioning of deformation within several external thrust zones of the Appalachian orogen. *Journal of Geology* 96: 351–361.

Wu, Q., Chapman, M., and Chen, X. (2018). Stress-drop variations of induced earthquakes in Oklahoma. *Bulletin of the Seismological Society of America* 108 (3A): 1107–1123. https://doi.org/10.1785/0120170335.

Yadav, A., Bansal, B.K., and Pandey, A.P. (2016). Five decades of triggered earthquakes in Koyna-Warna Region, western India – a review. *Earth-Science Reviews* 162: 433–450. https://doi.org/10.1016/j.earscirev.2016.09.013.

Index

An Integrated Framework for Structural Geology: Kinematics, Dynamics, and Rheology of Deformed Rocks, First Edition. Steven Wojtal, Tom Blenkinsop, and Basil Tikoff.